HEALTH AND SAFETY ENFORCEMENT
Law and Practice

HEALTH AND SAFETY ENFORCEMENT

Law and Practice

SECOND EDITION

Richard Matthews
James Ageros

OXFORD
UNIVERSITY PRESS

Great Clarendon Street, Oxford OX2 6DP

Oxford University Press is a department of the University of Oxford.
It furthers the University's objective of excellence in research, scholarship,
and education by publishing worldwide in

Oxford New York

Auckland Cape Town Dar es Salaam Hong Kong Karachi
Kuala Lumpur Madrid Melbourne Mexico City Nairobi
New Delhi Shanghai Taipei Toronto

With offices in

Argentina Austria Brazil Chile Czech Republic France Greece
Guatemala Hungary Italy Japan Poland Portugal Singapore
South Korea Switzerland Thailand Turkey Ukraine Vietnam

Oxford is a registered trade mark of Oxford University Press
in the UK and in certain other countries

Published in the United States
by Oxford University Press Inc., New York

© Richard Matthews and James Ageros 2007

The moral rights of the authors have been asserted
Database right Oxford University Press (maker)

Crown copyright material is reproduced under Class Licence
Number C01P0000148 with the permission of OPSI
and the Queen's Printer for Scotland

First published 2007

All rights reserved. No part of this publication may be reproduced,
stored in a retrieval system, or transmitted, in any form or by any means,
without the prior permission in writing of Oxford University Press,
or as expressly permitted by law, or under terms agreed with the appropriate
reprographics rights organization. Enquiries concerning reproduction
outside the scope of the above should be sent to the Rights Department,
Oxford University Press, at the address above

You must not circulate this book in any other binding or cover
and you must impose the same condition on any acquirer

British Library Cataloguing in Publication Data
Data available

Library of Congress Cataloging in Publication Data

Matthews, Richard A., barrister.
 Health and safety enforcement: law and practice / Richard Matthews and
James Ageros. — 2nd ed.
   p. cm.
 Includes bibliographical references and index.
 ISBN 978–0–19–921286–6 (hardback : alk. paper) 1. Industrial safety—
Law and legislation—Great Britain. 2. Industrial hygiene—Law and
legislation—Great Britain. I. Ageros, James. II. Title.
 KD3168.M38 2007
 344.4104'65—dc22
                                                              2007031898

Typeset by Cepha Imaging Private Ltd., Bangalore, India
Printed in Great Britain
on acid-free paper by
Antony Rowe Ltd., Chippenham, Wiltshire

ISBN 978–0–19–921286–6

1 3 5 7 9 10 8 6 4 2

CONTENTS—SUMMARY

Table of Cases	xxiii
Table of Statutes	xxxiii
Table of Statutory Instruments	xlvii
1. The Enforcement Framework	1
2. The Reporting and Investigation of Health and Safety Incidents and Offences	25
3. Advising Clients: Interviews Under Caution, Representation, and Privilege	67
4. Improvement and Prohibition Notices	91
5. Legal Personality and Secondary Liability	123
6. Health and Safety Offences: Breach of the General Duties	141
7. Other Health and Safety Offences: HSWA 1974 s 33(1)(b)–(o)	195
8. Proceedings and Criminal Procedure	213
9. Judicial Review and Abuse of Process	263
10. Sentencing	277
11. Work-Related Deaths	311
12. Principal Health and Safety Regulations	345
Appendices	405
Index	667

CONTENTS

Table of Cases xxiii
Table of Statutes xxxiii
Table of Statutory Instruments xlvii

1. The Enforcement Framework

   A. Legislation, Regulations, and Guidance — 1.01
      The Health and Safety at Work etc. Act 1974 — 1.01
      Health and safety regulations — 1.07
      Approved codes of practice and guidance — 1.10

   B. Institutions — 1.13
      Introduction — 1.13
      The Health and Safety Commission — 1.16
      The Health and Safety Executive — 1.24
      Local authorities — 1.32
      Health and Safety (Enforcing Authority) Regulations 1998 — 1.36
      The CPS and the DPP: the protocol for liaison — 1.38
      The Crown — 1.41

   C. The European Revolution — 1.45
      The Framework Directive — 1.45
      The daughter Directives — 1.60

   D. Principles of Enforcement Action — 1.62
      Introduction — 1.62
      The Enforcement Policy Statement — 1.65
      The Code for Crown Prosecutors — 1.68
      The Enforcement Management Model — 1.72

   E. Methods of Enforcement — 1.75
      Introduction — 1.75
      Early intervention — 1.76
      Verbal warnings — 1.77
      Letters of advice/recommendation — 1.78
      Improvement notices and prohibition notices — 1.79
      Permissioning and licensing regimes — 1.80

Formal cautions	1.81
Crown censure	1.88
F. Prosecution and Particular Categories of Person	1.89
Prosecution of individuals	1.89
Employees: HSWA 1974 s 7	1.92
Directors and managers: HSWA s 37	1.95
Other individuals: HSWA ss 8 and 36	1.100
Prosecuting under HSWA s 3	1.102
Action to be taken when a decision is made to prosecute	1.104

2. The Reporting and Investigation of Health and Safety Incidents and Offences

A. Reporting Health and Safety Incidents	2.05
Notification and reporting of injuries and dangerous occurrences	2.10
'Accident' and 'dangerous occurrence'	2.17
Occupational diseases	2.19
Responsible person	2.21
Records	2.22
Breaching a duty under RIDDOR	2.23
B. The Service of Notices and Documents	2.25
C. The Investigation of Health and Safety Incidents and Offences	2.32
The ambit of the duty to investigate	2.34
Duty to make adequate arrangement for enforcement	2.36
The decision to investigate	2.41
The duty to pursue all reasonable lines of enquiry	2.46
Investigatory powers	2.53
Police powers and the new power of arrest	2.59
The likely exercise of the new powers of arrest	2.65
Police powers of search and entry	2.69
Inspectors and the Police and Criminal Evidence Act 1984	2.71
The Police and Criminal Evidence Act 1984 Codes of Practice	2.72
Searches under the Police and Criminal Evidence Act 1984, Code B	2.77
Cautioning	2.78
Interviews	2.81
Cautioning and the power under HSWA 1974 s 20(2)(j)	2.85
D. Inspectors' Powers Under HSWA 1974 s 20	2.87
Consequences of improper use of HSWA 1974 s 20 powers	2.96
Exclusion of evidence	2.96
Civil action for damages	2.100
The power to enter premises: HSWA 1974 s 20(2)(a)	2.101
Power to take a constable: HSWA 1974 s 20(2)(b)	2.105

Power to take with others and items of equipment: HSWA 1974 s 20(2)(c)	2.108
Powers to 'examine and investigate': HSWA 1974 s 20(2)(d)	2.109
Direction to leave undisturbed: HSWA 1974 s 20(2)(e)	2.113
Taking measurements, photographs, and recordings: HSWA 1974 s 20(2)(f)	2.115
Articles and substances: the power to take samples: HSWA 1974 s 20(2)(g)	2.117
The power to dismantle or subject dangerous articles or substances to testing: HSWA 1974 s 20(2)(h)	2.121
The power to take possession and detain dangerous articles or substances: HSWA 1974 s 20(2)(i)	2.128
The power to require answers: HSWA 1974 s 20(2)(j)	2.131
The status of the statement	2.135
Self-incrimination: HSWA 1974 s 20(7)	2.137
Ambit and purpose of power	2.139
Compelling answers from company officers	2.146
The presence of a nominated person	2.149
The power to require production, inspection, and to take copies: HSWA 1974 s 20(2)(k)	2.150
The legal privilege exception: HSWA 1974 s 20(8)	2.155
The power to require facilities: HSWA 1974 s 20(2)(1)	2.157
Any other power which is necessary: HSWA 1974 s 20(2)(m)	2.159
Powers to deal with immediate danger: HSWA 1974 s 25	2.163

3. Advising Clients: Interviews Under Caution, Representation, and Privilege

A. Interviews Under Caution	3.01
Advising suspects prior to interviews	3.05
Interviews under caution: advising clients	3.07
Pre-interview disclosure	3.11
Adverse inferences from a failure to mention facts	3.13
The effect of declining an invitation to be interviewed	3.19
Interviews with company directors and interviews by letter	3.21
The decision to be interviewed	3.26
Prepared statements	3.27
Representations	3.28
Interviews and the Police and Criminal Evidence Act 1984	3.31
Access to legal advice	3.35
The caution	3.39
Conduct of the interview	3.42
Limitations on the power to interview	3.46
Written statements under caution	3.50

B. Advising Clients in Health and Safety Investigations	3.51
Representation	3.55
Insurance and funding	3.55
Representing or advising more than one party: conflict of interest	3.59
Incidents involving a fatality	3.67
Cooperation with inspector and the use of s 20 powers	3.68
Privilege, compulsory production, and reports into causes	3.74
Privilege	3.74
Reports into causes: advising clients	3.77
The operation of legal privilege	3.81
Pre-existing documents	3.85
Admissibility of documents: the privilege against self-incrimination	3.86

4. Improvement and Prohibition Notices

A. Improvement and Prohibition Notices	4.01
Summary	4.01
Consequences of serving a notice	4.10
Contravening a notice	4.11
The inspector's decision to issue a notice	4.18
Improvement notices	4.25
Prohibition notices	4.34
Serving the notice	4.46
Withdrawal and extension	4.48
Liability of inspectors in negligence for misstatement	4.51
B. Appealing Improvement and Prohibition Notices	4.53
Procedure	4.59
Commencing an appeal	4.62
Suspension of notices pending appeal	4.68
Withdrawing an appeal	4.71
Case management	4.72
The hearing	4.75
Preliminary arguments	4.83
The burden of proof	4.84
Challenging the inspector's opinion	4.88
Did the inspector genuinely hold the opinion?	4.89
Did the activities complained of involve a breach of a statutory provision or a risk of serious personal injury?	4.95
Insignificant contravention	4.98
Fundamental flaw and modification	4.100
Example decisions	4.111
The tribunal's decision	4.120
Costs	4.123
Appeal from the decision of an employment tribunal	4.126

5. Legal Personality and Secondary Liability

A. Legal Personality and HSWA 1974 — 5.01
- Natural persons and corporations — 5.03
- Unincorporated bodies — 5.08
- Partnerships: England and Wales — 5.11
- Partnerships: Scotland — 5.15
- Limited liability partnerships — 5.16
- Trusts — 5.17

B. Secondary Liability — 5.27
- Common law — 5.27
- Section 36: the liability of others who cause the offence — 5.30
  - The 'commission by any person of an offence' — 5.34
  - 'Is due to': causation — 5.37
  - 'The act or default of' — 5.38
  - HSWA 1974 s 36(2): secondary liability and acts of the Crown — 5.45

C. Section 37: The Duties and Liability of Company Directors, Managers, and Officers — 5.46
- The secondary liability of company directors, managers, and officers — 5.46
  - Persons liable under HSWA 1974 s 37 — 5.48
  - Basis of liability — 5.52
  - Proof of the company's offence — 5.54
  - Consent — 5.56
  - Connivance — 5.60
  - Attributable to any neglect — 5.63
- Company directors' duties — 5.73
  - The continuing debate and calls for reform — 5.76

6. Health and Safety Offences: Breach of the General Duties

A. Section 33(1)(a): The Offence of Failing to Discharge one of the General Duties — 6.01

B. The Duties Under HSWA 1974 ss 2 and 3 — 6.04
- Introduction — 6.04
- HSWA 1974 s 2: the employer's duty to employees;
  - the elements of the offence — 6.10
  - The duty — 6.12
  - Failing to discharge the duty under HSWA 1974 s 2(1):
    - the elements of the offence — 6.14
  - HSWA 1974 s 2(3)–(7) — 6.16
- 'At work' — 6.18
- HSWA 1974 s 3: the duty to non-employees; the offence of
  - failing to discharge the duty — 6.24

The duty	6.25
Failing to discharge the duty under HSWA 1974 s 3(1): the elements of the offence	6.27
HSWA 1974 s 3(2): the duty on the self-employed	6.29
Undertaking	6.33
C. Elements Common to ss 2 and 3 HSWA 1974	6.43
Employment: employer and employee	6.43
Agency workers	6.52
Identifying an employer and the scope of his undertaking—companies without employees	6.53
Work experience	6.55
Crown servants, police officers, and domestic service	6.56
Employment: proof and practicalities	6.58
Failure to ensure safety/exposure to risks to safety: the possibility of danger	6.61
D. Reasonable Practicability and the General Duties: Meaning and Development to 1999	6.66
Introduction	6.66
The identification doctrine, vicarious liability, delegation, and the isolated acts of junior employees	6.68
British Steel and the 'identification doctrine'	6.68
Associated Octel and vicarious liability	6.75
The isolated act of a junior employee: *Gateway Foodmarkets Ltd*	6.80
R v Nelson Group Services (Maintenance) Ltd	6.91
Summary: *British Steel* to *Nelson Group Services*	6.96
E. Reasonable Practicability: *HTM Ltd* and the Current Position	6.101
Introduction	6.101
The Management of Health and Safety at Work Regulations 1999 reg 21	6.102
Reasonable practicability and 'fault of junior employee': summary of current position	6.109
Likely future development	6.115
Foreseeability	6.116
The current position	6.117
The civil case: *Dugmore v Swansea NHS Trust*	6.118
The criminal case: *R v HTM Ltd*	6.127
The reverse burden in HSWA 1974 s 40	6.135
F. HSWA 1974 s 4: The General Duty of Persons in Control of Premises to Persons other than their Employees	6.147
HSWA 1974 s 4(1): non-domestic premises used by non-employees as a place of work	6.148
HSWA 1974 s 4(2): by whom the duty is owed	6.155

'To take such measures as it is reasonable for a person in his position to take'	6.158
That the relevant premises, plant, or substance is or are safe and without risks to health	6.160
'So far as is reasonably practicable'	6.162
G. HSWA 1974 s 6: General Duties of Manufacturers etc as Regards Articles and Substances for Use at Work	6.165
Limitations on liability	6.168
HSWA 1974 s 6(1): designers, manufacturers, and suppliers of articles for work	6.173
Supply	6.174
Articles for use at work	6.176
The duty under HSWA 1974 s 6(1)(a)	6.180
The duty under HSWA 1974 s 6(1)(b)	6.183
The duty under HSWA 1974 s 6(1)(c)	6.184
The duty under HSWA 1974 s 6(1)(d)	6.185
Written undertakings	6.188
Imports	6.189
Fairground equipment	6.190
The duty under HSWA 1974 s 6(2): research	6.193
HSWA 1974 s 6(3): installation	6.196
HSWA 1974 s 6(4): substances	6.198
The duty under HSWA 1974 s 6(4)(a)	6.200
The duty under HSWA 1974 s 6(4)(b)	6.203
The duty under HSWA 1974 s 6(4)(c)	6.204
Substances and research: HSWA 1974 s 6(5)	6.206
H. HSWA 1974 s 7: The Duty of Employees at Work	6.209
Prosecution for a breach of HSWA 1974 s 7	6.217
7. Other Health and Safety Offences: HSWA 1974 s 33(1)(b)–(o)	
A. HSWA 1974 s 33(1)(b)–(o)	7.01
B. HSWA 1974 s 33(1)(b):Breach of HSWA 1974 ss 8 and 9	7.03
C. HSWA 1974 s 33(1)(c):Breach of Regulations	7.07
'To contravene any health and safety regulations'	7.10
Interpretation	7.13
Strict liability	7.15
Qualified standards of duty: reasonable practicability	7.18
Regulatory duties qualified by 'reasonable practicability'	7.21
Regulatory duties with a defence involving 'reasonable practicability'	7.30
Regulations in which measures to be taken are qualified by 'reasonably practicable'	7.35
Other qualifications in regulations	7.39

D. Other Offences	7.40
Health and safety inquiries: HSWA 1974 s 33 (1)(d)	7.40
Contravening requirements by inspectors: HSWA 1974 s 33(1)(e)	7.44
'Requirement'	7.47
HSWA 1974 s 25 powers	7.50
Preventing a person from attending or answering questions: HSWA 1974 s 33(1)(f)	7.52
Contravening improvement or prohibition notices: HSWA 1974 s 33(1)(g)	7.54
No 'reasonably practicable' defence	7.55
Obstructing an inspector: HSWA 1974 s 33(1)(h)	7.56
Obstructing an inspector	7.57
Intentionally	7.58
Provision of information to the commission: HSWA 1974 s 33(1)(i)	7.59
Disclosure of information: HSWA 1974 s 33(1)(j)	7.61
False statements: HSWA 1974 s 33(1)(k)	7.67
Knowledge or recklessness	7.69
Making a statement	7.72
A requirement to furnish any information imposed by or under any of the relevant statutory provisions	7.73
The offence and statements obtained under HSWA 1974 s 20(1)(j)	7.74
For the purpose of obtaining the issue of a document	7.78
Making false entries: HSWA 1974 s 33(1)(l)	7.80
Any register, book, notice, or other document required	7.82
Using a document with intent to deceive: HSWA 1974 s 33(1)(m)	7.83
'A document issued or authorised to be issued or required for any purpose'	7.85
'With intent to deceive'	7.86
'Calculated to deceive'	7.87
Impersonating an inspector: HSWA 1974 s 33(1)(n)	7.88
Failure to comply with a court order: HSWA 1974 s 33(1)(o)	7.89
8. Proceedings and Criminal Procedure	
A. Instituting Proceedings	8.01
Introduction	8.01
Arrestable offences	8.03
Future abolition of informations and summonses	8.05
Who may commence proceedings?	8.07
Independent Legal Oversight	8.13
Time limits	8.14
Bringing proceedings against a company in liquidation	8.17
Compulsory liquidation	8.21
Voluntary liquidation	8.24

Jurisdiction of the magistrates' court	8.26
Laying the information: obtaining and serving the summons	8.26
Content of a summons	8.36
Amending the information or summons	8.40
Withdrawal and Substitution of Informations	8.42
B. Conducting a Health and Safety Case in the Magistrates' Court	8.43
The initial appearance	8.45
Advance information in the magistrates' court	8.48
Friskies statements	8.54
The *Friskies* recommendations	8.57
Drawing up *Friskies* statements	8.59
Consequences of not providing a *Friskies* schedule	8.62
Relationship between advance information and the *Friskies* statements	8.66
Plea before venue	8.69
Effect of entering a guilty plea at the plea before venue stage	8.72
Benefits and drawbacks of the plea before venue procedure	8.75
Corporations entering pleas in the magistrates' court	8.78
Procedure for determining venue for trial or sentence	8.81
Deciding whether to commit for sentence after an indication of a guilty plea	8.82
Mode of trial guidelines	8.90
The *Howe* factors	8.91
Relevance of guilty plea	8.93
Previous convictions	8.97
Disputed basis of fact	8.98
Changing the decision to commit	8.99
Powers of the Crown Court when dealing with committals	8.100
Procedure for determining mode of trial	8.102
Previous convictions and committal	8.108
The individual's right to elect	8.111
Corporations and election	8.113
Orders on committal	8.115
Material to be sent to the Crown Court on committal	8.116
Judicial review of the decision to commit	8.117
C. Conducting a Health and Safety Case in the Crown Court	8.118
The Criminal Procedure Rules 2005	8.118
Lodging the indictment	8.121
The rule against duplicity	8.123
Effect of proceeding on a duplicitous charge or count	8.131
Selecting the appropriate charges	8.133
Plea negotiations: accepting lesser pleas	8.144

Progression of cases: case management	8.147
Plea and case management hearings (PCMH)	8.149
Complex cases	8.150
Plea bargaining: *R v Goodyear*	8.154
D. Disclosure	**8.157**
General legislative framework	8.159
Disclosure in the magistrates' court	8.162
Procedure for making disclosure	8.163
Defence disclosure	8.166
Third party disclosure	8.169
The investigation stage	8.170
Obtaining material from third parties	8.181
The disclosure officer	8.183
The prosecutor	8.184
Schedules of unused material	8.186
Effect of non-compliance with the Code	8.189
E. Expert Evidence	**8.190**
The 'ultimate issue rule'	8.193
Independence of an expert	8.195
The HSE Enforcement Guide and experts	8.196
The Criminal Procedure Rules	8.199
Rule 24 of the Criminal Procedure Rules	8.201
Calling the expert	8.205
Future development of disclosure duty	8.207
F. Miscellaneous Areas of Evidence	**8.208**
Hearsay in criminal cases	8.208
Bad character	8.214

9. Judicial Review and Abuse of Process

A. Judicial Review of the Decision to Prosecute or not to Prosecute	**9.01**
Judicial review of the decision to prosecute	9.03
Judicial review of the decision not to prosecute	9.09
B. Abuse of Process	**9.17**
The scope of the jurisdiction	9.17
Categories of abuse	9.22
Delay	9.23
Delay and Article 6 of the European Convention on Human Rights	9.28
Double jeopardy: two or more prosecutions on the same facts	9.34
Developments since *Beedie*	9.41
Breach of promise	9.46
Reneging on a plea bargain	9.49
Oppressive decision to prosecute contrary to policy	9.50

10. Sentencing

A. Sentence — 10.04
- The range of punishments — 10.04
- Compensation orders — 10.05
- Venue for sentence — 10.09
- Financial information to be provided by a defendant — 10.12
- Obtaining an indication of sentence: *R v Goodyear* — 10.15
- Determining the level of fine—established principles — 10.17
  - *R v Howe & Son (Engineers) Ltd* — 10.17
- Tariff: the issue of the consequence of any breach; and the importance of systematic fault — 10.29
  - Systemic fault — 10.36
- Commonly encountered situations — 10.41
  - Employees acting outside of their instructions — 10.41
  - Being 'let down' by others — 10.46
  - The relevance of a good safety record — 10.54
- Fines for breach of the general duties and regulations — 10.57
- Sentencing public bodies and the 'public element' — 10.59
- Sentencing a company which is part of a wider group — 10.62
- Examples of fines imposed by the Crown Court — 10.71
- Sentencing individuals — 10.87
  - Sentencing individuals for gross negligence manslaughter — 10.89
  - Disqualification from acting as the director of a company — 10.102
- Allowing defendants time to pay — 10.106
- Section 42 orders — 10.118
- Sentencing in the magistrates' court — 10.121

B. Costs — 10.130
- Prosecution costs — 10.131
- Defence costs — 10.137
- Examples of the approach taken — 10.148
- Appealing the decision for costs — 10.150
- Apportioning costs between a number of defendants — 10.152

11. Work-Related Deaths

A. The Investigation of Work-Related Deaths — 11.01
- The decision in *Beedie* — 11.05

B. The Work-Related Death Protocol — 11.08
- 'Work-related' deaths — 11.12
- Scope of the protocol — 11.15
  - Decision-making — 11.16
  - Advice prior to charge and the decision to prosecute — 11.19
  - Serious criminal offences: manslaughter — 11.20

C. Manslaughter	11.21
Individual manslaughter by gross negligence	11.25
Risk of death	11.29
Causation	11.31
The duty of care	11.32
R v Willoughby	11.35
R v Wacker	11.38
Health and safety duties and manslaughter	11.45
The *Caparo* test	11.57
Statutory duties and the duty of care	11.62
Company directors and statutory duties owed by the company as employer	11.63
Corporate manslaughter	11.66
The common law	11.67
The Corporate Manslaughter and Homicide Act 2007	11.69
Potential impact of the new corporate manslaughter offence	11.72
The provisions of the Act	11.75
'Relevant duty of care'	11.81
Senior management	11.85
Causation	11.86
Gross breach	11.87
D. Coroner's Inquests and Work-Related Deaths	11.89
Proposed reform	11.89
The provisions of the draft Bill	11.93
The current system	11.99
General principles	11.100
Jury inquests	11.103
The scope of an inquest	11.106
Matters to be ascertained at an inquest	11.117
Persons whose conduct may be called into question	11.122
Entitlement to examine witnesses	11.123
Witness statements and evidence at an inquest	11.125
The privilege against self-incrimination	11.127
Disclosure and documentary evidence	11.129
Coroner's disclosure and work-related deaths	11.133
HSWA 1974 s 20(2)(j) statements and inquests	11.140
Verdicts	11.142
Coroners Act 1988 s 16 and adjourned inquests	11.145

12. Principal Health and Safety Regulations

A. Health and Safety Regulations	12.01
Introduction	12.01
B. The Management of Health and Safety at Work Regulations 1999	12.09

Risk assessment	12.12
Particular requirements for risk assessments under other regulations	12.16
What does a risk assessment entail?	12.20
'Suitable and sufficient'	12.28
Relevance of the size of the undertaking in question	12.29
Model or generic risk assessments	12.30
Identifying who might be affected by the activities in question	12.31
Young people	12.34
Pregnant women	12.35
Principles of prevention	12.36
Continuing review and surveillance	12.37
Information and training	12.41
Procedures for evacuation in the event of imminent danger	12.45
The operation and effect of the Management of Health and Safety at Work Regulations 1999 reg 21	12.46
C. The Workplace (Health, Safety and Welfare) Regulations 1992	**12.47**
Introduction	12.47
Who owes the duty?	12.49
The extent of the duty	12.50
To whom is the duty owed?	12.52
What is a workplace?	12.53
The nature of the duty	12.57
D. Manual Handling Operations Regulations 1992	**12.60**
Introduction	12.60
Relationship between Manual Handling Operations Regulations 1992 and other provisions	12.62
The primary duties under Manual Handling Operations Regulations 1992	12.63
What is a manual handling operation?	12.70
E. Work Equipment: Provision and Use of Work Equipment Regulations 1998	**12.77**
Introduction	12.77
Suitability of work equipment	12.80
Risk assessment and the duty under Provision and Use of Work Equipment Regulations 1998 reg 4	12.84
Ergonomics	12.87
Maintenance of work equipment	12.88
Inspection of work equipment	12.97
Keeping records of inspections	12.100
Extent and frequency of the inspections	12.103
Competence of those who carry out inspections	12.106
Information and instructions	12.108

	Training	12.110
	Dangerous parts of machinery	12.112
	Dangerous part	12.115
	Protection against specified hazards	12.116
F.	Substances Hazardous to Health: the COSHH Regulations 2002	12.120
	The Control of Substances Hazardous to Health Regulations	12.120
	'Substance hazardous to health'	12.122
	Who owes duties under the Regulations?	12.123
	The duties	12.124
	Risk assessment	12.125
	The duty to prevent or control exposure	12.129
	Other duties	12.133
	Exemptions	12.138
G.	Construction	12.139
	Introduction	12.139
	Work at Height Regulations 2005	12.140
	Construction (Design and Management) Regulations 2007	12.146
	General provisions for health and safety	12.153
	Competence	12.154
	Cooperation and coordination	12.157
	Clients: changed and increased duties	12.158
	Notifiable projects	12.166
	The coordinator	12.169
	Designers	12.175
	The principal contractor	12.180
	Contractors	12.187
H.	Asbestos	12.192
	Introduction	12.192
	Duty-holders in premises	12.194
	Duties on employers	12.199
I.	Gas: Gas Safety (Installation and Use) Regulations 1998	12.211
	Gas Safety (Installation and Use) Regulations 1998	12.211
	Competent person	12.215
	Work	12.216
	Gas Safety (Installation and Use) Regulations 1998 reg 26	12.219
	Landlords	12.220
	Landlord's inspection certificate	12.223
J.	Electricity: Electricity at Work Regulations 1989	12.225
	Systems, work activities, and protective equipment	12.228
	'Construction'	12.230
	'Maintenance'	12.231

Protective equipment under the Electricity at Work Regulations 1989 reg 4(4)	12.233
Electricity at Work Regulations 1989 regs 5–12	12.234

Appendix A: Relevant extracts from the Health and Safety at Work etc. Act 1974	407
Appendix B: Enforcement Concordat; Enforcement Policy Statement (HSC15); Work Related Death Protocol	455
Appendix C: The Health and Safety (Enforcing Authority) Regulations 1998 (as amended)	475
Appendix D: The Employment Tribunals (Constitution and Rules of Procedure) Regulations 2004	483
Appendix E: Health and Safety: The Management of Health and Safety at Work Regulations 1999 (as amended)	525
Appendix F: Health and Safety: The Workplace (Health, Safety and Welfare) Regulations 1992 (as amended)	537
Appendix G: Health and Safety: The Manual Handling Operations Regulations 1992 (as amended)	547
Appendix H: Health and Safety: The Provision and Use of Work Equipment Regulations 1998 (as amended)	551
Appendix I: Health and Safety: The Control of Substances Hazardous to Health Regulations 2002 (as amended)	567
Appendix J: Health and Safety: The Construction (Design and Management) Regulations 2007	583
Appendix K: Health and Safety: The Work at Height Regulations 2005 (as amended)	605
Appendix L: Health and Safety: The Control of Asbestos Regulations 2006	619
Appendix M: Health and Safety: The Gas Safety (Installation and Use) Regulations 1998 (as amended)	639
Appendix N: Health and Safety: The Electricity at Work Regulations 1989 (as amended)	659
Index	667

TABLE OF CASES

Adams v Cape Industries Plc [1990] Ch 433 (CA)................................. 10.69
Albazero, The [1977] AC 774... 10.69
Amos v Worcester City Council (15 July 1996; CO 3693/95) (DC).................. 6.214
Andrews v DPP [1937] AC 576 (HL)... 11.28
Associated Dairies v Hartley [1979] IRLR 171................................... 4.114
Associated Provincial Picture Houses Ltd v Wednesbury Corporation [1948]
    1 KB 223... 9.12
Attorney-General's Reference (No 1 of 1990) (1992) 95 Cr App R 296; [1992]
    QB 630... 9.24, 9.31
Attorney-General's Reference (No 1 of 1995) [1996] 2 Cr App R 320 (CA).......... 5.56
Attorney-General's Reference (No 2 of 1999) [2000] 2 Cr App R 207.............. 11.67
Attorney-General's Reference (No 7 of 2000) [2001] 2 Cr App R 19............... 3.89
Attorney-General's Reference (No 2 of 2001) [2001] 1 WLR 1869 (CA).............. 9.30
Attorney-General's Reference (No 134 of 2004) [2004] EWCA Crim 3286............ 10.92
Austin Rover Group Ltd v HM Inspector of Factories [1990]
    1 AC 619 (HL)....................... 6.128, 6.130, 6.155, 6.159, 6.161, 6.163, 6.164

Bailey v Command Security Services Ltd [2001] All ER 352 (QB)............. 12.20, 12.50
Barnfather v London Borough of Islington [2003] EWHC 418 (Admin)........... 7.16, 7.17
Belhaven Brewery Co v McLean [1975] IRLR 370................................... 4.115
Beresford, Re (1952) 36 Cr App R 1....................................... 11.149, 11.150
Bernard v Dudley MBC and Dudley Magistrates' Court [2003]
    EWHC 147 (Admin)... 8.62, 8.67
Bilton v Fastnet Highlands Ltd 1998 SLT 1323................................... 6.125
Briggs Amasco Ltd v Thurgood *The Times* 30 July 1984 (DC)............ 6.194, 6.207, 7.37
British Airways Board v Henderson [1979] IC 77 (DC)............................ 4.106
Brook Street Bureau (UK) Ltd v Dacas [2004] EWCA Civ 217;
    [2004] ICR 1437 (CA)... 6.52
Brown v Stott [2003] 1 AC 681.. 3.89
BT Fleet Ltd v McKenna [2005] EWCA 387 (Admin)........................... 4.57, 4.108
Bunn v British Broadcasting Corporation [1998] 3 All ER 552................... 11.138

Cable & Wireless plc v Muscat [2006] EWCA Civ 220 (CA).......................... 6.52
Cadger v Vauxhall Motors, 28 March 2000 (Cty Ct)............................... 12.90
Campbell v Wallsend Slipway and Engineering Co Ltd [1978] ICR 1015.............. 8.08
Campion v Hughes (HM Inspector) [1975] IRLR 291................................ 4.100
Cantabrica Coach Holdings Limited v Vehicle Inspectorate [2001]
    UKHL 60 (HL)... 2.153
Canterbury City Council v Howletts & Port Lympne Estates Ltd [1997]
    ICR 925.. 4.117
Caparo Industries plc v Dickman [1990] 1 All ER 568...................... 11.57, 11.59
Carmichael v National Power plc [1999] ICR 1226 (HL)........................... 6.47
Carrington Carr Ltd v Leicestershire County Council [1993]
    Crim LR 938... 8.125

Chrysler (UK) Ltd v McCarthy [1978] ICR 939 . . . . 4.83, 4.103
City Equitable Fire Insurance Co Ltd, Re [1925] Ch 407 (CA) 5.65, 5.66, 11.64
City of Oxford Tramway Co v Sankey (1890) 54 JP 564 . . . . 8.41
Clode v Barnes [1974] 1 All ER 1176 (DC); [1974] Crim LR 268. . . . . 5.14
Coates v Jaguar Cars [2004] EWCA Civ 337. . . . . 12.58
Cockle pickers case *see* R v Lin Liang Reng and Others
Connelly v DPP [1964] AC 1254 (HL) 9.17, 9.34, 9.36, 9.38,
9.39, 9.41, 9.43
Consolidated Criminal Practice Direction [2002]
    2 Cr App R 35 . . . . 8.90, 8.115,
8.136, 8.137, 8.141, 8.226
    Pt V
        para 51 . . . . 8.106
        para 51.3 . . . . 8.107
        para 51.3(d) 8.106
Coult v Szuba [1982] ICR 380 (DC) 6.23, 6.212
Coupe v Guyett [1973] 2 All ER 1058 (DC). 5.34, 5.35
Cullen v Jardine [1985] Crim LR 668 . . . . 8.125

Davies (David Janway) v Health and Safety Executive [2002]
    EWCA Crim 2949 . . . . 5.59, 6.138
Deary v Mansion Hide Upholstery [1983] ICR 610 . . . . 4.14, 7.55
Deeley v Effer, COIT 1/72, Case/25354/77 . . . . 4.116
Den Norske Bank ASA v Antonatos [1999] QB 271 (CA (Civ Div)) 3.89
Direct Holidays Plc v Wirral Metropolitan Borough Council,
    28 April 1998 (DC) 3.22, 3.25
Donaldson v Hays Distribution Services Ltd 2005 SLT 733 . . . . 12.52
Douglas v Phoenix Motors [1970] SLT 57 . . . . 5.15
DPP v Humphrys (1976) 63 Cr App R 95 . . . . 9.42
DPP v Merriman [1973] AC 584 . . . . 8.125
DPP v Short (2002) 166 JP 474 . . . . 8.41
Dugmore v Swansea NHS Trust [2003]
    1 All ER 333 (CA) 6.117, 6.118, 6.123,
6.125, 6.128, 6.134, 7.18, 7.19, 7.25, 7.26, 12.129
Dyer v Watson [2002] UKPC D1 . . . . 9.32

Environmental Agency v Medley & Medley, September 1999
    (Leeds Crown Court) 9.45
Environmental Agency v Stanford [1998]
    DC (CO/4625/97) (DC) 9.20, 9.55
Express Ltd v The Environment Agency [2004] EWHC (Admin)
    [2005] JPL 242. . . . . 5.44

Field v Leeds City Council [2000] 17 EG 165; [2000] 1 EGLR 54;
    32 HLR 618 (CA, Civ). 8.195, 8.197
Foremans Relocatable Building Systems v Fuller, ET-3200213/99/S 4.91, 4.96
Franks v Reuters Ltd [2003] EWCA Civ 417; [2003] ICR 1166 (CA) 6.52
Frankson v Home Office; Johns v Home Office [2003] EWCA Civ 655 . . . . 11.137

Galashiels Gas Co Ltd v O'Donnell [1949] AC 275 (HL) 12.58
Garfield v Maddocks [1974] QB 7 . . . . 8.40
Garrard v AE Southey & Co and Standard Telephones and Cables Ltd
    [1952] 2 QB 174 . . . . 6.52
Gateshead Justices *ex p* Tesco Stores Ltd [1981] QB 470 . . . . 8.27

Table of Cases

Gateway Foodmarkets [1997] 3 All ER 78; [1997]
2 Cr App R 40 .... 6.84, 6.86, 6.90, 6.113
Gipps v Gipps and Hume (1861) 11 HL Cas 1 .... 5.62
Gissing v Walkers Smith Snack Foods Ltd [1999] CLY 3983 .... 12.75
Grovehurst Energy Ltd v Strawson, COIT 5035/90, reported in Tolleys
(HSIB 180, P 9) 4.45, 4.99
Guildford Justices *ex p* Healy [1983] 1 WLR 108 .... 9.19

Hampstead Heath Swimming Club v Corporation of London [2005]
EWHC 713 (Admin) 6.130
Harris v Evans & Health & Safety Executive [1998]
1 WLR 1285 .... 4.51
Harrison (1993) 14 Cr App R (S) 419 .... 10.154
Hellewell v Chief Constable of Derbyshire [1995] 4 All ER 473 .... 11.138
Hewlett Packard v O'Murphy [2002] IRLR 4 (EAT) 6.52
Highgrade Traders, Re [1984] BCLC 151 (CA) 3.82, 3.83
Hills v Ellis [1983] 2 WLR 234 .... 7.58
Hinchcliffe v Sheldon [1955] 1 WLR 1207 .... 7.57
Hirschler v Birch [1987] RTR 13 .... 5.66
Hoggetts v Chiltern District Council [1983] 2 AC 120 .... 8.125
HSE v George Tancocks Garage (Exeter) Ltd [1993]
Crim LR 605 .... 2.30, 4.47
HSE v Spindle Select Ltd *Times Law Reports* 9 December 1996 .... 8.129, 8.130
Huckerby v Elliot [1970] 1 All ER 189 (DC) 5.60, 5.64, 5.66, 11.64
Hughes v the Architects Registration Council of the United Kingdom
(1957) 1 QB 550 .... 4.93
Hui Chi-Ming v R [1992] 1 AC 34 (PC) 5.54, 9.17

Ikarian Reefer, The [1993] 2 Lloyd's Rep 455 .... 8.195
Interlink Express Parcels Ltd v Night Trunkers Ltd [2001]
IRLR 224 (CA) 6.50

J Armour v J Skeen (Procurator Fiscal, Glasgow) [1977] IRLR 310
(High Ct of Judiciary) 5.69
J.B. v Switzerland [2001] Crim. LR 748 .... 3.89
Jemmison v Priddle [1972] 1 QB 489; 56 Cr App R 229 .... 8.124
John v Matthews [1970] 2 QB 443 .... 5.39
Jordan, Re [1904] 1 Ch 260 .... 5.17
Joy v Federation against Copyright Theft Ltd [1993]
Crim LR 588 (DC) 2.74

King v RCO Support Services Ltd [2001] ICR 608 (CA) 12.51, 12.75
Kitching v Gateway Foodmarkets, 24 October 1988,
CO/171/88 (DC) 4.83, 4.104
Koeppler's Will Trusts, Re [1985] 2 All ER 869 .... 5.21
Koonjul v Thameslink Healthcare Services Ltd [2000] PIQR 123 (CA);
[2000] CLY 2983 .... 12.65, 12.72

Lamb v Sunderland and District Creamery Ltd [1951] 1 All ER 923 (DC) 5.41
Lane v Shire Roofing Co [1995] IRLR 493 (CA) 6.51
Larner v British Steel plc [1993] 4 All ER 102 .... 6.122, 6.123, 6.128
Ledger v DPP [1991] Crim LR 439 .... 7.57
Lee Ting Sang v Chung Chi-Keung [1990] 2 AC 374 .... 6.46
Lewis v Avidan Ltd [2005] EWCA Civ 670 .... 12.54, 12.58

Lewis v Cox [1984] 3 WLR 875 .... 7.58
Litster v Forth Dry Dock and Engineering Co Ltd [1989] ICR 341.................... 7.14
Liverpool Roman Catholic Archdiocese Trustees Incorporated v Goldberg (No 2)
    [2001] 4 All ER 950.. 8.195
London Borough of Wandsworth v South Western Magistrates' Court [2003]
    EWHC 1158 (Admin) (DC) 2.93, 2.110, 2.116, 2.139,
                                                                                     2.140, 2.152, 2.161, 7.49
Lonrho Ltd v Shell Petroleum Co Ltd (No 2) [1982] AC 173 .... 11.46

McFarlane v Tayside Health Board [2000] 2 AC 59................................ 11.58
McKay v Unwin Pyrotechnics Ltd [1991] Crim LR 547 .... 6.178
McLaughlin v East and Midlothian NHS Trust, 2000 Rep LR 87
    (OH (Scot))... 12.58
Mains v Uniroyal Englebert Tyres Ltd [1995] SC 518
    (Scot. Ct of Session) 6.122, 6.123, 6.124
Malcolm v Metropolitan Police Commissioner [1999] CLY 1494.................... 12.58
Mallon v Norman Hill Plant Hire Ltd [1993] Crim LR 605 .... 2.30, 4.47
Manifest Shipping v Uni-Polaris Insurance Co Ltd 2003 1AC 469 .... 5.72
Manning v Manning [1950] 1 All ER 602 (CA) 5.62
Marcel v Metropolitan Police Commissioner [1992] 1 All ER 72.................... 11.138
Mark and Nationwide Heating Services [2004] EWCA Crim 2490................... 11.37
Miller Mead v Minister of Housing and Local Government [1963]
    2 QB 196... 4.109
Montgomery v Johnson Underwood Ltd [2001] EWCA Civ 318;
    [2001] IRLR 269 .... 6.52
Mooney v Cardiff Magistrates' Court (1999) 164 JP 220 (DC) 10.141
Motorola Ltd v Davidson [2001] IRLR 4 (EAT)................................... 6.52
Munroe v Crown Prosecution Service [1998] Crim LR 823......................... 8.101

Neath RDC v Williams [1951] 1 KB 115 .... 5.44
Neil v Greater Glasgow Health Authority (OH) 1996 SLT 1260..................... 6.124
New Southgate Metals v London Borough of Islington
    [1996] CLR 334... 8.41
Nico Manufacturing Co Ltd v B Hendry [1975] IRLR 225......................... 4.99
Noss Farm Products Ltd v Lilico [1945] 2 All ER 609 (DC) 5.42

Ofterburn Mill Ltd v JW Bullman [1975] IRLR 223............................... 4.100
O'Hara v Chief Constable of the Royal Ulster Constabulary [1997]
    AC 286 (HL).. 2.64
Olgeirsson v Kitching [1986] 1 All ER 746 (DC) 5.39
Olliver [1989] 11 Cr App R (S) 10 .... 10.63
O'Neil v DSG Retail Ltd [2002] EWCA Civ 1139 .... 12.71
Osborne v Bill Taylor of Huyton Ltd [1982] ICR 168............................... 6.16

Paris v Stepney Borough Council [1951] AC 367 .... 6.75
Parsons v Barnes [1973] Crim LR 537.. 5.14
Peach v Metropolitan Police Commissioner (1986) 2 All ER 129.................... 11.126
Pickstone v Freemans plc [1988] ICR 697... 7.14
Post Office v Footitt [2000] IRLR 243 (DC)....................................... 4.107
Postle v Norfolk and Norwich NHS Healthcare Trust [2000] CLY 2970 .... 12.73
Practice Direction [2004] 2 All ER 1070.. 10.130
    Pt II.1.1 .... 10.141
    Pt II.2.2 .... 10.143
    Pt V, para 1.1(b) 10.142

Table of Cases

Pt V1, paras 1.4–1.5 .... 10.133

R v Aceblade Ltd [2001] 1 Cr App R (S) 366 .... 10.115
R v Adaway [2004] EWCA Crim 2831; [2004] 168 JP 645 .... 9.50, 9.51, 9.53
R v Adomako [1995] 1 AC 171 .... 11.26, 11.27, 11.28
R v Argent [1997] 2 Cr App R 27 (CA) 3.17, 3.25
R v Associated Octel Co Ltd [1994]
 4 All ER 1051 (CA) 6.33, 6.64, 6.69, 6.77, 6.82
R v Associated Octel Co Ltd [1996] 4 All ER 846–8 (HL);
 [1996] 1 WLR 1543 (HL) 6.12, 6.25, 6.33, 6.34, 6.35, 6.36,
           6.79, 6.84, 6.88, 6.93, 6.96, 6.98, 7.15
R v Associated Octel Ltd (2) [1997] 1 Cr App R (S) 435 (CA) 10.134
R v Avon Lippiatt Hobbs (Contractors) Ltd [2003]
 2 Cr App R (S) 427 .... 10.42
R v B & Q Plc [2005] EWCA Crim 2297 (CA) 10.116, 10.117
R v Bailey (1961) 66 Cr App R 31 .... 8.191
R v Balfour Beatty Civil Engineering Ltd & Geoconsult GES, 15 February 1999
 (Central Crim. Ct) 10.77
R v Balfour Beatty Infrastructure Services Ltd and Network Rail, 7 October 2005
 (Central Crim. Ct) 10.73
R v Balfour Beatty Rail Infrastructure Services Ltd [2006]
 EWCA Crim 1586 .... 10.35, 10.38, 10.39
R v Barber [2002] 1 Cr App R (S) 548 (CA) 8.74, 8.93
R v Bateman (1925) 19 Cr App R 8 (CCA) 11.28
R v Beckford [1996] 1 Cr App R 94 .... 9.18
R v Beedie [1997] 2 Cr App R 167 (CA) 9.16, 9.35, 9.36, 9.37, 9.38, 9.39,
          9.40, 9.41, 9.48, 11.05, 11.06, 11.149
R v Belmarsh Magistrates' Court *ex p* Watts [1999] 2 Cr App R 188 (DC) 9.21
R v Bennett [1994] 1 AC 42 (HL) 127 .... 9.19, 9.20, 9.57
R v Blackburn JJ *ex p* Holmes [2000] CLR 300 .... 8.16
R v Boal (1992) 95 Cr App R 272 (CA) 3.66, 5.49, 5.51
R v Board of Trustees of the Science Museum [1993]
 3 All ER 853 (CA) 5.17, 6.06, 6.62, 6.63, 6.69, 6.160
R v Board of Trustees of the Science Museum [1994]
 IRLR 25 .... 6.78
R v Brentford Justices *ex p* Wong [1981] 1 All ER 884 .... 8.16
R v Brintons Ltd, 22 June 1999 (CA) 10.57, 10.65
R v British Steel Plc [1998] 4 All ER 331; [1995] 1 WLR 1356 (CA); [1995]
 ICR 586; [1995] IRLR 310 .... 4.111, 6.12, 6.25, 6.68, 6.69, 6.70, 6.80,
          6.86, 6.93, 6.96, 6.98, 6.110
R v Brown (K) (1984) 79 Cr App R 115 .... 8.130, 8.140
R v Cappagh Public Works Ltd [1999] 2 Cr App R (S) 301 .... 10.56
R v Cardiff City Transport Services [2001] 1 Cr App R (S) 41 .... 10.54
R v Cato (1975) 62 Cr App R 41 .... 11.31
R v Central Criminal Court (Respondent) *ex p* Francis & Francis (Appellants)
 (1989) 88 Cr App R 213 .... 3.75
R v Ceri Davies [1999] 2 Cr App R (S) 356 .... 10.49
R v Chief Constable of the Royal Ulster Constabulary *ex p* Begley [1997]
 1 WLR 1475 (HL) 3.37
R v Colchester Justices *ex p* North East Essex Building Co Ltd [1977]
 1 WLR 1109 .... 8.108
R v Colthrop Board Mills Ltd [2002] 2 Cr App R (S) 80 .... 10.31, 10.32
R v Condron & Condron [1997] 1 Cr App R 185 .... 3.18
R v Connolly and Kennett [2007] EWCA Crim 270 .... 10.98

R v Conocophillips, 29 June 2005 (Grimsby Crown Ct) 10.75
R v Corus UK Ltd, 12 November 2000 (Cardiff Crown Ct) 10.82
R v Croydon JJ *ex p* Dean [1993] QBD 769 . . . . 9.07, 9.47, 9.48
R v Croydon Justices *ex p* W H Smith Ltd [2001]
    EHLR 12; (2000) 97 (46) LSG 39 . . . . 8.10
R v Curley (1909) 2 Cr App R 96 (CCA) 109 . . . . 11.31
R v Davies (David Janway) [2002] IRLR 170 (CA) 6.107, 7.16
R v Davison (1973) 57 Cr App R 113 (CA). 7.87
R v Derby Crown Court *ex p* Brooks [1985]
    80 Cr App R. 164 (DC) 9.20
R v Dickson and Wright 94 Cr App R 7 . . . . 8.22
R v Dover Magistrates' Court *ex p* Pamment [1994] Crim. LR 471 . . . . 8.99
R v DPP *ex p* C [1995] 1 Cr App R 136. . . . . 9.12
R v DPP *ex p* Kebilene [2000] 2 AC 326 (HL). 9.04, 9.21
R v DPP *ex p* Lee [1999] 2 Cr App R 304 . . . . 8.162
R v DPP and others *ex p* Jones [2000] IRLR 373 . . . . 9.13
R v Dudley Magistrates' Court *ex p* Power City Stores (1990) 154 JP 654 . . . . 10.148
R v Dunn 91 Cr App R 237 (CA) 2.84
R v F Howe & Son (Engineers) Ltd [1999] 2 All ER 249 (CA); [1999]
    2 Cr App R (S) 37 (CA) 3.69, 8.20, 8.56, 8.57, 8.59, 8.91, 8.106, 8.142,
                       10.03, 10.12, 10.17, 10.25, 10.27, 10.28, 10.29, 10.33, 10.36,
                       10.54, 10.60, 10.90, 10.122, 10.123, 10.124, 10.126, 10.132, 463
R v Faithful [1950] 2 All ER 1. . . . . 8.101
R v Flax Bourton Magistrates *ex p* Customs and Excise [1996]
    Crim. LR 907. . . . . 8.84
R v Fresha Bakeries Ltd [2003] 1 Cr App R (S) 202. . . . . 10.152
R v Friskies Petcare (UK) Ltd [2002]
    2 Cr App R 401 (CA) 3.20, 8.43, 8.52, 8.54, 8.55, 8.56, 8.59, 8.61,
                     8.62, 8.64, 8.65, 8.66, 8.67, 8.68, 8.108, 10.30, 10.32
R v Gateway Foodmarkets Ltd [1997] 3 All ER 78 (CA); [1997]
    2 Cr App R 40 . . . . 6.08, 6.12, 6.25, 6.26, 6.80, 6.81, 6.94,
                       6.99, 6.103, 6.106, 6.111, 7.15
R v Goodman [1993] 2 All ER 789. . . . . 10.103
R v Great Western Trains Company Limited, 27 July 1999
    (Central Crim. Ct) 10.76
R v Greenfield 57 Cr App R 279 (CA). 8.126
R v Guildhall Magistrates' Court *ex p* Primlaks Holdings Co
    (Panama) Inc (1989) 89 Cr App R 215 . . . . 3.75
R v Hall and Co Ltd [1999] 1 Cr App R (S) 306 . . . . 10.55
R v Hanson; R v Gilmore; R v P [2005] 2 Cr App R 21 (CA). 8.221
R v Hertfordshire County Council *ex p* Green Environmental
    Industries Ltd [2000] 2 AC 412 (HL); [2000] 2 WLR 412 (CA). 2.142, 2.143,
                       2.148, 3.87, 7.59
R v HM Coroner for North Humberside and Scunthorpe
    ex p Jamieson (1994) 3 All ER 972 . . . . 11.106, 11.112, 11.120, 11.125
R v Horseferry Road Magistrates' Court *ex p* Bennett [1994] 1 AC 42 (HL). 9.20
R v Horseferry Road Magistrates' Court *ex p* DPP [1999] COD 441 (DC) 9.21
R v Horsham Justices *ex p* Reeves (Note) [1980] 75 Cr App R 236. . . . . 9.19
R v HTM Ltd [2006] EWCA Crim 1156 (CA). 5.36, 6.09, 6.12, 6.25,
               6.67, 6.100, 6.101, 6.105, 6.106, 6.107, 6.113, 6.117, 6.127, 6.130,
                   6.132, 6.164, 7.18, 7.19, 7.21, 7.25, 7.26, 7.30, 7.33,
                   8.143, 8.151, 12.28, 12.46
R v Hughes [1988] Crim LR 519 (CA) 2.98
R v Hundal and Dhaliwal [2004] 2 Cr App R 19 (CA) 3.89

Table of Cases

R v Isleworth Crown Court *ex p* Buda [2000] 1 Cr App R (S) 538 .... 8.101
R v Jarvis Facilities Ltd [2006] Cr App R (S) 44 .... 10.34, 10.35, 10.61
R v Jarvis Fastline Ltd, 25 August 2000 (York Crown Ct) 10.83
R v Jefferson 99 Cr App R 13 (CA) 5.27
R v JF Alford Transport (1997) 141 Sol Jo LB 73 (CA) 5.14
R v Jisl and Tekin *Times Law Reports* 19 April 2004 (CA) 8.153
R v John Doyle Construction Ltd and Exterior International Plc,
    9 May 2006 (Central Crim. Ct) 10.85
R v Jones (1870) 11 Cox 544 .... 11.31
R v Karl Goodyear [2005] 3 All ER 117 (CA); [2006]
    Cr App R (S) 6 (CA) 8.154, 10.15, 10.16
R v Kearns [2003] 1 Cr App R 7 .... 3.87, 3.88
R v Kebeline [2000] 2 AC 326 .... 6.141
R v Keltbray Ltd [2001] Cr App R (S) 39 .... 8.142, 10.58, 10.66
R v Kew and Jackson (1872) 12 Cox 355 .... 11.31
R v Khan and Khan [1998] Crim LR 830 (CA) 11.34, 11.36
R v Kite [1996] 2 Cr App R (S) 295 .... 10.93
R v Klargester Environmental, 22 March 2002 (Aylesbury Crown Ct) 10.84
R v Knight [2004] 1 Cr App R 9 (CA) 3.27
R v Kvaerner Cleveland Bridge Ltd and Costain Ltd, 30 November 2001
    (Bristol Crown Ct) 10.80
R v Leeds Stipendiary Magistrate *ex p* Yorkshire Water Services
    (No 2) (CO/4834/99) 9.20
R v Leighton and Town and Country Refuse Collections,
    (CA 26 May 1995) 5.53
R v Lightwater Valley Ltd (1990) 12 Cr App R (S) 328 (CA) 10.03
R v Lin Liang Reng and Others 10.101
R v Litchfield, CA Ref No. 97 05973 W2, 17th December 1997 .... 10.89, 10.91, 10.99
R v London Borough of Hammersmith and Fulham,
    7 November 2001 (Blackfriars Crown Ct) 10.81
R v Looseley; Attorney-General's Reference (No 3 of 2000) [2002]
    1 Cr App R 29 (HL) 2.99
R v Mara [1987] 1 All ER 478 .... 6.40
R v Mersey Docks and Harbour Co (1995) 16 Cr App R (S) 806 .... 10.47
R v Milford Haven Port Authority [2000] 2 Cr App R (S) 423 (CA) 10.59, 10.61
R v Miller [1998] Crim LR 209 (CA) 2.84
R v Misra and Srivastava [2005] Crim LR 234 .... 11.26, 11.30
R v Mullen [1999] 2 Cr App R 143 (CA) 9.20
R v Nelson Group Services (Maintenance) Ltd [1998]
    4 All ER 331 (CA) 6.91, 6.92, 6.93, 6.99, 6.100, 6.102, 6.106, 6.108,
                                      6.110, 6.111, 6.112, 6.113, 12.46
R v Network Rail and others, Transcript p 93, paras 11–12 .... 5.71
R v Newton (1983) 77 Cr App R 13 .... 8.98
R v North East Suffolk Magistrates' Court *ex p* DPP [1998]
    2 Cr App R 307 .... 8.88, 8.97
R v North Essex Justices *ex p* Lloyd [2001] 2 Cr App R (S) 86 .... 8.84
R v North Humberside and Scunthorpe Coroner *ex p* Jamieson [1994]
    3 All ER 972 .... 11.112
R v North Sefton Magistrates' Court *ex p* Marsh [1994] Crim LR 865 .... 8.99
R v Northallerton Magistrates' Court *ex p* Dove [2001] 1 Cr App R (S) 137 (CA) 10.136
R v Nottingham Magistrates' Court *ex p* Davidson [2000] 1 Cr App R (S) 167 .... 8.99
R v O'Meara *Times Law Reports* 15 December 1989 (CA) 8.137
R v Otis, 7 April 2006 (Southampton Crown Ct) 10.86
R v Oxford City Justices *ex p* Smith [1982] 75 Cr App R 200 .... 9.19

R v P and another [2007] All ER (D) 173 (Jul) (CA) 5.70
R v Patchett [2001] 1 Cr App R (S) 138 .... 10.43
R v Pennine Acute Hospital NHS Trust etc (2003) 147 SJ 1426 .... 8.150, 8.151
R v Perry Wacker [2003] Cr App R (S) 92 .... 10.90, 10.100
R v Piggot and Litwin [1999] 2 Cr App R 320 (CA) 9.41
R v Pitts (1842) C & Mar 248 .... 11.31
R v Police Commissioner of the Metropolis *ex p* Blackburn [1968] 2 QB 118 (CA) 2.39
R v Port of Ramsgate and others, 28 February 1997 (Central Crim. Ct) 10.78
R v Puddick (1865) 4 F & F 497 .... 8.145
R v Rafferty [1999] 1 Cr App R 235 .... 8.74
R v Railtrack Plc & others, 1 September 2004, (Central Crim. Ct) 11.57, 11.60
R v Rhone-Poulenc Rorer Ltd [1996] ICR 1054 (CA) 6.194, 6.207, 7.37
R v Riebold [1967] 1 WLR 674 .... 9.41
R v Rimac Ltd [2000] 1 Cr App R (S) 468 .... 10.44
R v Roble [1997] Crim LR 449 (CA) 3.11
R v Rollco Screw and Rivet Co Ltd [1999] 2 Cr App R (S) 436 .... 10.88, 10.106,
                                                                           10.115, 10.117
R v S *Times Law Reports* 29 March 2006 .... 9.26
R v Sanyo Electrical Manufacturing (UK) Ltd (1992) 13 Cr App R (S) 657 .... 10.45
R v Secretary of State for Trade and Industry *ex p* UNISON, GMB and
    NASUWT [1996] ICR 1003 .... 1.09
R v Seelig and Lord Spens 94 Cr App R 17 (CA) 2.75
R v Shaw [2006] EWCA Crim 2570 .... 10.95
R v Sheffield Crown Court and Sheffield Stipendiary Magistrate *ex p* DPP
    [1994] Crim LR 470 .... 8.99
R v Sinclair (1998) 148 NLJ 1353 (CA) 11.34, 11.36
R v Singh (Gurphal) [1999] Crim LR 582 (CA): transcript (9803983 Z4),
    19 February 1999 .... 11.29, 11.30, 11.34, 11.50
R v Smith (WD) 99 Cr App R 223 (CA) 2.75
R v South West Magistrates' Court *ex p* James [2000] Crim LR 690 .... 10.138
R v Southampton Magistrates' Court *ex p* Sansome [1999] 1 Cr App R (S) 112 .... 8.109
R v Sparks [1991] Crim LR 128 (CA) 2.84
R v Staines Magistrates' Court *ex p* DPP [1998] 2 Cr App R 307 .... 8.88, 8.97
R v Staton [1983] Crim LR 190 .... 8.137
R v Stockwell (1993) 97 Cr App R 264 .... 8.190, 8.193, 8.194
R v Supremeplan Ltd [2001] 1 Cr App R (S) 244 .... 10.27
R v Surrey Coroner *ex p* Campbell [1982] QB 661 .... 11.119
R v Thamesway Homes Ltd, 24 August 2001 (Central Crim. Ct) 10.79
R v Thomas [1995] CLR 938 .... 9.49
R v Townsend, Dearsley and Bretscher [1997] 2 Cr App R 540 .... 9.46
R v Transco Plc [2006] EWCA Crim 838 (CA) 8.20, 10.13, 10.16, 10.25,
                                                                     10.36, 10.37, 10.72
R v Tropical Express Limited [2001] 1 Cr App R (S) 115 .... 10.28
R v Turner [1970] 2 QB 321 .... 8.154
R v Turner [1975] QBD 841 .... 8.190
R v Turner [1995] 2 Cr App R 94 (CA) 8.205
R v Vivian (1979) 68 Cr App R 53 .... 10.08
R v Wacker [2003] 1 Cr App R 22 (CA) 11.37, 11.38, 11.41, 11.43, 11.44
R v Ward, 98 Cr App R 337 (CA) 2.84
R v Warley Magistrates' Court *ex p* DPP [1998] 2 Cr App R 307;
    [1999] 1 WLR 216 .... 8.88, 8.97, 8.117
R v Willoughby [2005] 1 Cr App R 29 .... 11.35, 11.37, 11.44
R v Wilson (1979) 69 Cr App R 83 .... 8.124, 8.127
R v Woolmington [1935] AC 462 .... 10.138

Table of Cases

R v Z [2000] 2 Cr App R 281 (HL)... 9.42
R (on the application of Brenda Rowley) v DPP [2003] EWHC 693 (Admin)............ 9.16
R (on the application of Factortame & Others) v The Secretary of State for
    Transport [2002] EWCA Civ 932.. 8.195
R (on the application of Junttan Oy) v Bristol Magistrates' Court [2004]
    2 All ER 555 (HL)... 6.167
R (on the application of Middleton) v West Somerset Coroner [2004] UKHL 10;
    [2004] 2 All ER 465; [2004] 2 AC 182 .... 2.35, 11.108, 11.109, 11.110,
                                                                    11.113, 11.116, 11.121
R (on the application of Pullen) v Health & Safety Executive [2003]
    EWHC 2934 (Admin)... 2.45, 9.11
R (on the application of Stanley) v Inner North London Coroner [2003]
    EWHC 1180 (Admin).. 11.151
R (Hale) v Southport Justices *Times Law Reports* 29 January 2002.................. 10.149
R (Kent Pharmaceuticals Ltd) v Director of Serious Fraud Office [2005]
    1 WLR 1302 (CA, Civ Div).. 3.89
R (Social Security Secretary) v South Central Division Magistrates *Daily Telegraph*
    28 November 2000... 2.76
Railtrack Plc v Smallwood [2001] ICR 714.......... 4.44, 4.55, 4.83, 4.88, 4.92, 4.105, 4.106
Ralux NV/SA v Spencer Mason *The Times* 18 May 1989................................ 8.33
Readmans and Another v Leeds City Council [1992] COD 419 .... 4.82, 4.84, 4.96
Ready Mixed Concrete Ltd v Minister of Pensions and National
    Insurance [1968] 2 QB 497 .... 6.49
Redbridge Justices *ex p* Sainty [1981] RTR 13...................................... 8.42
Registrar of Restrictive Trading Agreements v WH Smith & Son Ltd [1969]
    3 All ER 1065... 5.50
Rhondda Waste Disposal Ltd, Re [2001] Ch 57 (CA) 8.23
Ricketts v Torbay Council [2003] EWCA Civ 613 (CA) 12.52
Ronson and Parnes (1992) 13 Cr App R (S) 153...................................... 10.154
Rubin v DPP [1990] 2 QB 80... 8.09

Saunders v Edwards [1987] 1 WLR 1116 .... 11.40
Shropshire County Council v Simon Dudley Ltd [1996] *Trading Law Reports* 69......... 9.51
Skinner v HM Advocate [1994] SCCR 316 .... 6.213
Skinner v John McGregor (Contractors) Ltd 1977 SLT (Sh Ct) 83 .... 2.96
Smith v Director of Public Prosecutions and Morris [2000] RTR 36 .... 11.150
South Surbiton Cooperative Society Ltd v Wilcox [1975] IRLR 292................... 4.98
Southwark Water Co v Quick (1878) 3 QBD 315....................................... 3.84
Stark v Post Office [2000] ICR 1013... 12.90
Stephenson v Delphi Diesel Systems Ltd [2003] ICR 471............................. 6.52
Surujpaul v R [1958] 3 All ER 300; 42 Cr App R 266 (PC)........................... 3.22
Swain v Denso Martin Ltd [2000] ICR 1079 (CA) 12.68, 12.74

Tarleton Engineering Co Ltd v Nattrass [1973] All ER 669 (DC) 5.37
Tate Access Floors Inc v Boswell [1990] 3 All ER 303 .... 2.147
Taylor v SFO [1998] 4 All ER 801 .... 11.138
TC Harrison (Newcastle-under-Lyme) Ltd v Ramsay [1976] IRLR 135 .... 4.99
Tesco v Kippax (COIT No 7605, HSIB 180, P 8)...................................... 4.35
Tesco Stores v Edwards [1977] IRLR 120 .... 4.113
Tesco Supermarkets Ltd v Nattrass [1972] AC 153; [1971]
    2 All ER 127... 5.34, 5.43, 5.69, 6.70
Tesco UK Ltd v The London Borough of Harrow 4.101

Walkers Snack Foods Ltd v Coventry City Council, Case no CO/2899/97 .... 2.86

Watson v Cammell Laird & Co (Shipbuilders & Engineers) Ltd [1959] 1 WLR 701 . . . . 3.85
Waugh v British Railways Board [1980] AC 521 (HL). 3.81, 3.82
Wells v West Hertfordshire HA, 5 April 2000 (QBD) 12.76
West Bromwich Building Society v Townsend [1983] ICR 257 . . . . 4.112
Westcott v Structural and Marine Engineers Ltd [1960] 1 All ER 775 . . . . 7.37
Westminster City Council v Select Management Ltd [1985]
    1 All ER 897 (CA) 6.152, 6.153
Wheeler v Le Marchant (1881) 17 Ch D 675 . . . . 3.81
Williams v Farne Salmon & Trout Ltd 1998 SLT 1329 . . . . 6.125
Willoughby [2005] 1 Cr App R 29 (CA). 11.61
Wilmott v Atack [1977] QB 498 . . . . 7.58
Wilson v Spindle Select Ltd, 28 November 1996, CO-2177-96 (DC); *The Times*
    9 December 1996. . . . . 6.14
Wilsons and Clyde Coal Co Ltd v English [1938] AC 57. . . . . 6.75
Woodhouse v Walsall MBC [1994] 1 BCLC 435 (DC) 5.51
Woolfson v Strathclyde Regional Council [1978] SLT 159 . . . . 10.67
Wotherspoon v HM Advocate 1978 JC 74 (Scot. Ct of Judiciary) 5.63, 5.67, 5.71

Yewens v Noakes (1880) 6 QBD 530 . . . . 6.51
Yorkshire Traction Co Ltd v Searby [2003] EWCA Civ 1856. . . . . 12.81
Young & Woods Ltd v West [1980] IRLR 201 CA 6.47

EUROPEAN UNION

Case C-127/05 EEC. 1.53
Commission v Austria, Case C-428/04, [2006] ECR I-3325, para 98 . . . . 1.57
Commission v Belgium, Case C-287/03, [2005] ECR I-3761, para 27 . . . . 1.57
Eckle v Germany (1982) 5 EHRR 1 . . . . 9.32
Funke v France, 16 EHRR 29 (ECtHR) 3.86, 3.87
Heaney and McGuinness v Ireland, (application/36887/97, 21 December 2000) 3.87
Jordan v UK (2001) 11 EHRC 1. . . . . 11.111
Keenan v UK (2001) 10 EHRC 319 . . . . 11.111, 11.113
Mattocia v Italy (2003) 36 EHHR 47 (ECtHR) 8.39
Osman v United Kingdom [2000] 29 EHRR 245. . . . . 7.17
Salabiaku v France (1988) 13 EHRR 379 (ECtHR) 6.138
Saunders v United Kingdom (1996) 23 EHRR 313 . . . . .3.87, 7.76
Telfner v Austria (2002) 34 EHRR 7. . . . . 3.89
Von Colson v Land Nordrhein-Westfalen [1984] ECR 1891 (ECJ). 7.14
Z v United Kingdom [2002] 34 EHRR 97 (ECtHR) 7.17

NEW ZEALAND

North v Commissioner of Inland Revenue (1985) 7 NZTC 5192 (HC, NZ). 5.20

TABLE OF STATUTES

Access to Justice Act 1999 .... 10.130
Accessories And Abettors Act
    1861, s 8 .... 5.28
Administration of Justice
    (Miscellaneous Provisions)
    Act 1933, s 2(1) 8.122
Agriculture (Poisonous Substances)
    Act 1952 448
Agriculture (Safety, Health and Welfare
    Provisions) Act 1956 448
Alkali, etc Works Regulation
    Act 1906 447
Alkali, etc Works Regulation
    (Scotland) Act 1951 447
Anti-terrorism, Crime and Security
    Act 2001, s 69 .... 5.08, 5.11
Atomic Energy
    Act 1946 436
Atomic Energy
    Act 1989 448

Banking
    Act 1987 .... 2.75
    s 3 .... 5.56
    s 96(1) 5.47, 5.56
Banking and Financial Dealings
    Act 1971 425
Boiler Explosions
    Act 1882 447
Boiler Explosions
    Act 1890 447
Building
    Act 1984 453
    s 133(2) 444
    Sch 7 444
Building (Scotland) Act 1959 407
    s 3 424
    s 10 423
    s 11 424
    s 16 423
Building (Scotland)
    Act 2003 453

Celluloid and Cinematograph Films
    Act 1922 447

Channel Tunnel
    Act 1987, s 1(7) 478
Charities
    Act 2006 .... 5.24
Children
    Act 1989, s 3 529
Children and Young Persons Act 1933,
    s 18(1)(f) 547
Children and Young Persons at Work
    Act 1920 .... 9.45
Children and Young Persons
    (Scotland) Act 1937,
    s 28(1)(f) 547
Civil Evidence
    Act 1972, s 3 .... 8.193
Civil Partnership
    Act 2004 422
Coal Industry Nationalisation
    Act 1946, s 42(1)–(2) 447
Commission for Revenue and
    Customs Act 2005 426
Companies
    Act 1985
    s 434 .... 2.75
    s 733 .... 5.47
    s 734 .... 5.10
    s 734(5) 5.11
Companies
    Act 1989
    s 83 .... 5.11
    s 85 .... 5.11
    s 90 .... 5.11
Companies Act 2006
    s 172 .... 5.73
    s 172(1) 5.73
Companies Consolidation
    (Consequential Provisions)
    Act 1985 444
Company Directors Disqualification
    Act 1986 .... 10.04, 463
    s 1 465
    s 2 .... 10.102, 465
    s 2(3) 10.105
Constitutional Reform Act 2005,
    s 109(4) 486

Table of Statutes

Consumer Protection Act
  1987 . . . . 6.165, 410, 411, 412, 422,
        425, 426, 429, 442, 443, 444
  s 40 . . . . 5.40
Contracts of Employment
  Act 1972 445
Control of Pollution
  Act 1974
  s 87 . . . . 5.51
  s 87(1) 5.47
  s 87(2) 5.31
Coroners Act 1988 . . . . 11.99, 11.110
  s 8(1) 11.100
  s 8(3)(c) 11.103
  s 8(3)(c)(1) 11.104
  s 8(3)(d) 11.105
  s 11(5)(b)(ii) 11.113
  s 16 . . . . 11.145, 11.150
  s 16(1) 11.145, 11.147
  s 16(2)–(3) 11.145
  s 16(7) 11.147
Coroners Bill (2006 draft) 11.91, 11.92,
        11.93, 11.96, 11.97
Corporate Manslaughter and Homicide
  Act 2007 . . . . .2.65, 5.79, 11.01,
        11.21, 11.66, 11.69, 11.70, 11.71,
        11.75, 11.77, 11.82, 11.83, 11.91
  s 1 . . . . 11.66, 11.76
  s 1(2) 11.81
  s 1(3) 11.85
  s 1(4)(b) 11.87
  s 2 . . . . 11.76, 11.81
  s 2(1) 11.71, 11.81
  s 2(1)(d) 11.81
  s 2(5) 11.84
  ss 3–7 . . . . 11.76
  s 8 . . . . 11.76, 11.87
  s 8(2)–(3) 11.87
  s 8(4) 11.88
  s 9 . . . . 11.76
  s 10 . . . . 11.76
  ss 11–16 . . . . 11.78
  s 17 . . . . 11.79
  s 19 . . . . 11.77
  s 20 . . . . 11.66
  s 27 . . . . 11.66
  Sch 1 . . . . 11.75, 11.79
  Schs 11.80
Courts and Legal Services Act 1990
  s 71 . . . . .486, 487, 511
  s 119(1) 11.123
Crime and Disorder Act 1998
  s 51 . . . . 8.102
  s 51(3) 8.112

Criminal Appeals Act 1968
  s 2(1) 6.39
  s 50(1) 10.151
Criminal Justice Act 1925, s 33 . . . . 5.10
Criminal Justice Act 1967, s 9 . . . . 2.134,
        2.135, 7.72
Criminal Justice Act 1982 430
Criminal Justice Act 1987
  s 2 . . . . 3.76
  s 2(9) 3.76
Criminal Justice Act 1988
  s 23 . . . . 8.116
  s 24 . . . . 8.116
  s 30 . . . . 8.205
Criminal Justice Act 1991 . . . . 10.127
Criminal Justice Act 2003 . . . . 2.136,
        8.159, 8.160, 8.163, 8.207, 8.208,
        8.214, 8.218, 10.10, 10.25
  s 29 . . . . 8.05, 8.26
  s 32 . . . . 8.164
  s 35 . . . . 8.207
  s 39 . . . . 8.168
  s 41 . . . . 10.10
  s 101(1)(d) 8.219
  s 101(3) 8.221
  s 103 . . . . 8.218
  s 103(1)(a) 8.221
  s 103(3) 8.221, 8.224
  s 103(4) 8.225
  s 114 . . . . 8.210
  s 114(1) 8.211
  s 114(2) 8.211
  s 114(2)(d) 8.211
  s 118 . . . . 8.210
  s 142 . . . . 10.39
  s 142(1)(d) 10.39
  s 144 . . . . 8.73
  s 162 . . . . 10.14
  s 164(1)–(4) 10.26
  s 331 . . . . 8.167
  Sch 3 . . . . .8.26, 8.102, 8.108, 8.156
  Pt 1
    para 1 . . . . 10.10
    para 6 . . . . 10.10
  Sch 36, Pt 3 . . . . 8.168
Criminal Justice and Police Act 2001,
  Sch 3, para 14(1) 441
Criminal Justice and Public Order
  Act 1994 . . . . 3.13
  Pt III 3.43
  s 34 . . . . 3.13, 3.15, 3.17
  s 34(1) 3.14
  s 34(1)(a) 3.19, 3.25, 3.27
  s 34(2) 3.14, 3.15

Table of Statutes

s 34(2A) 3.16	Education (Scotland) Act 1980,
s 34(4) 3.13, 3.17	s 31 525
s 38(3) 3.15	Electronic Communications Act 2000,
Criminal Law Act 1967, s 4(1) 9.47	s 15(1) 484
Criminal Law Act 1977 430	Employers Liability (Compulsory
Criminal Procedure (Attendance of	Insurance) Act 1969, s 5 . . . . 5.31
Witnesses) Act 1965 . . . . 8.169	Employment Act 1989 . . . . 425, 426
s 2 . . . . 2.136, 8.182	Employment Act 2002 484
Criminal Procedure and Investigations	s 32 . . . . 493, 494
Act 1996 . . . . 2.47, 2.52, 8.69,	s 32(2)–(4) 493, 494
8.76, 8.159, 8.162	s 32(6) 493
Pt I 2.48	Employment Agencies Act
Pt II 2.48, 2.49, 2.50, 8.170	1973 . . . . 492, 517
s 3 . . . . 8.213	s 3C 517
s 3(1) 8.164	Employment Medical Advisory Service
s 5(1) 8.167, 8.168	Act 1972 448
s 6D 8.207	s 1 448
s 8 . . . . 8.161	s 6 448
s 11 . . . . 8.168	Sch 1 448
s 22(1) 8.171	Employment Protection Act
s 23 . . . . 2.49, 8.170	1975 . . . . 407, 408, 413, 415, 416,
s 23(1) 8.171	417, 419, 428, 429, 430, 433, 434,
s 25 . . . . 2.49	435, 438, 439, 440, 442, 443,
s 26(1) 8.172	444, 445, 446, 453
s 26(4) 8.189	Employment Protection
s 29 . . . . 8.151	(Consolidation) Act 1978 445
s 35 . . . . 5.70, 8.151	Employment Rights Act
s 35(1) 6.105, 6.127	1996 . . . . 6.52, 445, 484, 525
Criminal Procedure (Scotland)	s 11 489
Act 1975 431	ss 13–28 503
s 331(3) 431	s 47B 503
Crown Proceedings Act 1947 413	s 50 503
s 38(3) 1.43, 439	ss 52–53 503
Crown Suits (Scotland) Act 1857 413	ss 55–56 503
Customs and Excise Management	s 64 503
Act 1979 . . . . 442, 451, 570, 632	s 67 532
	s 68 503
Dentists Act 1984 481	s 95(1)(c) 484
s 53(1) 569	s 103A 503
Disability Discrimination	s 105(6A) 503
Act 1995 . . . . 484, 515, 520	s 111 518
Pt II 503	s 128 501
s 1 537	ss 163–164 503
s 17A 503, 516	s 170 519
s 25(8) 503, 516	s 203(3) 484
Disability Rights Commission Act	s 230 . . . . 6.46
1999, Sch 3, para 10 490	Employment Rights (Dispute
Domestic Violence, Crime and Victims	Resoluton) Act 1998 424
Act 2004, s 5 . . . . 11.145	Employment and Training
	Act 1973 434
Education Act 1996	Employment Tribunals Act 1996 484
s 8 525	s 3 503
s 444(1) 7.16	s 4 488
Education Reform Act 1988 428	s 4(1) 501, 506

Table of Statutes

Employment Tribunals Act 1996 (cont.)
   s 4(1)(a)–(b) 488
   s 4(2) 501, 506
   s 4(5) 488
   s 4(6)–(6A) 483
   s 5(2)–(3) 510, 515
   s 6(1) 4.79
   s 7(1) 483
   s 7(3)–(3A) 483
   s 7(3ZA) 483
   s 7(4) 498
   s 7(5) 483
   s 7A(1)–(2) 483
   s 9(1)–(2) 483
   s 9(4) 483
   s 10(2) 483
   s 10(5)–(7) 483
   s 10A(1) 483
   s 11(1)–(2) 483
   s 13 483
   s 13A(1)–(2) 483
   s 19 483
   s 24(2) 483
   s 41(4) 483
Employment of Women, Young Persons and Children Act 1920 447
   s 1(2) 547
Environment Act 1995 . . . . 2.107, 426, 427, 432
Environmental Protection Act 1990 . . . . . . 2.107, 2.132, 2.142, 2.143, 2.145, 2.148, 3.87, 407, 409
   s 157(1) 5.47
Equal Pay Act 1970 . . . . 484, 520
   s 2(1) 503
   s 2A(1)(b) 491
European Communities Act 1972 . . . . . . 1.46
   s 2 525
   s 2(1) 1.09
   s 2(2) 525, 567, 619
European Economic Area Act 1993 567
Explosives Act 1875 447
   s 23 479
   ss 30–32 447
   s 80 447
   ss 116–121 447

Factories Act 1961 . . . . 12.115, 448, 540, 543, 641
   s 13(1) 4.115
   s 14(1) 4.115
   s 17 . . . . 12.81
   s 24 612
   s 29(1) 6.76, 6.121, 6.125, 7.28
   ss 123–126 641
   s 127(6) 5.69
   s 135 448
   s 175(5) 539
Fatal Accidents Inquiry (Scotland) Act 1895 431
Fatal Accidents and Sudden Deaths Inquiry (Scotland) Act 1906 431
Fatal Accidents and Sudden Deaths Inquiry (Scotland) Act 1976 416
Finance Act 1966 . . . . 5.64
   s 305(3) 5.64, 11.64
Financial Services and Markets Act 2000, s 177 . . . . 5.11
Fire Precautions Act 1971 . . . . 5.49, 444
Fire Precautions Act 2000 428
Fire and Rescue Services Act 2004 . . . . 423, 478
   s 1 621
Fire (Scotland) Act 2005
   Pt 3 599
   s 6 . . . . 478, 621
   s 61 599
Food and Drugs Act 1938 . . . . 5.42
   s 3 . . . . 5.41
   s 83 . . . . 5.31
   s 83(3) 5.41, 5.42
Food Safety Act 1990 . . . . 2.86, 7.51
Forgery and Counterfeiting Act 1981 . . . . 429, 430
Freedom of Information Act 2000 428
Friendly Societies Act 1992, s 5 . . . . 5.25
Gas Act 1986 . . . . 12.211, 12.212, 431, 640
   s 18(1) 431
   s 48 476
   Sch 2B
      para 17(1)–(2) 657
      para 17(5)–(6) 657

Gas Act 1995 . . . . 12.211
Government of Wales Act 1998
   Sch 8
      para 1 518
      para 36 483
Greater London Authority Act 1999 428

Health Protection Agency Act 2004 . . . . 417, 440, 444

Table of Statutes

Health and Safety at Work etc.
    Act 1974 . . . . .1.01, 1.02, 1.03,
      1.04, 1.05, 1.06, 1.11, 1.13, 1.14,
      1.18, 1.34, 1.43, 1.45, 1.79, 1.100,
      2.25, 2.30, 2.31, 2.36, 2.38, 2.40,
      2.55, 2.57, 2.103, 2.107, 2.111,
      3.71, 3.76, 3.87, 4.01, 4.02, 4.26,
      4.34, 4.47, 4.51, 4.52, 4.111,
      4.116, 4.118, 4.119, 5.02, 5.08,
      5.10, 5.11, 5.12, 5.15, 5.27, 5.33,
      5.36, 6.03, 6.05, 6.06, 6.56, 6.63,
      6.76, 6.116, 6.137, 6.143, 6.153,
      6.167, 6.174, 6.187, 6.205, 6.214,
      7.21, 7.22, 7.29, 7.51, 7.61, 7.82,
      8.03, 8.07, 8.14, 8.112, 9.40,
      10.42, 10.123, 11.46, 11.47, 11.63,
      11.67, 12.02, 12.03, 12.05, 12.10,
      12.46, 12.61, 12.62, 12.89, 12.111,
      407, 464, 471, 472, 475, 479,
      521, 551, 567, 605
    Pt I 1.03, 1.07, 1.41, 1.42,
      2.90, 2.123, 4.26, 4.34, 4.117, 7.06,
      7.41, 7.73, 407, 534, 549, 580
    Pts II–IV 444
    s 1 . . . . 2.36, 4.26, 4.34, 534, 549,
      552, 580, 585, 606, 634, 665
    s 1(1) 1.03, 6.06, 407
    s 1(1)(a) 6.07, 6.61, 6.76
    s 1(1)(b) 6.07
    s 1(2) 1.07, 407
    s 1(3) 1.04, 408
    s 1(4) 408
    s 2 . . . . 1.05, 1.07, 1.88, 2.53,
      4.26, 4.34, 4.113, 4.116, 5.36,
      5.70, 6.02, 6.03, 6.04, 6.05, 6.06,
      6.09, 6.10, 6.14, 6.17, 6.21, 6.43,
      6.44, 6.45, 6.58, 6.66, 6.67, 6.81,
      6.96, 6.97, 6.100, 6.107, 6.109,
      6.117, 6.128, 6.132, 6.134,
      6.143, 6.146, 6.159, 6.210, 7.01,
      7.24, 7.25, 7.26, 8.128, 8.220,
      10.85, 10.121, 11.46, 12.02,
      12.47, 12.79, 12.109, 464, 534,
      549, 552, 580, 585, 606, 634, 665
    s 2(1) 4.112, 4.115, 4.117,
      6.04, 6.08, 6.10, 6.11, 6.12, 6.14,
      6.15, 6.26, 6.28, 6.43, 6.44, 6.53, 6.61,
      6.64, 6.70, 6.80, 6.81, 6.82, 6.83, 6.84,
      6.85, 6.86, 6.99, 6.105, 6.113, 6.127,
      8.55, 8.129, 9.14, 10.55, 10.82, 10.108,
      10.115, 408
    s 2(2) 6.11, 6.13, 6.14,
      6.61, 8.129, 408
    s 2(2)(a) 4.117

    s 2(3)–(7) 6.16, 408
    s 3 . . . . .1.05, 1.07, 1.88,
      1.102, 2.34, 2.45, 2.53, 4.26, 4.34,
      5.36, 5.58, 5.70, 6.02, 6.03, 6.04, 6.05,
      6.06, 6.09, 6.24, 6.35, 6.38, 6.43, 6.58,
      6.66, 6.67, 6.70, 6.77, 6.79, 6.96, 6.97,
      6.100, 6.107,
      6.109, 6.117, 6.128, 6.132, 6.143, 6.146,
      6.159, 6.166, 6.210, 7.01, 7.24, 7.25,
      8.128, 8.220, 9.11,
      10.38, 10.39, 10.72, 10.85, 10.86, 10.121,
      11.46, 11.48, 12.02,
      12.79, 464, 534, 549, 552, 580,
      585, 606, 634, 665
    s 3(1) 5.32, 6.04, 6.08, 6.15,
      6.24, 6.25, 6.26, 6.27, 6.28, 6.29,
      6.32, 6.33, 6.37, 6.38, 6.40, 6.43, 6.44,
      6.54, 6.61, 6.65, 6.68, 6.69, 6.70, 6.73,
      6.82, 6.84, 6.85, 6.86, 6.91, 6.92, 6.93,
      6.94, 6.112,
      6.160, 9.14, 9.16, 10.47, 10.48, 10.82,
      10.108, 11.48, 409
    s 3(2) 5.32, 6.04, 6.29, 6.31,
      6.32, 9.35, 11.06, 11.48, 409, 580
    s 3(3) 409
    s 4 . . . . 1.05, 1.07, 1.88, 2.53,
      4.26, 4.34, 6.02, 6.03, 6.117,
      6.128, 6.132, 6.143, 6.146, 6.147, 6.154,
      6.155, 6.159, 6.161, 6.164, 6.198, 6.200,
      6.210, 7.01, 7.24,
      7.25, 8.128, 10.121, 12.02, 12.47, 12.79,
      464, 534, 549, 552, 580,
      585, 606, 634, 665
    s 4(1) 6.148, 409
    s 4(2) 6.155, 6.156, 6.158, 6.163, 409
    s 4(3) 6.156, 409
    s 4(4) 6.157, 409
    s 5 . . . . 1.05, 1.07, 1.88, 2.53,
      4.26, 4.34, 6.02, 6.03, 6.143,
      6.146, 6.210, 7.01, 8.128, 10.121, 12.02,
      12.79, 409, 464, 534, 549, 552, 580, 585,
      606, 634, 665
    s 6 . . . . .1.05, 1.07, 1.88, 2.53, 4.26,
      4.34, 6.02, 6.03, 6.143, 6.146,
      6.165, 6.166, 6.167, 6.168, 6.169, 6.175,
      6.178, 6.179, 6.190, 6.210, 7.01, 8.128,
      10.121, 12.02, 12.79, 464, 479, 534, 549,
      552, 580, 585, 606, 634, 665
    s 6(1) 6.173, 6.190, 6.192, 409
    s 6(1)(a) 6.170, 6.179, 6.180,
      6.183, 6.184, 6.188, 6.196
    s 6(1)(b) 6.183, 6.203
    s 6(1)(c) 6.184, 6.185
    s 6(1)(d) 6.185, 6.186

Health and Safety at Work etc.
Act 1974 (*cont.*)
- s 6(1A) 6.190, 6.196, 410
- s 6(1A)(a) 6.170, 6.188
- s 6(2) 6.193, 6.194, 6.195, 410
- s 6(3) 6.196, 410
- s 6(4) 6.198, 410
- s 6(4)(a) 6.170, 6.200, 6.203, 6.204, 6.206
- s 6(4)(b) 6.203
- s 6(4)(c) 6.204, 6.205
- s 6(4)(d) 6.205
- s 6(5) 6.206, 6.207, 6.208, 411
- s 6(6) 6.172, 6.183, 6.193, 6.203, 6.206, 411
- s 6(7) 6.168, 6.180, 6.183, 6.184, 6.189, 6.200, 6.203, 6.204, 411
- s 6(8) 6.188, 6.189, 411
- s 6(8A) .6.189, 411
- s 6(9) 411
- s 6(10) 6.169, 6.170, 6.180, 6.183, 6.200, 6.203, 412
- s 7 . . . . 1.07, 1.88, 1.89, 1.92, 1.99, 1.101, 2.53, 4.26, 4.34, 5.32, 6.02, 6.03, 6.23, 6.209, 6.210, 6.212, 6.213, 6.217, 6.218, 6.219, 6.220, 7.01, 10.09, 10.121, 12.02, 12.79, 412, 534, 549, 552, 580, 585, 606, 634, 665
- s 7(b) 6.216
- s 8 . . . . 1.89, 1.100, 1.101, 4.26, 4.34, 7.03, 7.04, 412, 534, 549, 552, 580, 585, 606, 619, 634, 665
- s 9 . . . . . 4.26, 4.34, 7.03, 7.05, 412, 534, 549, 552, 580, 585, 606, 634, 665
- s 10 . . . . .1.13, 4.26, 4.34, 448, 534, 549, 552, 580, 585, 606, 634, 665
- s 10(1)–(4) 412
- s 10(5)–(8) 413
- s 11 . . . . 4.26, 4.34, 534, 549, 552, 580, 585, 606, 634, 665
- s 11(1) 413
- s 11(2) 413
- s 11(2)(d) 475, 525, 537, 547, 551, 567, 583, 605, 619, 639
- s 11(2A) 413
- s 11(3) 413
- s 11(4)–(6) 414
- s 12 . . . . 4.26, 4.34, 414, 534, 549, 552, 580, 585, 606, 634, 665
- s 13 . . . . 4.26, 4.34, 534, 549, 552, 580, 585, 606, 634, 665
- s 13(1) 414
- s 13(2) 415
- s 14 . . . . 1.20, 1.21, 4.26, 4.34, 4.113, 7.41, 7.42, 7.43, 8.15, 534, 549, 552, 580, 585, 606, 634, 665
- s 14(1) 415
- s 14(2) 7.64, 415
- s 14(3)–(6) 415
- s 14(7) 416
- s 15 . . . . .1.07, 1.08, 4.26, 4.34, 7.10, 12.02, 450, 534, 549, 552, 580, 585, 606, 634, 665
- s 15(1) 1.07, 416, 475, 525, 537, 547, 551, 567, 583, 605, 619, 639, 659
- s 15(2) 416, 525, 537, 547, 551, 567, 583, 605, 619, 639, 659
- s 15(3) 416, 619
- s 15(3)(a) 475, 525, 537, 547, 551, 583, 605, 659
- s 15(3)(b) 567, 659
- s 15(3)(c) 475, 583
- s 15(4) 416, 567, 619
- s 15(4)(a) 639, 659
- s 15(5) 416, 525, 551, 619, 639
- s 15(5)(a) 547, 583
- s 15(5)(b) 537, 567, 605, 659
- s 15(6) 416
- s 15(6)(a) 551, 583, 605
- s 15(6)(b) 567, 583, 619, 639, 659
- s 15(7) 417
- s 15(8) 417, 583, 659
- s 15(9) 417, 525, 547, 567, 583, 619, 659
- s 15(10) 417
- s 16 . . . . .1.10, 4.26, 4.34, 12.05, 534, 549, 552, 580, 585, 606, 634, 665
- s 16(1)–(1A) 417
- s 16(2) 417
- s 16(3)–(8) 418
- s 17 . . . . 1.11, 4.26, 4.34, 12.05, 12.20, 534, 549, 552, 580, 585, 606, 634, 665
- s 17(1)–(3) 418
- s 18 . . . . 1.35, 2.33, 4.26, 4.34, 467, 534, 549, 552, 580, 585, 606, 634, 665
- s 18(1) 2.37, 419
- s 18(2) 419, 475, 479, 619
- s 18(3) 419
- s 18(4) 1.32, 419
- s 18(5)–(7) 419

Table of Statutes

s 19 . . . . 2.54, 2.91, 2.92, 4.26,
    4.34, 11.103, 534, 549, 552,
    580, 585, 596, 606, 634, 665
s 19(1) 1.34, 420, 521
s 19(2) 2.92, 420
s 19(3)–(4) 420
s 20 . . . . 2.56, 2.87, 2.89, 2.92,
    2.93, 2.94, 2.96, 2.100, 2.110,
    2.124, 3.51, 3.60, 3.68, 3.69,
    3.71, 4.26, 4.34, 7.46, 7.47, 7.51, 7.64,
    7.65, 11.139, 534, 549, 552, 580, 585,
    606, 634, 665
s 20(1) 2.88, 420
s 20(1)(j) 7.74
s 20(2) 2.103, 2.112, 2.161,
    2.163, 7.48, 7.52, 420
s 20(2)(a) 2.101, 2.108, 2.120,
    2.122, 2.139
s 20(2)(b) 2.68, 2.105
s 20(2)(c) 2.108
s 20(2)(d) 2.109, 2.110, 2.111,
    2.112, 2.115, 2.133,
    2.150, 2.152
s 20(2)(e) 2.110, 2.113, 2.114
s 20(2)(f) 2.115
s 20(2)(g) 2.117, 2.119, 2.120
s 20(2)(h) 2.121, 2.122, 2.123,
    2.125, 2.127, 2.128, 2.129
s 20(2)(i) 2.96, 2.120, 2.126,
    2.128, 2.130
s 20(2)(j) 2.85, 2.86, 2.89, 2.93,
    2.110, 2.131, 2.133, 2.134, 2.135,
    2.136, 2.137, 2.138, 2.139, 2.140,
    2.141, 2.142, 2.145, 2.152, 2.162,
    3.51, 3.68, 3.70, 7.47,7.49, 7.53,
    7.62, 7.75, 7.77, 11.140, 11.141
s 20(2)(k) 2.111, 2.150, 2.154,
    2.162, 3.68, 3.72, 3.78, 3.90, 7.47
s 20(2)(k)(ii). 2.152
s 20(2)(l) 2.157, 7.47
s 20(2)(m) 2.95, 2.159, 2.162
s 20(3) 2.118, 421
s 20(4) 2.125, 421
s 20(5) 2.127, 421
s 20(6) 2.96, 2.120, 2.130, 421
s 20(7) 2.85, 2.137, 2.140,
    2.141, 2.142, 2.148, 2.149,
    7.75, 7.77, 11.141, 422
s 20(7)(l) .11.141
s 20(8) 2.155, 3.73, 422
s 21 . . . . 1.41, 2.56, 2.128, 4.02,
    4.25, 4.26, 4.34, 4.90, 5.01, 6.07,
    422, 423, 464, 521, 534, 549, 552,
    580, 585, 606, 634, 665

s 21(a)–(b) .4.105
s 22 . . . . 1.41, 2.56, 2.128, 4.02,
    4.26, 4.34, 4.44, 4.90, 5.01, 6.07,
    464, 521, 534, 549, 552, 580,
    585, 606, 634, 665
s 22(1) 4.43, 422
s 22(2) 422
s 22(3) 422
s 22(3)(a)–(d). 4.35
s 22(4)(a) 4.36, 4.100
s 22(4)(b) 4.36
s 23 . . . . 1.41, 4.02, 4.26, 4.34, 534,
    549, 552, 580, 585, 606, 634, 665
s 23(1) 423
s 23(2) 4.30, 4.32, 4.39,
    4.41, 4.108, 423
s 23(3) 4.31, 4.40, 423
s 23(4) 423
s 23(5) 4.48, 423
s 23(6) 423
s 24 . . . . .1.41, 4.02, 4.26,
    4.34, 4.93, 4.94, 4.100, 4.103,
    4.105, 490, 534, 549, 552, 580,
    585, 606, 634, 665
s 24(1) 424
s 24(2) 4.53, 4.104, 424
s 24(3) 4.69, 424
s 24(3)(a) 4.68
s 24(3)(b) 522
s 24(4) 424, 522
r 8 . . . . .4.81
s 25 . . . . 1.41, 2.56, 2.104, 2.163,
    4.26, 4.34, 7.50, 534, 549, 552,
    580, 585, 606, 634, 665
s 25(1) 2.104, 424
s 25(2) 2.164, 424
s 25(3) 2.165, 424
s 25A(1)–(3). 425
s 26 . . . . .2.100, 4.26, 4.34, 425, 534,
    549, 552, 580, 585, 606, 634, 665
s 27 . . . . .1.23, 4.26, 4.34, 534, 549,
    552, 580, 585, 606, 634, 665
s 27(1) 2.142, 2.145, 7.59, 7.62, 425
s 27(2) 425
s 27(3) 426
s 27(4) 7.61, 426
s 27A(1)–(4). 426
s 28 . . . . .1.23, 1.43, 4.26,
    4.34, 7.61, 7.62, 11.139, 534, 549,
    552, 580, 585, 606, 634, 665
s 28(1) 7.62, 426
s 28(2) 7.62, 7.63, 426
s 28(3) 7.62, 426
s 28(3)(e) 7.63, 11.139

Table of Statutes

Health and Safety at Work etc.
Act 1974 (*cont.*)
s 28(4)–(5) 7.62, 427
s 28(6) 7.62, 428
s 28(7) 7.64, 7.66, 11.139, 428
s 28(8) 428
s 28(9) 7.66, 428
s 28(9A) 428
s 28(10) 428
ss 29–30 . . . . 4.26, 4.34, 428, 534,
  549, 552, 580, 585, 606, 634, 665
ss 31–32 . . . . 4.26, 4.34, 429, 534,
  549, 552, 580, 585, 606, 634, 665
s 33 . . . . 1.23, 1.41, 1.42, 2.61, 4.26,
  4.34, 5.45, 6.01, 6.56, 464, 534,
  549, 552, 580, 585, 606, 634, 665
s 33(1) 2.53, 5.01, 6.01,
  7.01, 8.103, 429
s 33(1)(a) 2.53, 5.01, 6.02,
  6.07, 6.84, 6.94, 7.01
s 33(1)(b) 2.53, 5.01, 7.03
s 33(1)(c) 2.23, 5.01, 7.07, 7.11
s 33(1)(d) 1.22, 5.01, 7.40, 8.103
s 33(1)(e) 2.89, 2.138, 2.151,
  3.71, 5.01, 7.44, 7.47,
  7.51, 8.103
s 33(1)(f) 2.89, 5.01, 7.52, 8.103
s 33(1)(g) 4.11, 5.01, 7.54
s 33(1)(h) 2.89, 2.105, 5.01,
  7.47, 7.56, 8.103
s 33(1)(i) 5.01, 7.59
s 33(1)(j) 5.01, 7.61
s 33(1)(k) 2.138, 5.01, 7.67, 7.77
s 33(1)(k)(i) 7.72
s 33(1)(l) 5.01, 7.80
s 33(1)(m) 5.01, 7.83, 7.85
s 33(1)(n) 5.01, 7.88, 8.103
s 33(1)(o) 5.01, 7.89
s 33(1A) 429
s 33(2) 430
s 33(2A) 10.87, 430
s 33(3) 430
s 33(3)(b)(i) 7.09
s 33(4) 7.09, 10.87, 430
s 33(5) 430
s 33(6) 430
s 34 . . . . 1.41, 1.42, 4.26,
  4.34, 8.15, 534, 549, 552,
  580, 585, 606, 634, 665
s 34(1) 430
s 34(2)–(6) 431
s 35 . . . . 1.41, 1.42, 4.26, 4.34,
  8.31, 534, 549, 552, 580,
  585, 606, 634, 665

s 36 . . . . 1.41, 1.42, 1.89, 1.100,
  1.101, 4.26, 4.34, 5.30, 5.31, 5.32,
  5.33, 5.34, 5.36, 5.37, 5.38, 5.41,
  5.43, 5.44, 465, 534, 549, 552,
  580, 585, 606, 634, 665
s 36(1) 5.30, 432
s 36(2) 5.45, 6.56, 432
s 36(3) 5.33, 432
s 37 . . . . . . 1.41, 1.42, 1.89, 1.95, 1.96,
  1.99, 1.101, 3.66, 4.26, 4.34,
  5.15, 5.32, 5.47, 5.48, 5.51, 5.53,
  5.54, 5.56, 5.59, 5.63, 5.64, 5.69,
  5.70, 5.71, 5.72, 10.88, 11.64,
  11.65, 465, 534, 549, 552, 580,
  585, 606, 634, 665
s 37(1) 5.46, 5.48, 5.67, 5.68,
  5.69, 5.72, 10.53, 10.108, 432
s 37(2) 5.16, 5.48, 432
s 38 . . . . 1.38, 1.41, 1.42, 2.54,
  4.26, 4.34, 8.07, 432, 534, 549,
  552, 580, 585, 606, 634, 665
s 39 . . . . 1.41, 1.42, 3.47, 4.26, 4.34,
  534, 549, 552, 580,
  585, 606, 634, 665
s 39(1) 8.11, 432
s 39(2) 432
s 40 . . . . 1.41, 1.42, 4.14, 4.26,
  4.34, 5.59, 6.64, 6.65, 6.107,
  6.115, 6.135, 6.138, 6.142, 6.144,
  6.163, 6.169, 6.181, 6.186, 6.195,
  6.197, 6.201, 6.205, 6.208, 6.210,
  7.16, 7.39, 7.55, 432, 534, 549,
  552, 580, 585, 606, 634, 665
s 41 . . . . 1.41, 1.42, 4.26, 4.34, 534,
  549, 552, 580, 585,
  606, 634, 665
s 41(1) 432
s 41(2) 433
s 42 . . . . 1.41, 1.42, 4.26, 4.34,
  7.89, 10.118, 10.120, 464, 534,
  549, 552, 580, 585, 606, 634, 665
s 42(1) 10.04, 433
s 42(2)–(5) 433
s 43 . . . . 4.26, 4.34, 534, 549,
  552, 580, 585, 606, 634, 665
s 43(1)–(5) 433
s 43(6) 434
s 43(8)–(9) 434
s 43A(1)–(7) 434
s 43A(8)–(9) 435
s 44 . . . . 4.26, 4.34, 534, 549, 552,
  580, 585, 606, 634, 665
s 44(1)–(7) 435
s 44(8) 436

Table of Statutes

s 45 . . . . 4.26, 4.34, 534, 549, 552, 580, 585, 606, 634, 665	s 52(3) 442, 525, 567
s 45(1)–(7) 436	s 53 . . . . .2.102, 4.26, 4.34, 534, 549, 552, 580, 585, 606, 634, 665
s 45(8)–(12) 437	
s 46 . . . . 2.25, 4.26, 4.34, 4.46, 4.47, 5.07, 5.09, 534, 549, 552, 580, 585, 606, 634, 665	s 53(1) 2.90, 2.109, 2.119, 2.122, 4.03, 4.26, 4.34, 4.35, 6.43, 6.46, 6.149, 6.151, 6.176, 8.07, 442
s 46(1) 2.26, 4.46, 437	s 53(2)–(6) 444
s 46(2) 2.27, 437	s 53(9) 6.175
s 46(3) 437	s 54 . . . . 444, 534, 549, 552, 580, 585, 606, 634, 665
s 46(3)(b) 2.28, 4.46	
s 46(4) 2.27, 2.28, 2.29, 437	s 55 . . . . 534, 549, 552, 580, 585, 606, 634, 665
s 46(4)(a)–(b) 2.28, 4.46	
s 46(5) 437	s 56 . . . . 534, 549, 552, 568, 580, 585, 606, 620, 634, 665
s 46(6) 2.30, 4.47, 438	
s 46(7)–(8) 438	ss 57–59 . . . . 534, 549, 552, 580, 585, 606, 634, 665
s 47 . . . . 4.26, 4.34, 534, 549, 552, 580, 585, 606, 634, 665	
	s 77 444
s 47(1) 438	s 78 . . . . .2.98, 444
s 47(1)(a) 6.06, 11.46	s 79 444
s 47(2) 11.46, 438, 525, 583	ss 80 . . . . 534, 549, 552, 580, 585, 606, 634, 665
s 47(3) 438, 583	
s 47(4)–(6) 438	s 80(1) 444, 525, 547, 583, 619
s 48 . . . . 1.41, 4.26, 4.34, 6.56, 534, 549, 552, 580, 585, 606, 634, 665	s 80(2) 445, 583
	s 80(2)(a) 547
s 48(1) 438	s 80(2A) 445
s 48(2) 1.42, 438	s 80(3) 445
s 48(3) 1.42, 6.56, 438	s 80(4) 445, 525, 547
s 48(4) 438	s 80(5) 445
s 48(5) 439	s 81 . . . . 445, 534, 549, 552, 580, 585, 606, 634, 665
s 48(6) 1.43, 439	
s 49 . . . . 4.26, 4.34, 534, 549, 551, 552, 580, 585, 606, 634, 665	s 82 . . . . 7.10, 7.45, 534, 549, 552, 580, 585, 606, 634, 665
s 49(1)–(4) 439	s 82(1) 445
s 50 . . . . .1.07, 4.26, 4.34, 534, 549, 552, 580, 585, 606, 634, 665	s 82(1)(c) 4.100
	s 82(2) 445
s 50(1)–(1A) 439	s 82(3) 446, 619
s 50(2)–(5) 440	s 82(3)(a) 475, 525, 537, 551, 567, 583, 605, 639, 659
s 50(3) 475, 525, 537, 547, 551, 567, 583, 605, 619, 639	
	s 82(4) 446
s 51 . . . . .4.26, 4.34, 6.57, 440, 534, 549, 552, 580, 585, 606, 634, 665	s 83 446
	s 84(1)–(4) 446
s 51(3) 6.30, 6.174, 6.191	s 84(5)–(6) 447
s 51A 6.56	s 85(1)–(3) 447
s 51A(1) 440	Sch 1 . . . . . 2.90, 4.26, 4.34, 7.06, 7.73, 447
s 51A(2)–(2C) 440	Sch 2 . . . . 1.13, 448
s 51A(2D)–(2F) 441	Sch 3 450
s 51A(3)–(4). 441	para 1(1). 551, 567, 583, 605, 619, 639
s 52 . . . . 4.26, 4.34, 6.23, 6.148, 12.55, 534, 549, 552, 580, 585, 606, 634, 665	
	para 1(1)(a) 547, 659
	para 1(1)(c) 547, 659
s 52(1) 6.18, 6.212, 441	para 1(2). 537, 551, 567, 583, 605, 619, 639, 659
s 52(2) 441, 525, 567	

Health and Safety at Work etc.
Act 1974 (cont.)
Sch 3 (cont.)
    para 1(3). 551, 567, 605,
                                           619, 639, 659
    para 1(4). 567, 619
    para 2 . . . . 567, 619
    para 3(2). 619
    para 4 619
    para 4(1). 639
    para 6 . . . . 583, 619
    para 6(1). 525, 567
    para 6(2). 659
    para 7 . . . . 525, 583
    para 8 . . . . .547, 567, 619
    para 8(1). 525, 583
    para 9 . . . . 537, 551, 567, 583,
                                           605, 619, 659
    para 10 . . . . .525, 537, 583, 619
    para 11 . . . . . . .567, 583, 605, 619, 659
    para 12. . . . . .583, 639, 659
    para 13(1). 619
    para 13(3). 619
    para 14 . . . . 525, 551, 567, 583,
                                           605, 619, 659
    para 15 525
    para 15(1). 551, 567, 583, 605,
                                           619, 639, 659
    para 16. . . . . 525, 551, 567,
                                       583, 605, 619, 639, 659
        para 18 583
        para 20. . . . . .567, 583, 619
        para 21 583
        para 21(b). 659
    Sch 4 453
        r 8 . . . . 4.81
    Schs 5–10. 453
Highways Act
    1980, s 329. . . . . 537, 569
Hours of Employment (Conventions)
    Act 1936 447
    s 5 447
House of Commons Disqualification
    Act 1975 446
House to House Collections
    Act 1939, s 5. . . . . .7.87
Human Rights Act 1998. . . . . 9.30, 9.31
    s 6(6) 2.35
    Sch 1 . . . . .2.35
Hydrogen Cyanide (Fumigation)
    Act 1937 447

Industrial and Provident Societies
    Act 1965, s 1. . . . . .5.26

Industrial Training Act 1982, s 12. 489
Insolvency Act 1986
    s 112(1) 8.25
    s 130(2) 8.21, 8.22
    s 236 . . . . .3.82
International Headquarters and Defence
    Organisations Act 1964 478
    Sch, para 1(1) 534, 549
Interpretation Act 1889, s 26 437
Interpretation Act 1978 . . . . 10.63
    s 7 . . . . .2.31
    Sch 1 . . . . 5.02, 8.14

Law Reform (Parent and Child)
    (Scotland) 1986, s 8 529
Law Reform (Personal Injuries)
    Act 1948, s 1. . . . . .6.75
Limited Liability Partnership
    Act 2000 .5.16
Local Government Act 1972
    s 250(1) 436
    s 250(2)–(5) 436, 437
Local Government Act 1985. . . . . . 428, 443
Local Government etc (Scotland)
    Act 1994, s 2 443
Local Government (Scotland) Act
    1947, s 355(2)–(9) 437
Local Government (Scotland) Act
    1973, s 210(2)–(8) 437
Local Government (Wales) Act
    1994 443

Magistrates' Courts Act 1980 . . . . 8.79,
                                   8.86, 10.151, 430
    ss 1–2. . . . . .8.28
    s 5A 2.135, 8.116
    s 5B 2.135
    s 5D(2). 8.116
    s 17A 8.81
    s 17A(4) 8.72
    s 17A(6) 8.72
    s 18 . . . . .8.81
    s 19 . . . . .8.81
    s 19(2)–(3) 8.104
    ss 20–21. . . . . .8.81
    s 44(1)–(2) .5.29
    s 46 . . . . .8.78
    s 97 . . . . .8.182
    s 100(1)–(2) 8.40
    s 123 . . . . .8.40
    s 123(1)–(2) 8.40
    s 127 . . . . 8.15, 8.16
    Sch 3 . . . . .5.10
        para 1 . . . . .8.114

para 3(1). 8.114	ss 33–34 448
r 2 . . . . 8.78, 8.114	s 42 448
r 4 . . . . 8.80	Sch 5 448
Medicines Act 1968, s 58 571	Police Act 1996
Merchant Shipping Act 1995,	s 64 441
s 313(1) 553	s 88(1) 441
Mines Management Act 1971 . . . . . 448, 665	s 89(2) 7.57, 7.58
Mines and Quarries Act 1954 . . . . 2.107,	s 97(9) 441
448, 537, 641, 660, 662, 665	Police Act 1997, Sch 8, para 7(1). 441
s 64(2) 664	Police and Criminal Evidence
s 151 448	Act 1984 . . . . .1.82, 2.61, 2.64,
s 157 664	2.71, 2.144, 2.145, 3.11,
s 180 . . . . 476, 568	3.23, 3.31, 3.32,
s 180(5) 476	3.37, 3.53, 458
Mines and Quarries (Tips)	s 8 . . . . .2.70
Act 1969 665	s 10 . . . . .3.75
ss 1–10 448	ss 11–14 . . . . .2.70
Ministry of Fuel and Power Act 1945 . . . 447	s 17 . . . . .2.70
s 1(1) 447	s 18 . . . . .2.70
	s 18(2)–(4) .2.70
National Heritage Act 1983 . . . . 5.05	s 18(5)–(5A). 2.70
Sch 1 . . . . 5.17	s 19 . . . . .2.70
National Minimum Wage Act 1998	s 19(2)–(4) .2.70
s 19 489	s 20 . . . . .2.71
s 20 489	s 23 . . . . .2.70
s 22 489	s 24 . . . . .2.59, 2.60, 2.106,
Norfolk and Suffolk Broads Act	3.03, 3.07, 8.03
1988. 428	s 24(5) 2.60
Nuclear Installations Act 1965 438	s 24(5)(c)(iv) 2.60
s 1 448	s 24(5)(e) 2.66
s 1(1)(b) 436	s 24(6) 2.60
ss 3–6 448	s 24A 2.61, 2.67, 2.106, 8.03
s 12 438	s 24A(3) 2.62
s 22 448	s 24A(4) 2.62, 2.63, 2.67
s 24A 448	s 25(5)(e) 3.02
Sch 2 448	s 32 . . . . .2.70
	s 58 . . . . .3.35
Offences Against the Person Act 1861	s 67(9) 2.72, 2.74, 2.75
s 18 . . . . .9.49	s 67(10) 2.73
s 20 . . . . .9.49	s 74 . . . . .5.55
Offices, Shops and Railway Premises	s 78 . . . . .2.97, 2.148, 3.78,
Act 1963 . . . . 4.98, 448, 545	3.90, 11.141
s 19 563	s 78(1) 2.96
s 90(4) 545	Police (Health and Safety) Act
Offshore Safety Act	1997. . . . . 6.19, 440, 441, 442
1992. . . . . .429, 430, 447	Police (Scotland) Act 1967,
s 1(1) 417	s 39 441
	Powers of Criminal Court
Perjury Act 1911, s 5. . . . . 2.138, 7.76	(Sentencing) Act 2000
Petroleum (Consolidation)	s 3(2) 8.83
Act 1928 447	s 4 . . . . .8.83
Petroleum (Transfer of Licences)	s 5 . . . . .8.100
Act 1936 447	s 12 . . . . .10.04
Pipe-lines Act 1962	ss 130–131 .10.05
ss 20–26 448	s 151 . . . . .8.97

Prevention of Crime Act 1953,
    s 1(4) 553
Prosecution of Offences Act 1985 . . . . 10.130
    s 16(1)–(3) 10.138
    s 16(6) 10.139, 10.142, 10.149
    s 16(7) 10.142, 10.146
    s 16(9) 10.141
    s 18 . . . . 10.131
    s 18(1) 10.131
    s 21(1) 10.140
Public Health Act 1961, s 73 448
Public Health (Smoke Abatement)
    Act 1926 447

Race Relations
    Act 1976 . . . . 485, 520
    Pt II 503
    s 54 503
    s 59 490
Radioactive Substances Act 1948,
    s 5(1)(a) 447
Railways Act 1993, s 117(4) 417
Railways Act 2005 . . . . 413, 415, 417,
                            419, 434, 435, 439
    Sch 3 . . . . 434, 444
    para 1(3) 417, 435
    para 2(5) 439
Railways and Transport Safety
    Act 2003 . . . . 426, 434, 435, 446
Redundancy Payments Act 1965 445
Rent Act 1977 653
Rent (Agriculture) Act 1976 653
Revenue Act 1909, s 11 447
Road Traffic Act 1988
    ss 1–2 . . . . 11.145
    s 3A 11.145
    s 3ZB 11.145
    s 172(5) 5.47
Roads (Scotland) Act 1984 477
    s 151 . . . . 537, 569

Scotland Act 1998
    Sch 6
        para 1 518
        para 37 483
Serious Organised Crime and Police
    Act 2005 . . . . 2.58, 8.03, 440, 441
    s 28 441
    s 110 . . 2.59, 2.61, 2.69, 2.70, 2.106, 8.03
    s 111 . . . . . . 2.59, 2.61, 2.69, 2.70, 2.106
    Sch 7 . . . . 2.69, 2.70, 2.106
Sex Discrimination Act
    1975 . . . . 448, 485, 520
    Pt II 503

s 63 503
s 68 490
Sex Discrimination Act 1986 . . . . . 448, 520
    s 6(4A) 517
Statistics of Trade Act 1947 425
    s 9 425
Statute Law (Repeals) Act 1993 . . . . 446, 453
Statutory Instruments Act 1946 . . . . 1.07
Suicide Act 1961, s 2(1) 11.145
Supreme Court Act 1981
    s 28 . . . . 9.05
    s 29(3) 9.04, 9.06

Territorial Waters Jurisdiction
    Act 1878, s 3 446
Theft Act 1968, s 18 . . . . 5.47, 5.61
Trade Descriptions Act
    1968 . . . . 5.15, 5.39, 5.47
    s 1 . . . . 5.39
    s 9 . . . . 5.39
    ss 12–14 . . . . 5.39
    s 20 . . . . 5.47
    s 20(1) 5.15
    s 23 . . . . . 5.31, 5.34, 5.35, 5.37, 5.39, 5.43
    s 24 . . . . 5.35, 5.43
    s 24(1) 5.35
Trade Union and Labour Relations
    Act 1974 445
Trade Union and Labour Relations
    (Consolidation) Act 1992 . . . . 445, 485
    s 68 504
    s 161 501
    ss 168–170 504
    s 178(2) 500
    s 192 504
    s 219 518
    s 238A 518
Transport Act 1968
    s 91(2)(bb) 2.153
    s 99(1)(bb) 2.153
Tribunals and Inquiries Act 1992 435
    s 8(1) 483
    s 11 . . . . 4.126
    s 11(1) 4.122

Vehicles (Excise) Act 1971 538
Veterinary Surgeons Act 1966, s 27 477
Visiting Forces Act 1952
    Pt I 534, 545, 562, 579
    s 12 478
    s 12(1) 533, 545, 549, 562

Water Act 1989 427
Water Resources Act 1991, s 217(2) 5.31

Table of Statutes

Youth Justice and Criminal Evidence
    Act 1999
    s 59 .... 7.76
    Sch 3 .... 7.76

Zoo Licensing Act 1981, s 1(2) 477

EUROPEAN UNION

EC Treaty
    art 137 .... 1.48
    art 226 .... 1.57
Single European Act 1987............ 1.47
    art 118A.................. 1.47, 1.48
Treaty Establishing the European
    Community, art 234............ 519
Treaty of Rome.................... 1.46

EUROPEAN UNION—SECONDARY LEGISLATION

Directive 76/769/EEC 634
Directive 82/130/EEC 663
Directive 88/35/EEC 663
Directive 89/391/EEC
    (Framework) 1.48, 1.49, 1.53,
        1.59, 1.60, 12.11, 12.153
    s 1 .... 1.49
    s 2 .... 1.50
    arts 1–3 .... 1.49
    art 4(1)....................... 1.49
    art 5 .... 6.104
    art 5(1) 1.50, 1.53, 1.55,
        1.57, 6.104, 6.139
    art 5(2)....................... 1.57
    art 5(3)........ 1.57, 6.103, 6.104, 6.140
    art 5(4)...... 1.51, 1.53, 1.56, 1.57, 1.58
    art 6 .... 1.52, 1.57
    art 6(1)....................... 1.52
    art 6(2)............... 1.52, 12.36, 536
    art 6(3)....................... 1.52
    arts 7–12 .... 1.57
Directive 89/654/EEC
    (Workplace) 12.48
Directive 89/655 (Work
    Equipment) 12.77, 12.115
Directive 90/269/EEC 12.60
Directive 91/269/EEC 663
Directive 92/85/EEC, Annexes I–II 532
Directive 94/33/EC, Annex 527
Directive 94/44/EC 663
Directive 95/63/EC (Work Equipment
    Amendment) 12.77
Directive 98/24/EC, art 9 579
Directive 98/65/EC 663

NEW ZEALAND

Interpretation Act 1924, s 4 .... 5.20

INTERNATIONAL

Agreement on the European Economic
    Area 1992................... 567
Brussels Protocol 1993 567
European Convention on Human
    Rights............ .2.35, 6.138, 9.05,
        9.33, 11.112
    art 2 .... .2.35, 11.108, 11.109,
        11.110, 11.111, 11.114,
        11.116, 11.121, 11.132
    art 6 .... 3.86, 3.87, 8.119, 9.09,
        9.28, 9.29, 10.138
    art 6(1).......... .3.79, 9.28, 9.30, 9.31
    art 6(2)............. 6.138, 7.16, 7.17
    art 6(3)(a)..................... 8.39
    art 7 .... 11.26

TABLE OF STATUTORY INSTRUMENTS

Active Implantable Medical Devices
Regulations 1992
(SI 1992/3146). 564
Air Navigation (Dangerous Goods)
Regulations 1994
(SI 1994/3187). 10.28
Asbestos (Licensing) Regulations 1983
reg 4(1). 634
reg 7(1). 634
Asbestos (Prohibitions) Regulations 1992
reg 7. . . . . 632, 633
reg 8(1). 634
reg 8(3). 635
Asbestos Regulations 1969,
reg 13. 633

Building (Safety, Health and Welfare)
Regulations 1948, reg 24(1) 7.37

Cableway Installations Regulations
2004 (SI 2004/129) 564
Carriage of Dangerous Goods and
Use of Transportable Pressure
Equipment Regulations
2004. . . . . 631, 636
Chemicals (Hazard Information and
Packaging for Supply) Regulations
1994, reg 2(1). 477
Chemicals (Hazard Information and
Packaging for Supply) Regulations
1999. 580
Chemicals (Hazard Information and
Packaging for Supply) Regulations
2002 (SI 2002/1689) 12.122,
568, 569, 636
reg 2(1). 567, 569
reg 4. 568
reg 5. 569
Civil Procedure Rules 1998
(SI 1998/3132). 4.125, 512, 523
Coal Mines (Respirable Dust)
Regulations 1975 570
Coal and Other Mines (Electricity)
Regulations 1956
reg 20. 664
reg 21A. 664
Confined Spaces Regulations 1997
(SI 1997/1713), reg 7 . . . . .7.31

Construction (Design and
Management) Regulations
1994 (SI 1994/3140) 5.14,
12.146, 12.147, 12.150, 12.152,
12.153, 12.169, 12.175, 12.180
reg 2(1). 475, 641
reg 4. 600
reg 6. 600
reg 7. 600
reg 7(1). 482
reg 14. .12.169
Construction (Design and
Management) Regulations
2007 (SI 2007/320) 12.146–191,
583, 605
Pt 1 . . . . .12.151
Pt 2 . . . . 12.151, 12.153
Pts 3–4. . . . . .12.151
reg 2(1). 605
regs 4–5 . . . . 12.161, 12.176
reg 6. . . . . 7.23, 12.161
reg 7. . . . . 12.153, 12.161
reg 9. . . . . 12.160, 12.161, 12.163
reg 9(1)–(2) 12.159
reg 10(1). 12.162
reg 11. . . . . 12.160, 12.168
reg 11(3)–(4) 12.176
reg 13(2). 12.187
reg 13(3). 12.188
reg 13(7). 12.187
reg 14(1). 12.166, 12.171
reg 14(2). 12.166, 12.180
reg 14(4). 12.166
reg 14(5). 12.146
regs 15–17 . . . . .12.166
reg 18. .12.177
reg 19(1). 12.189
reg 19(2)(a)(i)–(ii) 12.189
reg 19(2)(b) 12.190
reg 19(2)(d) 12.190
reg 20(1)–(2) 12.171
reg 22. .12.181
reg 22(1)(a) 12.180
reg 22(2)(a)–(b) 12.183
reg 23(1)–(2) 12.186
reg 33(1)(b) 602
apps 4–6. . . . . .12.155
app 7 . . . . .12.153

Table of Statutory Instruments

Construction (Head Protection)
  Regulations 1989 557
Construction (Health, Safety and
  Welfare) Regulations 1996
  (SI 1996/1592) 7.35, 12.140,
           12.142, 12.146, 12.150,
           12.151, 535, 538
  reg 2(1) 538
  reg 3(2) 538
  reg 6 . . . . 10.58, 10.66
  reg 20(2) 535
  reg 29 554
Construction Plant and Equipment
  (Harmonisation of Noise Emission
  Standards) Regulations 1985
  (SI 1985/1968) 563
Construction Plant and Equipment
  (Harmonisation of Noise Emission
  Standards) Regulations 1988
  (SI 1988/361) 563
Construction Products Regulations
  1991 (SI 1991/1620) 563
Construction (Working Place)
  Regulations 1966 . . . . 7.36
  reg 36(2) 7.36, 7.37
Control of Asbestos at Work
  Regulations 1987 556
Control of Asbestos at Work
  Regulations 2002 571
  reg 2(1) 482
  reg 25(1) 634
  reg 27(1) 635
Control of Asbestos Regulations 2006
  (SI 2006/2739) 12.118,
           12.193–210, 619
  Pt 2 . . . . 12.193, 12.194
  Pt 3 . . . . 12.193
  reg 4 . . . . 12.194
  reg 4(1)(a) 12.194
  reg 4(3) 12.195
  reg 4(7) 12.195
  reg 4(8) 12.196
  reg 4(9)(c) 589
  reg 5 . . . . 12.199
  reg 6 . . . . 12.200
  reg 7 . . . . 12.201
  reg 8 . . . . 7.09
  reg 10 . . . . 12.202
  reg 11 . . . . 12.203
  reg 11(1)(a)–(b) 12.203
  reg 11(2) 12.204
  reg 11(2)(b) 12.204
  reg 11(3) 12.205
  reg 11(5)(a)–(b) 12.206
  reg 12(1) 12.207
  reg 13 . . . . 12.208
  reg 13(3) 12.209
  reg 14(4) 636
  reg 20(4) 619
  reg 24(2)–(3) 636
  reg 30(1)–(2) 636
Control of Lead at Work Regulations
  1998 557
Control of Lead at Work Regulations
  2002 571
Control of Noise at Work Regulations
  2005 557
Control of Substances Hazardous to
  Health Regulations 1994 557
Control of Substances Hazardous to
  Health Regulations
  1999 . . . . 6.117, 7.19
  reg 6 . . . . 6.119
  reg 7 . . . . 7.26, 7.27, 7.28, 7.29
  reg 7(1) 6.118, 6.119, 6.125,
           6.134, 7.25, 7.26, 7.28
Control of Substances Hazardous to
  Health Regulations 2002
  (SI 2002/2677) 567-81, 6.118,
           7.19, 7.25, 12.118,
           12.120–38
  reg 2(1) 12.122, 12.135
  reg 3(1) 12.123, 12.126
  reg 4 . . . . 12.124
  reg 5 . . . . 12.138
  reg 6 . . . . 12.18, 12.138
  reg 6(1) 12.125
  reg 6(2) 12.127
  reg 6(3) 12.128
  reg 7 . . . . 12.127, 12.134, 12.138
  reg 7(1)–(2) 12.129
  reg 7(3) 12.129, 12.130
  reg 7(4) 12.130
  reg 7(5)–(6) 12.132
  reg 7(7) 12.131
  reg 7(11) 12.131
  reg 8 . . . . 12.123,
           12.133, 12.138
  reg 9 . . . . 12.134, 12.138
  reg 10 . . . . 12.127, 12.134, 12.138
  reg 11 . . . . 12.135, 12.138
  reg 12 . . . . 12.138
  reg 12(1)–(2) 12.136
  reg 13 . . . . 12.137, 12.138
  Sch 2 . . . . 12.124
  Sch 6 . . . . 12.135
Control of Vibration at Work
  Regulations 2005 557

Table of Statutory Instruments

Coroners Rules 1984
    (SI 1984/552)........ 11.99, 11.110
    r 4.3....................... 11.105
    r 20(1) 11.123
    r 20(2) 11.123
    r 20(2)(d) 11.124
    r 20(2)(h) 11.124
    r 22(1)–(2) 11.127
    r 24 .... 11.122
    r 36(1) 11.117
    r 36(1)(b) 11.113
    r 36(2) 11.117
    r 37 .... 11.132, 11.140
    r 37(1)–(2) 11.130
    r 37(3) 11.130, 11.131
    r 37(4)–(6) 11.130
    r 42 ..... 11.118, 11.119, 11.120, 11.121
    r 57 .... 11.129
Costs in Criminal Cases (General)
    Regulations 1986
    (SI 1986/1335)............. 10.130,
                         10.144, 10.147
    reg 7(1)(a) 10.145
    reg 9....................... 10.150
    reg 10(1)................... 10.150
    reg 14...................... 10.131
CPIA 1996 Code of Practice 1997 ..... 2.52
CPIA 1996 Code of Practice 2005
    Preamble 2.49
    para 1.1 .... 2.50, 8.173
    para 2.1 .... 2.52, 8.181
    para 3.4 .... 8.177
    para 3.5 .... 2.51, 8.181
    para 4.1 .... 8.177
    paras 5.6–5.7 .... 8.179
    para 14..................... 8.186
Criminal Procedure and Investigations
    Act 1996 (Code of Practice) Order
    1997 (SI 1997/1033) 2.49
Criminal Procedure and Investigations
    Act 1996 (Code of Practice) Order
    2005 (SI 2005/985) 2.49, 2.50,
                    2.51, 2.52, 8.159, 8.173,
                    8.177, 8.179, 8.181, 8.186
Criminal Procedure Rules 2005
    (SI 2005/384)....... 8.02, 8.44, 8.118,
                    8.120, 8.133
    Pt 4 .... 8.26
    Pt 7 .... 8.26
    Pt 33 .... 8.199
    r 1.1(e)...................... 8.68
    r 3.4........................ 8.148
    r 3.5(3)...................... 8.44
    r 4 .... 8.33
    r 4.1........................ 8.33
    r 4.1(2)...................... 8.34
    r 7(1) 8.27
    r 7.2(1)–(2) 8.36
    r 7.7..................... 8.36, 8.123
    r 7.7(3)...................... 8.37
    r 10.5....................... 8.116
    r 14.2....................... 8.121
    r 21 .... 8.48
    rr 21.2–21.3................... 8.48
    rr 21.5–21.6................... 8.52
    r 24 .... 8.201
    r 24.1....................... 8.202
    r 25.6....................... 8.161
    r 35.1....................... 8.226
    r 53(3) 8.195
Crown Court (Advance Notice of
    Expert Evidence) Rules 1987
    (SI 1987/716)
    r 3 .... 8.201
    r 5 .... 8.203

Defence Disclosure Time Limit Rules
    1997, r 2.................... 8.166
Docks Regulations 1988
    reg 2(1)................ 475, 476, 538
    reg 7(6)...................... 606

Electrical Equipment (Safety)
    Regulations 1994
    (SI 1994/3260)............ 563, 564
Electricity at Work Regulations 1989
    (SI 1989/635)....... 12.225–239, 659
    reg 2(1)...................... 12.229
    reg 3(1)...................... 12.225
    reg 4(1)............... 12.228, 12.233
    reg 4(2)–(3) 12.228, 12.231, 12.233
    reg 4(4)......... 12.226, 12.228, 12.233
    reg 5................... 12.226, 12.234
    regs 6–7 .... 12.234, 12.235
    regs 8–11 .... 12.226, 12.234
    reg 12.......... 12.226, 12.231, 12.234
    reg 13.......... 12.226, 12.231, 12.236
    reg 14.......... 12.226, 12.231, 12.236,
                        12.237, 12.238
    regs 15–16 .... 12.226, 12.231
    reg 25....................... 12.226
    reg 29.......... 12.233, 12.235, 12.238
Electricity (Factories Act) Special
    Regulations 1908 .... 10.45
Electricity (Factories Act) Special
    Regulations 1944 .... 10.45
Electricity Supply Regulations
    1988, reg 3(1)................. 476

Table of Statutory Instruments

Electro-medical Equipment (EEC
    Requirements) Regulations 1988
    (SI 1988/1586)............... 563
Electromagnetic Compatibility
    Regulations 1992
    (SI 1992/2372)................ 563
Employer's Health and Safety Policy
    Statement (Exceptions)
    Regulations 1975
    (SI 1975/1584)................6.16
Employment Equality (Age)
    Regulations 2006.............. 503
Employment Equality (Religion or
    Belief) Regulations 2003 503
Employment Equality (Sexual
    Orientation) Regulations
    2003..................... 503
Employment Tribunals (Constitution
    and Rules of Procedure)
    Regulations 2001 .... 483, 484
    Sch 1 491
    r 1 491
    r 2 491
    r 2(2) 491
    r 2(4)–(5)..................... 491
    r 3 491
    r 8 491
    r 14 491
    Sch 2, r 1 491
    Schs 3–6...................... 491
Employment Tribunals (Constitution
    and Rules of Procedure) Regulations
    2004 (SI 2004/1861) 483
    reg 3.......................4.61
    reg 8(3)(a)–(c) 4.76
    reg 9........................4.76
    reg 15.......................4.64
    reg 16(3)(b) 521
    Sch 1 .... 4.59, 492
    r 10 ....4.73
    r 10(2) 4.72
    r 10(2)(g) 4.72
    r 10(2)(i) 4.72
    r 10(2)(k)–(l) 4.72
    r 10(2)(r) 4.72
    r 14(4) 4.78
    r 16 ....4.75
    r 26 ....4.75
    r 27(3)–(4)....................4.80
    r 27(5)–(6)....................4.79
    r 28 ....4.77
    r 29 ....4.120
    r 30 ....4.121
    r 30(3) 4.121
    r 30(5) 4.121
    r 34 .... 4.70, 4.122
    r 34(1)(b) 4.122
    r 34(3) 4.70
    rr 35–36................ 4.70, 4.122
    r 60 ....4.60
    r 61 ....4.64
    r 61(2) 4.64
    Sch 4 .... 4.59, 4.62, 4.123, 521
    r 3 ....4.62
    r 4 ....4.63
    r 4(2) 4.63, 4.66
    r 5 ....4.67
    r 6 ....4.69
    r 6(1) 4.69
    r 6(3) 4.69
    r 6(4)–(5).....................4.70
    r 7 ....4.72
    r 9 ....4.71
    r 10 ....4.123
    r 10(1) 4.123
    r 10(2) 4.124
    r 10(3) 4.125
    r 10(4) 4.123
    r 11 ....4.72
Employment Tribunals (Constitution
    and Rules of Procedure) (Scotland)
    Regulations 2001 485
    reg 20........................ 483
    Sch 1 491
    r 1 491
    r 2 491
    r 2(2) 491
    r 2(4)–(5)..................... 491
    r 3 491
    r 8 491
    r 14 491
    Sch 2, r 1 491
    Sch 3 491
    r 13(3) 491
    Schs 4–6...................... 491
Employment Tribunals Extension of
    Jurisdiction (England and Wales)
    Order 1994, art 4 496
Employment Tribunals Extension of
    Jurisdiction (Scotland) Order 1994,
    art 4 496
Employment Tribunals (Health and
    Safety—Appeals Against
    Improvement and Prohibition
    Notices) Rules of Procedure
    2004........................ 483
Employment Tribunals (Levy Appeals)
    Rules of Procedure 2004......... 483

Employment Tribunals (National Security) Rules of Procedure 2004 483
Employment Tribunals (Non-Discrimination Notices Appeals) Rules of Procedure 2004. 483
Equipment and Protective Systems Intended for Use in Potentially Explosive Atmospheres Regulations 1996 (SI 1996/192) 564, 663

Fire Precautions (Workplace) Regulations 1997, Pt III. 535

Gas Appliances (Safety) Regulations 1992 (SI 1992/711) 563
Gas Appliances (Safety) Regulations 1995 (SI 1995/1629) 564
Gas Safety (Installation and Use) (Amendment) Regulations 1996. 657
Gas Safety (Installation and Use) (Amendment) (No 2) Regulations 1996 657
Gas Safety (Installation and Use) Regulations 1994 . . . . 6.91, 657
reg 2(1). 477
Gas Safety (Installation and Use) Regulations 1998 (SI 1998/2451). 11.51, 11.53, 12.211–224, 639
reg 6(1)–(6) 12.218
reg 26. . . . . 12.219
reg 34. . . . . 12.214
reg 36. . . . . 12.220, 12.221
reg 38. 657
Gas Safety (Management) Regulations 1996 (SI 1996/551) 12.211
reg 2. 656

Health and Safety at Work etc. Act 1974 (Application outside Great Britain) Order 1989 549
Health and Safety at Work etc. Act 1974 (Application outside Great Britain) Order 1995 . . . . 534, 552, 553, 665
Health and Safety at Work etc. Act 1974 (Application outside Great Britain) Order 2001 . . . . 580, 606, 634
art 8(1)(a). 585
Health and Safety (Display Screen Equipment) Regulations 1992 . . . . 1.61
Health and Safety Enforcement Regulations 1999 . . . . 1.32
Health and Safety (Enforcing Authority for Railways and Other Guided Transport Systems) Regulations 2006 620
reg 3(1). 580, 591
reg 3(2). 477
Health and Safety (Enforcing Authority) Regulations 1989 (SI 1989/1903). 2.91
reg 3 . . . . 2.91
Sch 1 . . . . 2.91
Health and Safety (Enforcing Authority) Regulations 1998 (SI 1998/494). 1.36, 2.06, 2.38, 475, 620
reg 3(1). 1.35
reg 6. 480
Schs 1–2. 481
Health and Safety (First-Aid) Regulations 1981 534
reg 6. 534
Health and Safety Inquiries (Procedure) Regulations 1975 (SI 1975/335). 1.22, 7.43
Health and Safety (Miscellaneous Amendments) Regulations 2002. . . . . 12.60
Health and Safety (Safety Signs and Signals) Regulations 1996. 615
Health and Safety (Training for Employment) Regulations 1990 (SI 1990/1380). 6.20, 6.55, 12.55
Health and Safety (Young Persons) Regulations 1997 (SI 1997/135). 12.34, 535
Housing (Scotland) Act 1987 654
Housing (Scotland) Act 1988 654

Ionising Radiations Regulations 1985. 556
Ionising Radiations Regulations 1999 (SI 1999/3232) 482
reg 2(1). 476, 482
Sch 1 482

Lawnmowers (Harmonisation of Noise Emission Standards) Regulations 1992 (SI 1992/168) 563
Lifting Operations and Lifting Equipment Regulations 1998
reg 9. 610
reg 9(4). 610
reg 10. 610

Lifts Regulations 1997
  (SI 1997/831) 564
Loading and Unloading of Fishing Vessels
  Regulations 1988, reg 5(3) 606
Low Voltage Electrical Equipment
  (Safety) Regulations 1989
  (SI 1989/728) 563

Magistrates' Courts (Advance Notice
  of Expert Evidence) Rules 1997
  (SI 1997/705) 8.204
  r 3 . . . . 8.201
Management and Administration of
  Safety and Health at Mines
  Regulations 1993, reg 29(5) 662
Management of Health and Safety at
  Work and Fire Precautions
  (Workplace) (Amendment)
  Regulations 2003
  (SI 2003/2457) 12.11, 526, 533
Management of Health and Safety at
  Work (Amendment) Regulations
  1994 535
Management of Health and Safety at
  Work Regulations 1992 535
Management of Health and Safety at
  Work Regulations 1999
  (SI 1999/3242) 6.16, 7.32,
            12.03, 12.09–46, 12.28, 12.62,
                      525, 578, 605, 627
  reg 3 . . . . . 8.139, 546, 588, 607, 617, 618
  reg 3(1) 12.13, 12.14, 12.35,
                12.62, 12.114, 12.119
  reg 3(2) 12.13
  reg 3(5) 12.34
  reg 3(6) 12.15
  reg 4 . . . . 12.36
  reg 5(1)–(2) 12.38
  reg 6 . . . . 12.39, 548
  reg 7(5) 12.40
  reg 8 . . . . 12.45
  reg 10 . . . . 12.41
  reg 11 . . . . 12.111
  reg 11(1)(a)–(c) 12.27
  reg 13(2) 12.42
  reg 13(2)(b) 588
  reg 14 . . . . 12.44
  reg 15 . . . . 12.43
  reg 16(1) 12.35
  reg 18 . . . . 12.35
  reg 21 . . . . 6.100, 6.101, 6.102, 6.103,
        6.106, 6.108, 6.115, 7.21, 7.22, 7.23,
              7.30, 7.32, 7.33, 12.46, 581, 635
  Sch 1 . . . . 12.36, 12.153, 584

Manual Handling Operations
  Regulations 1992
  (SI 1992/2793) 1.61, 12.60–76, 547
  reg 2 . . . . 12.75
  reg 2(2) 12.61
  reg 4 . . . . 12.72
  reg 4(1) 12.76
  reg 4(1)(a) 12.63
  reg 4(1)(b) 12.68
  reg 4(1)(b)(i) 12.16, 12.66, 549
  reg 4(1)(b)(ii) 12.67
  reg 4(1)(b)(iii) 12.74
  reg 4(2) 12.69
  Sch 1 . . . . 12.66, 549
Manufacture and Storage of Explosives
  Regulations 2005 479
  reg 2(1) 479
  reg 6 479
  reg 11 479
  reg 13 479
  reg 13(3) 479
  reg 25 479
  reg 27(19) 479
  Sch 1, para 1 479
Medical Devices Regulations 1994
  (SI 1994/3017) 563, 564
Merchant Shipping (Guarding of
  Machinery and Safety of
  Electrical Equipment) Regulations
  1988, regs 3–4 553
Merchant Shipping (Hatches and
  Lifting Plant) Regulations 1988,
  regs 5–10 553
Mines Miscellaneous Health and Safety
  Provisions Regulations 1995 534
  reg 4(2)(b) 535
  reg 4(5) 535
Mines (Shafts and Winding) Regulations
  1993 554

Noise at Work Regulations 1989 557
Noise Emission in the Environment by
  Equipment for Use Outdoors
  Regulations 2001 (SI 2001/1701) . . 564

Offshore Installations and Pipeline
  Works (First-Aid) Regulations
  1989 534
  reg 7(1) 534
  reg 7(3) 534
Offshore Installations and Pipeline
  Works (Management and
  Administration) Regulations
  1995, reg 3 478

Offshore Installations and Wells (Design
   and Construction etc) Regulations
   1996, reg 12 606

PACE Codes of Practice 2.97
   Code A....................... 2.72
   Code B.................. 2.72, 2.77
   Code C......... 2.72, 2.79, 2.82, 2.83,
         2.84, 3.11, 3.33, 3.35, 3.40,
               3.43, 3.45, 3.50
      para 10A..................... 2.79
      para 10D 3.40
      para 10.1 .... 2.78, 2.79, 2.81,
                 2.84, 2.86, 3.31
      para 11................. 2.84, 3.34
      para 11.1 .... 2.83
      para 11.1A 2.81, 3.11, 3.31, 3.41
      para 11.2 .... 2.76, 2.83, 3.36
      para 11.3 .... 2.83
      para 11.4 .... 2.83, 3.42
      para 11.5 .... 2.83, 3.44
      para 11.6 .... 2.83, 3.46
      para 11.7 .... 2.83
      para 11.7(a)–(c) 3.34
      paras 11.8–11.13 .... 2.83, 3.34
      paras 11.14–11.20 .... 2.83
      Note 11A 3.42
      para 12.5 .... 3.32
      para 16.5 .... 3.48
      Annex D..................... 3.50
   Code D 2.72
   Code E.................. 2.72, 3.33
   Codes F–G.................... 2.72
Personal Protective Equipment at
   Work Regulations 1992
   (SI 1992/2966), reg 6 ..... 1.61, 12.17
Personal Protective Equipment
   (EC Directive) Regulations 1992
   (SI 1992/3139)................ 564
Personal Protective Equipment
   Regulations 2002 .... 573, 626
Pipelines Safety Regulations 1996
   (SI 1996/825)............... 12.211
   reg 3......................... 482
   reg 28....................... 7.31
Police and Criminal Evidence Act 1984
   (Codes of Practice) Order 2005
   (SI 2005/3503).......2.76, 2.77, 2.78,
             2.79.2.81.2.83.2.84,
             3.11, 3.31, 3.32, 3.33, 3.35,
             3.40, 3.41, 3.42, 3.45, 3.46,
                 3.48, 3.50
Pressure Equipment Regulations 1999
   (SI 1999/2001)................ 564

Pressure System Safety Regulations 2000
   (SI 2000/128), reg 16 .... 7.31
Provision and Use of Work
   Equipment Regulations 1998
   (SI 1998/2306)...... 1.61, 12.27–119,
              12.77–119, 551, 612
   reg 4 .... 12.80, 12.84, 12.85
   reg 4(1)............. 12.80, 12.81
   reg 4(2)............. 12.80, 12.82
   reg 4(3)............. 12.80, 12.83
   reg 4(4)...................... 12.81
   reg 5 .... 12.208
   reg 5(1)........ 12.88, 12.89, 12.91
   reg 5(2).............. 12.88, 12.90
   reg 6......................... 12.97
   reg 6(2)...................... 12.99
   reg 6(3)..................... 12.100
   reg 6(4)..................... 12.101
   reg 6(5)....................... 612
   reg 8(1)..................... 12.108
   reg 9 .... 12.111
   reg 9(1)..................... 12.110
   reg 10 563
   reg 11(1).................... 12.112
   reg 11(1)(a) 10.84
   reg 12 .... 12.116
   reg 12(5)..................... 12.118
   reg 13(3)..................... 12.209
   regs 31–33 564
   reg 34 564
   reg 34(1)(b) 565
   reg 35 564
   Sch 1 563
   Sch 2 564
   Sch 3 565

Quarries Regulations 1999 .... 537, 641
   reg 3..................... 477, 660
   reg 45(1)..................... 481

Regulatory Reform (Fire Safety)
   Order 2005 .... 423, 599
   art 25 .... 423, 599
Rent (Scotland) Act 1984 654
Reporting of Injuries, Diseases and
   Dangerous Occurances Regulations
   1985 (SI 1985/2023) 2.05
Reporting of Injuries, Diseases and
   Dangerous Occurances Regulations
   1995 (SI 1995/3163) 2.01, 2.05,
           2.08, 2.09, 2.10, 7.73, 9.16,
              11.104, 12.190, 590
   reg 2..................... 2.17, 2.21
   reg 2(2)(c) 2.17

Reporting of Injuries, Diseases and
    Dangerous Occurances Regulations
    1995 (SI 1995/3163) (cont.)
    reg 3 . . . . .2.13, 2.14, 2.21, 2.22
    reg 3(1) 2.10, 2.11
    reg 3(1)(i)–(ii) 2.12
    reg 3(2) 2.11
    reg 4 . . . . 2.13
    reg 5 . . . . 2.19, 2.21
    reg 5(1) 2.22
    reg 5(3) 2.20
    reg 6 . . . . 2.10
    reg 7 . . . . 2.10, 2.22, 2.23
    regs 8–9 . . . . 2.10
    reg 10 . . . . 2.10
    reg 10(1) 2.14
    reg 10(3) 2.16
    reg 10(5) 2.12
    reg 11 . . . . 2.23
    Sch 1 . . . . 2.09
    Sch 2 . . . . 2.09, 2.18
    Sch 3 . . . . 2.09, 2.19, 2.20
    Pt 1 . . . . 2.19
    Sch 4 . . . . 2.09
Road Vehicles (Brake Linings Safety)
    Regulations 1999 637

Simple Pressure Vessels (Safety)
    Regulations 1991
    (SI 1991/2749) 563
Supply of Machinery (Safety)
    Regulations 1992
    (SI 1992/3073) 6.167, 563

Transfer of Undertakings (Protection
    of Employment) Regulations 2006,
    reg 15(10) 504

Visiting Forces and International
    Headquarters (Application of Law)
    Order 1965,
    art 3(2) 545, 549, 562
Visiting Forces and International
    Headquarters (Application of Law)
    Order 1999, Sch 2 . . . . 533, 579

Work at Height Regulations 2005 (SI
    2005/735) 7.36, 12.140–5, 541,
    554, 605
    reg 3(2)–(3) 12.140
    regs 4–5 . . . . 12.141
    reg 6(2) 12.142
    reg 6(3) 7.38, 12.142
    reg 6(4)(a) 612

reg 6(4)(a)(i) 12.143
reg 6(4)(b) 12.143
reg 6(5) 12.143
reg 7 . . . . 12.145
reg 8 . . . . 12.145
reg 8(b) 613
reg 8(c)–(d) 615
regs 9–11 . . . . 12.145
reg 12 . . . . 12.145, 554
reg 12(7) 618
reg 14 . . . . 12.144
reg 14(2) 12.144
Schs 1–3 612
Sch 4 615
Working Time Regulations 1998
    (SI 1998/1833)
    reg 13 504
    reg 14(2) 504
    reg 16(1) 504
Workplace (Health, Safety and
    Welfare) Regulations 1992
    (SI 1992/3004) 1.61, 12.47–59,
        12.48, 12.159, 12.175,
        537, 587, 588
    reg 2(1) 12.53, 585
    reg 3 . . . . 12.56
    reg 3(1) 12.56
    reg 3(3)–(5) 12.56
    reg 4(2) 12.49, 12.50
    reg 4(4) 12.49
    reg 5 . . . . 12.52
    reg 5(1) 12.57, 12.58
    reg 12(1) 12.59
    reg 24 . . . . 4.107
    Sch 1 545
SI 1989/1127 563
SI 1992/488 563
SI 1993/3074 564
SI 1994/2063 563
SI 1994/2326 564
SI 1994/3051 563
SI 1994/3080 563
SI 1994/3098 563
SI 1995/1671 564
SI 1995/2005 662
SI 1995/2036 . . . . 537, 538, 539
SI 1995/2357 563
SI 1996/3039 564
SI 1997/1993 665
SI 1998/81 564
SI 1999/860 553
SI 1999/2024 . . . . 477, 481, 537,
        641, 660
SI 1999/2550 663

Table of Statutory Instruments

SI 2001/3766 564
SI 2002/794 . . . . 416, 434, 439,
                              442, 445
SI 2002/2174 . . . . 537, 538, 539, 540,
                 544, 545, 548, 555, 556, 558, 562
SI 2002/2675 482
SI 2003/978 . . . . .568, 572, 578
SI 2004/2351 . . . . . 483, 484, 485, 489, 490,
                      492, 496, 507, 515, 521
SI 2004/3363 . . . . 427, 428
SI 2004/3386 . . . . 568, 569, 570, 571,
                           573, 574, 577
SI 2005/228 447
SI 2005/435 . . . . 492, 494
SI 2005/830 564
SI 2005/831 . . . . 563, 564
SI 2005/1082 . . . . .447, 448, 479
SI 2005/1093 557
SI 2005/1541 . . . . 423, 453, 478, 526,
                      527, 529, 530, 535
SI 2005/1643 557
SI 2005/1865 . . . . 483, 484, 485, 497,
                 498, 500, 501, 503, 504, 507,
                      510, 511, 513, 517, 523
SI 2005/2060 478
SI 2006/438 534
SI 2006/557 . . . . 475, 476, 477,
                           478, 482, 580
SI 2006/680 . . . . .484, 485, 486
SI 2006/2405 504
SI 2007/114 . . . . .606, 607, 611, 616
SI 2007/825 503

1

THE ENFORCEMENT FRAMEWORK

A. **Legislation, Regulations, and Guidance**	1.01	The Code for Crown Prosecutors	1.68
		The Enforcement Management Model	1.72
The Health and Safety at Work etc. Act 1974	1.01	E. **Methods of Enforcement**	1.75
Health and safety regulations	1.07	Introduction	1.75
Approved codes of practice and guidance	1.10	Early intervention	1.76
		Verbal warnings	1.77
B. **Institutions**	1.13	Letters of advice/recommendation	1.78
Introduction	1.13	Improvement notices and prohibition notices	1.79
The Health and Safety Commission	1.16	Permissioning and licensing regimes	1.80
The Health and Safety Executive	1.24	Formal cautions	1.81
Local authorities	1.32	Crown censure	1.88
Health and Safety (Enforcing Authority) Regulations 1998	1.36	F. **Prosecution and Particular Categories of Person**	1.89
The CPS and the DPP: the protocol for liaison	1.38	Prosecution of individuals	1.89
The Crown	1.41	Employees: HSWA 1974 s 7	1.92
C. **The European Revolution**	1.45	Directors and managers: HSWA s 37	1.95
The Framework Directive	1.45	Other individuals: HSWA ss 8 and 36	1.100
The daughter Directives	1.60	Prosecuting under HSWA s 3	1.102
D. **Principles of Enforcement Action**	1.62	Action to be taken when a decision is made to prosecute	1.104
Introduction	1.62		
The Enforcement Policy Statement	1.65		

A. Legislation, Regulations, and Guidance

The Health and Safety at Work etc. Act 1974

The cornerstone of the legal structure for the management and enforcement of health and safety law in the UK is the Health and Safety at Work Act 1974 (HSWA 1974). It was enacted following the publication of the Robens report into 'the **1.01** safety and health of persons in the course of their employment' in 1972.[1] The Act represented a watershed in the history of the regulation and enforcement of health and safety in the UK, and was arguably the single most significant legislative event in the field since the introduction of the Factories Acts during the Industrial Revolution. HSWA 1974 created new law and new institutions for the development and administration of that law. It brought together a series of disparate and industry-specific provisions into an instrument of general application; and a number of different inspectorates under the aegis of the HSC. With clearer definition of the duties incumbent on employers came extended powers of enforcement for inspectors.

1.02 Health and safety at work and the provision under the 1974 Act of bodies for its administration and enforcement are considered UK-wide matters and have not been devolved to the Scottish Parliament or Welsh Assembly. Rather, concordats with the Scottish Executive and the Cabinet of the Welsh Assembly exist, setting out arrangements for liaison with HSC/E on matters of common interest.

1.03 HSWA 1974 created the legal framework which regulates virtually all work activity in the UK. The general purposes of HSWA 1974 are set out in s 1, which provides, at s 1(1), that the provisions of HSWA 1974 Pt I:

> shall have effect with a view to:
> (a) securing the health, safety and welfare of persons at work;
> (b) protecting persons other than persons at work against risks to health or safety arising out of or in connection with the activities of persons at work;
> (c) controlling the keeping and use of explosive or highly flammable or otherwise dangerous substances, and generally preventing the unlawful acquisition, possession and use of such substances.

1.04 The approach of HSWA 1974 to achieving these purposes is exemplified by s 1(3), which provides:

> For the purposes of this Pt risks arising out of or in connection with the activities of persons at work shall be treated as including risks attributable to the manner of conducting an undertaking, the plant or substances used for the purposes of an undertaking and the condition of premises so used or any part of them.

1.05 The major provisions of HSWA 1974 are the general duties contained in ss 2–6. These are non-delegable duties placed on persons, whether human, corporate, or unincorporate, by virtue of their status of employer, self-employed, person in control of works, or occupier. The Act also contains provisions relating to the making of health and safety regulations and the preparation and approval of codes of practice; provisions defining the investigative and enforcement powers of health and

---

[1] Cmnd. 5034, 1972.

safety inspectors and provisions dealing with formal inquiries and powers of gathering information.

1.06 In terms of the development and administration of the law, HSWA 1974 provided for the creation of the Health and Safety Commission (HSC) and the Health and Safety Executive (HSE).

Health and safety regulations

1.07 If the primary tier of health and safety legislation contains the general duties at HSWA 1974 ss 2–7, the next tier down contains health and safety regulations.[2] In general terms, health and safety regulations are subordinate legislation created under HSWA 1974, as the parent or empowering Act. Section 15 of the Act gives the Secretary of State the power to make regulations for any of the general purposes of Pt 1 HSWA 1974.[3] One of the general purposes is set out in HSWA 1974 s 1(2); this provides that existing health and safety enactments in force at the time of its passage into law are to be '... progressively replaced by a system of regulations and approved codes of practice operating in combination with the other provisions of this Pt and designed to maintain or improve the standards of health, safety and welfare established by or under those enactments'. In practice, regulations are generally made on the proposal of the HSC, or by the Secretary of State after consultation with the HSC; but always after consultation with interested parties.[4]

1.08 At the time the Act came into force there were some thirty statutes and 500 sets of regulations. At the time of publication, some 150 sets of regulations have been made under the auspices of s 15 HSWA 1974, including all of the regulations created to give effect to the health and safety Directives emanating from Europe during the 'European Revolution' in the 1980s and 1990s. As of today, HSE 'owns'—ie is responsible for the enforcement of—some 200 sets of regulations.

1.09 In *R v Secretary of State for Trade and Industry ex parte UNISON, GMB and NASUWT* [1996] ICR 1003, it was held that s 2(1) of the European Communities Act 1972 enabled the repeal of a primary statute by regulations where the subject matter of such regulations 'relates to' a European law obligation.

Approved codes of practice and guidance

1.10 The tier beneath regulations contains approved codes of practice (ACOPs). By s 16 HSWA 1974 the HSC is empowered to approve codes of practice with the consent of the appropriate Secretary of State. According to the HSE, ACOPs offer

---

[2] Which are statutory instruments: see the Statutory Instruments Act 1946.
[3] HSWA 1974 s 15(1).
[4] See HSWA s 50.

practical examples of good practice, and give advice on how to comply with the law. For example, they may provide a guide to what is reasonably practicable for a work activity within a particular industry.

1.11 Though the codes of practice do not themselves impose legally enforceable duties, HSWA 1974 s 17 provides that, while a failure on the part of any person to observe any provision of an approved code of practice shall not of itself render him liable to any civil or criminal proceedings, failure to comply with the provisions may be taken by a court in criminal proceedings as evidence of a failure to comply with the requirements of HSWA 1974 or of regulations to which the ACOP relates, unless it can be shown by the defendant that those requirements were complied with in some equally effective way.

1.12 Beneath ACOPs comes guidance. HSE publishes guidance[5] on a range of subjects, its primary purpose being to assist in the understanding of and compliance with health and safety law. It is not compulsory to follow guidance; but HSE expresses the view that if duty-holders do follow guidance they will normally be doing enough to comply with the law.

B. Institutions

Introduction

1.13 As was stated above, the HSC and HSE, the institutions which develop, administer, and enforce the system of health and safety regulation within the UK, were the creation of HSWA 1974. Both are non-departmental public bodies with Crown status. Their constitutions are governed principally by s 10 and Sch 2 of the Act.

1.14 This two-tier system has remained unchanged since 1974 and, despite some criticism, has been generally considered successful in achieving the standards set by HSWA 1974. Despite this success, and the virtue of its flexibility, consideration is being given towards the merger of the two institutions.

1.15 Whether this planned merger takes place or not, a significant change in the regulatory landscape occurred in April 2006 when the Office of the Rail Regulator, which had hitherto only been the economic regulator for the railways, took over the HSE's responsibility for regulating safety within the industry. With this assumption of responsibility, the ORR became the body responsible for the development and enforcement of health and safety within the railway industry.[6]

---

[5] See paras 1.26–1.27 for HSE publications.
[6] The Rail Accident Investigation Board also assumed the responsibility for the investigation of major incidents.

B. Institutions

The Health and Safety Commission

1.16 HSC is a non-departmental body with specific statutory functions in relation to health and safety. It is made up of ten Commissioners who are appointed by the Secretary of State for Work and Pensions following consultation with various bodies, including employers' organizations, trade unions, and local authorities' representatives. In turn, the HSC appoint the three-person Executive.

1.17 In general terms the Commission oversees the work of the Executive and has extensive powers to delegate operational matters to it.

1.18 HSWA confers on the Commission a series of statutory duties and powers. The foremost of its duties is to make arrangements to secure the health, safety, and welfare of people at work and of the health and safety of the public in respect of the risks posed by the conduct of undertakings. Its other statutory functions include submitting proposals for regulations to Ministers after consultation with appropriate government departments and other bodies; arranging for the provision of information and advice to Ministers; arranging for the operation of an information and advisory service; arranging for research to be carried out and published and encouraging research by others; and arranging for the provision of training and information and encouraging their provision by others. It also has the responsibility of paying to the Executive sums considered appropriate for it to perform its functions.

1.19 Among its powers are the power to approve and issue codes of practice, with the consent of the relevant Secretary of State, subject to consultation with appropriate government departments and other bodies; to make Agency Agreements with government departments or others that enable them to perform functions on the HSC/E's behalf; to give guidance to local authorities on enforcement, and to appoint committees to advise it in relation to particular kinds of hazard and health.

1.20 Section 14 of HSWA 1974 provides that in respect of 'any accident, occurrence, situation or other matter whatsoever which the Commission thinks it necessary or expedient to investigate for any of the general purposes of this Pt or with a view to the making of regulations for those purposes', whether or not the HSE is the relevant enforcing authority concerning the matter in question, the Commission may at any time:

(a) direct the Executive or authorise any other person to investigate and make a special report on any matter to which this section applies; or
(b) with the consent of the Secretary of State direct an inquiry to be held into any such matter.

1.21 Section 14 was invoked following the explosion and fire at the Buncefield oil farm in 2005. The section was used to establish, for the first time, a Board under independent chairmanship—in this case Lord Newton. His investigation has wider terms of reference than the criminal investigation, and includes a review of the role of the HSE and its involvement in the inspection of the site. It has been argued that this has avoided the need for a lengthy public inquiry, while ensuring that there is independent oversight of the way in which the HSE discharged its duties prior to this major incident.

1.22 The Health and Safety Inquiries (Procedure) Regulations 1975 SI 1975/335 govern procedure and the powers of such inquiries. Section 33(1)(d) of HSWA 1974 makes it an offence to contravene any requirement imposed by or under regulations made under this section, or intentionally to obstruct any person in the exercise of his powers under the section.

1.23 By HSWA 1974 s 27, for the purpose of obtaining either any information which the Commission needs for the discharge of its functions, or any information which an enforcing authority needs for the discharge of the authority's functions, the Commission may, with the consent of the Secretary of State, serve on any person a notice requiring that person to furnish to the Commission, or, as the case may be, to the enforcing authority in question, such information about such matters as may be specified in the notice. HSWA 1974 s 28 contains detailed provisions relating to the restrictions on the use and the passing on of any information received. HSWA 1974 s 33 creates various offences connected with non-compliance with and disclosure of information obtained pursuant to s 27.

The Health and Safety Executive

1.24 The HSE is the main operational arm of the HSC. The three-person Executive advises and assists the HSC in carrying out its functions. It consists of a Director General (DG) and two other members. The DG is appointed by the Commission with the approval of the Secretary of State; the other two members are appointed by the Commission with the approval of the Secretary of State after consultation with the DG.

1.25 The Executive employs approximately 4,000 staff to undertake its functions and those delegated to it by the Commission. Its staff includes inspectors, specialist inspectors, scientists, technicians, medical experts, policy advisors, and solicitors. In 2004–2005 HSE inspectors instituted some 712 prosecutions, of which more than 95 per cent resulted in a conviction. HSE inspectors also issued a total of 8,445 improvement and prohibition notices during the same period.

1.26 The HSE publishes advisory literature on most aspects of health and safety which is divided into five main areas: Chemical Safety (CS), Environmental Hygiene (EH), General Series (GS), Medical Series (MS), and Plant or Machinery (PM). HSE publications are categorized by series and include the important Health and Safety (Guidance) Series (HS(G)), the Health and Safety (Regulations)

B. Institutions

Series (HS(R)), and the Legal Series (L). Other publications include Best Practicable Means (BPM) leaflets, Emission Test Methods (ETM) leaflets, Health and Safety Commission (HSC) leaflets, Health and Safety Executive (HSE) leaflets, Industry General (IND(G)) leaflets, Industry Safety (IND(S)) leaflets; and similarly, Methods for the Determination of Hazardous Substances (MDHS), Toxicity Reviews (TR), and Agricultural Safety (AS) leaflets.

1.27 The HSE website <http://www.hse.gov.uk> provides extensive information concerning health and safety and the HSE, and <http://www.hsebooks.com/Books> provides a full catalogue of HSE publications.

1.28 Enforcement is carried out by Health and Safety Inspectors, most of whom work in the Field Operations Directorate (FOD) of the HSE, based in Bootle, Merseyside. FOD is the largest operational inspectorate in the HSE and covers many employment sectors: these include construction, agriculture, general manufacturing, engineering, food and drink, quarries, entertainment, education, health services, local and central government, and domestic gas safety.

1.29 The Employment Medical Advisory Service is part of FOD. It supports all of HSE's front-line activities and provides occupational health advice direct to employers and employees.

1.30 Inspectors also work in the Hazardous Installations Directorate (HID), which is responsible for enforcement in the 'upstream' petroleum and diving industries, at chemical and explosive manufacturing, at processing and large-scale storage sites, in undertakings involving the transportation of hazardous substances, and in the mining industry. The Nuclear Safety Directorate (NSD) similarly has inspectors involved in regulating and enforcing safety legislation for the nuclear industry.

1.31 The Health and Safety Laboratory (HSL) operates as an agency of the Health and Safety Executive (HSE). As well as providing a technical service to the HSE, it also services other public sector organizations and the private sector. It operates from a main laboratory in Buxton and employs over 350 people, including scientists, engineers, psychologists, social scientists, health professionals, and technical specialists. Its work includes research and development, forensic investigation into the causes of accidents, and environmental and biological monitoring.

Local authorities

1.32 The 410 local authorities in the UK have statutory responsibilities for enforcement of health and safety law in certain premises, mainly those concerned with distribution, retail, office, leisure, and catering. It is estimated that this amounts to some 1.4 million premises. They work in partnership with the HSE and under the general direction of the HSC. The division of responsibility between the HSE and local authorities is governed by the Health and Safety Enforcement Regulations 1999; s 18(4) of HSWA 1974 provides:

> It shall be the duty of every local authority—
> (a) to make adequate arrangements for the enforcement within their area of the relevant statutory provisions to the extent that they are by any of those provisions or by (the Health and Safety Enforcement Regulations 1999) made responsible for their enforcement; and
> (b) to perform the duty imposed on them by the preceding paragraph and any other functions conferred on them by any of the relevant statutory provisions in accordance with such guidance as the Commission may give them.

1.33 Cooperation between HSE and the local authorities is pursued through the Health and Safety/Local Authority Enforcement Liaison Committee (HELA), which was set up in 1975 to provide effective liaison between the two. Its stated aims are to ensure that health and safety legislation is enforced in a consistent way among local authorities, and between local authorities and the HSE. It also provides a national forum for discussion and exchange of information on enforcement of legislation.

1.34 HSWA 1974 specifically provides for an enforcing authority to appoint health and safety inspectors 'for carrying into effect the relevant statutory provisions within its field of responsibility'.[7] In practice, local authorities usually appoint Environmental Health Officers with regulatory powers and responsibilities over various matters including health and safety.

1.35 In respect of local authorities, HSWA 1974 s 18 provides that the duty for enforcement of health and safety extends to 'their area'; and reg 3(1) of the Health and Safety (Enforcing Authority) Regulations 1998 SI 1998/494 defines the ambit of the enforcement duty in terms of specified activity at premises which rests upon 'the local authority for the area in which those premises are situated'.

Health and Safety (Enforcing Authority) Regulations 1998

1.36 The Health and Safety (Enforcing Authority) Regulations 1998 SI 1998/494[8] govern the division of responsibility for enforcement between the HSE and local authorities. The regulations contain detailed provisions concerning various activities, premises, industries, organizations, and bodies, specifying whether the HSE or the local authority is responsible for enforcement in respect of each. In general the regulations provide that the HSE is the 'enforcing authority' in the case of all industrial premises and that in the case of commercial premises within its area, the local authority is the enforcing authority.

---

[7] HSWA 1974 s 19(1).
[8] Reproduced at App C.

B. Institutions

The regulations provide that the HSE cannot enforce provisions in respect of its own premises and that similarly in respect of local authorities' premises and activities, enforcement is undertaken by the HSE. Enforcement can be transferred (though not in the case of Crown premises), by prior agreement, from the HSE to the local authority, and vice versa. The Health and Safety Commission is also empowered to effect such a transfer, without the necessity of such agreement. In either case, parties affected by such a transfer must be notified. However, transfer occurs by operation of the regulations when an event or activity changes so as to cause the responsibility to transfer, whether the affected party is notified or not. Where there is uncertainty as to with whom responsibility lies, the regulations also permit responsibility to be assigned by the HSE and the local authority jointly to either body. **1.37**

The CPS and the DPP: the protocol for liaison

HSWA 1974 s 38 provides that proceedings for an offence under any of the relevant statutory provisions shall not be instituted in England and Wales except by an inspector, or by the Environment Agency, or by or with the consent of the Director of Public Prosecutions. **1.38**

Thus the Director of Public Prosecutions has the power to institute proceedings for a health and safety offence; and the CPS often institutes proceedings for manslaughter with additional/alternative health and safety offences alleged. **1.39**

Where, following a work-related death, evidence indicates that a serious criminal offence other than a health and safety offence may have been committed, the investigation should be progressed in accordance with the protocol for liaison for work-related deaths. This has been agreed between the Health and Safety Executive (HSE), the Association of Chief Police Officers (ACPO), the British Transport Police (BTP), the Local Government Association and the Crown Prosecution Service (CPS) and now recently, the Office of the Rail Regulator (ORR). It sets out the principles for effective liaison between the parties in relation to work-related deaths in England and Wales, and is available to the public. **1.40**

The Crown

HSWA 1974 s 48 provides that HSWA 1974 Pt I applies to the Crown with the exception of ss 21–25 (improvement and prohibition notices) and ss 33–42 (criminal offences, liability, and court orders). Thus, the Crown cannot be either prosecuted or served with a notice requiring improvement or prohibition. **1.41**

However, HSWA 1974 s 48(2) provides that although they do not bind the Crown, HSWA 1974 ss 33–42 shall apply to persons in the public service of the Crown as they apply to other persons; and s 48(3) provides that for the purposes of HSWA 1974, Pt I and health and safety regulations, persons in the service of **1.42** the Crown shall be treated as employees of the Crown whether or not they would be so treated apart from the subsection.

1.43 HSWA 1974 s 48(6) specifies that nothing in HSWA 1974 s 28 shall authorize proceedings to be brought against Her Majesty in her private capacity, and that the subsection is to be construed as if s 38(3) of the Crown Proceedings Act 1947 (interpretation of references in that Act to Her Majesty in her private capacity) were contained in HSWA 1974.

1.44 Where Crown bodies have breached health and safety legislation, then the procedure adopted is to consider issuing non-statutory Crown prohibition or improvement notices and/or Crown Censure, which is considered below. Between 2001 and 2005, seventeen Crown Censures and five non-statutory improvement notices were issued by the HSE. The bodies censured included HM Prison Service, the Ministry of Defence, and the Royal Mint.

C. The European Revolution

The Framework Directive

1.45 Some commentators have argued that the raft of Directives which came from the European Commission during the 1980s and 1990s represented a greater revolution in health and safety management than the enactment of HSWA 1974.[9] Certainly, in the last two decades it has been Brussels and not Westminster which has been the driving force for regulation in the field of health and safety.

1.46 On 1 January 1973 the UK joined the European Community (now called the European Union), having signed the Treaty of Rome the previous year. It discharged its treaty obligations by enacting the European Communities Act 1972 which embodied the principle that European law was now valid and binding in the UK, and should override domestic law.

1.47 However, from 1973 to 1987 only six Directives relating to health and safety emanated from the European Commission. It is likely that the paucity of Directives is connected with the fact that until 1986, with the advent of the Single European Act (SEA), unanimous agreement amongst Member States was required to introduce minimum standards for health and safety. By contrast, the SEA (which came into force in 1987) required only a 'qualified majority' of Member States. Moreover, the SEA embodied a more widespread and progressive social agenda, providing at Art 118A, 'Member States shall pay particular attention to encouraging improvements, especially in the working environment,

---

[9] See, for example, *Munkman on Employer's Liability*, 14th edn (London, Lexis Nexis Butterworths).

C. The European Revolution

as regards the health and safety of workers, and shall set as their objective the harmonization of conditions in this area, while maintaining the improvements made'. This initiative came to fruition with the approval by the Council of Ministers in December 1987 of the Third Community Action Programme on safety, hygiene, and health at work, which contained fifteen new Health and Safety Directives.

1.48 The first of these Directives to be issued was the Framework Directive (89/391/EEC). It was adopted on the basis of Art 118A which was subsequently replaced by Art 137 of the EC Treaty. It lays down general rules concerning the prevention of occupational risks and the protection of the safety and health of workers, and seeks to achieve the harmonization of safety rules within the Community.

1.49 The Framework Directive consists of four sections. Section One is entitled 'General Provisions', and consists of four articles. Articles 1 and 2 define the object and scope of the Directive; while Art 3 defines the terms 'worker', 'employer',[10] 'workers' representative', and 'prevention'. Article 4(1) provides that 'Member States shall take the necessary steps to ensure that employers, workers and workers' representatives are subject to the legal provisions necessary for the implementation of this Directive'.

1.50 Section Two of the Directive consists of eight articles and contains the 'Employer's Obligations'. Article 5(1) sets out the employer's duty to ensure safety as follows: 'The employer shall have a duty to ensure the safety and health of workers in every aspect related to the work.'

1.51 Article 5(4) provides that the Directive 'shall not restrict the option of Member States to provide for the exclusion or the limitation of employers' responsibility where occurrences are due to unusual or unforeseeable circumstances beyond the employers' control, or to exceptional events the consequences of which could not have been avoided despite the exercise of all due care'.

1.52 Article 6, is entitled 'General obligations on employers' and provides as follows:

1. Within the context of his responsibilities, the employer shall take the measures necessary for the safety and health protection of workers, including prevention of occupational risks and provision of information and training, as well as provision of the necessary organisation and means.
   The employer shall be alert to the need to adjust these measures to take account of changing circumstances and aim to improve existing situations.
2. The employer shall implement the measures referred to in the first subparagraph of paragraph 1 on the basis of the following general principles of prevention:
   (a) avoiding risks;
   (b) evaluating the risks which cannot be avoided;
   (c) combating the risks at source;

---

[10] 'Employer' is defined as 'any natural or legal person who has an employment relationship with the worker and has responsibility for the undertaking and/or establishment'.

(d) adapting the work to the individual, especially as regards the design of work places, the choice of work equipment and the choice of working and production methods, with a view, in particular, to alleviating monotonous work and work at a predetermined work-rate and to reducing their effect on health;
(e) adapting to technical progress;
(f) replacing the dangerous by the non-dangerous or the less dangerous;
(g) developing a coherent overall prevention policy which covers technology, organization of work, working conditions, social relationships and the influence of factors related to the working environment;
(h) giving collective protective measures priority over individual protective measures;
(i) giving appropriate instructions to the workers.

3. Without prejudice to the other provisions of this Directive, the employer shall, taking into account the nature of the activities of the enterprise and/or establishment:
(a) evaluate the risks to the safety and health of workers, inter alia in the choice of work equipment, the chemical substances or preparations used, and the fitting-out of work places.
Subsequent to this evaluation and as necessary, the preventive measures and the working and production methods implemented by the employer must:
— assure an improvement in the level of protection afforded to workers with regard to safety and health;
— be integrated into all the activities of the undertaking and/or establishment and at all hierarchical levels;
(b) where he entrusts tasks to a worker, take into consideration the worker's capabilities as regards health and safety;
(c) ensure that the planning and introduction of new technologies are the subject of consultation with the workers and/or their representatives, as regards the consequences of the choice of equipment, the working conditions and the working environment for the safety and health of workers;
(d) take appropriate steps to ensure that only workers who have received adequate instructions may have access to areas where there is serious and specific danger.

1.53 The very compatibility of the system for the management of health and safety in the UK with the Framework Directive was challenged in Case C-127/05 EEC. Here the Commission of the European Communities sought a declaration from the Court that, by restricting the duty upon employers to ensure the safety and health of workers in all aspects related to work to a duty to do this only 'so far as is reasonably practicable', the UK had failed to achieve the result intended by Art 5(1) when transposing Directive 89/391/EEC of 12 June 1989; even if that provision was read in conjunction with the exception provided for in Art 5(4).

1.54 The Commission had taken the view that by inclusion of the phrase 'so far as is reasonably practicable' in domestic legislation the UK had thereby limited the employer's liability in the event of an accident; also that it had the capacity to affect the scope of the general duty of safety placed on the employer.

C. The European Revolution

They argued that Art 5(1) of Directive 89/391 made employers liable for the consequences of any event detrimental to workers' health and safety, regardless of whether that event or the consequences could be attributed to the employer's negligence. The effect of the Commission's argument was that the Directive imposed a no-fault liability, in both civil and criminal contexts. In terms of the duty itself, the Commission accepted that Art 5(1) of Directive 89/391 does not impose a duty upon employers to ensure an absolutely safe working environment; but acknowledged that, as a result of carrying out a risk assessment, the employer could conclude that the risks were so small that no preventive measures would be necessary. **1.55**

Despite this they said that it did imply that the employer remains responsible for the consequences of any event detrimental to the health and safety of workers occurring in his undertaking. They contended that the only derogation possible from this responsibility was in the circumstances expressly laid down in Art 5(4) of Directive 89/391,[11] arguing that the provision, which is an exception to the general principle that the employer is responsible, must be interpreted strictly. **1.56**

The Court rejected the Commission's arguments, ruling in favour of the UK. They held that the Commission had not 'sufficiently clarified' its interpretation of the duty set out in Art 5(1)—ie that it imposed an absolute duty on employers—apart from any question of civil or criminal liability in the event of accident, and irrespective of the obligations stemming from Art 5(2) and (3) and Arts 6 to 12 of Directive 89/391. Consequently, they found that the Commission had failed to establish how—in the light of the case-law cited by both parties—the reasonable practicability exception infringes Art 5(1) and (4) of Directive 89/391.[12] **1.57**

They held that the first subparagraph of Art 5(4) of Directive 89/391 was intended to clarify the scope of certain provisions of the Directive by explaining the margin of manoeuvre available to the Member States in transposing those provisions into national law; but did not accept that the presence of the clause implied that the Community legislature intended to create a no-fault liability regime for employers. **1.58**

---

[11] Which allows Member States the option of limiting employers' responsibility 'where occurrences are due to unusual and unforeseeable circumstances, beyond the employers' control, or to exceptional events, the consequences of which could not have been avoided despite the exercise of all due care'.

[12] They pointed out in an action brought on the basis of Art 226 EC, it is for the Commission to prove the existence of the alleged infringement and to provide the Court with the information necessary for it to determine whether the infringement is made out, and the Commission may not rely on any presumption for that purpose (see Case C-287/03 *Commission v Belgium* [2005] ECR I-3761, para 27, and the case law cited, and Case C-428/04 *Commission v Austria* [2006] ECR I-3325, para 98).

1.59 Further the Court held that the Commission had not shown in what respect the objective of Directive 89/391, consisting in 'the introduction of measures to encourage improvements in the safety and health of workers at work', could not be attained by means other than the setting up of a no-fault liability regime. They stated that although the reasonable practicability exception was a proviso to the employer's duty to ensure the safety and health of workers in every aspect related to the work as regards what is 'reasonably practicable', the significance of the proviso depended on the precise content of that duty.

The daughter Directives

1.60 Following on from the Framework Directive came six 'daughter' Directives, all of which were implemented in UK law by regulations supported by ACOPs and guidance notes. The first six of these new regulations were commonly known as the six-pack. The first six daughter Directives were followed by six further Directives, which were in turn implemented in the UK through the enactment of regulations.

1.61 The daughter Directives were: Workplace (Health Safety and Welfare) Regulations 1992 (as amended in 2003); Manual Handling Operations Regulations 1992 (as amended in 2002); Provision and Use of Work Equipment Regulations 1998; Health and Safety (Display Screen Equipment) Regulations 1992 (as amended in 2002); and Personal Protective Equipment at Work Regulations 1992 (as amended in 2002).

D. Principles of Enforcement Action

Introduction

1.62 The stated aim of the HSC[13] is to ensure that duty holders manage and control risks effectively and so prevent harm. The intended purpose of enforcement action is to:

(a) ensure that duty holders take action to deal immediately with serious risks;
(b) promote and achieve sustained compliance with the law;
(c) ensure that those who breach health and safety requirements, including directors and managers, are held to account.

1.63 Plainly, enforcement in the field of health and safety encompasses a great deal more than prosecution; which should properly be viewed as a measure of last resort.

---

[13] See Enforcement Policy Statement (HSC 15) produced at App B.

Accordingly, inspectors have a wide range of enforcement options available to them and are given substantial discretion in deciding which, if any, to take. Particular enforcement measures may be taken alone, or in combination with others—for example, prohibition notices and improvement notices are widely imposed with or without prosecution for the substantive offence.

1.64 Enforcement decisions should be made in accordance with the principles set out in a number of important documents. One of these is the Enforcement Concordat,[14] a statement of policy agreed between the Cabinet and Home Office and the local authority associations. The Concordat sets out the general principles to be followed by the HSE and local authorities when deciding what, if any, enforcement action is necessary.[15] They are:

(a) proportionality; which is defined as the need to relate the enforcement action to the level of risk created by the activity in question, the extent of the breach and the actual or potential for harm;
(b) targeting; which involves directing the available resources towards the regulation of high-risk activities and concentrating attention on those who are either responsible for creating risk or are in a position to control it;
(c) consistency; which is not the same as uniformity, but involves taking a similar approach in similar circumstances to achieve similar ends;
(d) transparency; which involves letting the duty holder know what is expected of him and what he should expect from the regulator; and
(e) accountability; which means that the enforcing authorities must have a settled set of policies and standards setting out their general approach and which are readily available for inspection.

The Enforcement Policy Statement

1.65 The Health and Safety Commission's attitude towards enforcement action is set out in its publication, 'HSE Enforcement Policy Statement'[16] (EPS). The Enforcement Policy Statement sets out the general principles which HSE inspectors and local authorities should apply when making decisions about taking enforcement action. Enforcing authorities are expected to use their discretion by observing the principle that the action taken should be proportionate to the risks created and the nature of the breach.

1.66 The five factors set out in the Enforcement Concordat are reiterated in the Enforcement Policy Statement. In terms of prosecution, the Enforcement Policy

---

[14] See Enforcement Concordat (Cabinet Office), reproduced at App B.
[15] These are reiterated in the Enforcement Policy Statement (HSC 15).
[16] Enforcement Policy Statement (HSC 15) reproduced at App B.

Statement says that the public interest would normally require enforcing authorities to prosecute where one or more of the following circumstances apply:

(a) Death resulted from the breach of legislation.
(b) The offence alleged is a grave one and gave rise to serious actual or potential harm.
(c) The general record and approach of the offender warrants it.
(d) The duty holder's standard of managing health and safety is found to be far below what is required by law and gave rise to serious risk.
(e) Reckless disregard of health and safety requirements has taken place.
(f) Repeated breaches which give rise to significant risk have occurred.
(g) There has been persistent and significant poor compliance.
(h) Work has been carried out without or in serious non-compliance with an appropriate licence or safety case.
(i) There has been a failure to comply with an improvement notice or prohibition notice.
(j) There has been a repetition of a breach that was the subject of a formal caution.
(k) False information has been supplied wilfully or there has been an intent to deceive in relation to a matter which gives rise to a serious risk.
(l) Inspectors have been intentionally obstructed in the lawful course of their duty.

1.67 It continues to say that prosecution should also be considered where it is appropriate in all of the circumstances to draw general attention to the need for compliance with the law, and the standards required by law, and where a conviction may deter others from committing similar offences.

The Code for Crown Prosecutors

1.68 Even though the Code refers only to Crown Prosecutors and is not specifically extended to cover those who have a duty to investigate beyond the CPS, inspectors should have regard to the principles laid out in the Code for Crown Prosecutors when deciding whether prosecution within the criminal courts is justified. The Code provides that there are always two steps in the decision to prosecute. The first is the evidential test: that there is enough evidence to provide a realistic prospect of conviction. In determining whether the evidential test is met, the enforcing authority must apply an objective standard and evaluate whether the available evidence is reliable and admissible. Consideration should also be given to the defence case and how it is likely to affect the prosecution case.

1.69 This last requirement has particular significance in the context of health and safety law since, if the defendant is able to show on the balance of probabilities that he did what was reasonably practicable in all of the circumstances, the prosecution will fail. Thus, while the prosecution need only prove a limited amount in

D. Principles of Enforcement Action

health and safety cases—that the defendant was an employer and that he created risk to his employees through the conduct of his undertaking—the prosecution will usually obtain and seek to lead by way of 'advance rebuttal' evidence that the precautions taken fell short of what was reasonably practicable.

1.70 If a case does not pass the evidential test then it founders immediately. And even if the case does pass the evidential test, the enforcing authority must go on to consider whether the public interest requires prosecution.

1.71 The Code sets out the general principles for determining whether it is in the public interest to pursue a prosecution. While the principles are intended to apply to conventional criminal prosecutions, some are of general application and may apply equally to health and safety prosecutions. The following factors, where present, may militate in favour of prosecution:

(a) the defendant's previous convictions or cautions are relevant to the present offence;
(b) there are grounds for believing that the offence is likely to be continued or repeated, for example, by a history of recurring conduct;
(c) the offence, although not serious in itself, is widespread in the area where it was committed.

The Enforcement Management Model

1.72 The Enforcement Management Model[17] (EMM) is a document providing a detailed framework intended to assist inspectors to take decisions in accordance with the Enforcement Policy Statement. Although inspectors are expected to have regard to the principles contained within it whenever they take enforcement decisions, it is not intended to fetter their overall discretion, and need only be formally applied in certain defined circumstances, such as when a work-related fatal accident has occurred. It is also intended to enable those who may be affected to understand the principles which are applied when enforcement action is taken.

1.73 It is also intended to provide an empirical framework for auditing the consistency and correctness of decisions across a wide range of activities, sectors, and enforcement regions, and to assist in achieving accountability.

1.74 The EMM requires an inspector to make an assessment of the risk arising from the activities as they are carried on; to establish what level of risk would be tolerable if all necessary controls were in place, and to use the gap between the two—in one instance or across a number of instances—to make an initial determination of what level of enforcement action is expected. This is known as the

---

[17] Available at <http://www.hse.gov.uk/enforce/emm.pdf>.

Initial Enforcement Expectation (IEE). Factors which are peculiar to the duty holder, such as previous enforcement record etc, should then be taken into account. This may have the effect of turning a situation in which the IEE was not for prosecution into one where prosecution is justified, or vice versa. Strategic factors, ie whether the action proposed is in the public interest or is necessary to protect vulnerable groups, should then be applied.

E. Methods of Enforcement

Introduction

1.75 The Enforcement Policy Statement[18] sets out the range of enforcement measures available to inspectors to achieve these ends. They may offer duty holders information and advice (including an opinion that they are failing to comply with the law), orally or in writing; they may serve improvement notices and prohibition notices, withdraw approvals, vary licence conditions, or issue formal cautions. Beyond saying what powers an inspector has, the Enforcement Policy Statement is relatively silent on the question of the circumstances in which they should be used.

Early intervention

1.76 The HSC has stated that it will attempt to influence health and safety compliance through early intervention. An example of this is within the construction industry where the HSE has sought to influence duty-holders in the design and concept stages of projects, before the construction processes begin.

Verbal warnings

1.77 An inspector may give a verbal warning to a duty holder when he encounters some instance of non-compliance with health and safety law, but the circumstances are not so serious as to justify a written letter of advice. A verbal warning is effectively the lowest in the hierarchy of possible enforcement measures.

Letters of advice/recommendation

1.78 The main means of securing compliance with health and safety law is through written advice by an inspector. Written advice is generally given following a visit to premises and contains a statement of where it is thought that the duty holder

---

[18] Paras 1.65–1.67.

E. Methods of Enforcement

is falling short, and recommendations as to the remedial measures which are required to secure compliance with the law. It may be appropriate to offer written advice when the breaches, if there are any, do not suggest that the duty holder is falling far short of the standards expected of him, or where the risks of harm are not excessive. An inspector may, if asked, distinguish between what amounts to best practice and what is a legal requirement; but the inspector should make it clear what is legal advice and what is best practice.

Improvement notices and prohibition notices

1.79 These were an innovation of HSWA 1974 and have proved to be one of the most effective and commonly used of an inspector's tools. For a detailed account of the operation of improvement notices and prohibition notices, see Chapter 4.

Permissioning and licensing regimes

1.80 Where a particular work activity involves significant hazards, levels of risk, or gives rise to wider public concern—for instance where there is a significant risk of multiple fatalities from a single incident, and a high degree of regulatory control is therefore required—a permissioning regime may be imposed. A permissioning regime broadly describes a situation whereby a particular work activity is not allowed to commence or continue unless a consent, licence, letter of conclusion, or acceptance of a safety case or safety report has been given or accepted. Their purpose is to ensure that hazardous activities are managed in a systematic way throughout all stages of their operation, and so by controlling risk effectively, to secure public confidence without compromising the efficient working of industry. The HSC's and HSE's attitude towards such regimes is set out in the policy document which is to be found on their website. The following areas are subject to some form of permissioning regime:

(a) the nuclear industry;
(b) railways;
(c) offshore installations;
(d) onshore major hazard sites;
(e) work with genetically modified organisms;
(f) gas distribution;
(g) work with asbestos;
(h) explosive manufacture and storage;
(i) keeping of petrol.

Formal cautions

1.81 An inspector has the power to formally caution a person in respect of an offence instead of instituting proceedings in respect of it.

1.82 Formal cautions in the field of health and safety enforcement are distinct from cautions under the Police and Criminal Evidence Act 1984. When deciding whether it may be an appropriate disposal, the enforcing authorities should take account of the guidance given in Home Office Circular 18/1994 (HOC 18/1994) and the specific practical guidance given in HSE Operational Circular 130/7.

1.83 The purposes of a formal caution are to deal quickly and simply with less serious offenders; to divert them from unnecessary appearances in the criminal courts and to reduce the chances of their re-offending. They should not be considered a 'let off' where there are some mitigating circumstances, where there is doubt about the 'public interest', or where either the prosecutor's office or the court is 'too busy'.

1.84 The guidance provides that they may only be used where the evidence available is sufficient to give a reasonable prospect of conviction, where the duty holder admits the offence and when the public interest does not require prosecution. The Board of the HSE has confirmed that inspectors may consider a formal caution in a case which ordinarily would meet the public interest test but where personal and/or exceptional circumstances weigh against it. Such circumstances may include: situations in which a court appearance would be likely to have a seriously adverse effect on a victim's health, or when the accused is elderly, or was suffering significant physical or mental ill health at the time of the offence.

1.85 Formal cautions should not generally be considered where the offender has already received a caution, since repeated cautioning brings the disposal into disrepute. A second caution should only be administered when there has been a substantial lapse of time since the first, in circumstances where it may be inferred that it had a preventative effect. Nor should cautions be administered to an offender in circumstances where there can be no reasonable expectation that they will curb further offending. The formal caution may also be referred to in court if there is a further prosecution within the next five years. Details of the formal caution may also be made publicly available.

1.86 If there should be a 'breach' of the caution, in the sense of a repetition of the conduct giving rise to it, the Enforcement Policy Statement para 39 provides that it should be treated in the same way as failure to comply with an Improvement Notice or Prohibition Notice, ie normally by prosecution.

1.87 In terms of procedure, a caution may be carried out by addressing it to the managing director or company secretary and delivering it to the registered offices in the case of UK companies, or to any place in the UK where the corporation trades or conducts its business in the case of non-UK companies. However, the Home Office National Standards set down a preference for administration of cautions in person, in the case of both a natural person and a body corporate.

Crown censure

Crown bodies are subject to the requirements of health and safety law, including the general duties under HSWA 1974 ss 2–7, but may not be prosecuted or made the subject of any other enforcement action, whether the imposition of improvement notices or prohibition notices or the use of inspectors' powers. Crown bodies may, however, be made the subject of Crown censure. This is an administrative procedure whereby the HSE sets out a written account of the breaches alleged and the evidence relied on in support of them. The appropriate Crown body may accept the censure or not. **1.88**

F. Prosecution and Particular Categories of Person

Prosecution of individuals

Detailed guidance on the circumstances in which it may be appropriate to proceed against individuals, whether under HSWA 1974 ss 7, 8, 36, or 37, is given in HSE Operational Circular 130/8 (OC 130/8). While different criteria apply when considering whether to prosecute employees under s 7, and directors under s 37, some of the guidance is relevant to both categories of person. **1.89**

In general, the prosecution of individuals will be warranted where there are substantial failings by them: for example where they have shown wilful or reckless disregard for health and safety requirements, or there has been a deliberate act or omission that seriously imperilled their health/safety or that of others. **1.90**

The guidance provides that since a body corporate operates only by and through the actions of its employees and officers (including directors and managers), when a body corporate commits an offence there is likely also to be some personal failures by directors, managers, or employees. Be this as it may, it does not follow that there should always be prosecution of individuals. **1.91**

Employees: HSWA 1974 s 7

The guidance provides that expressions such as 'reasonable care' and 'necessary to enable' should be considered in the context of the employer's provisions. The expression 'so far as is necessary' does not require employees to compensate for employers' failure to make adequate provisions, as this remains the responsibility of the employer. **1.92**

Some practical examples are supplied: **1.93**

> ... a machine operator who has received inadequate training might be considered to have acted reasonably in all the circumstances if he/she removes a guard from a machine and continues to use it, and this is the generally accepted and condoned

practice in the company. In other circumstances the same act might be considered unreasonable, if the employee has received proper training, if the guard in question is sufficient, and if removal of guards is neither accepted nor condoned in the company.

1.94 In determining whether the public interest requires the prosecution of employees, inspectors should consider the following:

(a) whether the company had done all it reasonably could to ensure compliance;
(b) whether the offence was solely the result of the actions/inactions of the individual;
(c) whether employees, as a matter of general practice, followed the systems of work alleged by the employer to be in force;
(d) any previous warnings to the employee, from whatever source;
(e) whether the offence by the employee was flagrant;
(f) the risks to health and safety arising from the offence by the employee; and
(g) whether prosecution would be seen by others as fair, appropriate, and warranted.

Directors and managers: HSWA s 37

1.95 The circular provides that prosecution is intended to bring home to directors/managers the extent of their responsibilities, and to bring them to public account for their failings where appropriate. Therefore the prosecution should be seen by others—particularly by other directors/managers with knowledge of the industry concerned—as justified not only in legal terms, but also as a matter of practical judgment.

1.96 The circular provides that the following factors should be taken into consideration when deciding whether the public interest requires prosecution under s 37:

(a) The matter was, in practice, clearly within the director/manager's effective control—where the steps that could reasonably have been taken to avoid the offence fall properly and reasonably within their duties, responsibilities, and scope of functions.
(b) The director/manager had personal awareness of the circumstances surrounding or leading to the offence.
(c) The director/manager failed to take obvious steps to prevent the offence.
(d) The director/manager had previous advice/warnings regarding matters relating to the offence. (This may also include whether previous advice to the company had meant that he/she had the opportunity to take action. In such a case you would need to show that he/she knew, or ought reasonably to have known, about the advice/warning.)
(e) The director/manager was personally responsible for matters relating to the offence—eg when the manager had personally instructed, sanctioned, or

positively encouraged activities that significantly contributed to or led to the offence.
(f) Prosecution would be seen by others as fair, appropriate, and warranted.
(g) The individual knowingly compromised safety for personal gain, or for commercial gain of the body corporate, without undue pressure from the body corporate to do so.

1.97 However, prosecutions should be avoided in cases that cause directors or managers to refuse explicit responsibility for oversight of occupational health and safety, or that lead to safety policies and job descriptions being written defensively or to excessive delegation of responsibility. It is important that these points are seen in context and that they are not considered disproportionately.

1.98 Section 37 cases should not be taken against directors/managers just because a company has closed down. The circumstances of the closure should be scrutinized, but prosecution may follow if there is evidence that the closure was motivated by a deliberate attempt to avoid prosecution.

1.99 Directors or managers liable to be prosecuted under s 37 may also be employees and therefore also subject to action under s 7. Generally, the appropriate charge depends on the role being fulfilled at the time: if the individual is acting as a director of the company and directing its affairs then s 37 should be used; if acting as an employee and carrying out the company's procedures in the same way as other employees, then s 7 may be more appropriate.

Other individuals: HSWA ss 8 and 36

1.100 Sections 8 and 36 allow cases to be taken against persons who did not commit the original offence but who nevertheless caused the offence. This includes people who are not specific duty-holders under HSWA, such as members of the public or elected members of local authorities. Employees and directors/managers may fall within ss 8 and 36.

1.101 As far as the public interest is concerned, similar principles and considerations apply to ss 8 and 36 as to ss 7 and 37. This means that prosecution should be a proportionate response to the offence by the 'other person' and take account of the risk created by the offence, the role played by the 'other person', whether remedial measures were truly under their control, and whether other duty-holders were more at fault.

Prosecuting under HSWA s 3

1.102 Acknowledging the broad scope of s 3 HSWA 1974, the HSC has published a policy with the intention of assisting enforcing authorities properly to exercise their discretion when investigating or taking action for suspected breaches of the section. The guidance provides that s 3 will continue to be enforced in major and high-hazard activity, such as the nuclear industry and construction, and also in cases where there is a high level of risk, or where enforcing authorities need to act/investigate in the interests of justice.

1.103 The guidance also provides that every effort will be made to ensure that the appropriate authority investigates the activity in question, and that there is no time-consuming and costly duplication of activity. When deciding who should investigate, it provides that the following considerations should be borne in mind:

(a) Which authority is best equipped, including appropriate powers, to investigate the alleged risks?
(b) Is the other body capable of ensuring public safety?
(c) Does it have those enforcement powers necessary to do so?
(d) Which body knows most about the risks concerned and the effective control measures?
(e) Is either body already inspecting/visiting the premises or activity in question?
(f) Can duplication of visits be avoided?
(g) Is health and safety enforcing authority involvement a good use of resources when considered against the scale of risk or level of public concern?

Action to be taken when a decision is made to prosecute

1.104 It is the practice of the HSE, especially where there has been no interview under caution with the duty holder or where there has been subsequent delay, to write to a particular duty holder in the event that a preliminary decision has been taken to prosecute, seeking his representations as to why a prosecution ought not to be brought. Such a request is certainly consistent with the application of the general principles to the decision-making process, not least because the duty holder is thereby given the opportunity to raise any matters which may affect the assessment of the evidential and public interest tests. It is also a step which is specifically referred to in the Enforcement Concordat. The opportunity to make final representations is particularly important in cases where the duty holder has not been interviewed under caution as part of the investigation, and might therefore may not have had a full say.

1.105 It is submitted that those whom it is intended to prosecute should avail themselves fully of the opportunity to make such representations. What representations ought to be made obviously depends on the circumstances of the particular case, but in general terms it is worthwhile to make reference to:

(a) previous good health and safety record;
(b) positive action towards improving health and safety following complaints.

2

THE REPORTING AND INVESTIGATION OF HEALTH AND SAFETY INCIDENTS AND OFFENCES

A. Reporting Health and Safety Incidents	2.05	Powers to 'examine and investigate': HSWA 1974 s 20(2)(d)	2.109
Notification and reporting of injuries and dangerous occurrences	2.10	Direction to leave undisturbed: HSWA 1974 s 20(2)(e)	2.113
B. The Service of Notices and Documents	2.25	Taking measurements, photographs, and recordings: HSWA 1974 s 20(2)(f)	2.115
C. The Investigation of Health and Safety Incidents and Offences	2.32	Articles and substances: the power to take samples: HSWA 1974 s 20(2)(g)	2.117
The ambit of the duty to investigate	2.34	The power to dismantle or subject dangerous articles or substances to testing: HSWA 1974 s 20(2)(h)	2.121
Investigatory powers	2.53		
Police powers and the new power of arrest	2.59		
Police powers of search and entry	2.69	The power to take possession and detain dangerous articles or substances: HSWA 1974 s 20(2)(i)	2.128
Inspectors and the Police and Criminal Evidence Act 1984	2.71		
D. Inspectors' Powers Under HSWA 1974 s 20	2.87	The power to require answers: HSWA 1974 s 20(2)(j)	2.131
Consequences of improper use of HSWA 1974 s 20 powers	2.96	The power to require production, inspection, and to take copies: HSWA 1974 s 20(2)(k)	2.150
The power to enter premises: HSWA 1974 s 20(2)(a)	2.101		
Power to take a constable: HSWA 1974 s 20(2)(b)	2.105	The power to require facilities: HSWA 1974 s 20(2)(l)	2.157
Power to take with others and items of equipment: HSWA 1974 s 20(2)(c)	2.108	Any other power which is necessary: HSWA 1974 s 20(2)(m)	2.159
		Powers to deal with immediate danger: HSWA 1974 s 25	2.163

Chapter 2: The Reporting and Investigation of Health and Safety Incidents

2.01 Many health and safety investigations are commenced as a result of a duty-holder's compliance with the statutory scheme for reporting health and safety incidents, created by the Reporting of Injuries, Diseases and Dangerous Occurrences Regulations 1995 SI 1995/3163. The ambit and operation of these duties is considered in the first section of this chapter.

2.02 There is a range of health and safety duties placed on persons, requiring them to serve notices upon health and safety inspectors. Equally, there exists a range of powers, duties, and situations in which an inspector is required to serve a notice or document on a duty-holder. The statutory provisions relating to the service of such notices are considered in the second section of this chapter.

2.03 The third section of the chapter considers the investigation of health and safety incidents and offences. Such investigations are undertaken by health and safety inspectors and, on occasions, by the police. Health and safety inspectors have distinct and different powers from police officers, including those of compulsory questioning and production. These are backed up by criminal sanctions for non-compliance.

2.04 Section D relates to the conduct of interviews under caution, which feature in nearly every investigation by health and safety inspectors that precedes a prosecution. Section E is concerned with some important areas of advice to be given to those who are the subject of a health and safety investigation. It includes consideration of any conflict of interest, the operation of legal professional privilege, and comments on advising clients in relation to interviews under caution.

A. Reporting Health and Safety Incidents

2.05 The Reporting of Injuries, Diseases and Dangerous Occurrences Regulations 1995 SI 1995/3163[1] (RIDDOR) created far-ranging duties. These are imposed on employers, those responsible for work premises, and those in control of various industrial processes (such as mines, quarries, and offshore installations). RIDDOR contains the duty to report deaths, major injuries, and dangerous occurrences to either, or both, the Health and Safety Executive (HSE) and the local enforcing authority. In short, RIDDOR provides for enforced notification of health and safety incidents.

2.06 Notification must be made to the 'relevant enforcing authority'. This is determined by reference to activity and location, through the operation of the Health

---

[1] These replaced the Reporting of Injuries, Diseases and Dangerous Occurrences Regulations 1985 SI 1985/2023.

A. Reporting Health and Safety Incidents

and Safety (Enforcing Authority) Regulations 1998 SI 1998/494.[2] Generally, the local authority is the relevant authority, in respect of leisure, retail, and hotel premises/activities. The HSE is the relevant authority in respect of construction, manufacturing industry, health, and local government/the Crown.

2.07 It is through the operation of these regulations that the HSE obtains the data on workplace safety from which it compiles the health and safety statistics for the UK. The HSE devotes a department to these statistics, and a website to their publication, at <http://www.hse.gov.uk/statistics/index.htm>.

2.08 Equally importantly, the duties operate so that health and safety inspectors are notified of potential offences. Often, it transpires that notification is made by the very person who ends by being investigated for a suspected health and safety offence. A RIDDOR return is, in many situations, the cause for a visit by an inspector.

2.09 Various of the terms employed by RIDDOR[3] are defined in detail in schedules, as follows:

(a) sch 1: 'Major Injuries';
(b) sch 2: 'Dangerous Occurrences';
(c) sch 3: 'Reportable Diseases';
(d) sch 4: requirements in relation to the keeping of records.

Notification and reporting of injuries and dangerous occurrences

2.10 The principal duty[4] created by RIDDOR[5] is set out in reg 3(1). This provides that a 'responsible person'[6] shall, forthwith, notify the relevant enforcing authority, by the quickest practicable means. Within ten days, the 'responsible person' must send[7] a report to the relevant enforcing authority, on an approved form.[8] This obligation operates where any of the following occurs:

(a) any person dies as a result of an accident arising out of or in connection with work;
(b) any person at work suffers a major injury as a result of an accident arising out of or in connection with work;

---

[2] See App C.
[3] Reporting of Injuries, Diseases and Dangerous Occurrences Regulations 1995 SI 1995/3163.
[4] RIDDOR contains special and specific provisions related to offshore installations and workplaces, mines, quarries, and to the reporting of gas incidents: Reporting of Injuries, Diseases and Dangerous Occurrences Regulations 1995 SI 1995/3163 rr 6–10.
[5] Reporting of Injuries, Diseases and Dangerous Occurrences Regulations 1995 SI 1995/3163.
[6] Para 2.21.
[7] 'Send' see paras 2.31 and 2.26.
[8] Unless within that period he makes a report to the HSE by some other approved means.

(c) any person not at work suffers an injury as a result of an accident arising out of or in connection with work and that person is taken from the site of the accident to a hospital for treatment in respect of that injury;

(d) any person not at work suffers a major injury as a result of an accident arising out of or in connection with work at a hospital; or

(e) there is a dangerous occurrence.

2.11 Regulation 3(2) provides for situations in which a person at work is incapacitated, for work of a kind which he might reasonably be expected to do. The regulation envisages that incapacitation occurred either under his contract of employment, or, if there is no such contract, in the normal course of his work, for more than three consecutive days.[9] Incapacitation is equated with an injury resulting from an accident arising out of, or in connection, with work (other than one reportable under reg 3(1)). The responsible person shall, as soon as practicable, and, in any event, within ten days of the accident, send a report to the relevant enforcing authority, on an approved form.[10]

2.12 Regulation 3(1)(i) does not apply to a self-employed person who is injured at premises of which he is the owner or occupier; but reg 3(1)(ii) does apply to such a self-employed person (other than in the case of death). However, it is deemed sufficient compliance if the self-employed person makes arrangements for the report to be sent to the relevant enforcing authority by some other person.[11]

2.13 Regulation 4[12] further provides where an employee has suffered an injury, as a result of an accident at work which is reportable under reg 3, and which is a cause of his death, within one year of the date of that accident, the employer has to inform the relevant enforcing authority in writing of the death, as soon as it comes to his knowledge. This duty rests with the employer, whether or not the accident has been reported under reg 3.

2.14 Regulation 10[13] restricts the operation of the duties. The requirements of reg 3, as they relate to the death or injury of a person as a result of an accident, do not apply to an accident causing death or injury to a person arising out of the conduct of any operation on, or any examination or other medical treatment of, that person. 'Treatment' must be administered by, or conducted under the supervision of, a registered medical practitioner or a registered dentist.

---

[9] Excluding the day of the accident, but including any days which would not have been working days, ie weekends.
[10] Unless, within that period, he makes a report to the HSE by some other approved means.
[11] Reporting of Injuries, Diseases and Dangerous Occurrences Regulations 1995 SI 1995/3163 reg 10(5).
[12] Reporting of Injuries, Diseases and Dangerous Occurrences Regulations 1995 SI 1995/3163.
[13] ibid reg 10(1).

Furthermore, the requirements of reporting, as they relate to the death or injury **2.15** of a person as a result of an accident, only apply to an accident arising out of, or in connection with, the movement of a vehicle on a road, if that person:

(a) was killed, or suffered an injury as a result of exposure to a substance being conveyed by the vehicle; or
(b) was himself either engaged in, or was killed or suffered an injury as a result of the activities of another person who was at the time of the accident engaged in, work connected with the loading or unloading of any article or substance on to, or off, the vehicle; or
(c) was himself either engaged in, or was killed or suffered an injury as a result of the activities of another person who was at the time of the accident engaged in, work on or alongside a road; that work being concerned with the construction, demolition, alteration, repair, or maintenance of:
  i. the road or the markings or equipment thereon;
  ii. the verges, fences, hedges, or other boundaries of the road;
  iii. pipes or cables on, under, over, or adjacent to the road; or
  iv. buildings or structures adjacent to, or over the road; or
(d) was killed or suffered an injury as a result of an accident involving a train.

The notification requirements relating to any death, injury, or case of disease, do **2.16** not apply to a member of the armed forces, of the Crown, or of a visiting force on duty at the relevant time.[14]

'Accident' and 'dangerous occurrence'

The term 'accident' is not defined other than as to include 'an act of non-consensual **2.17** physical violence done to a person at work', and 'an act of suicide which occurs on, or in the course of the operation of, a relevant transport system'.[15] 'Accident' is a word in ordinary English usage that is wide enough to encompass any unfortunate occurrence involving injury or pain. Additionally, reg 2(2)(c)[16] provides that:

> an accident or a dangerous occurrence which arises out of or in connection with work shall include a reference to an accident, or as the case may be, a dangerous occurrence attributable to the manner of conducting an undertaking; the plant or substances used for the purposes of an undertaking; and the condition of the premises so used or any part of them.

---

[14] Reporting of Injuries, Diseases and Dangerous Occurrences Regulations 1995 SI 1995/3163 reg 10(3).
[15] ibid reg 2.
[16] ibid reg 2(2)(c).

2.18 A 'dangerous occurrence' is defined as 'an occurrence which arises out of or in connection with work' and is of a class specified, depending on where it occurs, in RIDDOR, Sch 2.[17] Schedule 2 is divided into five parts, defining 'general' situations properly reported as 'dangerous occurrences', those arising in relation to mines, quarries, relevant transport systems (railway, tramway, trolley vehicle system, or guided transport system), and offshore workplaces. The general part includes paragraphs concerned, variously, with lifting machinery, pressure systems, freight containers, electrical short circuit, explosives, biological agents, malfunction of radiation generators, breathing apparatus, diving operations, collapse of scaffolding, train collisions, wells, pipelines or pipeline works, fairground equipment, carriage of dangerous substances by road, collapse of building or structure, explosion or fire, escape of flammable substances, and escape of substances.

Occupational diseases

2.19 Regulation 5[18] creates a duty to report an incidence of certain occupational diseases, by sending[19] a report forthwith on an approved form to the relevant enforcement authority. 'Certain occupational diseases' are ones that arise when a person at work suffers from any of the occupational diseases specified in column 1, RIDDOR, Sch 3, Pt 1.[20] If the person's work involves one of the activities specified in the corresponding entry, in column 2, the case falls within the contemplation of Sch 3. The diseases or conditions (in column 1) are specific, but many of the corresponding work activities (in column 2) are quite general. Two examples, from Sch 3, Pt 1, are given in the table below:

	Column 1—Diseases	Column 2—Activities
12	Traumatic inflammation of the tendons of the hand or forearm or of the associated tendon sheaths.	Physically demanding work, frequent or repeated movements, constrained postures or extremes of extension or flexion of the hand or wrist
19	Legionellosis	Work on, or near, cooling systems which are located in the workplace and use water; or work on hot water service systems located in the workplace, which are likely to be a source of contamination.

---

[17] Reporting of Injuries, Diseases and Dangerous Occurrences Regulations 1995 SI 1995/3163 Sch 2.
[18] ibid reg 5.
[19] 'Send': see paras 2.26 and 2.31.
[20] Reporting of Injuries, Diseases and Dangerous Occurrences Regulations 1995 SI 1995/3163, Sch 3, Pt 1, col 1.

A. Reporting Health and Safety Incidents

The duty only arises if, in the case of an employee, the responsible person has received a written statement, prepared by a registered medical practitioner, diagnosing the disease as one of those specified in RIDDOR, Sch 3. In the case of a self-employed person, the duty arises where the responsible person has been informed, by a registered medical practitioner, that the employee is suffering from such a disease.[21] **2.20**

Responsible person

Regulation 2,[22] of RIDDOR, defines a 'responsible person' in relation to various specific activities and industries. In general, the responsible person is his employer, in the case of the death of, or other injury to, an employee, which is reportable under reg 3, and of a disease suffered by an employee reportable under reg 5. In any other case, the person for the time being having control of the premises, in connection with the carrying on by him of any trade, business, or other undertaking (whether for profit or not) at which, or in connection with the work at which, the reportable accident, dangerous occurrence, or disease happened. **2.21**

Records

Regulation 7 of RIDDOR[23] requires the responsible person to keep records of any event which he is required to report, under reg 3, or any disease which must be reported, under reg 5(1). **2.22**

Breaching a duty under RIDDOR

Failure to comply with a duty under the regulations is an offence, contrary to HSWA 1974 s 33(1)(c). Regulation 11[24] of RIDDOR, provides for a defence in proceedings against any person for an offence under the regulations. It shall be a defence 'for that person to prove that he was not aware of the event requiring him to notify or send a report to the relevant enforcing authority, and that he had taken all reasonable steps to have all such events brought to his notice'. **2.23**

The burden placed upon such a defendant is to show, on a balance of probabilities, that he was unaware of the event. It is for him to show that he took all reasonable steps. **2.24**

---

[21] It is deemed sufficient compliance if the self-employed person makes arrangements for the report to be sent to the relevant enforcing authority, by some other person: Reporting of Injuries, Diseases and Dangerous Occurrences Regulations 1995 SI 1995/3163 reg 5(3).
[22] Reporting of Injuries, Diseases and Dangerous Occurrences Regulations 1995 SI 1995/3163 reg 2.
[23] ibid reg 7.
[24] ibid.

B. The Service of Notices and Documents

2.25 The Health and Safety at Work etc. Act 1974 (HSWA 1974), and the various health and safety regulations, create a wealth of provisions that require the service of notices or documents by, or upon, health and safety inspectors. HSWA 1974 s 46 provides possible means for service of notices and documents.

2.26 Section 46(1) provides that any notice or document required, or authorized, by any of the relevant statutory provisions to be served on, or given to, an inspector, may be served, or given, by delivering it to him. Service may also be effected by leaving it at, or sending it by post to, his office.

2.27 Section 46(2) provides that any such notice or document required, or authorized, to be served on, or given to, a person other than an inspector, may be served, or given, by delivering it to him, or by leaving it at his proper address, or by sending it by post to him at that address. Other than in respect of a corporation or partnership, the proper address of any person on, or to whom, any such notice or document is to be served, or given, shall be his last known address.[25]

2.28 In the case of a body corporate, the notice may be served on, or given to, the secretary or clerk of the body. The proper address shall be the address of the registered or principal office of that body.[26] In the case of a partnership, the notice may be served on, or given to, a partner or the person having control or management of the partnership business,[27] at the principal office of the partnership.[28] The principal office of a company registered outside the UK, or of a partnership carrying on business outside the UK, shall be their principal office within the UK.[29]

2.29 If the person to be served with, or given, any such notice or document has specified an address within the UK other than his proper address,[30] as the one at which he or someone on his behalf will accept notices or documents of the same description as the notice or document to be served, then that address is also to be treated as his proper address.

2.30 Section 46(6) provides that, where service is authorized on the owner or occupier of premises, it may be sent by post to those premises. It may also be addressed by name to the person on whom it is to be served, and delivered to some responsible person who appears to be resident or employed in the premises. In *HSE v George*

---

[25] HSWA 1974 s 46(4).
[26] ibid s 46(4)(a).
[27] ibid s 46(3)(b).
[28] ibid s 46(4)(b).
[29] ibid s 46(4).
[30] Within the meaning given in HSWA 1974 s 46(4).

Tancocks Garage Ltd,[31] the Divisional Court ruled that there is only a narrow scope to serve a notice in this way, namely where one was required or authorized to be served on, or given to, an owner or occupier of premises, where it was alleged that the person concerned was in breach of his duty as owner or occupier of those premises. If a company is the occupier, and a notice does not relate to it as the occupier, then service should be sent to its registered address. It was further held that s 46 did not contain a complete code of all the ways in which a notice might be served. The subsections were permissive only. If it is proved that a notice came to the attention of the proper officer of a company, there would be sufficient service under HSWA 1974.

2.31 In general, the terms 'serve' or 'send' are used in relation to duties in respect of notices and documents by the HSWA 1974, and health and safety provisions. The Interpretation Act 1978 s 7, provides that, where an Act authorizes or requires any document to be served by post (whether the expression 'serve' or the expression 'give' or 'send', or any other expression is used) then, unless the contrary intention appears, the service is deemed to be effected. The envelope containing the notice must be properly addressed and prepaid. Posting a letter containing the document is taken to have effected service of the notice at the time at which the letter would be delivered in the ordinary course of post, unless the contrary is proved.

C. The Investigation of Health and Safety Incidents and Offences

2.32 In 2004/2005 210 workers were killed in the UK; this translates to a rate of 0.7 fatal injuries per 100,000 employed. A total of 363,000 non-fatal injuries were reported to enforcement authorities—a rate of 1,330 per 100,000 workers. In the same year, the HSE alone issued a total of 8,445 enforcement notices. It instigated 1,267 prosecutions for alleged breaches of health and safety duties.

2.33 Clearly, the HSE does not have the resources to investigate every potential breach, nor would the deployment of all its resources to investigation and enforcement amount to a means of fulfilling the HSE's statutory objectives. Thus, the HSE, other enforcing authorities, and individual inspectors, are not under a duty to investigate every suspected breach of a health and safety duty. However, the HSE does have a statutory obligation 'to make adequate arrangements for the enforcement of the relevant statutory provisions'.[32]

---

[31] *Mallon v Norman Hill Plant Hire Ltd* [1993] Crim LR 605.
[32] HSWA 1974 s 18.

The ambit of the duty to investigate

2.34 The issue of whether an enforcing authority is under a duty to investigate a suspected breach of a health and safety duty has most often arisen in relation to the HSE and the investigation of fatalities to members of the public, where the death arises from a breach of HSWA 1974 s 3.[33] In some situations, lawyers acting for injured parties pursuing a civil action, or the family of a deceased person, often press for an HSE investigation into an incident. The most common allegation is that it had a work-related management failure as its cause.

2.35 The ambit of any duty to investigate is founded upon principles of public law which govern the exercise of the functions of any public authority. The following principles, in particular, apply:

(a) A public authority must perform its legal duties.
(b) Where a function is properly analysed as a power, and not a duty, it must still be exercised lawfully and rationally. In particular, consideration must be given to exercising it without any fetter—by way of rigid policy or otherwise. Additionally, the function cannot be abdicated to another person or body.
(c) A public authority cannot use extra-statutory policies or practices to frustrate, impede, or otherwise contradict, the statutory scheme which governs its functions.
(d) A public authority must act[34] in a way which is compatible with the Convention rights set out in Sch 1 to the Human Rights Act 1998. (In particular, where a death has arisen, the Convention rights include the right to life, as enshrined in Art 2. This, in turn, may impose a procedural obligation to investigate the circumstances in which a death took place. Especially if state agents—for instance, a police force, a prison, or a health authority—may have contributed to the death.)[35]
(e) A public authority is entitled to have policies as to how its discretionary powers will *normally* be exercised. In particular, it can set priorities by reference to resources, to take account of the seriousness of the subject-matter. However, it is impermissible to have a policy which purports to rule out consideration of whole classes of incident/breach. Reference must first be made to such criteria based upon the particular facts.

Duty to make adequate arrangement for enforcement

2.36 Section 1 of HSWA 1974 expresses the wide protective intent of the Act. Section 1 provides that it 'shall have effect with a view to: securing the health, safety and

---

[33] The HSE's Operational Circular OC 130/10 gives general guidance to inspectors on the selection for investigation of incidents that may have arisen as a result of breaches of HSWA 1974 s 3.
[34] 'Act', for this purpose, includes a failure to act: Human Rights Act 1998 s 6(6).
[35] *R (Middleton) v West Somerset Coroner* [2004] UKHL 10, [2004] 2 AC 182.

C. The Investigation of Health and Safety Incidents and Offences

welfare of persons at work and protecting persons other than persons at work against risks to health or safety arising out of or in connection with the activities of persons at work'.

Section 18(1) of the Act provides: 2.37

> It shall be the duty of the Health and Safety Executive to make adequate arrangements for the enforcement of the relevant statutory provisions, except to the extent that some other authority or class of authorities is by any of those provisions or by regulations under subsection (2) below made responsible for their enforcement.

Regulations have, and may, pass some responsibility for enforcement of the relevant statutory provisions to authorities other than the HSE, such as the Health and Safety (Enforcing Authority) Regulations 1998 SI 1998/494.[36] In order for the HSE to fulfil its duty, however, to make 'adequate arrangements for the enforcement of the relevant statutory provisions', such arrangements must give effect to the intent of the Act. 2.38

In *R v Police Commissioner of the Metropolis ex parte Blackburn* [1968] 2 QB 118 (CA), the Court of Appeal was concerned with a challenge to the lawfulness of a policy of the Commissioner of the Metropolitan Police. The policy was not to take proceedings against gaming clubs for breach of the gaming laws, unless there were complaints of cheating or the clubs had become the haunts of criminals. The Court of Appeal held that it was the duty of chief constables to enforce the law. Although chief constables had discretions, including the discretion whether to prosecute in a particular case, nonetheless a policy decision amounting to a failure of the duty to enforce the law of the land would be unlawful. 2.39

In this way, the discretion of chief officers of police, over prosecution and enforcement of the statutory criminal law, must be so exercised as to give effect to the true intention of Parliament. The legislature's intention is what appears in, and what can be gleaned from the circumstances of, the relevant statutes. In respect of the HSE's enforcement responsibilities, the relevant Act is the HSWA 1974. 2.40

The decision to investigate

The HSC's Enforcement Policy Statement 'sets out the general principles and approach which HSC expects the health and safety enforcing authorities (mainly HSE and the local authorities) to follow'.[37] Inspectors are required to follow the terms of the Statement. This provides that, 'the appropriate use of enforcement powers, including prosecution, is important, both to secure compliance with the law and to ensure that those who have duties under it may be held to account for failures to safeguard health, safety and welfare'. 2.41

---

[36] See App C.
[37] App B: HSC, Enforcement Policy Statement, Introduction.

2.42 In para 7, the Statement provides that:

> Investigating the circumstances encountered during inspections, or following incidents or complaints, is essential before taking any enforcement action. In deciding what resources to devote to these investigations, enforcing authorities should have regard to the principles of enforcement set out in this statement and the objectives published in HSC and HELA strategic plans. In particular, in allocating resources, enforcing authorities must strike a balance between investigations and mainly preventive activity.

2.43 In para 10, the Statement sets out the principles which inspectors should follow, when arriving at judgments concerning both the exercise of discretion in deciding when to investigate and what enforcement action may be appropriate. The Enforcement Policy Statement deals specifically with investigation, at paras 30 to 33. At para 31, the Statement provides that investigations are undertaken in order to determine:

- causes;
- whether action has been taken, or needs to be taken, to prevent a recurrence and to secure compliance with the law;
- lessons to be learnt and to influence the law and guidance;
- what response is appropriate to a breach of the law.

2.44 The paragraph continues:

> [T]o maintain a proportionate response, most resources available for investigation of incidents will be devoted to the more serious circumstances. HSC's Strategic Plan recognizes that it is neither possible nor necessary for the purposes of the Act to investigate all issues of non-compliance with the law which are uncovered in the course of preventive inspection, or in the investigation of reported events.

2.45 At para 33, the Enforcement Policy Statement provides:

> In selecting which complaints or reports of injury or occupational ill health to investigate, and in deciding the level of resources to be used, the enforcing authorities should take account of the following factors:
> - the severity and scale of potential or actual harm;
> - the seriousness of any potential breach of the law;
> - knowledge of the dutyholder's past health and safety performance;
> - the enforcement priorities;
> - the practicality of achieving results;
> - the wider relevance of the event, including serious public concern.

A person with a sufficient interest in a matter can seek to judicially review a decision by an inspector, the HSE, or an enforcing authority, not to investigate a suspected breach of a health and safety duty. A number of such actions have been threatened against the HSE in previous years, particularly in relation to alleged breaches of HSWA 1974 s 3. No case involving a refusal to investigate has reached a hearing in the High Court, but one has in relation to a challenged decision not

C. The Investigation of Health and Safety Incidents and Offences

to prosecute following such an investigation: *R (on the application of Pullen) v Health & Safety Executive* [2003] EWHC 2934 (Admin).

The duty to pursue all reasonable lines of inquiry

Whether or not an inspector is under a duty to commence a particular investigation into an incident or suspected breach of duty, if such an investigation is commenced, he is thereafter subject to statutory duties concerning the ambit and extent of his duty. **2.46**

The Criminal Procedure and Investigations Act 1996 created a new procedure and new duties in respect of prosecution and defence disclosure during the trial process. The 1996 Act also placed duties upon investigators in the conduct of a criminal investigation, both in respect of the retention of material and in relation to the duty to investigate. **2.47**

The definition of a criminal investigation under Pt I of the Criminal Procedure and Investigations Act 1996 (concerned with prosecution and defence disclosure) is different to that under Pt II (concerned with criminal investigations). Part I defines a criminal investigation as an investigation which police officers or other persons have a duty to conduct; Pt II defines it as extending only to investigations conducted by the police. **2.48**

Two codes of practice have been issued, pursuant to the Criminal Procedure and Investigations Act 1996 ss 23 and 25. The first code came into force on 1 April 1997 (SI 1997/1033). A revised code of practice was brought into force by the Criminal Procedure and Investigations Act 1996 (Code of Practice) Order 2005 SI 2005/985, in respect of criminal investigations conducted by police officers which began on or after 4 April 2005. The Codes were issued under Pt II of the Act, 'setting out the manner in which police officers are to record, retain and reveal to the prosecutor material obtained in a criminal investigation and which may be related to the investigation, and related matters'.[38] The Codes govern the action the police must take in recording and retaining material obtained in the course of a criminal investigation. Further, the Codes regulate its supply to the prosecutor, in lieu of a decision on disclosure. **2.49**

While concerned with criminal investigations under the Criminal Procedure and Investigations Act 1996, Pt II (ie those conducted by the police), para 1.1 of the Code[39] provides: 'Persons other than police officers, who are charged with the duty of conducting an investigation, as defined in the Act, are to have regard **2.50**

---

[38] Preamble to Criminal Procedure and Investigations Act 1996 (Code of Practice) Order 2005 SI 2005/985.
[39] Criminal Procedure and Investigations Act 1996 (Code of Practice) Order 2005 SI 2005/985.

to the relevant provisions of the Code, and should take these into account in applying their own operating procedures.'

2.51 Under 'General Functions', the Code sets out some important provisions defining the functions of the 'investigator' and the 'disclosure officer'. In respect of the investigator, the Code, at para 3.5, makes a further observation: that 'In conducting an investigation, the investigator should pursue all reasonable lines of inquiry, whether these point towards or away from the suspect. What is reasonable in each case will depend on the particular circumstances.'

2.52 This duty—to pursue 'all reasonable lines of inquiry'—is one to which a health and safety inspector, being a person under a duty to investigate, as defined in CPIA 1996, must 'have regard' when conducting any investigation. 'Criminal Investigation', under the Code, is an investigation with a view to it being ascertained whether a person should be charged with an offence, or whether a person charged with an offence is guilty of it.[40] The Code provides that this will include:

(a) investigations into crimes that have been committed;
(b) investigations whose purpose is to ascertain whether a crime has been committed, with a view to the possible institution of criminal proceedings; and
(c) investigations which begin in the belief that a crime may be committed.

Investigatory powers

2.53 Section 33(1) of HSWA 1974 creates fifteen health and safety offences. Most importantly s 33(1)(a) creates the offence of breaching one of the general duties in ss 2 to 7 of the Act. Section 33(1)(b) of HSWA 1974 makes it an offence to breach any provision in a health and safety regulation. Both these offences, and most health and safety offences, can be tried either in the magistrates' court or the Crown Court: they are indictable offences.

2.54 Proceedings for a health and safety offence, in England and Wales, can only be instituted by a health and safety inspector,[41] by the Environment Agency, or by, or with, the consent of the Director of Public Prosecutions, through the Crown Prosecution Service.[42]

2.55 Health and safety inspectors are a creation of statute, namely HSWA 1974. Inspectors have only those powers given to them by that Act, and none through the operation of the common law.

---

[40] Criminal Procedure and Investigations Act 1996 (Code of Practice) Order 2005 SI 2005/985 para 2.1.
[41] Appointed pursuant to HSWA 1974 s 19.
[42] HSWA 1974 s 38.

C. The Investigation of Health and Safety Incidents and Offences

2.56 The principal investigatory powers of a health and safety inspector are contained within HSWA 1974 s 20. These include compulsory powers of questioning, the power to have documents produced, and to take copies, and a power to seize certain articles. In addition to the s 20 powers, HSWA 1974 s 25 empowers inspectors to deal with imminent causes of danger. Sections 21 and 22 of HSWA 1974 give inspectors the power to issue prohibition and improvement notices.

2.57 No power of arrest is given by virtue of HSWA 1974 to inspectors, nor is a power given to search. Prior to 1 January 2006, the health and safety offences created by HSWA 1974 were not arrestable offences (meaning that neither the police, an inspector, nor any other person had any power to arrest a person in respect of, or on suspicion of, having committed such an offence). Not being arrestable offences, the power of the police to obtain a warrant and to search premises or persons, in respect of a health and safety investigation, was also limited. Effectively, this was limited only to where there were grounds for suspecting the commission of gross negligence manslaughter by an individual.

2.58 The Serious Organized Crime and Police Act 2005 has, from 1 January 2006, removed the distinction between serious arrestable, arrestable, and non-arrestable offences. It introduces a distinction in powers, based upon whether offences are indictable, as is the case with most health and safety offences, or simply summary. Police officers now have a power to arrest in respect of any offence. In certain circumstances, a person other than a police officer has a power to arrest anyone whom he has reasonable grounds for suspecting is committing, or is guilty of, an indictable offence, such as a health and safety offence.

Police powers and the new power of arrest

2.59 Section 24 of the Police and Criminal Evidence Act 1984 (PACE) (as amended)[43] now provides that a police constable may arrest without a warrant:

(a) anyone who is about to commit an offence;
(b) anyone who is in the act of committing an offence;
(c) anyone whom he has reasonable grounds for suspecting to be about to commit an offence;
(d) anyone whom he has reasonable grounds for suspecting to be committing an offence.

2.60 Section 24 provides that if an offence has been committed, or a constable has reasonable grounds for suspecting that an offence has been committed, he may arrest without a warrant anyone who is guilty of the offence. He may also arrest

---

[43] From 1 January 2006 amended by the Serious Organized Crime and Police Act 2005 ss 110–111.

without a warrant anyone whom he has reasonable grounds to suspect of being guilty of it. This power of summary arrest is exercisable only if the constable has reasonable grounds for believing that, for any of the reasons mentioned in s 24(5), it is necessary to arrest the person in question. Those reasons are:

  (a) to enable the name of the person in question to be ascertained (in the case where the constable does not know, and cannot readily ascertain, the person's name, or has reasonable grounds for doubting whether a name given by the person as his name is his real name);
  (b) correspondingly, as regards the person's address;
  (c) to prevent the person in question—
     (i) causing physical injury to himself or any other person;
     (ii) suffering physical injury;
     (iii) causing loss of, or damage to, property;
     (iv) committing an offence against public decency;[44] or
     (v) causing an unlawful obstruction of the highway.
  (d) to protect a child or other vulnerable person from the person in question;
  (e) to allow the prompt and effective investigation of the offence or of the conduct of the person in question;
  (f) to prevent any prosecution for the offence from being hindered by the disappearance of the person in question.

2.61 Section 24A of PACE 1984 (as amended)[45] has created the power for a person other than a constable to arrest without warrant, in respect of an indictable offence. This will include health and safety inspectors. Many of the health and safety offences under HSWA 1974[46] fall within this ambit. Section 24A provides that a person other than a constable may arrest, without a warrant:

(a) anyone who is in the act of committing an indictable offence;
(b) anyone whom he has reasonable grounds for suspecting to be committing an indictable offence;
(c) where an indictable offence has been committed, anyone who is guilty of the offence; and
(d) anyone whom he has reasonable grounds for suspecting to be guilty of it.

2.62 Section 24A(3) provides that these powers of summary arrest are exercisable only if:

  (a) the person making the arrest has reasonable grounds for believing that, for any of the reasons mentioned in s 24A(4), it is necessary to arrest the person in question; and
  (b) it appears to the person making the arrest that it is not reasonably practicable for a constable to make it instead.

---

[44] Subject to subs (6), which provides that s 24(5)(c)(iv) applies only where members of the public going about their normal business cannot reasonably be expected to avoid the person in question.
[45] From 1 January 2006, amended by the Serious Organized Crime and Police Act 2005 ss 110–111.
[46] HSWA 1974 s 33.

C. The Investigation of Health and Safety Incidents and Offences

The reasons, provided by s 24A(4), are to prevent the person in question: **2.63**

 (i) causing physical injury to himself or any other person;
 (ii) suffering physical injury;
 (iii) causing loss of, or damage to, property; or
 (iv) making off before a constable can assume responsibility for him.

'Reasonable grounds' for suspicion or belief are not defined in PACE 1984. In *O'Hara v Chief Constable of the Royal Ulster Constabulary* [1997] AC 286 (HL), it was held that the test as to whether reasonable grounds for the suspicion to justify an arrest existed is partly subjective. The arresting officer must have formed a genuine suspicion that the person being arrested was guilty of an offence. The Court found that the test is also partly objective, in that there had to be reasonable grounds for forming such a suspicion. Such grounds could arise from information received from another (even if it subsequently proves to be false), provided that a reasonable man, having regard to all the circumstances, would regard them as reasonable grounds for suspicion. A mere order from a superior officer to arrest a particular individual, however, could not constitute reasonable grounds for such suspicion. **2.64**

The likely exercise of the new powers of arrest

Currently, the police rarely become involved with suspected health and safety offences unless during the course of an investigation into potential gross negligence manslaughter, conducted jointly with a health and safety inspector pursuant to the Work Related Death Protocol.[47] Following an investigation where it has been decided not to bring manslaughter charges, there have been rare instances when the police have pursued health and safety offences, rather than the HSE taking primary responsibility for the investigation. Otherwise, such offences will appear as alternative or joined allegations on a manslaughter indictment. With the passing of the Corporate Manslaughter Act 2007,[48] police involvement in health and safety-related inquiries is likely to increase. The pursuit by the Crown Prosecution Service of health and safety offences may increase, either as joined offences or following a decision not to prosecute for manslaughter. **2.65**

It is possible, in the future, that there will be arrests by the police of individuals on suspicion of having committed a health and safety offence, pursuant to s 24 Police and Criminal Evidence Act 1984. Such individuals may be arrested following a work-related death which is the subject of a joint police/HSE investigation. The justification for any such arrest may be 'to allow the prompt and effective investigation of the offence or of the conduct of the person in question'.[49] This will **2.66**

---

[47] See, in full, para 6, 'Special Inquiries'.
[48] See paras 11.66 *et seq*. in relation to Corporate Manslaughter.
[49] Police and Criminal Evidence Act 1984 s 24(5)(e).

permit the police to take swift action, particularly in respect of seizure of documents and interviewing under caution.

2.67 It is unlikely that health and safety inspectors will have cause to use their power of arrest that derives from the Police and Criminal Evidence Act 1984 s 24A. The criteria required before an arrest can be made, in particular the conditions in s 24A(4), will rarely be established.

2.68 More significant is likely to be the effect upon the position of a health and safety inspector who fears serious obstruction from a person. An inspector in this position may, pursuant to HSWA s 20(2)(b),[50] be accompanied by a police constable. The offence of obstructing a health and safety inspector is triable only by magistrates and is non-imprisonable. However, it may be that either an accompanying police officer or the inspector himself has cause to exercise his powers of arrest, in respect of serious obstruction, fearing physical injury or damage to property.

Police powers of search and entry

2.69 Police powers of search and seizure under the Police and Criminal Evidence Act 1984 are varied and, to some extent, co-existent with one another. With the principal health and safety offences becoming arrestable offences[51] and the test for the more extensive police search powers now being related to indictable offences,[52] these powers will be more important to, and more likely to be used during, joint police/HSE investigations following a work-related death.

2.70 In summary, the relevant entry, search, and seizure powers are set out in the following sections of the Police and Criminal Evidence Act 1984:

(a) Section 17 (as amended):[53] a constable may enter and search any premises for various purposes, including for arresting a person for an indictable offence, if the constable has reasonable grounds for believing that the person whom he is seeking is on the premises.

(b) Section 32 (as amended):[54] a constable may enter any premises in which a person arrested for an indictable offence was present when arrested, or immediately before he was arrested, and search for evidence relating to the offence for which he has been arrested.

---

[50] s 20(2)(b) provides for a threshold test: '... if he [the inspector] has reasonable cause to apprehend any serious obstruction in the execution of his duty.'
[51] Serious Organised Crime and Police Act 2005 ss 110–111.
[52] ibid and Sch 7.
[53] From 1 January 2006, amended by Serious Organized Crime and Police Act 2005 ss 110–111 and Sch 7.
[54] ibid.

C. The Investigation of Health and Safety Incidents and Offences

(c) Section 18 (as amended):[55] a constable can enter and search any premises[56] occupied or controlled by a person who is under arrest for an indictable offence, if he has reasonable grounds for suspecting that there is, on the premises, evidence other than items subject to legal privilege, that relates to that offence, or to some other indictable offence which is connected with, or similar to, that offence. Thereafter he may seize and retain anything for which he may search, under s 18.[57] The power is to search to the extent that is reasonably required for the purpose of discovering such evidence;[58] the powers conferred by this section cannot be exercised unless an officer of the rank of inspector or above has authorized them in writing,[59] except where the search is made before the person is taken to a police station or released on bail, and the presence of the person at a place (other than a police station) is necessary for the effective investigation of the offence.[60]

(d) Section 8: on an application made by a constable, a magistrate will grant a warrant to enter and search premises if he is satisfied that there are reasonable grounds for believing that an indictable offence has been committed; that there is material on the premises specified in the application which is likely to be of substantial value (whether by itself or taken together with other material) to the investigation of the offence; that the material is likely to be relevant evidence;[61] that it does not consist of, or include, items subject to legal privilege, excluded material[62] or special procedure material;[63] and that any of the following conditions apply:

  (i) that it is not practicable to communicate with any person entitled to grant entry to the premises;

---

[55] From 1 January 2006, amended by Serious Organized Crime and Police Act 2005 ss 110–111 and Sch 7.
[56] Premises 'includes any place and, in particular, includes—(a) any vehicle, vessel, aircraft or hovercraft; (b) any offshore installation; (ba) any renewable energy installation; and (c) any tent or movable structure': Police and Criminal Evidence Act 1984 s 23.
[57] Police and Criminal Evidence Act 1984 s 18(2).
[58] ibid s 18(3).
[59] ibid s 18(4).
[60] ibid ss 18(5), (5A).
[61] 'Relevant evidence', in relation to an offence, means anything that would be admissible in evidence at a trial for the offence.
[62] Police and Criminal Evidence Act 1984 s 11 defines 'excluded material' as: (a) personal records which a person has acquired or created in the course of any trade, business, profession or other occupation or for the purpose of any paid or unpaid office which he holds in confidence; (b) human tissue or fluid which has been taken for the purposes of diagnosis or medical treatment and which a person holds in confidence; (c) journalistic material which a person holds in confidence and which consists—(i) of documents; or (ii) of records other than documents. 'Personal records' are defined by s 12; 'Journalistic material' is defined by s 13.
[63] 'Special procedure material' is defined by s 14.

(ii) that it is practicable to communicate with a person entitled to grant entry to the premises, but it is not practicable to communicate with any person entitled to grant access to the evidence;
(iii) that entry to the premises will not be granted unless a warrant is produced;
(iv) that the purpose of a search may be frustrated, or seriously prejudiced, unless a constable arriving at the premises can secure immediate entry to them.

In such circumstances, a constable may seize and retain anything for which a search has been authorized (and he will be lawfully on the premises, for the purposes of the Police and Criminal Evidence Act s 19).

(e) Section 19: general powers of seizure exercisable by a constable who is lawfully on any premises. The constable may seize anything which is on the premises, if he has reasonable grounds for believing that either it has been obtained in consequence of the commission of an offence, or that it is evidence in relation to an offence which he is investigating, or any other offence; and, in any event, that it is necessary to seize it in order to prevent it being concealed, lost, damaged, altered, or destroyed.[64] In the same circumstances, the constable may require any information, which is stored in any electronic form and is accessible from the premises, to be produced in a form in which it can be taken away, and in which it is visible and legible, or from which it can readily be produced in a visible and legible form.[65] The section also provides that no power of seizure conferred on a police constable under any enactment is to be taken to authorize the seizure of an item which the constable exercising the power has reasonable grounds for believing to be subject to legal privilege.

Inspectors and the Police and Criminal Evidence Act 1984

2.71 HSWA 1974 s 20, affords a health and safety inspector extensive powers to obtain evidence. Crucially, though, it does not include any general power of search and seizure, nor does it give inspectors the power to apply for, and obtain, a warrant. It follows that any search of premises, conducted by a health and safety inspector, could only properly be conducted with the consent of the occupier of the premises.

The Police and Criminal Evidence Act 1984 Codes of Practice

2.72 Section 67(9) of the Police and Criminal Evidence Act 1984 provides, 'Persons other than police officers who are charged with the duty of investigating offences or charging offenders shall in the discharge of that duty have regard to any relevant

---

[64] Police and Criminal Evidence Act 1984 s 19(2), (3).
[65] ibid s 19(4).

provision of such a code.' Revised Codes A to F, and new Code G, came into force on 1 January 2006: Police and Criminal Evidence Act 1984 (Codes of Practice) Order 2005 SI 2005/3503.

Section 67(10), of the Police and Criminal Evidence Act 1984, further provides: 2.73

A failure on the part of any person other than a police officer who is charged with the duty of investigating offences or charging offenders to have regard to any relevant provision of such a code in the discharge of that duty, shall not of itself render him liable to any criminal or civil proceedings.

In *Joy v Federation against Copyright Theft Ltd* [1993] Crim LR 588 (DC) it was held that the duty referred to in the Police and Criminal Evidence Act 1984 s 67(9) was any type of legal duty—whether imposed by statute, common law, or by contract. A health and safety inspector is clearly charged with the duty of investigating health and safety offences, and thus is subject to 'any relevant provision' of the codes of practice. 2.74

For the purpose of 'investigating offences', within the meaning of Police and Criminal Evidence Act 1984 s 67(9), it has been held that neither Department of Trade and Industry inspectors, when asking questions under the Companies Act 1985 s 434,[66] nor a Bank of England manager, when exercising supervisory powers over a bank under the Banking Act 1987,[67] are contemplated. Thus, when performing duties and functions other than investigating offences, a health and safety inspector will not be subject to the codes of practice. In nearly every situation in which an inspector is concerned with a potential breach of a health and safety provision, he will be investigating an offence. 2.75

The duty on investigators other than police officers, such as health and safety inspectors, is 'to have regard to any relevant provision' of the codes. Compliance with the letter of the codes will sometimes be impossible, because they refer explicitly to police officers and police stations. In this way in *R (Social Security Secretary) v South Central Division Magistrates*,[68] it was held that there had been no breach by a social security investigator of what is now Code C: 11.2,[69] which provides that, 'immediately prior to the commencement or re-commencement of any interview at a police station or other authorised place of detention, the interviewer should remind the suspect of their entitlement to free legal advice and that the interview can be delayed for legal advice to be obtained'. The social security investigator had not told an interviewee who attended a social security office voluntarily of his entitlement to free legal advice, because there was no such 2.76

---

[66] *R v Seelig and Lord Spens* 94 Cr App R 17 (CA).
[67] *R v Smith (WD)* 99 Cr App R 223 (CA).
[68] *Daily Telegraph* (28 November 2000).
[69] Police and Criminal Evidence Act 1984 (Codes of Practice) Order 2005 SI 2005/3503; C Code of Practice for the Detention, Treatment and Questioning of Persons by Police Officers para 11.2.

provision of free legal advice and thus no entitlement. The position is the same when interviews under caution are conducted by health and safety inspectors, where there is no provision for free legal advice.

Searches under the Police and Criminal Evidence Act 1984, Code B

2.77 In order for an inspector properly to conduct a search with consent, he would be required to comply with the provisions of Code B.[70] To conduct such a search, in accordance with Code B, an inspector must first inform the person of the purpose of the proposed search. He must then make clear that there is no obligation to give consent to a search, and that anything seized may be produced in evidence. A further obligation is to obtain the written consent of a person entitled to grant entry to the premises; and thereafter, a complete written record of the search must be made, in accordance with the code.

Cautioning

2.78 During the course of enquiring into the circumstances of an accident, injury, death, or dangerous occurrence or some suspected breach of a health and safety provision, an inspector may suspect a person of involvement in contravening such a provision and thus have grounds to suspect an offence has been committed. Paragraph 10.1 of Code C,[71] the Code of Practice for the detention, treatment, and questioning of persons by police officers, provides that, 'A person whom there are grounds to suspect of an offence, must be cautioned before any questions about an offence, or further questions if the answers provide the grounds for suspicion, are put to them if either the suspect's answers or silence, (i.e. failure or refusal to answer or answer satisfactorily) may be given in evidence to a court in a prosecution'.

2.79 The Code[72] continues with the following: 'A person need not be cautioned if questions are for other necessary purposes.' It goes on to give various examples of when such a caution is not necessary, including, 'to obtain information in accordance with any relevant statutory requirement'. The guidance at Code C, para 10A, provides: 'There must be some reasonable, objective grounds for the suspicion, based on known facts or information which are relevant to the likelihood that the offence has been committed and the person to be questioned committed it.'

---

70 Police and Criminal Evidence Act 1984 (Codes of Practice) Order 2005 SI 2005/3503; B Code of Practice for Searches of Premises by Police Officers and the Seizure of Property Found by Police Officers on Persons or Premises.
71 ibid; C Code of Practice for the Detention, Treatment and Questioning of Persons by Police Officers para 10.1.
72 ibid.

C. The Investigation of Health and Safety Incidents and Offences

In short, if the answers to questions asked by the inspector of a person who is suspected of involvement in an offence may be sought to be relied upon in any subsequent prosecution of the person, then the person must have been first cautioned. The obligation to caution remains, no matter what course of enquiries the inspector was conducting at the time the questions were asked, nor for what purpose the questions were asked. **2.80**

Interviews

The conduct of formal interviews under caution, in relation to the person 'invited' to answer such questioning, is considered at para 10.1. However, para 11.1A of Code C[73] defines an 'interview' in wide terms: 'the questioning of a person regarding their involvement or suspected involvement in a criminal offence or offences which, under paragraph 10.1, must be carried out under caution. Whenever a person is interviewed they must be informed of the nature of the offence, or further offence.' **2.81**

This provision has been interpreted strictly by the courts. Conversations with police officers, in which suspects have allegedly incriminated themselves, have been ruled inadmissible, as amounting to interviews not conducted in full compliance with the provisions of Code C. A conversation with a person, by a health and safety inspector, during the course of an investigation into a dangerous occurrence, workplace injury, or even during the course of a routine visit, may involve the asking of a question concerning that person's involvement in a contravention of a provision. Potentially, this can amount to an interview. Unless the inspector has complied with the relevant provisions of Code C, any answer given to his question is likely to be ruled inadmissible, if relied upon by the prosecution at any trial of the person for the alleged contravention. **2.82**

Many of the provisions of Code C apply specifically to interviews conducted in police stations, and the need for such interviews to be conducted at a police station, following a decision to arrest. Importantly, however, others are of universal application.[74] These include the need for an accurate record to be made of each interview, which should state the following: **2.83**

(a) the place of interview;
(b) the time it begins and ends;
(c) any interview breaks, and the names of all those present;

---

[73] Police and Criminal Evidence Act 1984 (Codes of Practice) Order 2005 SI 2005/3503; C Code of Practice for the Detention, Treatment and Questioning of Persons by Police Officers para 11.1A.
[74] ibid paras 11.1–11.20.

(d) the need for any written record to be made and completed during the interview, unless this would not be practicable or would interfere with the conduct of the interview (in which case it must be made as soon as practicable after its completion), and;

(e) that such a record must constitute either a verbatim record of what has been said, or, failing this, an account of the interview which adequately and accurately summarizes it.

2.84 In *R v Sparks* [1991] Crim LR 128 (CA) it was held that any questioning, formal or informal, of a person regarding his involvement, or suspected involvement, in an offence, which, by virtue of Code C[75] para 10.1, should be under caution, amounted to an interview. The provisions of Code C para 11, were held to apply. A single question can constitute 'any questioning' and, therefore, may amount to an interview,[76] for the purpose of Code C.

Cautioning and the power under HSWA 1974 s 20(2)(j)

2.85 It is neither necessary, nor a proper course of action, for an inspector to caution the person being questioned, when exercising the power under HSWA 1974 s 20(2)(j), 'to require any person whom he has reasonable cause to believe to be able to give any information relevant to any examination or investigation . . . to answer . . . questions and to sign a declaration of the truth of his answers'. HSWA 1974 s 20(7) provides that no answer given by a person in pursuance of a requirement imposed under s 20(2)(j) 'shall be admissible in evidence against that person or the husband or wife of that person in any proceedings'. The power under HSWA 1974 s 20(2)(j) should not be exercised in order to question a person regarding his involvement or suspected involvement in an offence.

2.86 The Divisional Court case of *Walkers Snack Foods Ltd v Coventry City Council*[77] concerned a virtually identical provision to HSWA 1974 s 20(2)(j), under the Food Safety Act 1990. Environmental Health Officers had not cautioned an employee of the applicant company, who had been questioned under the compulsory power during the early stages of the investigation, when inspectors were attempting to discover the circumstances of a reported incident. Rose LJ concluded: 'Bearing in mind that Code C: 10.1 expressly excludes from the requirement of a caution where questions are put to obtain information in accordance with any relevant statutory requirement, I am of the view that the Justices were

---

[75] Police and Criminal Evidence Act 1984 (Codes of Practice) Order 2005 SI 2005/3503; C Code of Practice for the Detention, Treatment and Questioning of Persons by Police Officers para 10.1.
[76] *R v Miller* [1998] Crim LR 209 (CA); *R v Ward* 98 Cr App R 337 (CA); *R v Dunn* 91 Cr App R 237 (CA).
[77] Case no CO/2899/97.

correct in reaching the conclusion which they did as to the absence of any need to caution.'

D. Inspectors' Powers Under HSWA 1974 s 20

2.87 While HSWA 1974 does not provide an inspector with a power of arrest, nor the power to enter premises to conduct a search, nonetheless HSWA 1974 s 20, affords the inspector substantial powers of investigation. In some important respects, these investigatory powers vested in inspectors are far more potent than those enjoyed by police or customs officers.

2.88 Section 20(1), of HSWA 1974, provides that, 'Subject to the provisions of section 19 and this section, an inspector may, for the purpose of carrying into effect any of the relevant statutory provisions within the field of responsibility of the enforcing authority which appointed him, exercise the powers set out in subsection (2) . . .'

2.89 The various powers created by HSWA 1974 s 20 are supported by a number of potential criminal sanctions:

(a) obstructing an inspector in the exercise of any of these powers amounts to an offence under s 33(1)(h);
(b) contravening any requirement imposed by an inspector under s 20 amounts to an offence under s 33(1)(e);
(c) preventing, or attempting to prevent, any other person from appearing before an inspector or from answering any question to which an inspector may, by virtue of s 20(2) require an answer, is an offence under s 33(1)(f);
(d) s 33(1)(h) provides that it is an offence for a person to make a statement which he knows to be false, or recklessly to make a statement which is false, where the statement is made in various circumstances, including when required by an inspector, pursuant to s 20(2)(j).

2.90 The 'relevant statutory provisions' means the provisions of HSWA 1974 Pt I, and any health and safety regulations, along with the existing statutory provisions set out in HSWA 1974 Sch 1. These should be read in conjunction with, and include, any regulation, order, or other instrument of a legislative character made or having effect under any of the provisions.[78]

2.91 HSWA 1974 s 19 gives enforcing authorities power to appoint inspectors, and stipulates how the appointment is to be effected. The division of responsibility, as between the HSE and local authorities, is set out in the Health and Safety

---

[78] HSWA 1974 s 53(1).

(Enforcing Authority) Regulations 1989 SI 1989/1903.[79] Regulation 3 provides that, where the main activity carried on in non-domestic premises is specified in Sch 1 of the regulations, 'the local authority for the area in which the premises are situated is the enforcing authority for those premises'.

2.92 HSWA 1974 s 19 provides that every enforcing authority may appoint as inspectors such persons having suitable qualifications as it thinks necessary for carrying into effect the relevant statutory provisions, within its field of responsibility. An inspector is to be appointed by an instrument, in writing, specifying which of the powers conferred on inspectors by the relevant statutory provisions are to be exercisable by the person appointed and that he shall exercise only such of those powers as are so specified.[80] In practice, inspectors are appointed with all of the powers provided by HSWA 1974 s 20. Moreover, s 19 provides that an inspector shall 'be entitled to exercise the powers so specified only within the field of responsibility of the authority which appointed him'. Thus, the power of local authorities to enforce health and safety extends only to their local boundary, although local authority inspectors' powers may in some respects extend beyond such boundaries. HSE inspectors are not appointed with powers specific to, or contained within, a geographical location. Instead, their powers are triggered solely by the conditions set out in HSWA 1974 s 20, and, in such circumstances, can be exercised anywhere within the jurisdiction.

2.93 In the case of *London Borough of Wandsworth v South Western Magistrates' Court* [2003] EWHC 1158 Admin, the Divisional Court had cause to consider the extent and ambit of the powers afforded to inspectors by HSWA 1974 s 20. In particular, the court elaborated on the power to require answers from persons, under s 20(2)(j), and commented on whether that extended to a power to require answers, in writing, to questions supplied by letter. Lord Justice Baker Scott drew especial attention to the limitation on the geographical extent of the powers afforded to local authority inspectors. He picked out the dichotomy that exists, where such an inspector may wish to seek information from persons far removed from the locality where any incident occurred. Developing this, Lord Justice Scott Baker commented: 'An inspector has no power to enter premises, except in the area of the local authority that employs him. Thus on a narrow interpretation, requiring face-to-face meetings, he could travel miles to see someone whom he regards as relevant, only to be refused entry to the premises when he gets there.'

2.94 This, the Court ruled, was one of the reasons why s 20, taken as a whole, was to be construed widely, as 'obviously intended to contain wide powers'.

---

[79] Reproduced at App C.
[80] HSWA 1974 s 19(2).

D. Inspectors' Powers Under HSWA 1974 s 20

2.95 The Court described the power under HSWA 1974 s 20(2)(m), as an 'important sweeping-up provision which clothes the inspector with any other power that is necessary for him to carry into effect any statutory provisions within the field of responsibility of the enforcing authority'.

Consequences of improper use of HSWA 1974 s 20 powers

Exclusion of evidence

2.96 If an inspector obtains evidence improperly, by exceeding any of the powers vested in him by virtue of HSWA 1974 s 20, a court may, in a subsequent prosecution, rule that the evidence so obtained should be excluded.[81] The test for admissibility of evidence garnered in these circumstances is a threshold one. Section 78(1), of the Police and Criminal Evidence Act 1984 provides:

> In any proceedings the court may refuse to allow evidence on which the prosecution proposes to rely to be given if it appears to the court that, having regard to all the circumstances, including the circumstances in which the evidence was obtained, the admission of the evidence would have such an adverse effect on the fairness of the proceedings that the court ought not to admit it.

2.97 This provision has proved to be the cornerstone of the courts' discretion to exclude evidence. Each case has to be decided upon its own facts. The Police and Criminal Evidence Act 1984 s 78, has been invoked in relation to the widest range of evidential material. Most frequently, this is in connection with evidence said to have been obtained following a breach of the Codes, under the Police and Criminal Evidence Act 1984.

2.98 There are two stages in the application of HSWA 1974 s 78. First, the circumstances in which the evidence came to be obtained, and, second, whether admission of the evidence would have an adverse effect upon the fairness of the proceedings. In considering fairness, a balance has to be struck between that which is fair to the prosecution and that which is fair to the defence.[82] The final aspect of the fairness test relates only to the defendant. An issue is whether the admission of the evidence would have 'such an adverse effect on the fairness of the proceedings that the court ought not to admit it'.

2.99 In the House of Lords case of *R v Looseley; Att-Gen's Reference (No 3 of 2000)* [2002] 1 Cr App R 29, Lord Scott stated that evidence could be excluded under HSWA 1974 s 78, if it had been obtained by unfair means. More far-reaching, Lord Nicholls said that the concept of fairness was not limited to procedural fairness.

---

[81] See *Skinner v John McGregor (Contractors) Ltd* 1977 SLT (Sh Ct) 83 for a good example of where the exclusion of evidence has been obtained under HSWA 1974 s 20(2)(i), through non-compliance with the terms of s 20(6).
[82] *R v Hughes* [1988] Crim LR 519 (CA).

Civil action for damages

2.100 An unlawful or unjustified purported exercise of power, in excess of the terms of HSWA 1974 s 20, could expose an inspector to a possible civil claim for damages. In such circumstances s 26 of the Act provides for the possibility of the inspector's indemnity, by the enforcing authority, where honest belief on his part is established:

> Where an action has been brought against an inspector in respect of an act done in the execution or purported execution of any of the relevant statutory provisions and the circumstances are such that he is not legally entitled to require the enforcing authority which appointed him to indemnify him, that authority may, nevertheless, indemnify him against the whole or part of any damages and costs or expenses which he may have been ordered to pay or may have incurred, if the authority is satisfied that he honestly believed that the act complained of was within his powers and that his duty as an inspector required or entitled him to do it.

The power to enter premises: HSWA 1974 s 20(2)(a)

2.101 Under HSWA 1974 s 20(2)(a), an inspector is empowered to enter any premises, without first having to seek permission, if certain conditions are satisfied. He may, 'at any reasonable time (or, in a situation which in his opinion is, or may be, dangerous, at any time) . . . enter any premises which he has reason to believe it is necessary for him to enter for the purpose' of carrying into effect any of the relevant statutory provisions within the field of responsibility of the enforcing authority which appointed him.

2.102 The word 'premises' is defined widely, in HSWA 1974 s 53 to include 'any place' and, in particular:

(a) any vehicle, vessel, aircraft, or hovercraft;
(b) any installation on land (including the foreshore and other land intermittently covered by water); any offshore installation; and any other installation (whether floating, or resting on the seabed or the subsoil thereof, or resting on other land covered with water or the subsoil thereof); and
(c) any tent or movable structure.

2.103 Unlike a police officer, the inspector's power to enter premises is not related to either arrest or the obtaining of a warrant from a court. However, an inspector has no power to search premises nor, it follows, to effect an entry in order to search. The power to enter arises where an inspector 'has reason to believe it is necessary'. This appears to impose an objective standard, requiring the existence of an identifiable reason capable of supporting the inspector's subjective belief of necessity. Entry must be necessary in order either to exercise any of the other powers under HSWA 1974 s 20(2), or any other provision of the Act, or in order to effect any provision of a health and safety regulation.

D. Inspectors' Powers Under HSWA 1974 s 20

2.104 The term 'any reasonable time' is not defined, but it does not appear to import any necessity for an inspector to give notice of intended entry. Where an inspector forms the opinion that a situation is, or may be, dangerous, he can enter a premises at any time. If such circumstances obtain, the inspector enjoys additional powers, under HSWA 1974 s 25.[83]

Power to take a constable: HSWA 1974 s 20(2)(b)

2.105 By virtue of HSWA 1974 s 20(2)(b) the inspector is entitled to take with him a constable, if he has reasonable cause (an objective standard) to apprehend 'any serious obstruction' in the execution of his duty. While any obstruction of an inspector in such circumstances would amount to an offence, under s 33(1)(h),[84] it is only the apprehension of a 'serious obstruction' that gives an inspector the power to take along a constable.

2.106 Since 1 January 2006,[85] police officers and all other persons have new powers of arrest, pursuant to the Police and Criminal Evidence Act 1984 ss 24 and 24A. The offence of obstructing a health and safety inspector is triable only by magistrates, and is non-imprisonable. Nevertheless, it may be that either an accompanying police officer, or the inspector himself, may have cause to exercise his powers of arrest in respect of obstruction, if one of them fears physical injury or damage to property.

2.107 'Serious obstruction' is not defined by HSWA 1974. The same expression appears, without definition, in a similar context in a raft of other legislation, including: the Mines and Quarries Act 1954; the Environment Act 1995, and the Environmental Protection Act 1990. It would appear that 'serious obstruction requires, and is constituted by, an apprehension of an act calculated, or likely, to hinder the inspector in some significant way'.

Power to take with others and items of equipment: HSWA 1974 s 20(2)(c)

2.108 Whenever the inspector exercises his power under HSWA 1974 s 20(2)(a) to enter premises, s 20(2)(c) provides that he may take with him any other person duly authorized by the inspector's enforcing authority. In addition, he may take with

---

[83] Per s 25(1): 'Where, in the case of any article or substance found by him in any premises which he has the power to enter, an inspector has reasonable cause to believe that, in the circumstances in which he finds it, the article or substance is a cause of immediate danger of serious personal injury, he may seize it and cause it to be rendered harmless (whether by destruction or otherwise).' Certain conditions attach to the exercise of this power, and the other powers outlined in s 25.
[84] 'It is an offence for a person—[s 33(1)(h)] intentionally to obstruct an inspector in the exercise or performance of his powers or duties . . .'
[85] Serious Organised Crime and Police Act 2005 ss 110–111 and Sch 7.

Powers to 'examine and investigate': HSWA 1974 s 20(2)(d)

2.109 Section 20(2)(d) of HSWA 1974 vests a wide power in an inspector. It provides for him to make such examination and investigation as may, in any circumstances, be necessary for the purpose of carrying into effect any of the relevant statutory provisions[86] within the field of responsibility of the enforcing authority which appointed him.

2.110 In *London Borough of Wandsworth v South Western Magistrates' Court* [2003] EWHC 1158, the Divisional Court held that, 'most but not all of the powers' given to an inspector, by HSWA 1974 s 20, 'envisage the inspector's attendance at premises'. In particular, the court scrutinized the power under s 20(2)(j), 'to require any person whom he has reasonable cause to believe to be able to give any information relevant to any examination or investigation under paragraph (d)' to answer questions. It held that the power did not require that the inspector first attend at the premises, in order to be exercised. Neither s 20(2)(d), nor subs (2)(j), use the term 'premises'. It follows that, consistent with the interpretation of the Divisional Court, the power to examine and investigate goes beyond a power limited to the inspector's attendance at premises.[87] Furthermore, the language of HSWA 1974 s 20(2)(e), which begins 'as regards any premises', is suggestive that such an interpretation of s 20(2)(d) is correct.

2.111 Neither the term 'examination', nor 'investigation', is defined by HSWA 1974. The limits of a power to 'make such examination and investigation' are, accordingly, difficult to ascertain. HSWA 1974 s 20(2)(k), gives the inspector a power to 'require the production of, inspect and take copies of, any document which it is necessary for him to see, for the purposes of any examination or investigation under paragraph (d)'. It therefore appears that the power under s 20(2)(d) would not itself, without the operation of s 20(2)(k), extend to including a power to require production of such items.

2.112 The backdrop to all of this is the plethora of detailed statutory safeguards and limitations on the power of the police to conduct a search of premises. Also significant is the absence of a paragraph in HSWA 1974 s 20(2), referring to a power to search. Having regard to these factors, the power to 'make such examination and/or investigation', under s 20(2)(d), cannot be interpreted to include the power to search, and must, at least, be limited in this way.

---

[86] Defined by HSWA 1974 s 53(1).
[87] [2003] EWHC 1158 Admin para 20.

Direction to leave undisturbed: HSWA 1974 s 20(2)(e)

Section 20(2)(e), of HSWA 1974, provides that, **2.113**

> as regards any premises which he has power to enter, [an inspector may] . . . direct that those premises or any part of them, or anything therein, shall be left undisturbed (whether generally or in particular respects) for so long as is reasonably necessary for the purpose of any examination or investigation under paragraph (d) above.

This power is often exercised to allow for subsequent examination by a specialist inspector, or expert, of machinery or plant, following an accident. The power is usually exercised through the issuance of a written notice to the occupier of premises. The notice must refer to s 20(2)(e), and give a fixed time during which no disturbance or interference should be made. Such a fixed time can then be extended by further notice. **2.114**

Taking measurements, photographs, and recordings: HSWA 1974 s 20(2)(f)

Section 20(2)(f) of HSWA 1974 gives an inspector the power 'to take such measurements and photographs and make such recordings as he considers necessary' for the purpose of any examination or investigation under s 20(2)(d). **2.115**

The power is not expressed to be limited in its exercise in relation to, or upon, premises: *London Borough of Wandsworth v South Western Magistrates' Court* [2003] EWHC 1158 Admin.[88] **2.116**

Articles and substances: the power to take samples: HSWA 1974 s 20(2)(g)

Section 20(2)(g), of HSWA 1974, gives an inspector the power 'to take samples of any articles or substances found in any premises which he has power to enter, and of the atmosphere in, or in the vicinity of, any such premises'. **2.117**

HSWA 1974 s 20(3) gives the Secretary of State the power to make regulations dealing with the procedure to be followed in connection with the taking of samples. To date, none has been enacted. **2.118**

The term 'Articles for use at work' is defined in HSWA 1974,[89] but 'articles' is not. It is such a wide term that it appears to include, in the context of s 20(2)(g), any item—whether plant, equipment, a component, or indeed anything else. 'Substance' is defined by HSWA 1974 s 53(1) to mean any natural or artificial substance (including micro-organisms), whether in solid or liquid form, or in the form of a gas or vapour. **2.119**

Section 20(2)(g) of HSWA 1974 is clearly expressed to create a power exercisable in relation to premises which an inspector has a power to enter. Specifically, this **2.120**

---

[88] As spoken.
[89] HSWA 1974 s 53(1).

refers to the power under HSWA 1974 s 20(2)(a). However, HSWA 1974 s 20(2)(g) appears to envisage the taking of samples for testing or analysis, as opposed to their use in court as evidence. The wording of the section suggests—but does not require—that only a sample of articles, or a sample of a substance, and thus not every article or all the substance, will be taken. Such an interpretation is consistent with the fact that the paragraph requires the leaving of no notice. This is unlike, and in contradistinction to, the power under s 20(2)(i) to take possession of and detain a dangerous article and substance, when, pursuant to HSWA 1974 s 20(6), a written notice must be left.

The power to dismantle or subject dangerous articles or substances to testing: HSWA 1974 s 20(2)(h)

2.121 HSWA 1974 s 20(2)(h) creates a power for an inspector, governing the dismantling and testing of what may be dangerous articles or substances:

> In the case of any article or substance found in any premises which he has power to enter, being an article or substance which appears to him to have caused or to be likely to cause danger to health or safety, to cause it to be dismantled or subjected to any process or test (but not so as to damage or destroy it unless this is in the circumstances necessary for the purpose of carrying into effect any of the relevant statutory provisions within the field of responsibility of the enforcing authority which appointed him).

2.122 It was stated above, at paragraph 2.120, that 'Articles for use at work' is a term defined in HSWA 1974,[90] whereas 'articles' is not. 'Articles' is such a wide term that it must include, in the context of s 20(2)(h), any item—whether plant, equipment, a component or anything else. 'Substance' is defined by HSWA 1974 s 53(1), to mean any natural or artificial substance (including micro-organisms), whether in solid or liquid form, or in the form of a gas or vapour. This power is exercisable in relation to premises which an inspector has a power to enter, ie the power under HSWA 1974 s 20(2)(a).

2.123 Section 20(2)(h) of HSWA 1974 requires the inspector to have formed the opinion that the article or substance is one which has had certain effects. Either it has caused danger to health or safety, ie someone has been put at risk, or else the article or substance is 'likely to cause danger to health and safety'. This second phrase is unusual, in the context of HSWA 1974 Pt I. Elsewhere, it employs the standard of a 'risk to health and safety' or 'without risks'—ie it engenders a possibility of danger. 'Likely' is akin to 'probable': it appears that something is 'likely' to happen when there is more than a 50 per cent chance of its occurring. (The inspector's powers to issue an improvement or prohibition notice, preventing the use of the

---

[90] HSWA 1974 s 53(1).

article or substance, do not require him to have formed such a view as to the likelihood for danger to be posed.) However, the absence of a qualifying word, such as 'reasonable', in HSWA 1974 s 20(2)(h), in respect of what 'appears' to the inspector to cause danger, creates an extremely wide power. This power is exercisable without limitation: there is no need for the existence of objective grounds to be established.

2.124 The inspector must avoid damaging, or destroying, the article or substance which is to be dismantled, subjected to any process, or tested. The exception to this general rule is where, in the circumstances, damage or destruction is necessary for the purposes of any of the relevant statutory provisions. 'Relevant statutory provisions' must be read in the complete sense of including the inspector's powers, under HSWA 1974 s 20, to investigate and examine. In short, where such process or testing to be carried out requires damage or destruction of the article or substance, this can be done.

2.125 Section 20(4), HSWA 1974, governs situations in which an inspector proposes to exercise the power conferred by s 20(2)(h). It provides that:

> he shall, if so requested by a person who at the time is present in, and has responsibilities in relation to, those premises, cause anything which is to be done by virtue of that power to be done in the presence of that person, unless the inspector considers that its being done in that person's presence would be prejudicial to the safety of the State.

2.126 The duty to allow such a person who has 'responsibilities in relation to the premises' to be present only arises if the inspector carries out the processing or testing on the premises. Conversely, the duty does not apply where he exercises the power under s 20(2)(i). The inspector is not required to await the arrival at the premises of such a person, and need only comply if requested by such a person 'who at the time is present'. 'The safety of the State' refers to national security.

2.127 Section 20(5) of HSWA 1974 requires that, before exercising the power conferred by HSWA 1974 s 20(2)(h), 'an inspector shall consult such persons as appear to him appropriate for the purpose of ascertaining what dangers, if any, there may be in doing anything which he proposes to do under that power'.

The power to take possession and detain dangerous articles or substances: HSWA 1974 s 20(2)(i)

2.128 Section 20(2)(i) of HSWA 1974 gives an inspector a power in respect of an article or substance to which HSWA 1974 s 20(2)(h), applies. This refers to an article or substance found in any premises which the inspector has power to enter, being an article or substance which appears to him to have caused or to be likely to cause danger to health or safety. Section 20(2)(i) permits that the inspector take possession of such an article or substance, and detain it, for so long as is necessary for all or any of the following purposes, namely:

(a) to examine it and do to it anything which he has power to do under HSWA 1974 s 20(2)(h);

(b) to ensure that it is not tampered with before his examination of it is completed;

(c) to ensure that it is available for use as evidence in any proceedings for an offence under any of the relevant statutory provisions or any proceedings relating to an improvement or prohibition notice, under s 21 or 22.

2.129 In most circumstances in which an inspector would exercise his power under HSWA 1974 s 20(2)(h), it will be necessary for the inspector to remove the article or substance.

2.130 Section 20(6) of HSWA 1974 requires that, where an inspector exercises the power conferred by HSWA 1974 s 20(2)(i):

> he shall leave there, either with a responsible person or, if that is impracticable, fixed in a conspicuous position, a notice giving particulars of that article or substance sufficient to identify it and stating that he has taken possession of it under that power; and before taking possession of any such substance under that power an inspector shall, if it is practicable for him to do so, take a sample thereof and give to a responsible person at the premises a portion of the sample marked in a manner sufficient to identify it.

The power to require answers: HSWA 1974 s 20(2)(j)

2.131 Section 20(2)(j) of HSWA 1974 gives an inspector the power:

> to require any person whom he has reasonable cause to believe to be able to give any information relevant to any examination or investigation under paragraph (d) above to answer (in the absence of persons other than a person nominated by him to be present and any persons whom the inspector may allow to be present) such questions as the inspector thinks fit to ask and to sign a declaration of the truth of his answers.

2.132 While this power is replicated in other regulatory legislation, such as the Environmental Protection Act 1990, neither the police nor Revenue & Customs officers enjoy any equivalent power. Furthermore, the subsection operates in a different fashion to the compulsory questioning powers created by the financial regulatory Acts and the insolvency legislation.

2.133 The inspector's power pursuant to HSWA 1974 s 20(2)(j) extends to requiring an answer to such questions as he 'thinks fit' of any person about whom there exists a reasonable cause for the belief that the person is able to give 'any' information relevant to an examination or investigation[91] that the inspector is conducting.

---

[91] HSWA 1974 s 20(2)(d): 'examination or investigation as may in any circumstances be necessary for the purpose mentioned in subsection (1) above.'

D. Inspectors' Powers Under HSWA 1974 s 20

An inspector should make it plain to a person, when he is exercising his power to **2.134** obtain answers under compulsion. It would be wholly inappropriate to record answers obtained under this power on a standard witness statement form containing a declaration of the type embodied by the Criminal Justice Act 1967, in s 9. However, there are situations in which an inspector, having exercised his powers under HSWA 1974 s 20(2)(j), thereafter approaches the person and requests him to sign a Criminal Justice Act 1967 s 9-type statement. In such a situation, it should be made clear to the person that he is not compelled to cooperate, nor is he obliged to sign such a statement.

The status of the statement

By giving a power to require the person to sign a declaration 'of the truth of his **2.135** answers', HSWA 1974 s 20(2)(j), envisages that such answers will be recorded in writing. It thus assumes that these will form a statement in writing. However, such a statement does not require that the maker sign a declaration consistent with the provisions of the Criminal Justice Act 1967 s 9 and the Magistrates' Court Act 1980 ss 5A, 5B. It cannot, therefore, itself be relied on, as evidence, for the purposes of committal proceedings.[92] The terms of a declaration consistent with these provisions are, first, that it is true to the best of that person's knowledge and belief; and additionally, a declaration that he made the statement knowing that if it were tendered in evidence, he would be liable to prosecution if he wilfully stated in it anything which he knew to be false or did not believe to be true.

Such a s 20(2)(j) statement would be sufficient for the purposes of an application **2.136** to obtain a witness summons, pursuant to the Criminal Procedure (Attendance of Witnesses) Act 1965 s 2. Furthermore, providing 'any information relevant to any examination or investigation', as is the requirement in HSWA 1974 s 20(2)(j), would amount to a statement for the purposes of the hearsay provisions in the Criminal Justice Act 2003. Thus, application could be made to have the statement read as evidence of the truth of its contents.

Self-incrimination: HSWA 1974 s 20(7)

Section 20(7) of HSWA 1974 provides that 'No answer given by a person in pursuance of a requirement imposed under subsection (2)(j) above shall be admissible in evidence against that person or the husband or wife of that person in any proceedings'. **2.137**

This apparent blanket prohibition of the use in evidence of any answer in **2.138** 'any proceedings' abrogates the right of a person not to incriminate himself, but

---

[92] The maker of such a statement, taken pursuant to the compulsory power under HSWA 1974 s 20(2)(j), would have to be called as a witness, to give evidence on oath at committal proceedings or at trial (unless agreed by all parties).

protects such a person from the consequences of any self-incrimination. The words 'any proceedings' appear to go so far as to prevent the prosecution of a person who gives false answers to an inspector exercising his power under s 20(2)(j) for an offence of either making a false declaration to an inspector, contrary to HSWA 1974 s 33(1)(k), or of making a false declaration other than on oath, contrary to the Perjury Act 1911 s 5. There appears to be no bar to a prosecution of a person who simply fails to answer the questions an inspector requires him to answer, under HSWA 1974 s 20(2)(j), for an offence of contravening a requirement by an inspector, contrary to HSWA 1974 s 33(1)(e).

Ambit and purpose of power

2.139 In *London Borough of Wandsworth v South Western Magistrates' Court* [2003] EWHC 1158 Admin, the Divisional Court ruled that the power under HSWA 1974 s 20(2)(j) should be construed widely. Accordingly, that power permitted an inspector to seek and obtain information in writing, as well as at a face-to-face meeting. The Court confirmed that the power was independent of that under HSWA1974 s 20(2)(a), and thus did not require an inspector's attendance at any premises before, or in order to be, exercised.

2.140 The Court held that the power was 'used to gather information in carrying out both formal and informal enforcement action under the Health and Safety at Work Act 1974'. The terms of HSWA 1974 s 20(7) were regarded by the Court as, 'important'. The operation and effect of s 20(7) was also considered significant 'because it emphasises that [HSWA 1974 s 20(2)(j)] is directed towards information gathering, rather than criminal or indeed civil proceedings'. Lord Justice Scott Baker stated that, to his mind, this 'points towards a wide, rather than a narrow, interpretation of [s 20(2)(j)]'.[93]

2.141 The Divisional Court was left in no doubt that such 'information gathering', using the power under HSWA 1974 s 20(2)(j), was often a prelude to enforcement action. However, bearing in mind the terms of HSWA 1974 s 20(7), the Court relied heavily on the fact that no answers could ever feature in a criminal prosecution of the maker of such a statement.

2.142 In *Regina v Hertfordshire County Council ex parte Green Environmental Industries Ltd* [2000] 2 AC 412, the House of Lords ruled upon the status of the provisions in the Environmental Protection Act. The opinion of the House of Lords was the culmination of a case that was first heard by the Divisional Court, and, on appeal, by the Court of Appeal. The case concerned the exercise, by an enforcing authority, of a power framed in almost identical terms to the compulsory questioning power in HSWA 1974 s 20(2)(j). The power was limited by an identical provision,

---

[93] [2003] EWHC 1158, at para 5.

D. Inspectors' Powers Under HSWA 1974 s 20

concerning the admissibility of such answers to that contained in HSWA 1974 s 20(7). The exercise of the compulsory questioning power was then followed by further requests for information by the enforcing authority, using powers in the Environmental Protection Act 1990. These powers were similar to those contained in HSWA 1974 s 27(1).

2.143 In the Court of Appeal,[94] Lord Justice Waller gave the judgment of the Court. He ruled that both the sections creating compulsory questioning powers, and those powers creating the notice requiring information, provide powers to enable an authority to discharge its 'functions' under the Environmental Protection Act 1990. These 'functions' include the obtaining of evidence during a criminal investigation. Provided the information is reasonably needed by the authority, it was the intention of Parliament that requisitions demanding such information should be answered. This is even if, in the result, evidence would be provided against the person or entity that provided it.[95]

2.144 In the House of Lords, the argument of the appellant focused not on the purpose for which the authority used the powers, but on whether a person could rely upon the right against self-incrimination not to answer the requests. Lord Hoffman ruled:

> Those powers have been conferred not merely for the purpose of enabling the authorities to obtain evidence against offenders but for the broad public purpose of protecting the public health and the environment. Such information is often required urgently and the policy of the statute would be frustrated if the persons who knew most about the extent of the health or environmental hazard were entitled to refuse to provide any information on the ground that their answers might tend to incriminate them. Parliament is more likely to have intended that the question of whether the obligation to provide potentially incriminating answers has caused prejudice to the defence in a subsequent criminal trial should be left to the judge at the trial, exercising his discretion under the 1984 Act.[96]

2.145 Clearly, in respect of the power under HSWA 1974 s 20(2)(j), like the equivalent under the Environmental Protection Act 1990, the statute itself prevents admission of the answers in evidence. However, in respect of the power under HSWA 1974 s 27(1), as with its equivalent under the Environmental Protection Act 1990, the position is different. Any incriminating answers so obtained are likely to be ruled inadmissible at any trial of the person who supplied them, as having been unfairly obtained in breach of the protections offered by the Police and Criminal Evidence Act 1984.

---

[94] [2000] 2 WLR 412.
[95] ibid at 414.
[96] Police and Criminal Evidence Act 1984.

Compelling answers from company officers

2.146 Any compelled statement taken from a company director could not be used in any proceedings against that director for any offence. However, it does appear that such witness evidence would be potentially admissible in proceedings against the company.

2.147 In *Tate Access Floors Inc v Boswell* [1990] 3 All ER 303, the directors of a company sought to argue that the company was effectively the same entity as themselves. The second limb of the argument was that, therefore, any disclosure by the company was a disclosure by them. The court rejected the argument. It reasoned that if people conducted their business through a corporation, and took advantage of the separate legal entity, they could not then claim that they were not separate legal entities from the corporation.

2.148 The Court of Appeal proceedings, in *Regina v Hertfordshire County Council ex parte Green Environmental Industries Ltd* [2000] 2 WLR 412, avail here. Lord Justice Waller there expressed the view that the identical provision to HSWA 1974 s 20(7), as contained in the Environmental Protection Act 1990, could only prevent a compelled answer from being admitted against a director. It would, subject to any argument to exclude the evidence pursuant to the Police and Criminal Evidence Act 1984 s 78, be admissible in proceedings against the company. Such a statement may not provide much assistance to any prosecution. During contested proceedings against a company, the prosecution would have to call as a witness any director from whom it has obtained such a compelled statement. Rather than being able to rely upon the statement itself (the defendant would most likely require the prosecution to call such a witness), the prosecution would have to rely upon the director as a witness of truth. He, in turn, would be able to provide the defendant company with an excellent source for helpful cross-examination.

The presence of a nominated person

2.149 HSWA 1974 s 20(7), gives the person answering questions, if this is being done in a face-to-face situation, the right to nominate another person to be present during the questioning. It also gives the inspector discretion to conduct his questioning in the presence of any other persons.

The power to require production, inspection, and to take copies: HSWA 1974 s 20(2)(k)

2.150 Section 20(2)(k), of HSWA 1974, gives an inspector a power to require the production of, inspect, and take copies of, any entry, in:

(a) any books or documents which, by virtue of any of the relevant statutory provisions, are required to be kept; and

D. Inspectors' Powers Under HSWA 1974 s 20

(b) any other books or documents which it is necessary for him to see, for the purposes of any examination or investigation.[97]

2.151 A failure to comply with such a requirement, imposed by an inspector, under this paragraph, amounts to an offence contrary to HSWA 1974 s 33(1)(e).

2.152 The power is not dependent upon the inspector's attendance at premises. By analogy with the interpretation of HSWA 1974 s 20(2)(j) by the Divisional Court in *London Borough of Wandsworth v South Western Magistrates' Court* [2003] EWHC 1158 Admin, the power can be exercised by the inspector simply setting out his requirements in writing. The power can be, and often is, exercised for the purpose of obtaining evidence for a contemplated prosecution. As such, the terms of HSWA 1974 s 20(2)(k)(ii) extend the power to the widest class of documents. Included here is any document necessary for the inspector to see for the purposes of any 'examination or investigation', under HSWA 1974 s 20(2)(d). An inspector is thus armed with the most powerful of investigative powers. While the power does not extend to a power for the inspector to search for documents himself, he is able to require a person to produce to him documents that he can specify in broad terms, by class or relevance to a particular project. Effectively, this places a burden upon such a required person to sort through documentation, in order to find all those documents that meet the specified criteria.

2.153 An inspector can require the production of documents, inspect them, and take copies of them. He is not given a power to take the originals into his possession. The Transport Act 1968 ss 99(1)(bb) and 91(2)(bb), provides a similar power for vehicle inspectors to inspect and copy various classes of documents and records. These provisions were the subject of an appeal considered by the House of Lords in *Cantabrica Coach Holdings Limited v Vehicle Inspectorate* [2001] UKHL 60. Lord Slynn ruled that it was:

> essential to bear in mind that the power is given so that the officer can 'inspect and copy'. If the documents are few and copying facilities are available at the company's premises, so that it is not reasonably necessary to take them away, production to the officer for inspection and copying there is in my opinion a compliance with the section. If, on the other hand, the documents are many or such that a proper inspection and copying is only possible with other equipment, which is not immediately available then to enable the officer to inspect and copy, he must be able to take them away but only for such period as is reasonably required for their inspection and copying.

2.154 Thus, where it is 'necessary' for an inspector to inspect documents, in respect of an investigation pursuant to his powers in HSWA 1974 s 20(2)(k), if it is not reasonably practicable for him to have them copied in situ, he has the power to take the documents away in order to have the documents copied.

---

[97] 'Examination or investigation' conducted under HSWA 1974 s 20(2)(d).

Chapter 2: The Reporting and Investigation of Health and Safety Incidents

The legal privilege exception: HSWA 1974 s 20(8)

2.155 Section 20(8) of HSWA 1974 provides for the protection of legal privilege and how it might remain intact:

> Nothing in this section shall be taken to compel the production by any person of a document of which he would, on grounds of legal professional privilege, be entitled to withhold production on an order for discovery in an action in the High Court or, as the case may be, on an order for the production of documents in an action in the Court of Session.

2.156 Legal privilege is considered in detail at paragraphs 3.81 to 3.84.

The power to require facilities: HSWA 1974 s 20(2)(1)

2.157 By virtue of HSWA 1974 s 20(2)(1), an inspector has the power to coerce cooperation. He can require any person to afford him such facilities and assistance with respect to any matters or things within that person's control, or in relation to which that person has responsibilities, as are necessary to enable the inspector to exercise any of the powers conferred on him by HSWA 1974 s 20.

2.158 The facilities and assistance must be necessary in order to enable the inspector to exercise any of the other powers under the section. The wording of this power does not limit it to premises. Thus, it does not require an inspector's attendance at any premises before being, or in order to be, exercised.

Any other power which is necessary: HSWA 1974 s 20(2)(m)

2.159 Section 20(2)(m), of HSWA 1974, gives the inspector 'any other power which is necessary for the purpose' of carrying into effect any of the relevant statutory provisions within the field of responsibility of the enforcing authority which appointed him.

2.160 The ambit of the power is drafted in remarkably uncertain terms. It is limited and circumscribed only by the requirement that the power must be 'necessary for the purpose' of carrying into effect any of the relevant statutory provisions. In the absence of any judicial scrutiny of the limits of this power, any discussion of its ambit is necessarily speculative.

2.161 In *London Borough of Wandsworth v South Western Magistrates' Court* [2003] EWHC 1158 Admin, Lord Justice Scott Baker drew particular attention to this power. Forming the conclusion that the powers under HSWA 1974 s 20(2) were to be construed widely, Scott Baker LJ described it as 'an important sweeping-up provision'.

2.162 One particular use of the power under HSWA 1974 s 20(2)(m) may be in relation to information held on computers. (Such electronic data is not specifically catered for in HSWA s 20(2)(k).) Section 20(2)(m) may afford an inspector the power to require a person to print relevant information that an inspector knows is contained on a computer. Such knowledge may potentially have been obtained through compulsory questioning of the computer operator, using the inspector's power under HSWA 1974 s 20(2)(j).

Powers to deal with immediate danger: HSWA 1974 s 25

2.163 Section 25, of HSWA 1974, provides an inspector with particular powers to deal with the cause of imminent danger. The powers complement those granted by s 20(2), in particular, those under paragraphs (h) and (i). In the case of any article or substance found by an inspector, in any premises which he has power to enter, if he has reasonable cause to believe that, in the circumstances in which he finds it, the article or substance is a cause of imminent danger of serious personal injury, he may seize it and cause it to be rendered harmless (whether by destruction or otherwise).

2.164 By virtue of HSWA 1974 s 25(2), before any article that forms part of a batch of similar articles, or any substance, is rendered harmless under the section, the inspector is required, if it is practicable for him to do so, to take a sample thereof. The inspector must then ensure that he gives to a responsible person at the premises where the article or substance was found by him, a portion of the sample, marked in a manner sufficient to identify it.

2.165 HSWA 1974 s 25(3) provides that as soon as may be after any article or substance has been seized and rendered harmless, the inspector is required to prepare and sign a written report. This must give particulars of the circumstances in which the article or substance was seized and so dealt with by him, and, further:

(a) give a signed copy of the report to a responsible person at the premises where the article or substance was found by him; and

(b) unless that person is the owner of the article or substance, also serve a signed copy of the report on the owner.

2.166 In relation to para (b), if the inspector cannot, after reasonable enquiry, ascertain the name or address of the owner, the copy may be served on the responsible person. This may be done by giving it to the person to whom a copy was given under para (a).

3

ADVISING CLIENTS: INTERVIEWS UNDER CAUTION, REPRESENTATION, AND PRIVILEGE

A. Interviews Under Caution	3.01	B. Advising Clients in Health and Safety Investigations	3.51
Advising suspects prior to interviews	3.05	Representation	3.55
Interviews under caution: advising clients	3.07	Cooperation with inspector and the use of s 20 powers	3.68
Interviews and the Police and Criminal Evidence Act 1984	3.31	Privilege, compulsory production, and reports into causes	3.74
Access to legal advice	3.35		
The caution	3.39		
Conduct of the interview	3.42		
Limitations on the power to interview	3.46		
Written statements under caution	3.50		

A. Interviews Under Caution

An investigation conducted by a health and safety inspector which produces evidence of a suspected breach of a health and safety duty, will usually result in the inspector inviting the person suspected of having breached the duty, or a representative (in the case of a company), to be interviewed under caution. Such an invitation will invariably be extended if the circumstances are such that the inspector is contemplating the instigation of criminal proceedings in respect of the breach. Significantly, though, there is no express legal requirement that a person suspected of having committed an offence must be interviewed under caution before the prosecution decision is taken. Even where the inspector is of the opinion that the existing evidence is sufficient to provide a realistic prospect of conviction, the invitation will be extended. Inspectors are advised[1] that it is desirable that

3.01

---

[1] See the Enforcement Concordat, reproduced at App B. See, also, HSE Enforcement Guide: <http://www.hse.gov.uk/enforce/enforcementguide/investigation/witness/questioning.htm>.

persons who are suspected of committing offences are interviewed under caution, because:

(a) the interview may provide important evidence against the suspect;
(b) the interview may provide important information revealing further lines of inquiry (particularly concerning the reasonable practicability of steps that were, or were not, taken);
(c) the interview may provide relevant information to be considered in the prosecution decision;
(d) it is fair and proper to allow a potential defendant an opportunity to answer the allegations and give their own account;
(e) an interview under caution may satisfy the provisions of the Enforcement Concordat.

3.02 Since 1 January 2006, health and safety offences have attracted a police power of arrest, and, coupled with this, a general power of arrest exercisable by any person. Therefore, where the police are investigating health and safety offences, or acting as the lead investigator in a work-related death investigation, a suspected person may be arrested. The rationale is that this 'allow[s] the prompt and effective investigation of the offence or of the conduct of the person in question'.[2] Suspected persons can thus be brought to a police station and interviewed under caution.

3.03 While health and safety inspectors do now have a technical power of arrest, it is not as wide as a police officer's, but is limited to the general power given to any person in respect of an indictable offence.[3] As the power does not allow an inspector to arrest to facilitate '. . . the prompt and effective investigation of the offence . . .', he may not arrest a suspect in order to interview him.

3.04 Local authority health and safety inspectors often attempt to interview those suspected of a breach of a health and safety duty, through 'cautioned correspondence', ie a letter that sets out the caution and poses a number of questions. HSE inspectors are discouraged from adopting this course, and ought instead to invite suspects or representatives to attend for a taped interview under caution. In both situations, suspected persons should be written to by the inspector, and advised in terms that they should seek legal advice before responding.

Advising suspects prior to interviews

3.05 Legal advice given to a suspected person prior to interview under caution may have far-reaching procedural and substantive implications during any future prosecution. Accordingly, it should be given with care.

---

[2] Police and Criminal Evidence Act 1984 s 25(5)(e).
[3] ibid s 24.

A. Interviews Under Caution

The legal representation of corporate suspects, directors, and employees in health and safety investigations has been a cause of tension in the past between solicitors acting for such persons and the HSE. The Law Society, at the invitation of the HSE, has recently issued guidance to solicitors, which has gone some way towards identifying the relevant matters to which any legal representative should advert before representing both employee and employer. The guidance recommends that these matters should be addressed before advising both employees and directors, or the company itself. Also see below in section B, 'Advising clients in health and safety investigations'. **3.06**

Interviews under caution: advising clients

Health and safety inspectors do not have power to arrest, beyond the general power given to any person in respect of an indictable offence.[4] This general power does not extend to a power to arrest so as to allow the prompt and effective investigation of the offence, for example, to conduct an interview. **3.07**

Local authority health and safety inspectors often attempt to interview those suspected of a breach of a health and safety duty through 'cautioned correspondence', ie a letter that sets out the caution and poses a number of questions. Conversely, HSE inspectors are discouraged from adopting this course, and may instead invite suspects or their representatives to attend for a taped interview under caution. **3.08**

In both situations, suspected persons should be written to by the inspector, and advised in terms that they should seek legal advice before responding. **3.09**

There are a number of factors to consider before advising a duty-holder to accept, or decline, an invitation to attend for an interview under caution from a health and safety inspector. These include the nature and extent of any pre-interview disclosure, the effect of declining an invitation, whether a duty-holder can benefit from offering any explanation, whether to supply a pre-prepared statement, and whether to simply supply representations. **3.10**

Pre-interview disclosure

Paragraph 11.1A, of Code C[5] requires that 'whenever a person is interviewed they must be informed of the nature of the offence, or further offence'. Despite this, the Code and PACE 1984 are otherwise silent as to the extent of the disclosure that should be made. Before conducting the interview, police officers tend to disclose some particulars of the allegation and evidence to a detained person's legal **3.11**

---

[4] Police and Criminal Evidence Act 1984 s 24.
[5] Police and Criminal Evidence Act 1984 (Codes of Practice) Order 2005 SI 2005/3503; C Code of Practice for the Detention, Treatment and Questioning of Persons by Police Officers para 11.1A.

representative unless some good reason for not so disclosing exists. To do other than tend towards disclosure is likely to result in the representative interrupting the interview when a matter arises which has not been previously disclosed, and to ask for a break, in order that he might advise his client and take instructions. On occasions, lack of disclosure is a reason cited on tape in interview for clients being advised to remain silent and not to answer questions. In *R v Roble* [1997] Crim LR 449 (CA) it was said that good reasons underpinning such advice to remain silent could include the following scenarios: the interviewing officer has disclosed little or nothing of the nature of the case against the defendant, and the solicitor cannot usefully advise his client; or the nature of the offence, or the material in the hands of the police is so complex, or relates to matters so long ago, that no sensible response may immediately be feasible.

3.12 A suspected person, or his legal representative, is entitled to make a request for disclosure of the inspector. Health and safety inspectors should consider requests for pre-interview disclosure seriously and be prepared to disclose sufficient particulars to enable a person to obtain legal advice concerning, and prior to, any interview.

Adverse inferences from a failure to mention facts

3.13 The Criminal Justice and Public Order Act 1994 introduced a principled basis for courts to draw adverse inferences from silence, in interview and when giving evidence. Section 34 of the Act concerns the drawing of adverse inferences from a failure to mention facts in interview. Section 34(4) provides that the section applies in relation to questioning by persons (other than constables) charged with the duty of investigating offences, ie health and safety inspectors. The section is intended to apply to inspectors as it applies in relation to questioning by constables.

3.14 The relevant parts of the Criminal Justice and Public Order Act 1994 s 34(1) and (2), provide that subsection (2) applies, in any proceedings against a person for an offence, in the following circumstances. Where evidence is given that the accused:

> at any time before he was charged with the offence, on being questioned under caution by a constable trying to discover whether, or by whom, the offence had been committed, failed to mention any fact relied on in his defence in those proceedings . . . being a fact which in the circumstances existing at the time the accused could reasonably have been expected to mention when so questioned.

3.15 Section 34(2) provides that, in such circumstances, a court may draw such inferences from the failure as appear proper. This includes a magistrates' court enquiring into the offence as examining justices; the court, in determining whether there is a case to answer; and the court or jury, in determining whether the accused is guilty of the offence charged. The section provides that a court, in deciding

A. Interviews Under Caution

whether there is a case to answer, may draw such inferences from a failure to mention a fact relied upon as appear proper. Counter-balancing the rigour of this provision, the Criminal Justice and Public Order Act 1994 s 38(3), stipulates that a finding of a case to answer shall not be based solely on such an inference. It appears that s 34 will seldom arise in practice, in determining whether there is a case to answer. The adverse inference is restricted to a fact relied on by the defendant in his defence, as opposed to his failure to explain a fact relied on by the prosecution.

3.16 The Criminal Justice and Public Order Act 1994 s 34(2A), now makes provision for situations in which the accused was at an authorized place of detention at the time of the failure, ie a police station. The section does not apply if he had not been allowed an opportunity to consult a solicitor prior to being questioned.

3.17 In *R v Argent* [1997] 2 Cr App R 27 (CA) at 32–33, Lord Bingham CJ found that there were six formal conditions that had to be met, before s 34 of the Criminal Justice and Public Order Act 1994 could operate:

(a) there had to be proceedings against a person for an offence;
(b) the failure to answer had to occur before a defendant was charged;
(c) the failure had to occur during questioning under caution by a constable or other person, within s 34(4);
(d) the questioning had to be directed to trying to discover whether, or by whom, the offence had been committed;
(e) the failure had to be to mention any fact relied on in the person's defence in those proceedings; and
(f) the fact the defendant failed to mention had to be one which, in the circumstances existing at the time of the interview, he could reasonably have been expected to mention when so questioned.

3.18 In *R v Condron & Condron* [1997] 1 Cr App R 185, the court held that it would seldom be appropriate to rule on the question of whether a jury should be invited to draw an adverse inference before the conclusion of all the evidence. Further, where a 'no comment' interview has been made by an accused, the prosecution should simply adduce the fact that the accused did not answer questions, or made no comment. This is the usual course followed. If, and when, the accused gives evidence, and mentions a fact which, in the view of prosecuting counsel, he could reasonably have been expected to mention in interview, he may be asked why he did not mention it in interview.

The effect of declining an invitation to be interviewed

3.19 The Criminal Justice and Public Order Act 1994, in s 34(1)(a), clearly does not require an arrest to have been made. Instead, the requirement is that a suspect must have been cautioned before the section applies. The failure to mention has to be in response to questioning. A person's refusal to be interviewed or, rather, declining a health and safety inspector's invitation, would not satisfy the terms of the section.

3.20 It follows that the declining of an invitation to be interviewed cannot properly found any adverse inference against a person at trial. This must be right, since the drawing of adverse inferences is a statutory exception to what was a common law right. It appears, then, that such a refusal to be interviewed would not be admissible, nor should it form the basis of any adverse criticism levelled by the prosecution. (On occasions, such a refusal has been cited, on a *Friskies* schedule, as evidence of a failure to cooperate with an investigation.)

Interviews with company directors and interviews by letter

3.21 Local authority trading standards officers routinely attempt to conduct interviews under caution by letter. This is a practice yet to be adopted by the HSE. In any event, it is only suited to matters concerning straightforward issues of fact, where there are limited questions to be asked.

3.22 In *Direct Holidays Plc v Wirral Metropolitan Borough Council* (DC 28 April 1998), the Divisional Court was concerned with the admissibility against a company of a letter responding to the questions posed. The letter had been signed by the managing director of the company. The Court considered the case of *Surujpaul v R* [1958] 3 All ER 300, 42 Cr App R 266, a Privy Council case. Lord Tucker, in *Surujpaul*, formulated what has become a familiar and often-cited statement of principle:

> A voluntary statement made by an accused person is admissible as a 'confession'. He can confess as to his own acts, knowledge or intentions, but he cannot 'confess' as to the acts of other persons which he has not seen and of which he can only have knowledge by hearsay.[6]

3.23 The letter written by the Trading Standards Office, to the company, detailed the complaint that the office was investigating. In addition, it attached a document setting out a number of specific questions of fact, and the letter ended by drawing to the attention of the company the PACE 1984 caution. This caution was set out in full. The letter went on to remind the appellants that they could seek legal advice before answering, suggesting that this may be an advisable course of action. The Divisional Court ruled that an admission of knowledge of a fact in the letter, signed by the managing director, was admissible against the company. It constituted evidence that the company had such knowledge.

3.24 The HSE routinely invite company directors to be interviewed under caution 'on behalf of the company'. Answers given by a director, in such an interview, are clearly admissible as evidence against the company.

---

[6] 42 Cr App R 266 at 273.

A. Interviews Under Caution

In *Direct Holidays Plc v Wirral Metropolitan Borough Council* (DC 28 April 1998), **3.25** the court was not concerned with any issue surrounding the drawing of an adverse inference from silence. Nor was the court troubled by a failure to mention a fact in answer to the questions posed, pursuant to the Criminal Justice and Public Order Act 1994 s 34(1)(a). Instead, the court appears to have treated the response to the questions as a 'voluntary statement' which contained an admission. The issue remains an open one as to whether any refusal to answer questions in such an interview (conducted by letter), or a failure to mention a relevant fact in response, would satisfy the terms of s 34(1)(a) and the conditions set out in *R v Argent* [1997] 2 Cr App R 27.

The decision to be interviewed

While a person suspected of a health and safety offence by an inspector commits **3.26** no offence by declining an invitation to be interviewed, nor is liable to suffer any adverse inference at trial, his failure to provide an explanation may still affect his case adversely. If he intends to fight the case his failure to provide an explanation means that he thereby fails to set out any of the reasonably practicable steps he may have taken; and equally, that he fails to place his defence on record at an early stage in a form which will be admissible at trial. Conversely, if he ultimately pleads guilty, his failure to admit the offence in interview may lead to an eventual reduction in credit for the guilty plea.[7]

Prepared statements

It often happens that a person suspected of committing an offence is advised to **3.27** attend the interview, read out a pre-prepared statement, but thereafter refuse to answer any questions. By voluntarily attending at interview, a suspected person places himself within the territory of the potential operation of the Criminal Justice and Public Order Act 1994 s 34(1)(a). In *R v Knight* [2004] 1 Cr App R 9 (CA), it was observed that the objective sought to be achieved by s 34(1)(a) is early disclosure of a suspect's account and not—separately and distinctly—to subject it to police cross-examination. Accordingly, there was no place for any adverse inference, where the defendant gave the police his full account in a pre-prepared statement from which he did not depart in evidence. This remained the position, even though it was not given in response to questioning and the defendant had said 'No comment' to all subsequent police questions. Crucially, though, it does not follow that no adverse inference may ever be drawn where there is a prepared statement. What is said at trial may be fuller, or it may be inconsistent. In any event, it does not necessarily follow in a case where there is no possibility of an

---

[7] See Ch 11 on credit for guilty pleas.

adverse inference, that the prosecution are obliged to adduce a wholly self-serving prepared statement as part of their case.

Representations

3.28 The Enforcement Concordat, agreed by the Cabinet Office, sets out the principles and policies that central and local government should follow, when performing regulatory enforcement. The document provides, as one of the 'Principles of Good Enforcement':

> Before formal enforcement action is taken, officers will provide an opportunity to discuss the circumstances of the case, and, if possible, resolve points of difference unless immediate action is required (for example, in the interests of health and safety or environmental protection or to prevent evidence from being destroyed).

3.29 The HSE, in attempting to put this policy into practice, encourages inspectors,[8] where no interview under caution has occurred, to write to duty-holders whom it is intended to prosecute. Inspectors should inform duty-holders of the decision, and the nature of the proposed offences, and invite them to make any representations as to why proceedings should not be instituted. Only upon receipt and consideration of any representations the party wishes to make, is a final decision whether to prosecute then taken.

3.30 Thus, a suspected person who is invited to be interviewed under caution can decline to attend; instead, he may either inform the inspector of his wish to make representations in writing before any final decision to prosecute is taken, or he can simply make such written representations at that time. Not being under caution, and being likely to be wholly self-serving in nature, such representations will not necessarily be adduced by the prosecution at any trial. Nonetheless, such an approach gives a suspected person an opportunity to put his explanation before the inspector, without having to attend for interview under caution.

Interviews and the Police and Criminal Evidence Act 1984

3.31 Paragraph 11.1A of Code C[9] of the Police and Criminal Evidence Act 1984 defines an interview as 'the questioning of a person regarding their involvement or suspected involvement in a criminal offence or offences which, under paragraph 10.1 [of the Code], must be carried out under caution'. It is a proper course to conduct an interview under caution prior to a decision to prosecute—both as a legitimate

---

[8] <http://www.hse.gov.uk/enforce/enforcementguide/investigation/approving/report.htm>, at para 13.
[9] Police and Criminal Evidence Act 1984 (Codes of Practice) Order 2005 SI 2005/3503; C Code of Practice for the Detention, Treatment and Questioning of Persons by Police Officers para 11.1A.

attempt to obtain evidence against a person suspected of an offence, and to afford that person an opportunity to offer an explanation.

3.32 A suspect, whose detention by the police without charge has been authorized under the Police and Criminal Evidence Act 1984, because the detention is necessary for an interview to obtain evidence of the offence for which he has been arrested, may choose not to answer questions. The police do not require the suspect's consent or agreement to interview him for this purpose.[10] However, a suspect who has not been arrested and detained at a police station cannot be compelled to attend an interview.

3.33 The conduct of interviews under caution, by health and safety inspectors, is subject to the relevant provisions of Code C, the Code of Practice for the Detention, Treatment and Questioning of Persons by Police Officers.[11] Alongside this is Code E of the Police and Criminal Evidence Act 1984, the Code of Practice on Audio Recording Interviews with Suspects.[12] The HSE provides inspectors with the facilities to conduct tape-recorded interviews, and to comply with the provisions of Code E, relating to access to tapes and record-keeping.

3.34 Paragraph 11 of Code C of PACE sets out the requirements concerning interview records. These are more onerous if such an interview is not tape recorded. The principal requirements are:

(a) an accurate record must be made of each interview: para 11.7(a);
(b) the record must state the place of interview, the time it begins and ends, any interview breaks and the names of all those present: para 11.7(b);
(c) any written record must be made and completed during the interview, unless this would not be practicable or would interfere with the conduct of the interview. It must constitute either a verbatim record of what has been said, or, failing this, an account of the interview which adequately and accurately summarizes it: para 11.7(c);
(d) if a written record is not made during the interview, it must be made as soon as practicable after its completion: para 11.8;
(e) written interview records must be timed and signed by the maker: para 11.9;
(f) if a written record is not completed during the interview, the reason must be recorded in the interview record: para 11.10;

---

[10] Police and Criminal Evidence Act 1984 (Codes of Practice) Order 2005 SI 2005/3503; C Code of Practice for the Detention, Treatment and Questioning of Persons by Police Officers para 12.5.
[11] Police and Criminal Evidence Act 1984 (Codes of Practice) Order 2005 SI 2005/3503; C Code of Practice for the Detention, Treatment and Questioning of Persons by Police Officers.
[12] ibid; E Code of Practice on Audio Recording Interviews with Suspects.

(g) unless it is impracticable, the person interviewed shall be given the opportunity to read the interview record, and to sign it as correct, or to indicate how they consider it inaccurate. If the person interviewed cannot read, or refuses to read the record or sign it, the senior interviewer present shall read it to them. He shall go on to ask whether they would like to sign it as correct, or make their mark, or indicate how they consider it inaccurate. The interviewer shall certify on the interview record itself what has occurred: para 11.11;

(h) if the appropriate adult, or the person's solicitor, is present during the interview, they should also be given an opportunity to read and sign the interview record, or any written statement taken down during the interview: para 11.12;

(i) a written record shall be made of any comments made by a suspect, including unsolicited comments, which are outside the context of an interview but which might be relevant to the offence. Any such record must be timed and signed by the maker. When practicable, the suspect shall be given the opportunity to read that record and sign it as correct, or to indicate how they consider it inaccurate: para 11.13.

Access to legal advice

3.35 Section 58 of the Police and Criminal Evidence Act 1984, and the terms of Code C,[13] are concerned with a person's right to be granted access to legal advice, following arrest and detention. Code C requires, subject to certain exceptions, that as soon as is practicable after arrest, such a person will be taken to a police station. Following a decision to arrest a person, any interview must be conducted at a police station. The effect of the different provisions is that a person has a right to legal advice prior to being interviewed at a police station. This right encompasses having a solicitor present at interview.

3.36 Paragraph 11.2 of Code C further elaborates on conditions at interview:

> Immediately prior to the commencement or re-commencement of any interview at a police station or other authorized place of detention, the interviewer should remind the suspect of their entitlement to free legal advice and that the interview can be delayed for legal advice to be obtained ... It is the interviewer's responsibility to make sure all reminders are recorded in the interview record.

3.37 It follows that, where an interview is conducted with a person who is not under arrest or at a police station, the Police and Criminal Evidence Act 1984 does not expressly grant a right to have a solicitor present.[14] Such a person, not being either

---

[13] Police and Criminal Evidence Act 1984 (Codes of Practice) Order 2005 SI 2005/3503; C Code of Practice for the Detention, Treatment and Questioning of Persons by Police Officers.

[14] In *R v Chief Constable of the Royal Ulster Constabulary ex p Begley* [1997] 1 WLR 1475 (HL) at 1480, the House of Lords ruled on the involvement of solicitors. The court found that the

A. Interviews Under Caution

under arrest or detained, cannot be prevented from seeking legal advice, or refusing to be interviewed.

Where a health and safety inspector intends to invite a person to be interviewed under caution, he should inform the person that he can obtain legal advice prior to interview. The inspector should also make clear that that person also has the right to have a solicitor present at interview. Moreover, the inspector should give the person an opportunity to arrange this before proceeding to interview. HSE inspectors invariably write to suspected persons, inviting them to attend an interview under caution and advising them in terms that they should obtain legal advice and informing them of their right to have a solicitor present at the interview. 3.38

The caution

The terms of the relevant caution to be given during every interview are as follows. 'You do not have to say anything. But it may harm your defence if you do not mention when questioned something which you later rely on in Court. Anything you do say may be given in evidence.' 3.39

Paragraph 10D of Code C[15] provides that if it appears a person does not understand the caution, the person giving it should explain it in their own words. Minor deviations from the words of the caution do not constitute a breach of the Code, provided the sense of the relevant caution is preserved. After any break in questioning under caution, the person being questioned must be made aware that they remain under caution. If there is any doubt, the relevant caution should be given again, in full, when the interview resumes. 3.40

Paragraph 11.1A of Code C[16] not only requires that a person must be cautioned but requires that, 'Whenever a person is interviewed they must be informed of the nature of the offence, or further offence.' HSE inspectors, when writing to a person inviting them to be interviewed under caution, usually set out the suspected breach of duty/offence or breaches/offences, on which it is proposed to interview. 3.41

---

common law recognized a general right, vesting in an accused person, to communicate and consult privately with his solicitor outside the interview room. Crucially, that right did not necessarily extend to recognition of a suspect's right to be accompanied by his solicitor in a police interview.

[15] Police and Criminal Evidence Act 1984 (Codes of Practice) Order 2005 SI 2005/3503; C Code of Practice for the Detention, Treatment and Questioning of Persons by Police Officers para 10D.

[16] ibid para 11.1A.

Conduct of the interview

3.42 Paragraph 11.4 of Code C[17] requires that:

> At the beginning of an interview the interviewer, after cautioning the suspect, see section 10, shall put to them any significant statement or silence which occurred in the presence and hearing of a police officer or civilian interviewer before the start of the interview and which have not been put to the suspect in the course of a previous interview. See Note 11A. The interviewer shall ask the suspect whether they confirm or deny that earlier statement or silence and if they want to add anything.

3.43 The Code defines a significant statement as one which appears capable of being used in evidence against the suspect. In particular, a direct admission of guilt is envisaged here. A significant silence, which is far less likely to arise during the course of an investigation by a health and safety inspector, is defined as a failure or refusal to answer a question, or answer satisfactorily when under caution. Either finding might give rise to an adverse inference at trial, under the Criminal Justice and Public Order Act 1994 Pt III.

3.44 Paragraph 11.5 of Code C provides that no interviewer may try to obtain answers or elicit a statement by the use of oppression. The paragraph further provides that:

> no interviewer shall indicate, except to answer a direct question, what action will be taken ... if the person being questioned answers questions, makes a statement or refuses to do either. If the person asks directly what action will be taken if they answer questions, make a statement or refuse to do either, the interviewer may inform them what action [it is] proposed to take, provided that action is itself proper and warranted.

3.45 Special considerations and provisions apply when interviewing persons under 17 years of age, or those who may be mentally disordered or otherwise mentally vulnerable. These provisions are set out in Code C to the Police and Criminal Evidence Act 1984.[18]

Limitations on the power to interview

3.46 Paragraph 11.6 of Code C[19] requires that the interview, or further interview, of a person, about an offence with which that person has not been charged, or for

---

[17] Police and Criminal Evidence Act 1984 (Codes of Practice) Order 2005 SI 2005/3503; C Code of Practice for the Detention, Treatment and Questioning of Persons by Police Officers para 11.4.

[18] Police and Criminal Evidence Act 1984 (Codes of Practice) Order 2005 SI 2005/3503; C Code of Practice for the Detention, Treatment and Questioning of Persons by Police Officers.

[19] ibid para 11.6.

which they have not been informed they may be prosecuted, must cease when the officer in charge of the investigation:

(a) is satisfied all the questions they consider relevant to obtaining accurate and reliable information about the offence have been put to the suspect: this includes allowing the suspect an opportunity to give an innocent explanation and asking questions to test if the explanation is accurate and reliable, for example, to clear up ambiguities or clarify what the suspect said;
(b) has taken account of any other available evidence; and
(c) the officer in charge of the investigation . . . reasonably believes there is sufficient evidence to provide a realistic prospect of conviction for that offence if the person was prosecuted for it.

3.47 In general, a person may not be interviewed about an offence after they have been charged with it, or informed that they may face prosecution (ie by way of information and summons). Proceedings instituted by the HSE, or enforcing authorities in respect of health and safety offences, are commenced by inspectors.[20] This is done only by way of information and summons, not charge. Therefore, the equivalent time must be when a decision has been made by the inspector to prosecute.

3.48 An interview can take place after this time only if the interview is necessary for one of three reasons, as set out in para 16.5, of Code C,[21] namely:

(a) to prevent or minimize harm or loss to some other person, or the public;
(b) to clear up an ambiguity in a previous answer or statement;
(c) when it is in the interests of justice for the person to have put to them, and have an opportunity to comment on, information concerning the offence which has come to light since they were charged or informed they might be prosecuted.

3.49 In these circumstances, the person being interviewed must be told of his rights to have legal advice. Further, he must be cautioned in terms that preclude an adverse inference from silence, namely: 'You do not have to say anything, but anything you do say may be given in evidence.'

Written statements under caution

3.50 Whenever a person who is suspected of a criminal offence wishes to say something, concerning his involvement in the alleged offence or the investigation,

---

[20] Health and Safety at Work etc. Act 1974 s 39.
[21] Police and Criminal Evidence Act 1984 (Codes of Practice) Order 2005 SI 2005/3503; C Code of Practice for the Detention, Treatment and Questioning of Persons by Police Officers para 16.5.

unless he is being interviewed under caution on tape, he should be invited to write the matter down. This should take the form of a statement under caution. Such a written statement, made under caution by a person, will be admissible in proceedings against him. Annex D[22] to Code C provides that 'a person shall always be invited to write down what they want to say'. The Code continues to provide detailed provisions concerning the different situations in which a written statement under caution can be made.

B. Advising Clients in Health and Safety Investigations

3.51 Those who are the subject of—or concerned with—a health and safety investigation may seek legal advice in a variety of different situations, and at different times. Common situations include: following the service of an enforcement notice; following a fatal accident; where an inspector has exercised his HSWA 1974 s 20, powers; following a request for a person to provide a witness statement, or to answer questions pursuant to HSWA 1974 s 20(2)(j), and following an invitation to be interviewed under caution.

3.52 As has been stated above, a number of issues have recently arisen surrounding potential conflicts of interest, and these have implications for legal representation or advice for companies and employees.

3.53 The changes to the Police and Criminal Evidence Act 1984—and police powers in general—will add a new dimension to advising in health and safety investigations. Even more so with the expected involvement of the police in the investigation of the proposed health and safety-based new offence of corporate manslaughter.

3.54 The following part of this chapter attempts to provide a background and guidance to the most common issues met by those who are asked to advise in such situations.

Representation

Insurance and funding

3.55 Most often, following a health and safety incident, an employer will be the subject of an investigation. That employer should have compulsory public and employer's liability insurance. This cover can include provision for legal advice and representation, both during a health and safety investigation and through

---

[22] Police and Criminal Evidence Act 1984 (Codes of Practice) Order 2005 SI 2005/3503; C Code of Practice for the Detention, Treatment and Questioning of Persons by Police Officers Annex D.

B. Advising Clients in Health and Safety Investigations

the course of any criminal proceedings arising; it may well cover simply the latter. Additionally, the directors of an employing company may have a separate or associated insurance policy, in respect of directors' personal liability. Again, this may include provision for legal advice and representation, both during a health and safety investigation and through the course of any criminal proceedings arising, or simply the latter.

3.56 The route for initially determining the extent of cover is usually through the insured's broker, and by recourse to the insurance policy. Many insurers have 'panel' solicitors—whom, to a greater or lesser extent, they encourage an insured person to consult—and fixed hourly rates for remuneration. Such 'panel' solicitors will simultaneously represent the insurer in respect of civil liability and any civil claim, and thus the insured will be encouraged to seek the advice of such a solicitor. An insured person who instructs independent solicitors may find that advice and representation prior to commencement of any proceedings is not met by the insurer. In the past, instructed solicitors have had recourse to specialist insurance counsel, to advise on the extent of cover provided by a policy.

3.57 A number of insurance policies describe the cover for legal expenses as 'related and limited to proceedings for health and safety offences'. Arguably, this does not cover proceedings for gross negligence manslaughter. However, insurers have previously extended cover under such policies where proceedings have been instituted in respect of manslaughter. It is not uncommon for insurance cover in respect of legal representation to be withdrawn. This may happen where the ultimate criminal allegation includes, or the evidence discloses, material suggesting that the insurer had been misled as to the state of risk assessment/control measures that had been required under the policy. In the event of a conviction at trial it is common for the insurer to withdraw cover at that point, and not to fund any appeal.

3.58 Legal aid representation is not available for companies. Only limited legally aided representation is available for any natural person, prior to the commencement of criminal proceedings, and it is not offered in respect of interviews under caution, where a person is not detained at a police station.

Representing or advising more than one party: conflict of interest

3.59 In recent years, solicitors advising employers during health and safety investigations have found themselves in conflict with health and safety inspectors, when seeking to advise employees. The Law Society Rules and Ethics Committee issued Guidance in March 2006. After much discussion and debate, the Guidance was revised, in August 2006. It is specifically concerned with whether it is ethically appropriate for solicitors, who act for an employer (either in-house solicitors or solicitors in private practice), to be present at the Health and Safety Executive's Inspector's interview of an employee, following a work-related incident.

3.60 The Guidance does not distinguish between interviews under caution, compulsory statement-taking, under HSWA 1974 s 20, and the obtaining of a voluntary statement. The Guidance advises that if a solicitor is asked to act for both the employer and the employee, 'very careful consideration' should be given before deciding whether to accept instructions, due to the risk of a conflict of interests. The Law Society's Solicitors' Practice Rule 16D(2)(b)(i) details a circumstance when a conflict of interest arises. This will be established where the solicitor or his practice owes 'separate duties to act in the best interests of two or more clients in relation to the same or related matters, and those duties conflict, or there is a significant risk that those duties may conflict'. The Guidance provides that in assessing this risk, the solicitor should take account of the duty of disclosure and the duty of confidentiality, and the likelihood of these coming into conflict. The Guidance gives the example of how the duty to tell the employer client about information, provided by the employee to the HSE inspector in the course of the interview, may conflict with the duty of confidentiality owed to the employee client.

3.61 The Guidance provides that if a solicitor is acting only for the employer and not for the employee, the solicitor may provide to the employee information about the process being undertaken by the inspector. In such circumstances, 'The solicitor must make sure the employee understands that he or she is not acting for them, owes duties only to the employer, and cannot advise them and that anything they say will not be treated as confidential and will probably be reported to the employer'. However, the Guidance warns:

> But even if these precautions are taken, it is difficult to justify the employer's solicitor accompanying the employee to the interview. The employer's solicitor's conduct obligation is to pass on to the employer all information which is material to the client's business regardless of the source of that information (Practice Rule 16E(3)). This means that the solicitor will be obliged to pass material information to the employer, whether or not to do so is detrimental to the interests of the employee. The solicitor would be present at the interview at the employee's behest, in the knowledge that this situation might arise. The solicitor would be exposing himself to the possibility of obtaining information detrimental to the employee which he/she would be duty bound to pass on to the employer. The solicitor could even be perceived as taking unfair advantage of the employee, in that the employee's right to be accompanied would have been exercised in favour of a solicitor whose duty was owed to the employer and not to the employee (Principle 17.01).

3.62 Furthermore, the Guidance warns that there are other, more general, public interest issues to be considered: namely, whether the presence of the employer's solicitor might inhibit the employee from making a full and proper disclosure of facts relevant to the enquiry. The Guidance strikes a cautionary note: 'An employee may understandably be reluctant to say anything which may have an adverse effect on their continuing employment or prospects of promotion, in the presence of an employer's solicitor, whose presence may have the effect of constraining the employee in what they will say (Practice Rule 1).'

It advises that, for all these reasons, 'it is generally inappropriate for the employer's **3.63**
solicitor to attend such interviews as the employee's nominee, or to seek to obtain
the employee's consent to being present at the interview'. Notwithstanding that
statement, the Guidance advises that an employer's solicitor's role in a health and
safety investigation can be:

- To advise and represent their employer client;
- To inform employees as to their general legal rights and obligations, while making clear that they neither act for the employee nor owe the employee any of the duties which solicitors owe to their clients;
- To advise and represent the employer and one or more employees if it is absolutely clear that there is no conflict or potential for conflict, for example, where the entire responsibility for an accident clearly rests with a third party outside the organization, then there would not seem to be any difficulty in the solicitor representing both parties. The solicitor should consider carefully whether at the early, investigative stage that conclusion can safely be reached, and make a record of the reasons for their decision. The question of conflict should be kept under constant review throughout the investigation;
- To attend an HSE interview with an employee where the employee's interest is so co-incident with the employer's that the solicitor is able to conclude that, if so requested, he/she would be able to accept formal instructions to act for both of them, without any breach of the solicitor's professional duties, as set out in this guidance. Such a situation could arise, for example, where the employee is the controlling mind of the company, or is authorised to represent the employer in connection with the matter. Again, the question of conflict should be kept under constant review throughout the investigation.

Where an employer has engaged a solicitor to advise and represent him in a health **3.64**
and safety investigation, it may be in the employer's interest to arrange for, and
perhaps fund, the services of an independent solicitor to be available to advise
employees during the investigation. Obviously, such an independent solicitor will
have to be mindful of a potential conflict of interest if he advises more than one
employee. Solicitors regularly instructed on behalf of employers in health and
safety investigations increasingly have referral arrangements in place. They may
then consider it appropriate to recommend to the employer such an independent
solicitor capable of advising employees.

There is a much narrower potential for conflict when advising and representing a **3.65**
company and individual directors, or very senior management. In respect of liability for manslaughter and health and safety offences, an act or omission on the part
of a director will amount to such an act or omission on the part of the company.

Section 37 of HSWA 1974 creates a personal liability on the part of senior com- **3.66**
pany officers.[23] This liability attaches where such a person has consented to, or connived at, an offence by the company, or such an offence is attributable to neglect on

---

[23] *R v Boal* (1992) 95 Cr App R 272.

his part. Thus, there is potential for one company director to assert that the relevant neglect was on the part of another company director, and not himself.

Incidents involving a fatality

3.67 Work-related deaths are investigated pursuant to the Work Related Death Protocol.[24] Whenever a fatality has occurred, there will be, at the very least, consideration by the police of whether a homicide investigation should be undertaken. Consideration will include, and focus on, the grounds for suspecting gross negligence manslaughter and, in the future, corporate manslaughter. Employers, directors, and employees need to be advised of the police powers of arrest, search, and seizure, in respect of manslaughter and, since January 2006, health and safety offences. Equally, consideration must be given to the almost inevitable intervention of an inquest, following a decision by the police and CPS that there is no basis for a homicide prosecution and prior to the completion of an investigation into health and safety breaches.

Cooperation with inspector and the use of s 20 powers

3.68 Health and safety inspectors have a battery of powers, pursuant to HSWA 1974 s 20. These include a power to obtain information by way of a statement (s 20(2)(j)), and a power to compel production of a document, and take a copy of the same (s 20(2)(k)). During the course of an investigation, an inspector is likely to request information and documentation, before exercising a power of compulsion. Those who are the subject of such a request often require, and are entitled to receive, legal advice prior to responding.

3.69 Careful consideration should be given to the question of when, and in what circumstances, it is appropriate for a person to cooperate with an inspector who is not purporting to exercise his s 20 powers.[25] In *R v Howe*, the Court of Appeal set out the leading guidance in respect of sentencing for health and safety offences, including the aggravating and mitigating features that may be relevant.[26] Mitigating features include whether, and to what extent, the defendant has cooperated with the inspector's investigation. For this reason and others—not least a desire to ensure a continuing good relationship with the inspector—an employer may wish to cooperate fully with requests for information. However, it may very well not be in the interests of a person to provide answers to questions concerning his suspected involvement in a breach of health and safety legislation; and employers should be advised that requests for information from a person suspected of

---

[24] Reproduced at App B.
[25] When advising, it must always be borne in mind that it is an offence to obstruct an inspector.
[26] *R v Howe and Sons (Engineers) Ltd* [1992] 2 All ER 249.

a criminal offence can amount to an 'interview', which should therefore be conducted under caution.

3.70 However, it should be remembered that if such information is not volunteered, an inspector may compel answers to be given under HSWA 1974 s 20(2)(j). Most importantly, such a statement cannot be used in any criminal proceedings as evidence against its maker.

3.71 Once an inspector has used his HSWA 1974 s 20 powers, the duty to comply with the requirement needs to be clearly understood, by both the affected person and the legal adviser. It is an offence, under HSWA 1974 s 33(1)(e), 'to contravene any requirement imposed by an inspector under s 20'. The offence requires no proof of intent against a defendant; nor does HSWA 1974 create any defence of reasonable excuse for such failure, upon which a defendant might rely.

3.72 The power, under HSWA 1974 s 20(2)(k), to compel a person to produce a document, and for the inspector to take a copy of the same, should not be used to force a person to create a document: it should only be used to provide a pre-existing document. Thus, a refusal to provide voluntary information should not result in a requirement to create a document, pursuant to HSWA 1974 s 20(2)(k), with the very requested information within.

3.73 The powers of production are subject to a limitation in relation to production of material subject to legal privilege. Section 20(8) of HSWA 1974 explicitly provides for this. 'Nothing in this section shall be taken to compel the production by any person of a document of which he would on grounds of legal professional privilege be entitled to withhold production on an order for discovery in an action in the High Court or, as the case may be, on an order for the production of documents in an action in the Court of Session.'

Privilege, compulsory production, and reports into causes

Privilege

3.74 Legal professional privilege extends to communications, statements, reports and documents created during the course of a relationship between a solicitor and a client. These are created for the purpose of the obtaining and giving of legal advice (the 'advice privilege'), or in contemplation, or furtherance of, anticipated or ongoing legal proceedings (the 'litigation privilege'). The privilege extends to such communications with barristers, their clerks, and agents.

3.75 Much of the case law that forms the common law rules of legal professional privilege has been concerned with disclosure in civil cases, or with questions of admissibility of evidence. Section 10 of the Police and Criminal Evidence Act 1984 contains a definition of legally privileged material, for the purposes of that Act. Broadly, this follows the common law. The police powers of search and seizure are limited by a constraint in terms of legal privilege, as is the power to obtain a warrant. This legislation has generated a number of decided cases upon the ambit of the privilege. Mostly, these have been in respect of the exclusion from legal privilege of documents and communication, in furtherance of a criminal enterprise. That privilege has been held not to extend to communications held by a solicitor, which were intended to further any criminal purpose, where the client has such a criminal intention, even though his intention is unknown to his solicitor.[27]

3.76 HSWA 1974 provides no procedure for resolving any dispute that could arise between an inspector, requiring the production of a document, and a person asserting that the document is protected by legal professional privilege. The problem is regularly met by the Serious Fraud Office, which enjoys a power under Criminal Justice Act 1987 s 2 to serve a notice requiring a person to produce documents. Section 2(9) of the Criminal Justice Act 1987 excludes from the operation of a notice any material which is subject to legal privilege. In cases of dispute, the Serious Fraud Office has adopted a procedure of appointing a mutually agreed independent barrister to arbitrate on the status of a document.

Reports into causes: advising clients

3.77 A common area of potential difficulty in relation to legal privilege in health and safety investigations arises in respect of reports commissioned by an employer to examine the immediate or underlying causes of a workplace accident or dangerous occurrence. The report will clearly be covered by legal professional privilege if it was commissioned by the employer following consultation with a solicitor in anticipation of potential enforcement proceedings. Similarly, if the report was prepared in anticipation of a potential civil action for damages by an injured employee, the report will also be legally privileged. However, if the report is commissioned before the employer has instructed a solicitor, the issue may be more difficult, although it may still be covered by legal professional privilege. Clearly, it is important that an employer receives legal advice on this issue at an early stage.

3.78 Such reports, and any materials obtained during the investigation—such as witness statements—may become the subject of a request or requirement by the inspector for production, usually under HSWA 1974 s 20(2)(k). Unless the report and material attracts legal privilege, the duty-holder can be required to hand the material over to the inspector; although the fact that an inspector may obtain possession of the report does not automatically render it admissible at trial, and there may be a basis for objecting to its admission into evidence under the admissibility provisions of s 78 PACE 1984.

---

[27] *R v Central Criminal Court (Respondent) ex p Francis & Francis (Appellants)* (1989) 88 Cr App R 213, and *R v Guildhall Magistrates' Court ex p Primlaks Holdings Co (Panama) Inc* (1989) 89 Cr App R 215.

B. Advising Clients in Health and Safety Investigations

Some duty-holders are under statutory duties to undertake and supply to the HSE investigations and reports into the cause of particular health and safety incidents. While such duty-holders must comply, it is arguable that reliance in a prosecution upon such a report would offend a defendant's right against self-incrimination under Art 6(1) of the European Convention on Human Rights. **3.79**

Similarly, in some highly regulated industries—such as the railway industry— regulations impose obligations on certain industry bodies to conduct inquiries into major incidents or accidents. Such inquiries are generally intensive and far-reaching, and often address the very matters at issue in a later criminal trial. The Rail Safety and Standards Board (RSSB), which is charged with the duty of leading formal inquiries, has shown itself to be reluctant to disclose the fruits of its inquiries—whether the final report or the transcripts of the evidence of witnesses testifying before it—to either defence or prosecution, for fear that such disclosure will have the effect of inhibiting the attendance of witnesses in the future. **3.80**

The operation of legal privilege

The important limit upon the scope of the advice privilege is that it extends only to communications between the client and his lawyer. Privilege cannot be claimed for a communication between either the solicitor and the client, and/or a third party, unless the communication is made in respect of litigation privilege.[28] In terms of practical application, it is very difficult to specify with any precision the extent of the connection which is required between the preparation of a document and the anticipation of the litigation in relation to which it was made, in order for the document to be protected by litigation privilege. The leading authority is the opinion of the House of Lords in *Waugh v British Railways Board* [1980] AC 521. There, it was decided that in order for a claim to litigation privilege to succeed, the claimant must show that the qualifying communication is made with the 'dominant' purpose of being used in, or in connection with, litigation in which the maker of the document is or is likely to be a party. Use in such litigation need not be the sole purpose for which the communication was made. Where a document has other uses or purposes, however, its use in the litigation must be the dominant purpose behind its preparation. **3.81**

In *Re Highgrade Traders* [1984] BCLC 151, the Court of Appeal clarified an area of some confusion. The appellate court rejected the argument that *Waugh v British Railways Board* [1980] AC 521 established that it was only if documents were brought into existence for the dominant purpose of actually being used as evidence, in the anticipated proceedings, that privilege could attach, such that the purpose of taking advice on whether or not to litigate was some separate purpose **3.82**

---

[28] *Wheeler v Le Marchant* (1881) 17 Ch D 675.

which did not qualify for privilege. In that case, the stock and premises of a company were destroyed by fire shortly after a substantial increase in insurance cover had been taken out. The insurers suspected foul play, and had several reports prepared into the cause of the fire by external investigators. From the time they received the first report, it was clear that any claim on the policy might be disputed. As a result, solicitors were instructed—after the insurers had received their first report—and, in due course, the insurers decided not to honour the policy. Shortly afterwards, the company entered into a members' voluntary liquidation. The only hope the creditors had of making any realistic recoveries was if the insurance claim was met. The liquidator attempted to negotiate a resolution with the insurers, in the course of which he requested copies of their reports. These were withheld on the grounds of privilege, whereupon he applied for an examination, under insolvency legislation, of a responsible officer of the insurers, coupled with an order for production of the reports (now contained within Insolvency Act 1986 s 236).

3.83 These orders were made and the dispute was appealed to the Court of Appeal, which ruled that the reports were indeed privileged. Oliver LJ stated that there was one specific purpose apparent: 'The purpose was, and was only, to determine aye or no were they to litigate, and it was clearly in order to enable the solicitors to advise them on that matter that the relevant documents were obtained.'[29]

3.84 Furthermore, privileged materials, created for the purposes of litigation, do not actually need to be used in connection with the anticipated litigation for which they were created. In *Southwark Water Co v Quick* (1878) 3 QBD 315, it was held that a document prepared by a client which contained information, procured for the purposes of consulting his solicitor in relation to litigation, was privileged. This was the case even though the document was ultimately not sent to the solicitor. In such cases, the intention of the document's creator is examined, for the purpose of the privilege, at the time of creation.

Pre-existing documents

3.85 The position—both in respect of advice and litigation privilege—is more complicated when considering the question of pre-existing documents. In relation to copies of pre-existing documents, acquired by a solicitor as part of the evidence-gathering process for litigation, the position is now clear. The Court of Appeal has held that privilege can, in certain circumstances, be asserted in the litigation for which they were acquired, principally where to disclose them would reveal the trend of advice being given by the solicitor. The decisions on this topic are inconsistent and apparently without logical coherence. This has a practical knock-on effect.

---

[29] [1984] BCLC 151 at 174.

B. Advising Clients in Health and Safety Investigations

While a pre-existing document, whether a copy or an original, obtained by a solicitor from his client, will not, without more, be privileged from production even in any litigation in which it is relevant, where the solicitor obtains a copy of a pre-existing document from a third party, it may be privileged.[30]

Admissibility of documents: the privilege against self-incrimination

In *Funke v France* 16 EHRR 29 the judgment of the European Court was that the right to a fair trial guaranteed by Art 6 of the European Convention on Human Rights included 'the right of anyone charged with a criminal offence to . . . remain silent and not to contribute to incriminating himself'. **3.86**

The Court of Appeal, in the case of *R v Kearns* [2003] 1 Cr App R 7, reviewed the nature and effect of the privilege against self-incrimination in the light of Art 6 of the European Convention on Human Rights. The Court conducted a review[31] of the principal UK cases, and the jurisprudence of the European Court of Human Rights. Concluding that Art 6 was concerned with the fairness of a judicial trial, the appellate court found that it was not concerned with extra-judicial enquiries as such. The Court adopted the analysis of Lord Hoffmann, in the House of Lords' decision of *R v Hertfordshire County Council ex parte Green Environmental Industries Ltd* [2000] 2 AC 412. That case concerned the powers of compulsory questioning and information-gathering, under the Environmental Protection Act 1990, which are identical to those under HSWA 1974. These powers, it was observed, did not themselves offend Art 6 of the European Convention on Human Rights. Neither the right to silence, nor the right not to incriminate oneself, both of which are implicit in Art 6 of the European Convention on Human Rights, were found not to be absolute. Both can be qualified and restricted, where to do so was proportionate to the social or economic problem with which the measure in question was intended to deal. **3.87**

Thereafter, the Court in *R v Kearns* [2003] 1 Cr App R 7, found that there was a distinction between, on the one hand, the compulsory production of documents, or other material which had an existence independent of the will of the suspect or accused person, and, on the other, statements made under compulsion. In the former case, the rights in question were not infringed. In the latter case, those rights might be infringed, depending on the circumstances. The Court of Appeal found that a further distinction was to be drawn. This was between a situation in which a person was required to produce information for an administrative **3.88**

---

[30] *Watson v Cammell Laird & Co (Shipbuilders & Engineers) Ltd* [1959] 1 WLR 701.
[31] The Court of Appeal considered, in particular, the European Court judgments in *Funke v France* 16 EHRR 29, *Saunders v United Kingdom* (1996) 23 EHRR 313, *Heaney and McGuinness v Ireland* (application/36887/97, 21 December 2000), and the Court of Appeal decision in *Att-Gen's Reference (No 7 of 2000)* [2001] 2 Cr App R 19.

purpose, or in the course of an extra-judicial enquiry; and a situation where a person was compelled to give information which could be used in criminal or civil proceedings. In the former situation, it was unlikely that the right to silence or the right not to incriminate oneself would be infringed. In the latter situation, the use of the information in subsequent judicial proceedings could breach those rights and so make that trial unfair.

3.89 In that latter situation, the question of whether a person's rights have been breached would depend on all the circumstances of the case. In particular, a determining issue is whether the information demanded was factual or an admission of guilt. Also relevant is whether the demand for the information and its subsequent use in proceedings was proportionate to the particular social or economic problem that the relevant law was intended to address.[32]

3.90 Thus, the admission of documents into evidence, during the course of any prosecution against a person, obtained from him under compulsion, by an inspector using his power under HSWA 1974 s 20(2)(k), may be the subject of a challenge pursuant to the Police and Criminal Evidence Act 1984 s 78. Where the material obtained is evidence that a defendant has been compelled to create, for example, a compelled statement, it is unlikely to be admissible if it offends the person's right not to incriminate himself. Where the material obtained was already in existence, so that the effect of the compulsory power was to make such evidence available to the court, it is likely to be admissible.

---

[32] See also *Den Norske Bank ASA v Antonatos* [1999] QB 271 (CA (Civ Div)); *J.B. v Switzerland* [2001] Crim. LR 748; *Telfner v Austria* (2002) 34 EHRR 7; *Att-Gen's Reference (No 7 of 2000)* [2001] 2 Cr App R 19; *Brown v Stott* [2003] 1 AC 681; *R v Hundal and Dhaliwal* [2004] 2 Cr App R 19 (CA); *R (Kent Pharmaceuticals Ltd) v Director of Serious Fraud Office* [2005] 1 WLR 1302 ((CA) (Civ Div)).

4

IMPROVEMENT AND PROHIBITION NOTICES

A. Improvement and Prohibition Notices	4.01	Suspension of notices pending appeal	4.68
Summary	4.01	Withdrawing an appeal	4.71
Consequences of serving a notice	4.10	Case management	4.72
Contravening a notice	4.11	The hearing	4.75
The inspector's decision to issue a notice	4.18	Preliminary arguments	4.83
		The burden of proof	4.84
Improvement notices	4.25	Challenging the inspector's opinion	4.88
Prohibition notices	4.34		
Serving the notice	4.46	Insignificant contravention	4.98
Withdrawal and extension	4.48	Fundamental flaw and modification	4.100
Liability of inspectors in negligence for misstatement	4.51	Example decisions	4.111
		The tribunal's decision	4.120
B. Appealing Improvement and Prohibition Notices	4.53	Costs	4.123
Procedure	4.59	Appeal from the decision of an employment tribunal	4.126
Commencing an appeal	4.62		

A. Improvement and Prohibition Notices

Summary

The 1972 Robens Report,[1] which framed the provisions of the Health and Safety at Work etc. Act 1974 (HSWA 1974), recommended that inspectors should be vested with broad-ranging and comprehensive powers: 'Inspectors should have

4.01

---

[1] Cmnd 5034, 1972. The Robens Committee was set up by the Government to determine the effectiveness of health and safety law existing at that time. The Committee concluded that the fragmented and patchy statutory instruments that had governed health and safety should be swept away, and replaced by a single statute (an 'enabling' Act). The Health and Safety at Work etc. Act 1974 directly arose from the observations and recommendations of the Committee. The Act was to be the centrepiece of health and safety law, supplemented by Regulations and Codes of Practice.

the power, without reference to the courts, to issue a formal improvement notice to an employer requiring him to remedy particular faults or to institute a specified programme of work within a stated time limit.'[2] Moreover, the report made a further recommendation: 'The improvement notice would be the inspector's main sanction. In addition, an alternative and stronger power should be available to the inspector for use where he considers the case for remedial action to be particularly serious. In such cases he should be able to issue a prohibition notice.'[3]

4.02 An inspector's power to issue both improvement notices, and prohibition notices, derives from ss 21 and 22 of the HSWA 1974, respectively. Section 23 makes supplementary provisions in relation to notices; s 24 deals with appeals against notices. The relevant parts of the HSWA 1974, including ss 21 to 24, are reproduced at Appendix A.

4.03 An inspector may serve an improvement notice if the circumstances warrant it. If an inspector is of the opinion that a person is contravening one or more health and safety provisions,[4] an improvement notice may be served. Similarly, a notice may be issued in circumstances where one or more of those provisions outlined above has been contravened in circumstances that make it likely that the contravention will continue or be repeated. An improvement notice specifies a period of time within which the person is required to remedy the contravention, or remedy the matters occasioning such contravention.

4.04 An inspector may also serve a prohibition notice on a person. One may be served if the inspector is of the opinion that an activity is being carried on which either does involve or will involve a risk of serious personal injury if carried on. The prohibition notice directs that the activities shall not be carried on by the person unless matters specified in the notice have been remedied. The prohibition notice can be immediate or deferred.

4.05 The decision to issue an improvement or prohibition notice, and its service on the affected person, is very often taken by an inspector during the course of his attendance at an incident, inspection, or visit. Since 1997, there have been no procedures requiring inspectors to notify a person of an intention to serve a notice. At that time, the Health and Safety Commission expressed a view that it expected that informal communication between inspectors and employers would continue: 'Inspectors should discuss the alleged breaches of the law and the action required to remedy them before they issue the notice, in an attempt to resolve any points of difference before the notice is served.'[5] However, an inspector is not required to, and rarely will, consult the affected person, hear representations, or otherwise delay, prior to taking

---

[2] Cmnd 5034, 1972 para 269.
[3] ibid para 276.
[4] The 'relevant statutory provisions' are defined by HSWA 1974 s 53(1), on which, see below.
[5] HSC Press Release C11/98, 31 March 1998.

A. Improvement and Prohibition Notices

the decision to issue a notice. The criteria for intervention are that either continuing contravention is likely, or there is deemed to be a risk of serious personal injury.

4.06 An affected person will almost always be in the position of seeking legal advice following receipt of, and not prior to, service of the notice. A notice cannot be withdrawn once a time period set by the notice has expired; a prohibition notice of immediate effect cannot subsequently be withdrawn. An inspector can extend, or further extend, a time period set in a notice at any time, unless an appeal against the notice has been lodged. Any representations concerning the correctness of the terms of a notice need to be made prior to the period specified in the notice. Such representations, if accepted, are likely to be met with a withdrawal of the notice, coupled with the service of a fresh notice in amended terms. In such circumstances, the issue of appeal against the improvement or prohibition notice is the most important potential recourse open to the affected person.

4.07 Inspectors are advised to give, or send, leaflet ITL 19 ('Appeal to the Employment Tribunal against an Enforcement, Improvement or Prohibition Notice in Health and Safety and related matters') when serving a notice. This leaflet provides useful advice, and includes a form for submitting an appeal.

4.08 Appeal against a notice lies to the Employment Tribunal. A person in receipt of an improvement or prohibition notice must lodge an appeal within 21 days from the date of service of the notice. The Employment Tribunal can extend this period. This may be done where it is satisfied, on an application made in writing (either before or after expiry of the 21-day period), that it was not reasonably practicable for the appeal to be brought within the 21-day period. The lodging of an appeal against an improvement notice has the effect of suspending its operation. A separate application can be made to suspend a prohibition notice, prior to the substantive hearing of an appeal against such a notice. On appeal, the Employment Tribunal may either affirm or cancel a notice. If it affirms a notice, it may so do with modifications in the form of additions, omissions, or amendments.

4.09 The HSE publishes statistics which include the annual totals of notices issued by local authority and HSE inspectors. The most recent figures available are those for 2004/2005:

Notices issued in 2004/2005:	Improvement Notices	Deferred Prohibition Notices	Immediate Prohibition Notices	Total Enforcement Notices
HSE	5,186	49	3,236	8,471[6]
Local authorities	5,100	50	1,260	6,420
Total	10,286	99	4,496	14,891

[6] This is substantially lower than for the four previous years.

Consequences of serving a notice

4.10 Issuing a notice amounts to enforcement action against a person. An individual notice can be issued in respect of a number of breaches. Notices have potentially serious consequences. HSE inspectors most often enforce health and safety standards by providing advice on how to comply with the law. The HSE publishes a public register of issued notices.[7] Persons subject to a notice can suffer, as a consequence, harm to reputation, increased insurance costs, and real handicap in successfully tendering for contracts.

Contravening a notice

4.11 To contravene any requirement of an improvement notice or a prohibition notice—including a notice which has been modified on appeal—amounts to a serious criminal offence;[8] and it is the stated policy of the HSE to prosecute for breaches of a notice. There are, however, occasions when no further action is taken for reasons of public interest.[9] Nonetheless, the HSE will continue to monitor and enforce health and safety standards in these cases.

4.12 The HSE's approach to health and safety enforcement aims to represent fair and effective enforcement. In particular, enforcement is proportionate to the risk: thus 'serious breaches will get a tough response'.[10]

4.13 When the offence is committed by an individual, it is punishable in the magistrates' court by a fine of £20,000 or six months' imprisonment. An unlimited fine, or two years' imprisonment, may also be given in the Crown Court. In May 2002 a gas fitter became the fourth person to be imprisoned for breaching the terms of a prohibition notice. The magistrates imposed a term of four months' imprisonment, although the sentence was subsequently varied on appeal to a community penalty.[11]

4.14 In the event of a prosecution for contravening a notice, no challenge to the terms of an improvement or prohibition notice can be raised. Neither do the provisions of HSWA 1974—s 40, the reasonable practicability defence—apply. The case of

---

[7] Available at <http://www.hse.gov.uk/notices>. The site is updated on a weekly basis; notices appear for up to five years. HSE policy is that 'steps are taken to ensure that notices which are withdrawn or subject to appeal are not included'. There are plans afoot, in 2008, to also include local authorities' enforcement data: HSE Offences and Penalties Report 2004/2005.

[8] Contrary to HSWA 1974 s 33(1)(g).

[9] HSE give examples of where prosecution would not be in the public interest; the client who has ceased trading, for instance, would not attract prosecution. Prosecution is also not envisaged where a person has carried out sufficient work to essentially comply, but there remains a minor deficiency which does not give rise to a health and safety risk. For further, see the HSC's Enforcement Policy Statement, which inspectors follow to decide on the most appropriate action to take.

[10] Foreword to the HSE Offences and Penalties Report, 2004/2005.

[11] Health and Safety Bulletin 309, June 2002.

A. Improvement and Prohibition Notices

Deary v Mansion Hide Upholstery [1983] ICR 610 deals with the issue; the respondent company was served with an improvement notice requiring it to provide fire-resistant storage for polyurethane foam. It did not comply with the notice, nor did it appeal against it. When it was prosecuted for failure to comply with the notice, it contested the matter at the magistrates' court. The justices applied HSWA 1974 s 40, and found that it had done all that was reasonably practicable. It was held that, in doing so, they had erred, as the section had no bearing on the offence of failing to comply with an improvement notice.

4.15 Notices are not necessarily issued as either an alternative, or a precursor, to prosecution for a contravention of a health and safety duty. Often, a notice will be issued and prosecution for a health and safety offence will follow: although, in many situations, a decision will subsequently be taken that prosecution is not warranted.

4.16 The Health and Safety Commission's Enforcement Policy Statement[12] provides that:

> 6. Formal cautions and prosecution are important ways to bring duty holders to account for alleged breaches of the law. Where it is appropriate to do so in accordance with this policy, enforcing authorities should use one of these measures in addition to issuing an improvement or prohibition notice.

4.17 Neither a decision on prosecution, nor the commencement of such criminal proceedings, is likely to occur until well after the expiry of the period permitted for lodging an appeal against a notice. The fact that a notice was issued will be irrelevant in a criminal trial for breach of the substantive offence.

The inspector's decision to issue a notice

4.18 Both the Health and Safety Commission's Enforcement Policy Statement,[13] and the Enforcement Management Model[14] (a matrix that HSE inspectors use to make and record enforcement decisions), should inform an HSE inspector's decision to issue a notice. These also govern whether an improvement or prohibition notice should be issued. The Enforcement Policy Statement describes the purpose of enforcement as being to:

(a) ensure that duty-holders take action to deal immediately with serious risks;
(b) promote and achieve sustained compliance;
(c) ensure that duty-holders who breach health and safety requirements, and directors and managers who fail in their responsibilities, may be held to

---

[12] Available at <http://www.hse.gov.uk/pubns/hsc15.pdf>.
[13] Reproduced at App B.
[14] Available at <http://www.hse.gov.uk/enforce/emm.pdf>.

account. This may include bringing the alleged offenders before the courts in England and Wales, or recommending prosecution in Scotland.

4.19 The Enforcement Management Model outlines the range of tools available to inspectors to secure compliance: 'Inspectors use various enforcement techniques to deal with risks and secure compliance with the law, ranging from the provision of advice, to enforcement notices. They can also initiate or recommend prosecution where the circumstances warrant punitive action. Making decisions about appropriate enforcement is fundamental to the role of an inspector.'[15]

4.20 The Enforcement Management Model describes how:

> During a regulatory contact, inspectors collect information about hazards and control measures. This is used to make an initial assessment of the health and safety risks posed by the various activities and to determine the actual risk . . . They should compare this to the risk accepted by the law or guidance and decide the benchmark (the level of risk remaining once the actions required of the dutyholder by the relevant standards, enforceable by law, are met). The difference between where the dutyholder is and where they should be, is the risk gap.[16]

4.21 The concept of a so-called 'risk gap' is fundamental to the decision-making process. Risk gap analysis is used in two ways: first, to assess what enforcement is necessary to secure compliance with the law; and secondly, to determine whether prosecution should be considered.

4.22 Inspectors are advised to assess the level of actual risk arising from the dutyholder's activities. This judgment should be based on 'information about hazards and control measures informed by their training, experience, guidance and other relevant sources of information'.[17] The risk gap principle can be used for events which have already happened, for example, during investigation of an accident or dangerous occurrence. Crucially, however, 'it is the potential for harm which should inform the decision: not what actually happened'.[18]

4.23 Thus the decision to issue an improvement or prohibition notice, rather than merely to give informal advice, may be based upon many factors, such as perceived attitude and previous conduct of the affected person. The starting-point is likely to be an inspector's judgment of the risk to health and safety arising from the situation. First, the inspector must form the opinion that a state of affairs exists that fulfils the statutory criteria for the issuance of a notice. The judgment would then appear to be whether such a notice amounts to a proportionate and effective response to the situation.

---

[15] s 1 para 2.
[16] s 3 para 13.
[17] ibid para 17.
[18] ibid para 18.

A. Improvement and Prohibition Notices

The HSE's Enforcement Guide[19] (which is drafted by HSE Solicitor's Office and provides advice to inspectors in relation to all aspects of enforcement) provides: **4.24**

> Where there is a risk of serious personal injury, a prohibition notice is more appropriate than an improvement notice. For example, a prohibition notice might be appropriate to cover a defective scaffold or an unguarded power press, but improvements in a maintenance system should be dealt with by an improvement notice. In some cases, both may be issued to deal with the same set of circumstances.

Improvement notices

Section 21 HSWA 1974 sets out the power of an inspector to issue an improvement notice. It also prescribes what particulars must be included in such a notice. Improvement notices may be issued in two circumstances: an inspector may serve an improvement notice if he is of the opinion that a person is contravening one or more of the 'relevant statutory provisions'; and a notice may also be served where it appears a person has contravened one or more of those provisions in circumstances that make it likely that the contravention will continue or be repeated. **4.25**

The 'relevant statutory provisions' are defined by HSWA 1974 s 53(1), to comprise: **4.26**

(a) the provisions of HSWA 1974, pt I (ss 1–53) (which includes the general health and safety duties);
(b) the health and safety regulations passed under HSWA 1974; and
(c) the 'existing statutory provisions'; those enactments specifically preserved in HSWA 1974 Sch 1 which have not subsequently been repealed.

The 'relevant statutory provisions', which are the health and safety provisions, create different duties on many different classes of person. Among these are: employers, employees, the self-employed, suppliers of machinery, occupiers of premises, those in control of various activities, premises and hazardous undertakings, and those who exercise some control over how machinery or operatives work, in various fields. **4.27**

If the inspector forms such an opinion, then he may serve an improvement notice on the person. In the improvement notice, the inspector must: **4.28**

(a) state that he has formed the opinion (that a person is contravening one or more of the 'relevant statutory provisions', or has contravened one or more of those provisions in circumstances that make it likely that the contravention will continue or be repeated);
(b) specify the provision(s), which in his opinion have been, or are being contravened;
(c) give particulars of the reasons why he is of that opinion; and

---

[19] Available at <http://www.hse.gov.uk/enforce/enforcementguide/index.htm>.

(d) specify the period within which the person is required to remedy the contravention or the matters occasioning such a contravention.

4.29 The period specified in the notice within which the requirement must be carried out must be at least 21 days. This is the time limit for lodging an appeal against the notice with an employment tribunal.

4.30 Section 23(2) of HSWA 1974 provides that a notice may (but need not) include directions as to the measures to be taken to remedy any contravention or matter to which the notice relates. Any such directions may be framed to any extent by reference to any approved code of practice. They may be framed so as to afford the person on whom the notice is served a choice between different ways of remedying the contravention or matter.

4.31 Section 23(3) of HSWA 1974 provides for where any of the relevant statutory provisions apply to a building. The inspector should not direct any measures to be taken which are more onerous than any current building regulation. This is the case unless the relevant statutory provision imposes specific requirements which are more onerous.

4.32 Inspectors, both HSE and those employed by local authorities, use a standard form (LP/1) improvement notice, filling in particulars and deleting non-applicable words. Directions as to the measures to be taken, pursuant to HSWA s 23(2), are detailed in a schedule attached to, and as part of, the notice. The directions may refer to an Approved Code of Practice. Where this is the case, these must still allow the recipient of the notice to choose an alternative way of remedying the contravention that achieves the same standard. The actual terms of the notice should not include guidance on good practice, or other advice that goes beyond the duty imposed by the particular health and safety provision. The HSE advises inspectors to include such recommendations in a covering letter. Further, it cautions inspectors to specify that such information is guidance. They should state that it does not form part of the notice.

4.33 An inspector should only serve an improvement notice in a situation which can be remedied within the period specified. An improvement notice should not be used as a means of imposing an obligation which has no attainable end. Thus, in the example given in the HSE's Enforcement Guide,[20] a notice can properly require the provision of a guard. It should not, however, require that such a guard be maintained in good condition. Instead, a notice can properly require that, by a given date, there should be in place a system for maintenance. It should be made clear that this is a requirement to remedy the matters that led to the breach.

---

[20] Available at <http://www.hse.gov.uk/enforce/enforcementguide/index.htm>.

A. Improvement and Prohibition Notices

Prohibition notices

Section 24 of HSWA 1974 sets out the power of an inspector to serve a prohibition notice. It prescribes what particulars must be included in the notice. The power extends to any activity which is under the control of any person. Importantly, this must be one to which the relevant statutory provisions do apply or will apply. Again, the 'relevant statutory provisions' are defined by HSWA 1974 s 53(1) to comprise: **4.34**

(a) the provisions of Pt I (ss 1–53) of HSWA 1974 (which includes the general health and safety duties);
(b) the health and safety regulations passed under HSWA 1974; and
(c) the 'existing statutory provisions'; those enactments specifically preserved in HSWA 1974 Sch 1 which have not subsequently been repealed.

An inspector may serve a prohibition notice if he is of the opinion that such an activity is being carried on which either does, or will, involve a risk of serious personal injury. The risk of injury need not be imminent.[21] Personal injury includes any disease and impairment of a person's physical or mental condition.[22] In the prohibition notice, the inspector must: **4.35**

(a) state that he has formed the opinion that such an activity is being carried on which either does involve, or will involve, a risk of serious personal injury;
(b) specify the matters which give rise to the risk, or will give rise to the risk;
(c) where, in his opinion, any of those matters involves, or will involve, a contravention of any statutory provision, specify the provision;
(d) state why he is of that opinion and give particulars of his reasons;
(e) direct that the activities to which the notice relates shall not be carried on by, or under the control of, the person on whom the notice is served, unless the matters specified in the notice, and any associated contraventions of statutory provisions, have been remedied.[23]

The prohibition notice can be immediate or deferred. Compliance must take place immediately, or at the end of a specified period.[24] **4.36**

The HSE Enforcement Guide advises inspectors when deciding whether a prohibition notice should apply immediately or be deferred. An inspector should consider the relative risks. The only reason a prohibition should not be made to have immediate effect is that the risk posed by stopping immediately is greater than the risks of deferring the prohibition until later.[25] **4.37**

---

21 *Tesco v Kippax* (COIT No 7605, HSIB 180, P 8).
22 HSWA 1974 s 53(1).
23 ibid s 22(3)(a)–(d).
24 ibid s 22(4)(a), (b).
25 Available at <http://www.hse.gov.uk/enforce/enforcementguide/notices/notices/intro.htm>.

4.38 Provided that an inspector forms an opinion that there is a risk of serious personal injury, he need not be of the opinion that a contravention of a statutory provision has occurred, or will occur. The person operating a process or piece of equipment that poses a risk of serious personal injury may not be subject to any statutory duty. He may even be unaware of the risk. Nevertheless, he can still be served with a prohibition notice requiring him to close down the process or equipment. The HSE Enforcement Guide advises inspectors on this. 'In such circumstances there may be a need to serve a notice on the employee, as well as the employer.'

4.39 As with improvement notices, HSWA 1974 s 23(2) provides that a notice may (but need not) include directions as to the measures to be taken to remedy any contravention or matter to which the notice relates. Any such directions may be framed to any extent by reference to any approved code of practice. Directions may be framed so as to afford the person on whom the notice is served a choice between different ways of remedying the contravention or matter.

4.40 As with improvement notices, s 23(3) governs situations where one of the relevant statutory provisions applies to a building. It provides that the inspector should not direct any measures to be taken which are more onerous than any current building regulation, unless the relevant statutory provision imposes specific requirements which are more onerous.

4.41 Inspectors—both HSE and those employed by local authorities—use a standard form (LP/2) prohibition notice, filling in particulars and deleting non-applicable words. Directions as to the measures to be taken, pursuant to HSWA 1974 s 23(2), are detailed in a schedule attached to the notice. The schedule then forms part of the notice. Where the directions refer to an Approved Code of Practice, these must still allow the recipient of the notice to choose an alternative way of remedying the contravention that achieves the same standard.

4.42 As with improvement notices, the actual terms of a prohibition notice should not include guidance on good practice, or other advice, that goes beyond the duty imposed by the particular health and safety provision. The HSE advises inspectors to include such recommendations in a covering letter, to specify that such information is guidance. Crucially, it should be stated that it does not form part of the notice.

4.43 If the activity has not previously been carried on, the inspector must be of the opinion that it is likely to be carried on in a way that will involve a risk of serious personal injury. If the activity has been carried on, but has temporarily stopped, the inspector must be of the opinion that it is likely that the activity will recommence. The fact that an activity may have been suspended or interrupted, following a major accident or incident, does not necessarily mean that it is not being carried on for the purposes of HSWA 1974 s 22(1). Neither does anything hinge on an assurance from the duty-holder that the prohibited activity will

A. Improvement and Prohibition Notices

not recommence or continue. However earnestly the assurance might be offered, it will not prevent the service of a prohibition notice.

4.44 The case of *Railtrack Plc v Smallwood* [2001] ICR 714 is relevant here. In that case, an assurance was given by the appellant, in the aftermath of the Ladbroke Grove rail crash, in which 31 people were killed and 259 injured. The assurance stated that it was 'unthinkable' for passenger trains to run in the area where the signal had been passed at danger, but was not held to mean that activities had ceased for the purposes of HSWA 1974 s 22. The court held that it was a question of fact and degree whether the activities had ceased. It applied a purposive approach to the implementation of that section, so as to render it effective in its role of protecting public safety.

4.45 Similarly, in *Grovehurst Energy Ltd v Strawson*,[26] it was held that the undertaking by the managing director of the company was insufficient. The undertaking, in that case, was to the effect that the managing director would take additional precautions to guard against the risk of serious injury from unsafe plant. It was held to be insufficient reason to suspend a prohibition notice. This was despite the considerable loss of profit occasioned to both the appellant and his customer. Where an improvement notice has been served, the notice may be withdrawn by an inspector at any time before the end of the period specified within the notice. That period may be extended or further extended by an inspector at any time unless an appeal against the notice has been lodged.

Serving the notice

4.46 Section 46 of HSWA 1974 provides possible means for service of notices. Section 46(1) provides that service of any notice may be effected by delivering it to the person in question, leaving it at his proper address, or by sending it to him by post at that address. In the case of a body corporate, the notice may be served on, or given to, the secretary or clerk of the body; the proper address shall be the address of the registered or principal office of that body.[27] In the case of a partnership, the notice may be served on, or given to, a partner or the person having control or management of the partnership business,[28] at the principal office of the partnership.[29]

4.47 Section 46(6) provides that the notice may be deposited with a person in a position of authority. Where service is authorized on the owner or occupier of premises, it may be sent by post to those premises. If addressed by name, to the person on

---

[26] CO.5035/90, reported in *Tolleys* (HSIB 180 P 9).
[27] HSWA 1974 s 46(4)(a).
[28] ibid s 46(3)(b).
[29] ibid s 46(4)(b).

whom it is to be served, the notice may be delivered to some responsible person who appears to be resident or employed in the premises. In *HSE v George Tankcocks Garage Ltd*,[30] the Divisional Court ruled specifically on this point. Where it was alleged that the person concerned was in breach of his duty as owner or occupier of those premises, only a notice that was required or authorized to be served on, or given to, an owner or occupier of premises, could be served in this way. If a company is the occupier, and a notice does not relate to it as occupier, then service should be sent to its registered address. It was further held that s 46 did not contain a complete code of all the ways in which a notice might be served. The subsections were permissive only. If it is proved that a notice came to the attention of the proper officer of a company, there would be sufficient service under HSWA 1974.

Withdrawal and extension

4.48 Where an improvement or prohibition notice has been served, it may be withdrawn by an inspector at any time before the end of the period specified within the notice. That period may be extended, or further extended, by an inspector, at any time, unless an appeal against the notice has been lodged.[31] Thus a prohibition notice of immediate effect cannot subsequently be withdrawn.

4.49 The standard forms for both types of enforcement notice advise that the period may be extended, if the need arises. Any extension will normally be given in writing, by the inspector, on a standard form (LP4).

4.50 Inspectors very often seek check compliance with a notice by visiting the premises prior to the expiry of the period for compliance.

Liability of inspectors in negligence for misstatement

4.51 In *Harris v Evans & Health & Safety Executive* [1998] 1 WLR 1285, the liability inspectors might incur for negligent misstatement was examined. It was held that an inspector who was acting pursuant to a duty, under HSWA 1974, could not be liable in negligence for alleged misstatements made about a business which caused it to suffer economic loss. The inspector had given advice, in the course of his duties for the HSE, to certain local authorities about the safety of a mobile telescopic crane used by the plaintiff, for the purposes of his bungee jumping business. As a result, an improvement notice and then a prohibition notice had been served on the owner of the equipment. The issue was whether the inspector owed a duty of care to the plaintiff for the content of his advice.

---

[30] *Mallon v Norman Hill Plant Hire Ltd* [1993] Crim. LR 605.
[31] HSWA 1974 s 23(5).

It was held that the duty of enforcing officers was to have regard to the health and safety of members of the public. If steps which they thought should be taken would have an adverse economic effect on the business enterprise in question, so be it. A tortious duty would be very likely to engender untoward caution and reluctance on the part of an inspector. It was implicit, under HSWA 1974, that improvement notices and prohibition notices may cause economic loss or damage to the business enterprise in question. It would be seriously detrimental to the proper discharge by enforcing authorities of their responsibilities, in respect of public health and safety, if they were to be exposed to potential liability in negligence at the suit of the owners of the businesses affected by their decisions. HSWA 1974 itself provided remedies against errors or excesses on the part of inspectors and enforcing authorities.

B. Appealing Improvement and Prohibition Notices

Section 24(2), HSWA 1974, provides:

> A person on whom a notice is served may within such period from the date of its service as may be prescribed appeal to an [employment tribunal]; and on such an appeal the tribunal may either cancel or affirm the notice and, if it affirms it, may do so either in its original form or with such modifications as the tribunal may in the circumstances think fit.

The ambit of this apparently wide discretion to cancel, affirm, and modify, as the tribunal may 'think fit', remains uncertain. There is no authoritative judgment on the provision. The most that can be identified is principles apparent in previously decided cases, and some judicial opinion can be ascertained. Much of it is, though, conflicting. In particular, there is a real issue as to whether the tribunal conducts simply a 'review' of the inspector's decision, or holds an 'appeal' by way of a fresh hearing (a hearing *de novo*) of the issues expressed in the notice.

The potential grounds of appeal are not capable of being reduced to an exhaustive list. In practice, grounds against a notice can be identified as having fallen into the following categories:
(a) the inspector wrongly interpreted the law or exceeded his powers;
(b) the notice is fundamentally flawed in some way;
(c) the inspector did not hold the necessary opinion, or the opinion was perverse/based on unreasonable grounds;
(d) the inspector's opinion, whilst genuinely held, was not one that the tribunal should endorse (following the judgment of Sullivan J in *Railtrack Plc v Smallwood*,[32] considered below, at paras 4.89–4.94);

---

[32] [2001] ICR 714.

(e) no contravention of a relevant statutory provision has occurred/will occur (there is no risk to health and safety, or not the risk identified), or there is no risk of serious personal injury (in respect of a prohibition notice);

(f) the risk to health and safety exists/is admitted, but the steps identified as necessary to remedy the contravention, or required in order to avert the risk, are not 'reasonably practicable', or there are no such steps that will reduce the risk.

(g) the contravention is so insignificant that the notice should be cancelled (see below, at paras 4.98–4.99).

4.56 An appeal may raise grounds that can be cured by modification. Before the hearing of the appeal, the inspector may accept the criticism made of the notice and withdraw it, and then serve a fresh notice in amended terms. Or, at the hearing of the appeal, the tribunal affirming the notice can modify the notice to meet the criticism. This may not be the outcome sought by an appellant who wishes to avoid the consequent harm to reputation arising from service of a notice.

4.57 The seriousness of harm to reputation that may ensue from the service of a notice on a person should be acknowledged. Instructive here is the successful Administrative Court appeal launched by BT Fleet Ltd, following an appeal to the employment tribunal. In that case, the appeal against an improvement notice resulted in the notice being affirmed, but with substantial modifications. At the time of the Administrative Court challenge, there were no remaining issues between the appellant company and the HSE, since the notice had been complied with. However, the Court noted that:

> BT retains an interest in challenging the Tribunal's findings . . . This is because the fact that BT have been served with an Improvement notice by an inspector of HSE will be found on HSE's website. BT's primary business is servicing British Telecom vehicles but it does contract to provide services to outside commercial organisations. In bidding for any contract exceeding £50,000 it would be bound to disclose that it had been served with an Improvement notice in respect of its operations.[33]

4.58 Appeals of improvement or prohibition notices form a small—and entirely discrete—part of an employment tribunal's caseload. As a consequence, the particular tribunal that is hearing an appeal is likely to have very little—if any—previous experience of such an appeal, let alone knowledge of health and safety law and provisions. At appeals, HSE inspectors are usually legally represented; local authority inspectors are often represented by an in-house lawyer. The HSE usually instruct from a small pool of counsel who have extensive experience of enforcement notice appeals.

---

[33] [2005] EWCA 387 (Evans-Lombe J) at para 9.

B. Appealing Improvement and Prohibition Notices

Procedure

Improvement and prohibition notice appeals are now subject to a set of rules found in Employment Tribunals (Constitution and Rules of Procedure) Regulations 2004 SI 2004/1861 Sch 4. This schedule is reproduced at Appendix D. These rules of procedure both apply and exclude many of the general rules, the Employment Tribunals Rules of Procedure, contained in Sch 1 to the same regulations. **4.59**

Schedule 1 r 60[34] provides that, subject to the rules of procedure, a tribunal or chairman may regulate its, or his, own procedure. To a very large extent, tribunals do exactly this and proceedings are less formal than in a criminal court. In particular, the tribunal is not concerned with the rules of hearsay. It will admit all relevant evidence, assessing what weight seems appropriate. **4.60**

The Employment Tribunals (Constitution and Rules of Procedure) Regulations 2004 SI 2004/1861 reg 3 gives an overview of the purpose of the regulations. It describes how the overriding objective of the regulations and the rules in the schedules is to enable tribunals and chairmen to deal with cases justly. Moreover, it provides that dealing with a case justly includes, so far as is practicable, ensuring that the parties are on an equal footing. This involves dealing with the case in ways which are proportionate to the complexity or importance of the issues, ensuring that it is dealt with expeditiously and fairly, and saving expense. The regulation further provides that a tribunal or chairman shall seek to give effect to the overriding objective when exercising any power pursuant to the regulations or the rules in the schedules. This objective should be furthered when interpreting the same. The regulation also requires that the parties shall assist the tribunal or the chairman to further the overriding objective. **4.61**

Commencing an appeal

Rule 3 of Sch 4[35] provides: **4.62**

> A person wishing to appeal an improvement notice or a prohibition notice (the appellant) shall do so by sending to the Employment Tribunal Office two copies of a notice of appeal which must include the following:
> (a) the name and address of the appellant and, if different, an address to which he requires notices and documents relating to the appeal to be sent;
> (b) the date of the improvement notice or prohibition notice appealed against and the address of the premises or the place concerned;
> (c) the name and address of the respondent;

---

[34] Employment Tribunals (Constitution and Rules of Procedure) Regulations 2004 SI 2004/1861 Sch 4 r 60.
[35] ibid Sch 4.

(d) details of the requirements or directions which are being appealed; and
(e) the grounds for the appeal.

4.63 Rule 4 of Sch 4 provides that a notice of appeal against an improvement or prohibition notice, 'must be sent to the Employment Tribunal Office within 21 days from the date of the service on the appellant of the notice appealed against'. Rule 4(2) provides that the tribunal may extend the time, where it is satisfied on an application made in writing to the Secretary, either before or after the expiration of that time period, that 'it is or was not reasonably practicable for an appeal to be brought within that time'.

4.64 Regulation 15[36] provides that, where any act must be done within a certain number of days of or from an event, the date of that event is not to be included in the calculation. Thus, if the date of service of an improvement or prohibition notice is stated to be 1 October, the last day for any appeal to have been sent is 22 October. Rule 61[37] of Sch 1 provides that any notice given or document sent under the rules shall (unless a chairman or tribunal orders otherwise) be in writing. It may be given in person, or sent by post, sent by fax or other means of electronic communication, or by personal delivery. According to Sch 1 r 61(2), such notice or documentation sent in this way is, unless the contrary is proved, taken to have been received by the party to whom it is addressed:

(a) in the case of a notice or document given, or sent by post, this will be taken as the day on which the notice or document would be delivered in the ordinary course of post;
(b) in the case of a notice or document transmitted by fax or other means of electronic communication, this will be taken as the day on which the notice or document is transmitted;
(c) in the case of a notice or document delivered in person, this will be taken as the day on which the notice or document is delivered.

4.65 When serving a notice, inspectors should give an affected person a copy of booklet ITL 19 ('Appeal to the Employment Tribunal against an Enforcement, Improvement or Prohibition Notice in Health and safety and related matters'). The booklet contains a list of the addresses of the Employment Tribunals and a standard form for appeal. The addresses of Employment Tribunals can also be found at the employment tribunal website: <http://www.employmenttribunals.gov.uk/>.

4.66 Rule 4(2)[38] permits a tribunal to allow an extension of the 21-day period. This appears to be a limited discretion to allow out-of-time appeals: the power is

---

[36] Employment Tribunals (Constitution and Rules of Procedure) Regulations 2004 SI 2004/1861.
[37] ibid Sch 1 r 61.
[38] ibid Sch 4 r 4(2).

B. Appealing Improvement and Prohibition Notices

not founded on 'an interests of justice' test. Instead, it requires a party to demonstrate that it was not practicable to comply. Lack of awareness of a notice by a relevant company officer within the 21 days appears to be the type of situation envisaged.

Once a notice of appeal has been received by the tribunal, the Secretary to the tribunal is obliged to send a copy of the notice of appeal to the respondent inspector. The Secretary must then inform the parties, in writing, of the case number of the appeal, and of the address to which notices and other communications to the Employment Tribunal Office shall be sent.[39] This number should be referred to in all correspondence relating to the appeal. **4.67**

Suspension of notices pending appeal

In the case of an improvement notice, bringing an appeal has the effect of suspending the operation of the notice until the appeal is fully disposed of, or withdrawn,[40] whichever the case may be. **4.68**

Section 24(3) HSWA 1974 provides that, in the case of a prohibition notice, the tribunal can direct that the notice be suspended until determination of the appeal, on application by the appellant. Rule 6[41] of Sch 4 sets out the procedure for making a preliminary application for a direction suspending the operation of a prohibition notice. Rule 6(1) provides that the application must be presented to the Employment Tribunal Office in writing. It shall include the case number of the appeal, or, if there is no case number, sufficient detail to identify the appeal. Importantly, it must set out the grounds on which the application is made. Following receipt of such an application, the Secretary to the tribunal is required to send a copy to the respondent inspector. This should be as soon as practicable upon receipt of the application. The Secretary must then inform the inspector that he has the opportunity to submit representations in writing, if he so wishes, within a specified time, but not less than seven days later. Thereafter, the tribunal chairman considers the application and any representations submitted by the inspector. The chairman may:[42] **4.69**

(a) order that the application should not be determined separately from the full hearing of the appeal;

---

[39] Employment Tribunals (Constitution and Rules of Procedure) Regulations 2004 SI 2004/1861 Sch 4 r 5.
[40] HSWA 1974 s 24(3)(a).
[41] Employment Tribunals (Constitution and Rules of Procedure) Regulations 2004 SI 2004/1861 Sch 4 r 6.
[42] ibid Sch 4 r 6(3).

(b) order that the operation of the prohibition notice be suspended, until the appeal is determined or withdrawn;
(c) dismiss the appellant's application; or
(d) order that the application be determined at a hearing.

4.70 The chairman must give reasons for his decision, and any decision made following a hearing of the application.[43] The rules[44] provide that any such decision shall be treated as a decision which may be reviewed upon the application of a party under rr 34–36 of Sch 1.[45] This procedure permits a party to apply to the chairman to consider whether there are grounds for the tribunal to review its decision. The grounds for review[46] are:

(a) that the decision was wrongly made as a result of an administrative error;
(b) a party did not receive notice of the proceedings leading to the decision;
(c) the decision was made in the absence of a party;
(d) new evidence has become available since the conclusion of the hearing to which the decision relates, provided that its existence could not have been reasonably known of or foreseen at that time; or
(e) that the interests of justice require such a review.

Withdrawing an appeal

4.71 Rule 9[47] of Sch 4 provides that an appellant may withdraw all, or part, of a notice appeal at any time. This may be done either orally while at a hearing, or in writing to the Employment Tribunal Office.

Case management

4.72 Rule 7 of Sch 4[48] provides for a party to the appeal to apply to the chairman to make pre-hearing orders. The rule provides that, in such circumstances, the chairman may, 'make an order in relation to any matter which appears to him to be appropriate'. It is further provided that, subject to various exclusions,[49]

---

[43] Employment Tribunals (Constitution and Rules of Procedure) Regulations 2004 SI 2004/1861 Sch 4 r 6(4).
[44] ibid Sch 4 r 6(5).
[45] ibid Sch 1 rr 34–36.
[46] ibid Sch 1 r 34(3).
[47] ibid Sch 4 r 9.
[48] ibid Sch 4 r 7.
[49] Sch 4 r 11: excluding those orders at Sch 1 r 10(2)(g), (i), (k), (l), and (r).

B. Appealing Improvement and Prohibition Notices

such orders may be those listed in r 10(2) of Sch 1[50] (the general rules). Additional or such other orders as the chairman thinks fit may be included.

The potential orders contained in r 10 of Sch 1 include: **4.73**

(a) those relating to the manner in which the proceedings are to be conducted, including any time limit to be observed;
(b) that a party provides additional information. It requires the attendance of any person in Great Britain either to give evidence, or to produce documents or information, or to disclose documents or information to a party. This allows a party to inspect such material as might be ordered by a county court;
(c) requiring the provision of written answers to questions put by the tribunal or chairman;
(d) that a witness statement be prepared or exchanged; and
(e) the use of experts or interpreters in the proceedings.

The orders may be issued as a result of a chairman considering the papers before him in the absence of the parties. An order may also be issued at a hearing. Contrary to the general rules, in an enforcement notice appeal, a tribunal can only make such orders on the application of one of the parties. If a party applies, then the chairman may make such orders as he sees fit. If the parties agree, in writing, upon the terms of any decision to be made by the tribunal or chairman, the chairman may, if he thinks fit, decide accordingly. **4.74**

The hearing

Rule 26 of Sch 1[51] provides that a hearing is held for the purpose of determining outstanding procedural or substantive issues, or disposing of the proceedings. It provides that, in any proceedings, there may be more than one hearing. There may also be different categories of hearing (such as a hearing in relation to an application to suspend the operation of a prohibition notice pending determination of an appeal). The hearing takes place in public, subject to national security considerations. Limited confidentiality matters, as set out in r 16 of Sch 1,[52] are also relevant. **4.75**

A tribunal normally consists of three members, with one chosen from each of the three panels of members. The chairman is chosen from the panel of chairmen, and must be legally qualified and of seven years' experience.[53] The other members are **4.76**

---

[50] Employment Tribunals (Constitution and Rules of Procedure) Regulations 2004 SI 2004/1861 Sch 1 r 10(2).
[51] ibid Sch 1 r 26.
[52] ibid Sch 1 r 16.
[53] ibid reg 8(3)(a).

selected from the panels chosen by the Secretary of State. This is following consultation with employees' representative organizations[54] and employers' representative organizations,[55] respectively. If all parties consent, the tribunal may sit with only two members. One of the two must be the chairman.[56] If the parties give their written consent, the tribunal may consist of the chairman alone. If he wishes, he may himself require that he is joined by other members in certain circumstances.

4.77 Where a tribunal is composed of three persons, any order or judgment may be made or issued by a majority. If a tribunal is composed of only two persons, the chairman has a second or casting vote.[57]

4.78 Unless the parties agree to shorter notice, the secretary to the employment tribunal will send notice of any hearing to the parties. This must be done not less than 14 days before the date fixed for the hearing.[58]

4.79 A person may appear before a tribunal in person. He may also be represented by counsel or a solicitor, a representative of a trade union or an employers' association, or by any other person he chooses.[59] If a party fails to attend, or is not represented, at the time and place fixed for the hearing, the tribunal may dismiss or dispose of the proceedings. This may be done in the absence of that party, or a decision may be made to adjourn the hearing to a later date.[60] Importantly, the tribunal must first consider any information in its possession which has been made available to it by the parties.[61]

4.80 At the hearing of the appeal, each party is entitled to give evidence, call witnesses, question witnesses, and address the tribunal.[62] Witnesses who attend to give their evidence do so on oath or affirmation. The tribunal has the power to exclude from the hearing any person who is to appear as a witness in the proceedings, until such time as they give evidence. It may do this if it considers that it is in the interests of

---

[54] Employment Tribunals (Constitution and Rules of Procedure) Regulations 2004 SI 2004/1861 reg 8(3)(b).
[55] ibid reg 8(3)(c).
[56] ibid reg 9.
[57] ibid Sch 1 r 28.
[58] ibid Sch 1 r 14(4).
[59] Employment Tribunals Act 1996 s 6(1).
[60] Employment Tribunals (Constitution and Rules of Procedure) Regulations 2004 SI 2004/1861 Sch 1 r 27(5).
[61] ibid Sch 1 r 27(6).
[62] ibid Sch 1 r 27(3).

B. Appealing Improvement and Prohibition Notices

justice to do so.[63] In practice, this is rarely done and, unlike in criminal courts, all witnesses remain to hear each others' evidence.

4.81 The President of the Employment Tribunal, the Vice President, or a Regional Chairman, may appoint persons with special knowledge or experience. This knowledge will amount to expertise of the subject matter of an improvement or prohibition notice appeal, and competence to sit with the tribunal as assessor.[64] There is, however, no recorded occasion when this has occurred.

4.82 Notwithstanding that it is not his appeal, it is common practice for the respondent to present his case first. This is for the reasons which were identified by Roch J in *Readmans and Another v Leeds City Council* [1992] COD 419.

Preliminary arguments

4.83 In *Chrysler (UK) Ltd v McCarthy* [1978] ICR 939, it was held that it is inappropriate for the tribunal to decide a preliminary point without first investigating the facts of the matter.[65] The appellant had argued, as a preliminary point, that the notice was invalid because it was too imprecise and vague. The court held that this argument was inappropriate, in the light of the fact that the purpose of the tribunal was to give effect to the requirements of safety. Further, the court observed that, in any event, it had far-reaching powers to amend or modify the notice if it chose. Even though, in that case, Forbes J doubted that the tribunal had the power to decide a preliminary point, in *Railtrack plc v Smallwood* [2001] ICR 714, and Sullivan J stated that:[66]

> ... in my view the tribunal does have power to determine a preliminary issue, if it considers that such a course is desirable for regulating its own procedure. But the power must be used very sparingly ... an appeal under section 24 of the 1974 Act is concerned with issues of public safety, the tribunal is there to find the facts, and has very wide powers to modify the notice.

The burden of proof

4.84 The judgment of Roch J in *Readmans and Another v Leeds City Council* [1992] COD 419 is, again, instructive here; it contains useful guidance on the issue of the burden and standard of proof. In that case, the first appellant ran a large cash and carry store which dealt in clothing. The second appellant manufactured the shopping trolleys which were used in the store, and which had seats for babies and toddlers.

---

[63] Employment Tribunals (Constitution and Rules of Procedure) Regulations 2004 SI 2004/1861 Sch 1 r 27(4).
[64] HSWA 1974 s 24(4) and Sch 4 r 8.
[65] See also *Kitching v Gateway Foodmarkets Ltd* (24 October 1988, CO/766/86). See below, at para 4.104, for a synopsis of the facts in *Kitching*.
[66] [2001] ICR 714 at [56].

An accident occurred involving an 11-month-old child, caused when the trolley tipped forward, resulting in distressing injuries. A prohibition notice was served and the employment tribunal upheld the notice.

4.85 The appellant appealed to the High Court, arguing that the tribunal had placed the burden of proof wrongly on them to show that the trolleys were not dangerous. Roch J said that, in the majority of cases, the burden of proof would not be determinative of the issues which the tribunal had to decide. Nevertheless, he went on to say that a correct approach to decision-making was important. The right starting-point was found to be that the issue of a prohibition notice was a substantial interference in the freedom of an individual to carry on his business as he thought fit. There could also be serious financial implications. He concluded that it would be inelegant, if not absurd, that the burden of proof in appeals against a prohibition notice should be any different from that in criminal proceedings. Therefore, it was for the inspector to show, on the balance of probabilities, that there was a risk to health and safety. If that could be shown, it was for the duty-holder to prove that he had done all that was reasonably practicable.

4.86 Roch J observed that an appeal to a tribunal was not an appeal in the true sense. The 'entrepreneur' had played no part in the process leading up to the issue of the notice. Since the inspector had issued the notice, it was for him to satisfy the tribunal that the opinion which gave rise to it was justifiable. In the light of those matters, it was not surprising that cases at the tribunal were usually commenced by the respondent calling its evidence first. Roch J did, however, go further. He acknowledged that, where it was accepted that there was a prima facie breach of the legislation, the only issue might be whether what was done was reasonably practicable. It might then be appropriate for the appellant to call his evidence first.

4.87 Improvement and prohibition notices may not include an expression by the inspector of an opinion that the affected person has contravened one of the duties, subject to the reasonably practicable qualification. For example, in the former case, it was considered a contravention of an absolute duty; in the latter, no contravention was found, merely a risk of serious personal injury. Nonetheless, the practice is that the respondent inspector calls his evidence first. Roch J's observation provides a justification for this course. It lays down that, to the effect that the inspector had issued the notice, it was for him to satisfy the tribunal that the opinion which gave rise to it was a justifiable one.

Challenging the inspector's opinion

4.88 The conventional approach to deciding appeals has been for the tribunal to pose two questions:

(a) At the date of the notice, did the inspector genuinely hold the view stated in the notice?

B. Appealing Improvement and Prohibition Notices

(b) If so, at the date of the notice, did the activities complained of involve or, as the case may be, would they involve, a contravention of the statutory provisions; or do or will they involve a risk of serious personal injury?

The first of these questions must now be considered in the light of the judgment in *Railtrack v Smallwood* [2001] ICR 714.

Did the inspector genuinely hold the opinion?

The tribunal considers whether the inspector genuinely held the view that there was either a contravention of statutory procedures and/or activities were being carried out which might involve a risk of serious personal injury. Whether that view was based on reasonable grounds is highly relevant. **4.89**

In the case of both HSWA 1974 s 21 and s 22, the inspector must show that the activities in question were ones to which the statutory provisions applied. Some care must be exercised here. It is only in the former case that the inspector must prove that the relevant legislative provision has been broken. In the latter, he need not. All that must be established is that the activities complained of involve, or will involve, a risk of serious personal injury. **4.90**

The burden of proof is on the inspector, on the balance of probabilities, to satisfy the tribunal that he did indeed hold that view. It must be based on reasonable grounds. If the inspector fails to satisfy the tribunal that he did genuinely hold that view, and that it was based on reasonable grounds, then the notice must be cancelled. He will not otherwise have acted in compliance with the section. This approach is founded upon the unreported Employment Tribunal decision of *Foremans Relocatable Building Systems v Fuller*.[67] This is known as the '*Foremans* test'. **4.91**

In *Railtrack v Smallwood* [2001] ICR 714, Sullivan J expressed a provisional view, in argument. He said that a tribunal hearing an appeal against a notice was not limited to reviewing whether the view was genuinely held, and/or the reasonableness of the inspector's opinions. Rather, it was required to form its own view. In coming to its view, the tribunal must pay due regard to the inspector's expertise. The position remains, though, that even where the inspector has genuinely formed the opinion, expressed on reasonable grounds, the tribunal can form a different opinion based on the same facts. **4.92**

During proceedings in the Administrative Court, on appeal from the tribunal, counsel for both sides had initially agreed with the *Foremans* test. Upon reflection, counsel for the appellant agreed with Sullivan J's formulation. Counsel for the respondent inspector, observing that such a decision would have far-reaching **4.93**

---

[67] ET-3200213/99/S.

implications, preferred not to make any submissions on the issue. In the absence of full argument, but observing that the tribunal had, in any event, fully endorsed the inspector's view, Sullivan J said only that, 'this decision should not be regarded as endorsing the "*Foremans* test"'. This provisional view was founded upon analogy with other statutory appeals against the exercise of discretionary powers,[68] coupled with the wide nature of the appeal envisaged by HSWA 1974 s 24. However, a salient argument against Sullivan J's view should be considered. The wide power to 'modify', when affirming a notice, provides the tribunal with the means to do justice without forming any view on the correctness, as opposed to genuineness, of the inspector's opinion.

4.94 The present position is that it remains unclear whether the issue of notice appeal proceedings under HSWA 1974 s 24, should be limited to a review of the inspector's opinion. It might, alternatively be argued that any such appeal proceedings should be treated as an appeal proper, with a fresh hearing of the matters which gave rise to that opinion, with the power to substitute an alternative view. What parts of the inspector's decision should remain intact, and the extent to which exercisable powers of review should be circumscribed, remains unclear.

Did the activities complained of involve a breach of a statutory provision or a risk of serious personal injury?

4.95 The second question the tribunal should consider is whether the activities complained of involve, or, as the case may be, will involve, a contravention of the statutory provisions; or do or will involve a risk of serious personal injury.

4.96 In the case of the alleged breach of a statutory provision, it will be for the inspector to prove that there was a breach of that provision. If the provision was subject to a qualification of reasonable practicability, then it will then be for the appellant to show that he had done all that was required of him by that provision. As was said by Roch J in *Readmans v Leeds City Council* [1992] COD 419, this approach is consistent with the approach adopted in criminal proceedings. If the appellant could show that he had done all that was required by statute, then the notice would either be cancelled, or affirmed with modifications.

4.97 When considering whether there has been a breach of statutory provision, the tribunal will clearly have regard to the standard of compliance required under the provision in question. Usually, this will be whether the taking of a particular step was reasonably practicable. It is on this issue, rather than the first, that appeals most often turn.

---

[68] *Hughes v The Architects Registration Council of the United Kingdom* (1957) 1 QB 550, in particular, the dicta of Lord Goddard CJ at 557 and 558.

B. Appealing Improvement and Prohibition Notices

Insignificant contravention

However, if the contravention relied upon by the inspector is an absolute or strict requirement, and not subject to any qualification, then no recourse can be had to such an argument. All that can be suggested is that the contravention was so insignificant that the notice should be cancelled. In such circumstances, in *South Surbiton Cooperative Society Ltd v Wilcox* [1975] IRLR 292, a company unsuccessfully attempted to appeal against a notice. The notice had been issued in respect of a cracked wash-hand basin, this being a breach of an absolute duty under the Offices, Shops and Railway Premises Act 1963. **4.98**

Neither the loss of profit,[69] nor the fact that employees' jobs will be endangered,[70] amount to sufficient grounds for the suspension of a prohibition notice. An argument to the effect that the company could not afford to take the precautions required by law is irrelevant to the question of whether a notice should or should not be affirmed. **4.99**

Fundamental flaw and modification

If the tribunal affirms the inspector's opinion (or substitutes its own view, under the formulation promulgated by Sullivan J), it has wide powers thereafter. Section 24 of HSWA, provides that the tribunal may either cancel or affirm the notice. If it affirms the notice, it may do so either in its original form, or with such modifications as it thinks fit in the circumstances of the case. 'Modifications' include additions, omissions, and amendments.[71] Such modification clearly includes the power to extend the time allowed for correcting the contravention in question.[72] When determining the length of the time for compliance with a deferred prohibition notice under s 22(4)(a), the tribunal is entitled to take into account the time taken to perform a particular safety operation.[73] **4.100**

In other respects, the ambit of the tribunal's power to affirm and modify, so as to cure a defect rather than cancel because of such defect, is unclear.[74] **4.101**

In respect of prohibition notices, a tribunal may find that no contravention of a statutory provision has occurred. If the inspector has stated he is of such an opinion, provided that the inspector shows that he had reasonable grounds for his **4.102**

---

[69] *Grovehurst Energy Ltd v Strawson* (COIT 5035/90); *Nico Manufacturing Co Ltd v B Hendry* [1975] IRLR 225.
[70] *TC Harrison (Newcastle-under-Lyme) Ltd v Ramsay* [1976] IRLR 135.
[71] HSWA 1974 s 82(1)(c).
[72] *Campion v Hughes (HM Inspector)* [1975] IRLR 291.
[73] *Ofterburn Mill Ltd v JW Bullman* [1975] IRLR 223.
[74] In *Tesco UK Ltd v The London Borough of Harrow*, the tribunal held at first instance that they would have had the power to modify the notice by including a regulation not cited by the inspector in the original notice.

opinion, that there was a risk of serious personal injury, the prohibition notice will still be valid.

4.103 In *Chrysler United Kingdom Limited v McCarthy* [1978] ICR 939, Eveleigh J considered the provisions concerned with enforcement notice appeals, and the power of the tribunal in HSWA s 24. He considered how the tribunal might rehabilitate a defective enforcement notice: 'The whole scheme, in my view, envisages a review of the situation, if necessary, by the Industrial Tribunal, who, if the notice is not sufficiently precise, will be in a position to re-draft it, as it were, and make it precise.'[75]

4.104 In *Kitching v Gateway Foodmarkets*,[76] the Divisional Court quashed a decision of an employment tribunal. The tribunal had cancelled a notice as a result of a preliminary challenge. This was on the basis that the notice and schedule had failed to include the statutory qualification to the duty of reasonable practicability. Remitting the matter back to the tribunal, the court directed that it should consider the exercise of its powers under s 24(2) HSWA 1974. These powers are to cancel, or vary, or affirm the notice, having regard to, first, the evidence it has heard and, second, the qualification in the statute 'so far as reasonably practicable'.

4.105 It is likely that the tribunal also has the power to amend an improvement notice by substitution of a finding under HSWA 1974 s 21(b). This section provides that the person 'has contravened in circumstances that make it more likely that the contravention will continue or be repeated'. An opinion may be substituted that he 'is contravening the relevant statutory provisions' under HSWA 1974 s 21(a). Sullivan J, in *Railtrack plc v Smallwood*,[77] accepted that the scope of HSWA 1974 s 24 was broad enough to encompass such a modification when the facts permitted it, though they did not in that particular case.

4.106 However, in *British Airways Board v Henderson* [1979] IC 77 it was held that the tribunal does not have the power to amend an improvement notice. The Divisional Court, in that case, ruled that the power did not extend to permit a tribunal to affirm and modify a notice so as to allow the inspector to rely upon a provision of the Act not specified in the notice.[78]

4.107 In *Post Office v Footitt* [2000] IRLR 243, the Divisional Court upheld the decision of an employment tribunal to affirm the service of an improvement notice requiring the provision of separate cubicles within same-sex changing rooms. The improvement notice was made pursuant to reg 24 of the Workplace

---

[75] At 942.
[76] 24 October 1988, CO/171/88.
[77] [2001] ICR 714.
[78] See also the comments of Sullivan J in *Railtrack*.

B. Appealing Improvement and Prohibition Notices

(Health, Safety & Welfare) Regulations 1992 SI 1992/3004. Regulation 24 provides that suitable and sufficient facilities shall be provided for any person at work in the workplace to change clothing. Such facilities shall not be suitable unless they include separate facilities for, or separate use of facilities by, men and women where necessary 'for reasons of propriety'. Neither the notice nor the schedule of requirements made clear the inspector's contention as to what was required. In Ognall J's judgment, there was no authority to the effect that a tribunal must view the wording of the notice in isolation. Instead, there was scope to view it in its overall context of the correspondence sent by the inspector.

4.108 A different, and far more restrictive, approach was taken by the Administrative Court in *BT Fleet Ltd v McKenna* [2005] EWCA 387. In that case, a tribunal had found that the schedule attached to an improvement notice identifying the steps required of the appellant to remedy the contravention was defective, in that it was not sufficiently specific. The tribunal had affirmed the notice and modified the schedule. In the Administrative Court, Evans-Lombe J noted how HSWA s 23(2), did not make it compulsory for the notice to contain 'directions as to the measures to be taken to remedy any contravention'. Despite this, he held otherwise, stating: 'If the provisions of the relevant statute provide an option to proscribe how the recipient can comply with the notice, and that option is taken, then the specification of how compliance can be effected form part of the notice and, if confusing, may operate to make it an invalid notice.'

4.109 Evans-Lombe J referred to the judgment of Upjohn J in *Miller Mead v Minister of Housing and Local Government*.[79] In that case, Upjohn J had taken a purposive approach, holding: 'The function of the court is not to introduce strict rules not justified by the words of the section ... the test must be: does the notice tell him fairly what he has done wrong and what he must do to remedy it?'

4.110 In Evans-Lombe J's judgment, depending on the circumstances of the case, the tribunal should have realized that the notice was fatally flawed. It should not have attempted to put matters right by amendment. The notice did not properly enable the recipient to know what was wrong, why it was wrong, and how the giver of the notice intended that what was wrong should be put right.

Example decisions

4.111 The following are examples of the approach taken by tribunals and the Administrative Court, on appeal from such cases. Many of the cases pre-date the series of defining Court of Appeal and House of Lords decisions concerning the

---

[79] [1963] 2 QB 196.

interpretation of the general duties under HSWA 1974. This line began with *R v British Steel* [1995] 1 WLR 1356.

4.112 In *West Bromwich Building Society v Townsend* [1983] ICR 257, an inspector had served a notice requiring the erection of anti-bandit screens to separate employees from the general public. The decision of a tribunal to uphold a notice, issued under HSWA 1974 s 2(1), was overturned. McNeill J held that the tribunal had erred in considering the question of the general desirability of screens. It ought to have concentrated on the specific offence in question, and to have weighed the degree of risk against the onerousness of the measures required to avert that risk.

4.113 In *Tesco Stores v Edwards* [1977] IRLR 120, Tesco appealed against improvement notices which claimed that the seating provided for its check-out operators contravened s 14 of the Offices, Shops and Railways Premises Act 1963. It was held that, as the Act did not require it, reasonable practicability under HSWA 1974 s 2, was irrelevant to the issue for the tribunal.

4.114 In *Associated Dairies v Hartley* [1979] IRLR 171, an improvement notice was issued following an accident. The notice required the company to provide protective footwear free of charge to all employees involved in operating hydraulic trolley jacks. It was held that the cost of providing protective footwear free of charge was disproportionate to the risks involved. The test of reasonable practicability was satisfied by the existing arrangements, whereby footwear could be purchased at cost price and by way of instalments.

4.115 In *Belhaven Brewery Co v McLean* [1975] IRLR 370, an appeal against an improvement notice requiring that interlocking guards should be fitted to prevent access to transmission areas and dangerous machinery was rejected. It was held that, without the guards, the dangerous machinery would not be securely fenced within the meaning of the Factories Act 1961 ss 13(1) and 14(1). It was not beyond what was reasonably practicable under HSWA 1974 s 2(1), the court further observed. The argument that a high level of supervision was provided, and that the operatives were of sufficient integrity not to open the gates when the power was on, was rejected.

4.116 In *Deeley v Effer*,[80] it was held by an employment tribunal that an inspector had misinterpreted the statute in question. The category of people he had in mind to protect, namely customers/members of the public, by issuing a notice citing s 2 of the HSWA, were not in fact protected under the particular provision of the HSWA 1974. The provision was concerned with risks to employees.

---

80 COIT 1/72, Case/25354/77.

B. Appealing Improvement and Prohibition Notices

In *Canterbury City Council v Howletts & Port Lymne Estates Ltd* [1997] ICR 925, **4.117** a zoo-keeper employed by the respondents was killed by a Siberian tiger, after he had entered the tiger's enclosure. The Council's environmental health officer served a prohibition notice stating that he was of the opinion that the respondents were in breach of HSWA 1974 s 2(1) and (2)(a), by failing to ensure a safe system of work. He had considered that permitting employees to enter the enclosures of tigers, while the animals roamed freely inside, failed to ensure a safe system. Accordingly, the inspector directed that the activities should cease forthwith. The notice was overturned on appeal to the employment tribunal, and the council appealed against that decision. The court elaborated on the construction and purpose of Pt 1 of the Health and Safety at Work Act 1974. The proper interpretation, it held, has to be that Pt 1 HSWA is concerned with the requirements an employer must meet to see that his employees are safe in the work which it is the employer's business to carry out.

The court was 'quite unable' to accept the primary submission advanced on behalf **4.118** of the appellant, which was to the effect that HSWA 1974 was concerned with the measures which an employer was required to take for the purposes of conducting his business. The court ruled that HSWA 1974 is not seeking to legislate what work could or could not be performed, but is properly concerned with the manner of its doing, ie whether the risk arising from such work is reduced so far as is reasonably practicable.

It was said that the tribunal should consider the nature of the business being carried **4.119** on, in terms of its ethos or idiosyncrasy. It was the work done in pursuit of that business which the Act sought to regulate, not the doing of the business itself. If work, which by ordinary standards could be described as dangerous, was necessary, it was permissible for risk arising from it to persist as long as it was properly controlled.

The tribunal's decision

A tribunal may issue its judgment orally at the end of the hearing. If it wishes, the **4.120** tribunal can also reserve its judgment to be given in writing at a later date. When judgment is reserved, a written judgment is sent to the parties as soon as practicable thereafter. All judgments are be recorded in writing and signed by the chairman.[81]

Rule 30 of Sch 1[82] provides that a tribunal or chairman must give reasons (either **4.121** oral or written) for its judgment. (Any order must also be reasoned if a request for

---

[81] Employment Tribunals (Constitution and Rules of Procedure) Regulations 2004 SI 2004/1861 Sch 1 r 29.
[82] ibid Sch 1 r 30.

reasons is made before or at the hearing at which the order is made.) Reasons may be given orally at the time of issuing the judgment or order, or they may be reserved to be given in writing at a later date. Written reasons are only provided in relation to judgments if requested within 14 days by one of the parties.[83] Rule 30(6) prescribes various matters which must be included in the written reasons. These include, 'how the relevant findings of fact and applicable law have been applied in order to determine the issues'.

4.122 The determination of a notice appeal by a tribunal is neither subject to the review procedure set out in r 34(1)(b) of Sch 1,[84] nor capable of appeal to the Employment Appeals Tribunal. The only recourse open to a dissatisfied party is the very limited rights afforded by judicial review, or case stated proceedings in the High Court.[85]

Costs

4.123 Schedule 4[86] provides for 'costs and expenses' in improvement and prohibition notice appeals. Rule 10(1)[87] provides that a tribunal or chairman may make an order ('a costs order') that a party ('the paying party') make a payment in respect of the costs incurred by another party ('the receiving party'). Rule 10(4) provides that the tribunal or chairman shall have regard to the paying party's ability to pay, when considering whether it shall make a costs order, or how much that order should be.

4.124 'Costs' is defined so as to mean 'fees, charges, disbursements, expenses or remuneration incurred by or on behalf of a party in relation to the proceedings'.[88] It is arguable that the cost of the time during which an inspector is engaged, while investigating an incident or contravention, does not fall within this definition.

4.125 Rule 10(3)[89] provides that the amount of a costs order against the paying party can be determined in the following ways:

(a) the tribunal may specify the sum which the party must pay to the receiving party, provided that sum does not exceed £10,000;

---

[83] Employment Tribunals (Constitution and Rules of Procedure) Regulations 2004 SI 2004/1861 Sch 1 rr 30(3), 30(5).
[84] ibid Sch 1 rr 34–36.
[85] Tribunals and Inquiries Act 1992 s 11(1).
[86] Employment Tribunals (Constitution and Rules of Procedure) Regulations 2004 SI 2004/1861 Sch 4 r 10.
[87] ibid Sch 4 r 10.
[88] ibid Sch 4 r 10(2).
[89] ibid Sch 4 r 10(3).

B. Appealing Improvement and Prohibition Notices

(b) the parties may agree on a sum to be paid by the paying party to the receiving party and if they do so the costs order shall be for the sum so agreed (which may exceed £10,000);

(c) the tribunal may order the paying party to pay the receiving party the whole or a specified part of the costs of the second party, with the amount to be paid (which may exceed £10,000) being determined by way of detailed assessment in a county court in accordance with the Civil Procedure Rules.

Appeal from the decision of an employment tribunal

4.126 There is no appeal from a decision of an employment tribunal in a notice appeal to the Employment Appeals Tribunal. The only appeal is by operation of s 11 of the Tribunals and Inquiries Act 1992. Section 11 of that statute provides that if any party to an improvement or prohibition notice appeal 'is dissatisfied in point of law with a decision of the tribunal he may . . . either appeal from the tribunal to the High Court or require the tribunal to state and sign a case for the opinion of the High Court'. Thus, the only basis for appealing against a finding on the facts would be that no reasonable tribunal could have properly reached such a decision.

43,685

5

LEGAL PERSONALITY AND SECONDARY LIABILITY

A. Legal Personality and HSWA 1974	5.01
B. Secondary Liability	5.27
Common law	5.27
Section 36: the liability of others who cause the offence	5.30
C. Section 37: The Duties and Liability of Company Directors, Managers, and Officers	5.46
The secondary liability of company directors, managers, and officers	5.46
Company directors' duties	5.73

A. Legal Personality and HSWA 1974

5.01 The Health and Safety at Work etc. Act 1974 (HSWA 1974) s 33(1) defines and creates criminal liability for health and safety offences by stating that 'it is an offence for a person' to do any of the various acts listed under paras (a) to (o) of the subsection. Similarly, an inspector is given the power to serve improvement and prohibition notices, pursuant to HSWA 1974 ss 21 and 22, upon any 'person'.

5.02 'Person' is not defined by HSWA but is defined by the Interpretation Act 1978 Sch 1 to include a 'body of persons corporate or unincorporate'. The use of the word 'person' in HSWA 1974 is clearly intended to have this wide meaning.

Natural persons and corporations

5.03 No difficulty lies in taking enforcement action, either by issuing a notice or instituting proceedings, against any natural person or any incorporated body such as a private limited liability company, a public limited company, or other corporation.

5.04 In addition to private limited and public limited companies there are other types of corporation, including companies limited by guarantee and corporations sole.

Corporations sole are the particular office, the holder of which undertakes, usually, a public function; examples include the holders of certain offices in the Church of England (vicars) and the Office of Commissioner of Police for the Metropolis.

5.05 There are a number of bodies required to be incorporated by statute, eg the various boards of trustees of museums,[1] and commonhold associations. Other bodies, such as housing associations, may be incorporated, unincorporated, or be formed as a trust. In any proceedings, the prosecution must be in a position, if necessary, to prove that the body is a 'person'.

5.06 The liability of such persons, natural or corporate, in respect of health and safety offences arises from their status as 'employer', 'self-employed person', 'employee', 'occupier', 'person in control', or any other of a number of duty-holder situations in respect of which health and safety legislation imposes obligations. Proceedings against such persons will require the prosecution to prove their status as a relevant duty-holder.

5.07 HSWA 1974 s 46 (considered at paragraphs 2.25–2.31), makes provision for the service of notices and documents upon persons and makes specific provision in respect of corporations.

Unincorporated bodies

5.08 Difficulties arise as a result of the absence of a definition of 'body of persons unincorporate' in HSWA, and the absence of any provision in respect of the conduct of proceedings against such a body. Unlike HSWA, recent legislation makes detailed provision in respect of the institution of proceedings, procedures and liability for fines,[2] but the decided common law provides little assistance.

5.09 Sporting clubs and associations most easily fit the description of a body of persons unincorporated, but when an inspector issues proceedings, he will need to obtain evidence capable of establishing the status of the body. The constitution of the club or association is likely to provide both the best evidence and the best indication of the status of the body. The institution of proceedings against such a body will require documents to be served upon it: once again no specific provision is made by HSWA 1974 s 46 and the best course for an inspector to take is to refer to the body's constitution or rulebook to establish whether any person, such as a secretary, treasurer, or even a committee, is identified as a person with power to transact business on behalf of the body, and upon whom such documentation can be served.

---

[1] See the National Heritage Act 1983.
[2] eg Anti-terrorism, Crime and Security Act 2001 s 69.

By contrast with HSWA 1974 s 734 Companies Act 1985 makes provision for proceeding against an unincorporated body for various offences under that Act; this provision is replicated in various other pieces of legislation. It provides: the proceedings shall be brought in the name of the unincorporated body (and not in that of any of its members); and that for the purposes of any such proceedings, any rules of court relating to the service of documents should apply as if that body were a corporation; that a fine imposed on an unincorporated body on its conviction of such an offence shall be paid out of the funds of that body and that in a case in which an unincorporated body is charged in England and Wales with such an offence s 33 of the Criminal Justice Act 1925 and Sch 3 to the Magistrates' Courts Act 1980 (procedure on charge of an offence against a corporation) shall have effect in like manner as in the case of a corporation so charged. **5.10**

Partnerships: England and Wales

A partnership does not enjoy a separate legal personality from its partners under English common law. While a number of more recent Acts[3] have created criminal liability for partnerships themselves, allowing proceedings to be instituted against the name of the partnership, and notwithstanding the significant number of professional partnerships engaged in activities which are subject to health and safety regulations, HSWA 1974 makes no such provision. **5.11**

Partners are jointly and severally liable for each other's acts: the business is carried on by each partner acting as an agent for all the partners. Therefore, proceedings for an offence under HSWA 1974 are instituted against the partners of the partnership in respect of offences committed by the partnership. **5.12**

Having said this, it sometimes happens that when a partnership accepts its guilt in respect of a particular charge, for reasons of expediency proceedings are brought against one partner only, on the understanding that he represents the whole of the partnership. **5.13**

Care needs to be taken when considering the basis upon which enforcement action is to be taken: where an offence has been committed by the partnership, for example, in its capacity as 'employer', then proceedings against all the partners may be appropriate. However, where a partnership undertakes professional work with each of the partners bearing responsibility for particular tasks, for example, with an individual partner being appointed as a planning supervisor under the Construction (Design and Management) Regulations 1994 SI 1994/3140, then it may be more appropriate to proceed against the partner with responsibility for **5.14**

---

[3] See Companies Act 1985 s 734(5); Companies Act 1989 ss 83, 85, 90; Financial Services and Markets Act 2000 s 177; Anti-terrorism, Crime and Security Act 2001 s 69.

that particular duty alone. Particular care needs to be taken when considering an offence involving *mens rea* (a guilty mind: intent or recklessness).[4]

Partnerships: Scotland

5.15 Scottish partnerships are a collection of persons recognized as a separate legal entity under Scottish law. In *Douglas v Phoenix Motors* [1970] SLT 57 it was held that for the purposes of the Trade Descriptions Act 1968, which like HSWA 1974 creates criminal liability for certain acts committed by any 'person' and which in s 20(1) creates identical secondary liability to that contained in HSWA 1974 s 37, a Scots partnership, although not a corporation in the true sense, was nonetheless a body corporate and, as such, the partnership could be proceeded against for an offence, as could a partner if he 'consented' or 'connived' etc at the partnership's offence.

Limited liability partnerships

5.16 Limited liability partnerships (LLP) were an entity created by the Limited Liability Partnership Act 2000. The key characteristic of an LLP is that it is a body corporate with its own legal personality separate from that of its members. Proceedings against an LLP can be instituted in the same way as with any other corporate body, with service of documentation on the LLP at the registered office, and the members of an LLP will be subject to potential personal liability for the criminal offences of the LLP under HSWA 1974 s 37(2).

Trusts

5.17 Trusts, under common law, are not legal persons even though for many fiscal purposes they are regarded as separate entities. Each trustee is, in general, jointly and severally liable for a breach of trust.[5] The individual trustees of any trust are likely to be the proper 'persons' to be proceeded against in relation to any breach of health and safety legislation arising on or from the management of trust property and are likely to be the 'employer' of staff engaged to act in respect of any property which is the subject of a trust.[6] The trust deed or documentation should identify the powers and responsibilities of the trustees.

5.18 Charitable trusts may take a number of different forms; some may have legal personality and some may not. They may be 'bare' trusts, unincorporated associations

---

[4] See *Parsons v Barnes* [1973] Crim LR 537; *Clode v Barnes* [1974] 1 All ER 1176 (DC), [1974] Crim LR 268; *R v JF Alford Transport* (1997) 141 Sol Jo LB 73 (CA).

[5] In a civil action for breach of trust it has been held that all trustees are a necessary party to an action: *Re Jordan* [1904] 1 Ch 260.

[6] The well-known and significant decision of *R v Board of Trustees of Science Museum* [1993] 3 All ER 853 (CA) involved a board of trustees that, by operation of statute (National Heritage Act 1983 Sch 1), was a corporate body.

A. Legal Personality and HSWA 1974

with a charitable purpose, incorporated trusts, a friendly society, or an industrial and provident society.

5.19 A bare trust is one where the trustees hold the legal title to the trust property, and are under a duty to carry out the charitable objects of the trust. It is likely that this type of trust has no legal personality; rather it is merely a collection of obligations surrounding the subject matter of the trust which are enforceable in equity by the Attorney-General.

5.20 In *North v Commissioner of Inland Revenue* (1985) 7 NZTC 5192 Holland J, giving the judgment of the New Zealand High Court in response to an argument that a trust was a 'person' for the purposes of the New Zealand tax laws, considered s 4 of the Act Interpretation Act 1924, which reads: '"Person" includes a corporation sole, and also a body of persons, whether corporate or unincorporate.' He went on to hold that:

> . . . the trust is not a body of persons either corporate or unincorporate. A trust is a legal concept from which persons may benefit and have duties but it is not a legal entity.

5.21 An unincorporated association was defined in the case of *Re: Koeppler's Will Trusts* [1985] 2 All ER 869 as 'an association of persons bound together by identifiable rules and having an identifiable membership'.

5.22 An unincorporated association is formed by the agreement between the members. This is usually in the form of a written constitution or set of rules, although the fact the agreement is oral will not prevent there being an unincorporated association. If a constitution has not been formally adopted, a set of rules which has been used consistently can be adopted as the constitution by long usage.[7]

5.23 The question arises: is an unincorporated association a body unincorporate for the purposes of the Interpretation Act? While there is no direct authority on the point, it is likely that it is.

5.24 Charitable trusts have increasingly adopted the legal structure of a company limited by guarantee. In this way, the personal liability of the trustees may be limited. As they are both trusts and companies, they are governed not only by trust and charity law, but also by company law. A trust of this sort is likely to have legal personality in its own right. When the relevant sections of the Charities Act 2006 come into force, a new structure called a Charitable Incorporated Organisation (CIO) will be created. While this is likely to be a structure well suited to the particular demands of charities, the precise implications for the health and safety liabilities of a charity incorporated in this way remain to be seen.

---

[7] The Charity Commissioners have provided a Model Constitution for a Charitable Unincorporated Association, as has the Charity Law Association.

5.25 A charitable mutual assurance association, ie one which restricts its members to those who are poor, may adopt the legal structure of a friendly society. Formation is by registration with the 'central office' in accordance with s 5 of the Friendly Societies Act 1992, after submission by at least seven members of a memorandum and rules which comply with that Act. A society registered under the Act will have corporate status and therefore legal personality.

5.26 A charity may adopt the structure of an Industrial and Provident Society if it is established for the purpose of carrying on an industry, or trade business intended to be conducted for the benefit of the community. This form is often adopted by charitable housing associations. It is formed by registration with the Financial Services Authority Mutual Societies Section after a minimum of seven members have submitted a set of rules containing the matters specified in the Industrial and Provident Societies Act 1965 s 1. If so registered, the society has corporate status and thus legal personality.

B. Secondary Liability

Common law

5.27 Under the common law a secondary party to a criminal offence is liable to be tried, convicted, and sentenced as a principal offender. Common law secondary liability has been held to be of general application to all offences, unless expressly or impliedly excluded by statute.[8] Thus, at least technically, there is no reason why a person may not in appropriate circumstances be charged with aiding and abetting another in the commission of a health and safety offence. In practice, however, the secondary liability established under HSWA is usually relied on.

5.28 Section 8 of the Accessories and Abettors Act 1861 provides that:

> Whosoever shall aid, abet, counsel or procure the commission of any indictable offence, whether the same be an offence at common law or by virtue of any Act passed or to be passed, shall be liable to be tried, indicted and punished as a principal offender.

5.29 In relation to proceedings in magistrates' courts, s 44 of the Magistrates' Court Act 1981 provides:

> (1) A person who aids, abets, counsels or procures the commission by another person of a summary offence shall be guilty of the like offence and may be tried (whether or not he is charged as a principal) either by a court having jurisdiction to try that other person or by a court having by virtue of his own offence jurisdiction to try him.

---

[8] *R v Jefferson* 99 Cr App R 13 (CA).

B. Secondary Liability

(2) Any offence consisting in aiding, abetting, counselling or procuring the commission of an offence triable either way... shall by virtue of this subsection be triable either way.

Section 36: the liability of others who cause the offence

5.30 HSWA 1974 s 36 creates a statutory form of secondary liability for those whose act or default causes another to commit a health and safety offence. Subsection (1) provides:

> Where the commission by any person of an offence under any of the relevant statutory provisions is due to the act or default of some other person, that other person shall be guilty of the offence, and a person may be charged with and convicted of the offence by virtue of this subsection whether or not proceedings are taken against the first-mentioned person.

5.31 Various other Acts contain a similar provision in respect of secondary liability; most notably s 23 of the Trade Descriptions Act 1968 and the now repealed s 83 of the Food and Drugs Act 1938.[9] In both those statutes the provision is more directly concerned with the situation that arises with a chain of onward supply of either misleadingly described or defective articles by manufacturers and vendors. However, identically termed provisions also appear in various environmental protections statutes, which create strict forms of liability for pollution offences in a similar way to health and safety offences.[10] The nature and scope of the liability created by HSWA 1974 s 36 is not without difficulty.

5.32 In the vast majority of situations, potential defendants, whose 'act or default' may make them liable under HSWA 1974 s 36, will themselves owe duties under HSWA 1974 ss 3(1), 3(2), or 7; or they will be liable as company officers under HSWA 1974 s 37. Thus, they are liable to prosecution without reliance upon this section.

5.33 The liability created by HSWA 1974 s 36 extends to an offence created by any of the 'relevant statutory provisions' meaning the offences created by HSWA 1974 and by any health and safety regulations. However, s 36(3) provides that the provisions of this section are subject to what may be specified in health and safety regulations.

The 'commission by any person of an offence'

5.34 While the principal offender need not be prosecuted for liability to arise against the secondary party, HSWA 1974 s 36 does appear to require proof against the secondary party that an offence was committed by the principal offender. The same

---

[9] See also s 5 Employers Liability Compulsory Insurance Act 1969.
[10] eg Control of Pollution Act 1974 s 87(2); Water Resources Act 1991 s 217(2).

has been held to be the position under s 23 of the Trade Descriptions Act 1968, which is in identical terms to HSWA 1974 s 36.[11]

5.35 The Trade Descriptions Act 1968 s 24 creates a statutory defence of due diligence. In *Coupe v Guyett* [1973] 2 All ER 1058 (DC) it was explained by Lord Widgery CJ that where the principal offender had a defence on the merits and without reference to this statutory defence, then the secondary offender, whose act or default gave rise to the matter in complaint, could not be convicted. Conversely, if the principal offender would have been liable to conviction but for Trade Descriptions Act 1968 s 24(1), then the secondary offender could properly be regarded as having 'committed an offence' for the purposes of the prosecution under Trade Descriptions Act 1968 s 23.

5.36 By analogy, under HSWA 1974 it could be argued that if the prosecution was to establish a case of a breach of a health and safety duty against a principal offender, who subsequently established that he had taken all reasonably practicable steps, then this would nonetheless amount to the 'commission by any person of an offence' for the purposes of a prosecution against the secondary offender under HSWA 1974 s 36. However, the Court of Appeal in *R v HTM* [2006] EWCA Crim 1156 has ruled that 'reasonable practicability' in the general duties of HSWA 1974 ss 2 and 3 is not a defence, but rather a qualification on the duty, which militates against such an interpretation.

'Is due to': causation

5.37 The wording of HSWA 1974 s 36 appears to require proof of a direct causal connection between the offence by the principal offender and the act or default alleged against the secondary offender. The identical terms of Trade Descriptions Act 1968 s 23 have been strictly interpreted in this regard.[12]

'The act or default of'

5.38 HSWA 1974 s 36 contains no requirement that an act alleged to have caused the offence must be shown to have been either unlawful or in any sense wrongful, for liability to arise. The principal health and safety offences are committed by the breach of personal duties imposed upon persons either as a result of a particular status—ie employer, employee, self-employed person, person in control of premises etc—or a particular activity being undertaken by that person, eg principal contractor, designer, etc. The secondary party may not be under the same or any duty but his actions may nonetheless cause the principal party to commit an offence. Potentially, a lawful act by a person who was not subject to any health and safety

---

[11] *Tesco Supermarkets Ltd v Nattrass* [1972] AC 153 and *Coupe v Guyett* [1973] 2 All ER 1058 (DC).
[12] *Tarleton Engineering Co Ltd v Nattrass* [1973] All ER 669 (DC).

B. Secondary Liability

duty could found liability under s 36 and lead to that person's conviction for an offence involving breach of that duty or regulation.

5.39 This is the apparent position under the identically worded Trade Descriptions Act 1968 s 23. Under that Act, primary liability for misdescriptions arises only for acts done in the course of a trade or business.[13] The Divisional Court in *Olgeirsson v Kitching* [1986] 1 All ER 746 (DC) upheld the conviction under s 23 of a private individual, not acting in the course of any trade or business, who had falsely represented to a garage that the odometer on his vehicle represented the car's true mileage, the garage salesman having then made the same representation during the onward sale of the car. The Divisional Court queried whether liability would arise if the act or default had been done neither falsely, recklessly, nor carelessly.

5.40 The more recently drafted Consumer Protection Act 1987 s 40 creates similarly worded secondary liability, but only for an act or default done 'in the course of a business' of the secondary party.

5.41 *Lamb v Sunderland and District Creamery Ltd* [1951] 1 All ER 923 was a Divisional Court judgment concerned with s 83(3) of the Food and Drugs Act 1938, which was a similar provision to HSWA 1974 s 36. Milk had been bought by the respondents, who were wholesale milk dealers, under a warranty as to its quality. It was bottled and delivered to a retailer who, without tampering with it in any way, then sold the milk to the appellant inspector, whereupon it was then found to be deficient in fat. Justices dismissed an information against the respondents under s 83(3) of the Food and Drugs Act 1938, on the ground that there was no evidence of actual adulteration of the milk by them and that they had given a reasonable explanation of their failure to test the milk before selling it. The Court ruled that as the principal offence of selling food not of the quality demanded by the purchaser under the Food and Drugs Act 1938 s 3 did not require proof of *mens rea* or negligence, then the mere selling by the manufacturer of the milk when it was deficient in fat constituted an 'act or default' within s 83(3) and, therefore, the respondents should have been convicted.

5.42 However, in *Noss Farm Products Ltd v Lilico* [1945] 2 All ER 609, another Divisional Court decision concerned with s 83(3) of the Food and Drugs Act 1938, a manufacturer of food had sold an article of food with a label attached to it and which complied, at the time of the sale, with the then existing regulations. By the time the retailer sold the item on to a customer, the regulations had changed and the label represented an infringement. The Divisional Court held that the earlier sale of the manufacturer did not infringe the Food and Drugs Act 1938, and he was, therefore, not liable to prosecution under s 83(3) of that Act, if, due to

---

13 See Trade Descriptions Act 1968 ss 1, 9, 12, 13, and 14 and *John v Matthews* [1970] 2 QB 443.

a subsequent alteration of law, the sale of such article with the label attached becomes unlawful at the time when the retailer sells it. The Court held that the words 'due to his act or default' in s 83(3) had to be construed as meaning 'due to his wrongful act or default' and, therefore, a manufacturer cannot be made responsible under the section for a sale which takes place after he has parted with the article of food, on the basis that it would be wrong for him, after the alteration of the law, to sell an article which he then is not proposing to sell and does not sell.

5.43 In *Tesco Ltd v Nattrass*,[14] in relation to the Trade Descriptions Act 1968 ss 23 and 24, which are in similar terms to HSWA 1974 s 36, Lord Diplock said:

> In the expression 'act or default' . . . the word 'act' is wide enough to include any physical act of the other person which is causative of the offence. But the use of the word 'default' instead of the neutral expression 'omission' connotes a failure to act which constitutes a breach of a legal duty to act. A legal duty to act may arise independently of any contract or it may be a duty owed to another person arising out of a contract with him.[15]

5.44 Thus, as the meaning of default is a failure to act in accordance with a duty,[16] it is difficult to conceive of a 'default' within the terms of HSWA 1974 s 36 that would not involve a substantive breach of a health and safety duty or regulation sufficient to found primary liability.

HSWA 1974 s 36(2): secondary liability and acts of the Crown

5.45 HSWA 1974 s 36(2) provides that where there would be or have been the commission of an offence under HSWA 1974 s 33 by the Crown, but for the circumstance that s 33 does not bind the Crown, and that fact (ie the offence) is due to the act or default of a person other than the Crown, that person shall be guilty of the offence which, but for that circumstance, the Crown would be committing or would have committed, and may be charged with and convicted of that offence accordingly.

C. Section 37: The Duties and Liability of Company Directors, Managers, and Officers

The secondary liability of company directors, managers, and officers

5.46 Section 37(1) of HSWA 1974 provides:

> Where an offence under any of the relevant statutory provisions committed by a body corporate is proved to have been committed with the consent or connivance of,

---

[14] [1972] AC 153, [1971] 2 All ER 127.
[15] ibid at 196G.
[16] See also *Neath RDC v Williams* [1951] 1 KB 115; but see the very wide meaning given to 'duty' in *Express Ltd v The Environment Agency* [2004] EWHC (Admin) [2005] JPL 242.

C. Section 37: Duties and Liability of Company Directors, Managers, Officers

or to have been attributable to any neglect on the part of, any director, manager, secretary or other similar officer of the body corporate or a person who was purporting to act in any such capacity, he as well as the body corporate shall be guilty of that offence and shall be liable to be proceeded against and punished accordingly.

5.47 Many regulatory and criminal Acts contain a similar provision to HSWA 1974 s 37.[17] The Trade Descriptions Act 1968 has produced a number of decided cases upon the equivalent section that are of particular relevance to the interpretation of HSWA 1974 s 37.

Persons liable under HSWA 1974 s 37

5.48 In addition to the directors, secretary, or other officers of a corporate body the section extends liability to a person 'who was purporting to act in any such capacity'. HSWA 1974 s 37(2) provides that where the affairs of a body corporate are managed by its members, then s 37(1) shall apply in relation 'to the acts and defaults of a member in connection with his functions of management as if he were a director of the body corporate'.

5.49 In *R v Boal* (1992) 95 Cr App R 272, a Court of Appeal case concerning the identical provision in the Fire Precautions Act 1971, Simon Brown LJ, having reviewed earlier authorities concerned with similar statutory provisions, held that the intention of the section was to fix the criminal liability only on those who were in a position of real authority: the decision-makers within the company who had both the power and responsibility to decide corporate policy and strategy. Its purpose was to catch those responsible for putting proper procedures in place; 'it was not meant to strike at underlings'.

5.50 The Court cited with approval the judgment of Lord Denning in *Registrar of Restrictive Trading Agreements v WH Smith & Son Ltd* [1969] 3 All ER 1065 and in particular a passage at p 1069:

> The word 'manager' means a person who is managing the affairs of the company as a whole. The word 'officer' has a similar connotation . . . the only relevant 'officer' here is an officer who is a 'manager'. In this context it means a person who is managing in a governing role the affairs of the company itself.

5.51 In *Woodhouse v Walsall MBC* [1994] 1 BCLC 435 the Divisional Court was concerned with the identical provision to HSWA 1974 s 37 contained within the Control of Pollution Act 1974 s 87. The Court confirmed that whether a person came within the ambit of someone who is managing the affairs of the company was a question of fact; but stressed the importance of applying the full words of

---

[17] Trade Descriptions Act 1968 s 20; Banking Act 1987 s 96(1); Control of Pollution Act 1974 s 87(1); Environmental Protection Act 1990 s 157(1); Theft Act 1968 s 18; Companies Act 1985 s 733; Road Traffic Act 1988 s 172(5).

Simon Brown LJ in *R v Boal* (1992) 95 Cr App R 272, which were the proper test. That test was whether such a person was in a position of 'real authority', and that phrase meant: 'The decision makers within the company who have both the power and the responsibility to decide corporate policy and strategy.'

Basis of liability

5.52 Whilst liability arises from proof of any one of three different bases, namely consent, connivance, or being attributable to neglect, they are not all mutually exclusive alternatives. For example a director may consent to or connive at the commission of an offence, while it may also be attributable to his neglect.

5.53 An allegation containing all three bases, in the alternative, is neither duplicitous nor improper although where the prosecution puts its case on one basis only then the allegation should reflect this: *R v Leighton and Town and Country Refuse Collections* (CA 26 May 1995). Furthermore, in the same way as between 'knowledge' and 'belief' in a charge of handling stolen goods, it is permissible for a conviction under HSWA 1974 s 37 to be on the basis that a defendant was of either state of mind.

Proof of the company's offence

5.54 The prosecution bear the burden of proving all the elements of HSWA 1974 s 37 against a person. In this way s 37 requires proof, against the person charged, that the company committed the offence. While in most situations the company will be prosecuted at a joint trial with the director, the section requires neither the prosecution nor the conviction of the company. Under the common law an acquittal of an alleged principal at an earlier trial is no bar to the subsequent conviction of an accessory and is even inadmissible at the trial of the accessory since, being merely evidence of the opinion of the first jury, it is irrelevant.[18]

5.55 Where the company has been convicted of the offence, s 74 of the Police and Criminal Evidence Act 1984 provides a convenient procedure for proving the conviction.

Consent

5.56 In *Att-Gen's Reference (No 1 of 1995)* [1996] 2 Cr App R 320 the Court of Appeal was asked to rule upon what state of mind was required to be proved against a director to show 'consent', pursuant to the Banking Act 1987 s 96(1), which is in identical terms to HSWA 1974 s 37, in relation to a strict liability offence committed by a company contrary to the Banking Act 1987 s 3.

---

[18] *Hui Chi-Ming v R* [1992] 1 AC 34 (PC).

C. Section 37: The Duties and Liability of Company Directors, Managers

5.57 The Court concluded that a director must be proved to have known the material facts which constituted the offence by the company and to have agreed to its conduct of its business on the basis of those facts. The fact that a director may be ignorant that the conduct of the business in that way will involve a breach of the law can be no defence.

5.58 Thus it appears that in relation to the commission of an offence by a company involving a breach of HSWA 1974 s 3 this will require proof that the director was aware that the possibility of danger existed in relation to the company's undertaking, that he was aware that there was a reasonably practicable step that the company could have taken to reduce the risk from the danger and that he agreed to the company carrying on its undertaking in that way.

5.59 The provisions of HSWA 1974 s 40 (reversal of burden) do not apply to HSWA 1974 s 37.[19]

Connivance

5.60 In *Huckerby v Elliot* [1970] 1 All ER 189 (DC) at 194, Ashworth J commented in passing that the formulation of connivance as a state of mind in which a director is 'well aware of what is going on but his agreement is tacit, not actively encouraging what happens but letting it continue and saying nothing about it', was one with which he did not disagree.

5.61 Connivance is a word considered to be in ordinary usage and thus, perhaps, not requiring a detailed definition for a jury. The Law Commission Report,[20] which proposed what eventually became the Theft Act 1968 s 18, described the liability created by the clause in the following terms:

> [Section 18] makes directors and other officers of a body corporate, and persons in similar positions, liable for an offence . . . committed by the body corporate if the offence is proved to have been committed with their consent or connivance. The clause follows a form of provision commonly included in statutes where an offence is of a kind to be committed by bodies corporate and where it is desired to put the management under a positive obligation to prevent irregularities, if aware of them. Passive acquiescence does not, under the general law, make a person liable as a party to the offence, but there are clearly cases (of which we think this is one) where the director's responsibilities for his company require him to intervene to prevent fraud and where consent or connivance amounts to guilt.

5.62 Connivance appears to be a mental state not amounting to actual consent to the commission of the offence in question, involving a failure to take any step to prevent or discourage the commission of that offence. It has been described as 'wilful

---

[19] See *David Janway Davies v Health and Safety Executive* [2002] EWCA Crim 2949.
[20] Law Commission's Eighth Report (Cmnd 2977) at para 104.

blindness' or an intentional shutting of the eyes to something which the person would, in his own interests, prefer to remain unaware.[21]

Attributable to any neglect

5.63 Unlike consent or connivance, the third basis of liability set out in HSWA 1974 s 37 requires proof of a causal connection between the negligence and the commission of the offence. The offence must be proved to have been attributable in some way or to some extent and not necessarily wholly attributable to the director's neglect.[22]

5.64 In *Huckerby v Elliot* [1970] 1 All ER 189 (DC) the Court of Appeal was concerned with a prosecution in relation to failure to hold the requisite gaming licence under the Finance Act 1966, which contained a provision similarly worded to HSWA 1974 s 37. The appellant, a director of the company, was charged with an offence against the Finance Act 1966 s 305(3), in that the offence was attributable to her neglect.

5.65 Lord Parker CJ stated that a director of a company is not under a general duty to exercise some degree of control over the company's affairs, nor to acquaint himself with all the details of the running of the company. Nor is a director under a duty to supervise the running of the company or his co-directors. Lord Parker CJ quoted the judgment of Romer J in *Re City Equitable Fire Insurance Co Ltd* [1925] Ch 497 at 428–430, where it was held that amongst other things, it is perfectly proper for a director to leave matters to another director or to an official of the company, and that he is under no obligation to test the accuracy of anything that he is told by such a person, or even to make certain that he is complying with the law.

5.66 The decision demonstrates that 'neglect' requires proof of more than a mere failure to see that the law is observed and requires the identification of a duty resting upon an individual to do a specific act and failure so to do.[23]

5.67 In the Scottish Court of Judiciary decision in *Wotherspoon v HM Advocate* 1978 JC 74, the managing director of a company had been convicted together with the company of health and safety offences relating to machinery guarding. The Court held that, in considering in a given case whether there has been neglect within the meaning of HSWA 1974 s 37(1) on the part of a particular director or other

---

[21] See the construction given to the word in connection with the old matrimonial offences: *Gipps v Gipps and Hume* (1861) 11 HL Cas 1 at 14; *Manning v Manning* [1950] 1 All ER 602 (CA).
[22] See *Wotherspoon v HM Advocate* 1978 JC 74.
[23] However, see *Hirschler v Birch* [1987] RTR 13, where *Huckerby v Elliot* [1970] 1 All ER 189 DC and *Re City Equitable Fire Insurance Co Ltd* [1925] Ch 407 (CA) were distinguished and it was held that, 'in every case where delegation was pleaded or relied on, it was necessary to look at the facts of the particular case' (per MacPherson J at p 21F). In that case, there appears to have been no argument or analysis of what duty it was that the director had neglected.

C. Section 37: The Duties and Liability of Company Directors, Managers

particular officer charged, the search must be to discover whether the accused has failed to take some steps to prevent the commission of an offence by the corporation to which he belongs if the taking of those steps either expressly falls or should be held to fall within the scope of the functions of the office which he holds.

However, the court added: **5.68**

> In all cases accordingly the functions of the office of a person charged with a contravention of section 37(1) will be a highly relevant consideration for any Judge or jury and the question whether there was on his part, as the holder of his particular office, a failure to take a step which he could and should have taken will fall to be answered in light of the whole circumstances of the case including his state of knowledge of the need for action, or the existence of a state of fact requiring action to be taken of which he ought to have been aware.

In *J Armour v J Skeen (Procurator Fiscal, Glasgow)* [1977] IRLR 310, the High Court of Justiciary was concerned with an appeal against conviction by the Director of Roads[24] for Strathclyde Regional Council against his conviction pursuant to HSWA s 37 in relation to the corporation's breach of the Factory Act s 127(6). Rejecting the appellant's arguments, the Court held: **5.69**

> This argument proceeds on the basis that the neglect on the part of the person in one of the categories mentioned in s 37(1) must be a neglect in relation to a duty which the legislation has placed on that person. In my opinion that is a misconception. Section 37(1) refers to any neglect, and that seems to me to relate to any neglect in duty, however constituted, to which the contravention of the safety provisions was attributable.

In *R v P and another*,[25] the Court of Appeal was concerned with a preliminary ruling of a judge upon the meaning of neglect in s 37 HSWA in a trial where a director of a company was charged as co-defendant to the company, which faced allegations of having breached ss 2 and 3 of the Act. At a preparatory hearing, the trial judge had ruled that, where neglect was relied upon against the director, the prosecution had to prove (i) that the company had committed a breach of ss 2 and/or 3 of the Act; (ii) that the defendant had had a duty to inform himself of the facts giving rise to the breach and had known of the material facts; (iii) that the defendant had been under a duty to act in relation to those facts; (iv) that the defendant had been neglectful of his duty because he had either known or should have known, but had shut his eyes to, the facts, or because reasonably practicable steps which he could have taken had not been taken; and (v) that his neglect of duty had caused or contributed to the company breaching its duty under ss 2 and/or 3. **5.70**

---

[24] The Court in *J Armour v J Skeen (Procurator Fiscal, Glasgow)* [1977] IRLR considered *Tesco Supermarkets Ltd v Nattrass* [1972] AC 153, but had 'no difficulty in holding' that the Director of Roads, 'came within the ambit of the class of persons referred to in s 37(1)'.
[25] [2007] All ER (D) 173 (Jul).

The prosecution appealed, pursuant to s 35 of the Criminal Procedure and Investigations Act 1996, against the judges' ruling in so far as he had ruled that it had to establish that the defendant had actual knowledge of the material facts.

5.71 Latham LJ, giving the judgment of the Court in allowing the appeal, endorsed the approach set out in *Wotherspoon v H M Advocate*[26] and described as wrong comments of McKay J, in his first instance decision in *R v Network Rail* and others ('the Hatfield Rail Disaster'), which had equated neglect in s 37 to 'wilful neglect', requiring actual knowledge of the material facts of a company's breach. Latham LJ held that s 37 did not refer to 'wilful neglect' and neither, where the question was whether the defendant should have been aware of the material facts of a breach, did that involve a determination of whether the defendant had 'turned a blind eye'. To so require, would equate the test of neglect with that to be applied where the allegation was connivance, whereas Parliament had chosen to apply a distinction between the words consent, connivance, and neglect. The question was whether, if there had not been actual knowledge of the relevant state of facts, nevertheless the officer of the company should have, by reason of the surrounding circumstances, been put on enquiry.

5.72 The Court stressed how the extent of any company officer's duty would depend on the evidence in every case. In this way, the functions of the office of a person charged with a contravention of s 37(1) would be a highly relevant consideration for any judge or jury and the question whether there was on his part, as the holder of his particular office, a failure to take a step which he could and should have taken would fail to be answered in the light of the whole circumstances of the case, including his state of knowledge of the need for action, or the existence of a state of a fact requiring action to be taken of which he should to have been aware. The word 'neglect' in its natural meaning presupposed the existence of some obligation or duty on the part of the person charged with neglect. It was clear that s 37 as a whole was concerned primarily to provide a penal sanction against those persons charged with functions of management who could be shown to have been responsible for the commission of a relevant offence by an artificial persona or a body corporate.

Company directors' duties

5.73 The Companies Act 2006 s 172 creates a duty upon company directors, 'to promote the success of the company'. The section, which at the date of publication of this work is not in force, provides:

> (1) A director of a company must act in the way he considers, in good faith, would be most likely to promote the success of the company for the benefit of its members as a whole, and in doing so have regard (amongst other matters) to—
>   (a) the likely consequences of any decision in the long term,

---

[26] 1978 JC74.

C. Section 37: The Duties and Liability of Company Directors, Managers

    (b) the interests of the company's employees,
    (c) the need to foster the company's business relationships with suppliers, customers and others,
    (d) the impact of the company's operations on the community and the environment,
    (e) the desirability of the company maintaining a reputation for high standards of business conduct, and
    (f) the need to act fairly as between members of the company.

5.74 There are no statutory duties placed upon directors of companies in relation to health and safety, nor does a director of a company owe any duty, statutory or otherwise, to employees or others in respect of the company's compliance with any health and safety duties it owes, simply by holding the office of director.

5.75 The current guidance issued by the Health and Safety Commission, entitled *Directors' Duties for Health and Safety* (INDG343) details only the general responsibilities on the board of directors of a company and advises that certain actions are followed. The document itself, under the heading, 'What the guidance does', provides:

> The guidance sets out the roles and responsibilities of the board and its members in respect of health and safety risks arising from the organization's activities. It recommends that every board should appoint one of their number to be a 'health and safety director'. It is also important that directors, in carrying out their responsibilities, set out their expectations of senior managers with health and safety responsibilities and the arrangements for keeping the board informed and advised of all relevant matters concerning performance.

The continuing debate and calls for reform

5.76 The current guidance has been subject to review for some time and fresh guidance has long been promised. More importantly, over recent years there has been a series of reports and recommendations calling for the introduction of specific directors' duties in respect of health and safety.

5.77 Back in June 2000, the Government published its strategy for 'revitalizing' health and safety[27] which included an 'action point' that the Health and Safety Commission was to advise Ministers on how the law needed to be changed in order to place directors' responsibilities with respect to health and safety on a statutory footing.[28]

5.78 However, by 2004, the House of Commons Work and Pensions Committee noted in its report into the work of the Health and Safety Commission and Executive how the Government appeared to have changed its mind on this issue. The Committee

---

[27] Department of the Environment, Transport and the Regions, Revitalising Health and Safety: Strategy Statement, June 2000.
[28] *Health and Safety: Strategy Statement*, June 2000 para 68.

recommended that the Government should, 'reconsider its decision not to legislate on directors' duties and that it should bring forward proposals for pre-legislative scrutiny in the next Parliament'.[29] In response, the Government said that it had 'asked HSC to undertake further evaluation to assess the effectiveness and progress of the current measures in place, legislative and voluntary, and to report its findings and recommendations by December 2005'.[30]

5.79 The House of Commons Home Affairs and Work and Pensions Joint Committee in its first report into the Government's proposals for the Corporate Manslaughter and Homicide Bill, in December 2005, stated:

> We acknowledge that statutory health and safety duties could be introduced outside the Bill, but believe that since they might help clarify directors' duties with regard to corporate manslaughter law the Government should aim to introduce them either in the Bill, alongside the Bill, or as closely as possible afterwards.

5.80 However, even by April 2006, the HSE was reporting to the HSC how, 'There is agreement amongst stakeholders on the value of director leadership, the need for credible guidance, and effective enforcement of the current legislation. There is not agreement on the need for new legislation.'[31]

5.81 The report further stated that, 'The Commission is invited to agree a proposed package of actions for improving director leadership on health and safety.'[32]

5.82 In Autumn 2006, the HSC invited the Institute of Directors to form a steering group and devise proposals for non-statutory guidance to directors by way of revision of *Directors' Duties for Health and Safety (INDG343)*. The Institute of Directors, jointly with the HSC, published a draft guidance document for consultation in May 2007, with a subheading, 'Action and Good Practice for Board Members'. At the date of publication of this work, no final document had been issued following consultation. However, it is apparent that no statutory responsibility or duty upon company directors in respect of health and safety is proposed and none seems likely in the near future.

---

[29] Work and Pensions Committee, Fourth Report of Session 2003–04, The Work of the Health and Safety Commission and Executive, HC 456-I para 60.
[30] ibid, Third Special Report of Session 2003–04, Government Response to the Committee's Fourth Report into the Work of the Health and Safety Commission and Executive, HC 1137, p 5.
[31] Health and Safety Commission Paper HSC/06/44 para 28.
[32] ibid para 29.

6

HEALTH AND SAFETY OFFENCES: BREACH OF THE GENERAL DUTIES

A. Section 33(1)(a): The Offence of Failing to Discharge one of the General Duties	6.01	F. HSWA 1974 s 4: The General Duty of Persons in Control of Premises to Persons other than their Employees	6.147
B. The Duties Under HSWA 1974 ss 2 and 3	6.04	HSWA 1974 s 4(1): non-domestic premises used by non-employees as a place of work	6.148
HSWA 1974 s 2: the employer's duty to employees; the elements of the offence	6.10	HSWA 1974 s 4(2): by whom the duty is owed	6.155
'At work'	6.18	'To take such measures as it is reasonable for a person in his position to take'	6.158
HSWA 1974 s 3: the duty to non-employees; the offence of failing to discharge the duty	6.24	That the relevant premises, plant, or substance is or are safe and without risks to health	6.160
The duty	6.25	'So far as is reasonably practicable'	6.162
Undertaking	6.33	G. HSWA 1974 s 6: General Duties of Manufacturers etc as Regards Articles and Substances for Use at Work	6.165
C. Elements Common to ss 2 and 3 HSWA 1974	6.43	Limitations on liability	6.168
Employment: employer and employee	6.43	HSWA 1974 s 6(1): designers, manufacturers, and suppliers of articles for work	6.173
Failure to ensure safety/exposure to risks to safety: the possibility of danger	6.61	Supply	6.174
D. Reasonable Practicability and the General Duties: Meaning and Development to 1999	6.66	Articles for use at work	6.176
Introduction	6.66	The duty under HSWA 1974 s 6(1)(a)	6.180
The identification doctrine, vicarious liability, delegation, and the isolated acts of junior employees	6.68	The duty under HSWA 1974 s 6(1)(b)	6.183
E. Reasonable Practicability: *HTM Ltd* and the Current Position	6.101	The Duty under HSWA 1974 s 6(c)	6.184
Introduction	6.101	The duty under HSWA 1974 s 6(1)(d)	6.185
The Management of Health and Safety at Work Regulations 1999 reg 21	6.102	Written undertakings	6.188
Foreseeability	6.116	Imports	6.189
The reverse burden in HSWA 1974 s 40	6.135	Fairground equipment	6.190

The duty under HSWA 1974 s 6(2): research	6.193	The duty under HSWA 1974 s 6(4)(c)	6.204
HSWA 1974 s 6(3): installation	6.196	Substances and research: HSWA 1974 s 6(5)	6.206
HSWA 1974 s 6(4): substances	6.198		
The duty under HSWA 1974 s 6(4)(a)	6.200	H. HSWA 1974 s 7: The Duty of Employees at Work	6.209
The duty under HSWA 1974 s 6(4)(b)	6.203	Prosecution for a breach of HSWA 1974 s 7	6.217

A. Section 33(1)(a): The Offence of Failing to Discharge one of the General Duties

6.01 The Health and Safety at Work etc. Act 1974 (HSWA 1974) s 33 created a range of health and safety offences. Section 33(1) is subdivided into fifteen paragraphs, each one creating a separate health and safety offence.

6.02 The first subparagraph, s 31(1)(a), creates the offence of failing to discharge one of the general duties under HSWA 1974, namely those duties set out in ss 2–7 of the Act which are placed variously upon employers, the self-employed, those in control of premises, manufacturers, and employees. These duties form the core of the Act and this offence is often prosecuted where the most serious alleged health and safety breaches have occurred. The offence is triable either in the magistrates' court or, upon indictment, in the Crown Court.

6.03 In the magistrates' court, an offence of failing to discharge one of the duties under ss 2–6 carries an exceptional maximum fine of £20,000; in respect of the failure by an employee to discharge his duty under HSWA 1974 s 7, the magistrates' court maximum penalty is a fine up to Level 5 (currently £5,000); in the Crown Court, the maximum penalty in respect of a failure to discharge any of the duties under ss 2–7 is the imposition of an unlimited fine.

B. The Duties Under HSWA 1974 ss 2 and 3

Introduction

6.04 HSWA 1974 s 2(1) imposes a wide general duty upon employers to ensure the safety of their employees at work. Section 3(1) imposes a wide general duty upon employers, and s 3(2) the same duty on the self-employed, to conduct their undertaking so as to ensure the safety of those not in their employment who may be thereby affected. These duties form the cornerstone of health and safety legislation. An investigation following a workplace injury or death will often result in prosecution of an offence for failing to discharge the duty under either or both of these sections. A breach of either or both of these duties is considered a serious matter.

B. The Duties Under HSWA 1974 ss 2 and 3

HSWA 1974 created an entirely new health and safety regime: this included the creation of the Health and Safety Commission, the Health and Safety Executive, health and safety inspectors, the enforcement powers, and the general duties. The two general duties created by HSWA 1974 ss 2 and 3, while modelled on the common law duties that an employer owed to an employee in respect of his health and safety, have been interpreted as imposing both a higher and a different standard upon the employer. **6.05**

The ss 2 and 3 duties have been held to be part of the means by which effect is given to the preventative or protective aim of the Act. Breach of these duties is not dependent upon proof of harm or damage nor do the duties give rise to civil liability,[1] but only to a more strict criminal liability. It is this 'protective' intent[2] of the Act which, in any criminal prosecution, cannot be defeated by an argument that civil liability would not have arisen upon the same facts. **6.06**

Thus these principal general duties are intended to 'secure the health, safety and welfare of persons at work'[3] and to protect 'persons other than persons at work against risks to health or safety arising out of or in connection with the activities of persons at work'.[4] This purpose is achieved, in part, through enforcement of these duties under HSWA 1974 s 33(1)(a) (the provision which creates a criminal offence of breaching a general duty) and, importantly, through the powers of an inspector to enforce compliance through the service of notices under HSWA 1974 ss 21 and 22. **6.07**

The general duties placed upon an employer by HSWA 1974 ss 2(1) and 3(1) have been described as having the same structure and been held to be 'the same kind of duty—the company as employer is liable when the necessary conditions for liability are fulfilled'.[5] **6.08**

In the light of the similarity between the two duties, the components common to both are considered together later in this chapter.[6] Of the elements common to both HSWA 1974 ss 2 and 3, it is the concept of reasonable practicability whose ambit has proved the most difficult to define, and which has provoked the greatest debate. In a series of cases commencing in the late 1990s and culminating **6.09**

---

[1] HSWA 1974 s 47(1)(a) provides that nothing in the Act shall be construed as conferring a right of action in any civil proceedings in respect of any failure to comply with any duty imposed by ss 2 to 7.

[2] HSWA 1974 s 1(1) provides that the provisions shall have effect with a view to: '(a) securing the health, safety and welfare of persons at work; (b) protecting persons other than persons at work against risks to health or safety arising out of or in connection with the activities of persons at work'; see also Steyn LJ in *R v Board of Trustees of Science Museum* [1993] 3 All ER 853 (CA) at 857–859.

[3] HSWA 1974 s 1(1)(a), above.

[4] ibid s 1(1)(b), above.

[5] *Per* Evans LJ in *R v Gateway Foodmarkets* [1997] 2 Cr App R 40 (CA) at 44G.

[6] See paras 6.43–6.60.

in the Court of Appeal decision in *R v HTM Ltd* [2006] EWCA Crim 1156, the courts have considered the nature of the statutory duties imposed by ss 2 and 3 HSWA 1974. These authorities, and the development of the law, are considered under the headings 'Reasonable practicability and the general duties: meaning and development to 1999', below at paragraphs 6.66–6.100, and 'Reasonable practicability: HTM Ltd and the current position', below at paragraphs 6.101–6.147.

HSWA 1974 s 2: the employer's duty to employees; the elements of the offence

6.10 HSWA 1974 s 2(1) provides that it shall be the duty of every employer to ensure, so far as is reasonably practicable, the health, safety, and welfare at work of all his employees.

6.11 HSWA 1974 s 2(2) details matters which the duty under s 2(1) includes:

(a) the provision and maintenance of plant and systems of work that are, so far as is reasonably practicable, safe and without risks to health;

(b) arrangements for ensuring, so far as is reasonably practicable, safety and absence of risks to health in connection with the use, handling, storage, and transport of articles and substances;

(c) the provision of such information, instruction, training and supervision as is necessary to ensure, so far as is reasonably practicable, the health and safety at work of his employees;

(d) so far as is reasonably practicable as regards any place of work under the employer's control, the maintenance of it in a condition that is safe and without risks to health, and the provision and maintenance of means of access to and egress from it that are safe and without such risks;

(e) the provision and maintenance of a working environment for his employees that is, so far as is reasonably practicable, safe, without risks to health, and adequate as regards facilities and arrangements for their welfare at work.

The duty

6.12 HSWA 1974 s 2(1) sets out the general duty that is owed to all employees. The statute has been interpreted to impose a higher standard than the common law duties owed by an employer to an employee in respect of his health and safety: namely a duty, qualified only by reasonable practicability, to ensure the health, safety, and welfare of his employees at work.[7]

---

[7] *R v British Steel Plc* [1995] 1 WLR 1356 (CA), [1995] ICR 586, [1995] IRLR 310; *R v Associated Octel Co Ltd* [1996] 4 All ER 848 (HL); *R v Gateway Foodmarkets* [1997] 2 Cr App R 40 (CA); and *R v HTM Ltd* [2006] EWCA Crim 1156.

B. The Duties Under HSWA 1974 ss 2 and 3

The particularization in HSWA 1974 s 2(2) is expressly without prejudice to the wide terms of the general duty set out in s 2(1) and, as such, particularizes only part of the duty. **6.13**

Failing to discharge the duty under HSWA 1974 s 2(1): the elements of the offence

A breach of HSWA 1974 s 2 is a continuing offence; ie it can be committed continuously over a period of time but will remain a single offence. Unless truly alternative allegations are laid, or offences are alleged to have occurred at distinctly different times or places, a failure to discharge any number of the particulars in subs (2) could still properly be alleged as a single breach of the duty under s 2(1) with such particulars alleged in support.[8] **6.14**

As has been said above, a number of the elements of the offence of failing to discharge the duty under HSWA 1974 s 2(1) are in common with the offence of failing to discharge the duty under HSWA s 3(1); and are considered in detail later in this chapter. They are: **6.15**

(a) that the defendant was, at the material time, an employer (paragraphs 6.43–6.60); and
(b) there was the possibility of danger (paragraphs 6.61–6.62);
(c) to employees at work (paragraph 6.212);
(d) the defendant will have failed to have ensured the health, safety, and welfare of his employees at work *unless* he can show (paragraph 6.135):
  (i) that all that was reasonably practicable (paragraphs 6.135–6.146) was done to minimize or eliminate the risk to health and safety,
  (ii) by him, or on his behalf, (paragraphs 6.135–6.146).

HSWA 1974 s 2(3)–(7)

6.16

HSWA 1974 s 2(3) places a duty upon every employer to prepare, and as often as may be appropriate revise, a written statement of his general policy with respect to the health and safety at work of his employees and the organization and arrangements for carrying out that policy, and to bring the statement and any revision of it to the notice of all his employees. A written statement is not required if there are no more than five workers[9] present on premises at one time.[10] In reality, employers' general safety policies are often broad and generic documents. The Management of Health and Safety Regulations 1999[11] SI 1999/3242 place employers under a far more stringent duty in respect of the preparation of risk assessments.[12]

---

[8] See *Wilson v Spindle Select Ltd* (28 November 1996, CO-2177-96, DC and The Times 9 December 1996).
[9] Employer's Health and Safety Policy Statement (Exceptions) Regulations 1975 SI 1975/1584.
[10] See *Osborne v Bill Taylor of Huyton Ltd* [1982] ICR 168.
[11] Reproduced at App E.
[12] See paras 12.09–12.46.

6.17 The remaining part of HSWA 1974 s 2 is concerned with the appointment of and consultation with employee safety representatives and safety committees.

'At work'

6.18 HSWA 1974 s 52(1) provides that the terms 'work' and 'at work' shall be construed in whatever context according to the following definition:

'work' means work as an employee or as a self-employed person;

an employee is at work throughout the time when he is in the course of his employment, but not otherwise; and

a self-employed person is at work throughout such time as he devotes to work as a self-employed person.

6.19 This meaning is expressed to be capable of extension or adaptation by provisions of health and safety regulations. By way of example, the Police (Health and Safety) Act 1997 extended the definition to include police officers throughout the time they are on duty, but not otherwise.

6.20 Those engaged on certain training courses or undertaking work experience are 'at work' by operation of the Health and Safety (Training for Employment) Regulations 1990 SI 1990/1380.

6.21 The issue of whether a failure to ensure the health and safety of an employee or employees has occurred at work or otherwise, arises less frequently in criminal prosecutions for a breach of HSWA 1974 s 2 than in civil actions concerned with vicarious liability.

6.22 The fact that harm or the risk of harm arises in circumstances where an employee was not acting in accordance with his employer's instructions, will not generally enable an employer to argue, by way of defence, that the incident in question was not 'in the course of employment'. Rather, the issue in such a case is likely to be whether the employer did all that was reasonably practicable to ensure that his employee followed instructions, normally through supervision and training. In any event, even if the employer is able to show that he did all that was reasonably practicable to ensure that the disobedient employee followed instructions—and thus may escape liability for the risk of harm this employee created for himself—he may still incur liability by virtue of the fact that the conduct of the disobedient employee exposed other employees to the risk of harm.

6.23 The issue of interpreting the phrase 'in the course of his employment' in HSWA 1974 s 52 was considered by the Divisional Court in *Coult v Szuba* [1982] ICR 380 (DC), a case concerned with an allegation of breach of the duty under HSWA 1974 s 7. It was held that the phrase was not capable of bearing a precise definition, and that it had to be construed in the context of a penal statute on the basis of looking at the facts and considering the purpose of the legislation.

B. The Duties Under HSWA 1974 ss 2 and 3

HSWA 1974 s 3: the duty to non-employees; the offence of failing to discharge the duty

6.24 HSWA 1974 s 3(1) provides that it shall be the duty of every employer to conduct his undertaking in such a way as to ensure, so far as is reasonably practicable, that persons not in his employment who may be affected by it are not thereby exposed to risks to their health or safety.

The duty

6.25 HSWA 1974 s 3(1) defines the general duty upon an employer as being to ensure the safety of all those who are not employed by him and who may thereby be affected by the way in which he conducts his undertaking. The section has been interpreted both to impose a higher standard than the relevant common law duty owed by an employer and to encompass a potentially wider class of persons than such a common law duty of care.[13]

6.26 HSWA 1974 s 3(1) creates a duty, qualified only by reasonable practicability, to ensure that the employer does not conduct his undertaking in a way that exposes non-employees to risks to their health and safety. A breach of s 3(1) will often occur in tandem with a breach of s 2(1) and in the same manner; the difference being the class of person to whom the duty is owed under each section. Although the duties under ss 2(1) and 3(1) are of the same kind (*R v Gateway Foodmarkets Ltd* [1997] 2 Cr App R 40 (CA) at 44) and the law and decided authorities upon each are relevant to both, the scope of the duty under s 3(1) is defined through the conduct of the employer's undertaking, and the group of people to whom that duty is owed is those who may be affected by the conduct of that undertaking.

Failing to discharge the duty under HSWA 1974 s 3(1): the elements of the offence

6.27 A breach of the duty under HSWA 1974 s 3(1) is a continuing offence; ie it can be committed continuously over a period of time but will remain a single offence. The failure to ensure the health and safety of non-employees may arise in a number of different ways; and each instance may properly form a separate particular of a single allegation of a failure to discharge the duty.

6.28 A number of the elements of the offence of failing to discharge the duty under HSWA 1974 s 3(1) are common to the breach of HSWA 1974 s 2(1), and are considered in detail later. They are:

(a) that the defendant was, at the material time, an employer (paragraphs 6.43–6.60); and

---

[13] *R v British Steel Plc* [1995] 1 WLR 1356 CA, [1995] ICR 586, [1995] IRLR 310; *R v Associated Octel Co Ltd* [1996] 4 All ER 848 (HL); *R v Gateway Foodmarkets* [1997] 2 Cr App R 40 (CA) and *R v HTM Ltd* [2006] EWCA Crim 1156.

(b) there was the possibility of danger (paragraphs 6.61–6.62)
(c) to persons not employed by the defendant (paragraph 6.25)
(d) arising from the conduct of defendant's undertaking (paragraphs 6.33–6.42).
(e) The defendant will have failed to have ensured the health and safety of such persons not in his employment *unless* he can show (paragraph 6.135)
   (i) that all that was reasonably practicable (paragraphs 6.135–6.146) was done to minimize or eliminate the risk to health and safety,
   (ii) by him, or on his behalf (paragraphs 6.135–6.146).

HSWA 1974 s 3(2): the duty on the self-employed

6.29 HSWA 1974 s 3(2) places upon a self-employed person an identical duty to that created by s 3(1), namely a duty to ensure the health and safety of all persons who may be affected by the conduct of his undertaking. It follows that a self-employed person who employs others will owe two identical duties to non-employees in respect of the conduct of his undertaking under s 3(1) and (2).

6.30 HSWA 1974 s 51(3) defines 'self-employed person' in the following terms: '"self-employed person" means an individual who works for gain or reward otherwise than under a contract of employment, whether or not he himself employs others.'

6.31 In any prosecution in respect of a breach of HSWA 1974 s 3(2), the issue of whether a person is a self-employed person or an employee will involve questions of both law and fact (see below paragraphs 6.43–6.60).

6.32 As the duties under ss 3(1) and 3(2) are identical, save for the person owing them, they are not separately considered in the remainder of this chapter.

Undertaking

6.33 The word 'undertaking' is an ordinary English word that means 'enterprise' or 'business'.[14] The meaning of the term 'undertaking' was considered by Lord Hoffman in his single judgment in the leading House of Lords' authority upon HSWA 1974 s 3(1), *R v Associated Octel* [1996] 4 All ER 847.

6.34 The issue of whether an activity that has caused a risk amounts to part of the conduct by the employer of his undertaking is in each case strictly a question of fact, and not of law,[15] but is something that the prosecution must establish.

6.35 Lord Hoffman stressed that the duty under HSWA 1974 s 3 is imposed on the employer himself by reference to the conduct of his undertaking.[16] The question

---

[14] *R v Associated Octel Co Ltd* [1994] 4 All ER 1051 (CA) at 1062 per Stuart Smith LJ.
[15] [1996] 4 All ER 847 at 852.
[16] [1996] 1 WLR 1543 (HL) at 1547B.

B. The Duties Under HSWA 1974 ss 2 and 3

remains 'whether the activity in question can be described as part of the employer's undertaking'.[17]

6.36 The issue has consistently arisen when considering whether a breach has occurred by an employer who has engaged independent contractors. In *Associated Octel*, the facts concerned a large chemical plant operated by the appellant, who had used a small firm of specialist contractors for certain repairs. The contractors' eight employees were employed virtually full-time on the site and, like all other contractors on the site, they operated under a 'permit to work' system. This required them to obtain authorization from the appellant's engineers, who decided what safety precautions were needed and issued a 'safety certificate' imposing conditions under which the work was to be done. While undertaking repairs as part of the plant's annual maintenance programme, an employee of the contractors was in a tank cleaning the lining by the light of an electric light bulb attached to a lead when the light broke, causing a bucket of highly inflammable acetone, being used for the cleaning to ignite. The workman was badly burned in the ensuing flash fire.

6.37 The appellant was prosecuted for breach of HSWA 1974 s 3(1). The appellant submitted that there was no case to answer because the injury to the workman was not caused by the way in which it had conducted its undertaking within the meaning of s 3(1), arguing that since the workman was an employee of independent contractors, the cleaning of the tank was part of the conduct of their undertaking and the appellant had no right to control the way in which its independent contractors carried out their work.

6.38 The House of Lords ruled that whether an employer was under a duty to exercise control over an independent contractor he employed to carry out work was not the relevant test; the concept of control being one of the tests for determining the contractual relationship, something which is not decisive for HSWA 1974 s 3. Rather, they held that the important question of whether an employer may leave an independent contractor to do the work as the contractor thinks fit depends upon whether having the work done forms part of the employer's conduct of his undertaking: if it does, he owes a duty under s 3(1) to ensure that it is done without risk, albeit subject to reasonable practicability, which may limit the extent to which the employer can supervise the activities of a specialist independent contractor.[18]

---

[17] [1996] 1 WLR 1543 (HL) at 1547H.

[18] It is this final qualification which often becomes the focus of argument at trial; with the defendant company arguing that it ought not to have been expected to do more than it did to supervise a specialist contractor.

6.39 At the trial, the judge had been invited to make a ruling upon the issue as a matter of law, and the effect of his summing up was wrongly to direct the jury to find in favour of the prosecution upon this point as a matter of law. As a result, in determining the appeal the House of Lords had to consider the proviso to s 2(1) of the Criminal Appeal Act 1968, as it was then, and determine whether the appellant would undoubtedly have been convicted had the jury been properly directed. Lord Hoffman's judgment was that the question of fact that should have been left to the jury was simply whether having the tank repaired was part of the conduct of the appellant's chemical undertaking; and that in the circumstances of the case even a properly instructed jury could not have returned a negative answer, and that the appellant's appeal against conviction should therefore fail.

6.40 In the course of his judgment, Lord Hoffman drew attention to the decision in *R v Mara* [1987] 1 All ER 478 at 481, in which Parker LJ gave the following illustration:

> A factory, for example, may shut down on Saturdays and Sundays for manufacturing purposes, but the employer may have the premises cleaned by a contractor over the weekend. If the contractor's employees are exposed to risks to health or safety because machinery is left insecure, or vats containing noxious substances are left unfenced, it is, in our judgment, clear that the factory owner is in breach of his duty under s 3(1). The way in which he conducts his undertaking is to close his factory for manufacturing purposes over the weekend and to have it cleaned during the shut down period. It would clearly be reasonably practicable to secure machinery and noxious vats, and on the plain wording of the section he would be in breach of his duty if he failed to do so.

6.41 In this way, the employer must take all reasonably practicable steps to avoid risks to the contractors' employees that arise not merely from the physical state of the premises, but also from the inadequacy of the arrangements that the employer makes with the contractors for how they will do the work.

6.42 Location of the activity in question is likely to be highly relevant; but even where an activity is being carried out by independent contractors in a location remote from the employer's premises, the activity may nevertheless form part of the employer's undertaking and the resolution will depend upon the jury's determination of the facts.

C. Elements Common to ss 2 and 3 HSWA 1974

Employment: employer and employee

6.43 In any prosecution for an offence of failing to comply with the duty under HSWA 1974 s 2(1) or s 3(1), it will be for the prosecution to prove that the defendant was an employer and that there was a risk to the health and safety of an employee or

C. Elements Common to ss 2 and 3 HSWA 1974

employees, while at work. HSWA 1974 s 53(1) defines an 'employee' as someone who works under a contract of employment. This must be considered in contrast to an independent contractor working under a contract of service.

6.44 Because the duties under HSWA 1974 ss 2(1) and 3(1) only fall on employers, proving that a defendant is an employer of at least an employee is central to any prosecution for a breach of such a duty. Furthermore, showing whether the worker was an employee or an independent contractor will be important, as it will determine whether charges should be brought for failing to comply with the duty under HSWA 1974 s 2 (in respect of employees) or under s 3 (independent contractors).

6.45 In any trial concerned with HSWA 1974 s 2 in which employment is in issue, the matter is likely to involve questions of both fact and law. While it is simple to state that an employee is someone who works under a contract of service, in contrast to an independent contractor who works under a contract for service, proof of the position may be far more complex.

6.46 Both the Health and Safety at Work etc. Act 1974 s 53(1) and Employment Rights Act 1996 s 230 provide partial definitions of the term 'employee' related to the existence of a contract of employment. In *Lee Ting Sang v Chung Chi-Keung* [1990] 2 AC 374, Lord Griffiths, when delivering the advice of the Privy Council, described the question of what was the appropriate common law test by which to determine whether a workman was working as an employee or as an independent contractor as one which 'has proved to be a most elusive question', and how, 'despite a plethora of authorities the courts have not been able to devise a single test that will conclusively point to the distinction in all cases'.

6.47 The House of Lords confirmed in *Carmichael v National Power plc* [1999] ICR 1226 that the determination of the question whether there is a contract of employment is one of fact and law, stating that the question could only be an issue of law if all the terms of the contract were contained in documents which required to be construed. A written contract will not necessarily be solely determinative: the manner in which any contract categorizes the relationship is relevant, but not decisive, because the final determination is dependent upon the factual relationship between the parties: *Young & Woods Ltd v West* [1980] IRLR 201 CA.

6.48 Various tests have been described in a variety of cases, the perspective being different according to the nature of the issue in the case. In some, the question has been whether a worker was liable for tax or national insurance; in others the context has been the breach of an employer's statutory duties in respect of health and safety; others have examined whether there was a succession of pieces of work which might arguably be linked to form one employment under an 'umbrella' arrangement; others whether the choice between employee or independent contractor masked the fact that the essential contract was not one of service or for services, but of a different character altogether; or whether there had been continuity of employment for the purposes of claiming employment rights against an alleged employer.

6.49 The judgment of MacKenna J in *Ready Mixed Concrete Ltd v Minister of Pensions and National Insurance* [1968] 2 QB 497 at 515 contains the nearest to a classic definition. There it was said that a contract of service exists if these three conditions are fulfilled:

(a) the servant agrees that, in consideration of a wage or other remuneration, he will provide his own work and skill in the performance of some service for his master;
(b) he agrees, expressly or impliedly, that in the performance of that service he will be subject to the other's control in a sufficient degree to make that other master; and
(c) the other provisions of the contract are consistent with its being a contract of service.

6.50 The 'control test' has remained central to the determination of the issue of employment in civil cases and has been described by the Court of Appeal as 'the paramount test'.[19] However, it is only one of a cluster of tests adopted by courts, including the 'organizational test', the 'economic reality test', the 'mutuality of obligation', and the 'multiple test'.

6.51 The 'control test' stems from the concept of master and servant and involves the notion that a 'servant is a person subject to the command of his master as to the manner in which he shall do his work'.[20] In *Lane v Shire Roofing Co* [1995] IRLR 493, the Court of Appeal said that in the context of safety at work, there was a real public interest in recognizing the employer/employee relationship. Lord Justice Henry described how a great many relationships which are in truth 'master and servant', and thus 'employer and employee', are, to the advantage of both parties, dressed up as self-employment, and that courts must look to the realities.

Agency workers

6.52 Many recent employment rights cases have considered the position of agency workers, in circumstances where the agency worker has sought to assert rights of the kind conferred by the Employment Rights Act 1996.[21] Agency workers

---

[19] *Interlink Express Parcels Ltd v Night Trunkers Ltd* [2001] IRLR 224 CA.
[20] *Yewens v Noakes* (1880) 6 QBD 530.
[21] See *Garrard v AE Southey & Co and Standard Telephones and Cables Ltd* [1952] 2 QB 174; *Hewlett Packard v O'Murphy* [2002] IRLR 4 (EAT); *Montgomery v Johnson Underwood Ltd* [2001] EWCA Civ 318, [2001] IRLR 269; *Franks v Reuters Ltd* [2003] EWCA Civ 417, [2003]

C. Elements Common to ss 2 and 3 HSWA 1974

usually have a contract with the agency through whom they secure work, which is unlikely to be a contract of employment. The person for whom they work, the 'end-user', usually has a contract with the agency for the supply of the services of the agency worker, which the agency worker is not party to; because the agency worker has no contract with the end-user, he cannot be his employee.. Thus, on any conventional analysis, agency workers are employees of no one; in both *Franks v Reuters Ltd* [2003] EWCA Civ 417, [2003] ICR 1166 (CA) and *Brook Street Bureau (UK) Ltd v Dacas* [2004] EWCA Civ 217, [2004] ICR 1437 the possibility that a contract might be implied between the agency worker and the end-user, by virtue of his having provided the services integrally for the organization over a long period of time, was raised as a possible common law answer. In *Dacas* it was suggested that the agency and end-user might jointly be the employer. The approach taken in *Dacas* was endorsed by the Court of Appeal in *Cable & Wireless plc v Muscat* [2006] EWCA Civ 220.

Identifying an employer and the scope of his undertaking—companies without employees

6.53 There are a significant number of companies which directly employ no persons under contracts of service, but which nevertheless have undertakings that pose risks to the health and safety of many individuals. Some such businesses are formed through a complex corporate structure in which employees are employed by a 'service' company that acts as the group's employer. The employing service company will be subject to the duty under HSWA 1974 s 2(1) in respect of its employees; but in respect of non-employees, the question of the scope of the employing company's undertaking beyond its 'service' role may be more difficult to establish. However, an undertaking may form part of the undertaking of a number of different persons or companies. The issue will always be a question of fact.

6.54 HSWA 1974 s 3(1) only places a duty upon an employer. There are independent companies which employ nobody under a written contract of service but engage a significant number of workers styled as 'self-employed'. In these circumstances, a prosecution of such a company may be undertaken for a breach of s 3(1) with the company alleged to be an employer through the extent of its control over those performing the work.

Work experience

6.55 The Health and Safety (Training for Employment) Regulations 1990 SI 1990/1380 state that a person provided with relevant training (namely, work experience as

---

ICR 1166 (CA); *Stephenson v Delphi Diesel Systems Ltd* [2003] ICR 471; *Brook Street Bureau (UK) Ltd v Dacas* [2004] EWCA Civ 217, [2004] ICR 1437 (CA); *Motorola Ltd v Davidson* [2001] IRLR 4, EAT.

part of a training course or programme of training, other than at an educational establishment or under a contract of employment) is treated as being the employee of the person whose undertaking (whether carried on for profit or not) is for the time being the immediate provider to that person of the training.

Crown servants, police officers, and domestic service

6.56 Crown servants are treated as employees for the purposes of health and safety regulations,[22] as are police officers.[23] The Crown is not bound by HSWA 1974 s 33, but persons in the public service of the Crown are subject to the provisions of HSWA 1974 and deemed to be employees of the Crown.[24]

6.57 The general duties and offences are expressly excluded from applying to any person by reason only that he employs another, or is himself employed, as a domestic servant in a private household.[25]

Employment: proof and practicalities

6.58 There is no authoritative judgment on the meaning of employer and employee in relation to the general duties under HSWA 1974 ss 2 and 3, nor is there likely to be one. This is because the provisions do not create civil liability and, in the criminal context, the issue will rarely be controversial, and will only attract appellate judicial consideration if the trial judge misdirects a jury on the relevant law.

6.59 In practice the issue is usually addressed in one or more of the following ways:

(a) by witnesses describing their status;
(b) a copy contract of employment;
(c) copy wage slips;
(d) copy P60 certificate of national insurance deductions;
(e) copy of compulsory employer's liability insurance certificate.

6.60 The issue of who is the employer and whether a defendant company is an employer can be difficult in situations where operations are undertaken by a group of linked companies, or where an individual is concerned, who has avoided complying with any of the formalities required of an employer.

Failure to ensure safety/exposure to risks to safety: the possibility of danger

6.61 HSWA 1974 ss 2(1) and 3(1) refer to a similar duty but employ different language to express it: the former talks of a duty on employers to 'ensure . . . the health,

---

[22] HSWA 1974 s 48(3).
[23] ibid s 51A.
[24] See HSWA 1974 ss 36(2) and 48.
[25] ibid s 51.

C. Elements Common to ss 2 and 3 HSWA 1974

safety and welfare' of employees; the latter a duty on employers to ensure that non-employees are not 'exposed to risks to their health or safety'. Thus HSWA 1974 s 2(2), which particularizes part of the duty under s 2(1), employs the terms 'safe and without risks to health' and the protective intent of the Act, set out in s 1(1)(a), is to 'secure the health, safety and welfare of persons at work' who may or may not be employees.

6.62 A risk to health and safety occurs whenever there is a possibility of danger and does not require the danger to have eventuated. Were it to be otherwise, the provision would afford little protection to employees. In *R v Board of Trustees of Science Museum* [1993] 3 All ER 853 CA, this 'ordinary meaning'[26] of risk was firmly found to be the correct interpretation by Steyn LJ, who equally firmly rejected the alternative, stating '"Risks" conveys the idea of a possibility of danger. Indeed, a degree of verbal manipulation is needed to introduce the idea of actual danger.'[27]

6.63 Steyn LJ additionally relied upon the 'preventative aim' of the Act[28] to reinforce this meaning of 'risk', which was required in order to render the 'statutory provisions effective in their role of protecting public health and safety', and described the alternative as resulting in a 'substantial emasculation of a central part' of the Act.

6.64 Thus, it can be plainly stated that an employer will have failed to ensure the health, safety, and welfare of his employees, pursuant to HSWA 1974 s 2(1), if a risk to their health, safety, or welfare while at work can be shown to exist. It follows that all the prosecution need prove is that a possibility of danger existed to an employee while at work: thereafter, the burden will pass to the defendant, pursuant to HSWA 1974 s 40, to show that he took all reasonably practicable steps to avoid or minimize such risk. Where an injury to such an employee can be established, it follows that such a risk will have been proved.[29]

6.65 Similarly, in respect of HSWA 1974 s 3(1), an employer will have failed to ensure that persons not in his employment were not exposed to risks to their health and safety by the conduct of his undertaking if a possibility of danger can be proved to have existed. Proving the extent of the undertaking and the exposure to risk falls to the prosecution, but thereafter the burden will pass to the defendant, pursuant to HSWA 1974 s 40, to show that he took all reasonably practicable steps to avoid or minimize such risk.

---

[26] [1993] 3 All ER 853 CA at 858–859.
[27] ibid.
[28] ibid at 859.
[29] See Stuart Smith LJ in *R v Associated Octel* [1995] ICR 281 (CA) at 294 B to C.

D. Reasonable Practicability and the General Duties: Meaning and Development to 1999

Introduction

6.66 Of the elements common to both HSWA 1974 ss 2 and 3, it is the concept of reasonable practicability whose ambit has proved the most difficult to define, and which has provoked the greatest debate. In the large majority of contested health and safety cases in the criminal courts, the key question is whether the duty-holder did all that was reasonably practicable. Within this wider inquiry, the narrower issue is often whether the employer should be liable when it is alleged that injury or risk was caused by the isolated act of a junior employee.

6.67 In a series of cases commencing in the late 1990s and culminating in the recent Court of Appeal decision in *R v HTM Ltd* [2006] EWCA Crim 1156, the courts have considered the nature of the statutory duties imposed by ss 2 and 3 HSWA 1974. Although in the first of these cases the actual issue before the courts was whether employers were in breach of their duties through the act or omission of a junior employee—ie whether the identification doctrine or vicarious liability had any place when considering the general duties—they naturally fell to considering the ambit and scope of reasonable practicability. Indeed, in the later cases it became key to the understanding of how the duties applied in practice. In general terms, while in the early judgments the duties were interpreted in onerous and restrictive terms, and the scope of any defence limited, in the later cases a relaxation of approach appears to have occurred.

The identification doctrine, vicarious liability, delegation, and the isolated acts of junior employees

British Steel *and the 'identification doctrine'*

6.68 In *R v British Steel Plc* [1995] 1 WLR 1356 the Court of Appeal was concerned with a conviction for an offence of having breached HSWA 1974 s 3(1), following an incident where the employee of a sub-contractor had been killed and another injured, which the company alleged had occurred 'because the two men had disobeyed instructions and gone about the work in an extraordinary and unforeseeable manner'. Furthermore, the company's defence was that its employee supervisor, even if at fault concerning the standard and level of his supervision, was neither a 'directing mind' nor at senior management level, and the company had taken reasonable care to delegate supervision of the operation to a competent employee.

6.69 The Court first reviewed what were then the very recent decisions of the Court of Appeal in *R v Board of Trustees of the Science Museum* [1993] 3 All ER 853 and

D. Reasonable Practicability and the General Duties: Meaning and Development

R v Associated Octel Co Ltd [1994] 4 All ER 1051 (CA), the latter of which was subsequently upheld in the single opinion of Lord Hoffman in the House of Lords. Steyn LJ commented in *British Steel* on how these two decisions established 'that prima facie, and subject to the stated qualification (reasonable practicability), s 3(1) created an absolute prohibition'. He went on to state: 'If we had to consider the matter de novo we would have still concluded that the words of s 3(1) are in context capable of one interpretation only, namely that subject to the defence of reasonable practicability s 3(1) creates an absolute prohibition.'[30]

The Court in *British Steel* answered firmly in the negative the question as to whether the common law 'identification doctrine'—whereby a company can be liable only through its 'directing minds', essentially at board level (*Tesco Supermarkets v Nattrass* [1972] AC 153 (HL))—had any relevance or application in respect of the duty under HSWA 1974 s 3. Steyn LJ based this response upon 'the words of s 3(1) of the 1974 Act read in its contextual setting', and added that such a 'construction of s 3(1) must have relevance to the interpretation of s 2(1) which provides for the employer's duty to his own employees'.[31]

6.70

In the view of the Court, to have accepted the appellant company's submission would have resulted in an 'emasculation' of the legislation. Moreover, Steyn LJ stated 'If we accept British Steel's submission, it would be particularly easy for large industrial companies, engaged in multifarious hazardous operations, to escape liability on the basis that the company through its "directing mind" or senior management was not involved.'

6.71

The Court then considered a point which it described as 'troublesome': namely that the strict interpretation which it preferred 'would lead to manifestly absurd consequences'. It considered a number of examples which, according to this strict interpretation, would leave the employer without a defence of reasonable practicability. These included 'the employee who drops a spanner or an employee who drives without due care and attention'.

6.72

The Court's answer was that:

6.73

> The suggested absurdities are unlikely to arise in practice. An action such as the dropping of a spanner will only be relevant if it exposes a person not in the employer's employment to a risk to his health or safety. That will only occur if he is, as it were, in the danger zone. Thus he will only be exposed to the risk if the system (if any) designed to ensure his safety has broken down and it does not matter for the purposes of s 3(1) at what level in the hierarchy of employees that breakdown has taken place. Similarly, the driver's carelessness may have resulted from something for which his employer is to be regarded as responsible, such as trying to meet excessively tight

---

[30] [1995] 1 WLR 1356 CA at 1362.
[31] ibid at 1360, 1361.

delivery schedules or tiredness due to over-long hours of work. We do recognize that there may be circumstances in which it might be regarded as absurd that an employer should even be technically guilty of a criminal offence. An example might perhaps be the driver who is guilty of an error of judgment when driving his employer's lorry on his employer's business. But, in any event, so-called absurdities are not peculiar to this corner of the law: at the extremities of the field of application of many rules surprising results are often to be found. That circumstance is inherent in the adoption of general rules to govern an infinity of particular circumstances. Fortunately, the cases to which counsel referred will in practice cause no real difficulty in relation to s 3(1) of the 1974 Act. Nobody has suggested that there has ever been a prosecution in such a case, and it is most unlikely that there would in future be a prosecution in such cases. Moreover, if such prosecutions are brought, they are not likely to be viewed sympathetically by a judge and jury or by magistrates. In the most unlikely event of a conviction, the judge would be entitled to impose an absolute discharge and refuse to order costs in favour of the prosecution. Despite the intellectual difficulties created by counsel's examples, they do not deflect us from the firm conclusion at which we have arrived.

6.74 It is noteworthy that the Court viewed its interpretation as, 'a satisfactory measure for the attainment of the aim of the legislation. On this interpretation the law will also be simplified.' The Court expressed the view that long trials which examined where in a company's hierarchy an employee belonged should be consigned to history: 'On the basis of our judgment such enquiries will in future not be necessary. Employers and employees ought now to know where they stand. Furthermore, a culture of guarding against the risks to health and safety by virtue of hazardous industrial operations will be promoted.'

Associated Octel *and vicarious liability*

6.75 At common law, an employer's duty to his employees was developed into a special non-delegable duty, largely to ameliorate the effects of the doctrine of 'common employment', which held sway until 1948,[32] and which provided that an employee voluntarily assumed the risk of injury through the negligence of a fellow worker in the same employment. Thus in response an employer's duty was considered to include a duty to provide and maintain a safe place of work and equipment; to provide competent employees; and to establish and enforce a safe system of work. This became a personal, non-delegable duty of care. At common law, the primary duty arises by virtue of the employment relationship and exists 'whether or not the employer takes any share in the conduct of the operations'.[33] Thus the employer can delegate the performance of the duty to others, whether employees or independent contractors, but not responsibility for its negligent performance. This is the case even where a statute requires him to delegate the task to a suitably

---

[32] Abolished by the Law Reform (Personal Injuries) Act 1948 s 1.
[33] Lord Wright in *Wilsons and Clyde Coal Co Ltd v English* [1938] AC 57 at 84.

D. Reasonable Practicability and the General Duties: Meaning and Development

qualified person.[34] The duty is owed to each individual employee and must take account of his circumstances.[35]

6.76 The Factory Acts were introduced partly as a response to the inadequacies of the common law in terms of compensation for injury. The general duties under the Health and Safety at Work etc. Act 1974 were introduced as part of a wholly new regime with the aim of 'securing the health, safety and welfare of persons at work'.[36] Some of the terms in the 1974 Act, such as 'reasonable practicability', found their origins in the Factory Acts.[37]

6.77 In *R v Associated Octel Co Ltd* [1994] 4 All ER 1051, in the Court of Appeal, it was submitted on behalf of the appellant company that the duty imposed by HSWA 1974 s 3 was coterminous with the duty imposed by the law of tort in respect of the activity of a person to those who are not in his employment, which does not, save in exceptional cases, involve liability for the acts of independent contractors. It was submitted that:

> The effect of the section is merely to impose a criminal sanction in those cases where, under the law of tort, the defendant would be liable to the injured man with the sole modification that the onus of proving negligence is not on the prosecution, since the onus is on the defendant to show, so far as is reasonably practicable, that he had taken all proper care.

6.78 Stuart Smith LJ rejected this submission, noting that the concept of conducting an undertaking was not an expression used in the law of tort to define the scope of liability; and that *R v Board of Trustees of the Science Museum* [1994] IRLR 25 'does show very graphically that the section goes much further than common law liability in tort, since it is concerned with risk of injury and not the actuality of injury'.

6.79 In the House of Lords,[38] Lord Hoffman delivering the single opinion upheld the decision of the Court of Appeal. Lord Hoffman described how the appellant's submissions involved a confusion between two quite different concepts: vicarious liability, which is the employer's responsibility for the unlawful act of an employee and depends generally on the contractual relationship between the employer and the employee; and a duty imposed upon the employer himself. Lord Hoffman stated that HSWA 1974 s 3 was not concerned with vicarious liability, but: 'It imposes a duty upon the employer himself. That duty is defined by reference to a certain kind of activity, namely, the conduct by the employer of his undertaking.

---

[34] *Wilsons and Clyde Coal Co Ltd v English* [1938] AC 57.
[35] *Paris v Stepney Borough Council* [1951] AC 367.
[36] HSWA 1974 s 1(1)(a)—'securing the health, safety and welfare of persons at work'.
[37] See s 29(1) of the Factories Act 1961.
[38] [1996] 4 All ER 847.

The isolated act of a junior employee: Gateway Foodmarkets Ltd

6.80 Some of the arguments that had been advanced and rejected by the Court of Appeal in British Steel in respect of HSWA 1974 s 3(1) were revisited in *R v Gateway Foodmarkets Ltd* [1997] 3 All ER 78, [1997] 2 Cr App R 40. Here the appellant company had been convicted of an offence of having failed to discharge the duty under HSWA 1974 s 2(1), to ensure, so far as was reasonably practicable, the health and safety of employees. The charge arose out of a fatal accident at one of the company's supermarkets when a 22-year-old section manager fell to his death through a trapdoor in the floor of the lift control room. On the day in question the employee had been acting as duty manager in the absence of the regular store manager. Throughout the previous year, there had been a persistent problem with the lift cable stretching, so that a faulty electrical contact caused the lift to jam frequently. A firm of lift contractors was employed under contract with the company to provide regular maintenance and also a call-out service, with a required maximum response time of two hours. These arrangements were made by the company's head office in Bristol. However, a system had developed at the store in response to this recurring defect: the defect could be cured by freeing the contact manually, and the contractors had told the store personnel how to do this. The store manager had learned how to do this, and it had become the regular practice for him or another of the section managers to go to the control room and free the contact himself. There was no evidence that this had been authorized by the head office, nor that the head office was aware of it. On the day prior to the fatal incident, the contractors had carried out routine maintenance of the lift and had left open the trapdoor in the control room floor. When the lift jammed on the following morning, the duty manager went to the control room in order to re-set the contact. He went from sunshine into darkness and so did not see that the trapdoor was open. He fell about 26 feet to the floor of the lift shaft and died in hospital the same afternoon.

6.81 At trial the judge had been asked to rule whether the appellants could be held to have breached the duty under HSWA 1974 s 2(1) if the employee section manager was exposed to the risk of injury 'not by any act or omission of the appellants, meaning their head office personnel or senior management who could be identified with the company itself, but by those of their employees who were not in that category, specifically the store manager or the section managers who had allowed the irregular system to grow up and who had implemented it in contradiction of their instructions from head office'.[39] The judge had ruled that the offence was

---

[39] [1997] 3 All ER 78 at 81–82; [1997] 2 Cr App R 40 at 42.

D. Reasonable Practicability and the General Duties: Meaning and Development

one of strict liability, subject to the reasonable practicability defence, and he had concluded 'there will be a liability on the defendants under s 2 if there has been a breach of the duty created by the servants of these defendants'.[40] Having so ruled, the appellants had pleaded guilty and thereafter appealed against conviction, on the ground that the judge's ruling was wrong in law.

Evans LJ, giving the judgment of the Court of Appeal, dealt with the arguments of counsel concerning Lord Hoffman's opinion in *Associated Octel* and the impact of the interpretation of HSWA 1974, s 3(1). The Court of Appeal again considered 'vicarious liability' and the relevance of this to companies, this time in the context of HSWA 1974 s 2(1): **6.82**

> It is said first that section 2(1) imposes a 'strict' liability on the employer, subject only to the defence contained in the subsection itself ('so far as is reasonably practicable'). This means 'strict' as distinct from offences which require proof of mens rea, meaning broadly that the wrong-doer was blameworthy or at fault. It is said, secondly, that the issue is whether section 2(1) imposes 'vicarious liability' for the acts or omissions of another person. . . . The phrase is used to mean that one person, the defendant, is guilty of a criminal offence even though the relevant acts and omissions, including mens rea where appropriate, were those of another person. . . . These two concepts may overlap. A company can only act through persons who are its servants or agents. If the individuals concerned were those whose acts can be regarded as those of the company itself, then no question of vicarious liability is involved. But if the terms of the statute are such that the company has committed an offence, even though the acts or omissions which make it liable were those of persons outside that category, then it may be said that the company incurs a vicarious liability, although in our judgment that is not strictly correct. The liability, in such a case, is the company's own.

Thereafter the Court analysed the concept, introduced by counsel for the appellant company, of 'isolated' acts or omissions by individual employees, distinguishing these from situations where the employers could be said to have failed in their task of control or supervision of the employees concerned. The Court stated in terms 'we doubt whether this additional concept is of assistance in determining the scope of section 2(1) for the purposes of this appeal'. **6.83**

The appellants submitted that HSWA 1974 s 2(1) does not render an employer 'liable for the acts or omissions of an employee who cannot be regarded as the embodiment of the company itself' and that the persons to be so categorized were only the 'directing minds' of a company. Thus, unless the relevant failure to ensure the safety of employees could be attributed to the directing minds, there could be no breach of the duty. This, in effect, was the same argument that had been advanced and rejected by the Court of Appeal in British Steel in respect of HSWA 1974 s 3(1). In *Gateway Foodmarkets*, the Court of Appeal rejected this **6.84**

---

[40] [1997] 3 All ER 78 at 81; [1997] 2 Cr App R 40 at 42.

argument and, having considered Lord Hoffman's opinion in *Associated Octel*, expressly sought to 'adopt the same approach to section 2(1)', as had there been adopted in respect of HSWA 1974 s 3(1), finding:

> The structure of the two subsections is the same. The duty is imposed on the employer. It is a duty to 'ensure . . . the health, safety and welfare at work of all his employees'. If the duty is broken, the employer is guilty of an offence (section 33(a)). . . . There is no reference in section 2(1) to the conduct of the undertaking, which is the basis for liability under section 3(1), and so it is manifest that the content of the duty under section 2(1) is different from that under section 3(1). But in our judgment it is the same kind of duty—the company, as employer, is liable when the necessary conditions for liability are fulfilled.

6.85 Thus the Court found that:

> the general considerations referred to in the authorities, including the purpose and object of the legislation, make it overwhelmingly clear that section 2(1), like section 3(1), should be interpreted so as to impose liability on the employer whenever the relevant event occurs, namely, a failure to ensure the health, etc., of an employee.

6.86 However, as had occurred during the course of argument in *R v British Steel*, the Court of Appeal was addressed concerning what was suggested to be the 'absurd' consequence that 'the employer is criminally liable under the section for the acts or omissions of even its most junior employees, where these have put another employee, or even the same employee, at risk of injury to his health, safety or welfare', which, it was submitted, Parliament could not have intended. However, in *Gateway Foodmarkets* the Court of Appeal adopted a different approach to that taken in *British Steel*:

> We would suggest that a principled answer can be found. First, however, we would respectfully differ from the suggestion that these results, if they have been prescribed by Parliament, should be called 'absurd'. We agree that it is a somewhat extreme contention that the employer should be held criminally liable even for an isolated act of negligence by a junior employee, affecting the health, safety or welfare either of a fellow employee (section 2(1)) or of some other person (section 3(1)). The question is whether that extreme consequence is one which results from the proper construction of the two subsections. We would hold not, but we hasten to add that a conclusion on this issue is not necessary for the purposes of the present appeal.

6.87 Thus the Court, in *obiter dicta*, went on to doubt whether this allegedly extreme consequence did result:

> The answer lies, we suggest, in the application of the qualification or caveat contained in the statute itself. The duty under each section is broken if the specified consequences occur, but only if 'so far as is reasonably practicable' they have not been guarded against. So the company is in breach of duty unless all reasonable precautions have been taken, and we would interpret this as meaning 'taken by the company or on its behalf'. In other words, the breach of duty and liability under the section do not depend upon any failure by the company itself, meaning those persons who

D. Reasonable Practicability and the General Duties: Meaning and Development

embody the company, to take all reasonable precautions. Rather, the company is liable in the event that there is a failure to ensure the safety, etc., of any employee, unless all reasonable precautions have been taken—as we would add, by the company or on its behalf. If this is correct, then it follows that the qualification places upon the company the onus of proving that all reasonable precautions were taken both by it and by its servants and agents on its behalf. The concept of the 'directing mind' of the company has no application here. The further question is whether this includes all those persons for whose negligence the employer is vicariously liable to third parties for the purposes of the law of tort. If it does, then the employer is not able to rely on the statutory defence when any of his employees has been negligent, ie failed to take reasonable precautions 'in the course of his employment'. That phrase has been widely defined, and if the same test applies here then the statutory defence is limited to the rare case where the individual employee was on a frolic of his own, and where there was no failure to take reasonable precautions at any other level. It is possible that some narrower test should be defined, but as stated above we do not consider that it is necessary to decide this for the purposes of the present appeal.

6.88 The Court suggested that such an approach was consistent with that of the House of Lords in *Associated Octel*,[41] citing Lord Hoffmann's assertion that if an employer 'engages an independent contractor to do work which forms part of the employer's undertaking, he must stipulate for whatever conditions are needed to avoid those risks and are reasonably practicable'.[42] They suggested that this phrase implied that the principles of vicarious liability did have some application, so that the statutory qualification of reasonable practicability applied 'when all reasonable precautions have been taken by the employer and those for whom he is responsible in law'.[43]

6.89 The Court rejected an approach that the offence was only committed when there has been some failure to take reasonable precautions by 'management' as distinct from junior employees, citing two difficulties with such an interpretation:

> First, there is no clear legal basis for distinguishing between 'management' and (other) employees. Secondly, if the test is whether all reasonable precautions have been taken by the company or on its behalf, then it would not seem to be material to consider whether the individual concerned, who acted or was authorised to act on behalf of the company, was a senior or a junior employee.

6.90 In the event, the Court in *Gateway Foodmarkets* expressly only decided, 'that a failure at store management level is certainly attributable to the employer, while leaving open the question whether the employer is liable in circumstances where the only negligence or failure to take reasonable precautions has taken place at some more junior level'.[44]

---

[41] [1996] 4 All ER 848 (HL).
[42] ibid.
[43] ibid.
[44] [1997] 3 All ER 78 at 84; [1997] 2 Cr App R 40 at 47.

R v Nelson Group Services (Maintenance) Ltd

6.91 In *R v Nelson Group Services (Maintenance) Ltd* [1998] 4 All ER 331, the Court of Appeal was confronted with an appeal against convictions by a company from separate trials arising out of different facts. The company installed, serviced, and maintained gas appliances. In one of the indictments, the company had been convicted of an offence of failing to discharge the duty under HSWA 1974 s 3(1) and two offences of contravening absolute provisions in the Gas Safety (Installation and Use) Regulations 1994. In that case, one of the appellant's fitters had removed a defective and dangerous gas fire in a private house and the judge had directed the jury that if they found that the fitter had not capped the gas pipe, they could not find that the appellants had done all that was reasonably practicable to ensure that the occupier was not exposed to risk to her health and safety. The appellants were convicted and they appealed. The question arose whether the mere fact that the fitter had left the gas fittings in such a condition that the occupier of the house was exposed to risk to her health and safety rendered the appellants in breach of their duty under HSWA 1974 s 3(1); or whether they were entitled to show that they had, so far as reasonably practicable, conducted their undertaking in such a way as to ensure that persons not in their employment who might be affected thereby were not exposed to risk to their health by the fitter's negligent omission.

6.92 In *Nelson Group Services (Maintenance) Ltd*, the Court of Appeal first considered the *British Steel* decision, finding in that case, that there was 'clear evidence of a failure to lay down a safe system of work and by supervision to see that that system was applied' and that the present appeal was the first 'in which a court has had to decide whether the act or omission of a fitter during the course of his work inevitably involves the employer in criminal liability for failing to discharge the duty to which the employer is subject by virtue of s 3(1)'.[45]

6.93 In *Nelson Group Services (Maintenance) Ltd*, the appellants conceded that the duty under HSWA 1974 s 3(1) 'could rest on a person far down the chain if he was carrying out one of the duties of the company as an employer, for example a fitter supervising an apprentice'. However, the Court, relying on the judgment of Steyn LJ in *British Steel*,[46] rejected such an approach. The Court went on to consider *R v Associated Octel Co Ltd* [1996] 4 All ER 846 (HL) and reiterated that 'The employer's duty under s 3(1) cannot be delegated', finding that in the light of this, on the facts of the present appeal, there could be 'no doubt that the activities of the appellants' fitters are part of the conduct by the appellants of

---

[45] [1998] 4 All ER 331 at 348.
[46] ibid.

D. Reasonable Practicability and the General Duties: Meaning and Development

their undertaking. Moreover, it is clear that such acts and omissions of the fitters exposed the householders to risks to their health or safety.'[47]

Thus the remaining issue was whether the appellants 'were unable to establish a defence of reasonable practicability by the simple fact that an act or omission of one of their employees had exposed the householder to danger?'[48] Deriving considerable assistance from the judgment of the Court of Appeal in *R v Gateway Food Markets Ltd* [1997] 3 All ER 78, Evans LJ summarised the Court's view in the following way:

6.94

> First, if persons not in the employment of the employer are exposed to risks to their health or safety by the conduct of the employer's undertaking, the employer will be in breach of s 3(1) and guilty of an offence under s 33(1)(a) of the Act unless the employer can prove on the balance of probability that all that was reasonably practicable had been done by the employer or on the employer's behalf to ensure that such persons were not exposed to such risks. It will be a question of fact for the jury in each case whether it was the conduct of the employer's undertaking which exposed the third persons to risks to their health or safety. The question what was reasonably practicable is also a question of fact for the jury depending on the circumstances of each case. The fact that the employee who was carrying out the work, in this case the fitter installing the appliance, has done the work carelessly or omitted to take a precaution he should have taken does not of itself preclude the employer from establishing that everything that was reasonably practicable in the conduct of the employer's undertaking to ensure that third persons affected by the employer's undertaking were not exposed to risks to their health and safety had been done.[49]

The Court expressed the view that such an analysis was consistent with the distinction in the gas safety regulations between the duties of employers of persons and the duties of persons performing the work, and went on to state:

6.95

> It is not necessary for the adequate protection of the public that the employer should be held criminally liable even for an isolated act of negligence by the employee performing the work. Such persons are themselves liable to criminal sanctions under the Act and under the regulations. Moreover it is a sufficient obligation to place on the employer in order to protect the public to require the employer to show that everything reasonably practicable has been done to see that a person doing the work has the appropriate skill and instruction, has had laid down for him safe systems of doing the work, has been subject to adequate supervision, and has been provided with safe plant and equipment for the proper performance of the work.

Summary: British Steel *to* Nelson Group Services

Following *British Steel* [1995] 1 WLR 1356 and *Associated Octel* [1996] 4 All ER 847, the duties under HSWA 1974 ss 2 and 3 are to be considered personal duties

6.96

---

[47] [1998] 4 All ER 331 at 349.
[48] ibid at 350.
[49] ibid at 351.

placed upon an employer. They are not concerned with vicarious liability. In so far as an employer delegates the performance of these duties, he will remain liable, if a failure to ensure the health and safety of employees or others affected by the conduct of his undertaking occurs.

6.97 In any prosecution for a breach of the general duties under HSWA 1974 ss 2 and 3, the issue is not whether the risk to the safety of employees or others has been caused by the acts or omissions of persons whose acts can be said to be those of the company; the duty to ensure health and safety of employees and others falls upon the employer and the liability for a breach is the employer's.

6.98 This is subject to the qualification of the duty by the words 'so far as is reasonably practicable': Lord Hoffman in *Associated Octel*[50] explicitly recognized that the extent to which an employer is in a position to supervise a third party may be limited and that this is a matter falling for consideration under 'reasonable practicability'. However, the earlier decision of the Court of Appeal in *British Steel*[51] was made without the benefit of Lord Hoffman's analysis of the difference between vicarious liability and personal non-delegable duty. In *British Steel*,[52] at the same time as rejecting the application of the identification doctrine to the general duties, the Court appears to have accepted that in so far as an employee is 'at work', his acts and omissions were those of the employer and in such circumstances, 'the corporate employer' cannot have, 'a defence of reasonable practicability'.[53]

6.99 The judgment of the Court in *R v Gateway Foodmarkets Ltd* [1997] 3 All ER 78, went beyond considering the nature of the duty in HSWA 1974 s 2(1) and suggested that the qualification of 'reasonable practicability' may have scope for ameliorating the strictness of the duty to avoid 'absurdity'. The extent of the qualification created by the words 'so far as is reasonably practicable' became less certain following the decision of the Court of Appeal in *R v Nelson Group Services (Maintenance) Ltd* [1998] 4 All ER 331.

6.100 The debate as to what extent the decision in *Nelson Group Services (Maintenance) Ltd*[54] had affected the interpretation of the scope of the duties under HSWA 1974 ss 2 and 3, was halted almost before it began by the enactment of Management of Health and Safety at Work Regulations 1999 reg 21. This remained the position until the recent decision of the Court of Appeal, in *R v HTM Ltd* [2006] EWCA Crim 1156, when the law was substantially returned to that which it had been at the date of *Nelson Group Services (Maintenance) Ltd*.[55]

---

[50] *R v Associated Octel Co Ltd* [1996] 4 All ER 848 (HL).
[51] *R v British Steel Plc* [1995] 1 WLR 1356 (CA), [1995] ICR 586, [1995] IRLR 310.
[52] ibid.
[53] ibid.
[54] [1998] 4 All ER 331.
[55] ibid.

E. Reasonable Practicability: *HTM Ltd* and the Current Position

Introduction

In *HTM Ltd*[56] the Court revisited some old issues and sought to address ones, such as 'foreseeability' and the effect of Management of Health and Safety at Work Regulations 1999 SI 1999/3242 reg 21, which, having been largely unaddressed, sat uncertainly in the background of the debate. Arguably, the judgment has resulted in greater uncertainty as to the ambit of the general duties and the scope of the qualification of reasonable practicability. Each of the issues is analysed in detail below. **6.101**

The Management of Health and Safety at Work Regulations 1999 reg 21

Within months of the Court of Appeal's judgment in *Nelson Group Services (Maintenance) Ltd*,[57] the Management of Health and Safety at Work Regulations 1999 reg 21 was enacted, which provides: **6.102**

> **21 Provisions as to liability**
>
> Nothing in the relevant statutory provisions shall operate so as to afford an employer a defence in any criminal proceedings for a contravention of those provisions by reason of any act or default of—
> (a) an employee of his, or
> (b) a person appointed by him under regulation 7 (a health and safety advisor).

While it was widely assumed that the two events were connected, in reality reg 21 had been previously proposed and was most likely intended to deal in part with the perceived uncertainty created by certain dicta in the *Gateway Foodmarkets*[58] judgment and to ensure that the UK's law was compliant with Art 5(3) of the Framework Directive. **6.103**

Article 5 of the Council Directive (89/391/EEC), on the introduction of measures to encourage improvements in the safety and health of workers at work, is entitled 'General Provision' and sets out at 5(1) that 'The employer shall have a duty to ensure the safety and health of workers in every aspect related to the work.' Article 5(3) provides that 'The workers' obligations in the field of safety and health at work shall not affect the principle of the responsibility of the employer.' **6.104**

R v HTM Ltd [2006] EWCA Crim 1156 was an appeal under Criminal Procedure and Investigations Act 1966 s 35(1) against rulings made by a trial judge in a preparatory hearing. The defendant company faced an indictment which included a **6.105**

---

[56] [2006] EWCA Crim 1156.
[57] [1998] 4 All ER 331 (CA).
[58] [1997] 3 All ER 78 (CA).

count of failure to discharge the duty under HSWA 1974 s 2(1), and wished to assert that it had done everything that was reasonably practicable to ensure the safety of two employees who had been injured as a result of having ignored their training.

6.106 In *HTM Ltd* the Court of Appeal considered the impact of Management of Health and Safety at Work Regulations 1999 reg 21 on the general duties and the 'reasonable practicability' in this context. The Court ruled that, subject to the effect of reg 21, the decisions in *Gateway Foodmarkets*[59] and *Nelson Group Services*[60] represented the binding position in law and described how:

> Their effect is that the defendants will be entitled to put before the jury evidence to show that what happened was purely the fault of one or both of their employees. If the jury were persuaded that everything had been done by or on behalf of the company to prevent that accident from happening, it would be entitled to be acquitted.

6.107 The Court in *HTM Ltd* found that it was an overriding principle 'that secondary legislation can only have the effect of amending primary legislation, assuming all other tests are met, if the wording is clear and unambiguous and the intention to achieve that objective manifestly established'. The Court noted that there was no doubt that in a significant number of the judgments given in the cases concerned with the general duties under HSWA 1974, the effect of HSWA 1974 s 40 has resulted in judges referring to the duties under ss 2 and 3 as being subject to a 'defence', or a 'limited defence', of reasonable practicability. However, the Court found that the correct analysis was, as Tuckey LJ had stated in *Davies*, that 'The duty cast on the defendant is a "duty . . . to ensure as far as is reasonably practicable." It is a breach of this qualified duty which gives rise to the offence.'[61]

6.108 Notwithstanding that a legal burden of proof in relation to that aspect of the duty is imposed on a defendant, nonetheless the Court ruled that the phrase 'so far as reasonably practicable' is not a defence. For this reason they found that Management of Health and Safety at Work Regulations 1999 reg 21 'can have no application to it' and 'that it has not affected the decision of this Court in *Nelson*.'[62]

Reasonable practicability and 'fault of junior employee': summary of current position

6.109 The principles that apply to the duties under ss 2 and 3, but for the qualification of reasonable practicability, remain intact; namely that these are non-delegable duties, the breach of which the employer remains liable for whenever the relevant event, being the exposure to the possibility of danger, occurs.

---

[59] [1997] 3 All ER 78 (CA).
[60] [1998] 4 All ER 331 (CA).
[61] [2006] EWCA Crim 1156 para 32.
[62] ibid para 31.

E. Reasonable Practicability: HTM Ltd *and the Current Position*

6.110 The extent and operation of the reasonably practicable qualification is now uncertain; just as the principle established in *British Steel*,[63] that at what level in the hierarchy of employees a breakdown of failure has taken place is irrelevant, is irreconcilable with that set out in *Nelson Group Services (Maintenance) Ltd*,[64] namely that the employer should not be necessarily held criminally liable for an isolated act of negligence by his employee performing work.

6.111 There is a conflict between the approach of the Court in *Gateway Foodmarkets* and that in *Nelson Group Services (Maintenance) Ltd*, despite the former having apparently informed the latter. In *Gateway Foodmarkets* the Court suggested that the position may be that the statutory qualification of reasonable practicability applied only when all reasonable precautions have been taken by the employer and those for whom he is responsible in law, and this would certainly include every employee while at work. Importantly, in that case, the Court rejected an approach that the offence was only committed when there has been some failure to take reasonable precautions by 'management', as distinct from junior employees.

6.112 The judgment in *Nelson Group Services (Maintenance) Ltd* was concerned only with HSWA 1974 s 3(1), namely the duty to those not employed by the employer. The rationale for the ruling appears to have been based upon 'public policy', with the Court finding that 'It is not necessary for the adequate protection of the public that the employer should be held criminally liable even for an isolated act of negligence by the employee performing the work'.

6.113 However, according to the Court of Appeal in *HTM Ltd*,[65] the effect of the decisions in *Gateway Foodmarkets* and *Nelson Group Services* is to represent a binding position in law that a defendant, on a charge of having breached HSWA 1974 s 2(1) 'will be entitled to put before the jury evidence to show that what happened was purely the fault of one or both of their employees'.[66]

6.114 It is submitted that the following represents the best attempt at a general principle:

> An employer must show that all reasonably practicable steps were taken by him, or on his behalf. The duty cannot be delegated to any person, be they manager, employee or independent contractor. Should there be a failure to take all reasonably practicable steps by the employer, or on his behalf, then the employer will remain liable unless he can show that the only failing was by the person doing the work and all of the following applied: he had the appropriate skill and instruction; safe systems of work had been laid down; he had been subject to adequate supervision; he had been provided with safe plant and equipment for the proper performance of the work.

---

[63] [1995] 1 WLR 1356.
[64] [1998] 4 All ER.
[65] [2006] EWCA Crim 1156.
[66] ibid, para 28.

Likely future development

6.115 In the relatively near future a legislative attempt to reinstate Management of Health and Safety at Work Regulations 1999 reg 21's application to the general duties is likely; either through an amendment to the words of HSWA 1974 s 40, to include the term 'a defence', or to remove the same from reg 21 itself. When this happens it is to be anticipated that Parliament will express a clear intent to establish the extent of an employer's liability and responsibility for the acts or defaults of his employee within the context of the employer's health and safety duties.

Foreseeability

6.116 The case law from both UK jurisdictions which has sought to answer the question of what part if any foreseeability plays in either the general duties under HSWA 1974 or the qualification of 'reasonable practicability' is unsatisfactory and conflicting.

The current position

6.117 The position has recently been complicated by two apparently conflicting decisions of the Court of Appeal of England and Wales: one civil[67] and the other the decision of the Court of Appeal, Criminal Division, in *HTM Ltd*.[68] The three sub-paragraphs following represent a summary of what appears to be the current position. After these is an analysis of these two decisions on foreseeability:

(a) Foreseeability is no part of a test for the existence of a risk to health and safety: subject to reasonable practicability, an employer will be in breach of the duties under HSWA 1974 ss 2 and 3 if there exists a possibility of danger to a relevant class of person

(b) Foreseeability may be relevant to reasonable practicability, in relation to the general duties under HSWA 1974 ss 2, 3, and 4, but only in a closely confined way. Foreseeability is merely a tool with which to assess the likelihood of a risk eventuating. In asking a jury to consider whether he has established that he has done all that is reasonably practicable, a defendant can adduce evidence as to the likelihood of the incidence of the relevant risk eventuating, in support of his case that he had taken all reasonable means to eliminate it.

(c) Foreseeability plays no part in relation to risk or reasonable practicability in respect of other health and safety regulations, such as the Control of Substances Hazardous to Health Regulations 1999, which are expressed as a duty to prevent so far as reasonably practicable.

---

[67] *Dugmore v Swansea NHS Trust* [2003] 1 All ER 333.
[68] [2006] EWCA Crim 1156.

E. Reasonable Practicability: HTM Ltd *and the Current Position*

The civil case: Dugmore v Swansea NHS Trust

6.118 In *Dugmore v Swansea NHS Trust* [2003] 1 All ER 333 the Court of Appeal considered an appeal concerned with Control of Substances Hazardous to Health Regulations 1999[69] reg 7(1), which provides:

> Every employer shall ensure that the exposure of his employees to a substance hazardous to health is either prevented or, where this is not reasonably practicable, adequately controlled.

6.119 Giving the single judgment of the Court, Hale LJ stated:

> Many legislative provisions imposing duties upon employers to protect the health and safety of their employees impose strict liabilities different in kind from their duties at common law. Regulation 7(1) uses the language of strict liability in providing that an employer 'shall ensure' that exposure is either prevented or controlled. The primary duty is to prevent exposure altogether, unless this is not reasonably practicable. 'This' must refer to 'prevented' rather than both limbs of the duty. Where prevention is not reasonably practicable, the secondary duty is adequately to control the exposure. 'Adequately' is restrictively defined, the only relevant factors being the nature of the substance and the nature and degree of exposure generally. Nowhere is there any reference to the reasonable foreseeability of the risk. Nor is the duty dependent upon what a risk assessment would have revealed. It is therefore irrelevant whether or not a reg 6 assessment (the duty to carry out a risk assessment) would have revealed it.

6.120 The judgment then continued:

> The relevance of the common law concept of foreseeability in legislative provisions such as these has been the subject of a number of decisions, some going one way and some going the other. Much will depend upon the precise wording of the provision in question.

6.121 The Court then reviewed all of the principal authorities that had considered the matter, including those concerned with the interpretation of s 29(1) of the Factories Act 1961, which read 'every place at which any person has at any time to work . . . shall, so far as is reasonably practicable, be made and kept safe for any person working there'.

6.122 The Court cited the case of *Larner v British Steel plc* [1993] 4 All ER 102, which concerned the nature of the duty and not reasonable practicability, where the Court of Appeal held that in order to establish a case the claimant did not have to prove that the danger which had made his place of work unsafe was reasonably foreseeable. In that case, both members of the Court had pointed out that otherwise there would be little if any distinction between the common law duty of care and the duty under the statute. It also considered the case of *Mains v Uniroyal*

---

[69] Now repealed and replaced by the Control of Substances Hazardous to Health Regulations 2002.

Englebert Tyres Ltd [1995] SC 518, in which the Scottish Inner House of the Court of Session reached the same conclusion concerning the nature of the duty as in *Larner*, for essentially the same reasons, but in which reference to foreseeability of risk was made in the context of reasonable practicability.

6.123 The Court of Appeal in *Dugmore* noted that neither *Larner* nor *Mains* was directly concerned with how foreseeability of risk might come into the question of what it was reasonably practicable for the employer to do.

6.124 In *Mains v Uniroyal Englebert Tyres Ltd* [1995] SC 518, the Scottish Court had considered a long line of conflicting English and Scots cases on the point, including the judgment of Lord Abernathy in *Neil v Greater Glasgow Health Authority (OH)* 1996 SLT 1260. In *Mains* Lord Sutherland stated that because considerations of reasonable practicability involve weighing the degree and extent of the risk, on the one hand against the time, trouble, and expense of preventing it, on the other, therefore, it held, foreseeability clearly came into the matter as it is impossible to assess the degree of risk in any other way.

6.125 However, in *Dugmore*, Hale LJ agreed with two decisions[70] of Lord Nimmo Smith in the Scottish Outer House of the Court of Session, which she described as being more directly on point, where it was held that the duty was absolute and not couched in terms which allowed for reasonable foreseeability, stating:

> In our view, that analysis is correct. The duty in reg 7(1) is an absolute one: to ensure that exposure is prevented or controlled. [Counsel for the defendant] sought to persuade us that the words 'so far as is reasonably practicable' should be moved from their current position qualifying the duty to prevent exposure so as to qualify the duty to ensure that exposure is either prevented or controlled. There is no warrant for us to rewrite the regulation in this way. Its wording is even stricter than that in s 29(1) of the 1961 Act, where the phrase 'so far as is reasonably practicable' came between 'shall' and 'be made and kept safe'. If that was an absolute duty, then so must this be.

6.126 Having dealt with the narrow point in the case she then went on to consider the matter in broader terms, addressing the essence of the issue of 'reasonable practicability' and the part that foreseeability played. She stated: 'To import into the defence of reasonable practicability the same approach to foreseeability of risk as is contained in the common law of negligence would be to reduce the absolute duty to something much closer to the common law, albeit with a different burden of proof.'

The criminal case: R v HTM Ltd

6.127 *R v HTM Ltd* [2006] EWCA Crim 1156 was an appeal under Criminal Procedure and Investigations Act 1966 s 35(1) against rulings made by a trial judge in a

---

[70] *Bilton v Fastnet Highlands Ltd* 1998 SLT 1323 and *Williams v Farne Salmon & Trout Ltd* 1998 SLT 1329.

E. Reasonable Practicability: HTM Ltd *and the Current Position*

preparatory hearing. The defendant company faced an indictment which included a count of failure to discharge the duty under HSWA 1974 s 2(1). One of the rulings made by the trial judge was that evidence of foreseeability was not irrelevant and therefore inadmissible to the case alleged against the defendant company, particularly with regard to the reasonable practicability of their ensuring the health, safety, and welfare of their employees. The ruling of the trial judge did not particularize what it was that evidence could show was foreseeable, but the defendant company wished to assert that the behaviour of the deceased employees was unforeseeable in that they ignored their training.

Giving the judgment of the Court, Latham LJ first considered the judgments in *Dugmore* and *Larner v British Steel plc* [1993] 4 All ER 102 before considering *Austin Rover Group Ltd v HM Inspector of Factories* [1990] 1 AC 619 and the minority judgment of Lord Goff. *Austin Rover Group Ltd* is the defining authority on the duty under HSWA 1974 s 4, owed by those in control of premises that are used for work. In that case, the opinion of Lord Jauncey, with which the majority agreed except Lord Goff, held that the ambit of HSWA 1974 s 4 was far wider than that of ss 2 and 3 and that, as a result, Parliament had included in s 4 the words 'to take such measures as it is reasonable for a person in his position to take', in order to require consideration to be given not only to the extent to which the individual in question has control of premises, but also to his knowledge and reasonable foresight at all material times:[71] it was these words, not the term 'so far as is reasonably practicable', that incorporated a limitation to liability based on reasonable foreseeability. **6.128**

However, in that case, Lord Goff, having reviewed the authorities defining reasonable practicability, concluded at page 626H: **6.129**

> It follows from the passages which I have quoted that, for the purpose of considering whether the defendant has discharged the onus which rests upon him to establish that it was not reasonably practicable for him, in the circumstances, to eliminate the relevant risk, there has to be taken into account (inter alia) the likelihood of risk eventuating. The degree of likelihood is an important element in the equation. It follows that the effect is to bring into play foreseeability in the sense of likelihood of the incidence of the relevant risk, and that the likelihood of such risk eventuating has to be weighed against the means, including cost, necessary to eliminate it.

In *HTM Ltd*, Latham LJ held that Lord Jauncey and the other members of the Judicial Committee who had agreed with his opinion, did not in any way dissent from Lord Goff's analysis of reasonable practicability;[72] Latham LJ then cited how **6.130**

---

[71] [635C–636C]; pp 253–254.
[72] See also, *Hampstead Heath Swimming Club v Corporation of London* [2005] EWHC 713 (Admin), where Stanley Burnton J. stated: 'The meaning of the words "so far as is reasonably practicable" was authoritatively considered by the House of Lords in *Austin Rover Group Ltd v Her*

Lord Jauncey had dealt with the relationship between the phrase, 'to take such measures as it is reasonable for a person in his position to take', and the relevant phrase 'so far as is reasonably practicable', namely:

> Thus while only one yardstick determines whether premises are safe at any one time, the measures to ensure the safety required of each person having a degree of control may vary. Approaching the matter in this way, content may be given to the words 'so far as reasonably practicable'. It could, having regard to his degree of control and knowledge of likely use, be reasonable for an individual to take a measure to ensure the safety of premises, but it might not be reasonably practicable for him to do so having regard to the very low degree of risk involved and the very high cost of taking the measure.

6.131 Latham LJ held that 'in short form, that is precisely the meaning ascribed to the relevant phrase by Lord Goff'. The difficulty with this is that Lord Jauncey's 'short form' definition of reasonable practicability ('very low degree of risk involved and the very high cost of taking the measure') contains no reference to foreseeability. Furthermore, Lord Goff's longer definition ('It follows that the effect is to bring into play foreseeability in the sense of likelihood of the incidence of the relevant risk') while referring to foreseeability, does not use the term in any sense that incorporates foreseeability: likelihood of risk is not the same as foreseeability; the former is the chance of something occurring, while the latter is the ability of a person to know or find out the chance of that thing occurring; something may be very likely to happen but nonetheless impossible to foresee.

6.132 Thus, the judgment of the Court of Appeal in *HTM Ltd*,[73] in respect of foreseeability and reasonable practicability, was that:

> Lord Goff's analysis of what is the right approach, is the one which, on the authorities, correctly identifies the proper approach to the jury question posed by the relevant phrase. It is to be noted that he expresses the relevance of foreseeability in a closely confined way. Foreseeability is merely a tool with which to assess the likelihood of a risk eventuating. It is not a means of permitting a defendant to bring concepts of fault appropriate to civil proceedings into the equation by the back door; still less does it mean that the phrase 'reasonably foreseeable' in itself provides the answer to the jury question. But it seems to us that a defendant to a charge under section 2 or indeed section 3 or 4, in asking the jury to consider whether it has established that it has done all that is reasonably practicable, cannot be prevented from adducing evidence as to the likelihood of the incidence of the relevant risk eventuating in support of its case that it had taken all reasonable means to eliminate it.

6.133 It appears from this that the only sense in which foreseeability is relevant is in relation to 'the likelihood of the incidence of the relevant risk eventuating'.

Majesty's Inspector of Factories [1990] 1 AC 619. It was held that the relevant factors are the foreseeable risk of injury and the cost of the preventive measures.'

[73] [2006] EWCA Crim 1156, para 18.

E. Reasonable Practicability: HTM Ltd *and the Current Position*

The Court went on to state that it did not disagree with the analysis of the Court of Appeal in *Dugmore* as to the nature of the duty in Control of Substances Hazardous to Health Regulations 1999 reg 7(1), on the basis that 'The wording of that regulation is significantly different from the wording of section 2 of the 1974 Act.'[74] **6.134**

The reverse burden in HSWA 1974 s 40

HSWA 1974 s 40 provides: **6.135**

> **40 Onus of proving limits of what is practicable etc**
>
> In any proceedings for an offence under any of the relevant statutory provisions consisting of a failure to comply with a duty or requirement to do something so far as is practicable or so far as is reasonably practicable, or to use the best means to do something, it shall be for the accused to prove (as the case may be) that it was not practicable or not reasonably practicable to do more than was in fact done to satisfy the duty or requirement, or that there was no better practicable means than was in fact used to satisfy the duty or requirement.

Wherever a defendant is placed under a burden to 'prove' a matter in a criminal trial, the standard he must satisfy is the balance of probability, in other words, that something was more likely than not. **6.136**

Thus in proceedings for an offence of breaching one of the general duties under HSWA 1974 a defendant bears the burden of showing that he did all that was reasonably practicable to ensure safety. **6.137**

In *David Janway Davies v Health and Safety Executive* [2002] EWCA Crim 2949, the Court of Appeal was concerned with a challenge to the compatibility of the legal burden placed by HSWA 1974 s 40 on a defendant with the terms of the European Convention on Human Rights. The Court recognized that while HSWA 1974 s 40 did make some inroads into the presumption of innocence, nevertheless, it did not follow that the section was incompatible with the Convention. The Court of Appeal noted that the European Court of Human Rights had said in *Salabiaku v France* (1988) 13 EHRR 379 at para 28: **6.138**

> Presumptions of fact or law operate in every legal system. Clearly the Convention does not prohibit such presumptions in principle. It does however require the contracting States to remain within certain limits in this respect as regards criminal law . . . Article 6 (2) does not therefore regard presumptions of fact or law provided for in the criminal law with indifference. It requires States to confine them within reasonable limits which take into account the importance of what is at stake and maintain the rights of the defence.

---

[74] However, see paras 7.26–7.29 for an analysis of Latham LJ's reasoning for distinguishing the decision in *Dugmore*.

Chapter 6: Health and Safety Offences: Breach of the General Duties

6.139 The Court of Appeal further held that it was relevant to consider that Art 5.1 of Council Directive[75] 89/391/EEC made it possible to impose absolute duties on employers in the field of health and safety, by requiring Member States to provide that 'The employer shall have a duty to ensure the safety and health of workers in every aspect related to the work.'

6.140 The Directive, in Art 5(3), provides that Member States may, but are not required:

> . . . to provide for the exclusion or the limitation of employers' responsibility where occurrences are due to unusual and unforeseeable circumstances beyond the employer's control, or to exceptional events, the consequences of which could not have been avoided despite the exercise of all due care.

6.141 The Court of Appeal held that the approach to be adopted was to consider whether a fair balance had been struck between the fundamental right of the individual and the general interests of the community, paying due regard to the choice which the legislature has made when striking that balance, particularly where social or economic policy is involved. The Court cited *R v Kebeline* [2000] 2 AC 326 at 386, where Lord Hope had described how in considering where the balance lay, it was useful to consider the following questions:

(a) What does the prosecution have to prove in order to transfer the onus to the defence?
(b) What is the burden on the accused? Does it relate to something which is likely to be difficult for him to prove, or does it relate to something which is likely to be within his knowledge or to which he readily has access?
(c) What is the nature of the threat faced by society which the provision is designed to combat?

6.142 Finally, having considered the questions, the Court of Appeal concluded that the imposition of a legal burden of proof upon the defendant by HSWA 1974 s 40 was justified, necessary, and proportionate, so that it was not incompatible with the Convention.

6.143 The Court's reasons for this finding are important. It found that HSWA 1974 was 'regulatory' and that its purpose was to protect the health and safety of those affected by the activities referred to in ss 2–6. The need for such regulation was said to have been amply demonstrated by statistics showing that fatal injuries reported to the UK enforcing authorities by industry are running at an average of about 700 a year and non-fatal major injuries at nearly 200,000 a year. The Court heard that, following a survey in 1995–6, the Office of Statistics put the annual cost of accidents at work in the UK at between £14.5 and £18.1 billion. In the view of

---

[75] Para 9.

the Court, this therefore meant that the purpose of HSWA 1974 was both social and economic.

6.144 It was the opinion of the Court, expressed in the single judgment of Lord Justice Tuckey, that the reversal of the burden of proof in HSWA 1974 s 40 took into account the fact that:

> duty holders are persons who have chosen to engage in work or commercial activity (probably for gain) and are in charge of it. They are not therefore unengaged or disinterested members of the public and in choosing to operate in a regulated sphere of activity they must be taken to have accepted the regulatory controls that go with it. This regulatory regime imposes a continuing duty to ensure a state of affairs, a safety standard. Where the enforcing authority can show that this has not been achieved it is not unjustifiable or unfair to ask the duty holder who has either created or is in control of the risk to show that it was not reasonably practicable for him to have done more than he did to prevent or avoid it.

6.145 The Court noted that any facts relied on in support of a defence would be within the knowledge of the defendant and that whether the defendant should have done more would be judged objectively. In the Court's view, if all that the defendant had to do were to raise the defence to require the prosecution to disprove it, the focus of the statutory scheme would be changed. The trial would become focused on what it was that the enforcing authority maintained should have been done, rather than on what the defendant had done or ought to have done, as Parliament had intended.

6.146 It was material to the Court's decision that 'in cases where the reverse burden of proof applies', the defendant does not face potential imprisonment. In the Court's view, it was also relevant that:

> The offence involves failure to comply with an objective standard. The consequences of such failure may be newsworthy in some cases but the moral obloquy is not the same as that involved in truly criminal offences. The statistics we have been provided with show that only about 15% of those prosecuted under sections 2 to 6 of the Act are individuals. The rest are companies.

F. HSWA 1974 s 4: The General Duty of Persons in Control of Premises to Persons other than their Employees

6.147 The drafting of HSWA 1974 s 4 is complex. The section has been held to create an absolute duty to ensure that premises which are made available for use by non-employees are safe. The duty is owed by all those who have a degree of control over the relevant premises or of the relevant plant or substance. The duty can be divided into three parts:

(a) to take such measures as it is reasonable for a person in his position to take to ensure

(b) that the relevant premises, plant, or substance is or are safe and without risks to health,

(c) so far as is reasonably practicable.

HSWA 1974 s 4(1): non-domestic premises used by non-employees as a place of work

6.148 The safety duty is owed 'in relation to' non-employees who use non-domestic premises made available to them as a place of work[76] or as a place where they may use plant or substances provided for their use. HSWA 1974 s 4(1) provides that the duty applies to premises 'so made available and other non-domestic premises used in connection with them'.

6.149 HSWA 1974 s 53(1) provides that the term 'non-domestic premises' is to be construed by reference to the definition of domestic premises, namely:

> premises occupied as a private dwelling (including any garden, yard, garage, outhouse or other appurtenance of such premises which is not used in common by the occupants of more than one such dwelling).

6.150 The term 'premises' is itself defined by the same section in the broadest possible terms:

> 'premises' includes any place and, in particular, includes—
> (a) any vehicle, vessel, aircraft or hovercraft,
> (b) any installation on land (including the foreshore and other land intermittently covered by water), any offshore installation, and any other installation (whether floating, or resting on the seabed or the subsoil thereof, or resting on other land covered with water or the subsoil thereof), and
> (c) any tent or movable structure.

6.151 The term 'plant' is defined by HSWA 1974 s 53(1) as 'including any machinery, equipment or appliance'. The same section defines the term 'substance' as any natural or artificial substance (including micro-organisms), whether in solid or liquid form or in the form of a gas or vapour.

6.152 In *Westminster City Council v Select Management Ltd* [1985] 1 All ER 897, the Court of Appeal was concerned with an appeal from a decision of the Divisional Court. The Divisional Court had considered the meaning of non-domestic premises following a challenge to the finding of an industrial tribunal in relation to an improvement notice served upon a property management company, in respect of the lifts in a residential block of flats.

6.153 The Court ruled that the common parts of a block of flats were properly capable of being construed as 'non-domestic premises', as defined by HSWA 1974, since,

---

[76] See HSWA 1974 s 52 and paras 6.18–6.23 for definition of the term 'at work'.

first, they were either a 'place' or an 'installation on land' and, secondly, the common parts were not 'domestic' since they were used in common by the occupants of more than one private dwelling.[77]

The Court of Appeal considered that the common parts were made available as a 'place of work' to persons who were not employees of the company but who came to repair and maintain the premises, or, alternatively, that the lifts and electrical installations in the common parts were 'plant' provided for the use of such persons. Accordingly, for the purposes of HSWA 1974 s 4, the common parts of the flats were non-domestic premises made available to persons not employed by the company as a place of work or where they could use plant provided for their use, and therefore the company was under a duty in accordance with s 4 to ensure that they were kept safe.

6.154

HSWA 1974 s 4(2): by whom the duty is owed

HSWA 1974 s 4(2) creates a duty on a person who has, to any extent, control of relevant premises or of the relevant plant or substance. Lord Jauncey, in the House of Lords' decision of *Austin Rover Group Ltd v Her Majesty's Inspector of Factories* [1990] 1 AC 619, held that HSWA 1974 s 4 recognized that more than one person may have a degree of control of those premises at any one time and hence be under a duty in relation to them.

6.155

HSWA 1974 s 4(3) provides that where a person has, by virtue of any contract or tenancy, an obligation of any extent in relation to:

6.156

(a) the maintenance or repair of any relevant premises, or of any means of access/egress to/from such premises; or
(b) the safety/absence of risks to health arising from plant or substances in any such premises

then that person shall be treated, for the purposes of s 4(2), as being a person who has control of the matters to which his obligation extends.

HSWA 1974 s 4(4) provides that any reference to a person having control of any premises or matter is a reference to a person having control of the premises or matter in connection with the carrying on by him of a trade, business, or other undertaking[78] (whether for profit or not).

6.157

'To take such measures as it is reasonable for a person in his position to take'

HSWA 1974 s 4(2) employs the 'reasonableness' limitation in two different ways on an otherwise absolute duty, placed upon all those who have a degree of control

6.158

---

[77] [1985] 1 All ER 897 at 899b to h, 901b to d, and 904c to d.
[78] See paras 6.33–6.42 for a consideration of the term 'undertaking'.

over the relevant premises, to ensure the safety of premises that are made available for use by non-employees.

6.159 The first limitation is created by the term 'to take such measures as it is reasonable for a person in his position to take'. A majority of the House of Lords in *Austin Rover Group Ltd v Her Majesty's Inspector of Factories* [1990] 1 AC 619 considered that the ambit of HSWA 1974 s 4 was far wider than that of ss 2 and 3 and that, as a result, Parliament had included these words in order to require consideration to be given not only to the extent to which the individual in question has control of the premises, but also to his knowledge and reasonable foresight at all material times. Thus, when a person makes available premises for use by another, the reasonableness of the measures which he is required to take to ensure the safety of those premises must be determined in the light of his knowledge of the anticipated use for which the premises have been made available, and of the extent of his control and knowledge, if any, of the actual use thereafter. If the premises were not a reasonably foreseeable cause of danger to persons using the premises in a manner or circumstance that might reasonably be expected to occur, it is not reasonable to require any further measures to be taken to guard against unknown and unexpected events which might imperil their safety.

That the relevant premises, plant, or substance is or are safe and without risks to health

6.160 In *R v Board of Trustees of the Science Museum* [1993] 3 All ER 853 (CA), the Court of Appeal defined 'risk' as bearing its ordinary meaning in the context of the term 'risks to health or safety' in HSWA 1974 s 3(1), thereby denoting the possibility of danger rather than actual danger. Following this decision, 'safe and without risks to health' must equally mean that the premises, plant, or substance are free from the possibility of danger.

6.161 In the earlier decision of *Austin Rover Group Ltd v Her Majesty's Inspector of Factories* [1990] 1 AC 619, the House of Lords found that there was an absolute duty to ensure that premises which had been made available for use by others were safe, breach of which was to be determined by reference to the time when the premises were being used rather than when they were made available. Liability for breach of the duty was thereafter limited by the other parts of s 4.

'So far as is reasonably practicable'

6.162 The second 'reasonableness' limitation on the absolute duty relates to the measures that must be taken to ensure the relevant premises are safe and without risks to health, namely, all those that are 'reasonably practicable'.

6.163 Lord Jauncey's opinion, with which the majority of the Judicial Committee agreed in *Austin Rover Group Ltd v Her Majesty's Inspector of Factories* [1990] 1 AC 619, confirmed that once it was proved in a prosecution under HSWA 1974 s 4(2): (a) that premises which had been made available for use by others were unsafe and constituted a risk to health; (b) that the defendant had a degree of control over those premises; and (c) that having regard to his degree of control and his knowledge of the likely use, it would have been reasonable for him to take measures which would ensure that the premises were safe and without risks to health, then, by operation of s 40, the onus lay on the defendant to show that, weighing the risk to health against the means, including cost, of eliminating the risk, it was not reasonably practicable for him to take those measures.

6.164 The clarity with which the House of Lords in *Austin Rover Group Ltd v Her Majesty's Inspector of Factories* [1990] 1 AC 619 interpreted HSWA 1974 s 4 has been somewhat clouded by the decision of Latham LJ in the Court of Appeal in *R v HTM Ltd* [2006] EWCA Crim 1156. Latham LJ held that Lord Jauncey's opinion included a 'shorthand' agreement with the definition in Lord Goff's minority opinion of the phrase 'reasonable practicability'. Lord Goff's opinion answered the questions posed to the House of Lords differently to Lord Jauncey, with whom the other members of the Judicial Committee agreed. Lord Goff opined that foreseeability was part of reasonable practicability; Lord Jauncey that foreseeability came into HSWA s 4 only through the phrase, 'To take such measures as it is reasonable for a person in his position to take'. It is submitted that dicta in *HTM Ltd* to the effect that Lord Jauncey's opinion included any element of foreseeability in reasonable practicability in respect of HSWA 1974 s 4 cannot be right. (For a detailed consideration of the meaning and operation of the term 'so far as is reasonably practicable', see paragraphs 6.101–6.146.)

G. HSWA 1974 s 6: General Duties of Manufacturers etc as Regards Articles and Substances for Use at Work

6.165 HSWA 1974 s 6 contains a series of interrelated provisions of some complexity, which remain, it is submitted, poorly drafted. It owes its present form to amendments by the Consumer Protection Act 1987. The section places duties on designers, importers, manufacturers, and suppliers in respect of the safety relating to specific aspects of the use of articles for work, fairground equipment, and substances.

6.166 In many situations in which a breach of HSWA 1974 s 6 might be alleged, such a designer, importer, manufacturer, or supplier is likely to be demonstrably in breach of the more straightforward duty under s 3 to those 'who may be affected' by the conduct of such a person's undertaking.

6.167 In *R (on the application of Junttan Oy) v Bristol Magistrates' Court* [2004] 2 All ER 555, a bare majority of the House of Lords overturned a decision of the

Divisional Court[79] and ruled that nothing in the Supply of Machinery (Safety) Regulations 1992 SI 1992/3073 prevented a prosecution under HSWA 1974 s 6. The House of Lords held that HSWA 1974 and the 1992 Regulations functioned in parallel, at different levels of seriousness: a person found guilty of an offence under HSWA 1974 s 6 was liable on conviction on indictment to an unlimited fine; the 1992 Regulations prohibited manufacturers supplying machinery unless it satisfied the relevant health and safety requirements and was 'in fact safe'. Charges under the 1992 Regulations were triable summarily and were punishable only by a 'moderate' fine.

Limitations on liability

6.168 All of the duties imposed by the provisions of HSWA 1974 s 6 are limited by subs (7) which provides that the duties 'extend only to things done in the course of a trade, business or other undertaking' carried on by the person subject to the duty, whether for profit or not and, importantly, only 'to matters within his control'. Proof of these matters lies with the prosecution.

6.169 Similarly, HSWA 1974 s 6(10) provides that for the purposes of s 6, an absence of safety or a risk to health shall be disregarded 'in so far as the case in or in relation to which it would arise is shown to be one the occurrence of which could not reasonably be foreseen'. While this language seems to suggest that a burden is being placed on the defendant, the subsection does not do this and nor do the terms of HSWA 1974 s 40 apply, which place such a burden upon a defendant to show that he has taken all reasonably practicable steps.

6.170 HSWA 1974 s 6 (10) continues by providing that in determining whether any duty imposed by virtue of s 6(1)(a), (1A)(a) or (4)(a) (which are considered below) has been performed:

> regard shall be had to any relevant information or advice which has been provided to any person by the person by whom the article has been designed, manufactured, imported or supplied or, as the case may be, by the person by whom the substance has been manufactured, imported or supplied.

6.171 The rationale for this part of the subsection is unclear, as these words appear merely to state an obvious and inevitable position.

6.172 HSWA 1974 s 6(6) provides that nothing in s 6 shall be taken to require a person to repeat any testing, examination, or research which has been carried out otherwise than by him or at his instigation, in so far as it is reasonable for him to rely on the results thereof for the purposes of the provisions of the section.

---

[79] *R v Junttan Oy ex p Bristol Magistrates' Court* [2002] 4 All ER 965.

G. HSWA 1974 s 6: General Duties of Manufacturers etc

HSWA 1974 s 6(1): designers, manufacturers, and suppliers of articles for work

6.173 HSWA 1974 s 6(1) places a duty upon designers, manufacturers, importers, and suppliers of articles for use at work (and fairground equipment):

(a) to ensure, so far as is reasonably practicable, that the article is so designed and constructed that it will be safe and without risks to health at all times when it is being set, used, cleaned or maintained by a person at work;

(b) to carry out or arrange for the carrying out of such testing and examination as may be necessary for the performance of the duty imposed on him by the preceding paragraph;

(c) to take such steps as are necessary to secure that persons supplied by that person with the article are provided with adequate information about the use for which the article is designed or has been tested and about any conditions necessary to ensure that it will be safe and without risks to health at all such times as are mentioned in paragraph (a) above and when it is being dismantled or disposed of; and

(d) to take such steps as are necessary to secure, so far as is reasonably practicable, that persons so supplied are provided with all such revisions of information provided to them by virtue of the preceding paragraph as are necessary by reason of its becoming known that anything gives rise to a serious risk to health or safety.

Supply

6.174 'Design' and 'designer' are not defined under HSWA 1974, nor are the terms 'manufacture' or 'manufacturer', and thus these terms must bear their ordinary meaning. In relation to the supply of articles or substances s 51(3) provides that this means supplying them by way of sale, lease, hire or hire purchase, whether as principal or agent for another.

6.175 However, HSWA 1974 s 53(9) provides that where a person ('the ostensible supplier') supplies any article or substance to another ('the customer') under a hire-purchase agreement, conditional sale agreement, or credit-sale agreement, and the ostensible supplier:

(a) carries on the business of financing the acquisition of goods by others by means of such agreements; and

(b) in the course of that business acquired his interest in the article or substance supplied to the customer as a means of financing its acquisition by the customer from a third person ('the effective supplier');

(c) then the effective supplier and not the ostensible supplier is to be treated for the purposes of s 6 as supplying the article or substance to the customer so that any duty imposed by the section on suppliers falls on the effective supplier and not on the ostensible supplier.

Articles for use at work

6.176 HSWA 1974 s 53(1) defines 'article for use at work' to mean:

(a) any plant designed for use or operation (whether exclusively or not) by persons at work, and

(b) any article designed for use as a component in any such plant.

6.177 The term 'plant' includes any machinery, equipment, or appliance. (See paragraphs 6.18–6.23 for a consideration of the term 'at work'.)

6.178 The unusual case of *McKay v Unwin Pyrotechnics Ltd* [1991] Crim LR 547 concerned a prototype dummy mine that had been constructed by the respondent. A self-employed explosives consultant (S) was injured when it detonated as he was carrying out the first trials. The justices dismissed informations alleging a breach of HSWA 1974 s 6 against the respondent, concluding that the respondent had not, at the relevant date, manufactured an article for use at work, that S was conducting a trial of a mine which he had helped to develop, and that S was not using the article as a person at work.

6.179 It was argued on appeal that the dummy mine was an article for use at work, and that it had to be safe at all times when it was being used by a person at work, which included a self-employed person. Dismissing the appeal, the Divisional Court held that while it was not doubted that S was a person at work, to focus on that was to approach the problem from the wrong end. It had first to be ascertained whether the respondent had manufactured an article for use at work. HSWA 1974 s 6(1)(a) dealt with articles to be used in the work place that must be so designed and constructed to be safe when being 'set, used, cleaned or maintained' and the dummy mine could not on that basis be described as an article for use at work. The Court held that if the mine was 'plant' or a component of plant in the sense that it was an appliance, it could not be said that the prototype was designed or manufactured for use at work. It was designed and manufactured so that it might be determined whether it could be used at work. Accordingly s 6 was not contravened.

The duty under HSWA 1974 s 6(1)(a)

6.180 In any case alleging a breach of the duty under HSWA 1974 s 6(1)(a), the prosecution must establish against a defendant that the article was not safe when it was being set, used, cleaned, or maintained by a person at work and that it was a matter within the defendant's control (s 6(7)) to have ensured that the article was designed or constructed to be reasonably foreseeably (s 6(10)) safe and without reasonably foreseeable risks to health at all such times.

6.181 Thereafter, HSWA 1974 s 40 provides that in any proceedings for an offence consisting of a failure to comply with a duty or requirement to do something so far as is reasonably practicable, it shall be for the accused to prove that it was not reasonably practicable to do more than was in fact done to satisfy the duty or requirement.[80]

The term 'safe and without risks to health' means that there is no possibility of danger at the relevant times.[81] **6.182**

The duty under HSWA 1974 s 6(1)(b)

It is difficult to envisage a situation in which a breach of the duty under HSWA 1974 s 6(1)(b), the duty in relation to testing and examination, would be proven without first proving the breach of s 6(1)(a). Once a breach of the duty under para (a) has been established, then proof of the breach of para (b) must become dependent upon establishing the necessity of particular testing or research of the article. However, the words of the section strictly require such testing as 'may be necessary'—a much wider duty limited only by subss (6), (7), and (10). **6.183**

The duty under HSWA 1974 s 6(1)(c)

The absolute nature of the duty to 'take such steps as are necessary to secure' that persons supplied with the substance are provided adequate information relevant to the time and functions set out in HSWA 1974 s 6(1)(a) is limited by the terms of subs (7) which provides that duties extend only to matters within a person's control. **6.184**

The duty under HSWA 1974 s 6(1)(d)

The duty under HSWA 1974 s 6(1)(d) concerns the need to provide persons with revisions of information. Any case alleging a breach of s 6(1)(d) will involve the prosecution establishing that a defendant, within whose control the matter lay, failed to take such steps as were necessary to secure that a person, who had been supplied with information pursuant to the duty under s 6(1)(c), was not thereafter provided with all such revisions of information as were necessary by reason of it becoming known that something gave rise to a serious risk to health or safety. **6.185**

Thereafter, unlike the preceding duty, HSWA 1974 s 6(1)(d) (through the operation of HSWA 1974, s 40) provides that it shall be for the defendant to show, on a balance of probabilities, that it was not reasonably practicable to do more than was in fact done to satisfy the duty or requirement.[82] **6.186**

---

[80] Paras 6.135–6.146.
[81] Paras 6.160–6.161.
[82] Reasonable practicability—paras 6.101–6.146.

6.187 The term 'serious' risk to health or safety is not defined by HSWA 1974 and might prove difficult to establish, involving potentially both a question of degree of harm and likelihood of coming to pass.

Written undertakings

6.188 The provision created by HSWA 1974 s 6(8) in a single sentence of 123 words is set out in three parts below to aid easy comprehension. It provides:

(a) where a person designs, manufactures, imports, or supplies an article for use at work or an article of fairground equipment, *and*

(b) he does so for or to another on the basis of a written undertaking (by that other) to take specified steps sufficient to ensure, so far as is reasonably practicable, that the article will be safe and without risks to health, at all such times as are mentioned in s 6(1)(a) and/or s 6(1A)(a), *then*

(c) the undertaking shall have the effect of relieving the first-mentioned person from the duty imposed by virtue of that paragraph, ie s 6(1)(a) and or s 6(1A)(a), to such extent as is reasonable having regard to the terms of the undertaking.

Imports

6.189 HSWA 1974 s 6(8A) provides that nothing in s 6(7) (duties extend only to matters within a person's control—paragraph 6.200), or s 6(8) (written undertakings—paragraph 6.188), shall relieve any person who imports any article or substance from any duty in respect of anything which:

(a) in the case of an article designed outside the UK, was done by and in the course of any trade, profession or other undertaking carried on by, or was within the control of, the person who designed the article; or

(b) in the case of an article or substance manufactured outside the UK, was done by and in the course of any trade, profession, or other undertaking carried on by, or was within the control of, the person who manufactured the article or substance.

Fairground equipment

6.190 HSWA 1974 s 6 places 'articles of fairground equipment' into a special category, specifically applying the duties under s 6(1) to designers, suppliers, and manufacturers of such equipment and, by s 6(1A), placing such persons under an additional duty to:

(a) ensure, so far as is reasonably practicable, that the article is so designed and constructed that it will be safe and without risks to health at all times when it is being used for or in connection with the entertainment of members of the public;

(b) carry out or arrange for the carrying out of such testing and examination as may be necessary for the performance of the duty imposed on him by the preceding paragraph;
(c) take such steps as are necessary to secure that persons supplied by that person with the article are provided with adequate information about the use for which the article is designed or has been tested and about any conditions necessary to ensure that it will be safe and without risks to health at all times when it is being used for or in connection with the entertainment of members of the public; and
(d) to take such steps as are necessary to secure, so far as is reasonably practicable, that persons so supplied are provided with all such revisions of information provided to them by virtue of the preceding paragraph as are necessary by reason of its becoming known that anything gives rise to a serious risk to health or safety.

HSWA 1974 s 51(3) defines 'article of fairground equipment' as any fairground equipment or any article designed for use as a component in any such equipment. 'Fairground equipment' means any fairground ride, any similar plant designed to be in motion for entertainment purposes with members of the public on or inside it, or any plant that is designed to be used by members of the public for entertainment purposes either as a slide or for bouncing upon. The reference to 'plant which is designed to be in motion with members of the public on or inside it' is defined 'to include a reference to swings, dodgems and other plant which is designed to be in motion wholly or partly under the control of, or to be put in motion by, a member of the public'. **6.191**

The operation of the duty in other respects follows HSWA 1974 s 6(1). **6.192**

The duty under HSWA 1974 s 6(2): research

By HSWA 1974 s 6(2), a person who undertakes the design or manufacture of any article for use at work, or of any article of fairground equipment, is placed under a duty to carry out, or arrange for the carrying out, of any necessary research with a view to the discovery and, so far as is reasonably practicable, the elimination or minimization of any risks to health or safety to which the design or article may give rise. The limitation provided by s 6(6) is important and provides that a person is not required to repeat any testing, examination, or research which has been carried out otherwise than by him or at his instigation 'in so far as it is reasonable for him to rely on the results' for the purposes of this duty. **6.193**

The meaning of the term 'reasonably practicable' is considered at paragraphs 6.101–6.146. However, the use of this term in HSWA 1974 s 6(2) appears to qualify the standard necessary to eliminate or minimize risks. (See the cases concerned with similar wording in absolute provisions: *Briggs Amasco Ltd v* **6.194**

Thurgood Times, 30 July 1984 (DC) and *R v Rhone-Poulenc Rorer Ltd* [1996] ICR 1054 (CA).)

6.195 HSWA 1974 s 40 provides that in any proceedings for an offence consisting of a failure to comply with a duty or requirement to do something so far as is reasonably practicable, it shall be for the accused to prove that it was not reasonably practicable to do more than was in fact done to satisfy the duty or requirement.[83] However, the application of s 40 to this provision is far from straightforward as the wording of s 6(2) suggests that the burden would rest upon the prosecution to prove the existence of a risk and that research was necessary for the discovery of a risk to health or safety arising from the design or article, but that the burden would rest upon a defendant to show that such research was not necessary to eliminate or minimize the risk, so far as was reasonably practicable.

HSWA 1974 s 6(3): installation

6.196 HSWA 1974 s 6(3) provides that it shall be the duty of any person who erects or installs any article for use at work in any premises where that article is to be used by persons at work, or who erects or installs any article of fairground equipment, to ensure, so far as is reasonably practicable, that nothing about the way in which the article is erected or installed makes it unsafe or a risk to health at any such time as is mentioned in s 6(1)(a)[84] or (1A).[85]

6.197 By HSWA 1974 s 40, in any prosecution, once the prosecution has proven that a risk to safety existed at the relevant time, then the defendant will bear the burden of showing that it was not reasonably practicable to have done more than was in fact done.[86]

HSWA 1974 s 6(4): substances

6.198 HSWA 1974 s 6(4) provides that it shall be the duty of any person who manufactures, imports, or supplies any substance:

(a) to ensure, so far as is reasonably practicable, that the substance will be safe and without risks to health at all times when it is being used, handled, processed, stored or transported by a person at work or in premises to which HSWA 1974, s 4 applies;[87]
(b) to carry out or arrange for the carrying out of such testing and examination as may be necessary for the performance of the duty imposed on him by the preceding paragraph;

---

[83] Paras 6.135–6.146.
[84] Paras 6.180–6.182.
[85] Paras 6.190.
[86] Paras 6.135–6.146.
[87] HSWA 1974 s 4.

(c) to take such steps as are necessary to ensure that persons supplied by that person with the substance are provided with adequate information about any risks to health or safety to which the inherent properties of the substance may give rise, about the results of any relevant tests which have been carried out on or in connection with the substance, and about any conditions necessary to ensure that the substance will be safe and without risks to health at all such times as are mentioned in paragraph (a) above and when the substance is being disposed of; and

(d) to take such steps as are necessary to ensure, so far as is reasonably practicable, that persons so supplied are provided with all such revisions of information provided to them by virtue of the preceding paragraph as are necessary by reason of its becoming known that anything gives rise to a serious risk to health or safety.

6.199 'Substance' means any natural or artificial substance (including micro-organisms[88]), whether in solid, liquid, gas, or vapour form.

The duty under HSWA 1974 s 6(4)(a)

6.200 In any case alleging a breach of the duty under HSWA 1974 s 6(4)(a), the prosecution must establish against a defendant that the article was not safe at a time when it was being used, handled, processed, stored, or transported by a person at work or in premises to which HSWA 1974 s 4 applies[89] and that it was a matter within the defendant's control (s 6(7)) to have ensured that the article was designed or constructed to be reasonably foreseeably (s 6(10)) safe and without reasonably foreseeable risks to health at all such times.

6.201 Thereafter, HSWA 1974 s 40 provides that in any proceedings for an offence consisting of a failure to comply with a duty or requirement to do something so far as is reasonably practicable, it shall be for the accused to prove that it was not reasonably practicable to do more than was in fact done to satisfy the duty or requirement.

6.202 The term 'safe and without risks to health' means that there is no possibility of danger at the relevant times (paragraphs 6.160–6.164).

The duty under HSWA 1974 s 6(4)(b)

6.203 As with HSWA 1974 s 6(1)(b), it is difficult to envisage a situation in which a breach of the duty under s 6(4)(b) would be proven without first proving the breach of s 6(4)(a). Once a breach of the duty under para (a) has been established, then proof of the breach of para (b) must become dependent upon establishing the necessity of particular testing or examination of the substance. However, the words of the section strictly require such testing as 'may be necessary': a much wider duty limited only by s 6(6), (7), and (10).

---

[88] 'Micro-organism' includes any microscopic biological entity which is capable of replication.
[89] Paras 6.160–6.164.

The duty under HSWA 1974 s 6(4)(c)

6.204 The absolute nature of the duty to 'take such steps as are necessary to secure' that persons supplied with the substance are provided with information is limited by the terms of HSWA 1974 s 6(7), which provides that duties extend only to matters within a person's control.[90] The nature of the 'adequate information' concerns risks arising from the 'inherent properties' of the substance, 'relevant' tests that have been carried out and (the very wide term) about 'any conditions necessary to ensure' the safety of the substance at the times mentioned in s 6(4)(a) and during disposal of the substance.

6.205 The duty under HSWA 1974 s 6(4)(d) concerns the need to provide persons with revisions of information. Any case alleging a breach of s 6(4)(d) will involve the prosecution establishing that a defendant, within whose control the matter lay, failed to take such steps as were necessary to secure that a person so supplied by the defendant with information pursuant to the duty under s 6(4)(c) was not provided with all such revisions of information as were necessary by reason of its becoming known that something gave rise to a serious risk to health or safety. Thereafter, unlike the preceding duty, s 6(4)(d) (by operation of HSWA 1974 s 40) provides that it shall be for the defendant to show, on a balance of probabilities, that it was not reasonably practicable to do more than was in fact done to satisfy the duty or requirement. The term 'serious' risk to health or safety is not defined by HSWA 1974 and might prove difficult to establish, involving potentially both a question of degree of harm and likelihood of coming to pass.

Substances and research: HSWA 1974 s 6(5)

6.206 By HSWA 1974 s 6(5), a person who undertakes the manufacture of any substance is placed under a duty to carry out, or arrange for the carrying out of, any necessary research with a view to the discovery and, so far as is reasonably practicable, the elimination or minimization of any risks to health or safety to which the substance may give rise at all times mentioned in s 6(4)(a). The limitation provided by s 6(6) is important and provides that a person is not required to repeat any research which has been carried out otherwise than by him or at his instance 'in so far as it is reasonable for him to rely on the results' for the purposes of this duty.

6.207 The meaning of the term 'reasonably practicable' is considered at paragraphs 6.101–6.146. The use of this term in HSWA 1974 s 6(5) appears to qualify the standard necessary to eliminate or minimize risks. (See the cases concerned with

---

[90] Para 6.189.

similar wording in absolute provisions: *Briggs Amasco Ltd v Thurgood* Times, 30 July 1984 (DC) and *R v Rhone-Poulenc Rorer Ltd* [1996] ICR 1054 (CA)).

6.208 HSWA 1974 s 40 provides that in any proceedings for an offence consisting of a failure to comply with a duty or requirement to do something so far as is reasonably practicable, it shall be for the accused to prove that it was not reasonably practicable to do more than was in fact done to satisfy the duty or requirement.[91] However, the application of s 40 to this provision is far from straightforward as the wording of s 6(5) suggests that the burden would rest upon the prosecution to prove the existence of a risk and that research was necessary for the discovery of a risk to health or safety arising from the substance, but that the burden would rest upon a defendant to show that such research was not necessary to eliminate or minimize, so far as was reasonably practicable, the risk.

H. HSWA 1974 s 7: The Duty of Employees at Work

6.209 HSWA 1974 s 7 provides that it shall be the duty of every employee while at work:

(a) to take reasonable care for the health and safety of himself and of other persons who may be affected by his acts or omissions at work; and
(b) as regards any duty or requirement imposed on his employer or any other person by or under any of the relevant statutory provisions, to cooperate with him so far as is necessary to enable that duty or requirement to be performed or complied with.

6.210 Unlike HSWA 1974 ss 2–6, the provisions of s 40 (placing a legal burden on defendants in respect of showing the taking of all reasonably practicable steps) have no application to s 7. Furthermore, the magistrates' court maximum sentence for a failure to comply with the duty under s 7 remains a fine of £5,000 (but this is unlimited in the Crown Court).

6.211 The issue of employment and who is an employee is considered at paragraphs 6.43–6.60.

6.212 The definition of 'at work' can prove important in any prosecution for a breach of this section. HSWA 1974 s 52(1) provides that the term 'at work' shall be construed according to the following definition: 'an employee is at work throughout the time when he is in the course of his employment, but not otherwise.' The issue of interpreting the phrase 'in the course of his employment' in relation to HSWA 1974 s 7 was considered by the Divisional Court in *Coult v Szuba* [1982] ICR 380. It was held that the phrase was not capable of having a precise definition and

---

[91] Paras 6.135–6.146.

that it had to be construed in the context of a penal statute on the basis of looking at the facts and considering the purpose of the legislation.

6.213 What is 'reasonable care' will always be a question of fact for magistrates or a jury. In *Skinner v HM Advocate* [1994] SCCR 316, S, a supervisor with the Gas Board with overall responsibility for several sites, including a site where a new mains pipe was being laid on a public roadway, appealed against his conviction for failing to comply with the duty under HSWA 1974 s 7 by failing to properly warn motorists of roadworks. S argued that there was insufficient evidence for conviction as, on the evidence, the site was in the charge of a co-accused (K) and S could not be expected to be at that site all the time. His appeal was refused. The Court held that it was significant that S was not merely a superior of K, but a supervisor of the entire site, and had been put on notice of the danger at this particular location. It was therefore open to the jury to have concluded that S had failed to take reasonable care for the safety of others in that he should have taken steps to ensure that instructions he gave for properly signing the works were carried out.

6.214 *Amos v Worcester City Council* (15 July 1996; CO 3693/95) is a decision of the Divisional Court on the particular facts of a 'case stated' appeal. The Court in that case stressed that resolving an issue as to scope of such a duty involved the resolution of a question of fact and found that the justices had rightly concluded a danger to have been an 'obvious' one, commenting that 'it is just those dangers which an active store manager is expected to identify and to guard against'. A comment of general application is to be found in the judgment of Beldam LJ when he stated:

> The policy of the Health and Safety at Work Act is to require, from all at work, initiative to take reasonable care in their work in those aspects which are within the scope and sphere of their responsibility to see that danger is not caused to others.

6.215 The section is concerned with 'reasonable care' for the safety of those who may be affected by the employees' acts or omissions, not with causation or the mechanics of any accident.

6.216 The duty under HSWA 1974 s 7(b) upon an employee to 'cooperate with his employer so far as is necessary' to enable the employer to comply with the employer's duty appears to impose an onerous statutory standard of cooperation upon the employee.

Prosecution for a breach of HSWA 1974 s 7

6.217 Prosecution of employees for a breach of HSWA 1974 s 7 remains relatively rare. In 2004/5 just fifteen such prosecutions were brought by the HSE. A breach by an employee of HSWA 1974 s 7 cannot of itself provide a defence to an allegation of breach against an employer of his own duty or duties.[92] To a significant extent there is a duty upon an employer to anticipate that employees may make mistakes through carelessness or by taking short-cuts; and the taking of all reasonably practicable steps involves a consideration by the employer of just this.

6.218 The HSE guidance to inspectors on the prosecution of individuals[93] provides that the HSWA 1974 s 7 duty needs 'to be considered in the context of the employer's provisions', so that inspectors are advised, in assessing whether the conduct of an employee is reasonable or not, or whether the employee has done what was necessary or not, to give consideration to the extent to which the employer has been complying with his obligations to provide adequate training, supervision, and safe systems of work etc. The following example is given:

> For example, a machine operator who has received inadequate training might be considered to have acted reasonably in all the circumstances if he/she removes a guard from a machine and continues to use it, and this is the generally accepted and condoned practice in the company. In other circumstances the same act might be considered unreasonable, if the employee has received proper training, if the guard in question is sufficient, and if removal of guards is neither accepted nor condoned in the company.

6.219 The guidance provides that 'Some acts of horseplay and violence against people will come within the scope of section 7 if they arise out of or in connection with work and put people's health and safety at risk'. In relation to the meaning of 'so far as is necessary', the guidance provides that this 'does not require employees to compensate for employers' failure to make adequate provisions. This remains the responsibility of the employer.'

6.220 Similarly, when considering whether a public interest exists for the prosecution of such an employee for a breach of HSWA s 7, inspectors are advised to consider:

(a) whether the company had done all it reasonably could to ensure compliance;
(b) whether the offence was solely the result of the actions/inactions of the individual;
(c) whether employees, as a matter of general practice, followed the systems of work alleged by the employer to be in force;
(d) any previous warnings to the employee, from whatever source;
(e) whether the offence by the employee was flagrant;

---

[92] However, see the discussion of the meaning of 'reasonable practicability' at paras 6.101–6.146.
[93] HSE Operational Circular 130/8 (version 2, 18.05.06), 'Prosecuting individuals' App 1.

(f) the risks to health and safety arising from the offence by the employee; and
(g) whether prosecution would be seen by others as fair, appropriate, and warranted.

6.221 The guidance describes how:

> It is also possible, but probably less likely, that you may want to prosecute both the company and an individual employee. This might be where there were deficiencies in the company's arrangements/procedures and additional, separate actions/inactions by an individual—both of which warrant prosecution. However, we do not generally prosecute individuals whose actions arose from their employer's unsatisfactory working arrangements and procedures.

7

OTHER HEALTH AND SAFETY OFFENCES: HSWA 1974 s 33(1)(b)–(o)

A. HSWA 1974 s 33(1)(b)–(o)	7.01	Disclosure of information: HSWA 1974 s 33(1)(j)	7.61
B. HSWA 1974 s 33(1)(b): Breach of HSWA 1974 ss 8 and 9	7.03	False statements: HSWA 1974 s 33(1)(k)	7.67
C. HSWA 1974 s 33(1)(c): Breach of Regulations	7.07	The offence and statements obtained under HSWA 1974 s 20(1)(j)	7.74
D. Other Offences	7.40	Making false entries: HSWA 1974 s 33(1)(l)	7.80
Health and safety inquiries: HSWA 1974 s 33 (1)(d)	7.40	Using a document with intent to deceive: HSWA 1974 s 33(1)(m)	7.83
Contravening requirements by inspectors: HSWA 1974 s 33(1)(e)	7.44	Impersonating an inspector: HSWA 1974 s 33(1)(n)	7.88
Preventing a person from attending or answering questions: HSWA 1974 s 33(1)(f)	7.52	Failure to comply with a court order: HSWA 1974 s 33(1)(o)	7.89
Contravening improvement or prohibition notices: HSWA 1974 s 33(1)(g)	7.54		
Obstructing an inspector: HSWA 1974 s 33(1)(h)	7.56		
Provision of information to the commission: HSWA 1974 s 33(1)(i)	7.59		

A. HSWA 1974 s 33(1)(b)–(o)

The Health and Safety at Work etc. Act 1974 (HSWA 1974) s 33 creates a range **7.01** of health and safety offences. Section 33(1) is subdivided into fifteen paragraphs, each one creating a separate health and safety offence. Section 33(1)(a), creates the offence of breaching one of the general health and safety duties under ss 2–7 of HSWA 1974 and this offence is considered in the previous chapter.

Chapter 7: Other Health and Safety Offences: HSWA 1974 s 33(1)(b)–(o)

7.02 The health and safety offences created by s 33(1)(b)–(o) are:

(b) to contravene HSWA 1974 ss 8 or 9;
(c) to contravene any health and safety regulations . . . or any requirement or prohibition imposed under any such regulations (including any requirement or prohibition to which he is subject by virtue of the terms of or any condition or restriction attached to any licence, approval, exemption or other authority issued, given or granted under the regulations);
(d) to contravene any requirement imposed by or under regulations under HSWA 1974 s 14 or intentionally to obstruct any person in the exercise of his powers under that section;
(e) to contravene any requirement imposed by an inspector under HSWA 1974 s 20 or 25;
(f) to prevent or attempt to prevent any other person from appearing before an inspector or from answering any question to which an inspector may by virtue of HSWA 1974 s 20(2) require an answer;
(g) to contravene any requirement or prohibition imposed by an improvement notice or a prohibition notice (including any such notice as modified on appeal);
(h) intentionally to obstruct an inspector in the exercise or performance of his powers or duties [or to obstruct a customs officer in the exercise of his powers under HSWA 1974 s 25A];
(i) to contravene any requirement imposed by a notice under HSWA 1974 s 27(1);
(j) to use or disclose any information in contravention of HSWA 1974 s 27(4) or s 28;
(k) to make a statement which he knows to be false or recklessly to make a statement which is false where the statement is made—
  (i) in purported compliance with a requirement to furnish any information imposed by or under any of the relevant statutory provisions; or
  (ii) for the purpose of obtaining the issue of a document under any of the relevant statutory provisions to himself or another person;
(l) intentionally to make a false entry in any register, book, notice or other document required by or under any of the relevant statutory provisions to be kept, served or given or, with intent to deceive, to make use of any such entry which he knows to be false;
(m) with intent to deceive, to . . . use a document issued or authorised to be issued under any of the relevant statutory provisions or required for any purpose thereunder or to make or have in his possession a document so closely resembling any such document as to be calculated to deceive;
(n) falsely to pretend to be an inspector;
(o) to fail to comply with an order made by a court under HSWA 1974 s 42.

B. HSWA 1974 s 33(1)(b): Breach of HSWA 1974 ss 8 and 9

7.03 HSWA 1974 s 33(1)(b) creates an offence of contravening HSWA 1974 s 8 or s 9. The offence is triable either in the magistrates' court, where the maximum fine is £5,000, or at the Crown Court where an unlimited fine can be imposed.

HSWA 1974 s 8 creates a 'duty not to interfere with or misuse things provided' **7.04**
and requires that: 'No person shall intentionally or recklessly interfere with or
misuse anything provided in the interests of health, safety or welfare in pursuance
of any of the relevant statutory provisions.'

Section 9 creates a 'duty not to charge employees for things done or provided' and **7.05**
requires that: 'No employer shall levy or permit to be levied on any employee of
his any charge in respect of anything done or provided in pursuance of any specific
requirement of the relevant statutory provisions.'

The 'relevant statutory provisions' include HSWA 1974 Pt 1, the pre-1974 regu- **7.06**
lations listed at HSWA 1974 Sch 1 and the wealth of health and safety regulations
passed since 1974.

C. HSWA 1974 s 33(1)(c): Breach of Regulations

By HSWA 1974 s 33(1)(c): **7.07**

it is an offence for any person to contravene any health and safety regulations ... or
any requirement or prohibition imposed under any such regulations (including any
requirement or prohibition to which he is subject by virtue of the terms of or any
condition or restriction attached to any licence, approval, exemption or other authority issued, given or granted under the regulations).

In 2005/6 the HSE prosecuted 539 alleged offences of contravening a health and **7.08**
safety regulation, making this offence the most frequently prosecuted.

The offence is triable in either the magistrates' court or the Crown Court. The **7.09**
maximum sentence in the magistrates' court is a fine of level five, currently £5,000,
and in the Crown Court an unlimited fine. However, pursuant to the remarkably
complicated provisions of HSWA 1974 s 33(3)(b)(i) and (4), in the Crown Court
the offence can attract a sentence of imprisonment of up to two years in certain
circumstances. Namely, where the offence consists of:

(a) contravening any health and safety regulation 'by doing otherwise than under
the authority of a licence issued by the Executive ... something for the doing
of which such a licence is necessary under health and safety regulations';[1]
(b) contravening a term of or a condition or restriction attached to any such
licence as mentioned in the preceding paragraph;
(c) 'acquiring or attempting to acquire, possessing or using an explosive article or
substance ... in contravention of any health and safety regulations'.

---

[1] An example being a licence under the Control of Asbestos Regulations 2006 SI 2006/2739
reg 8.

'To contravene any health and safety regulations'

7.10 HSWA 1974 s 82 defines 'contravention' as including a failure to comply and provides that 'contravene' has a corresponding meaning. Health and safety regulations are those passed since 1974, normally pursuant to the power in HSWA 1974 s 15.

7.11 The effect of this provision and HSWA 1974 s 33(1)(c) is to create a criminal offence of very wide application. There are a wealth of health and safety regulations placing obligations upon those engaged in a wide range of activities in different capacities. Many of these regulations are drafted in absolute terms. A failure to comply with any such regulation will amount to a 'contravention' and thus a criminal offence.

7.12 While a number of matters of general relevance can be identified, the nature of the offence is determined by the terms of the particular regulation alleged to have been contravened.

Interpretation

7.13 While there are a great number of decided civil cases concerning the interpretation and application of health and safety regulations, care has to be taken when attempting to apply such decisions in a criminal context. Judgments in civil cases often involve findings of fact and law in which the distinction between the two may not be always apparent. In a criminal case, the jury will always remain the sole judges of fact and must apply the law as directed by the judge to the facts as they find them to be.

7.14 The majority of health and safety regulations have been passed to implement European Directives and there is persuasive English and European authority to the effect that a regulation or statute intended to implement a European Directive must be construed 'purposively' so as to give effect to the results envisaged by that Directive.[2] However, the relevance of such Directives to the interpretation of the regulations in a criminal context is very limited. Certainty is essential to the application of a criminal provision and such provisions have to be construed narrowly. The language of regulations must be given its ordinary and natural meaning in a criminal context; and so nothing but the plain terms of the regulations could be enforced by way of criminal prosecution.

Strict liability

7.15 Regulations imposing strict liability duties are not uncommon. The rationale and justification for the imposition of such strict liability in relation to health and

---

[2] *Von Colson v Land Nordrhein-Westfalen* [1984] ECR 1891 at 1909 and 1910–1911 (ECJ); *Pickstone v Freemans plc* [1988] ICR 697 at 722–723, 725 (HL); *Litster v Forth Dry Dock and Engineering Co Ltd* [1989] ICR 341 at 350, 353, 354, 357–358, 370.

C. HSWA 1974 s 33(1)(c): Breach of Regulations

safety having been affirmed in *R v Associated Octel Co Ltd* [1996] 4 All ER 846 and *R v Gateway Foodmarkets Ltd* [1997] 3 All ER 78.

7.16 Article 6(2) of the European Convention on Human Rights guarantees that 'everyone charged with a criminal offence shall be presumed innocent until proved guilty according to law'. The compatibility of strict criminal liability with this guarantee has been recently considered by the Administrative Court in *Barnfather v London Borough of Islington* [2003] EWHC 418 (Admin). The Court decisively ruled that Art 6(2) did not restrict Parliament from creating strict liability offences and that it was not incompatible for a parent to be convicted of the strict liability offence under s 444(1) of the Education Act 1996 for failing to ensure that her child attended school regularly. When considering the issue of justification for such liability, the Court referred to the decision of the Court of Appeal in *R v Davies (David Janway)* [2002] IRLR 170 concerning HSWA 1974 s 40.[3]

7.17 The Administrative Court in *Barnfather* stressed that in determining the essentials of a criminal offence, courts should keep in mind the distinction between what were procedural guarantees provided by Art 6(2) of the European Convention on Human Rights and the substantive elements of an offence, a distinction that the Strasbourg Court acknowledged in the civil sphere in *Z v United Kingdom* [2002] 34 EHRR 97, at 138, paras 100–101, when reviewing its decision in *Osman v United Kingdom* [2000] 29 EHRR 245, at paras 138–139. Lord Justice Kay, giving the judgment of the Court, quoted with approval from an article entitled 'The Presumption of Innocence Brought Home? Kebilene Deconstructed' [2002] 118 LQR 41 at 50, in which the author Paul Roberts stated:

> Article 6(2) has no bearing on the reduction or elimination of mens rea requirements, and is therefore perfectly compatible with offences of strict or even absolute liability.

Qualified standards of duty: reasonable practicability

7.18 The meaning of the term, 'so far as is reasonably practicable' is considered in detail at paragraphs 6.101–6.146 and the important decisions of the Court of Appeal in *R v HTM Ltd* [2006] EWCA Crim 1156 and *Dugmore v Swansea NHS Trust* [2003] 1 All ER 333 at paragraphs 6.127–6.134 and 6.118–6.126, respectively. However, in respect of health and safety regulations, three different categories of duty in which 'reasonable practicability' may occur, require consideration.

7.19 The drafting of health and safety regulations has evolved over the past fifteen years. Many of the principal regulations that were first enacted in 1992 have been replaced with fresh provisions. Many regulations drafted post-1997 follow the same form, incorporating a statutory defence that requires a defendant to show

---

[3] [2003] EWHC 418 (Admin) para 31.

'reasonable practicability' of steps taken and all 'due diligence', rather than qualifying the duty itself by the phrase 'so far as reasonably practicable'. Following the Court of Appeal's decision in *HTM Ltd*[4] the distinction between the first category of qualified duties and the second category of post-1997 statutory defence duties, is likely to prove important. As has been said previously, the position is further complicated by Latham LJ's interpretation in *HTM Ltd*[5] of the Court of Appeal's judgment in *Dugmore*,[6] which concerned the duty in Control of Substances Hazardous to Health 1999.[7]

7.20 A third category of health and safety regulations incorporate 'reasonable practicability' only as a qualification upon the standard of required measures.

Regulatory duties qualified by 'reasonable practicability'

7.21 Where the duties are qualified by the term 'reasonable practicability', the second limb of Latham LJ's judgment in *R v HTM*[8] is likely to apply. The Management of Health and Safety at Work Regulations reg 21 provides that, 'Nothing in the relevant statutory provisions shall operate so as to afford an employer a defence in any criminal proceedings for a contravention of those provisions by reason of any act or default of (a) an employee of his'. Latham LJ, in *HTM Ltd*,[9] ruled in terms that the phrase 'reasonably practicable' was properly to be viewed as a qualification of the general duties in HSWA 1974.

7.22 Thus, notwithstanding that a defendant will bear the burden of showing that he took all reasonably practicable measures, pursuant to the terms of HSWA 1974, 'reasonably practicable' is not a defence, and as such, reg 21 has no application.

7.23 A considerable number of duties created by health and safety regulations are similarly worded as 'qualified duties' and thus, by the same argument, are unaffected by the prohibition in Management of Health and Safety at Work Regulations reg 21. Some very recently drafted duties, such as the Construction (Design and Management) Regulations 2007 reg 6 are similarly worded. This provides:

> All persons concerned in a project on whom a duty is placed by these Regulations shall coordinate their activities with one another in a manner which ensures, so far as is reasonably practicable, the health and safety of persons—
> (a) carrying out the construction work; and
> (b) affected by the construction work.

---

[4] [2006] EWCA Crim 1156.
[5] ibid.
[6] [2003] 1 All ER 333.
[7] Now repealed and replaced by the Control of Substances Hazardous to Health Regulations 2002.
[8] See paras 6.127–6.134.
[9] [2006] EWCA Crim 1156.

C. HSWA 1974 s 33(1)(c): Breach of Regulations

7.24 The first limb of Latham LJ's judgment concerned the application of the concept of 'foreseeability' to 'reasonable practicability' and is considered in detail at paragraphs 6.127–6.134. Latham LJ held that 'foreseeability' may be relevant to reasonable practicability, in relation to the general duties under HSWA 1974 ss 2, 3, and 4, but only in a closely confined way: foreseeability being merely a tool with which to assess the likelihood of a risk eventuating.

7.25 This aspect of his ruling is far more difficult to apply to provisions other than those considered in *HTM Ltd*,[10] ie the general duties under HSWA 1974 ss 2–4. In *HTM Ltd*,[11] Latham LJ affirmed the decision in *Dugmore*, which held that foreseeability played no part in reasonable practicability in Control of Substances Hazardous to Health Regulations 1999[12] reg 7(1). That provision provides: 'Every employer shall ensure that the exposure of his employees to a substance hazardous to health is either prevented or, where this is not reasonably practicable, adequately controlled.'

7.26 Latham LJ in *HTM Ltd*,[13] sought to distinguish the general duties under HSWA 1974 from the terms of Control of Substances Hazardous to Health Regulations 1999 reg 7 but did so by holding that, 'The wording of that regulation is significantly different from the wording of section 2 of the 1974 Act'.[14] Latham LJ continued:

> As Hale LJ explained, the primary duty in the regulation is an absolute one. The defendant (in Dugmore) sought to avoid this by arguing that the phrase 'so far as is reasonably practicable' qualified, in effect, the word 'ensure'. The court rejected that argument for obvious reasons. Unlike regulation 7(1), however, the phrase does qualify the word 'ensure' in section 2 of the 1974 Act.

7.27 As has been stated in the previous chapter, the difficulty with Latham LJ's judgment in this respect is that Hale LJ in fact rejected a different submission, namely that, 'the words "so far as is reasonably practicable" should be moved from their current position qualifying the duty [in reg 7] to prevent exposure so as to qualify the duty to ensure that exposure is either prevented or controlled'.

7.28 In the Control of Substances Hazardous to Health 1999 reg 7(1), the words qualify the duty to ensure that exposure is prevented, ie as opposed to both prevented or controlled. As Hale LJ stated, 'the defence of reasonable practicability qualifies only the duty of total prevention'. Furthermore, Hale LJ immediately observed how in the Factory Act 1961 s 29(1), the phrase 'so far as is reasonably practicable'

---

[10] [2006] EWCA Crim 1156.
[11] ibid.
[12] Now repealed and replaced by the Control of Substances Hazardous to Health Regulations 2002.
[13] See paras 6.127–6.134.
[14] [2006] EWCA Crim 1156.

came between 'shall' and 'be made and kept safe', stating, if that was an absolute duty, then so must be Control of Substances Hazardous to Health 1999 reg 7.

7.29 Thus in short, Hale LJ's interpretation of 'reasonable practicability' in the Control of Substances Hazardous to Health 1999 reg 7, would appear to be capable of equal, if not more, general application to health and safety regulations than Latham LJ's interpretation of the same phrase in the general duties under HSWA 1974. Of course, Hale LJ's judgment was that foreseeability can play no part in 'reasonable practicability'.

Regulatory duties with a defence involving 'reasonable practicability'

7.30 There are two different provisions that occur in many health and safety regulations drafted post-1997 which expressly create a 'defence' involving 'reasonable practicability' and thus, despite the decision in HTM Ltd, are subject to the terms of the Management of Health and Safety at Work Regulations 1999 reg 21, which provides an employer is not to be afforded a defence in any criminal proceedings for a contravention of a health and safety provision by reason of any act or default of an employee of his.

7.31 The first form of provision[15] provides:

> Defence in proceedings
> ... it shall be a defence for the person charged to prove—
> (a) that the contravention was due to the act or default of another person not being one of his employees (hereinafter called 'the other person'); and
> (b) that he took all reasonable precautions and exercised all due diligence to avoid the contravention.

7.32 Provisions in later Regulations, drafted after the enactment of the Management of Health and Safety at Work Regulations provide:

> **Defence**
> Subject to regulation 21 of the Management of Health and Safety at Work Regulations 1999, in any proceedings for an offence consisting of a contravention of these Regulations it shall be a defence for any person to prove that he took all reasonable precautions and exercised all due diligence to avoid the commission of that offence.

7.33 Regulations which include either of the above provisions are unaffected by the decision in *HTM Ltd* concerning Management of Health and Safety at Work Regulations 1999 reg 21, and thus subject to the prohibition in that Regulation that nothing in any health and safety provision shall operate so as, 'to afford an employer a defence in any criminal proceedings for a contravention of those provisions by reason of any act or default of ... an employee of his'.

---

[15] See, for example, Confined Spaces Regulations 1997 SI 1997/1713 reg 7; Pressure System Safety Regulations 2000 SI 2000/128 reg 16; Pipelines Safety Regulations 1996 SI 1996/825 reg 28.

C. HSWA 1974 s 33(1)(c): Breach of Regulations

In the context of what 'all reasonable steps' involves in respect of health and safety, it is difficult to envisage a situation in which a defendant could succeed in demonstrating that he had taken all such steps but fail at the same time to demonstrate that he had shown 'all due diligence'. **7.34**

Regulations in which measures to be taken are qualified by 'reasonably practicable'

In a third category of regulation, 'reasonably practicable' is employed not as a qualification on compliance but as a qualification on the measures required. **7.35**

The Work at Height Regulations 2005 replaced and repealed the Construction (Health Safety and Welfare) Regulations 1996, which in turn had replaced the earlier, similarly worded, Construction (Working Place) Regulations 1966, reg 36(2) of which was drafted in the following terms: **7.36**

> Without prejudice to the provisions of Regulations 33 and 35, no person shall pass or work near [material which would be liable to fracture if his weight were to be applied to it and so situated that if it were to be so fractured he would be liable to fall a distance of 2 meters] unless provision is made by means of such one or more of all of the following, that is to say, suitable guard rails, suitable coverings and other suitable means as are necessary for preventing, so far as reasonably practicable, any person so passing or working from falling through the said material.

This Regulation has been held to have imposed an absolute duty to provide the suitable means referred to, and 'reasonably practicable' was held to qualify the standard necessary to prevent a person falling through the material (*Briggs Amasco Ltd v Thurgood* The Times 30 July 1984 (DC)). The case of *R v Rhone-Poulenc Rorer Ltd* [1996] ICR 1054 (CA) affirmed this point when it held that the words 'reasonably practicable' in reg 36(2) related specifically to the ability of the safety device to prevent an accident of the type envisaged.[16] **7.37**

Thus, where the Work at Height Regulations 2005 reg 6(3) provides: 'Where work is carried out at height, every employer shall take suitable and sufficient measures to prevent, so far as is reasonably practicable, any person falling a distance liable to cause personal injury', 'reasonably practicable' qualifies the standard of the measures; the employer is under an absolute duty to take suitable and sufficient measures. **7.38**

Other qualifications in regulations

Some regulations passed since 1974 have employed terms other than 'reasonably practicable' or 'practicable' (in particular, 'all reasonable steps'). Unless the regulation employs a term encompassed by HSWA 1974 s 40, then that section will **7.39**

---

[16] The predecessor reg 24(1) of the Building (Safety, Health and Welfare) Regulations 1948 was similarly held to create an absolute duty. (See *Westcott v Structural and Marine Engineers Ltd* [1960] 1 All ER 775.)

not apply to the regulation: the ambit of such a provision affecting criminal liability must be narrowly construed. However, most regulations employing terms other than those governed by HSWA 1974 s 40 contain a separate reverse burden clause.

D. Other Offences

Health and safety inquiries: HSWA 1974 s 33 (1)(d)

7.40 By HSWA 1974 s 33(1)(d) it is an offence 'to contravene any requirement imposed by or under regulations under section 14 or intentionally to obstruct any person in the exercise of his powers under that section'. The offence is triable only in the magistrates' court and is punishable with a maximum fine of £5,000.

7.41 HSWA 1974 s 14 concerns the Health and Safety Commission's power to order the Health and Safety Executive, or any other person, to investigate 'any accident, occurrence, situation or other matter' that it thinks necessary to investigate with a view to introducing new regulations or for any of the general purposes of HSWA 1974, Pt I, namely to secure the health, safety, and welfare of persons at work and other persons affected by their work. It also empowers the Commission to direct the Executive to hold an inquiry into any such matter, and to provide for reports of investigations and inquiries to be made public.[17]

7.42 The requirements 'imposed by or under' s 14 include the power of a person holding such an inquiry to enter and inspect, to summon witnesses to give evidence or produce documents, to take evidence on oath and administer oaths, to require the making of declarations and, with the permission of a Government minister, to require any such inquiry to be held otherwise than in public.

7.43 The Health and Safety Inquiries (Procedure) Regulations 1975 SI 1975/335 were introduced pursuant to HSWA 1974 s 14 and govern procedure at, and the powers of, Health and Safety Inquiries.

Contravening requirements by inspectors: HSWA 1974 s 33(1)(e)

7.44 It is an offence under HSWA 1975 s 33(1)(e) 'to contravene any requirement imposed by an inspector under section 20 or 25'. The offence is triable only in the magistrates' court and is punishable with a maximum fine of £5,000.

7.45 HSWA 1974 s 82 defines 'contravention' as including a failure to comply and provides that 'contravene' has a corresponding meaning.

---

[17] This power was first invoked in 2005 when an inquiry under the chairmanship of Lord Newton was ordered following the explosion and fire at the Buncefield Oil Farm. See para 1.21.

D. Other Offences

7.46 The extensive powers available to inspectors under HSWA 1974 s 20 are considered in Chapter 2.

'Requirement'

7.47 HSWA 1974 s 20 gives inspectors various powers to 'direct' and to 'require'. There seems no good reason to differentiate between the two terms but the failure to include the word 'direction' in the very definition of what constitutes a criminal offence under s 33(1)(e) arguably means that, bearing in mind the need to construe a criminal provision restrictively, this offence must be limited to a contravention of HSWA 1974 s 20(2)(j) (k) and (l), being the three subsections in which an inspector is given a power to 'require'. In any event, the offence created by s 33(1)(h), of wilful obstruction, seems apt to cover an intentional contravention of a direction issued by an inspector.

7.48 The relevant subsections of HSWA 1974 s 20(2) give an inspector the power to require:

(j) any person whom he has reasonable cause to believe to be able to give any information relevant to any examination or investigation . . . to answer (in the absence of persons other than a person nominated by him to be present and any persons whom the inspector may allow to be present) such questions as the inspector thinks fit to ask and to sign a declaration of the truth of his answers;
(k) the production of, inspect, and take copies of or of any entry in—
   (i) any books or documents which by virtue of any of the relevant statutory provisions are required to be kept; and
   (ii) any other books or documents which it is necessary for him to see for the purposes of any examination or investigation under paragraph (d) above;
(l) any person to afford him such facilities and assistance with respect to any matters or things within that person's control or in relation to which that person has responsibilities as are necessary to enable the inspector to exercise any of the powers conferred on him by this section.

7.49 The ambit of s 20(2)(j), the inspector's power to require a person to require a person to give information, is considered at paragraphs 2.132–2.150. No answers obtained from a person pursuant to this power can be used in a prosecution against him (or his spouse). In the case of *London Borough of Wandsworth v South Western Magistrates' Court* [2003] EWHC 1158 (Admin) the Divisional Court ruled that the power under s 20(2)(j) should be construed widely and that it permitted an inspector to seek and obtain information in writing, as well as face to face.

HSWA 1974 s 25 powers

7.50 Surprisingly, HSWA 1974 s 25 does not expressly give an inspector a power to 'require'. However, the section does give an inspector very wide power in respect of articles or substances that he believes are 'a cause of imminent danger of serious personal injury'. These powers include a power to 'cause it to be rendered harmless (whether by destruction or otherwise)': presumably 'cause' includes a power to require a person to do something.[18]

7.51 Similar powers to require information or assistance to that created by HSWA 1974 s 20 are given to other regulatory inspectors, for example under the Food Safety Act 1990. However, unlike offences created under the Food Safety Act 1990, the offence under s 33(1)(e) of contravening a requirement imposed by a health and safety inspector requires no proof of intent against a defendant; nor does HSWA 1974 create any defence of reasonable excuse for such failure upon which a defendant might rely. The effect is that, upon its face, HSWA 1974 creates a wide power of compulsory questioning backed up by a strict criminal sanction for any contravention.

Preventing a person from attending or answering questions: HSWA 1974 s 33(1)(f)

7.52 HSWA 1974 s 33(1)(f) makes it an offence 'to prevent or attempt to prevent any other person from appearing before an inspector or from answering any question to which an inspector may by virtue of section 20(2) require an answer'. The offence is triable only in the magistrates' court and is punishable with a maximum fine of £5,000.

7.53 This offence provides another criminal sanction to ensure compliance with an inspector's power to obtain information under HSWA 1974 s 20(2)(j). The wording of the offence is not limited by any defence of 'reasonable cause'.

Contravening improvement or prohibition notices: HSWA 1974 s 33(1)(g)

7.54 HSWA 1974 s 33(1)(g) creates an offence of contravening any requirement or prohibition imposed by an improvement notice or a prohibition notice (including any such notice as modified on appeal).[19] The offence is punishable by a fine of £20,000 or six months' imprisonment in the magistrates' court, and by an unlimited fine, two years' imprisonment, or both in the Crown Court.

No 'reasonably practicable' defence

7.55 The provisions of HSWA 1974 s 40 do not provide any defence to this offence and a defendant cannot seek to challenge the merits of the notice[20] to establish a defence: the offence creates strict liability for a failure to comply with a requirement or prohibition within the notice.

---

[18] See Ch 2.
[19] See Ch 4 for Improvement and Prohibition Notices generally.
[20] See *Dearey v Mansion Hide* [1983] ICR 610.

D. Other Offences

Obstructing an inspector: HSWA 1974 s 33(1)(h)

7.56 HSWA 1974 s 33(1)(h) creates an offence of intentionally obstructing an inspector in the exercise or performance of his powers or duties. The offence is triable only in the magistrates' court and is punishable with a maximum fine of £5,000.

Obstructing an inspector

7.57 The offence requires proof that an inspector was at the time of the alleged obstruction exercising or performing his powers or duties. The offence is similarly worded to that now under s 89(2) of the Police Act 1996 of wilfully obstructing a constable in the execution of his duty, which has established a considerable body of case law. It is apparent that 'obstructing' a police officer is not confined to physical obstruction, but includes anything which makes it more difficult for him to carry out his duties (*Hinchcliffe v Sheldon* [1955] 1 WLR 1207 and *Ledger v DPP* [1991] Crim LR 439).

Intentionally

7.58 It appears by analogy with s 89(2) of the Police Act 1996 that the prosecution must prove an intention that the deliberate act should in fact obstruct the inspector (*Wilmott v Atack* [1977] QB 498). The act of obstruction need not necessarily be aimed at or hostile to the inspector, but must in fact make it more difficult for the inspector to carry out his duty and the person must know that his conduct would have this effect (*Lewis v Cox* [1984] 3 WLR 875). A well-known example was where a defendant was held to have wilfully obstructed a police constable when, without hostility, he intervened because he thought the constable was arresting the wrong man (*Hills v Ellis* [1983] 2 WLR 234).

Provision of information to the commission: HSWA 1974 s 33(1)(i)

7.59 It is an offence under HSWA 1974 s 33(1)(i) 'to contravene any requirement imposed by a notice under section 27(1)'. Section 27(1) affords the Commission the power to serve a notice on any person requiring that person to provide the Commission with any information it, or any enforcing authority, needs to discharge its functions. A person receiving such a notice cannot rely upon the privilege against self-incrimination as a reason not to comply (see *R v Hertfordshire County Council ex parte Green Environmental Industries Ltd* [2000] 2 AC 412).

7.60 This offence is triable either in the magistrates' court, where the maximum sentence is a fine of £5,000, or at the Crown Court where a fine in any sum can be imposed.

Disclosure of information: HSWA 1974 s 33(1)(j)

7.61 HSWA 1974 s 33(1)(j) makes it an offence to use or disclose any information in contravention of HSWA 1974 s 27(4) or s 28. Section 27(4) prohibits the HSC or the HSE from using any information obtained under HSWA 1974 for any purpose other than a purpose of the HSC or HSE. This offence is triable in the magistrates' court, where the maximum sentence is a fine of £5,000, or at the Crown Court, where the maximum sentence is an unlimited fine, imprisonment of up to two years, or both.

7.62 HSWA 1974 s 28 imposes limits on the disclosure of information obtained by the Commission under powers by preceding sections. Section 28(1)–(6) provide that no information obtained under s 27(1) or in pursuance of a requirement imposed by the relevant statutory provisions may be disclosed by the recipient of the information without the consent of the person who gave it. The section details various bodies or individuals to whom disclosure may be made but specifies that where such information is disclosed it may only be used for certain purposes. The offence is presumably intended to principally apply to such recipients of information who use the information for a purpose other than that permitted by s 28(5) but importantly the definition of 'relevant information' is wide enough to include all material obtained by an inspector by the exercise of his s 20 powers, including a statement obtained by compulsion under s 20(2)(j).

7.63 The general prohibition exception contained in HSWA 1974 s 28(2) that no relevant information shall be disclosed without the consent of the person by whom it was furnished is subject to the broad exclusion in s 28(3)(e) relating to the permissible 'disclosure of information for the purposes of any legal proceedings'.

7.64 HSWA 1974 s 28(7) provides that a person shall not disclose any information obtained by him as a result of the exercise of any power conferred by s 20 except:

(a) for the purposes of his functions; or
(b) for the purposes of any legal proceedings or any investigation or inquiry held by virtue of section 14(2) or for the purposes of a report of any such proceedings or inquiry or of a special report made by virtue of section 14(2); or
(c) with the relevant consent.

7.65 'The relevant consent' means, in the case of information furnished in pursuance of a requirement imposed under s 20, the consent of the person who furnished it, and, in any other case, the consent of a person having responsibilities in relation to the premises where the information was obtained.

7.66 HSWA 1974 s 28(9) provides that notwithstanding anything in s 28(7), a person who has obtained such information as is referred to in that subsection may furnish to a person who appears to him to be likely to be a party to any civil proceedings arising out of any accident, occurrence, situation, or other matter, a written statement of relevant facts observed by him in the course of exercising any of the powers referred to in that subsection.

D. Other Offences

False statements: HSWA 1974 s 33(1)(k)

HSWA 1974 s 33(1)(k) provides that it is an offence for a person: **7.67**

to make a statement which he knows to be false or recklessly to make a statement which is false where the statement is made:
in purported compliance with a requirement to furnish any information imposed by or under any of the relevant statutory provisions; or
for the purpose of obtaining the issue of a document under any of the relevant statutory provisions to himself or another person.

This offence is triable either in the magistrates' court, where the maximum sentence is a fine of £5,000, or at the Crown Court where a fine in any sum can be imposed. **7.68**

Knowledge or recklessness

It is for the prosecution to prove that a defendant knew, at the time that the statement was made, that it was false or that he recklessly made a false statement. This is proved, either directly, by the evidence of the defendant, or circumstantially and will obviously involve the prosecution proving that the statement was false. **7.69**

Knowledge is a word in ordinary English usage and in this context it must import an awareness that the statement was false. **7.70**

Recklessness has been the subject of recent judicial consideration. **7.71**

Making a statement

The term 'statement' is not intended to be interpreted in the sense of one that complies with the requirements of s 9 of the Criminal Justice Act 1967, but rather should be understood as carrying its ordinary every day meaning of 'the action or an act of stating; the manner in which something is stated'.[21] Whilst most statements falling within the terms of HSWA 1974 s 33(1)(k)(i) and (ii) will be either made in writing or written down by the inspector and declared in writing to be true by the person making the statement, a statement can be made orally. **7.72**

A requirement to furnish any information imposed by or under any of the relevant statutory provisions

The 'relevant statutory provisions' include HSWA 1974, Pt I, the pre-1974 regulations listed at HSWA 1974, Sch 1 and the wealth of health and safety regulations passed since 1974. This body of legislation includes numerous provisions requiring a person to supply information either to the HSE or other enforcing authorities with one notable example being the duties under the Reporting of Injuries, Diseases and Dangerous Occurrences Regulations 1995 SI 1995/3163 (RIDDOR). **7.73**

---

[21] Shorter Oxford English Dictionary.

The offence and statements obtained under HSWA 1974 s 20(1)(j)

7.74 HSWA 1974 contains various provisions requiring the furnishing of information but the most significant is the power of the inspector under s 20(1)(j) 'to require any person ... to answer such questions or provide such information as he sees fit for the purpose of his investigations ... and to sign a declaration of the truth of his answers'.

7.75 However, HSWA 1974 s 20(7) abrogates the right of a person not to incriminate himself but protects such a person from the consequences of any self-incrimination, providing: 'No answer given by a person in pursuance of a requirement imposed under subsection (2)(j) above shall be admissible in evidence against that person or the husband or wife of that person in any proceedings.'

7.76 On its face, this provision goes far further than s 59 of and Sch 3 to the Youth Justice and Criminal Evidence Act 1999, which were enacted following the European Court's decision in the *Saunders*[22] case to prevent the prosecution relying on answers obtained as a result of compulsory questioning under various Acts in subsequent prosecutions, save in certain circumstances including in respect of offences of giving false information and offences contrary to s 5 of the Perjury Act 1911 (false declarations other than under oath).

7.77 HSWA 1974 s 20(7), by preventing the admission against the person of any answer in 'any proceedings', appears to prevent a prosecution of such a person for an offence under s 33(1)(k) in respect of making a false statement in answer to the inspector exercising his power under s 20(2)(j).

For the purpose of obtaining the issue of a document

7.78 Once again, the regulations abound with examples of licences, consents, and even acknowledgements that a person is required to obtain in differing circumstances before undertaking activities or as proof of notification from and to either the HSE or local enforcing authorities.

7.79 The prosecution is required to prove, directly or by inference, that the purpose of making the false statement was to obtain the document.

Making false entries: HSWA 1974 s 33(1)(l)

7.80 HSWA 1974 s 33(1)(l) provides that it is an offence 'intentionally to make a false entry in any register, book, notice or other document required by or under any of the relevant statutory provisions to be kept, served, or given or, with intent to deceive, to make use of any such entry which he knows to be false'.

---

[22] *Saunders v UK* (1997) 23 EHRR 313.

D. Other Offences

7.81 This offence is triable either in the magistrates' court, where the maximum sentence is a fine of £5,000, or at the Crown Court where a fine in any sum can be imposed.

Any register, book, notice, or other document required

7.82 The offence appears to encompass every conceivable document required for any purpose under HSWA 1974 and health and safety regulations.

Using a document with intent to deceive: HSWA 1974 s 33(1)(m)

7.83 Under HSWA 1974 s 33(1)(m), it is an offence:

> with intent to deceive to . . . use a document issued or authorised to be issued under any of the relevant statutory provisions or required for any purpose thereunder or to make or have in his possession a document so closely resembling any such document as to be calculated to deceive.

7.84 This offence is triable either in the magistrates' court, where the maximum sentence is a fine of £5,000, or at the Crown Court where a fine in any sum can be imposed.

'A document issued or authorised to be issued or required for any purpose'

7.85 HSWA 1974 s 33(1)(m) apparently encompasses all documents issued by the HSE or enforcing authorities and, with the term 'required for any purpose thereunder', those documents prepared by persons in order to comply with requirements under health and safety regulations.

'With intent to deceive'

7.86 The words 'with intent to deceive' apply to the whole of the subsection so that the prosecution is required to prove, directly or by inference, that a defendant used, made, or had in his possession such a specified document with intent to deceive.

'Calculated to deceive'

7.87 In *R v Davison* (1973) 57 Cr App R 113, the Court of Appeal was concerned with an offence of using a document resembling a prescribed certificate of authority that so closely resembled a certificate as to be 'calculated to deceive' contrary to s 5 of the House to House Collections Act 1939. The Court held that the words 'calculated to deceive' meant likely to deceive and not necessarily intended to deceive.

Impersonating an inspector: HSWA 1974 s 33(1)(n)

7.88 This section makes it an offence 'falsely to pretend to be an inspector'. The offence is triable only in the magistrates' court and is punishable with a maximum fine of £5,000.

Failure to comply with a court order: HSWA 1974 s 33(1)(o)

7.89 It is an offence under this section to fail to comply with an order made by a court under HSWA 1974 s 42. Under s 42, where a person is convicted of an offence under any of the relevant statutory provisions in respect of any matters which appear to the court to be matters which it is in his power to remedy, the court may, in addition to or instead of imposing any punishment, order him, within such time as may be fixed by the order, to take such steps as may be specified in the order for remedying the matters.

7.90 The offence is punishable by a fine of £20,000 or six months' imprisonment in the magistrates' court, and by an unlimited fine, two years' imprisonment, or both in the Crown Court.

8

PROCEEDINGS AND CRIMINAL PROCEDURE

A.	**Instituting Proceedings**	8.01		Plea negotiations: accepting lesser pleas	8.144
	Introduction	8.01		Progression of cases: case management	8.147
	Arrestable offences	8.03			
	Future abolition of informations and summonses	8.05		Plea bargaining: *R v Goodyear*	8.154
	Who may commence proceedings?	8.07	D.	**Disclosure**	8.157
	Time limits	8.14		General legislative framework	8.159
	Bringing proceedings against a company in liquidation	8.17		Disclosure in the magistrates' court	8.162
				Procedure for making disclosure	8.163
	Jurisdiction of the magistrates' court	8.26		Defence disclosure	8.166
	Amending the information or summons	8.40		Third party disclosure	8.169
				The investigation stage	8.170
B.	**Conducting a Health and Safety Case in the Magistrates' Court**	8.43		Obtaining material from third parties	8.181
	The initial appearance	8.45		The disclosure officer	8.183
	Advance information in the magistrates' court	8.48		The prosecutor	8.184
				Schedules of unused material	8.186
	Friskies statements	8.54		Effect of non-compliance with the Code	8.189
	Plea before venue	8.69			
	Procedure for determining venue for trial or sentence	8.81	E.	**Expert Evidence**	8.190
				The 'ultimate issue rule'	8.193
	Powers of the Crown Court when dealing with committals	8.100		Independence of an expert	8.195
				Calling the expert	8.205
	Procedure for determining mode of trial	8.102		Future development of disclosure duty	8.207
	Judicial review of the decision to commit	8.117	F.	**Miscellaneous Areas of Evidence**	8.208
				Hearsay in criminal cases	8.208
C.	**Conducting a Health and Safety Case in the Crown Court**	8.118		Bad character	8.214
	The Criminal Procedure Rules 2005	8.118			
	Lodging the indictment	8.121			

A. Instituting Proceedings

Introduction

8.01 As with many other areas of criminal procedure, recent legislation has effected major changes to the process for instituting proceedings for breaches of health and safety law. In turn, this has had significant implications for taking cases through both the magistrates' court and Crown Court. Although not all of the relevant provisions are yet in force, the majority are covered here, in anticipation of their introduction.

8.02 Practice and procedure regarding the institution of proceedings is now largely governed by the Criminal Procedure Rules 2005, which came into force on 4 April 2005. They substantially replace the patchwork of rules which regulated discrete areas in the magistrates' court and Crown Court. They represent something of a harmonization of the previously separate procedures operating in the two courts. That said, except for the expression of a number of 'overriding objectives' for the conduct of criminal litigation generally,[1] the Rules do not effect any radical changes; rather, they merely consolidate, and develop, to an extent, the existing rules and procedures.

Arrestable offences

8.03 Before the advent of the Serious Organized Crime and Police Act 2005, none of the offences created by health and safety law was an arrestable offence. Section 24 of the Police and Criminal Evidence Act 1984 has been amended by s 110 of the Serious Organized Crime and Police Act 2005, so that it now provides a police officer with a power of arrest in respect of all offences—whether summary, triable either way, or indictable only. Furthermore, a new s 24A, Police and Criminal Evidence Act 1984 has created the power for a person other than a police constable to arrest another without warrant in certain limited circumstances in respect of an indictable offence, ie for many of the health and safety offences under HSWA 1974. Therefore, at least technically, a person who is suspected of committing a health and safety offence may be arrested.

8.04 Thus, proceedings against individuals for breaching health and safety legislation, where instituted by the Crown Prosecution Service, may conceivably be commenced by arrest, charge, and bail to the magistrates' court. All proceedings instituted by enforcing authorities, and those commenced by the Crown Prosecution Service against companies, will continue to be instituted by way of laying an information and the issue of summons in the magistrates' court.

---

[1] See paras 8.119–8.120 for an elaboration of the objectives the court will seek to further.

A. Instituting Proceedings

Future abolition of informations and summonses

8.05 Although not yet in force, s 29 of the Criminal Justice Act 2003 will abolish the laying of an information and issue of a summons as a means of commencing criminal proceedings. Instead, a public prosecutor will be responsible for commencing proceedings by issuing a document called a written charge. This document must be accompanied by a further document—called a requisition—requiring the person facing the charge to attend a specified court at a specified time. These two documents must also be served on the court specified in the requisition at the same time.

8.06 In this way, the person facing the charge will be notified by the prosecutor, and not by the court, that he is charged with an offence. The necessity to attend court should also be made apparent to him. With the new procedure in place, the court will have no involvement at all with a given case until the defendant first appears at court.

Who may commence proceedings?

8.07 Proceedings under the Health and Safety at Work etc. Act 1974 (HSWA), or under health and safety regulations, may only be instituted by an inspector, or with the consent of the Director of Public Prosecutions.[2] Alongside these, the Environment Agency is also in some circumstances a competent enforcement authority. This amounts to a specific limitation on the right, which exists otherwise, of any individual to lay an information for the breach of a statutory provision, providing that the offence does not concern an individual grievance.

8.08 An inspector is appointed by an instrument in writing, specifying which of the relevant statutory powers conferred on inspectors he may exercise. This certificate of appointment does not need to specifically empower him to bring proceedings,[3] as he has the right to do so in any event.

8.09 Proceedings must be instituted by, and in the name of, an inspector and may not be instituted in the name of an unincorporated body, such as the HSE, or a particular local authority. In *Rubin v DPP* [1990] 2 QB 80, it was held that an information had to be laid by the constable who reported the commission of the offence. Alternatively, the chief constable himself, or some individual authorized by him might lay an information, but not—as applied in this case—Thames Valley Police. In spite of this, the court found that the fact that the informant was Thames Valley Police did not invalidate the information, as it was plain that it had been laid by a police officer whose identity could easily have been established. The same principles should apply to the laying of informations for health and safety offences.

---

[2] HSWA 1974 s 38. For definition of 'statutory provisions' see HSWA 1974 s 53(1).
[3] *Campbell v Wallsend Slipway and Engineering Co Ltd* [1978] ICR 1015.

The information should be laid by the inspector himself, or by some person who is authorized by the Director of the Health and Safety Executive to lay informations.

8.10 Further, an inspector's power cannot be delegated to a solicitor purporting to act on his behalf. In those circumstances, proceedings would not have been validly instituted,[4] with the effect that the laying of the informations would not be properly effected.

8.11 There is no statutory requirement that the proceedings should be instituted by the inspector who investigated the case, or even by the reviewing inspector. Inspectors have locus to prosecute cases in the magistrates' court, and frequently do so, even though they are not counsel or a solicitor.[5]

8.12 A large number of prosecutions are conducted in the magistrates' court by the inspector who was responsible for investigating the case, although, it is common practice for an HSE inspector to instruct a solicitor agent in private practice to undertake the prosecution.

Independent Legal Oversight

8.13 In April 2004 a new system of Independent Legal Oversight was instituted by the HSE, which required inspectors to refer the 'most serious, complex and sensitive criminal cases' to HSE's Legal Adviser's Office for conduct. The criteria used to identify cases which require such Independent Legal Oversight are set out in an Operational Circular, OC 168/11.[6] If a case is accepted for Independent Legal Oversight, then the HSE's lawyers assume the role of prosecutor whilst the inspectors retains the role of investigator. (The inspector remains the only person who can lawfully commence proceedings.)

Time limits

8.14 There is no time limit for making a complaint regarding an indictable offence.[7] An offence of breaching one of the general duties under HSWA, an offence of breaching a health and safety regulation and most of the other commonly prosecuted health and safety offences, are all triable on indictment: that is to say they may be tried either in the magistrates' court or the Crown Court.

8.15 For those offences which are summary only, the information must be laid within six months of the commission of the offence, or of the time when the matters complained of became apparent.[8] This six-month rule is subject to the important

---

[4] See *R v Croydon Justices ex p W H Smith Ltd* [2001] EHLR 12, (2000) 97 (46) LSG 39.
[5] HSWA 1974 s 39(1).
[6] http://www.hse.gov.uk/foi/internalops/fod/oc/100-199/168_11.pdf.
[7] Interpretation Act 1978 Sch 1.
[8] Magistrates' Courts Act 1980 s 127.

A. Instituting Proceedings

exception, in HSWA 1974 s 34, which extends the period in certain defined circumstances; namely, where, following an accident or other occurrence the HSC has directed the HSE to investigate and prepare a special report, where with the consent of the Secretary of State an inquiry is held;[9] or where a coroners' inquest has been held in relation to a work-related death, and it appears that any health and safety provision has been contravened.

8.16 With summary-only offences, the failure to lay an information within the six-month period will lead to the proceedings being time-barred. It is also undesirable to delay the laying of an information until the last possible moment.[10] In *R v Brentford Justices ex parte Wong* [1981] 1 All ER 884, the prosecutor laid an information the day before the six-month period was due to expire. This was done without having reached a conclusion or view as to whether the prosecution would be continued or not, as a means of gaining more time to make a decision. It was held that such conduct amounted to an abuse of process. The purpose of s 127 of the Magistrates' Courts Act 1980, the Court went on to reflect, was to ensure that prosecutions were brought within a reasonable time.

Bringing proceedings against a company in liquidation

8.17 It sometimes happens that a company which faces prosecution for health and safety offences goes into liquidation, or else is wound up before it is prosecuted or appears before the court. Companies may also go into administration, although the situation is less frequently encountered in the context of health and safety enforcement. This does not necessarily prevent prosecution of the company.

8.18 There are two types of liquidation: compulsory liquidation and voluntary liquidation. Compulsory liquidation occurs after an application is made to the court to wind up the company. Voluntary liquidation occurs after a resolution of the shareholders. In the first instance, leave of the court will be required if proceedings are to be brought; in the second it will not.

8.19 If proceedings against a company in liquidation are contemplated, it should be borne in mind that, in the event of a conviction, the imposition of a financial penalty will lead to the court being treated as an unsecured creditor in the usual way. This is potentially a relevant factor, when considering whether the public interest requires prosecution; and in practice companies which have gone into liquidation are often not prosecuted for this reason.

8.20 In the event of a guilty plea or conviction after trial, if the sentencing judge is likely to fine the company, it is prudent to obtain a report from the liquidators informing the court when the company's assets are likely to be realized, and how much

---

[9] HSWA 1974 s 14.
[10] *R v Blackburn JJ ex p Holmes* [2000] CLR 300.

unsecured creditors are likely to receive for each pound owed to them. Although such a report may be obtained by either the prosecution or the defence, the information is likely to be more readily available to the defence. Moreover, the provision of such information is not inconsistent with the obligation on the defence, as set out in *R v Howe*,[11] to furnish the court with information concerning its ability to meet any financial penalty, and with the dicta in *R v Transco Ltd*.[12]

Compulsory liquidation

8.21 If it is intended to commence or pursue proceedings against a company where a winding-up order has been made by the court, or a provisional liquidator has been appointed—ie a compulsory winding up—leave of the court with jurisdiction over the winding-up order must be sought.[13] It may be refused, and the court has the power to impose such terms as it thinks fit. The winding-up order is deemed to commence at the time the presentation of a petition for winding up is made at the court. Where the company has passed a resolution to wind up voluntarily, the order takes effect at the time of the passing of that resolution.

8.22 In *R v Dickson and Wright* 94 Cr App R 7, it was assumed that a criminal prosecution amounts to a 'proceeding' within the meaning of the Insolvency Act 1986 s 130 (2), and therefore the leave of the court is required before one may be brought. In that case, however, it was said that the fact that none had been obtained made no material difference to the course of the trial and did not render the conviction unsafe.

8.23 The factors to be taken into account when deciding whether a prosecution ought to be allowed or not were considered in *Re Rhondda Waste Disposal Ltd* [2001] Ch 57 CA. It was held that, when the public interest requires it, leave should be readily given. The interests of the creditors should not, the court proceeded to say, be taken to trump all other considerations.

Voluntary liquidation

8.24 There are two types of voluntary liquidation: members' voluntary or creditors' voluntary liquidation.

8.25 A members' voluntary winding up can only take place when the company is solvent, and a declaration of solvency must be filed with the court. A creditors' voluntary liquidation takes place when the directors call a meeting of the shareholders who agree (by formal resolution) to wind up the company. Although the winding up is initiated by a shareholders' resolution, it will be under the control

---

[11] *R v Howe & Son (Engineers) Ltd* [1999] 2 All ER 249.
[12] [2006] EWCA Crim 838.
[13] Insolvency Act 1986 s 130(2).

A. Instituting Proceedings

of the creditors—who will appoint a liquidator or liquidation committee. Unlike in the case of compulsory winding up, under English law there is no statutory provision which prevents actions and other proceedings being taken against a company which is being voluntarily wound up. However, the court has a discretion to stay such proceedings upon an application by the liquidator to the court under the Insolvency Act 1986 s 112(1).

Jurisdiction of the magistrates' court

Laying the information: obtaining and serving the summons

8.26 The rules of procedure which govern this area of practice are now largely consolidated in the Criminal Procedure Rules 2005 Pts 4 and 7. As has been stated above, though, the process for initiating proceedings is to be radically changed when Criminal Justice Act 2003 s 29 and Sch 3 come into force. The following sections state the law and practice at the date of publication.

8.27 An information may be laid orally: it does not need to be laid in writing or on oath.[14] Once the information is laid, the justice or justices' clerk should apply his mind in each case to the question of whether it is appropriate to issue a summons or warrant. The decision to issue involves the exercise of a judicial function and should not be undertaken without proper consideration.[15]

8.28 Any magistrates' court may issue a summons in respect of an offence, providing they have jurisdiction to try the offence. (A magistrates' court has the jurisdiction to try any summary offence and any either way offence.) Where they have accepted jurisdiction, they may technically issue a summons in respect of any offence, no matter in which part of the country it is alleged to have occurred.[16]

8.29 At least technically, therefore, HSE inspectors have a discretion as to where they issue proceedings. It is conventional to lay an information in the area where the offence was said to have been committed, though this is a matter of practice and there is no requirement to do so.[17] In the Greater London Area, the HSE presently have an agreement with the City of London Magistrates' Court that all cases from the London Area will commence there.

8.30 Indeed, the prosecution case in health and safety proceedings will generally encompass matters beyond the circumstances of the immediate accident. This will certainly be the case, for instance, when allegations are in connection with wider company procedures and policies. In this type of situation, at least, it is likely that

---

[14] Criminal Procedure Rules r 7.1.
[15] See for instance *Gateshead Justices ex p Tesco Stores Ltd* [1981] QB 470.
[16] See Magistrates' Courts Act 1980 ss 1 and 2.
[17] Within the metropolitan district of London, the HSE has reached an agreement with the City of London Magistrates' Court that cases should begin there.

the 'offence' may be said to have been committed at different locations, and, therefore, it may be appropriate to bring proceedings in one of a number of places.

8.31 It may be alleged that a health and safety offence has been committed in connection with any plant or substance. In these circumstances, the offence may be treated as having been committed at the place where that plant or substance is, for the time being, held. Proceedings may be instituted in the court covering that area.[18]

8.32 Once it is laid, a summons will be issued, and served by the court, on the individual or company concerned. The summons will include a date on which the matter is to be first heard at the magistrates' court.

8.33 The service of a summons, issued by the magistrates' court, is now governed by Criminal Procedure Rules 2005, r 4. In the case of a private individual, service of a summons may be effected by delivering it to the person to whom it is directed.[19] It may also be effected using the postal service. This can be done by leaving it for him with some person at his last known or usual place of abode, or by sending it by post in a letter addressed to him at his last known or usual place of abode.[20]

8.34 In the case of a corporation, service may be effected by delivering the summons to the registered office of the corporation in the UK, or alternatively sending it by post. If there is no registered office in the UK, it can be sent to any place in the UK where the corporation trades.[21]

8.35 Although there is no time limit for serving a summons after an information has been laid, it must not be delayed so as to prejudice the defendant. He must also be given reasonable notice of the hearing.

Content of a summons

8.36 A summons or information will be sufficient if it contains a description of the specific offence charged. It must be phrased in ordinary language, avoiding as far as possible the use of technical terms. It must give such particulars as are necessary for providing reasonable information of the nature of the charge.[22] If the offence is one that is created by statute or statutory instrument, there should be a reference to the section of the Act or regulation which creates the offence.[23] It is not necessary to state all the elements of the offence: for instance, there is no need to expressly negative any statutory exemption.[24] If the particulars given are

---

[18] HSWA 1974 s 35.
[19] See *Ralux NV/SA v Spencer Mason* The Times 18 May 1989 for a consideration of service by fax.
[20] Criminal Procedure Rules r 4.1.
[21] ibid r 4.1(2).
[22] ibid r 7.2(1).
[23] ibid r 7.2(1).
[24] ibid r 7.2(2).

A. Instituting Proceedings

considered to be inadequate, the court may order that further particulars are provided at any time after the information is laid.[25]

A single summons may be issued against a person in respect of a number of different informations, but the summons must separately state the detail of each information and act as though it were several summonses.[26] **8.37**

Each information should charge one offence only—ie it should not be duplicitous. Duplicity is dealt with in some detail in Section C Conducting a health and safety case in the Crown Court (below).[27] **8.38**

Article 6(3)(a) of the ECHR requires that everyone charged with a criminal offence has a right to be informed promptly in a language he understands and in detail, of the nature and cause of the accusation against him. The purpose of this requirement is to provide the defendant with the information necessary to enable him to start preparing his defence. In *Mattocia v Italy* (2003) 36 EHHR 47, the European Court of Human Rights held that the applicant's rights under Art 6(3)(a) had been violated. The Court held that it was an essential prerequisite that the accused should promptly and thoroughly be made aware of the material facts which formed the basis of the allegation. Although the extent of the detail required would vary according to the particular circumstances, he should be provided with sufficient material so as to enable him to understand the charges against him. Only if this scrupulous standard of detail is supplied can the preparation of an adequate defence be facilitated. If there were changes in the accusation as the matter progressed, he ought to be fully informed of any such changes. Further, he should be provided with adequate time and facilities to react to them and organize his defence. **8.39**

Amending the information or summons

An information may be amended at any time before final disposal of the case: ie before dismissal or sentence. Although the Magistrates' Courts Act 1980 s 123, which governs the issue, is drafted in wide terms and appears to allow any amendment to be made at any time, the model approach to be followed is that set out in *Garfield v Maddocks* [1974] QB 7. There it was said: **8.40**

> Those extremely wide words,[28] which on their face seem to legalize almost any discrepancy between the evidence and the information, have in fact always been given a more restricted meaning ... [I]f the variance is so substantial that it is unjust to the defendant to allow it to be adopted without a proper amendment of the information, then the practice is for the court to require the prosecution to amend in order to

---

[25] Criminal Procedure Rules r 7.7.
[26] ibid r 7.7(3).
[27] See paras 8.123–8.130 below.
[28] Magistrates' Courts Act 1952 s 100(1), which was drafted in identical terms to Magistrates' Courts Act 1980 s 123(1).

bring their information into line. Once they do that, of course, there is the provision in section 100(2)[29] whereby an adjournment can be ordered in the interests of the defence if the amendment requires him to seek an adjournment.

8.41 The general approach to be followed was set out in *New Southgate Metals v London Borough of Islington* [1996] CLR 334. In that case, three types of error that may occur in an information or summons were identified. These were:

(a) An error which was so fundamental that it could not be remedied by any appropriate and reasonable amendment and which would otherwise cause the prosecution to fail.[30]
(b) A defect which was substantial enough to require amendment, but not so grave as to be incapable of being remedied. In this case, magistrates have powers to allow, and should allow, amendment to take place, but should also allow an adjournment if the defence are placed at any disadvantage. A failure to detect such an error puts any conviction obtained on the evidence at risk of being quashed by the Divisional Court.
(c) Where the error is trivial, a conviction may be upheld even on an unamended information. In deciding whether to allow amendment or not, the fact that the full particulars of the offence were accurately set out in the information is to be taken into account. Applications to amend should normally be allowed unless to do so would be likely to cause injustice to the defendant.[31]

Withdrawal and Substitution of Informations

8.42 If no plea has been entered to an information, the prosecution may withdraw the summons, with the leave of the court.[32] However, if a plea has been entered, they are no longer entitled to withdraw the summons, and can only conclude proceedings by offering no further evidence. The prosecution may also prefer further information in substitution for, or in addition to, the withdrawn information. In such a case the defendant may seek an adjournment to consider the new information.

B. Conducting a Health and Safety Case in the Magistrates' Court

8.43 All health and safety cases start in the magistrates' court. A large number remain there for final disposal. Thus, it is important to have a clear understanding of the procedural and substantive requirements applicable. Some procedural requirements

---

[29] Now Magistrates' Courts Act 1980 s 123(2).
[30] Such as occurred in *City of Oxford Tramway Co v Sankey* (1890) 54 JP 564, where the information was laid against the wrong person.
[31] *DPP v Short* (2002) 166 JP 474.
[32] *Redbridge Justices ex p Sainty* [1981] RTR 13.

B. Conducting a Health and Safety Case in the Magistrates' Court

are common to both the magistrates' court and the Crown Court: for example, the requirements for the prosecution to draw up a *Friskies*[33] statement, and for the defence to respond. For the sake of convenience, *Friskies* statements have been dealt with in the context of the magistrates' court section, although the section applies equally to the Crown Court.

Since the advent of the Criminal Procedure Rules 2005 SI 2005/384, magistrates, like judges in the Crown Court, are required actively to manage at all stages cases which come before them. They should act in accordance with the overriding objectives contained in the Rules at all times. The Rules are wide in scope. They provide, for instance, that the magistrates' court may give a direction that will apply in the Crown Court, if the case is to continue there.[34] Being of wide general application, they should be kept in mind when considering any of the discrete procedural areas dealt with below.

8.44

The initial appearance

As was stated above, proceedings for health and safety offences are generally instigated by the service of a summons and information. These specify a date on which the defendant—whether human or corporate—must attend at the magistrates' court. Once the matter is before the court, if a second hearing is required, the case may simply be adjourned until a later date.

8.45

At this stage, there is no issue of bail for the human defendant. If he should fail to attend on the day appointed, there is no question of his having committed a bail act offence. Nevertheless, the magistrates may still be entitled to issue a warrant for the arrest of the non-attending defendant, in certain prescribed circumstances.

8.46

However, the magistrates are entitled to impose bail on a human defendant at any stage in the proceedings, at their discretion. Having imposed bail, they are also entitled to impose conditions on the bail, the breach of which may render the defendant liable for arrest. If the matter is committed to the Crown Court for sentence or trial, the committal will be on bail. In the event that the defendant fails to attend at the Crown Court, he may be said to be in breach of his bail, and therefore be liable for arrest.

8.47

Advance information in the magistrates' court

There is general requirement under Article 6 of the European Convention on Human Rights that 'all material evidence for or against the accused' is served before trial. Furthermore, when any person or company faces a summons for an offence triable either way, the prosecutor must inform them that they are entitled

8.48

---

[33] *R v Friskies Petcare (UK) Ltd* [2002] 2 Cr App R 401.
[34] Criminal Procedure Rules r 3.5(3).

to advance information,[35] as soon as practicable after the summons has been issued. If the accused requestsit, and before mode of trial is considered, the company or individual is entitledto certain information. The prosecutor is obliged to serve[36] on the accused either a copy of those parts of the statements which contain information as to all ofthe facts and matters on which the prosecution intend to rely, or a summary ofthe facts and matters of which the prosecutor proposes to adduce as evidencein the proceedings. If the written statements or summary make reference to adocument, the prosecutor should also provide the accused with a copy of it.[37]

8.49 Additional guidance on the service of advance information in the magistrates' court is contained in the Protocol for the Provision of Advance Information, Prosecution Evidence and Disclosure of Unused Material which was published in May 2006. Regard should also be had to the Attorney-General's Guidelines on Disclosure which provide, importantly, that the defendant 'should have sufficient time properly to consider the evidence before it is called'.[38]

8.50 The prosecutor may choose which form of advance information is given to the defendant.

8.51 In the case of a summary trial, the prosecutor must supply the defendant with all the evidence upon which he intends relying in sufficient time for him to consider it before it is called.[39]

8.52 The court is obliged to satisfy itself that the obligation to provide advance information has been complied with, prior to proceeding to mode of trial. If it finds that the obligation has not been complied with, it must adjourn proceedings pending compliance; although it has no power to order that the prosecution provide further material.[40] In any event, if the court forms the view that the accused has not been substantially prejudiced by the failure of the prosecution to comply with a request for advance disclosure, then it may proceed to mode of trial anyway.[41] (See below, for the courts' powers when the prosecution fails to serve a *Friskies* statement.)

8.53 If an adjournment is ordered as a result of the prosecutor's failure to comply with the obligation to provide advance disclosure, the costs of the wasted hearing may be awarded against him.

---

[35] Criminal Procedure Rules r 21.2.
[36] ibid r 21.
[37] ibid r 21.3.
[38] Att-Gen's Guidelines para 57.
[39] See the Att-Gen's Guidelines on Disclosure of Information in Criminal Proceedings para 57.
[40] Criminal Procedure Rules r 21.5.
[41] ibid r 21.6.

B. Conducting a Health and Safety Case in the Magistrates' Court

Friskies statements

In the case of *R v Friskies Petcare (UK) Ltd* [2002] 2 Cr App R 401, the Court of **8.54**
Appeal gave some general guidance on the documentation to be provided in
health and safety cases. Since the decision, its recommendations have been almost
universally adopted, and provide an effective framework for dealing with often
complex criminal prosecutions.

In *Friskies*, the appellants had pleaded guilty to one offence of failing to comply **8.55**
with the duty imposed under s 2(1), HSWA 1974, and had been committed to
the Crown Court for sentence. The case concerned the circumstances surrounding the electrocution of an employee of the appellant company, which manufactured pet food in a large plant which ran around the clock. The raw materials for
the pet food were mixed from frozen in large stainless steel silos using mechanical
stirring rods; these rods frequently broke during the early phase of the mixing and
had to be repaired. Rather than remove them from the silos and repair them in a
workshop, the procedure adopted was for them to be repaired *in situ* by a welder,
using an arc welder. This proved to be an extremely dangerous working method,
and an employee was electrocuted when attempting to change a welding rod in
the confined and conductive conditions of the silo. The employers had failed to
conduct any risk assessment of the process.

The difficulty which arose in *Friskies* was that the judge purported to sentence the **8.56**
defendant on the basis that it had deliberately sacrificed safety for profit's sake—
which according to *Howe*[42] was an aggravating feature—when the prosecution
had not specifically argued for it. As the Court of Appeal observed, in this case, no
party had sat down and deliberately placed safety considerations second to profit.
Rather, no proper consideration had been given to the risks associated with the
work. The appellate court observed that the aggravating feature identified in *Howe*
arose where a risk was identified but deliberately run for profit's sake.

The Friskies *recommendations*

The Court observed that problems can arise, on appeal, when there is some doubt **8.57**
as to the basis upon which the lower court passed sentence. It recommended,
therefore, that whenever the HSE commences proceedings, it should set down in
writing not merely the facts of the case, but also the aggravating features that are
alleged to exist in the case, as they were set out in *Howe*.

That document should then be served on both the court and the defendant, for **8.58**
their consideration. If the defendant pleads guilty, he should submit a similar
document, in writing, outlining the mitigating features it is claimed the court

---

[42] *R v F Howe & Son (Engineers) Ltd* [1999] 2 All ER 249.

Drawing up Friskies statements

8.59 Since the judgment in *Friskies* was delivered, it has become almost invariable practice for statements of the type recommended to be drawn up and served. The Court did not specify a precise format for the drafting of such statements, and it is possibly unsurprising that there is a considerable variation in the way they are drafted in practice. A certain amount of confusion has arisen. Whereas the court in *Friskies* referred to the need to set out the aggravating features, as elaborated on in the *Howe* case, in *Howe*, Scott Baker J had referred to both 'aggravating features' and to 'matters which are relevant to the offence'. The matters which he identified as being 'aggravating features' were:

(a) death or serious injury resulting from the breach of legislation;
(b) where the defendant has deliberately profited financially from a failure to take necessary health and safety steps or specifically run a risk to save money;
(c) failure to heed warnings.

8.60 The matters which he said were 'relevant to the offence' were:

(a) 'The degree of risk and extent of the danger created by the offence.'
(b) 'In assessing the gravity of the breach it is often helpful to look at how far short of the appropriate standard the defendant fell in failing to meet the reasonably practicable standard.'
(c) 'Whether it was an isolated incident or continued over a period of time.'

8.61 It is submitted that all of the matters listed above—both aggravating features and matters relevant to the offence—should be addressed in the *Friskies* statement as all relate to the overall seriousness of the offence. It may, ultimately, be of little significance, and a matter of individual preference, whether they are all included in a section entitled 'Aggravating Features', or those which are referred to as 'relevant to the offence' are subsumed in the main body of the statement.

Consequences of not providing a Friskies schedule

8.62 In *Bernard v Dudley MBC and Dudley Magistrates' Court* [2003] EWHC 147 (Admin), in a prosecution for health and safety offences, the applicant sought further particulars of the case advanced by the respondent. It was argued that: 'The service of a case-summary and information of any aggravating feature relied

B. Conducting a Health and Safety Case in the Magistrates' Court

upon is essential material for the assistance of the court and the defendant at a mode of trial hearing.'

8.63 Despite this, the respondent refused to provide any such materials. The applicant applied for further particulars before the magistrates' court. The court refused to order any, for reasons which Henriques J was 'unable to comprehend'. The applicant sought permission to judicially review the decision not to provide any further particulars. The Administrative Court ordered that a prosecution statement should be served, failing which permission would be granted. Having provided a statement, the prosecution declined to pay the applicant's costs for the judicial review; the matter was re-listed before the Administrative Court to hear the application for costs. After hearing argument, Henriques J found that the respondent had acted unreasonably in not providing further particulars. He ordered that they should pay the applicant's cost on an indemnity basis.

8.64 In a generally scathing judgment, Henriques J stated that, 'The defendants, that is the prosecuting authority, should at the very commencement of the criminal proceedings have provided the *Friskies* schedule. The judgment [in *Friskies*] makes that clear.'

8.65 By way of caution, however, if a *Friskies* schedule is provided to the magistrates' court before the mode of trial stage, there is a risk that they may thereby be provided with information that they should not have when deciding venue. A *Friskies* schedule may contain matters which refer to a defendant company's antecedent history, including the fact that it has previously been subject to enforcement action.

Relationship between advance information and the Friskies *statements* **8.66**

The advance information rules apply to health and safety cases in the same way that they apply to any other criminal case. However, since *Friskies*, the prosecution must also supply other particulars of their case. The *Friskies* requirements should be viewed as being additional and complementary to the advance information rules.

8.67 Both *Friskies* itself, and *Dudley MBC*, make it plain that the statement should be served when proceedings commence. Early service of a *Friskies* statement is also consistent with the wider purpose of the advance information rules. A defendant should be able to know the nature of the case against him, so as to be able to indicate a plea on an adequately informed basis. A properly defined case is particularly important in health and safety cases, where liability may arise for a number of different reasons.

8.68 Further, without a properly drafted *Friskies* schedule, the court may not be in a position to know which aggravating features are said by the prosecution to pertain in a given case. Without this information, it may not be able to fully assess whether its powers of punishment are adequate to retain jurisdiction. Indeed, as the overriding objective is for cases to be dealt with efficiently and expeditiously,[43] if a *Friskies* schedule has not been served at the outset, the court's powers of case management would entitle it to order that one is served forthwith.

Plea before venue

8.69 The Criminal Procedure and Investigations Act (CPIA) 1996 introduced significant changes to the existing mode of trial procedures, and, in particular, the provision for 'plea before venue'. Before CPIA 1996, the decision as to whether it was more suitable to hear a certain offence in the magistrates' court, or the Crown Court, was taken before the court had heard how a defendant was going to plead. The position is now that the court is required to invite the defendant to indicate how he would plead before it goes on to consider the appropriate venue, whether for sentence or trial.[44]

8.70 At the first or second hearing of the case, assuming that advance information has been provided, the court will usually expect the parties to deal with plea before venue.[45] If appropriate, mode of trial may also be determined. Plea before venue can only take place without advance information having been made if the defendant consents.

8.71 The defendant has three choices at the plea before venue stage: he may indicate a guilty plea, a not guilty plea, or he may indicate no plea. It is quite common in health and safety cases for a defendant to indicate no plea at the plea before venue stage; particularly, where the prosecution case has not been adequately particularized, or where the defendant has instructed an expert on the question of liability.

Effect of entering a guilty plea at the plea before venue stage

8.72 Although the defendant is invited only to indicate his plea, the effect, in practice, is the same as if a plea had been entered.[46]

8.73 Section 144 of the Criminal Justice Act 2003 requires the court to take into account the stage in the proceedings at which the offender indicated his intention to plead guilty, and the circumstances in which this indication was given.

8.74 Current sentencing guidelines state that a defendant will only receive maximum credit at the Crown Court where he has indicated a guilty plea at the plea before

---

[43] Criminal Procedure Rules r 1.1(e).
[44] Magistrates' Court Act 1981 s 17A(4).
[45] In this context see para 1.4 of the Protocol for the Provision of Advance Information, Prosecution Evidence and Disclosure of Unused Material in the Magistrates' Courts (May 2006), which provides that magistrates' courts should expect to deal with both plea and venue at the first hearing of the case.
[46] Magistrates' Court Act 1980 s 17A(4), (6).

B. Conducting a Health and Safety Case in the Magistrates' Court

venue stage. A guilty plea which is delayed in the Crown Court will attract less credit.[47] (See para 8.92, *Relevance of Guilty plea*.)

Benefits and drawbacks of the plea before venue procedure

The plea before venue procedure enables the defendant to enter a guilty plea at a very early stage. He therefore has opportunity to claim additional credit, as compared to a plea entered in the Crown Court. Counter-balancing this, though, he is being invited to enter his plea at an early stage of preparation, when he may not have seen all of the prosecution material available. Further, it is also highly unlikely that a defence expert report will have been obtained at this stage. **8.75**

This can be problematic in complex health and safety cases, where liability can arise for a number of different reasons. Practitioners should therefore be aware that they must be in a position to give important advice to their client at any early stage of the proceedings, and certainly earlier than was necessary before the advent of CPIA 1996. **8.76**

However, there may be cogent reasons why a guilty plea was not indicated at the plea before venue stage. It may, for instance, be properly argued that it was not the first reasonable opportunity in the terms envisaged by the Guidelines issued by the Sentencing Guidelines Council. A Crown Court judge may still be persuaded to give full credit. **8.77**

Corporations entering pleas in the magistrates' court

In the magistrates' court, a representative on behalf of a corporation may make the same representations as a non-corporate defendant. These may be concerned with fixing an appropriate venue for trial, consent to the matter being tried summarily, or entering a plea of guilty or not guilty to an information.[48] **8.78**

The term 'representative' is not defined in the Act. When a solicitor appears for a corporation, however, it has been held that he ought to be appointed as the corporation's representative. **8.79**

A director or company secretary may notify the court of the intention to enter a guilty plea without attendance at court.[49] The opportunity to enter such a plea only arises in the case of certain summary-only matters, and only then when the prosecution has served either a statement of facts or the relevant statements. A notice setting out any other information it intends to adduce before the court should also be laid. The court should not accept the plea of guilty, and convict the **8.80**

---

[47] See the guidelines issued by the Sentencing Guidelines Council in November 2004, and also *R v Rafferty* [1999] 1 Cr App R 235; *R v Barber* [2002] 1 Cr App R (S).
[48] Magistrates' Courts Act 1980 s 46, Sch 3 r 2.
[49] ibid Sch 3 r 4.

defendant, before first hearing any statement of mitigation submitted on behalf of the defendant.

Procedure for determining venue for trial or sentence

8.81 The procedure which must be followed for determining the venue for trial, or sentence, is laid down in Magistrates' Courts Act 1980 ss 17A–21. An annex to Home Office Circular 45/1997 sets out a suggested form of wording for use in the magistrates' court when inviting the defendant to indicate his plea. The following is an excerpt from that guidance, although a simpler version is often used.

> This offence(s) may be tried either by this court or by the Crown Court before a judge and jury.
>
> Whether or not this court can deal with your case today will depend upon your answers to the questions I am going to put to you. Do you understand?
>
> You will shortly be asked to tell the court whether you intend to plead guilty or not to the offence(s) with which you are charged. Do you understand?
>
> If you tell us that you intend to plead guilty, you will be convicted of the offence. We may then be able to deal with (part of) your case at this hearing. The prosecutor will tell us about the facts of the case, you (or your representative) will have the opportunity to respond, and we shall then go on to consider how to sentence you. Do you understand?
>
> We may be able to sentence you today, or we may need to adjourn the proceedings until a later date for the preparation of a pre-sentence report by the Probation Service. If we believe that you deserve a greater sentence than we have the power to give you in this court, we may decide to send you to the Crown Court, either on bail or in custody, and you will be sentenced by that court which has greater sentencing powers. Do you understand?
>
> If, on the other hand, you tell us that you intend to plead not guilty, or if you do not tell us what you intend to do, we shall go on to consider whether you should be tried by this court or by the Crown Court on some future date. If we decide that it would be appropriate to deal with your case in this court, we shall ask whether you are content for us to do so or whether you wish your case to be tried in the Crown Court.
>
> Before I ask you how you intend to plead; do you understand everything I have said or is there any part of what I have said which you would like me to repeat or explain?

Deciding whether to commit for sentence after an indication of a guilty plea

8.82 Although the plea before venue procedure outlined above must be carried out every time a person appears before a magistrates' court charged with an either-way offence, the procedure to be followed thereafter depends on whether a guilty plea is indicated, or either no plea or a not guilty plea is indicated.

8.83 The magistrates' court may commit the defendant to the Crown Court for sentence; either after conviction at summary trial or after a guilty plea has been indicated at

B. Conducting a Health and Safety Case in the Magistrates' Court

the plea before venue stage. To do so, it must be of the opinion that: 'The offence or the combination of the offence and one or more offences associated with it was so serious that greater punishment than they have the power to inflict should be imposed.'[50]

8.84 It has been held that, when performing this exercise, the court is also entitled to consider whether the level of fine it may impose is inadequate.[51]

8.85 If the accused pleads guilty, the prosecution will open the facts of the case in the usual manner. The court will, having heard mitigation, either sentence or commit the matter to the Crown Court, depending on whether they consider themselves to have adequate sentencing powers.

8.86 Both the prosecution and the defence are entitled to be heard on the question of whether the matter should be committed to the Crown Court, even though the Magistrates' Courts Act 1980 makes no express provision for it.

8.87 It is usual, in health and safety cases, for the defence to advance full mitigation before the magistrates, in an attempt to persuade them to retain jurisdiction for sentence. It is important, therefore, to ensure that sentencing bundles are prepared and available for the hearing; also that any authorities to be relied on are provided.

8.88 However, the court may take the view that the offence is so serious that the sentencing powers available are plainly insufficient. It may then be appropriate for the court to commit, without hearing full mitigation from the defence. If the court intends taking that course, it should indicate as much to the defendant or his representatives. This affords the defence to make some brief submissions, in an attempt to persuade the court to do otherwise. If, following defence submissions, the court is persuaded to alter its view, it should then give the prosecution the opportunity to make further submissions.[52]

8.89 The following are the factors which the magistrates should take account of when deciding whether their powers of punishment are adequate.

Mode of trial guidelines

8.90 The Consolidated Criminal Practice Direction[53] now contains the latest version of the Mode of Trial Guidelines. These provides guidance to magistrates in deciding whether defendants charged with either-way offences should be committed to

---

[50] Powers of Criminal Courts (Sentencing) Act 2000 ss 3(2), 4.
[51] *R v Flax Bourton Magistrates ex p Customs and Excise* [1996] Crim. LR 907; *R v North Essex Justices ex p Lloyd* [2001] 2 Cr App R (S) 86.
[52] See *R v Warley Magistrates' Court ex p DPP*; *R v Staines Magistrates' Court ex p DPP*; *R v North East Suffolk Magistrates' Court ex p DPP* [1998] 2 Cr App R 307.
[53] Consolidated Criminal Practice Direction [2002] 2 Cr App R 35.

the Crown Court for trial. Though not originally intended for the purpose, the original National Mode of Trial Guidelines have been held to be relevant in deciding whether or not to commit to the Crown Court for sentence, following the indication of a guilty plea at the plea before venue stage. In case law, they have been found to be more relevant than the Magistrates' Association Sentencing Guidelines, which are more concerned with sentence following a decision to retain jurisdiction. Although none of the guidelines is specific to health and safety offences, the most pertinent of the general observations are:

(a) The court should never make its decision on the grounds of convenience alone or expedition.
(b) The court should assume for the purposes of deciding mode of trial that the prosecution version of the facts is correct.
(c) Where two or more defendants were jointly charged with an offence, each has an individual right to elect his mode of trial.

The Howe *factors*

8.91 The ultimate question for the court, when deciding whether to commit or not, is whether their powers of sentencing are adequate. The court must necessarily have regard to the guidance given by the Court of Appeal, on the level of fine in health and safety cases, in *R v F Howe and Son (Engineers) Ltd* [1999] 2 Cr App R (S) 37. (See later, in Chapter 7.) The following observations from the *Howe* judgment are particularly relevant:

(a) Magistrates should be careful in accepting health and safety cases where it is arguable that the fine might exceed the limits of their jurisdiction, or where death or serious injury has resulted from the offence.
(b) The court observed, 'Disquiet has been expressed in several quarters that the level of fine for health and safety cases is too low'.
(c) In assessing the gravity of the breach, the court stated that it is helpful to see how far short the defendant fell from doing what was reasonably practicable.
(d) Regarding resulting death or serious injury, the court acknowledged that it is often a matter of chance whether death or serious injury results from even a serious breach. But it added that death resulting from any breach was to be regarded as an aggravating factor.
(e) The court held that an offence is seriously aggravated where the defendant has deliberately profited from a failure to take necessary health and safety measures or run a risk to save money.
(f) The court should advert to the degree of risk, and the extent of the danger created by the offence.
(g) The extent of the breach is a central feature. Whether it was an isolated incident, or whether it had been continuing for a period of time, is highly relevant.

B. Conducting a Health and Safety Case in the Magistrates' Court

(h) A failure to heed warnings is another aggravating factor. Although the court did not specify what warnings it had in mind, it is submitted that these might range from formal advice from the HSE to the occurrence of similar incidents in the past. Events which may have given warning of them may also be indicative.

(i) The court should also take account of whether any of the mitigating factors, outlined in *Howe*, are present. These include a prompt admission of responsibility and a timely plea of guilty, steps to remedy deficiencies after they were drawn to the defendants' attention, and a good safety record.

If a decision is taken to commit the case to the Crown Court, the case will either proceed directly to committal, or will be adjourned to a date when that committal is to take place. **8.92**

Relevance of guilty plea

When deciding whether their powers of punishment are adequate, magistrates must take the defendant's indication of a guilty plea into account. The maximum discount for a guilty plea is likely to be one third.[54] However, in *R v Barber* [2002] 1 Cr App R (S) 548, the Court of Appeal stated that the maximum possible credit should be given to those who indicated guilty pleas at the plea before venue stage; and that if a guilty plea was indicated at that stage on an either-way offence, a discount greater than one third might be appropriate. **8.93**

This level of reduction of sentence is only available if the defendant has indicated his willingness to plead guilty at the first reasonable opportunity.[55] Of course, the first reasonable opportunity may well be the plea before venue stage, but it is for the court to decide the time. It follows that the court would be entitled to conclude that guilt might reasonably have been admitted in interview.[56] This is something for a solicitor to be mindful of when advising a defendant whether an individual should be put forward for interview under caution following such a request by an inspector. **8.94**

It is conceivable that a defendant may informally indicate his guilt, at an earlier stage in the investigation. It is unlikely, however, that a failure informally to accept guilt at a pre-interview stage is likely to lead to any withholding of the discount. **8.95**

Magistrates are entitled to sentence a defendant to the maximum possible fine—ie £20,000 for a breach of one of the general duties—even though he pleads guilty, with the decision not to commit for sentence reflecting the discount. Assuming maximum credit is given, a fine of £20,000 represents an undiscounted fine of £30,000. **8.96**

---

[54] See para 4.2 of the Guidelines issued by the Sentencing Guidelines Council.
[55] See Guidelines Annex 2.
[56] ibid.

Chapter 8: Proceedings and Criminal Procedure

Previous convictions

8.97 When deciding whether the offence should be committed or not, the magistrates may take into account whether the defendant has previous convictions and all other relevant character and antecedents.[57]

Disputed basis of fact

8.98 It often happens, particularly in health and safety cases, that although liability is not disputed, the factual basis for it is not agreed between the parties. If a dispute about the factual basis for sentence becomes apparent, and that difference may directly affect whether the court's sentencing powers are adequate, the magistrates ought to hold a Newton hearing.[58] Conversely, the magistrates may take the view that their powers of punishment would be either adequate or inadequate, no matter in whose favour they resolve the factual dispute. They should then either commit or retain jurisdiction, as the case may be.

Changing the decision to commit

8.99 The decision to commit to the Crown Court may be made after the decision to accept jurisdiction has been determined, even if no new material has come to light.[59] Magistrates have what has been referred to as an 'open textured discretion'[60] to commit for sentence. This discretion is not tied to their decision to accept jurisdiction. This principle is subject to an important exception: if an accused is given a legitimate expectation that he will not be committed for sentence,[61] the court must honour that expectation. The court should be careful to explain to the accused that it has not ruled out the possibility of committing for sentence.

Powers of the Crown Court when dealing with committals

8.100 The Crown Court may deal with a matter that has been committed to it as though the defendant had been convicted on indictment before the court.[62] The question of sentence is entirely at large. The fact that a matter may have been committed to the Crown Court does not mean that the sentencing judge ought to impose a fine in excess of the magistrates' court limit.

---

[57] *R v Warley Magistrates' Court ex p DPP*; *R v Staines Magistrates' Court ex p DPP*; *R v North East Suffolk Magistrates' Court ex p DPP* [1998] 2 Cr App R 307. See also the Powers of Criminal Courts (Sentencing) Act (PCC(S)A) 2000 s 151.

[58] Following *R v Newton* (1983) 77 Cr App R 13. In that case, it was decided that a hearing should be held where evidence may be called to decide a disputed basis of fact.

[59] *R v Sheffield Crown Court and Sheffield Stipendiary Magistrate ex p DPP* [1994] Crim LR 470; *R v Dover Magistrates' Court ex p Pamment* [1994] Crim LR 471; *R v North Sefton Magistrates' Court ex p Marsh* [1994] Crim LR 865.

[60] Per Steyn LJ, in *R v North Sefton Magistrates' Court ex p Marsh* [1994] Crim LR 865.

[61] *R v Nottingham Magistrates' Court ex p Davidson* [2000] 1 Cr App R (S) 167.

[62] PCC(S)A 2000 s 5.

B. Conducting a Health and Safety Case in the Magistrates' Court

If, at the Crown Court, it becomes apparent that there is some dispute as to the factual basis for plea entered, the Crown Court has the power to resolve the dispute itself. It may also remit the matter. In deciding what course to take, the court should consider, *inter alia*, the stage at which the dispute became apparent.[63] The case of *R v Isleworth Crown Court ex parte Buda* [2000] 1 Cr App R (S) 538 is instructive here. In that case, Moses J held that where there has been a decision to commit on the wrong view of the facts, the proper approach would be for the defendant to make an application to change his plea. The matter should then be remitted to the magistrates' court. If the defendant had entered the plea at the plea before venue stage, there was generally no reason why he should not be allowed to change it. If the defendant indicates that he is appealing against his conviction in the magistrates' court, sentencing should be adjourned in the Crown Court until the appeal has been heard.[64] 8.101

Procedure for determining mode of trial

The following section contains the law at the time of publication. However, when Criminal Justice Act 2003 Sch 3 comes into force—possibly in the autumn of 2007—there will be wholesale changes in the way in which cases progress from the magistrates' court to the Crown Court. Committal proceedings will be abolished entirely. Under the terms of the Crime and Disorder Act 1998 s 51, cases which are triable either way—ie most health and safety cases—will now be sent to the Crown Court in the same way that indictable-only offences are sent at present. The decision as to whether an offence which is triable either way is more suitable for summary trial, or trial on indictment, will now be made by the Crown Court judge. The judge will remit cases which he considers suitable for summary trial. 8.102

The majority of offences created by s 33(1) of HSWA are triable either way. They may therefore be heard in either the magistrates' court or in the Crown Court.[65] At the plea before venue stage, if the defendant enters a not guilty plea, or fails to enter any plea, the court will proceed to consider mode of trial. There are some important differences between the criteria and procedure to be applied when considering mode of trial. These are linked to those for determining whether to commit for sentence. 8.103

The Court should proceed to determine mode of trial in the following way. The prosecutor and then the defendant should first be given the opportunity to make representations as to which mode of trial is more suitable (Magistrates' Courts 8.104

---

[63] *Munroe v Crown Prosecution Service* [1998] Crim LR 823.
[64] *R v Faithful* [1950] 2 All ER 1251.
[65] The exceptions are those created by s 33(1)(d), (e), (f), (h), and (n), which are summary-only offences. In the main, these are offences concerned with obstruction of inspectors.

Act 1980 s 19 (2)); and having heard representations from both parties, the court will indicate its view. The matters which the court should take into account when deciding the question are:[66]

(a) the nature of the case;
(b) whether the circumstances make the offence one of serious character;
(c) whether the punishment which a magistrates' court would have power to impose would be adequate;
(d) any other circumstance which may appear to the court to make it suitable to be tried one way or another;
(e) any representations made by the prosecution or the defence.

8.105 The most important of the above considerations is whether or not the magistrates' sentencing powers would be adequate in the event of conviction at summary trial. If the accused is charged with more than one offence the court should look at the totality of the offending; and should decline jurisdiction if their powers of punishment across the range of offences are not sufficient.

8.106 The court should also have regard to the mode of trial guidelines at para V51 in the Consolidated Criminal Practice Direction. The emphasis in these guidelines is for cases to be tried summarily, unless any of the particular aggravating features which are set out in relation to individual offences pertain. The list of offences included in the Practice Direction does not include health and safety offences. For these, magistrates will necessarily have to refer to the aggravating features set out in *Howe*. One of the most commonly invoked of the guidelines in the Practice Direction is V51.3(d) which provides that where cases involve complex questions of fact or difficult question of law, the court should consider Crown Court trial.

8.107 While the magistrates have a wide discretion as to the circumstances which may lead them to commit for trial, they should never make the decision simply on the grounds of convenience or expedition.[67]

Previous convictions and committal

8.108 The magistrates should not be informed of, nor have any regard to, any previous convictions which the defendant may have when deciding the question of venue.[68] Prosecutors should be aware of this when preparing *Friskies* statements. It is not uncommon for *Friskies* statements to contain details of a defendant's previous

---

[66] Magistrates' Courts Act 1980 s 19(3).
[67] Consolidated Criminal Practice Direction V51.3.
[68] See, for instance, *R v Colchester Justices ex p North East Essex Building Co Ltd* [1977] 1 WLR 1109. But this is subject to change, with the introduction of Sch 3 of the Criminal Justice Act 2003. The court will be entitled to hear about any previous convictions which the defendant may have.

B. Conducting a Health and Safety Case in the Magistrates' Court

convictions in the section concerned with previous safety record. This should be recognized and suitable caution exercised. It is possible that magistrates might inadvertently be provided with information they ought not to have, with the risk that a trial may be compromised.

If the court considers that summary trial is more appropriate, the defendant should be addressed as follows: **8.109**

> It appears to this court more suitable to be tried here. You may now consent to be tried by this court, but if you wish, you may choose to be tried by a jury instead. If you are tried by this court and are found guilty, this court may still send you to the Crown Court for sentence, if it is of the opinion that greater punishment should be inflicted for the offence than it has power to impose. Do you wish to be tried by this court or do you wish to be tried by jury?[69]

Therefore, even if the court accepts jurisdiction, the defendant may still elect to be tried at the Crown Court. The prosecution does not, however, have the right to determine where the case should be tried; the most they may usually do is to make representations that the matter is so serious as ought to be tried on indictment. **8.110**

The individual's right to elect

Where two or more defendants are charged together, each of them has an individual right to either elect trial on indictment or consent to summary trial. It is wrong for the court to ascertain, as a preliminary matter, whether any of the defendants intends to elect, and then to decline jurisdiction as a means of ensuring that there is only one trial. **8.111**

This principle of the right to individual election is subject to an important exception. When one of a number of defendants on linked charges is faced with an offence which is indictable only—ie one which may only be tried in the Crown Court—all matters linked to the indictable-only matter will automatically be committed to the Crown Court, and the party facing the linked matter has no individual right of election.[70] This will occur where an individual is charged with manslaughter as a result of a health and safety incident, and other individuals are charged with either way offences under the Health and Safety at Work Act 1974, or breach of other regulations. **8.112**

Corporations and election

A representative of a corporation may make the same representations as a non-corporate defendant about the appropriate venue. In other words, the appointed **8.113**

---

[69] This form of words was approved in *R v Southampton Magistrates' Court ex p Sansome* [1999] 1 Cr App R (S) 112.
[70] s 51(3) of the Crime and Disorder Act 1998.

representative may consent to the matter being tried summarily, or enter a guilty or not guilty plea to an information.

8.114 A magistrates' court may commit a corporation for trial by an order in writing. This empowers the prosecutor to prefer a bill of indictment in respect of the offence named in the order.[71]

Orders on committal

8.115 Once the magistrates have committed a case, they should direct that a Plea and Case Management Hearing (PCMH) occurs in the Crown Court. They will usually fix a date for the PCMH, by prior agreement with the Crown Court. The hearing should take place within seven weeks of committal, although it is usually fixed for an earlier date.[72]

Material to be sent to the Crown Court on committal

8.116 The magistrates must send the following material to the Crown Court officer as soon as practicable after committal:[73]

(a) the information, if it is in writing;
(b) the evidence tendered, in accordance with s 5A of the Magistrates' Courts Act 1980;
(c) any notification by the prosecutor under s 5D(2) of the 1980 Act regarding the admissibility of a statement under ss 23 or 24 of the Criminal Justice Act 1988 (first-hand hearsay; business documents);
(d) a list of the exhibits produced in evidence before the justices or treated as so produced;
(e) such of the exhibits referred to in paragraph (1)(f) as have been retained by the justices.

Judicial review of the decision to commit

8.117 The Divisional Court will not interfere lightly with the discretion of the justices to retain jurisdiction or commit. Nonetheless, it will not hesitate to do so in cases where the magistrates' decision can be described as being 'truly astonishing'.[74]

---

[71] Magistrates' Court Act 1980 Sch 3, paras 1–3(1).
[72] See, for example, Amendment No 11 to the Consolidated Criminal Practice Direction (Case Management), now in effect, amending the Consolidated Criminal Practice Direction 2002. In particular, see IV.41.6, on the case progression form to be used in the magistrates' court.
[73] Criminal Procedure Rules r 10.5.
[74] Per Kennedy LJ in *R v Warley Magistrates' Court ex p DPP* [1999] 1 WLR 216.

C. Conducting a Health and Safety Case in the Crown Court

The Criminal Procedure Rules 2005

8.118 Effective conduct of a health and safety case in the Crown Court will require knowledge of the Criminal Procedure Rules 2005. The Rules incorporate many of the important rules of practice and procedure for conduct of cases in the Crown Court. They govern wide and disparate areas—ranging from duplicity, to the procedure for the admissibility of expert evidence. Even when the Rules do not contain provisions governing a particular area of practice regard should be had to the underlying philosophy of the Rules—in particular, the ethos of the overriding objectives.

8.119 The Rules came into force on 4 April 2005 and apply to all criminal litigation in the Crown Court. They express a number of 'overriding objectives' for criminal cases, and outline the duties of all parties to criminal litigation in achieving those objectives. The overriding objective is stated to be that criminal cases are dealt with justly. Dealing with a case justly is said to involve, inter alia:

(a) acquitting the innocent and convicting the guilty;
(b) dealing fairly with the prosecution and the defence;
(c) recognizing the rights of a defendant, particularly those under Art 6 of the European Convention on Human Rights;
(d) respecting the interests of witnesses, victims, and jurors and keeping them informed of the progress of the case;
(e) dealing with the case efficiently and expeditiously;
(f) ensuring that appropriate information is available to the court when sentence is considered;
(g) and dealing with the case in ways that take into account the gravity of the offence alleged and the complexity of what is at issue.

8.120 The court is required to ensure that these overriding objectives are achieved by actively managing the case. To achieve this, the Rules admit of some flexibility and informality of approach. For instance, directions and orders need not be given in open court. The court may, for the purpose of giving directions, receive applications and representations by letter, by telephone, or by any other means of electronic communication. A hearing may be conducted by such means. The court is expected to give any direction appropriate to the needs of that case as early as possible, and the parties are expected to observe and honour these orders. It may also fix, postpone, bring forward, extend, or cancel a hearing and shorten or extend (even after it has expired) a time limit fixed by a direction.

Lodging the indictment

8.121 The prosecution must prefer the indictment, ie lodge it with the court within 28 days of the date of the committal.[75] This time may be extended before or after it has expired. There are a number of ways to effect this. An application to a judge of the Crown Court can be made, or a written application directed to an officer of the court; indeed, the officer of the court—of his own volition—may take action, in which case written reasons need not be provided. In contrast, an application to an officer of the court requires written reasons for the extension of the period, in support of the application. When an application is made before a judge, although the defendant has no right to make representations, he may be allowed to do so. Especially in circumstances where there has been a long delay and it is argued that prejudice may have occurred.

8.122 A distinction should be drawn between preferring an indictment and the signing of it. A bill of indictment does not become an indictment proper until it is signed in accordance with the provisions of s 2(1) of the Administration of Justice (Miscellaneous Provisions) Act 1933. A trial which takes place with an unsigned bill of indictment is likely to be rendered a nullity.

The rule against duplicity

8.123 Each count in the indictment must be for one offence only, so as not to fall foul of the so-called 'duplicity rule', in Criminal Procedure Rules r 7.7. The rule against duplicity applies to both informations and counts on an indictment, and therefore in the magistrates' court and the Crown Court.[76]

8.124 An information which charges two or more offences is bad for 'duplicity'. This has a practical knock-on effect. A magistrates' court cannot proceed on the basis of an information that charges more than one offence. It is legitimate to charge, in one count or information, an activity, even though that activity may comprise more than one act.[77] The question of whether a given activity is, in reality, one continuing offence, committed by a series of distinct acts, or a number of separate offences, can be a highly academic one. Whether one or more offence is disclosed is foremost a question of fact and degree.[78]

8.125 Separate informations can be contained in a single summons, but they must clearly be demarcated within the summons. A good rule of thumb is that a single activity

---

[75] Criminal Procedure Rules r 14.2.
[76] See Health and Safety Executive Drafting Informations on duplicity, available at <http://www.hse.gov.uk/enforce/enforcementguide/pretrial/preparing/drafting.htm>.
[77] See *Jemmison v Priddle* [1972] 1 QB 489, 56 Cr App R 229. See also *R v Wilson* (1979) 69 Cr App R 8.
[78] See *R v Wilson* (1979) 69 Cr App R 83, at 88.

C. Conducting a Health and Safety Case in the Crown Court

amounts to a single offence.[79] Separate informations are required where the acts of an accused have no common purpose. Generally, this will be established where the acts are separated in 'time and place'. As an example of where two offences are created, the HSE cites breach of a regulation requiring an appliance to be provided and used. If only one information were used, alleging that the appliance was not provided and not used, it would not survive application of the 'duplicity rule'.[80]

8.126 Duplicity relates to the form of the charge, rather than the evidence called in support of it. It will generally be unnecessary for the court to look beyond the words of the charge or count, to determine whether it is duplicitous. Nevertheless, in certain circumstances, the court is entitled to look at the particulars provided by the prosecution.[81]

8.127 Duplicity is a matter of form and not of substance. It is, therefore, immaterial that the particulars of the offence charge only one offence when the evidence called discloses the commission of more than one offence. This situation has been referred to as 'divergence or departure'[82] and is sometimes called 'quasi-duplicity'. However, even in this situation, it may be necessary for the prosecution to seek to amend the information or indictment. This may be done either by striking out certain particulars, or by applying to add a new charge. Otherwise, it may be successfully argued at appeal that the evidence adduced did not in fact support the offence charged.

8.128 For example, a count or information which alleges that an offence was committed 'on divers days' will be bad for duplicity if the offence charged is not a continuing one. Breaches of the general duties under HSWA 1972 ss 2–6 are usually continuing offences. There can, therefore, be no objection based on duplicity if the period is a long one, or the offence is said to have occurred on more than one day.

8.129 In *HSE v Spindle Select Ltd* Times Law Reports 9 December 1996, it was the prosecution case that the severing of an employee's right hand through coming into contact with the rotating cutters of a wood-turning machine was caused both by a failure to guard the machine and a failure to train the employee adequately. The Divisional Court rejected a submission that an information charging a breach of HSWA 1974 s 2(1) with both particulars was either bad for uncertainty or duplicity.

---

[79] *DPP v Merriman* [1973] AC 584.
[80] An excellent example of duplicity in practice is *Carrington Carr Ltd v Leicestershire County Council* [1993] Crim LR 938. The HSE also refer to, and outline in sum, supporting cases on this point: *Hoggetts v Chiltern District Council* [1983] 2 AC 120 (an information alleging that the defendant had on and since a particular day allowed land to be used in contravention of an enforcement notice was not bad for duplicity); *Cullen v Jardine* [1985] Crim LR 668 (an information alleging that the accused felled ninety trees over a three-day period was not bad for duplicity).
[81] *R v Greenfield* 57 Cr App R 279 (CA).
[82] See, in full, the speech of Browne LJ, in *R v Wilson* (1979) 69 Cr App R 83.

They held that even if the information itself, or the accompanying particulars, referred to more than one aspect of the overall duty, it would not be bad for duplicity; that the general duty was laid down by HSWA 1974 s 2(1) and the matters referred to in s 2(2) were no more than examples of that general duty, and that there was no need to refer to them specifically, nor were they exclusive in any event.

8.130 It follows, from *Spindle Select Ltd*, that there can be no objection to an information or count which alleges that a breach of duty occurred for a number of different reasons; nor would it be necessary for the jury to find that all of the matters alleged were proved in order to convict the defendant of the offence. However, in that situation the whole of the jury (or, where appropriate, a majority of them) must be satisfied that the same particular, at least, is proved and they should be directed accordingly.[83]

Effect of proceeding on a duplicitous charge or count

8.131 In the magistrates' court, if a summons is held to be duplicitous, the prosecution may elect with which charge to proceed. The same applies in the Crown Court, where a count which alleges more than one offence should be quashed before arraignment takes place. It is, however, open for the Crown to apply to amend the indictment by splitting the count into two. Conviction upon a duplicitous count may lead to that conviction being overturned on appeal. This will either be because the conviction was obtained on evidence which was not before the court, or because it was bad for uncertainty, unless it can be shown that the conviction is safe.

8.132 Duplicity can, however, be cured. If an Information is duplicitous, it must be amended before the hearing. To apply to the court to amend the Information, the acting solicitor is required to notify the defence in writing. This should be done promptly, so as to allow the defence sufficient time to consider the matter before the hearing. The hand of the prosecution is somewhat forced, however, in curing duplicity. On a defence application, if there is a finding by the court that the Information is duplicitous, the magistrates will require the prosecutor to *elect* the offence on which the prosecution wishes to proceed. The other allegation will then be struck out by the court. This rule of procedure is strictly upheld. Consequently, if the prosecutor refuses to make the election, the Information must be dismissed.

Selecting the appropriate charges

8.133 As was said above, all existing guidance about the selection of counts on indictments must be viewed in the light of the overriding objectives for the conduct of criminal litigation, as set out in the Criminal Procedure Rules 2005.

---

[83] See, for example, *R v Brown (K)* (1984) 79 Cr App R 115.

C. Conducting a Health and Safety Case in the Crown Court

The Code of Practice for Crown Prosecutors applies to prosecutions brought by the HSE or local authorities. It requires that crown prosecutors should select charges which: **8.134**

(a) reflect the seriousness of the offending;
(b) give the court adequate sentencing powers;
(c) enable the case to be presented in a clear and simple way.

It also provides that crown prosecutors should not always continue with the most serious charge where there is a choice, or continue with more charges than is necessary. **8.135**

Further guidance about selecting counts and framing indictments was given in the Consolidated Criminal Practice Direction [2002] 2 Cr App R 35, which provides, at para 34.3: **8.136**

> It is undesirable that a large number of counts should be contained in one indictment. Where defendants on trial have a variety of offences alleged against them, the prosecution should be put to their election and compelled to proceed on a certain number only, leaving a decision to be taken later whether to try any of the remainder.

The guidance given in the Practice Direction broadly reflects the guidance given in *R v Staton* [1983] Crim LR 190. In that case, the court acknowledged that there was a tendency to overload indictments. It suggested that the best approach was to identify the real offence or, if necessary, two or three offences alleged to have been committed by a defendant, and to indict him only for that offence or those offences. The court went on to state that the trial should not be complicated or lengthened by the addition of other offences, though there may have been evidence of them. The shorter and more direct the indictment, the better and swifter the process of justice. The principle in *Staton* was reiterated in *R v O'Meara* Times Law Reports 15 December 1989. The Court of Appeal observed that it is normally unwise to overburden a case with a multiplicity of counts in an indictment. The criminality could otherwise be embraced more satisfactorily if expressed in one count. **8.137**

The Code of Conduct of the Bar of England and Wales provides, at para 11.4: **8.138**

> In relation to cases tried in the Crown Court, prosecuting counsel . . . if he is instructed to settle an indictment . . . should bear in mind the desirability of not overloading an indictment with either too many defendants or too many counts, in order to present the case as simply and concisely as possible . . .

In applying the guidance above to the framing of the indictment in health and safety cases, careful consideration should be given to whether it is necessary to charge breaches of a general duty alongside breaches of regulations. Where the factual basis is common to both, overview and judgment are needed. An example of how these come into play follows. When framing an indictment, it is of course **8.139** desirable to reflect the circumstances surrounding an injury to an employee, for example, through contact with an unguarded part of machinery. But the wider allegation might involve a failure to conduct a suitable and sufficient risk assessment, and, in the event, a failure to guard against risk. On that analysis, it may not be necessary to charge offences of breaching reg 3 of the Management Health and Safety at Work Regulations 1999 (MHSWR) SI 1999/3242 alongside a count alleging a breach of general duty containing the same particulars.

8.140 Consideration should be given to whether it may be sufficient simply to charge a count alleging a breach of a general duty with separate particulars which reflect the wider components of the allegation. With such a count, the jury need only find one of the particulars proved to convict. Nonetheless, they must all be agreed on the same particular.[84] There are undoubtedly some advantages in proceeding in this way, although they may be offset by the inability to know the basis for conviction. This follows, given that the jury would be entitled to return a guilty verdict, upon proof of only one of a number of particulars.

8.141 It may be particularly undesirable to charge offences of failing to comply with general duties alongside failures to comply with specific duties. Where the factual basis is common to both, but the standard of proof varies between the two offences, there would seem no logic in charging the offences together. It may also invite unfairness. To require a jury to apply differing standards or burdens of proof to the same facts may be to place too great a strain on them, and is likely to infringe the letter and spirit of the Practice Direction.

8.142 It should also be remembered that the Crown Court's powers of punishment are identical for both breaches of the general duties and regulations. The Court of Appeal has upheld large fines in respect of breaches of regulations alone, and in such cases accepted that the factual basis for sentence may be broadly drawn.[85] For these reasons, there is no imperative to charge under the general duties so as, in some way, to 'mark the gravity of the case'.[86]

8.143 The judgment in the case of *R v HTM* has created other possible difficulties in this area. There, the Court of Appeal found that as the words 'so far as is reasonably practicable' in the general duties qualified the duty itself, and were not properly to be considered part of any defence. However, in certain health and safety regulations, it is stated that it shall be a *defence* for a person to show that he did all that was reasonably practicable.

---

[84] See *R v Brown (K)* (1984) 79 Cr App R 115.
[85] But see the dicta of Scott Baker in *Howe*.
[86] *R v Keltbray Ltd* [2001] Cr App R (S) 39.

C. Conducting a Health and Safety Case in the Crown Court

Plea negotiations: accepting lesser pleas

8.144 As was stated earlier, those who prosecute on behalf of the HSE or local authorities should have regard to the Code for Crown Prosecutors when prosecuting offences under health and safety legislation. Accordingly, the following principles should be foremost and clear, when deciding whether to accept lesser pleas. Under para 9.1 of the Code: 'Crown Prosecutors should only accept the defendant's plea if they think the court is able to pass a sentence that matches the seriousness of the offending, particularly where there are aggravating features. Crown Prosecutors must never accept a guilty plea because it is convenient.' The prosecution should be prepared to explain its reasons for accepting lesser pleas in open court.[87]

8.145 Counsel for the prosecution 'are to regard themselves as ministers of justice, and not to struggle for a conviction'.[88] In the same vein, the Code of Conduct of the Bar of England and Wales, at para 11.1 provides: 'Prosecution counsel should not attempt to obtain a conviction by all means at his command. He should not regard himself as appearing for a party.'

8.146 Developing this, the Farquharson Report ventures: '. . . great responsibility is placed upon prosecution counsel and although his description as "minister of justice" may sound pompous to modern ears it accurately describes the way in which he should discharge his function.'

Progression of cases: case management

8.147 With the perception that some criminal cases had become unmanageable came a plethora of rules and guidelines governing the conduct and progression of cases in the Crown Court. Some of the rules apply only to heavy fraud and other complex criminal cases; others are of universal application, and apply to health and safety cases in the same way that they apply to any other criminal case.

8.148 At the beginning of the case, each party must nominate an individual responsible for progressing the case. Known as the case progression officer, he must inform other parties and the court who he is and how to contact him. He has a variety of duties. These include ensuring that court orders are complied with, and marshalling witnesses and evidence.[89]

Plea and case management hearings (PCMH)

8.149 Each time the magistrates commit a case to the Crown Court, they must order a PCMH; this should be held within about seven weeks from the date of committal.[90]

---

[87] See *Att-Gen's Guidelines on the Acceptance of Pleas* [2001] 1 Cr App R 28.
[88] See *R v Puddick* (1865) 4 F & F 497.
[89] Criminal Procedure Rules r 3.4.
[90] Practice Direction (Criminal Proceedings: Consolidation) para IV.41.6 [2002] 1 WLR 2870 (as substituted by Practice Direction (Criminal Proceedings: Case Management) [2005] 1 WLR 1491).

PCMHs are intended to achieve active case management and so reduce the number of ineffective and disrupted trials. At the PCMH, the parties will be expected to complete a form (reproduced at Appendix E of the Practice Direction). The form is intended to deal with all the issues which might arise at trial, and to ensure that the trial is brought on and completed as expeditiously as possible. Judges are encouraged to list the PCMH with some flexibility, so as to allow the trial advocate to attend.

Complex cases

8.150 According to the provisions of s 29 Criminal Procedure and Investigations Act 1966, either the prosecutor or the judge may take the view that an indictment reveals that a case is likely to be of such complexity or length that substantial benefits are likely to accrue from the holding of a preparatory hearing. The judge may then order that such a hearing takes place. The purpose of such hearings is to identify the issues which are likely to be material to the verdict of the jury; to assist their comprehension of those issues; to expedite the proceedings before the jury, and to assist the judge's management of the trial.[91] Though the jury is not sworn at the time, they effectively mark the beginning of the trial.

8.151 One of the distinct advantages of preparatory hearings is that the judge may make a pre-trial ruling on the admissibility of evidence. He may also adjudicate on any question of law arising in the case,[92] either at his own motion or by application of either party. Appeal against any order or ruling of the judge lies with leave of the court to the Court of Appeal Criminal Division.[93] However, a preparatory hearing should only be held when the statutory provisions are fulfilled. When the statutory provisions are not otherwise met, it should not be used as vehicle to obtain a ruling on a point of law and interlocutory appeal.[94]

8.152 If the case is a complex one regard should also be had to the Protocol for the control and management of heavy fraud and other complex cases [2005] 2 All ER 429. The Lord Chief Justice has there given guidance as how judges and the parties should manage complex cases. While the guidance is specifically intended to apply to complex cases, it has been held to provide wider assistance as to the management of other criminal cases.

8.153 Regard should also be had to the judgment in *R v Jisl and Tekin* Times Law Reports 19 April 2004, where the Court of Appeal reiterated the need for active case management in complex cases.

---

[91] For an elaboration on this point, see *R v Pennine Acute Hospital NHS Trust etc* (2003) 147 SJ 1426.
[92] s 29 Criminal Procedure and Investigations Act 1996.
[93] s 35 Criminal Procedure and Investigations Act 1996; for an example of this in practice, see *R v HTM Ltd* [2006] EWCA Crim LR 1156.
[94] See *R v Pennine Acute Hospital NHS Trust etc.* (2003) 147 SJ 1426.

D. Disclosure

Plea bargaining: *R v Goodyear*

8.154 The 'vexed question of plea bargaining'[95] was addressed by a five-man Court of Appeal, in the case of *R v Goodyear* [2005] 3 All ER 117. There, the Court of Appeal addressed the issue of sentencing, as it relates to plea bargaining. The Court adverted to the scenario where a defendant personally instructs his counsel to seek an indication from the judge about his view on the maximum sentence which would be imposed on the defendant, if he were to plead guilty at that moment. The judge may give such an indication. However, the Court went on to impose a number of conditions. It stated that a judge should generally not give any indication unless it was specifically sought by the defendant. In any event, he may refuse to give one, or defer giving one, with or without giving reasons. Further, in the normal course of events, the judge should not give an indication of sentence until the basis for plea had been agreed between the parties. Any such agreement should be reduced to writing. The Court observed that the judge was most unlikely to be able to give any such indication in complex or difficult cases, unless the basis for plea had been reduced to writing. In such cases, no less than seven days' notice should be given to both the prosecution and the Court, of the intention to seek an indication.

8.155 By way of procedure, the Court stated that they anticipated that any sentence indication was likely to be sought at the plea and case management hearing. The hearing was to take place in open court, without the need for a full opening by the prosecution, or a full plea in mitigation by the defendant.

8.156 The Court also observed that it would be impracticable for the new arrangements to be extended to proceedings in the magistrates' court. Therefore, for the time being, magistrates should confine themselves to the statutory arrangements in the Criminal Justice Act 2003 Sch 3.

D. Disclosure

8.157 Effective and fair prosecution and defence of cases under health and safety legislation requires a particular approach. An understanding and correct application of the principles relating to disclosure of unused material are essential. The demands placed on an individual inspector may be particularly onerous, given that they may be—probably uniquely, in the field of criminal enforcement—investigator, prosecutor, and advocate in their own case.

8.158 Careful attention should be paid to the need for proper disclosure. As the Attorney-General's guidelines state: 'a failure to take action leading to proper disclosure

---

[95] See *R v Turner* [1970] 2 QB 321.

may result in a wrongful conviction or may alternatively lead to a successful abuse of process argument or an acquittal against the weight of the evidence.'

General legislative framework

8.159 Disclosure in criminal proceedings is governed by a number of different statutes, statutory instruments, and codes of practice. The primary legislative provision is the Criminal Proceedings and Investigation Act 1996 (CPIA), as amended by Criminal Justice Act 2003. CPIA 1996 applies to all cases which reach the Crown Court. A Code of Practice, which governs the preservation of material obtained during investigation and the provision of that material to the prosecutor, was originally produced under CPIA 1996. It has since been amended.[96] The amended Code applies in respect of investigations which began after 4 April 2005; the original code applies in respect of investigations conducted between 1 April 1997 and 3 April 2004.

8.160 Further, the Attorney-General issued guidelines on disclosure in November 2000, which augment and inform all aspects of the guidance above. The Guidelines themselves have been amended, in the light of the revisions made by the Criminal Justice Act 2003.

8.161 The Criminal Procedure Rules 2005 sets out rules which detail the procedure to be adopted when making applications for disclosure to the Crown Court.[97] A protocol for the control and management of unused material in the Crown Court was published under the authority of the Court of Appeal on 20 February 2006.

Disclosure in the magistrates' court

8.162 While the framework created by CPIA 1996, and associated provisions, is largely geared towards disclosure in the Crown Court, para 43 of the Attorney-General's Guidelines on Disclosure requires the prosecutor to provide all evidence on which the Crown seek to rely in summary proceedings. Guidance on pre-committal disclosure was also given in *R v DPP ex parte Lee* [1999] 2 Cr App R 304. There, the Court of Appeal identified a number of different types of material which it might be necessary for the prosecution to reveal prior to committal. In the main, this was material which may have been relevant to any decisions on bail, or material which was time-sensitive.

---

[96] Criminal Procedure and Investigations Act 1996 (Code of Practice) Order 2005 SI 2005/985.
[97] See, for example, Criminal Procedure Rules r 25.6 in relation to CPIA 1996 s 8 applications.

D. Disclosure

Procedure for making disclosure

8.163 The structure for disclosure of unused material has been substantially changed by Part 5 of the Criminal Justice Act 2003. The previous two-stage test—primary andsecondary disclosure—has been replaced. A single new objective test now governs the disclosure of unused material held by the prosecutor to the defence. He must disclose any material which might reasonably be considered capable of undermining the case for the prosecution against the accused, or of assisting the case for the accused.[98] If there is none, he should provide the accused with a written statement that there is no such material. The decision as to what should be disclosed is to be made by the prosecutor, and should not generally require the involvement of the judge.

8.164 While there is no specific time laid down in which the prosecutor must act, he must act as soon as is reasonably practicable after the accused has pleaded not guilty and the case has been committed (or sent) for trial.[99]

8.165 The prosecution should provide a schedule of the material which is in their possession, and which does not form part of their primary case. The schedule should contain detailed, clear, and accurate descriptions, possibly including summaries, of the material held. If the defence wishes to be provided with any of the material, they should justify why they seek it. The prosecution is not required to provide all the material on the schedule, or even make it available for inspection, unless a properly justified request is made.

Defence disclosure

8.166 Once the prosecutor has complied, or purported to comply, with the initial duty on him, the defendant is required to give a defence case statement within 14 days. If an application to extend the time for service of a defence is made prior to the expiry of the 14 days, the judge may extend the period at his discretion.[100]

8.167 If the investigation began after 4 April 2005, the defendant is now required to provide a more detailed defence case statement than before. In particular, he should set out the nature of his defence. He should also set out the matters on which he takes issue with the prosecution; stating why, in respect of each matter, he takes issue with the prosecution.[101]

8.168 If the defendant either fails to provide a defence case statement, or provides one late, or he puts forward at his trial a defence which was not mentioned in the

---

[98] See s 32 Criminal Justice Act 2003.
[99] Criminal Procedure and Investigations Act 1996 s 3(1); Criminal Justice Act 2003 s 32.
[100] Defence Disclosure Time Limit Rules 1997 r 2.
[101] Criminal Procedure and Investigations Act 1996 s 5(1); Criminal Justice Act 2003 s 331 Sch 36 Pt 3.

defence case statement, or which is different from the defence set out there, certain consequences follow. The trial judge, or any other party, may make such comments as may be appropriate, and the jury may draw such inferences as appear proper.[102]

Third party disclosure

8.169 It is sometimes necessary to obtain material from third parties. If the material is not in the possession of the prosecution, it will be necessary to obtain a summons under the Criminal Procedure (Attendance of Witnesses) Act 1965.

The investigation stage

8.170 Section 23 of Pt II of the Criminal Procedure and Investigation Act 1996 provides for the preparation, by the Secretary of State, of a code of practice. This governs the action that police officers must take in the recording of information and the retaining of material obtained in the course of a criminal investigation, and the preparation and revelation of that material to the prosecutor.

8.171 It contains 'provisions designed to secure'[103] that relevant material is gathered, recorded, and retained during the course of a criminal investigation. A 'criminal investigation'[104] is an:

> investigation conducted by police officers with a view to it being ascertained—
> (a) whether a person should be charged with an offence or
> (b) whether a person charged with an offence is guilty of it.

8.172 By s 26(1) of the Criminal Procedure and Investigation Act 1996

> a person other than a police officer who is charged with the duty of conducting an investigation with a view to it being ascertained—
> (a) whether a person should be charged with an offence or
> (b) whether a person charged with an offence is guilty of it
> shall in discharging that duty have regard to any relevant provision of a code which would apply if the investigation were conducted by police officers.

8.173 As inspectors are charged with a duty to conduct investigations, they must therefore have regard to the Code. Inspectors should take the above provisions into account when applying their own operating procedures.[105]

---

[102] Criminal Procedure and Investigations Act 1996 s 11; Criminal Justice Act 2003 s 39.
[103] Criminal Procedure and Investigations Act 1996 s 23(1).
[104] ibid s 22(1).
[105] ibid (Code of Practice) Order 2005 SI 2005/985) para 1.1.

D. Disclosure

8.174 The code envisages the creation of at least three different roles within a given investigation:

(a) the investigator;
(b) the officer in charge of the investigation;
(c) the disclosure officer.

8.175 There is no requirement that each role is undertaken by a different individual. Indeed, the code recognizes that the roles may be undertaken by one person only, according to the complexity of the case and the administrative arrangements within each police force. Specific limitations are made where there is likely to be a conflict of interest. It is relatively rare, in prosecutions for health and safety offences, that a separate disclosure officer is appointed. Rather, the function is usually carried out by the inspector leading the investigation. There is nothing wrong in this, though if an inspector assumes more than one role in an investigation, he should be careful to ensure that he is aware of the duties ascribed to each.

8.176 When discharging their duties, investigators must be fair and objective. They should always err on the side of recording and retaining material where they have any doubt as to whether it may be relevant.[106]

8.177 The investigator should pursue all reasonable lines of inquiry, whether they point towards the suspect or away from him.[107] What is reasonable in each case will depend on the particular circumstances.[108]

8.178 The failure to follow a line of enquiry which may seem to lead away from a defendant is often a source of adverse comment by a defendant during a trial. Although a judge will usually direct a jury not to speculate about evidence not before them, bearing in mind the burden of proof, experience has shown that the ability of a defendant to comment on the failure to follow a line of enquiry is often more damaging to a prosecution case, than the material which may have been obtained thereby.

8.179 The investigator must retain material obtained in an investigation which may be relevant.[109] This includes all material which is generated by him, such as interview records. It is acceptable to retain only copies of material or photographs of particular items if there are reasons for returning the original. The material must be retained at least until a decision is taken whether to institute proceedings, and in the event that they are, until they are concluded.

---

[106] Att-Gen's Guidelines, para 6.
[107] Criminal Procedure and Investigations Act 1996 (Code of Practice) Order 2005 SI 2005/985 para 3.4.
[108] ibid para 4.1.
[109] ibid paras 5.6 and 5.7.

8.180 Material which may be relevant to the investigation, but which is not recorded in any form, must be recorded in a durable or retrievable form. Acceptable forms include: writing, video, audio tape, or computer disk. 'Material', in this context, also includes negative information such as the fact that persons present at a relevant time say they saw nothing unusual.

Obtaining material from third parties

8.181 The Code requires the investigator to assume a proactive role. If he believes that others may be in possession of material which may be relevant to the investigation, and this material has not already been seized, he should inform them of the existence of the investigation. He should then advise them to retain the material in case they receive a request for its disclosure. However, there must be some reason to believe that a person is in possession of such material. The investigator has no obligation or remit to go on a fishing expedition or make speculative enquiries.[110]

> Material may be relevant to an investigation if it appears to an investigator, or to the officer in charge of the investigation or to the disclosure officer that it has some bearing on any offence under investigation, or any person being investigated, or on the surrounding circumstances of the case, unless it is incapable of having any impact on the case.[111]

8.182 In cases where it is suspected that a non-government agency or other third party has material which might be disclosable if it were in the prosecution's possession, consideration should be given to the question of whether it is appropriate to seek access to the material, particularly if it is thought that the material may undermine the prosecution case or assist the defence. If when requested to provide material the third party refuses, the prosecution should persist and may invoke s 2 of the Criminal Procedure (Attendance of Witnesses) Act 1965 or s 97 of the Magistrates' Courts Act 1980, where appropriate.[112] There should be consultation with the other agency lest there is any reason of public interest why the material is not disclosable.

The disclosure officer

8.183 The disclosure officer must inspect, view, or listen to, all the material that has been retained by the investigator.[113] The officer must, in addition, provide a personal

---

[110] Criminal Procedure and Investigations Act 1996 (Code of Practice) Order 2005 SI 2005/985 para 3.5.
[111] ibid para 2.1.
[112] Att-Gen's Guidelines on Disclosure para 31.
[113] ibid para 8.

D. Disclosure

declaration to the effect that this task has been done. This obligation does not apply in cases where the investigator has seized a large amount of material, out of an abundance of caution, which may not seem likely ever to be relevant. In those circumstances, the duty may extend only to retaining and informing the defence of the existence of the material, in order to afford an opportunity for inspection. Disclosure officers must specifically draw the prosecutor's attention to material under certain circumstances. Where there is any doubt as to whether it might undermine the prosecution case, or assist the defence case disclosed to date, the prosecutor's attention must be brought to the material.[114]

The prosecutor

Prosecutors, in turn, should be alert to the need to provide advice to disclosure officers, and to the possibility that material may exist which has not been revealed to them. They should also inform the investigator if, in their view, further lines of reasonable and relevant enquiry exist.[115] They should make a record of all the actions they make in discharging their disclosure responsibilities.[116] Similarly, prosecutors should err on the side of disclosure, if there is any doubt about whether material should be disclosed or not, unless the material is on the sensitive schedule.[117] **8.184**

Beyond the requirements of the above, the prosecutor is required at all times to act in the character of a minister of justice assisting in the administration of justice. **8.185**

Schedules of unused material

Schedules of both the non-sensitive unused, and the sensitive unused, material should be prepared by the disclosure officer. These should contain detailed, clear, and accurate descriptions, possibly including summaries, of the material held.[118] If no schedules have been provided, or there are apparent omissions from those provided, the prosecutor must take action to obtain properly completed schedules.[119] **8.186**

A decision must be made about what material does and does not go onto the schedules. In general terms all 'process' material—ie all material which relates to the conduct of the investigation—rather than the material obtained during the course of it need not be recorded on the schedules. Having said this, whether or **8.187**

---

[114] Att-Gen's Guidelines on Disclosure para 11.
[115] ibid para 17.
[116] ibid para 19.
[117] ibid para 20.
[118] ibid para 10.
[119] Criminal Procedure and Investigations Act 1996 (Code of Practice) Order 2005 SI 2005/985 para 14.

not a particular document is required to be listed on the schedule depends on its contents, and not on its nature or type. Thus consideration should be given to all sources of material generated during the conduct of a case; whether emails, internal memos, or otherwise.

8.188 Of course, consideration will have to be given to whether material which might otherwise fall to be disclosed is covered by privilege, or attracts public interest immunity.

Effect of non-compliance with the Code

8.189 The failure to comply with any provision of the Code will not render an inspector liable to any civil or criminal proceedings. Nonetheless, evidence of such a failure is admissible at trial. If non-compliance is shown, evidence may be excluded[120] in appropriate circumstances.

E. Expert Evidence

8.190 Expert evidence is admissible, as an exception to the general rule that evidence of opinion is inadmissible within court proceedings. The purpose of adducing expert evidence is to furnish the court with 'scientific information' which is likely to be outside the experience and knowledge of a judge or a jury.[121] It is for the court to decide whether the jury would, in fact, be assisted by expert evidence on a particular point. It is also for the court to determine whether the expert who is tendered does in fact have the necessary expertise.[122]

8.191 Where an expert is permitted to give an opinion, even on the ultimate issue, it should still be made clear to the jury that they must decide the issue. The point should be made to the jury that they are not bound by the expert's opinion. Expert evidence is not, in itself, determinative of the case—it is for the jury to decide what weight (if any) they wish to attach to it. However, the position remains that a conviction obtained in the face of clear and uncontradicted expert evidence suggesting innocence may be unsafe.[123]

8.192 In some cases, there may be a direct conflict of opinion between experts. The jury should then be directed that they should only accept the prosecution evidence if they are satisfied beyond reasonable doubt that it is correct.

---

[120] Criminal Procedure and Investigations Act 1996 s 26(4).
[121] *R v Turner* [1975] QBD 841.
[122] *R v Stockwell* (1993) 97 Cr App R 264.
[123] In this regard, see *R v Bailey* (1961) 66 Cr App R 31.

E. Expert Evidence

The 'ultimate issue rule'

If a rule ever existed to the effect that experts were unable to give an opinion on the ultimate issue in a criminal case, it has now been largely abandoned.[124] In *Stockwell*, the Lord Chief Justice stated that the rule that experts should not give evidence on the 'ultimate issue' was 'a matter of form rather than substance'. In the view of the court, if an expert was called to give his opinion, he should be allowed to do so. The Criminal Law Revision Committee, in its eleventh report,[125] expressed a similar view. The Committee found that the old common law rule—that a witness should not express an opinion on an ultimate issue—probably no longer existed. **8.193**

In health and safety cases, especially those concerned with breaches of the general duties, the ultimate issue is likely to be, 'did the defendant do all that was reasonably practicable in all of the circumstances?' Following the guidance in *Stockwell*, there will plainly be times when it will be appropriate for an inspector, specialist inspector, or an expert from industry, to give his opinion on that issue. He is then entitled to say that a given defendant did, or did not, take a particular step that could have been taken. Further, such a step would not have been disproportionately expensive, considering the level of risk involved: in other words, that the step was reasonably practicable. **8.194**

Independence of an expert

The overriding duty of the expert is to the court: '. . . he should never assume the role of an advocate.'[126] The court ruled on this point in *Liverpool Roman Catholic Archdiocese Trustees Incorporated v Goldberg (No 2)* [2001] 4 All ER 950. Where it can be shown that a relationship exists between the expert and the person calling him, and a reasonable observer may think the relationship capable of affecting the expert's views so as to make them unduly favourable to that party, his evidence should not be admitted. This is the case however unbiased the conclusions might have been.[127] The correctness of this conclusion was, however, doubted in *R (on the application of Factortame & Others) v The Secretary of State for Transport* [2002] **8.195**

---

[124] The common law position was formally abrogated in civil cases, by the Civil Evidence Act 1972 s 3.
[125] Cm 4991 para 268 *et seq.*
[126] *The Ikarian Reefer* [1993] 2 Lloyd's Rep 455 and cf, in the civil context, Criminal Procedure Rules r 53(3), which provides that the expert's duty is to 'help the court on matters within his expertise and this duty overrides any obligation to the person from whom he has received instructions or by whom he is paid'.
[127] Here, a Queen's Counsel specializing in tax law, who had been sued for professional negligence, called a Queen's Counsel from his own chambers, and who was a long-standing friend, to give expert evidence.

EWCA Civ 932. It was observed, by Lord Phillips MR (at [70]):[128] 'It is always desirable that an expert should have no actual or apparent interest in the outcome of the proceedings in which he gives evidence, but such disinterest is not automatically a precondition to the admissibility of his evidence.'

The HSE Enforcement Guide and experts

8.196 The HSE's Enforcement Guide now contains a section addressed to expert witnesses,[129] which is aimed at internal and external expert witnesses, inspectors, and all those who might be responsible for instructing such an expert. The Guide describes how specialist health and safety inspectors, 'might be involved in an investigation either as part of the investigation team (albeit with specialist knowledge) or as an independent expert witness. The specialist knowledge of inspectors involved in an investigation may allow them to be treated as an expert witness in the event of a prosecution.'

8.197 The Guide quotes how it has been held by the Court of Appeal, in the unreported civil case of *Field v Leeds City Council* (8 December 1999) that the fact that an expert may be the employee of one of the parties to litigation would not debar him from giving expert evidence and thus how the same principle applies to inspectors acting as experts in HSE prosecutions.

8.198 However, as the Guide describes, the roles of investigator and expert witness are usually mutually exclusive, and a specialist inspector will rarely be appointed as an independent expert if directly involved in the investigation. While a specialist inspector may advise the investigator on the gathering of evidence, such as questions to put to witnesses or to a suspect, if a specialist inspector is intended to act as a potential expert, then he should remain independent of the investigation and should be seen to do so: he should generally not take statements as part of the investigation or attend an interview of a suspect under caution.

The Criminal Procedure Rules

8.199 Rules relating to expert witnesses in criminal trials form Pt 33 of the Criminal Procedure Rules. In 2006 the Criminal Rules Committee published, and consulted upon, a series of draft rules. These closely followed the equivalent provisions in the Civil Procedure Rules. Principally, the proposals provide that an expert's report must:

(a) give details of the expert's qualifications and relevant experience;
(b) give details of any literature or other material which the expert has relied on in making the report;

---

[128] See also *Field v Leeds City Council* (CA 8 December 1999).
[129] http://www.hse.gov.uk/enforce/enforcementguide/investigation/expert/intro.htm.

E. Expert Evidence

(c) contain a statement setting out the substance of all facts and instructions given to the expert which are material to the opinions expressed in the report or upon which those opinions are based;
(d) make clear which of the facts stated in the report are within the expert's own knowledge;
(e) say who carried out any examination, measurement, test, or experiment which the expert has used for the report, give the qualifications of that person, and say whether or not the test or experiment has been carried out under the expert's supervision;
(f) where there is a range of opinion on the matters dealt with in the report, summarize the range of opinion, and give reasons for his own opinion;
(g) contain a summary of the conclusions reached;
(h) if the expert is not able to give his opinion without qualification, state the qualification;
(i) contain a statement that the expert understands his duty to the court, and has complied and will continue to comply with that duty;
(j) contain the same declaration of truth as a witness statement.

8.200 The proposed rules also include a power for the judge to direct that a pre-trial discussion occur between experts and a power to direct that evidence concerning an issue is to be given by one expert only.

Rule 24 of the Criminal Procedure Rules

8.201 Rule 24 of the Criminal Procedure Rules provides for disclosure of expert reports. It reproduces what was formerly r 3 of the Magistrates' Courts (Advance Notice of Expert Evidence) Rules 1997, and r 3 of the Crown Court (Advance Notice of Expert Evidence) Rules 1987.

8.202 Any party that proposes to adduce expert evidence in proceedings in the Crown Court, must, as soon as practicable after committal or transfer, furnish any other party to the proceedings with a statement in writing of any finding or opinion relied upon. Moreover, if he is requested to, he must provide that other party with a copy of the result of any observation, test calculation, or other procedure on which his finding or opinion is based. Any document, or object, in respect of which any such procedure has been carried out, must similarly be disclosed.[130]

8.203 A party who fails to comply with the above may not adduce the evidence without the leave of the court.[131]

---

[130] Criminal Procedure Rules r 24.1.
[131] Crown Court (Advance Notice of Expert Evidence) Rules 1987 SI 1987/716 reg 5.

8.204 Similarly, in the magistrates' court, a person intending to rely upon an expert opinion must serve a written statement of it on any other party. This must be as soon as practicable after a not guilty plea has been entered.[132]

Calling the expert

8.205 An expert can be compelled as a witness, like any other, and may also be required to bring documents. Counsel calling the expert witness should ask him in examination-in-chief to state the facts upon which his opinion is based.[133] An expert's report is admissible in criminal proceedings whether or not the person making it attends to give oral evidence, though the leave of the court must be obtained. When deciding whether to give leave, the court should take into account the contents of the report and the reasons why the person making it is not available to give expert evidence, and have regard to the risk of any unfairness which may arise.[134]

8.206 It is now more common in health and safety cases for experts to be called 'back to back'; rather than each in the course of their respective party's case. This has been common practice in civil cases for a long time, and is often appropriate in criminal cases. Indeed, even if the parties themselves do not ask for experts to be called one after the other, the trial judge may order it to take place if he considers it appropriate pursuant to his wide powers of case management.

Future development of disclosure duty

8.207 Section 6D of the Criminal Procedure and Investigations Act 1996 (which was inserted by s 35 Criminal Justice Act 2003) makes special provision for the instruction of expert witnesses. If the accused instructs an expert, with a view to his providing an expert opinion for possible use as evidence at trial, he must give to the prosecution and the court a notice specifying the person's name and address. At present, as with a significant number of provisions within the Criminal Justice Act 2003, no date has been set for this section to come into force. Ultimately, if it is brought into force, it may have a huge impact on criminal litigation, and will represent a radical departure from the present rules of disclosure.

F. Miscellaneous Areas of Evidence

Hearsay in criminal cases

8.208 The admissibility of hearsay, in criminal proceedings, is now governed by the Criminal Justice Act 2003. The 2003 Act made radical changes to the overall

---

[132] Magistrates' Courts (Advance Notice of Expert Evidence) Rules 1997 SI 1997/705.
[133] *R v Turner* [1995] 2 Cr App R 94 (CA).
[134] Criminal Justice Act 1988 s 30.

F. Miscellaneous Areas of Evidence

admissibility of hearsay in criminal cases. Previously, while it was the case that hearsay was only rarely admitted, now it is likely to be admitted into evidence more frequently. The thrust of the legislation is to make hearsay prima facie admissible, rather than prima facie inadmissible.

8.209 The effective date for commencement of the hearsay provisions is 4 April 2005. The sections will apply for any trial which starts on or after that date—regardless of when the investigation may have commenced, or when any plea or direction hearing may have commenced.

8.210 Section 114 provides that hearsay is admissible in criminal proceedings, but only if:

(a) it can be brought within a statutory exception, including the following provisions of this chapter of the Act;
(b) it is admissible under one of the common law exceptions preserved by s 118;
(c) all parties to the proceedings agree to it being admissible, or;
(d) 'the court is satisfied that it is in the interests of justice for it to be admissible'.

8.211 In making a decision, under para (d), a court is obliged to have regard to the matters listed in subs (2). Section 114(2) provides that, with the exception of the rules preserved by subs (1), the common law rules governing the admissibility of hearsay evidence in criminal proceedings are abolished.

8.212 The factors which must be taken into account, when deciding whether it is admissible, are the following:

(a) how much probative value the statement has (assuming it to be true), in relation to a matter in issue in the proceedings, or how valuable it is for the understanding of other evidence in the case;
(b) what other evidence has been, or can be, given on the matter or evidence mentioned in para (a);
(c) how important the matter or evidence mentioned in para (a) is, in the context of the case as a whole;
(d) the circumstances in which the statement was made;
(e) how reliable the maker of the statement appears to be;
(f) how reliable the evidence of the making of the statement appears to be;
(g) whether oral evidence of the matter stated can be given, and, if not, why it cannot;
(h) the amount of difficulty involved in challenging the statement;
(i) the extent to which that difficulty would be likely to prejudice the party facing it.

8.213 Procedurally, if the prosecution wish to rely upon hearsay evidence, they must give notice of their intention at the same time as they comply with s 3 of the Criminal Procedure and Investigations Act 1996 (ie disclosure by prosecutor). In the Crown

Court, this is to be not more than 14 days after the committal or transfer. A defendant must give notice of hearsay evidence not more than 14 days after the prosecutor has complied with, or purported to comply with, s 3 of the Criminal Procedure and Investigations Act 1996 (disclosure by prosecutor). A party who receives a notice of hearsay evidence may oppose it, by giving notice within 14 days, in the form set out in the Practice Direction. The notice should then be given to the court officer and all other parties.

Bad character

8.214　The Criminal Justice Act 2003 abolished the common law rules governing the admission of bad character in criminal trials. The relevant provisions, concerning bad character, came into force on 15 December 2004. The provisions apply to all trials commenced on, or after, that date, whether the preliminary hearing was held before or after.

8.215　'Bad character' is evidence of, or evidence of a disposition towards, misconduct, other than evidence bound up with the alleged facts of the offence with which the defendant is charged. It might also be evidence of misconduct in connection with the investigation, or prosecution, of that offence.

8.216　Evidence of bad character may be adduced against a person other than the defendant, ie witnesses and non-witnesses.

8.217　There are a number of different circumstances in which the evidence of a defendant's bad character may become admissible. These have been referred to as 'the seven gateway criteria'. The most significant ones are: that it is important explanatory evidence (c), or it is relevant to an important matter in issue between the defendant and the prosecution (d). In reality, important explanatory evidence is nothing more or less than motive.

8.218　The more significant paragraph is (d). This sets out the requirement that it is an important matter in issue between the defendant and the prosecution. Section 103 provides for the issues between the defendant and prosecution. This is further defined to include whether the defendant has a propensity to commit offences of the kind with which he is charged. The exception is where his having such a propensity makes it no more likely that he is guilty of the offence. The inclusion of propensity, as being a matter in issue between the defendant and the prosecution, is the major change wrought by this part of the Criminal Justice Act 2003.

8.219　Section 101(1)(d) requires that, before it is admitted, propensity must be an important issue in a case. It appears that it will be an important issue when the fact of such propensity may, when combined with the other evidence in the case, serve to eliminate doubt or found a convincing case.

F. Miscellaneous Areas of Evidence

8.220 It should be borne in mind that propensity suggests a predisposition to behave in a certain way. Breach of the duties, in ss 2 and 3 of the Health and Safety at Work etc. Act 1974, involves no *mens rea*. The duties are non-delegable: where performance is carried out by an employee or agent, the employer will still be liable for any breach. The behaviour, in the sense of acts or omissions of all persons at work or acting on behalf of the company, is attributable to the company.

8.221 The leading authority on the propensity provisions is contained in *R v Hanson*; *R v Gilmore*; *R v P* [2005] 2 Cr App R 21 (CA), where the Court of Appeal issued guidance. The Court laid down guidelines, when considering an application to adduce evidence of a defendant's previous convictions, to establish propensity to commit offences of the kind with which the defendant is charged. A judge should consider:

(a) whether the history of his convictions establishes a propensity to commit offences of the kind charged;
(b) whether that propensity made it more likely that the defendant had committed the offence charged (see the exception in s 103(1)(a));
(c) whether it was unjust to rely on convictions of the same description or category (see s 103(3)); and,
(d) in any event, whether the proceedings would be unfair if they were admitted (see s 101(3)).

8.222 Applications to adduce such evidence should not be made routinely. The simple fact that a defendant has previous convictions is not determinative. Consideration should, instead, be based on the particular circumstances of each case.

8.223 In ruling on any such application, judges should further bear in mind that:

(a) it is not necessarily sufficient, in order to show such propensity, that a conviction should be of the same description or category as that charged;
(b) there was no minimum number of events necessary to demonstrate such a propensity;
(c) though the fewer the number of convictions, the weaker was likely to be the evidence of propensity, a single previous conviction for an offence of the same description or category would often not show propensity. It might, where, for example, it showed a tendency to unusual behaviour;
(d) it would often be necessary to examine each individual conviction, rather than merely look at the name of the offence;
(e) where past events were disputed, the judge had to take care not to permit the trial to be unreasonably diverted into an investigation of matters not charged in the indictment.

8.224 However, the court must not admit evidence if, on an application by the defendant to exclude it, it appears to the court that the admission of the evidence would have such an adverse effect on the fairness of the proceedings that the court ought not to admit it (s 101(3)).

8.225 When considering whether to exclude the evidence, the court must have regard, in particular, to the length of time between the matters to which that evidence relates, and the matters which form the subject of the offence charged (s 101(4)).

8.226 Procedurally, a prosecutor who wants to introduce evidence of a defendant's bad character, or who wants to cross-examine a witness with a view to eliciting that evidence, must give notice. Notice must be given, in the form set out in the Practice Direction, to the court officer and all other parties to the proceedings. Notice must be given in a case to be tried in the Crown Court, not more than 14 days after the committal of the defendant for transfer.[135]

---

[135] See Criminal Procedure Rules r 35.1.

9

JUDICIAL REVIEW AND ABUSE OF PROCESS

A. Judicial Review of the Decision to Prosecute or not to Prosecute	9.01
Judicial review of the decision to prosecute	9.03
Judicial review of the decision not to prosecute	9.09
B. Abuse of Process	9.17
The scope of the jurisdiction	9.17
Categories of abuse	9.22

A. Judicial Review of the Decision to Prosecute or not to Prosecute

The doctrine of abuse of process has developed in the past years into a specialist area, an exhaustive study of which is beyond the scope of this book.[1] However, it is undoubtedly an important area for health and safety practitioners, with many of the complaints capable of giving rise to assertions of abuse occurring in health and safety cases. Indeed, some of the important case law on the topic originates from health and safety prosecutions, and accordingly it is covered in some depth here. **9.01**

Similarly, judicial review proceedings of decisions to prosecute, or of decisions not to either investigate or prosecute, while rare, have been contemplated and threatened against the HSE by aggrieved parties on many occasions in relation to work-related deaths and employee safety cases. **9.02**

Judicial review of the decision to prosecute

While it is possible to judicially review the decision of an enforcing authority to prosecute, such decisions being actions in relation to the exercise of a public function, the circumstances in which such applications have been entertained are extremely limited. **9.03**

---

[1] For further, see Colin Wells's *Abuse of Process: A Practical Approach* (London: the Legal Practice Group).

The Court's general approach has been that the trial process itself is well-equipped to deal with any complaint that may be made about the bringing of a prosecution—both through its ability to regulate what material is admitted into evidence, and through its inherent power to stay any prosecutions which are held to amount to an abuse of the process of the courts. (However, see below at paragraphs 9.19–9.21 in relation to the jurisdiction of the High Court in respect of abuse of process applications relating to magistrates' court proceedings.)

9.04 The issue was considered in *R v DPP ex parte Kebilene* [2000] 2 AC 326 (HL), where the House of Lords found that the DPP's decision to prosecute may be challenged in the High Court. The House went on to observe that any such challenge may only be brought in circumstances of mala fides or dishonesty of motive, or some other exceptional circumstance. Their lordships also proceeded to consider the ambit and effect of the Supreme Court Act 1981 s 29(3), which provides:

> In relation to the jurisdiction of the Crown Court, other than its jurisdiction in matters relating to trial on indictment, the High Court shall have all such jurisdiction to make orders of mandamus, prohibition or certiorari as the High Court possesses in relation to the jurisdiction of an inferior court.

9.05 Their Lordships then continued to examine the circumstances in which proceedings for judicial review may be brought, and reflected on a potential anomaly. The above provision would prevent proceedings for judicial review[2] being brought, if a Crown Court judge were to hold that a prosecution amounted to an abuse of process because of an alleged breach of a convention right. However, it would be 'curious' if the same issue could be raised in the Divisional Court by means of a challenge to the DPP's decision to institute or proceed with a prosecution. To allow such a course would be to 'outflank' the policy underlying the legislation—which was to prevent satellite litigation from clogging up and delaying the criminal justice system.

9.06 Their Lordships concluded that the Supreme Court Act 1981 s 29(3) may permit judicial review of the decision to prosecute in the limited circumstances described above. They asserted, too, that it would still be necessary to advert to, and consider, what Hobhouse LJ referred to as one of the 'well-established' principles that prevailed. That principle is that the remedy of judicial review only lies where there is no other legal remedy.[3]

9.07 Where the Administrative Court has made an order quashing the decision to prosecute, further observations on abuse arguments can be made. Almost invariably,

---

[2] The same applies in relation to applications for a case stated under SCA 1981 s 28.
[3] In this context, see De Smith, Woolf, and Jowell, *Judicial Review of Administrative Action* (5th edn) para 20–018, for clear elaboration on the discretion to refuse judicial review except in 'cases of last resort'.

A. Judicial Review of the Decision to Prosecute or not to Prosecute

matters which require the decision to be quashed would, if they had been argued, have amounted to an abuse of process in the Crown Court.[4]

In any event, the permission of the court is required before the claim for judicial review may proceed to a full hearing. The purpose of the provision is to winnow out claims which are frivolous, vexatious, or hopeless. The court will only grant permission when it considers that an arguable ground for seeking judicial review exists, and that it merits consideration at a full hearing. **9.08**

Judicial review of the decision not to prosecute

Conversely to the situation pertaining to review of decisions to prosecute, the Administrative Court has been more ready to entertain applications for judicial review of decisions not to prosecute in particular cases. This is on the basis that there is, *a priori*, no other forum for airing the complaint, and therefore no alternative remedy exists. Implications under European Convention on Human Rights, Art 6, and its guarantee of the right to a fair trial may, and are likely to, arise here. (The established common law principle of the right to be heard might also avail here, if the human-rights oriented threshold is not met.) **9.09**

Challenge may arise in respect of a refusal to investigate an incident, or a decision not to prosecute following such an investigation. The ambit of the duty on the HSE to investigate and the principles to be applied in any decision to investigate are considered at paragraphs 2.34–2.52. The relevant principles and the test in relation to a decision to take enforcement action are considered at paragraphs 4.01–4.24. **9.10**

Threatened judicial review of a decision by an HSE inspector not to investigate an incident as a potential breach of HSWA 1974 s 3, where a non-employee member of the public has been injured or killed, has occurred on a number of occasions in recent years; however, no such action has been pursued as far as a hearing at the High Court. A challenge to a decision not to prosecute following such an investigation by the HSE was heard, determined, and dismissed in *R (on the application of Pullen) v Health & Safety Executive* [2003] EWHC 2934 (Admin). **9.11**

Crucially, judicial review of a decision not to investigate or prosecute generally relates to the way in which the decision was taken, rather than the merits of the case itself. An important exception exists in the court's ability to find that a prosecuting authority acted irrationally in reaching such a decision. However, the threshold for success with this type of application is high, given the *Wednesbury*[5] formulation of irrationality. It may be possible to force the prosecuting authority **9.12**

---

[4] eg *R v Croydon JJ ex p Dean* [1993] QBD 769.
[5] *Associated Provincial Picture Houses Ltd v Wednesbury Corporation* [1948] 1 KB 223.

to review its decision not to investigate or prosecute on two bases: either by showing that it has not followed its own settled (and not irrational) policy for determining whether to prosecute or not; alternatively by arguing that it did not have regard to a material consideration. If either of these defective approaches features, the decision not to prosecute may be set aside.[6]

9.13 An example of the court's approach is found in *R v DPP and others ex parte Jones* [2000] IRLR 373. There, the deceased was working on the docks at Shoreham-on-Sea, when he was decapitated by the jaws of a grab-bucket which was being used to unload bags of cobblestones from a ship. The managing director of the company had altered the working of the bucket in such a way as to require employees to work close by, or directly beneath it. Employees were thereby put in danger if the bucket swung too close. Compounding this, the operator of the crane that was used to lift the bucket could not see into the hold area, but instead relied on a hatch-man to tell him when it was safe to close the jaws. The deceased had been sent to the company by an employment agency and it had been his first day of employment with them; he had received no formal training from either the agency or the company.

9.14 The HSE brought proceedings for breaches of HSWA 1974 s 2(1) and s 3(1). Nevertheless, the DPP, having taken advice from leading counsel, decided not to prosecute either the managing director or the company for manslaughter. In part, it was thought there was insufficient evidence to show causation. Proceedings to challenge the decision not to prosecute were initiated by the deceased's brother and leave was granted, whereupon the DPP decided to reconsider the decision not to prosecute. That review also concluded that the evidence was not sufficient to provide a reasonable prospect of satisfying a jury that the managing director or company were guilty of manslaughter by gross negligence. In reaching this conclusion, weight was placed on the managing director's lack of subjective recklessness, ie an awareness that the system for work was inherently dangerous, yet he made continued use of it.

9.15 The Divisional Court found that the DPP had not addressed the relevant law properly. He had placed too much emphasis on what he perceived to be the absence of subjective recklessness, rather than concentrating on objective liability for the dangerous system of work. The Court also concluded that they could not properly assess whether the decision not to prosecute had been taken irrationally. They had been provided with insufficient assistance about the reasoning which underpinned it; something which weighed in the balance as a further reason why the decision had to be reconsidered.

---

[6] See *R v DPP ex p C* [1995] 1 Cr App R 136.

9.16 In *R (on the application of Brenda Rowley) v DPP* [2003] EWHC 693 (Admin), the Administrative Court rejected the application for an order to quash the decision of the respondent not to prosecute. The potential defendants were the city council and a carer employed by them. At stake were charges of gross negligence manslaughter in respect of the death of the applicant's severely disabled son, who had been in their long-term care. In refusing an order to quash, the Court took into account the fact that the council had already been prosecuted under HSWA 1974 s 3(1), and the Reporting of Injuries, Diseases and Dangerous Occurrences Regulations 1995. A fine of £115,000 had been imposed. It was observed, in the judgment, that any prosecution for manslaughter was likely, in any event, to fall foul of the principle in *Beedie*[7] (on which, see para 9.35 *et seq*, below).

B. Abuse of Process

The scope of the jurisdiction

9.17 It has been said that the court has no role in either 'stifling' or 'instigating' prosecutions. As a basic proposition, this holds true. Nevertheless, it has been held that the court has an inherent power to protect its process from abuse. Moreover, this must include a power to safeguard an accused person from oppression or prejudice.[8] An abuse of process has been defined as 'something so unfair and wrong that the court should not allow a prosecutor to proceed with what is, in all other respects, a regular proceeding'.[9]

9.18 The jurisdiction to stay for abuse of process has been divided into two categories by the Court of Appeal:[10] namely, (a) where the defendant would not receive a fair trial; and/or (b) where it would be unfair for the defendant to be tried.

9.19 The power to stay proceedings as an abuse is one confined in magistrates' courts to matters directly affecting the fairness of the trial of the accused (ie category (a)—such as delay or unfair manipulation of court procedures—and has been held to be one that must only be exercised most sparingly).[11] Lord Griffiths in the House of Lords' decision of *Bennett* [1994] 1 AC 42 (HL) stated:

> I would accordingly affirm the power of the magistrates, whether sitting as committing justices or exercising their summary jurisdiction, to exercise control over their proceedings through an abuse of process jurisdiction. However, in the case of magistrates

---

[7] *R v Beedie* [1997] 2 Cr App R 167 (CA).
[8] *Connelly v DPP* [1964] AC 1254 (HL).
[9] *Hui Chi-Ming v R* [1992] 1 AC 34 (PC).
[10] See *R v Beckford* [1996] 1 Cr App R 94.
[11] See *R v Bennett* [1994] 1 AC 42 (HL), at 127; *R v Oxford City Justices ex p Smith* [1982] 75 Cr App R 200; and Ackner LJ in *R v Horsham Justices ex p Reeves (Note)* [1980] 75 Cr App R 236.

this power should be strictly confined to matters directly affecting the fairness of the trial of the particular accused with whom they are dealing, such as delay or unfair manipulation of court procedures. Although it may be convenient to label the wider supervisory jurisdiction with which we are concerned in this appeal under the head of abuse of process, it is in fact a horse of a very different colour from the narrower issues that arise when considering domestic criminal trial procedures. I adhere to the view I expressed in *Guildford Justices ex parte Healy* [1983] 1 WLR 108 that this wider responsibility for upholding the rule of law must be that of the High Court.[12]

9.20 The extent of this narrow category of abuse of process, involving oppression outside the trial process, in respect of which a magistrates' court has no jurisdiction,[13] remains undetermined. It includes where the prosecution have manipulated or misused the process of the court so as to deprive the defendant of a protection provided by the law;[14] and, where it would be contrary to the public interest in the integrity of the criminal justice system that a trial should take place.[15]

9.21 Furthermore, even in cases concerning complaints directed to the fairness or propriety of the trial process itself, a magistrates' court may decline jurisdiction and leave the matter to be challenged in the High Court;[16] something which would amount to an exception to the general rule explained in *R v DPP ex parte Kebilene* [2000] 2 AC 326 (HL).[17]

Categories of abuse

9.22 As case law has developed, the circumstances in which the courts have been willing to find abuse are extremely varied. Each case must inevitably turn on its own facts. Furthermore, the categories of abuse are not closed. What follows is an outline of the areas of abuse most commonly argued in health and safety cases.

Delay

9.23 Health and safety cases are often subject to delay, and take a long time to come to court. The investigation stage is often drawn out, exacting, and complicated; and in cases involving fatalities, the decision to prosecute will normally not be taken until after an inquest has taken place. It is by no means uncommon for a number of years to have elapsed between the alleged breaches and the trial itself.

---

[12] [1994] 1 AC 42 (HL) 127.
[13] *R v Bennett* [1994] 1 AC 42 (HL) 127; *Environment Agency v Stanford* (CO/4625/97) and paras 8.52–8.54; *R v Leeds Stipendiary Magistrate ex p Yorkshire Water Services (No 2)* (CO/4834/99)
[14] *R v Derby Crown Court ex p Brooks* [1985] 80 Cr App R 164 (DC); and *R v Horseferry Road Magistrates' Court ex p Bennett* [1994] 1 AC 42 (HL).
[15] *R v Mullen* [1999] 2 Cr App R 143 (CA).
[16] See *R v Belmarsh Magistrates' Court ex p Watts* [1999] 2 Cr App R 188 (DC); *R v Horseferry Road Magistrates' Court ex p DPP* [1999] COD 441 (DC).
[17] See paras 9.04–9.08.

B. Abuse of Process

When proceedings are against corporate defendants—as they frequently are—no issue of custody or bail arises.

9.24 The circumstances in which an indictment should be stayed for delay were set out in the case of *Att-Gen's Reference (No. 1 of 1990)* [1992] 1 QBD 630 (per Lord Lane CJ 643G–644C). It was held that a stay of a prosecution may be imposed for reasons of delay, but only in exceptional circumstances. Even when the delay can be said to be unjustifiable, the imposition of a stay should be the exception, rather than the rule. Rarer still should be cases where a stay is imposed in the absence of any fault on the part of the prosecution.

9.25 Delays which are occasioned by the complexity of the case may be justifiable. In any event, delay caused or contributed to by the actions of the defendant should never be the foundation for staying a case. No stay should be granted unless the defendant shows, on the balance of probabilities, that he will suffer serious prejudice owing to the delay such that no fair trial is possible.

9.26 The factors to be borne in mind when determining whether prejudice has arisen include the following:

(a) the power of the judge to regulate the admissibility of evidence at common law and under statute;
(b) the trial process itself, which should ensure that all relevant factual issues arising from the delay are placed before the jury for their consideration; and
(c) the power of the judge to give appropriate directions to the jury before they consider their verdict.

These principles were reiterated and endorsed in the judgment of Rose LJ, in *R v S* Times Law Reports 29 March 2006.

9.27 By way of example, such serious prejudice might arise when evidence which was available at an early stage is no longer so after the passing of time. 'Serious prejudice' might arise, for instance, on establishing the death of a key witness who might otherwise have given evidence in rebuttal of the prosecution case; although, where a written record or statement is available from such a witness, then the situation is likely to be one that will be held capable of being remedied by the trial process.

Delay and Article 6 of the European Convention on Human Rights

9.28 Article 6(1) of the European Convention on Human Rights (ECHR) deals with the right to a fair trial. It provides: 'In the determination of his civil rights and obligations or of any criminal charge against him, everyone is entitled to a fair and public hearing within a reasonable time by an independent and impartial tribunal established by law.'

9.29 It has been held that Art 6 creates a number of distinct rights, one of which is the right to a hearing within 'a reasonable time'. The nature and extent of that right has been considered in a number of cases over the recent years, and a substantial body of jurisprudence has developed.

9.30 In *Att-Gen's Reference (No 2 of 2001)* [2001] 1 WLR 1869, the Attorney-General sought clarification from the Court of Appeal on the effect of the advent of the Human Rights Act 1998 (HRA) concerning the common law position on delay. The two issues raised by the Attorney-General were:

(a) whether criminal proceedings could be stayed, on the ground that there had been a violation of the reasonable time requirement in Art 6(1) of the ECHR, in circumstances where the accused could not demonstrate any prejudice arising from the delay; and

(b) when, in the determination of whether, for the purposes of Art 6(1), a criminal charge had been heard within a reasonable time, the relevant time commenced.

9.31 The answer to the first question was a qualified 'no'. Woolf CJ (as he then was) elected to draw a careful distinction between, on the one hand, the conduct which constitutes the unlawful act for the purposes of Art 6(1), and, on the other, the remedy to be provided. The court found that even if it was shown that there had been an unreasonable delay—and therefore a breach of Art 6(1)—it did not follow that an effective remedy called for a stay of the proceedings for abuse of process. The Appellate Court went on to say that the approach which existed before incorporation into English law of the HRA 1998 continued to apply. Only when the conduct complained of amounted to an abuse, in the way described in *Att-Gen's Reference (No 1 of 1990)* (1992) 95 Cr App R 296, [1992] QB 630, should the proceedings be stayed for abuse of process. If there had been a breach of HRA 1998, which did not amount to an abuse and therefore justify a stay, that fact could be marked by a reduction in sentence if the defendant was convicted. Alternatively, an award of compensation could be made, if he was acquitted.

9.32 The position may conveniently be summarized as follows. The time in question is to be measured from the point of charge. More precisely, time runs from the point at which an individual is informed that he is being investigated for the commission of a criminal offence, to the point where the case comes for trial. The courts have consistently refused to define what amounts to 'a reasonable time', in terms of minimum or maximum periods. It has, however, been made plain that, for a delay to be unreasonable, it should be inordinate on its face and give rise to an obvious concern.[18] What amounts to a 'reasonable time' depends on the facts of any given case. Factors which the court should consider, when deciding whether a given period is reasonable or not, include primarily the complexity of the case, the conduct of the applicants, and the conduct of the judicial authorities.[19]

---

[18] See *Dyer v Watson* [2002] UKPC D1, paras 30–55.
[19] *Eckle v Germany* (1982) 5 EHRR 1 para 80.

B. Abuse of Process

9.33 The fact that it may be found that an unreasonable delay has occurred does not, of itself, mean that the defendant may not be tried or, if appropriate, sentenced. The ECHR details the right to a fair trial, but not the sanction to be imposed for the breach of it. In certain cases, the courts have reduced the sentence to be imposed, where it is found that there has been an unreasonable delay.

Double jeopardy: two or more prosecutions on the same facts

9.34 As a general (though not invariable) rule, it is likely to be an abuse to try a defendant for an offence which arises out of the same, or substantially the same, facts as an offence for which he has already been tried.[20]

9.35 The point was considered in *R v Beedie* [1997] 2 Cr App R 167. A young woman died through carbon monoxide poisoning, caused by the use of a faulty gas fire at her bed-sit. In proceedings instituted by the HSE, her landlord pleaded guilty at the magistrates' court to breaching his duty under HSWA 1974 s 3(2), and was fined. (He was also prosecuted and fined, in separate proceedings taken by the city council, for breaches of housing legislation.) After these proceedings had concluded, and an inquest had been held, the police notified the CPS about the case. After an interview, in which the defendant made no comment, he was charged with manslaughter and committed to the Crown Court. The trial judge rejected submissions that to proceed, in all of the circumstances, amounted to an abuse of the process of the court. The defendant pleaded guilty to the charge and was sentenced to a suspended sentence. The Court of Appeal held that an abuse of process had been established. The defendant had been prosecuted for the more serious offence of manslaughter, in the Crown Court, on the basis of substantially the same facts as gave rise to the prosecutions in the magistrates' court. Crucially, no new facts had emerged. This was an aspect of the case to which considerable weight was attached.

9.36 The judgment in *Beedie* amounted to a radical departure from the previous interpretation of the scope of the double jeopardy rule (derived from the majority decision in the House of Lords case of *Connelly v DPP* [1964] AC 1254). The decision in *Beedie* led to the creation of the Work Related Death Protocol. A uniform policy of liaison was instituted, in cases involving deaths, between the Crown Prosecution Service, HSE, and local authorities.

9.37 The decision in *Beedie* established the existence of a wide discretion to stay proceedings, where a second offence 'arises out of the same or substantially the same set of facts as the first'. Furthermore, it was held that the discretion should generally be exercised in favour of an accused, unless the prosecution establishes that there are special circumstances for not doing so.

---

[20] See *Connelly v DPP* [1964] AC 1254, per Lord Devlin at 1359.

9.38 The Court of Appeal, in *Beedie*, gave prominence to the speech of Lord Devlin, in *Connelly*. In that case, Lord Devlin stated (at 1359):

> As a general rule a judge should stay an indictment (that is, order that it remain on the file not to be proceeded with) when he is satisfied that the charges therein are founded on the same facts as the charges in a previous indictment on which the accused had been tried, or form part of a series of offences of the same or similar character as the offences charged in the previous indictment. He will do this because as a general rule it is oppressive to an accused for the prosecution not to use r 3 [of the Indictment Rules] where it can properly be used, but a second trial on the same facts or similar facts is not always and necessarily oppressive, and there may in a particular case be special circumstances which make it just and convenient in that case.

9.39 In *Beedie*, this passage was said to have been supported by the judgment of Lord Reid, in *Connelly*, and, in particular, where he said: 'the general rule must be that the prosecutor should combine in one indictment all the charges he intends to prefer . . . there must always be a residual discretion to prevent anything which savours of abuse of process.'

9.40 In terms of what was meant by 'special circumstances', the Court of Appeal in *Beedie* appears to have stressed that what may amount to good reasons may not correlate with what are special circumstances. There has to be something in the facts, or other relevant matter, which takes the case outside of the general rule. The court noted that in *Beedie*, prosecution counsel: 'did not seek to say that the fact that the Health and Safety Executive are an autonomous prosecuting authority gave rise to any special circumstances; for he accepted that manslaughter and the 1974 Act offences could have been joined in the same indictment' (at 175).

Developments since Beedie

9.41 The decision in *Beedie* has been cited with approval, and considered authoritative, on a number of occasions since it was laid down in 1997. In particular, the nature and ambit of the double jeopardy rule, as it was particularized in the passages of the speeches from *Connelly* as they are quoted in *Beedie*, has been confirmed in all subsequent authorities. (See the earlier cases of *R v Riebold* [1967] 1 WLR 674 and the later decision of *R v Piggot and Litwin* [1999] 2 Cr App R 320 (CA).)

9.42 In *R v Z* [2000] 2 Cr App R 281, the House of Lords again considered the ambit of the rule of double jeopardy. Lord Hutton analysed the authorities, including the House of Lords' decision in *DPP v Humphrys* (1976) 63 Cr App R 95, concerning issue estoppel. He considered (at 293):

> In my opinion, the speeches in the House recognized that as a general rule the circumstances in which a prosecution should be stopped by the court are where, on the facts, the first offence of which the defendant had been convicted or acquitted was founded on the same incident as that on which the alleged second offence is founded. This appears most closely in the speech of Lord Edmund-Davies.

Lord Hutton concluded: 9.43

> The principle of double jeopardy operates to cause a criminal court in the exercise of its discretion, and subject to the qualification as to special circumstances, stated by Lord Devlin in *Connelly*... to stop a prosecution where the defendant is being prosecuted on the same facts or substantially the same facts as gave rise to an earlier prosecution which resulted in his acquittal (or conviction).

The principle, as set out above, can raise particular difficulties for inspectors in relation to areas such as asbestos, where different aspects of regulation are shared between various enforcement authorities. Accordingly, they should exercise care in liaising with other authorities—such as the Environment Agency and local enforcement authorities—where multiple prosecutions in respect of one series of events may arise. 9.44

For example, in September 1999, Cockcroft J, sitting at Leeds Crown Court,[21] stopped the prosecution of two directors of a company, in respect of six allegations concerning the alleged unlawful disposal of asbestos. In April 1999 the two directors of the company had been prosecuted by the HSE for stripping out the same asbestos without a licence and for having unlawfully employed three teenagers to carry out the work.[22] They had been sentenced to perform community service orders. Cockcroft J ruled that the two prosecutions had arisen from broadly similar facts arising from one chain of events. As such, they should have been joined on one indictment. Specifically, Cockcroft J considered the prosecution concerning the teenage employees, which he viewed as the most serious aspect. He ruled that it would be unfair, and an abuse of the process, for a further prosecution to be allowed. 9.45

Breach of promise

It sometimes happens that a representation is made in correspondence, or orally, by an inspector that no prosecution will occur in relation to particular events; but later, that decision is reversed and a prosecution is started. It has been held that reneging on such a promise can, but does not necessarily, amount to an abuse of the process of the court. In turn, this may lead to proceedings being stayed on the ground that the defendant has a legitimate expectation that he will not be prosecuted. The longer he is given to thinking that he will not be prosecuted, the more likely it is that a prosecution will amount to an abuse. It will be inherently unfair to prosecute an individual when steps taken, as a result of the belief that no prosecution would arise, lead to manifest prejudice.[23] 9.46

---

[21] *Environmental Agency v Medley & Medley* (Leeds Crown Court, September 1999).
[22] In breach of the Children and Young Persons at Work Act 1920.
[23] See *R v Townsend, Dearsley and Bretscher* [1997] 2 Cr App R 540.

9.47 In *R v Croydon JJ ex parte Dean* [1993] QBD 7691, a juvenile arrested in the course of a murder enquiry admitted to doing acts with intent to impede the apprehension of other suspects. He was then released without charge, on the basis that he was to be a prosecution witness. He made witness statements to that effect, and assisted the police for a period of weeks, being treated as a prosecution witness only. Despite this, a decision had in fact been made by the Crown Prosecution Service to charge him with doing acts with intent to impede the arrest of another, contrary to s 4(1) of the Criminal Law Act 1967. At committal proceedings for that offence, counsel submitted that there was an abuse of process. Counsel applied, alternatively, for an adjournment, in order that he could make an application to the Divisional Court for a stay of proceedings. Both applications were refused. He sought, and was granted, leave to apply for judicial review of the magistrates' decision to sit as examining justices and to commit the case to the Crown Court. Staughton J quashed the committal of the defendant. He stated that the prosecution of a person who has received a promise, undertaking, or representation from the police that he will not be prosecuted, is capable of amounting to an abuse of process. Even in the absence of bad faith and lack of authority to make such a promise, Staughton J's reasoning prevails. Observing that the application ought to have been made to the Crown Court judge, under the principles of abuse of process, the learned judge then held that the case before him fell to be considered as an exceptional case. On that basis, the point could be decided upon disputed facts, taken together with any other facts he felt bound to accept as true, so as to reach a decision.

9.48 In the *Dean* case, the prosecution raised a 'constitutional' argument. Since the police did not have the authority to give the appellant an assurance that he would not be prosecuted, such a representation could not amount to an abuse. The argument was rejected, on the grounds that such authority was not an essential requirement. Rather, Staughton J suggested, the remedy 'must surely be a greater degree of liaison at an early stage'.[24]

Reneging on a plea bargain

9.49 *R v Thomas* [1995] CLR 938 is a first instance case, in which the prosecution reneged on a plea bargain. Before the pleas and directions hearing, the prosecution had written to a defendant charged with an offence under the Offences Against the Person Act 1861 (OAPA) s 18 (wounding with intent). It had been indicated that a plea to OAPA s 20 (wounding), would be sufficient. At the pleas and direction hearing, the defendant entered a guilty plea to s 20. Counsel for the prosecution, however, indicated that a trial was still sought on s 18, having disagreed with the initial view taken by the Crown Prosecution Service. The trial

---

[24] See *R v Beedie* [1997] 2 Cr App R 167.

judge concluded that there had been prejudice, and that proceedings should be stayed for abuse of process, observing: 'This case illustrates the need for careful consideration to be given when accepting or rejecting proposals on plea put forward by the defence and the prudence of seeking the advice of counsel in such instances.'

Oppressive decision to prosecute contrary to policy

9.50 A number of unsuccessful attempts have been made to argue that a prosecution for a health and safety offence undertaken by the HSE or a local enforcement authority is contrary to the stated enforcement policy of the body in case. Such arguments have been founded on the case of *R v Adaway* [2004] EWCA Crim 2831, [2004] 168 JP 645.

9.51 In *R v Adaway* the Court of Appeal was concerned with a prosecution of strict liability trading standards offences at the Crown Court which was said to have been oppressive. In that case the prosecuting local authority had a policy of only prosecuting in circumstances where fraud or persistent and repeated offences had been committed. As a result of the evidence called, the judge appeared satisfied that the decision to prosecute was not motivated by the presence of such circumstances, but solely by the insistence of the aggrieved party. The Court of Appeal had in mind the decision in *Shropshire County Council v Simon Dudley Ltd*,[25] and the judgment of Philips LJ, who had said:

> Trading standards officers must exercise discretion when deciding whether or not a particular case warrants the intervention of the criminal law . . . The Trade Descriptions Act is essentially concerned with consumer protection. It does not seem to me that this case falls within the type of mischief against which the Act is directed.

9.52 Philips LJ had added how it was easy to see that ' . . . a busy criminal court and a jury could become bogged down in a misplaced endeavour to resolve what are essentially civil disputes'.

9.53 In *R v Adaway*, despite finding that the prosecution had not identified a proper public interest for prosecuting, the trial judge had failed to stop the proceedings as oppressive and the Court of Appeal held that he should have done. The Court was concerned with the criminal law being mobilized to settle what was essentially a civil dispute between a customer and a supplier of goods.

9.54 Health and safety prosecutions are concerned with public safety. The Enforcement Policy Statement[26] requires inspectors to consider a broad range of public interest

---

[25] [1996] Trading Law Reports 69, at p 82.
[26] Reproduced at App B and considered at paras 1.65–1.67.

factors and has no single policy threshold that must be met before a decision to prosecute can be taken.

9.55 The Divisional Court decision of *Environmental Agency v Stanford* [1998] DC (CO/4625/97) involved a prosecutor's appeal by way of case stated against a decision of a magistrates' court that the prosecution of the respondent on four informations preferred against him should be stayed as an abuse of the process.

9.56 Having reviewed the facts before the justices and the basis upon which they arrived at their decision, the Lord Chief Justice pointed out how the magistrates regarded the case as falling within the category of cases where it would be unfair for the defendant to be tried, thereby treating the case as one falling outside the category over which magistrates enjoyed jurisdiction.

9.57 The Lord Chief Justice finally reiterated:

> The jurisdiction to stay, as has been repeatedly emphasized, is one to be exercised with the very greatest caution. Time and again parties, relying on the House of Lords' decision in *Bennett*,[27] invite the lower courts to exercise an ill-defined roving discretion to stay according to those courts' judgment of whether a prosecution is wise and well-advised or unwise and ill-advised. It is necessary to stress that that is not an issue for the court. The question whether or not to prosecute is for the prosecutor. Most of the points relied on in support of an argument of abuse are more properly to be relied on as mitigation. If a prosecutor obtains a conviction in a case which the court feels on reasonable grounds should never have been brought, the court can reflect that conclusion in the penalty it imposes. The circumstances in which it can intervene to stop the prosecution, however, are very limited indeed. At present many applications to stay are made in quite inappropriate circumstances. Sometimes courts are wrongly persuaded to accede and the present case is in my judgment an unfortunate example. It should be generally appreciated that a successful application to stay which leads to a successful appeal by the prosecutor renders the client no service.

---

[27] [1994] 1 AC 42 (HL).

10

SENTENCING

A. **Sentence**	10.04	Sentencing a company which is part	
The range of punishments	10.04	of a wider group	10.62
Compensation orders	10.05	Examples of fines imposed by the	
Venue for sentence	10.09	Crown Court	10.71
Financial information to be		Sentencing individuals	10.87
provided by a defendant	10.12	Allowing defendants time to pay	10.106
Obtaining an indication of sentence:		Section 42 orders	10.118
R v Goodyear	10.15	Sentencing in the magistrates'	
Determining the level of fine—		court	10.121
established principles	10.17	B. **Costs**	10.130
Tariff: the issue of the consequence		Prosecution costs	10.131
of any breach; and the		Defence costs	10.137
importance of systematic fault	10.29	Examples of the approach taken	10.148
Commonly encountered situations	10.41	Appealing the decision for costs	10.150
Fines for breach of the general		Apportioning costs between a	
duties and regulations	10.57	number of defendants	10.152
Sentencing public bodies and the			
'public element'	10.59		

Parliament has increasingly recognized the importance of protecting the health **10.01** and safety of people in the workplace through the enactment of relevant legislation. Courts have been encouraged to hand out substantial fines to those who flout their health and safety duties.

The level of financial penalty imposed by both the Crown and the magistrates' **10.02** courts, for health and safety offences, has shown a marked increase in recent years. In 2004/5, the average penalty was £18,765, up 31 per cent on the previous year, while fines of several hundred thousand pounds are no longer unusual. In October 2005 Balfour Beatty was fined £10 million, as a result of breaches that led to the Hatfield rail disaster in which four people died in October 2000. Previously, the largest fine ever imposed in the English courts was £2 million, on Thames Trains, following the 1999 Paddington rail crash. In August 2005 the gas supply company, Transco, was fined a record £15 million by the High Court in Edinburgh for numerous safety breaches that had led to the deaths of a family of four in an explosion in Larkhall, Lanarkshire, in 1999.

Chapter 10: Sentencing

10.03 In *R v Lightwater Valley Ltd* (1990) 12 Cr App R (S) 328, which predated the leading guideline decision of *R v F Howe & Son (Engineers) Ltd* [1999] 2 Cr App R (S) 37, the Court of Appeal considered the purpose of the legislation. This was a case in which three girls had suffered injuries on a fairground ride. Compounding this, an inexperienced employee had been crushed and sustained serious injuries as he started up the ride. Ian Kennedy J (as he then was) said:

> Parliament has enacted these provisions for the protection of the public and employees alike. Today the public themselves require that management should be seen to be responsible for the safety of their undertakings. It is the duty of the courts to enforce the standards of safety, and the means by which those standards are enforced is by means of criminal prosecutions. It is not simply a question of compensation for those who are injured.

A. Sentence

The range of punishments

10.04 The vast majority of health and safety offences are punished by way of a fine. However, it is also open to both the magistrates' court and the Crown Court to exercise other powers. They may hand out absolute and conditional discharges under PCC(S)A s 12. Orders can be made to disqualify directors under the Company Directors Disqualification Act 1986, breach of which is punishable by imprisonment for up to two years. Under HSWA 1974 s 42(1) the courts can order that a person convicted of an offence under any of the relevant statutory provisions must take steps to remedy any matters which it is in his power to remedy.

Compensation orders

10.05 Courts also have the power to make compensation orders against any convicted person, in respect of any personal injury resulting from the offence.[1] They may also require him to make payments for funeral expenses or bereavement. The limit for a compensation order in the magistrates' court is £5,000.[2]

10.06 Although the power to make an order for compensation exists, in practice it is rarely exercised. With many health and safety cases involving personal injury, there is likely to be an award of compensation—whether through the employer's insurer or otherwise. Often, this is for a substantial amount. If this is the case, it is generally unwise for the magistrates or a judge in the Crown Court to consider making a compensation order. Any such award will inevitably be clawed back in the final settlement of the claim.

---

[1] s 130 Powers of Criminal Courts (Sentencing) Act 2000.
[2] s 131 PCCA 2000.

In the usual course of events, a claim is likely to have been made by an injured party against the employer's compulsory insurance, in respect of employee and public liability. Often, liability has been accepted and, at most, only the issue of the quantum of damages is outstanding, with interim payment having been made. In the past, the imposition of a compensation order by a court has had the effect of further delaying settlement. Unsatisfactorily, this has been resolved by the employer's insurer deducting the sum of the compensation from the proposed award of damages, yet refusing to meet the compensation order. This is done on the basis that such an order amounts to part of the court's penalty payable by the employer. **10.07**

In any event, the Court of Appeal has held that no order for compensation should be made, unless the sum claimed by way of compensation order has either been agreed or proved,[3] and the criminal courts have been discouraged from conducting complicated investigations. **10.08**

Venue for sentence

As with all criminal offences, the most serious breaches of health and safety law will inevitably be dealt with by the Crown Court. The maximum fine that can be imposed by a magistrates' court—on either a corporate body or an individual—for breaches of the general duties, is £20,000. (The position is different in relation to a breach of the employee's HSWA 1974 s 7 duty, for which the maximum fine is £5,000.) For a breach of regulations, the limit is £5,000. **10.09**

Where a fine above these limits is likely to be warranted, the matter should be committed for sentence to the Crown Court, at the outset.[4] Following the implementation of the Criminal Justice Act 2003, a magistrates' court will no longer have the power to commit a case to the Crown Court for sentence once it has assumed jurisdiction at a mode of trial hearing, or following conviction at summary trial.[5] **10.10**

In the Crown Court, the fine that may be imposed in respect of a breach of either one of the general duties, or one of the regulations, is unlimited. **10.11**

Financial information to be provided by a defendant

For the purpose of determining a defendant's means, the Court of Appeal, in *R v F Howe & Son (Engineers) Ltd* [1999] 2 Cr App R (S) 37, has given guidance as to the financial information to be provided by the defendant, for the purpose of sentencing: **10.12**

> The starting point is its annual accounts. If a defendant company wishes to make any submission to the court about its inability to pay a fine, it should supply copies of its

---

[3] *R v Vivian* (1979) 68 Cr App R 53.
[4] In relation to the procedure for committal to the Crown Court for sentence see Ch 8 section B.
[5] Criminal Justice Act 2003 s 41, Sch 3 Pt 1 paras 1, 6, on a date yet to be appointed.

accounts and any other financial information on which it intends to rely in good time before the hearing both to the court and to the prosecution. This will give the prosecution the opportunity to assist the court should the court wish it. Usually, accounts need to be considered with some care to avoid reaching a superficial and perhaps erroneous conclusion. Where accounts or other financial information are deliberately not supplied, the court will be entitled to conclude that the company is in a position to pay any financial penalty it is minded to impose. Where the relevant information is provided late it may be desirable for sentence to be adjourned, if necessary at the defendant's expense, so as to avoid the risk of the court taking what it is told at face value and imposing an inadequate penalty.

10.13 However, in the case of *R v Transco Plc* [2006] EWCA Crim 838, the Court of Appeal clarified a particular issue. A court imposing a fine will always wish to have some information in relation to the means of the offender. It may not always be necessary, however, for the court to have detailed particulars of the financial position of a company, where it is made plain that the company's means are 'very substantial'.

10.14 Whilst the Criminal Justice Act 2003, s 162 provides a court with a power to make a 'financial circumstances order', requiring a convicted person to provide such a statement of his financial circumstances as the court may require, the provision applies only to an 'individual', being a term that does not include a company within its meaning.

Obtaining an indication of sentence: *R v Goodyear*

10.15 In the case of *R v Karl Goodyear* [2006] Cr App R (S) 6, the Court of Appeal introduced a procedure whereby a defendant may apply to the Crown Court for an indication of the level of sentence which the court would be minded to pass in the event of a guilty plea. If the court assents to the application, it will indicate in open court the maximum sentence it is minded to impose if a guilty plea is entered at that stage of the proceedings, or soon thereafter. If a guilty plea is entered, the court is bound to comply with its indication. This procedure does not apply to the magistrates' court.

10.16 The *Goodyear* procedure was adopted in a health and safety case and considered in this context by the Court of Appeal in *R v Transco PLC* [2006] EWCA Crim 838.

Determining the level of fine—established principles

R v Howe & Son (Engineers) Ltd

10.17 Wherever sentence takes place, the starting-point, and leading guideline case on determining the correct level of sentence in any health and safety case, remains *R v Howe & Son (Engineers) Ltd* [1999] 2 Cr App R (S) 37. The Court of Appeal there acknowledged that the circumstances of individual cases would vary almost infinitely. Elaborating on this, the Court gave some general guidance on the approach to be adopted when assessing culpability, and the appropriate level of fine.

A. Sentence

10.18 The case concerned a 20-year-old employee, who was electrocuted as he used a vacuum machine to clean the factory where he worked. The cable had become trapped beneath the machine, causing it to become live. A circuit-breaker, designed to trip the system in the event of a fault, had been deliberately overridden in a way that rendered it inoperable. Checks which would otherwise have revealed the fault had not taken place.

10.19 Having regard to the purpose of the statutory regime, Scott Baker J (as he then was) stated:

> The objective of prosecutions for health and safety offences in the work place is to achieve a safe environment for those who work there and for other members of the public who may be affected. A fine needs to be large enough to bring that message home where the defendant is a company, not only to those who manage it but also to its shareholders.

10.20 Regarding the purpose of the statutory regime, Scott Baker J stated:

> The objective of prosecutions for health and safety offences in the work place is to achieve a safe environment for those who work there and for other members of the public who may be affected. A fine needs to be large enough to bring that message home where the defendant is a company not only to those who manage it but also to its shareholders.

10.21 The Court observed that in recent years there had therefore been increasing recognition of the seriousness of health and safety offences and commented:

> Failures to fulfil the general duties under the Act are particularly serious as those duties are the foundations for prosecuting health and safety ... Disquiet has been expressed in several quarters that the level of fine for health and safety offences is too low. We think there is force in this and that the figures with which we have been supplied support the concern.

10.22 The Court set out a number of relevant factors to be taken into account when deciding the level of fine. Scott Baker J emphasized: 'It is impossible to lay down any tariff or to say that the fine should bear any specific relationship to the turnover or net profit of the defendant. Each case must be dealt with according to its own particular circumstances.'

10.23 The relevant factors are:

(a) How far short the defendant fell from doing what was reasonably practicable.
(b) Whether death or serous injury resulted from the breach. The Court conceded that it is often a matter of chance whether death or serious injury results from even a serious breach, but added that death resulting from any breach was to be regarded as an aggravating factor.
(c) Whether the defendant deliberately profited from a failure to take necessary health and safety measures or run a risk to save money. If so, it will be a seriously aggravating feature.

(d) The degree of risk and the extent of the danger created by the offence.
(e) The extent of the breach—whether it was an isolated incident, or whether it had been continuing for a period of time.
(f) A failure to heed warnings is another aggravating factor. Although the Court did not specify the nature of warning it had in mind, it is likely that it would include formal advice from the HSE resulting from the occurrence of similar incidents in the past, or warning of the danger of such an occurrence.

10.24 Mitigating factors will include: a prompt admission of responsibility and a timely plea of guilty; steps taken to remedy deficiencies after they were drawn to the defendant's attention; and a good safety record. When considering how much credit a defendant should receive for a plea of guilty, the Court should take into account the principles set out in the recent sentencing guidelines.

10.25 As is the case with any criminal offence, the Court of Appeal in *Howe* also confirmed that the fine should reflect not only the gravity of the offence, but also the means of the offender, whether corporate or otherwise. The principle that the level of fine should be proportionate to the means of the offender is now formally enshrined by the Criminal Justice Act 2003 (which came into force on 4 April 2005). It was referred to during the judgment of the Lord Chief Justice, in the case of *R v Transco Plc*.[6]

10.26 Section 164 of the Criminal Justice Act 2003, provides:

Fixing of Fines
(1) Before fixing the amount of any fine to be imposed on an offender who is an individual, a court must inquire into his financial circumstances.
(2) The amount of any fine fixed by a court must be such as, in the opinion of the court, reflects the seriousness of the offence.
(3) In fixing the amount of any fine to be imposed on an offender (whether an individual or other person), a court must take into account the circumstances of the case including, among other things, the financial circumstances of the offender so far as they are known or appear to the court.
(4) Subsection (3) above applies whether taking into account the financial circumstances of the offender has the effect of increasing or reducing the amount of the fine.

10.27 By way of example of the application of the principle, the case of *R v Supremeplan Ltd* [2001] 1 Cr App R (S) 244 is instructive. An employee, in that case, suffered serious burns in a gas explosion within a mobile food vehicle. A fine of £25,000 with a compensation order of £3,000 was reduced to a fine of £2,500 and a compensation order of £7,500. This was done so as to take into account the means of a company whose annual turnover was £140,000, with profits of £11,500. The court considered the case to be a serious one where many of the aggravating

---

[6] [2006] EWCA Crim 838 at para 25.

features in the case of *Howe*⁷ were present. Nevertheless, the court reduced the fine so as to take into account the limited means of the company.

In *R v Tropical Express Limited* [2001] 1 Cr App R (S) 115, contravention of the Air Navigation (Dangerous Goods) Regulations 1994 SI 1994/3187 was at issue. A fine of £20,000 was reduced to £5,000, in what was described as a 'wholly exceptional' case. The court, in reducing the fine to that level, adverted to *Howe*. It took into account the effect that the fine would have on the company's affairs, having heard argument that a fine at that level might have a devastating affect on the company's future. There was also an order for the payment of a substantial amount of costs.   **10.28**

Tariff: the issue of the consequence of any breach; and the importance of systematic fault

Unsurprisingly, the Court of Appeal has held that the highest fines should be reserved for those cases involving a major public disaster. However, while a death resulting from a breach is an aggravating feature to be reflected in arriving at the appropriate level of a fine, the absence of such an aggravating feature does not preclude a breach from being a serious one capable of attracting a substantial fine. As was noted in *Howe*,⁸ whether such a death or serious injury has occurred is often a matter of chance. The judgment in *Howe* expressly emphasized this point: 'It is impossible to lay down any tariff in respect of health and safety offences and nor can the extent to which a resultant death aggravates the offence be quantified in terms of a tariff.' Despite this, various dicta have suggested the existence of a tariff.   **10.29**

In *Friskies Petcare Ltd* [2000] 2 Cr App R (S) 401, the Court of Appeal approved of a purposive reading of the legislation, and examined the rationale for prosecuting. 'It must be remembered that the purpose of prosecutions under these regulations is not compensation and that the level of fine is not intended in any way to be a measure of the value of human life.'⁹ The Court was invited to consider reports of other decided cases. Having regard to these, the Court commented that, in particular, major public disasters tend to attract the highest fines: '. . . fines in excess of £500,000 tend to be reserved for those cases where a major public disaster occurs . . . that is to say, cases where the breaches of regulations put large numbers of the public at risk of serious injury or more.'¹⁰   **10.30**

---

⁷ *R v F Howe & Son (Engineers) Ltd* [1999] 2 Cr App Rep (S) 37.
⁸ ibid.
⁹ [2000] 2 Cr App Rep (S) 401 at 408.
¹⁰ ibid at 407.

10.31 The issue was revisited in *R v Colthrop Board Mills Ltd* [2002] 2 Cr App R (S) 359. There, total fines of £350,000 were reduced by the Court of Appeal to £200,000. A machine operative had his arm crushed, after it had been drawn into an unguarded part of a machine. Although it was not entirely clear what he had been doing before the accident, it was plain that he was attempting to assist his employees in keeping the machine running, albeit in a way which may have been unwise.

10.32 The Court was invited to reconsider or clarify what was meant in *Friskies*[11] when it was said that normally, financial penalties in excess of £500,000 were reserved for cases of major public disasters. Although it refused to do so, the Court did say: 'What is important is that companies in the position of this appellant can expect to receive financial penalties on a scale up to at least half a million pounds for serious defaults and proportionately lesser sums if the limitation upon means or some lesser blame justifies it.'[12] However, the Court noted:

> It appears from the authorities that financial penalties of up to around half a million pounds are appropriate for cases which result in the death even of a single employee, and perhaps of the serious injury to such a single employee. We would not wish the sum of £500,000 to appear to be set in stone or to provide any sort of maximum limit for such cases. On the contrary, we anticipate that as time goes on and awareness of the importance of safety increases, that courts will uphold sums of that amount and even in excess of them in serious cases, whether or not they involve what could be described as major public disasters.[13]

10.33 Courts have imposed a number of fines in excess of £500,000 on companies, in cases involving no death or serious injury. The courts have been prepared to do this where grave breaches have occurred. In these cases, in addition to the seriousness of the breach, the most important factor has been the means of the defendant. The judgment in *Howe* articulates that a fine is to be large enough 'to bring the message home . . . not only to those who manage it [a company] but also to its [a company's] shareholders'.[14]

10.34 However, in *R v Jarvis* [2006] Cr App R (S) 44, at para 7, it was noted that this may cause difficulty in practice:

> It will be seen at once that this gives rise to real difficulty, both in achieving consistency of sentence and in ensuring that some proportion is maintained between the quantum of the fine and the gravity of the specific case, given that offending companies may have vast disparities of economic strength. A fine that may hardly touch a multi-national might put a small company out of business, yet their offence may

---

[11] [2000] 2 Cr App Rep (S) 401 at 408.
[12] [2002] 2 Cr App R (S) 80 at para 22.
[13] [2002] 2 Cr App R (S) 80 [22].
[14] [1999] 2 Cr App R (S) 37 at 44.

have been the same. Consistency of level of fine may not therefore be a primary aim of sentencing in these cases.

10.35 The view espoused in *Jarvis*—that consistency need not be understood as a primary aim—was endorsed by the Lord Chief Justice, in *R v Balfour Beatty Rail Infrastructure Services Ltd* [2006] EWCA Crim 1586.[15]

Systemic fault

10.36 In *Transco Plc* [2006] EWCA Crim 838, the Lord Chief Justice endorsed one of the submissions advanced on behalf of the HSE. He cited with approval 'what was emphasised by the court in *Howe*,[16] namely how it is impossible to lay down any tariff or to say that the fine should bear any specific relationship to the turnover or net profit of the defendant and how each case must be dealt with according to its own particular circumstances. It is often a matter of chance whether death or serious injury results from even a serious breach.'[17]

10.37 However, the court distinguished the *Transco* case as unusual. It stated that *Transco* differed from most such prosecutions, 'in that it involved no systemic fault, but merely a mistake on the part of an individual or individuals managing an emergency situation on the ground'. Accordingly, where the defendant has been guilty of systemic managerial fault, the court decided that it would be inappropriate, on the facts of the case, to give further guidance on the approach to sentencing.

10.38 In *R v Balfour Beatty Rail Infrastructure Services Ltd* [2006] EWCA Crim 1586, the Court was concerned with an appeal from the fine imposed upon the company by Mr Justice Mackay. The fine constituted £10 million, and £300,000 costs, for the breach of the HSWA 1974 s 3 duty. This had been an operating cause of the Hatfield rail disaster on 17 October 2000, in which 102 passengers were injured and four lost their lives. The Court endorsed the sentencing judge's summary of the guidance, which he had reduced to thirteen propositions.[18]

10.39 The Lord Chief Justice referred to the objectives of sentencing, which are expressly set out in s 142 of the Criminal Justice Act 2003.[19] Noting the obvious difficulties

---

[15] Para 23 of the judgment.
[16] [1999] 2 Cr App Rep (S) 37.
[17] See paras 22–24 of the judgment.
[18] See para 22.
[19] Criminal Justice Act 2003
  '**142 Purposes of sentencing**
  (1) Any court dealing with an offender in respect of his offence must have regard to the following purposes of sentencing—
    (a) the punishment of offenders,
    (b) the reduction of crime (including its reduction by deterrence),
    (c) the reform and rehabilitation of offenders,
    (d) the protection of the public, and
    (e) the making of reparation by offenders to persons affected by their offences.'

in applying subs (1)(d), the Court adverted to the aim of 'the protection of the public', before stating:

> Section 3 of the 1974 Act requires positive steps to be taken by all concerned in the operation of the business of a company to ensure that the company's activities involve the minimum risk, both to employees and to third parties. Knowledge that breach of this duty can result in a fine of sufficient size to impact on shareholders will provide a powerful incentive for management to comply with this duty. This is not to say that the fine must always be large enough to affect dividends or share price. But the fine must reflect both the degree of fault and the consequences so as to raise appropriate concern on the part of shareholders at what has occurred. Such an approach will satisfy the requirement that the sentence should act as a deterrent. It will also satisfy the requirement, which will rightly be reflected by public opinion, that a company should be punished for culpable failure to pay due regard for safety, and for the consequences of that failure.
>
> A breach of the duty imposed by section 3 of the 1974 Act may result from a systemic failure, which is attributable to the fault of management. It may, however, be the result of negligence or inadvertence on the part of an individual, which reflects no fault on the part of the management or the system that they have put in place or the training that they have provided. In such circumstances a deterrent sentence on the company is neither appropriate nor possible. Where the consequences of an individual's shortcoming have been serious, the fine should reflect this, but it should be smaller by an order of magnitude than the fine for a breach of duty that consists of a systemic failure.[20]

10.40 In the event, the Court reduced the fine to £7.5 million, not because it was wrong in principle but purely because of what was found to be the disparity between the sentence on the company and that imposed upon the co-defendant, Railtrack. The Court also endorsed the sentencing judge's refusal to give any discount for the company's guilty plea, which was tendered on the ninety-third day of trial.

Commonly encountered situations

Employees acting outside of their instructions

10.41 Fines will not usually be reduced because an employee has acted inadvertently, or beyond instructions.

10.42 In *R v Avon Lippiatt Hobbs (Contractors) Ltd* [2003] 2 Cr App R (S) 427. In that case, the appellant argued that, in assessing how far short the company fell from its duties under the HSWA 1974, it was wrong to visit on it the consequences of wrongful acts committed by employees. The Court of Appeal said that this line of reasoning was 'at best far too high'. It pointed out that 'if it were correct, an employer could quite readily escape from the consequence of his failings in the discharge of his duties under the Act'.

---

[20] [2006] EWCA Crim 1586 at paras 42–43.

A. Sentence

10.43 The case of *R v Patchett* [2001] 1 Cr App R (S) 138 concerned the circumstances surrounding the death of an employee who was killed when he came into contact with an unguarded drive. It was submitted that the deceased had had no need to access, in the unorthodox way that he did, the area where a dangerously exposed drive shaft was located. Further, it was therefore not foreseeable that the fatal events would occur. The court rightly and properly rejected this submission. Commenting that it was well known that workers take short cuts from time to time, the court observed that workers often do not follow proper practice. It said that the particular part of the drive-shaft was not remote from any part of the machinery to which an operative would need access. That being the case, the need to guard the shaft should have been apparent to someone involved in the design of this machine. It would have been apparent to him, had he given sufficient thought to the question of safety and to the fact that some operatives do on occasions take short cuts. However, the fine of £75,000 was reduced to £20,000, in view of the company's financial situation.

10.44 In *R v Rimac Ltd* [2000] 1 Cr App R (S) 468, a 19-year-old trainee fell to his death through a non-load-bearing ceiling. The employer had failed to realize that employees might have to obtain access to the ceiling areas, and may have had cause to forget that they were not weight-bearing. Such a risk was obvious, foreseeable, and continuing. There were no signs to indicate the fragile nature of the ceilings. Nor, importantly, was the deceased given any instructions as to how he should carry out the task he was involved in, or any specific training or instruction about the dangers of the ceiling. A fine of £60,000, and costs of £9,273, were upheld against a company said to be well able to afford it.

10.45 In *R v Sanyo Electrical Manufacturing (UK) Ltd* (1992) 13 Cr App R (S) 657, the court took a purposive view of the governing health and safety statutory provision. An employee of the company had experienced a potentially fatal electrical shock while carrying out a test. The court examined the purpose of the Electricity (Factories Act) Special Regulations 1908 and 1944, which were then in force. The court held that it was clearly to protect employees against the consequences of doing things by reason of inadvertence, or inattention, which they would not normally do.

Being 'let down' by others

10.46 The Court of Appeal has not generally reduced a fine imposed on a company where it claims to have been failed by an independent contractor, or some other third party.

10.47 In *R v Mersey Docks and Harbour Co* (1995) 16 Cr App R (S) 806, a fine of £250,000 was upheld. It had been imposed on a harbour company for failing to take adequate precautions to avoid an explosion aboard a vessel which had previously carried a dangerous cargo. Two men were killed and others were injured. The appellant had pleaded guilty to one count of failing to discharge a duty imposed by HSWA 1974 s 3(1). The Court described the appellant's mitigation to the sentencing judge, and some of the 'tentative' submissions advanced before it, as 'surprising'. It had been argued that it was perfectly reasonable and appropriate that the port should be unaware that the vessel visiting its docks was a combination carrier (a carrier which is enabled and equipped to carry oil cargoes, bulk cargoes, or ore cargoes). Moreover, it was argued that it was perfectly proper, and indeed thoroughly reasonable, for it to delegate the responsibility for safety to the master of the vessel, by whom they had been badly let down.

10.48 In rejecting these submissions, Hobhouse LJ stressed that the duty under s 3(1) was non-delegable. It was no mitigation for the appellant to say that he had relied upon the master of the vessel to ascertain whether there were any dangerous spaces on his vessel. He added that if this were the appellant's attitude, the sentencing judge was entitled to impose a heavy fine in order to show how erroneous it was. The company's attitude towards establishing whether the visiting vessel did contain any dangerous spaces had been superficial and patently inadequate.

10.49 A similar argument was rejected, in *R v Ceri Davies* [1999] 2 Cr App R (S) 356. An employee of the defendant company, whose main business was the manufacture of plastic cases for cassettes and CDs, contracted legionnaires' disease and died, although it is unclear from the judgment whether this fact was taken into account. The legionella bacteria had built up in the cooling towers, which stored water used to dispel heat generated in the course of the manufacturing process. The appellant, who was the managing director of the company, had appointed one of his workforce to be responsible for health and safety matters. He had engaged a specialist contractor to institute a cleaning regime for the cooling towers. In the event, the regime was ineffective, with the result that dangerous levels of bacteria had built up in the towers.

10.50 It was argued, on the appellant's behalf, that he had properly relied upon the expertise of an apparently competent third party contractor for the cleaning of the water towers and had been let down by it. Further, it was argued that it was 'inapposite' for the third party contractors to be fined less than the appellant in those circumstances.

10.51 In rejecting the arguments, the Court stated:

> The short answer is, as we see it, that the purpose of the health and safety legislation is to ensure so far as possible that employees work in a healthy and safe place and atmosphere. The primary responsibility for ensuring their safety and their health must remain with their employers and it is not adequate for an employer simply to seek to transfer responsibility for all matters to an expert third party. At all stages it must be for the employer to supervise the independent contractor and not for the independent contractor to supervise the employer. We see no reason in principle why

in this case, or indeed any, the employer should not face up to the fact that it must carry the greater responsibility for the safety of the employees than those who engage on a part-time basis to assist them in that regard.

10.52 It is implicit from the judgment that the court found fault with not only the way in which the independent contractor carried out its investigations, but also the way in which the appellant and company supervised it, and indeed their overall approach to health and safety.

10.53 The appellant had pleaded guilty to an offence under HSWA 1974 s 37(1), namely, neglecting his duty as a director, whereby the body corporate committed an offence under HSWA 1974. Both the appellant's company, and two independent contractors, were prosecuted. Neither appealed. The appellant was fined £25,000, which was reduced to £15,000, for reasons unconnected with his reliance on third parties. His costs order, of £10,000, was not reduced. The third party contractors were fined £15,000, and ordered to pay £15,000 costs.

The relevance of a good safety record

10.54 A good safety record is one of the specific mitigating factors referred to in *R v Howe*. The relevance of a good safety record was considered in *R v Cardiff City Transport Services* [2001] 1 Cr App R (S) 41. This concerned a case in which an employee was struck and killed by a bus at the defendant's depot, whilst involved in moving buses. A fine of £75,000 was reduced to £40,000, on the court accepting that the death of the employee had not occurred as a result of the defendant's breach of duty. In conjunction with this the company had consistently shown a responsible attitude to health and safety.

10.55 This case is to be contrasted with *R v Hall and Co Ltd* [1999] 1 Cr App R (S) 306. There, a fine of £150,000 was upheld, where the appellant company pleaded guilty to breach of HSWA 1974 s 2(1). In this case, one of the company's employees had been crushed to death by a reversing lorry. There were a number of features indicative of a company that had not satisfactorily grasped, and implemented, adequate safety measures. There had been no safety area designated by hatching behind the lorry, the lorry was not fitted with reversing alarms or fixed mirrors, so as to enable drivers to see around corners, and the procedure of using lookouts was regularly flouted. The court took into account that the company had a substantial turnover, finding that it was well able to pay a fine of that order. In the event, the court thought it appropriate that the period for payment of six years and five months was reduced to five years and seven months.

10.56 In *R v Cappagh Public Works Ltd* [1999] 2 Cr App R (S) 301, the sentencing judge in the lower court had considered that the case involved 'a serious dereliction of duty leading to a death'. An employee was crushed to death after having been drawn into the unguarded workings of a concrete crushing machine, which he was cleaning.

It was argued, on appeal, that the sentencing court had placed insufficient weight on the company's thirty-year good safety record. Further, the process in question had been carried on for the same period of time without incident. Moreover, the deceased had failed to comply with his instructions to turn off the machine before entering it. It was also argued that the sentencing court had failed to have sufficient regard to the effect a fine of £40,000 would have on the company, whose net annual profits were said to be just under £150,000. Penry-Davey J upheld the level of fine, stating that it properly took into account the financial circumstances of the appellant. Adequate consideration had also been given all of the other matters which had been pleaded in mitigation.

Fines for breach of the general duties and regulations

10.57 In *R v Brintons Ltd* (CA 22 June 1999), it was argued that the fine imposed should reflect the fact that the HSE had charged only the breach of a regulation rather than a breach of one of the general duties. The submission was rejected, although the charge in this particular case was referred to as 'curious'.

10.58 The case of *R v Keltbray* [2001] Cr App R (S) 39 is a good example of where a substantial fine remained in place, despite the fact that the defendant was charged with breaching a single regulation. The Court of Appeal upheld a fine of £200,000, and costs of £8,000, in respect of a breach of reg 6 of the Construction (Health Safety and Welfare) Regulations 1996. The court commented: 'Though the charge was based on the regulation dealing with harnesses, the appropriate sentence must be considered on the basis of a breach of that regulation in the context of a lack of supervision and of a danger.'

Sentencing public bodies and the 'public element'

10.59 In *R v Milford Haven Port Authority* [2000] 2 Cr App R (S) 423, the Court of Appeal reduced a fine of £4m to £750,000. This was done for a number of different reasons. Among the submissions made, on behalf of the appellant, was that the status of the Port Authority, as a public trust port, was relevant to determining the appropriate level of fine. In argument, it was said that this was 'not because such bodies are subject to any lesser standard of duty or care in safety or environmental matters, but because the burden of paying the fine falls not on shareholders or directors or employees of the company but on either customers of the Port Authority ... or the public on whose behalf the Port Authority carries out its operations'.[21]

10.60 The Court held that whereas it would be wrong to suggest that public bodies are immune from criminal penalties because they have no shareholders and the

---

[21] [2000] 2 Cr App R (S) 433.

directors do not receive 'handsome annual bonuses', the sentencing judge does have to consider how any financial penalty will be paid, and the effect of a very substantial penalty on the proper performance by a statutory body was not something to be disregarded. This approach is not inconsistent with the dicta of Scott Baker J in *Howe*.[22]

10.61 The *Milford Haven* decision is to be viewed alongside, and contrasted with, *Jarvis Facilities Ltd* [2006] Cr App R (S) 44. In that case, the company, *Jarvis*, was responsible for aspects of the maintenance of the railway line in the vicinity of Sheffield. Having carried out some work on the line, it failed to take steps to close off a junction to traffic, or to inform the local signaller of the state of the line. In the event, a freight train being diverted off the main line was derailed at low speed. Despite reducing a fine of £400,000 to £275,000, the Court said that it was entitled to take a more severe view of breaches of health and safety legislation where there is 'a significant public element'. The Court observed:

> This is particularly so in cases (like the railway) where public safety is entrusted to companies in the work that they do . . . in our view public service cases will often be treated more seriously than those in which the breaches are confined within the private sector even where there is comparability between gravity of breach and economic strength of defendant.[23]

Sentencing a company which is part of a wider group

10.62 It is increasingly common for companies to be connected to a parent or sibling organization, or part of a wider group. The question therefore frequently arises as to who and what exactly a sentencing court is dealing with. Should it direct its punishment towards the single company convicted of the offence, or towards the wider group or corporate structure?

10.63 When imposing a fine, both the Crown Court and the magistrates' court must take into account the financial circumstances of the offender. It is wrong, in principle, to impose a fine beyond the means of the defendant's ability to pay.[24] However, difficulties arise when transposing principles, which are clearly designed to punish individual human beings, on to companies. An incorporated company is 'a person', within the meaning of the Interpretation Act 1978.

10.64 In conventional criminal cases, Parliament has made specific exception to the principle that a fine is to be imposed on the individual alone. The vehicle for this has been to allow the courts to order the parent or guardian of a juvenile convicted of a criminal offence to pay the fine or compensation themselves. There is no

---

[22] [1999] 2 Cr App R (S) 37.
[23] [2006] Cr App R (S) 44 at para 11.
[24] See, for example, Lord Lane CJ in *Olliver* [1989] 11 Cr App R (S) 10.

equivalent provision in the case of the corporate defendant. Arguably, it follows that the court should only impose a fine upon the company convicted of the offence and in cases involving a corporate defendant, the court should not be entitled to punish the family for the acts of the recalcitrant child.

10.65 In practice, however, sentencing courts frequently do look beyond the defendant, to the means of the wider corporate family. For example, in *R v Brintons Ltd*,[25] the appellant company bought premises which they were aware might contain asbestos, and which did in fact contain brown asbestos. The company allowed its employees to work on the premises, without ever testing to see whether asbestos was present or not. The Court of Appeal looked at the corporate group, of which the defendant company was a part, taking into account its annual turnover of some £91 million per year.

10.66 In *R v Keltbray Ltd* [2000] 1 Cr App R (S) 132, the defendant company was involved in the demolition of an eight-storey building. Two employees were killed when the floor, in which they had been drilling a hole through which to drop debris created in the process of demolition, collapsed beneath them. The company pleaded guilty to a breach of reg 6 of the Construction (Health Safety and Welfare) Regulations 1996, and was fined £200,000, by the judge at first instance. The company was owned by another company, and had a sister company working in the same field. The directors of the company were also its shareholders. The profits of the group were said to be in the region of £200,000 and £300,000, if one included the sister company. (It is assumed that this included the operating profits of the holding company.) The fine of £200,000 was upheld, despite the submission that business required a company to keep capital reserves for improvements or contingencies.

10.67 What is the basis for courts taking this approach? The question of whether the sentencing court is entitled to look at the wider corporate or group picture, when sentencing a company for breaching health and safety law, is closely connected with the question of when it is permissible to 'pierce the corporate veil'. Generally, the courts will only be entitled to pierce the corporate veil where it is found that the corporate structure is merely a façade, concealing the true facts.[26] This would be established, for example, if it can be shown that the corporate form is being used to conceal or facilitate fraud.

10.68 For the most part, the courts continue to show reluctance to pierce the corporate veil. The Victorian principle of limited liability, accordingly, prevails. For example, a holding company will not be made liable for the debts of an insolvent

---

[25] (CA 22 June 1999).
[26] *Woolfson v Strathclyde Regional Council* [1978] SLT 159, per Lord Wilberforce.

subsidiary, even if all or the majority of the company's shares are owned by that holding company.

10.69 This approach gained currency with the Court of Appeal, in the case of *Adams v Cape Industries Plc* [1990] Ch 433. There, it was argued that the companies linked with a group should be considered as a 'single economic unit'. In such cases, it would be permissible to pierce the corporate veil. The arguments, although sympathetically considered, were ultimately unsuccessful. The Court cited with approval the judgment of Roskill LJ in *The Albazero*,[27] to the effect that 'each company in a group of companies is a separate legal entity possessed of separate rights and liabilities'. Refraining from reversing this position, the appellate court was prepared to go no further than to suggest that a sentencing tribunal may have regard to the economic realities of a group of companies. This 'economic realities' approach could then be used as an aid in the interpretation of a statute or a contract.

10.70 A company's accounts are now required to reveal the nature of group ownership and inter-group transactions, thus, to a large extent, revealing the 'economic realities' of such a subsidiary company's affairs.

Examples of fines imposed by the Crown Court

10.71 The following represents a sample of recent larger fines imposed by the Crown Court, in health and safety cases.

10.72 In *R v Transco Plc*, the company was fined a record £15 million, at the High Court of Justiciary in Edinburgh, after a jury found it guilty of an offence contrary to s 3 of the HSWA. The charge arose from an explosion caused by a leaking gas main which destroyed a family home, killing its two adult and two child occupants, in December 1999.

10.73 In *R v Balfour Beatty Infrastructure Services Ltd and Network Rail* (Central Criminal Court 7 October 2005), Balfour Beatty and Network Rail were fined £10 million and £3 million, respectively, for failings that contributed to the Hatfield rail disaster, in which four people died and 102 were injured, on 17 October 2000. The sentencing judge referred to the incident as the 'worst piece of industrial negligence' he had encountered. The fine imposed upon Balfour Beatty was ultimately reduced by the Court of Appeal, principally on the basis of it being disproportionate to that imposed upon Network Rail. The reduction gave a final figure of £7.5 million.

10.74 On Friday, 30 March 2007 at the Blackfriars Crown Court, Network Rail Infrastructure Ltd was fined £4m and ordered to pay over £200,000 costs after

---

[27] *The Albazero* [1977] AC 774, 845.

pleading guilty to a health and safety offence arising out of the Ladbroke Grove rail disaster, where thirty-one were killed and hundreds injured following the head-on collision between a westbound Thames train and an eastbound First Great Western train at Ladbroke Grove Junction about two miles west of Paddington station shortly after 8 am on 5 October 1999.

10.75 In *R v Conocophillips* (Grimsby Crown Court 29 June 2005), two separate incidents resulted in a fine of £895,000, against gas giant Conocophillips, and an order to pay £218,854 costs. The sentence followed an investigation by the HSE, into a fire and explosion at the Humber Refinery on 16 April 2001, which led to large-scale damage to the property. A release of liquefied petroleum gas ensued, at the Immingham Pipeline Centre, on 27 September 2001. Neither incident led to injury or loss of life. The company pleaded guilty to seven breaches of the HSWA, and to contravention of the Pressure Systems and Transportable Gas Containers Regulations 1999.

10.76 In *R v Great Western Trains Company Limited* (Central Criminal Court 27 July 1999), Scott Baker J fined the defendant company £1.5m, in respect of charges arising out of the Paddington rail disaster, in which seven people died and 150 were injured.

10.77 In *R v Balfour Beatty Civil Engineering Ltd & Geoconsult GES* (Central Criminal Court 15 February 1999)—the 'Heathrow tunnel collapse' case—Cresswell J fined the first defendant £1.2m, and the second defendant £500,000. Each was ordered to pay costs of £100,000.

10.78 In *R v Port of Ramsgate and others* (Central Criminal Court 28 February 1997), the Port of Ramsgate walkway collapsed. As a result, six people died and seven were seriously injured. Clarke J imposed fines totalling £1.7m, on four separate defendants, the largest of which was £750,000.

10.79 In *R v Thamesway Homes Ltd* (Winchester Crown Court 24 August 2001), a young and inexperienced employee died, after the truck which he had been driving overturned. The defendant company (a wholly owned subsidiary of George Wimpey PLC), pleaded guilty to one count of breaching the duty under HSWA 1974 s 3(1). It was fined £350,000, with costs of some £14,000.

10.80 In *R v Kvaerner Cleveland Bridge Ltd and Costain Ltd* (Bristol Crown Court 30 November 2001), Owen J imposed fines totalling £500,000, and costs of £525,0000, in respect of offences arising out of the deaths of four workers, on the Avonmouth Bridge.

10.81 In *R v London Borough of Hammersmith and Fulham* (Blackfriars Crown Court 7 November 2001), the council was fined £350,000, in respect of offences connected with the deaths of two council tenants, through carbon monoxide poisoning, in 1998.

10.82 In *R v Corus UK Ltd* (Cardiff Crown Court 12 November 2000), the defendants were fined £300,000, after they had pleaded guilty to breaches of HSWA 1974 ss 2(1) and 3(1), in respect of an explosion at their plant in Llanwern. The explosion had left a contract worker paralysed from the waist down.

10.83 In *R v Jarvis Fastline Ltd* (York Crown Court 25 August 2000), a rail maintenance company was fined £500,000, in respect of two separate incidents where freight trains were derailed.

10.84 In *R v Klargester Environmental* (Aylesbury Crown Court 22 March 2002), an employee was killed when operating a machine which was not properly guarded. The defendant was fined £200,000, in respect of a breach of reg 11(1)(a) of the Provision and Use of Work Equipment Regulations 1998.

10.85 In *R v John Doyle Construction Ltd and Exterior International Plc* (Central Criminal Court 9 May 2006), the two companies were fined a total of £350,000. The fine followed the investigation into a fatal incident on a building site in Battersea, South London, on 6 August 2002. A man died after being struck by a large timber panel which was being lifted from the ground to the ninth floor. John Doyle Construction Ltd was fined £200,000 for contravening HSWA s 2, and £50,000 for contravening HSWA s 3. Exterior International Plc were fined £1,000, for an offence contrary to HSWA s 3.

10.86 In *R v Otis* (Southampton Crown Court 7 April 2006), the lift-operating company was fined £400,000, and ordered to pay £145,000 costs, as a result of an incident in which two young men were killed. The men had fallen down a lift shaft in a high-rise block of flats in Southampton. Otis Investments and Otis Ltd (sentenced as a single company) pleaded guilty to a single breach of HSWA s 3.

Sentencing individuals

10.87 Breach of the general duties, or of a regulation, is not punishable with imprisonment. However, pursuant to HSWA 1974 s 33(2A) and (4), the offence of contravening any requirement or prohibition imposed by an enforcement notice is punishable with up to two years' imprisonment. Similarly, any offence consisting of contravening any of the relevant statutory provisions—by doing otherwise than under the authority of a licence issued by the HSE or of contravening a term, condition, or restriction attached to such a licence—is also punishable with a maximum imprisonment term of two years. According to the HSE Enforcement Report for 2005, in total, six people have been imprisoned for such an offence, since 1996.

10.88 Where a director falls to be sentenced, pursuant to s 37 of the Health and Safety at Work Act 1974, in addition to the company, the court should be mindful that, with a small company whose directors are the shareholders, an element of 'double punishment' may arise: *R v Rollco Screw and Rivet Co Ltd* [1999] 2 Cr App R (S) 436.

Sentencing individuals for gross negligence manslaughter

10.89 Attempting to abstract principles of general application from the relatively few cases of gross negligence manslaughter to reach the appellate courts is neither easy nor particularly fruitful. No doubt this is in part because of the essentially accidental nature of the offence; and the almost infinite variety of circumstances in which it may occur. In *R v Litchfield* [28] Simon Brown LJ acknowledged the troubling nature of convictions based on findings of gross negligence where defendants are very often of previously good character and are being prosecuted for consequences they never intended or desired.

10.90 The following are examples of 'classic' gross negligence manslaughter cases: ie ones where the defendant's conduct was merely negligent, and not, for example, ones which arise directly from some illegal act, as in *R v Whacker* (below). In general terms, the Court of Appeal seems to have regard to factors analogous to those set out in *R v Howe*, ie the extent of the negligence, whether there had been prior warnings, the length for which the breaches had been going; whether safety was sacrificed for profit's sake; and whether, for instance, a safety feature was deliberately disabled or bypassed.

10.91 In *R v Litchfield* [29] the defendant was convicted for gross negligence manslaughter after a trial and sentenced to 18 months' imprisonment. The schooner of which he was owner and master foundered and broke up on rocks off the north Cornish coast after its engine failed, and three crew members were killed. The defendant was convicted on two bases: that he had sailed too close to the shore; and that he had sailed so as to have to rely on the vessel's engines when he knew that they might fail through fuel contamination. The prosecution alleged that the defendant was an experienced mariner and had been expressly told of the risks of using contaminated fuel by an expert.

10.92 In *Att-Gen's Reference (No 134 of 2004)* [2004] EWCA Crim 3286 the defendant had been sentenced to 12 months' imprisonment after pleading guilty to manslaughter by gross negligence. He had driven a car transporter weighing 35 tons, with brakes which he knew were defective, having been involved in a crash with a car at a zebra crossing two days before which had been caused as a result of the defective brakes. The Court did not consider it appropriate to interfere with the sentence imposed.

---

[28] CA Ref No. 97 05973 W2, 17 December 1997.
[29] ibid.

A. Sentence

In *R v Kite* [1996] 2 Cr App Rep (S) 295 the defendant was the managing director of a leisure centre which operated outward bound activities. In March 1993 four schoolchildren, aged 16 or 17, were drowned in a canoeing accident in the sea, having set out from his centre. He was prosecuted together with the manager of the leisure centre and the company, OLL Ltd. He was convicted after a trial and sentenced to three years' imprisonment. The prosecution case was that he had received express warning from two members of staff at the centre to the effect that they were close to having a very serious incident; also that staff were not supplied with a first aid kit and tow line. The wider case was that the operation was unsafe and disorganized and that the defendant foresaw the risk, but went on to run it nevertheless. Swinton Thomas LJ said:

10.93

> Cases such as this cause the court, so far as sentence is concerned, acute difficulty and very great anxiety. On one side, the lives of young people have been lost and, not only that enormous and acute distress has been suffered by the parents and other members of their family. Very often in cases such as this, and very naturally, the families of the deceased young people take the view that no sentence can be too long in the light of what has happened. On the other side, one has a man who was convicted of offences, in relation to which he had no criminal intent, and it may well be that he finds it difficult to understand why he was sent to prison at all. In some way the court has to hold a balance between those two extreme positions. The appellant was convicted in respect of his negligence. As we have said, and it is entirely and rightly accepted by the Crown, he had no criminal intent. Whether a particular case falls on the right or the wrong side of the line so as to amount to gross negligence may often be difficult. In this case the jury came to the conclusion that Mr Kite's conduct fell on the wrong side and in consequence he was criminally liable.

The Court of Appeal was persuaded that the sentence imposed by the judge was too long and substituted one of two years' imprisonment.

10.94

In *R v Shaw* [2006] EWCA Crim 2570 the Attorney-General appealed against a sentence of two years' imprisonment suspended for two years. The sentence was imposed following the defendant's guilty plea which was entered immediately after a trial where the jury had been unable to decide upon a verdict. The defendant was the founder and managing director of a company which manufactured work surfaces and fireplaces from stone and granite. The case concerned the death of a 21-year-old employee who was crushed to death by a stone-cutting machine with which he had been working. A light beam which when broken ought to have caused the machine to shut down had been deliberately disabled by engineers employed by the manufacturers of the machine at an earlier date, with the knowledge and consent of the defendant. In wider terms, it was the prosecution case that the defendant ran his business without proper regard for health and safety issues; and that safety was ignored for business convenience. The machine on which the deceased met his death and the other two machines (identical machines) had in fact been dangerous and potentially lethal from the date of when they were first installed some years earlier.

10.95

10.96 The sentencing judge imposed a suspended sentence solely through concern about the affect a sentence of imprisonment would have had upon his employees. It was said on behalf of the Attorney that this was an inappropriate consideration for the sentencing judge to have had in mind.

10.97 The Court held that while there will be occasions when it may well be relevant for a sentencing court to take account of the consequences of a sentence of imprisonment to a business run by the offender, with offences of this type, where the offence arose directly from the offender's misconduct of his business which unfolded disastrously on an employee, different considerations may apply. Otherwise, many small businesses which were dependent on the boss could operate without proper regard to the consequences of gross negligence resulting in death; as to take the boss away would almost certainly damage the business, and might therefore put innocent people out of work. But if that consideration were to produce a suspended sentence in each case like this, there would be no particular incentive on the bosses in most small organizations to see that health and safety measures were complied with. As this case shows, it is imperative that health and safety provisions should be observed. It can be, as we have said already, a matter of life and death. So in situations like the present, it seems to us that the consideration that the result of an immediate sentence of imprisonment would be the collapse of the business and unemployment for employees cannot normally be treated as an exceptional circumstance.

10.98 In *R v Connolly and Kennett*[30] the sentence of nine years' imprisonment imposed on the first defendant after a trial was reduced to seven years on appeal. The case concerned the Tebay Rail Disaster where four track workers were killed when they were hit by a runaway engineering trailer laden with scrap rail. The case for the prosecution was that the defendant had deliberately disabled the brakes on the trailer at an earlier date after they had seized; and not wishing to pay for the repair.

10.99 During the course of his judgment May LJ echoed the difficulties which Simon Brown LJ alluded to in *R v Litchfield* (above) and said:

> It is acknowledged, and this Court acknowledges, that sentencing cases of gross negligence manslaughter such as this is an extremely difficult and sensitive process. On the one hand, it is an offence which does not bear all the hallmark characteristics of, for instance, violent offences or sexual offences or offences of dishonesty. It is an offence which is not unique perhaps within the range of criminal offences in this jurisdiction but one which has distinctive characteristics of its own. One of those characteristics, of course, is that the state of mind of the accused, in this case the convicted defendant is not necessarily material to the offence of which he is convicted.

---

[30] [2007] EWCA Crim 270.

A. Sentence

10.100 By contrast with the above cases is *R v Perry Wacker* [2003] Cr App R (S) 92 where the defendant was convicted of manslaughter by gross negligence following the deaths of fifty-eight illegal immigrants who suffocated after being illegally imported in a container by the defendant. After a reference by the Attorney-General, the original sentence of six years' imprisonment was held to be unduly lenient, and was increased to fourteen years.

10.101 In *R v Lin Liang Reng and Others*—the 'Cockle pickers case'—where twenty-one people died on the sands of Morecambe Bay in illegal circumstances, the defendant was convicted of gross negligence manslaughter and was sentenced by Henriques J to nine years' imprisonment.

Disqualification from acting as the director of a company

10.102 In appropriate circumstances, a person convicted of health and safety offences may be disqualified from acting as a company director. Section 2 of the Company Directors Disqualification Act 1986, provides that:

> The court may make a disqualification order against a person where he is convicted of an indictable offence (whether on indictment or summarily) in connection with the promotion, formation, management or liquidation of a company, with the receivership of a company's property or with his being an administrative receiver of a company.

10.103 The question of whether an offence is 'in connection with the management of the company' is to be decided by looking at whether it has some relevant factual connection with the management of the company.[31]

10.104 A person who is the subject of a disqualification order 'shall not be a director of a company, act as a receiver of a company's property, or in any way, whether directly or indirectly, be concerned or take part in the promotion, formation or management of a company unless (in each case) he has the leave of the court'.

10.105 The maximum period of disqualification is five years when the order is made by a magistrates' court, and fifteen years in any other case.[32] A person who acts in contravention of a disqualification order, or disqualification undertaking, is liable, on conviction on indictment, to imprisonment for not more than two years or a fine, or both, and on summary conviction, to imprisonment for not more than six months or a fine not exceeding the statutory maximum, or both.

Allowing defendants time to pay

10.106 In *R v Rollco Screw and Rivet Co Ltd* [1999] 2 Cr App R (S) 436, the Lord Chief Justice considered the approach to be adopted when sentencing corporate and individuals together and the time to be allowed for the payment of fines.

---

[31] See *R v Goodman* [1993] 2 All ER 789.
[32] Company Directors Disqualification Act 1986 s 2(3).

Chapter 10: Sentencing

10.107 The prosecution arose out of what the Court described as an 'extremely serious' incident of asbestos contamination, in the Birmingham area, in late September 1997. The company operated a factory, the roof of which was in a poor condition. The directors of the company, father and son, were aware that the roof had asbestos in it, and arranged for contractors to remove it. Neither the contractors, nor an independent contractor employed by them, possessed an 'asbestos licence'. When they came to strip out the brown asbestos, they did it with inadequate equipment, and failed to take appropriate precautions. The result was that asbestos was disseminated over the factory premises, thereby exposing employees and members of the public to risk.

10.108 Later, a vehicle hired by the independent contractors was seen unloading bags of asbestos at a number of unauthorized sites. It was discovered that between 200 and 300 bags of brown asbestos had been fly-tipped on eight sites, to which the public had access. In this case, the company pleaded guilty to breaches of HSWA 1974 ss 2(1) and 3(1). Two of its directors entered pleas of guilty to conniving in the commission of those offences, contrary to HSWA 1974 s 37(1).

10.109 The company was ordered to pay a fine of £20,000 on each count, and £30,000 towards the cost of the prosecution. One director was fined £3,000 on each of the two counts to which he had pleaded guilty; the other was fined £2,000 on each count. Both were ordered to pay £2,000 towards the costs of the prosecution. Two partners from the contractor firm were each fined £6,000 and ordered to pay £1,000 costs. The independent contractor was sentenced to nine months' imprisonment. The judge had ordered that £5,000 of the total payable by the company should be paid within the first 12 months. The balance should be paid at a rate of £1,000 per month. The total period for payment, therefore, in the absence of accelerated payment, was six years and five months. The directors were ordered to pay the sums payable—£8,000 and £6,000 respectively—at a rate of £1,000 per month.

10.110 Upholding all of the fines, but reducing the company's costs, the Lord Chief Justice, Lord Bingham, stated that the correct approach in this type of case was first to determine what financial penalty the offence merited. Next, the court should go on to consider what the defendant, whether corporate or personal, could be reasonably expected to meet. It was this second question which involved a consideration of the time over which a fine was to be paid.

10.111 The court stated that, in the case of a personal defendant, there were strong and cogent arguments for keeping the period over which a fine was to be paid within reasonable bounds. Conversely, the same argument was weaker, and possibly inapplicable, in the case of a corporate defendant.

10.112 Bingham LJ added, referring to the corporate defendant:

> It . . . seems to us that it is not necessarily a more severe course to order a larger sum over a longer period than a smaller sum over a shorter period, since the former course may well give the company a greater opportunity to control its cash-flow and survive

difficult trading conditions. If, despite a long period, the company has the means or chooses to pay sooner than it need, then that is a course open to it.

10.113 The Court looked to the reality of the set-up of commercial enterprises where the company in question was small. It adverted to the fact that the directors are likely to be the shareholders, and therefore likely to lose out if fines were imposed on the company. The sentencing court should, in these cases, be careful to avoid what amounts to double punishment.

10.114 The Court then had regard to setting the level of fine for individual directors:

> It seems to us important in many cases that fines should be imposed which make it quite clear that there is a personal responsibility on directors and they cannot simply shuffle off their responsibilities to the corporation of which they are directors.

10.115 The decision in *Rollco Screw and Rivet Co Ltd*[33] was considered, and endorsed, in *R v Aceblade Ltd* [2001] 1 Cr App R (S) 366. There, the appellant had pleaded guilty to an offence of breach of HSWA 1974 s 2(1). A mobile crane, on which a safety device had been bypassed, overturned and killed the operator. The Court of Appeal upheld a fine of £20,000 and costs of £21,000. The judge at first instance had ordered the fine and costs to be paid at a rate of £1,000 per month. A submission, on appeal, to the effect that this would involve payment over too long a period, namely three and a half years, and that the fine should be reduced to one which could be paid over a year, namely £24,000, was rejected. The seriousness of the offence was said to merit financial penalties of that order. Drawing on what was held in *Rollco Screw and Rivet Co Ltd*, the Court said it was not inappropriate for a corporate defendant to be required to pay fines over a longer period than would have been appropriate for individuals.

10.116 However, in *R v B & Q Plc* [2005] EWCA Crim 2297, the Court of Appeal distinguished cases involving defendants with very substantial resources from those with far more limited resources. In that case, the appellant had sought to challenge the level of the fines totalling £550,000 and costs of £250,000 imposed upon it following conviction at trial of various health and safety offences arising from a fatal incident. The Court of Appeal held that:

> Considering the factors relating to the accident, the size of the appellants and their relative culpability and criminality in relation to the offence, there is no possible basis for contending that this fine was in any way manifestly excessive or wrong in principle.[34]

10.117 The Court of Appeal went on to state how:

> In considering this issue, we also have had some regard to the fact that the learned judge gave the appellants 28 days to pay the fine. That was the period formally

---

[33] [1999] 2 Cr App Rep (S) 436.
[34] [2005] EWCA Crim 2297, at para 46.

requested by the appellants, and the accession to that request was an indulgence to the appellants. Where a fine (which can only be viewed as modest when set against the appellants' overall turnover and profitability) is imposed on a company of anything approaching the size of the appellants, the seriousness of the offending and the impact of the penalty can be brought home to them by their being required to pay the fine within a much shorter period of time than 28 days; an undertaking of the size of the appellants does not need 28 days in which to pay a fine. Such fines ought, as a matter of course, to be paid either immediately or in a period to be measured in single figure days, unless very cogent evidence is provided to the court that more time is needed.[35]

By contrast, the principles applicable to very small companies, partnerships, or other forms of business enterprise are set out in the judgment of the Court of Appeal in *R v Rollco Screw & Rivet Co* [1999] 2 Cr App (S) 436. On the cogent evidence provided in that case a lengthy time was given to pay. But even in such a case, the first instalment should be made payable at a very early date—the Court maintained—so that the effects of the criminality are brought home.

Section 42 orders

10.118 Section 42 of HSWA 1974, provides, that where a person is convicted of a health and safety offence, in respect of any matters which appear to the court to be matters which it is in his power to remedy. The court may, in addition to, or instead of, imposing any punishment, order him, within such time as may be fixed by the order, to take such steps as may be specified in the order for remedying the matters.

10.119 Failure to comply with such an order is an offence. Default is punishable by a fine of £20,000, or six months' imprisonment in the magistrates' court. In the Crown Court, failure to comply is punishable by an unlimited fine, two years' imprisonment, or both.

10.120 Only in very rare circumstances will a sentencing court have resort to HSWA 1974 s 42. This is for the reason that all such matters will have been remedied, in the usual course of events, during the investigation. Remedy will have been secured either voluntarily, or through the service upon the defendant of improvement or prohibition notices by the inspector.

Sentencing in the magistrates' court

10.121 The magistrates' court has the power to impose a maximum fine of £20,000, in respect of a breach of one of the general duties under HSWA 1974 ss 2–6. A fine of up to £5,000 may be imposed for breach of HSWA 1974, s 7 and for breach of a health and safety regulation. There is no limit to the aggregate fine that can be imposed, in respect of a number of different offences. However, it would not be a

---

[35] [2005] EWCA Crim 2297, at para 47.

correct approach to assess the level of fine that might be warranted overall, and then spread it across all of the offences, when any offence, considered alone, would not merit the fine imposed on it.

10.122 The guidance in *Howe*[36] is as applicable to cases being dealt with in the magistrates' court as it is in the Crown Court. However, in dealing with offenders, lay magistrates generally rely heavily upon the Magistrates' Court Sentencing Guidelines, which are issued by the Magistrates' Association.[37] Issued only as guidelines, they are not intended to set tariffs. Nevertheless, the purpose of the Guidelines is to establish some measure of consistency, in sentencing for particular offences across the country. These Guidelines were devised in consultation with the Health and Safety Commission and followed the publication of advice from the Sentencing Advisory Panel to the Court of Appeal.

10.123 The Guidelines state that it is important for magistrates to seek guidance from their legal advisor, in all health and safety cases. They are specifically referred to in the case of *Howe*.[38] Magistrates are further reminded that offences under HSWA 1974 are serious. It is important that they consider carefully whether to accept jurisdiction, rather than commit to the Crown Court.

10.124 The Guidelines, though not exhaustive, list a number of factors which magistrates are encouraged to consider when assessing the serious of the offence. Largely mirroring the guidelines set out in *Howe*,[39] they include:

(a) whether the offence was a deliberate breach of law rather than carelessness;
(b) financial motive—profit or cost-saving, or neglecting to put in place preventative measures, or avoiding payment for relevant licence;
(c) considerable potential for harm to workers or public;
(d) regular or continuing breach, not isolated lapse;
(e) failure to respond to advice, cautions, or warning from regulatory authority;
(f) whether death or serious injury or ill-health has been a consequence of the offence;
(g) ignoring concerns raised by employees or others;
(h) having knowledge of risks but ignoring them;
(i) previous offences;
(j) extent of damage and cost of rectifying it;
(k) attitude to the enforcing authorities; and
(l) offending pattern.

---

[36] *F Howe & Sons (Engineers) Ltd* [1999] 2 Cr App R (S) 37, [1999] 2 All ER 249.
[37] The latest guidelines were issued in October 2003 for implementation on 1 January 2004.
[38] *F Howe & Sons (Engineers) Ltd* [1999] 2 Cr App R (S) 37; [1999] 2 All ER 249.
[39] ibid.

10.125 When computing the penalties for a company, magistrates are encouraged to look at its net turnover.

10.126 Supplementary advice, concerning the approach to be followed in deciding the level of fine in both environmental and health and safety cases, has been provided by the Magistrates Association. It is broadly in line with the guidance in *Howe*,[40] and other cases.

10.127 The guidance provides:

> A fine is considered by the Sentencing Advisory Panel to be the form of penalty which is usually appropriate for both companies and individuals for these offences. The normal principles of the Criminal Justice Act 1991 should apply and the seriousness of the offence and the financial circumstances of the defendant should be taken into account. The level of fine should reflect the extent to which the defendant's behaviour has fallen below the required standard. High culpability should be matched by a high fine even though actual damage turned out to be less than might reasonably have been anticipated.

10.128 In relation to the financial situation of the company, it provides:

> In all cases with corporate offenders the company's financial circumstances must be carefully considered. No single measure of ability to pay can apply in all cases. Turnover, profitability and liquidity should all be considered. It is not unusual for an expert accountant to be available in summary cases.

10.129 On the issue of a company's ability to pay a fine, it provides:

> A fine suited to the circumstances of a small local company would make no impact at all on a multi-national corporation with a huge turnover. The fine to any company should be substantial enough to have a real economic impact which, together with attendant bad publicity, would pressure both management and shareholders to tighten their regulatory compliance. Such fines on large companies might often be beyond the summary fines limit and in such circumstances the case should be transferred to the Crown Court for trial or sentence. Where the court does not transfer the case of a larger company to the higher court, magistrates should look to a starting point near the maximum fine level then consider the aggravating and mitigating factors.
>
> Care should be taken to ensure that fines imposed on smaller companies are not beyond their ability to pay. The court might not wish the payment of the fine to result in the company not being able to pay for improved procedures or cause the company to go into liquidation or make its employees redundant.

B. Costs

10.130 The correct approach to the making of orders for costs in criminal proceedings, whether in favour of the prosecution or the defendant, involves consideration of

---

[40] *F Howe & Sons (Engineers) Ltd* [1999] 2 Cr App R (S) 37; [1999] 2 All ER 249.

B. Costs

the Prosecution of Offenders Act 1985; the Access to Justice Act; the Lord Chief Justice's Practice Direction [2004] 2 All ER 1070; and the Costs in Criminal Cases (General) Regulations 1986 SI 1986/1335.

Prosecution costs

10.131 Where any person is convicted of an offence, before a magistrates' court or before the Crown Court, the court may make such order as to costs to be paid by the accused as the court considers just and reasonable.[41] This is also the position in relation to a convicted person whose appeal against sentence or conviction to the Crown Court or the Court of Appeal is dismissed. The Court may also require a person committed for sentence to the Crown Court to pay prosecution costs.[42]

10.132 Where a defendant is in a position to pay the whole of the prosecution costs, the court in *Howe*[43] made clear that, in addition to the fine, there was no reason in principle for the court not to make an order accordingly.

10.133 The power to order costs is to be exercised at the discretion of the Court, as a matter of general principle. Nonetheless, an order should be made where the Court is satisfied that the defendant or appellant has the means and the ability to pay[44] and the amount must be specified in the order by the court.[45]

10.134 In *R v Associated Octel Ltd* [1997] 1 Cr App R (S) 435, the Court of Appeal gave guidance on what costs may be claimed. The Court also elaborated on the procedure to be followed. It confirmed that the costs which a defendant is ordered to pay may include those incurred by a prosecuting authority, in carrying out investigations with a view to the prosecution of a defendant where a prosecution results and the defendant is convicted. The Court stated that:

(a) The prosecution should serve on the defence, at the earliest time, full details of its costs, so as to give the defence a proper opportunity to consider them and make representations on them if appropriate.
(b) If the defendant wished to dispute the whole or any part of the schedule after being served with a schedule of the prosecution's costs, he should give proper notice to the prosecution of the objections which he proposed to make. It should at least be made clear to the court what the objections were. It might in some cases be that a full hearing would need to be held for the objections to be resolved, as there was no provision for the taxation of the prosecution's costs in a criminal case.

---

[41] Prosecution of Offences Act 1985 (Prosecution of Offenders Act 1985) s 18(1).
[42] Costs in Criminal Cases (General) Regulations 1986 SI 1986/1335 reg 14; Prosecution of Offenders Act 1985 s 18.
[43] *R v F Howe & Son (Engineers) Ltd* [1999] 2 Cr App Rep (S) 37.
[44] See Pt VI. 1.4 Practice Direction [2004] 2 All ER 1070.
[45] Pt VI. 1.5 Practice Direction [2004] 2 All ER 1070.

10.135 The Court further held that the fact that a prosecution eventually modified its case is no reason for refusing or reducing costs.

10.136 In *R v Northallerton Magistrates' Court ex parte Dove* [2001] 1 Cr App R (S) 137, the Court of Appeal reviewed the relevant authorities. It set out a number of principles to be observed when requiring the defendant to pay the costs of the prosecution. Although the case concerned an order for costs within the magistrates' court, the principles are of general application:

(a) An order to pay the costs should never exceed the sum which the defendant was able to pay and it was reasonable for him to pay, having regard to his means and any other financial order imposed on him.

(b) Such an order should never exceed the sum which the prosecutor had actually and reasonably incurred.

(c) The purpose of such an order was to compensate the prosecutor and not to punish the defendant. Where the defendant had by his conduct put the prosecutor to avoidable expense he might, subject to his means, be ordered to pay some or all of that sum to the prosecutor. But he was not to be punished for exercising a constitutional right to defend himself.

(d) While there was no requirement that any sum ordered to be paid to a prosecutor by way of costs should stand in any arithmetical relationship to the fine imposed, the costs ordered to be paid should not in the ordinary way be grossly disproportionate to it. Justices should ordinarily begin by deciding on the appropriate fine to reflect the criminality of the defendant's offence, always bearing in mind his means and his ability to pay, and then to consider what, if any, costs he should be ordered to pay. If, when the costs sought by the prosecutor were added to the proposed fine, the total exceeded the sum which in the light of the defendants means and all other relevant circumstances the defendant could reasonably be ordered to pay, it was preferable to achieve an acceptable total by reducing the sum of costs which the defendant was ordered to pay rather than by reducing the fine.

(e) It was for the defendant facing a financial penalty by way of fine or an order to pay costs to a prosecutor to disclose to magistrates such data relevant to his financial position as would enable justices to assess what he could reasonably afford to pay. In the absence of such disclosure justices might draw reasonable inferences as to the defendant's means from evidence they had heard and all the circumstances of the case.

Defence costs

10.137 It is often the case that defendants in health and safety prosecutions are not in receipt of public funding.

10.138 In the event of a summons or indictment not being proceeded with, or in the event of an acquittal or discharge at committal, the court may make an order in favour of the defendant for a payment to be made out of central funds.[46]

---

[46] Prosecution of Offenders Act 1985 s 16(1)–(3).

B. Costs

Where proceedings are discontinued, a defendant is, prima facie, entitled to a defendant's costs order. The exception to that rule should be narrowly drawn, in order to respect the presumption of innocence both at common law[47] and under Art 6, guaranteeing the right to a fair trial, of the European Convention on Human Rights.[48]

10.139 Such an order is known as a 'defendant's costs order'. It should be for such an amount as the court considers reasonably sufficient to compensate the defendant for any expenses properly incurred by him in the proceedings.[49]

10.140 'Proceedings' includes the proceedings in any court below the court making the order.[50] Therefore, the Crown Court may make an order which includes provision for costs incurred in the magistrates' court. The Court of Appeal may make an order for costs incurred in the Crown Court and magistrates' court.

10.141 The court may either specify the amount payable out of central funds, if the person in whose favour it is made agrees the amount and the court considers it appropriate to do so, or put it off for taxation under the regulations.[51] In the circumstances described above, a defendant's costs order should normally be made unless there are positive reasons for not doing so.[52] Examples of such reasons are where the defendant has brought the proceedings onto himself or misled the prosecution into thinking that the case was stronger than it was.[53] (For the procedure to be adopted when such an assertion is made, see *Mooney v Cardiff Magistrates' Court* (1999) 164 JP 220 (DC).)

10.142 Equally, the court may take the view that there are circumstances which make it inappropriate for the defendant to recover the full amount, where he might otherwise be entitled to it, under the Prosecution of Offenders Act 1985 s 16(6). The court should then assess what amount it considers to be just and reasonable, and specify that amount in the order.[54] Costs claimed may be disallowed, if it is plain that those costs were not properly incurred. For instance, costs may be disallowed if it can be shown that the case has been conducted unreasonably, so as to incur unjustified expense. Where there has been a failure to conduct proceedings with reasonable competence and expedition,[55] costs may also be refused. When the court intends either to disallow costs itself, or to recommend to the taxing

---

[47] *R v Woolmington* [1935] AC 462.
[48] See *R v South West Magistrates' Court ex p James* [2000] Crim LR 690.
[49] Prosecution of Offenders Act 1985 s 16(6).
[50] ibid s 21(1).
[51] ibid s 16(9).
[52] Practice Direction Pt II.1.1.
[53] ibid.
[54] Prosecution of Offenders Act 1985 s 16(7).
[55] Practice Direction Pt V.1.1(b).

10.143 If a defendant is convicted of some counts, but acquitted on others, the court has discretion to make an order only for part of the costs.[56] The court should make whatever order seems to it to be just. Weighing in the balance, as relevant to its determination, is the relative importance of the charges faced, and the general conduct of the parties.

10.144 The regulations applicable to the assessment of costs are the Costs in Criminal Cases (General) Regulations 1986 SI 1986/1335. These provisions do not apply to taxation undertaken by a justices' clerk in the magistrates' court. In those cases, the remedy is to apply to the High Court for judicial review of the decision.

10.145 The approach to be adopted by the taxing authority is, first, to resolve whether work has actually and reasonably been done.[57] Then, the court must consider what sum is reasonably sufficient to compensate the appellant for expenses properly incurred. When deciding whether this has 'reasonably been done', the taxing authorities will no doubt have regard to the principles set out in Practice Direction 1, Costs in Criminal Cases (General) Regulations 1986 SI 1986/1335 reg 7(1)(a).

10.146 The court may come to the view that there are reasons for making a defendant's costs order for less than may otherwise be reasonably sufficient. It should then assess what amount would, in its opinion, be just and reasonable. The court must specify that amount in the order, and specify in writing the reasons why that view is taken.[58]

10.147 Detailed provisions as to what may and may not be claimed as part of a defendant's costs order are contained within the Costs in Criminal Cases (General) Regulations 1986 SI 1986/1335. For instance, there is no provision for claiming expenses such as loss of earnings.

Examples of the approach taken

10.148 In *R v Dudley Magistrates' Court ex parte Power City Stores* (1990) 154 JP 654, an appeal against the dismissal by the justices' clerk for payment of leading counsel's fees as part of a defendant's costs order, the appropriate question was held to be whether the defendant acted reasonably in instructing the counsel concerned, and not whether more junior counsel or a solicitor could have dealt with the case.

---

[56] Practice Direction Pt II.2.2.
[57] Costs in Criminal Cases (General) Regulations 1986 SI 1986/1335 reg 7(1)(a).
[58] Prosecution of Offences Act 1985 s 16(7).

B. Costs

In *R (Hale) v Southport Justices* Times Law Reports 29 January 2002, it was held that it was reasonable for a defendant, charged with common assault, to instruct a solicitor of more than four years' standing. It was not unreasonable to incur costs using a flat hourly rate. The justices' clerk had confused the test in the Prosecution of Offenders Act 1985 s 16(6). Mistakenly, he had approached the question of costs by asking himself whether a less experienced solicitor would have been reasonably sufficient. Rather, the words 'reasonably sufficient', as contained in the statute, were directed towards the level of compensation due to the claimant. **10.149**

Appealing the decision for costs

If an applicant is dissatisfied with the award of costs he may apply for a re-determination within 21 days of being notified of them.[59] Where an applicant is dissatisfied with the re-determination, he may appeal to the Chief Taxing Master within 21 days, specifying the grounds of his objection.[60] An applicant who is dissatisfied with the Taxing Master's decision may request him to certify a point of general importance, and if he agrees to do so, the applicant may appeal to the High Court. **10.150**

The order for payment of costs to a prosecutor has been held to be a sentence, within the terms of s 50(1) of the Criminal Appeals Act. It may, therefore, be appealed alone, or along with some other aspect of a sentence, to the Court of Appeal. There is no provision, within the Magistrates' Courts Act 1980, to appeal an order for costs to the Crown Court. **10.151**

Apportioning costs between a number of defendants

In *R v Fresha Bakeries Ltd* [2003] 1 Cr App R (S) 202, two employees of the defendant company had been killed after entering a large bread oven to conduct maintenance tasks. The oven had been insufficiently cooled after use. **10.152**

Both companies, and one of the individual defendants, had indicated guilty pleas at an early stage of the proceedings. The other three individual defendants—the chief executive, the chief engineer, and the production director—maintained their not guilty pleas. They then changed their pleas, but not until the day before the trial was due to start. At that time, two of them indicated guilty pleas, and the prosecution took the decision not to proceed against the third. **10.153**

At the time the companies entered their pleas, the costs of the prosecution amounted to £108,451. However, by the time the matter was ready for the trial of the three individual defendants, those costs had risen to £283,307. The court did **10.154**

---

[59] Costs in Criminal Cases (General) Regulations 1986 SI 1986/1335 reg 9.
[60] ibid reg 10(1).

not accept the submission on costs made by counsel for the companies. This was to the effect that their costs should be determined in accordance with the rule laid down in *Ronson and Parnes* (1992) 13 Cr App R (S) 153. That case established that the right general approach to the question of costs was to look to see what a reasonable estimate of costs of each defendant would be, if he had been tried alone. Instead, the court applied the principle in the case of *Harrison* (1993) 14 Cr App R (S) 419. There, the Court had held that it was not wrong in principle that the principal offender should bear 'the lion's share', or all, of the costs incurred. Moreover, the court in *Harrison* held that the sentencing court was entitled to conclude that any one particular defendant, amongst a number of defendants, was more responsible than the others for the criminal conduct leading to conviction. Then, it may be appropriate to order that the more blame-worthy defendant pay a greater share of the costs than he would otherwise have had to pay, if he had been tried alone.

10.155 Fines amounting to £350,000 were upheld against two related companies. However, the award of costs was ultimately reduced, from £250,000 to £150,000.

11

WORK-RELATED DEATHS

A. The Investigation of Work-Related Deaths	11.01	D. Coroner's Inquests and Work-Related Deaths	11.89	
The decision in *Beedie*	11.05	Proposed reform	11.89	
B. The Work-Related Death Protocol	11.08	The current system	11.99	
'Work-related' deaths	11.12			
Scope of the protocol	11.15			
C. Manslaughter	11.21			
Individual manslaughter by gross negligence	11.25			
Corporate manslaughter	11.66			

A. The Investigation of Work-Related Deaths

11.01 At the time of publication of this work, two very significant changes are imminent, each of which is likely to affect the investigation of work-related deaths. The long-awaited Corporate Manslaughter and Homicide Bill is currently stalled between the Houses of Parliament, with battle joined as to the extent of potential liability for deaths in custody. The Bill's central provisions seem very likely to become law and be brought into force sometime in 2008.

11.02 The other change is the promised reform of the coroners' system, which has been long-debated and awaited, with universal agreement as to the need for reform, but real disquiet particularly surrounding inadequate funding of the Government's proposals. Legislation is promised as soon as parliamentary time allows; any reforms might commence in 2008.

11.03 Currently, when a work-related death occurs, a number of agencies may be involved simultaneously in the investigation of the circumstances surrounding the death, not least the police and the HSE. Potentially, the police will be involved in investigating manslaughter, whilst the HSE (or other health and safety enforcement authority) will be investigating potential health and safety beaches. Whenever such a death occurs, an inquest before a jury is likely to be heard unless a criminal prosecution for manslaughter arises as a result of the police investigation.

11.04 The situation represents a complex fusion of often competing interests and whilst a series of principles has been developed to better ensure the effective investigation of such deaths, the whole process can nonetheless take a number of years to be completed.

The decision in *Beedie*

11.05 The law and practice in relation to work-related deaths has been greatly affected by the decision of the Court of Appeal in *R v Beedie* [1997] 2 Cr App R 167.

11.06 In *R v Beedie* the appellant had pleaded guilty in the magistrates' court to a health and safety offence brought by the HSE involving a breach of the Health and Safety at Work etc. Act 1974 (HSWA 1974) s 3(2). The offence arose out of the death of a young woman tenant from carbon monoxide poisoning caused by a defective gas fire in her flat. The appellant was her landlord. At the inquest into the death, which followed those proceedings, the coroner wrongly required the appellant to give evidence on the basis that, having already been prosecuted, there was no realistic possibility of a prosecution for manslaughter if a verdict of unlawful killing was to be returned. Such a verdict was returned and the appellant was subsequently charged with manslaughter. At his trial, he changed his plea to guilty, following an unsuccessful application to stay the indictment on the ground of a plea of autrefois convict, and thereafter appealed against conviction.

11.07 The Court of Appeal quashed his conviction for manslaughter and held that the rule against double jeopardy[1] meant that that no man should be punished twice for an offence arising out of the same, or substantially the same, set of facts and that there should be no sequential trials for offences on an ascending scale of gravity. The response to this decision was the development by the HSE and the Crown Prosecution Service (CPS) of the Work Related Death Protocol.[2]

B. The Work-Related Death Protocol

11.08 The Work Related Death Protocol represents an agreement between the HSE, the Association of Chief Police Officers, the British Transport Police, the Local Government Association, and the CPS upon the principles for effective liaison between the parties in relation to the investigation and prosecution of offences relating to work-related deaths.

---

[1] See paras 9.34–9.45 for the rule against double jeopardy.
[2] The Work Related Death Protocol is reproduced at App B.

B. The Work-Related Death Protocol

There is now a Work Related Death Investigators Guide,[3] which is intended as practical guidance and offers investigatory systems to be employed in furtherance of the protocol. **11.09**

The protocol's underlying principles are that: **11.10**

(a) an appropriate decision concerning prosecution will be made based on a sound investigation of the circumstances surrounding work-related deaths;
(b) the police will conduct an investigation where there is an indication of the commission of a serious criminal offence (other than a health and safety offence), and HSE, the local authority or other enforcing authority will investigate health and safety offences. There will usually be a joint investigation, but on the rare occasions where this would not be appropriate, there will still be liaison and cooperation between the investigating parties;
(c) the decision to prosecute will be co-coordinated, and made without undue delay;
(d) the bereaved and witnesses will be kept suitably informed; and
(e) the parties to the protocol will maintain effective mechanisms for liaison.

The protocol contains a statement of intent that recognizes that in the early stages of an investigation, whether any serious criminal offence has been committed is not always apparent. The protocol describes that should there be any issue as to who is to be involved in investigating any work-related death, then the parties will work together to reach a conclusion. **11.11**

'Work-related' deaths

The protocol defines a work-related death as a fatality resulting from an incident arising out of, or in connection with, work. The principles of the protocol are said to apply also to cases where the victim suffers injuries in such an incident that are so serious that there is a clear indication, according to medical opinion, of a strong likelihood of death. **11.12**

The protocol recognizes that there are cases in which it is difficult to determine whether a death is work-related within the terms set out above and gives examples of cases arising out of some road traffic incidents, or in prisons or health care institutions or following a gas leak. Another example, not cited in the protocol, is death from a disease, such as legionnaires' disease, which may have been caused by the conduct of an employer's undertaking and can potentially lead to the consideration of manslaughter charges. **11.13**

The protocol provides the principle that each fatality must be considered individually, on its own particular facts, according to organizational internal **11.14**

---

[3] http://www.hse.gov.uk/enforce/investigators.pdf.

guidance, and a decision made as to whether it should be classed as a work-related death. The protocol provides that in determining the question, the enforcing authorities will hold discussions and agree upon a conclusion without delay.

Scope of the protocol

11.15 The protocol contains principles dealing with initial action, the management of the investigation, decision-making, disclosure of material, special inquiries, advice prior to charge, the decision to prosecute, prosecution, dealings with HM Coroner, and liaison.

Decision-making

11.16 The protocol provides that where the investigation gives rise to a suspicion that a serious criminal offence (other than a health and safety offence) may have caused the death, the police will assume primacy for the investigation and will work in partnership with the HSE, the local authority, or other enforcing authority. Where it becomes apparent during the investigation that there is insufficient evidence that a serious criminal offence (other than a health and safety offence) caused the death, the investigation should, by agreement, be taken over by the HSE, the local authority, or other enforcing authority. Both parties should record such a decision in writing. The practical effect of this handover is to create what isan unavoidable delay in the decision making process of the HSE and will, despite the principles of the protocol, in almost all cases, cause a final decision on the prosecution of any health and safety offences to be left until after an inquest has been held.

11.17 Where the HSE, the local authority, or other enforcing authority is investigating the death, and new information is discovered which may assist the police in considering whether a serious criminal offence (other than a health and safety offence) has been committed, then the enforcing authority will pass that new information to the police.

11.18 The protocol makes clear that there will also be rare occasions where as a result of the coroner's inquest, judicial review or other legal proceedings, further consideration of the evidence and surrounding facts may be necessary. Where this takes place the protocol provides that the police, the enforcing authority with primacy for the investigation, and the CPS will work in partnership to ensure an early decision, but that there may also be a need for further investigation. In practice, such cases, whilst undoubtedly being rare, have occurred with regularity and have involved consideration and reconsideration at different stages by prosecuting authorities with a resultant delay in the final decision to prosecute.

C. Manslaughter

Advice prior to charge and the decision to prosecute

11.19 The protocol provides that the police should seek the advice of the CPS before charging an individual with any serious criminal offence (other than a health and safety offence) arising out of a work-related death and that the police must consult the CPS Casework Directorate for advice when there is any consideration of charging a company or corporation with any serious criminal offence (other than a health and safety offence).

Serious criminal offences: manslaughter

11.20 The protocol refers to 'serious criminal offences' arising from work-related deaths. In practice, the only serious criminal offence likely to arise in connection with a work-related death within the terms of the protocol is manslaughter.

C. Manslaughter

11.21 The common law offence of manslaughter, covering liability for intentional violent acts and liability for non-intentional gross negligence, has attracted much criticism, and proposals for its reform have been discussed for a number of years. At the time of publication of this work, the Corporate Manslaughter and Homicide Bill had reached a 'ping-pong' stage in Parliament, with the House of Lords insisting upon the extension of the proposed potential offence to cover deaths in custody, and the House of Commons resisting the same.

11.22 The central provisions of the offence are likely to become law in 2007 and brought into force in 2008. When this occurs, an organization will be liable for the offence of corporate manslaughter where it is grossly in breach of a relevant common law duty of care that is a substantial cause of a person's death.

11.23 The corporate manslaughter and homicide provisions will leave intact the common law offence of individual gross negligence manslaughter and it is intended that this will continue to be the position, with none of the many proposals for reform of the law in this area to be advanced.

11.24 Only the Director of Public Prosecutions, through the CPS, has, and will have, the power to institute proceedings against an individual or company for manslaughter. The Work Related Death Protocol provides that the decision to prosecute any serious criminal offence arising out of a work-related death will be taken by the CPS in accordance with the Code for Crown Prosecutors. Currently, such an offence may be prosecuted either with or without related health and safety offences, which the CPS also have the power to prosecute, but such cases are often jointly initiated and managed by the CPS and HSE, with the CPS formally instituting all proceedings. This position is unlikely to be materially affected by the advent of a new corporate manslaughter offence.

Individual manslaughter by gross negligence

11.25 Liability in respect of manslaughter can arise in various ways, but in respect of work-related death the form under consideration will be manslaughter by gross negligence.

11.26 The Court of Appeal, in *R v Misra and Srivastava* [2005] Crim LR 234, rejected an argument that manslaughter by gross negligence was an offence which lacked certainty, and as such that it offended against the terms of Art 7 of the European Convention on Human Rights, as had been posited in the Law Commission paper on Involuntary Manslaughter.[4] Lord Justice Judge said in terms, how 'The ingredients of the offence have been clearly defined, and the principles decided in the House of Lords in *Adomako*.[5] They involve no uncertainty.'[6]

11.27 *R v Adomako* [1995] 1 AC 171 authoritatively established that the ordinary principles of the law of negligence apply to determine whether a defendant was in breach of a duty of care towards the victim. Where a defendant can be proved to be in breach of a duty of care towards a deceased victim, which was a substantial cause of the death of the victim, then, if having regard to the risk of death involved, the defendant's conduct can be proved to have been 'so bad in all the circumstances as to amount to a criminal act or omission', he will be guilty of manslaughter by gross negligence.

11.28 How 'bad' the conduct is judged to be is ultimately a question for a jury: in *Adomoko*[7] it was held that it is always eminently for a jury to decide whether, having regard to the risk of death involved, the defendant's conduct was so bad in all the circumstances as to amount to a criminal act or omission.[8] Such allegations of manslaughter can, and often are, based upon an alleged omission to act in circumstances where a defendant is alleged to have been under a duty to act.

Risk of death

11.29 In *R v Singh (Gurphal)* [1999] Crim LR 582 (CCA) the Court of Appeal described as a model direction one in which a trial judge had stressed:[1]

> That the circumstances must be such that a reasonably prudent person would have foreseen a serious and obvious risk not merely of injury or even of serious injury but of death.

11.30 The decision in *R v Misra and Srivastava* [2005] Crim LR 234 clarified and confirmed the position at common law in relation to the nature of the relevant risk

---

[4] Law Com No 237.
[5] [1995] 1 AC 171.
[6] [2004] EWCA Crim 2375 at para 64.
[7] [1995] 1 AC 171.
[8] Following *R v Bateman* (1925) 19 Cr App R 8 (CCA) and *Andrews v DPP* [1937] AC 576 (HL).

C. Manslaughter

arising from the breach, being that described by the trial judge in *Singh (Gurphal)*, namely a risk of death alone.

Causation

In gross negligence manslaughter cases, as with all homicide cases, liability arises upon proof that the defendant's act, the grossly negligent breach of the duty of care, contributed significantly to the death or was a substantial cause of the death and thus it need not have been the sole or principal cause of the death.[9] Whilst it is no defence to an indictment for manslaughter, where the death of the deceased is shown to have been caused substantially by the negligence of the defendant, that the deceased was also guilty of contributory negligence in respect of his own death,[10] nonetheless, it may be relevant to whether the negligence of the defendant can properly be characterized as gross.

11.31

The duty of care

A number of uncertainties and difficulties remain in respect of individual gross negligence manslaughter in general and, in particular, where such an allegation arises in circumstances involving health and safety duties.

11.32

It remains uncertain whether with an offence of individual gross negligence manslaughter the existence of a duty of care is a matter solely of fact, for the jury, or involves both questions of law, for the judge, and questions of fact, for the jury.

11.33

In the judgment given by the Court of Appeal in *R v Khan and Khan* [1998] Crim LR 830 (CA) it was said that it is for the judge to decide, as a matter of law, whether the facts are capable of giving rise to a duty of care, and for the jury to decide in the light of the judge's directions whether the defendant owed such a duty of care to the victim. This approach was referred to with approval in *R v Sinclair* (1998) 148 NLJ 1353 (CA). However, in *R v Singh (Gurphal)* [1999] Crim LR 582 (CA), the trial judge had directed the jury firmly in terms that whether the appellant owed a duty of care was a question of law and that he had ruled that there was. In the Court of Appeal there was argument as to whether the facts gave rise to a duty of care but it was not submitted that it was inappropriate in principle for the judge to decide this issue and the summing-up was described by the Court as a 'model of its kind'.

11.34

R v Willoughby

The scope of the conflicting Court of Appeal authorities concerning this issue was considered in the recent judgment in *R v Willoughby* [2005] 1 Cr App R 29,

11.35

---

[9] See *R v Pitts* (1842) C & Mar 248; *R v Curley* (1909) 2 Cr App R 96 (CCA) 109; *R v Cato* (1975) 62 Cr App R 41 at 46.
[10] See *R v Jones* (1870) 11 Cox 544; *R v Kew and Jackson* (1872) 12 Cox 355.

which purported to definitively decide that, 'Whether a duty of care exists is a matter for the jury once the judge has decided that there is evidence capable of establishing a duty.'[11]

11.36 Lord Justice Rose described how the judgment in *Adomako* clearly envisaged that existence of duty, breach causing death, and judgment of criminality are all three usually matters for the jury and how this was the Court's interpretation in *R v Khan & Khan* [1998] Crim LR 83[12] which was followed in *R v Sinclair*.[13]

11.37 The judgment in *R v Willoughby* [2005] 1 Cr App R 29 then cited how in both *Wacker*[14] and *Mark and Nationwide Heating Services* [2004] EWCA Crim 2490, the issue of whether a duty of care existed had been left for the jury. Lord Justice Rose describes how, 'in the light of the illuminating analysis of principle in *Wacker*', the Court felt able to summarize a number of conclusions, including how whether a duty of care exists is normally a matter for the jury.[15]

R v Wacker

11.38 In *R v Wacker* [2003] 1 Cr App R 22 the Court of Appeal had been concerned with the infamous lorry driver who had been convicted of conspiracy to facilitate the entry into the UK of illegal immigrants and of fifty-eight offences of manslaughter by gross negligence, following the horrific discovery of the asphyxiated bodies of fifty-eight illegal immigrants within a container on his trailer at Dover. The vent that would have allowed air into the container, and potentially the noise of the occupants out, had been shut.

11.39 The appellant had submitted that one of 'the ordinary principles of the law of negligence', known by the Latin maxim of *ex turpi causa non oritur actio*, was that the law of negligence did not recognize the relationship between those involved in a criminal enterprise as giving rise to a duty of care owed by one participant to another.

11.40 The Court of Appeal held that that the civil law had introduced the concept of *ex turpi causa* as a matter of public policy as the courts will not 'promote or countenance a nefarious object or bargain which it is bound to condemn'[16] and that it was clear that the criminal law adopted a different approach to the civil law in this regard.

---

[11] [2005] 1 Cr App R 29 para 24.
[12] See transcript: (9702286 X3) 18 March 1998.
[13] (1998)148 NLJ 1353.
[14] [2003] 1 Cr App R 22.
[15] [2005] 1 Cr App R 29 para 24.
[16] Bingham LJ in *Saunders v Edwards* [1987] 1 WLR 1116 at 1134.

C. Manslaughter

Kay LJ, delivering the judgment of the Court of Appeal, stated: **11.41**

> Insofar as Lord Mackay referred to 'ordinary principles of the laws of negligence' we do not accept for one moment that he was intending to decide that the rules relating to *ex turpi causa* were part of those ordinary principles. He was doing no more than holding that in an 'ordinary' case of negligence, the question whether there was a duty of care was to be judged by the same legal criteria as governed whether there was a duty of care in the law of negligence.[17]

As to the question whether it is any answer to a charge of manslaughter for a defendant to say, 'we were jointly engaged in a criminal enterprise and weighing the risk of injury or death against our joint desire to achieve our unlawful objective, we collectively thought that it was a risk worth taking',[18] Lord Justice Kay's judgment firmly held, apparently as a matter of law, that it was not and that public policy considerations determine whether a duty of care exists. **11.42**

The implication of the Court's judgment in *R v Wacker* is that the jury should have been directed that if they were satisfied that the lorry driver was aware of the occupants of the container, and thus a party to the enterprise, then, as a matter of law, he owed them each a duty of care. **11.43**

Thus the import of the Court of Appeal's decision in *R v Wacker*[19] suggested the very opposite to that which the Court, apparently founding its judgment on *R v Wacker*, decided in *R v Willoughby*:[20] in *R v Wacker*, the Court held that the judge should have directed the jury that if the driver had been aware of the illegal immigrants, then he owed them a duty of care. In *R v Willoughby* the Court ruled that the existence of a duty of care was 'normally for the jury's determination'.[21] **11.44**

Health and safety duties and manslaughter

The second difficulty arises with the relationship and relevance of statutory health and safety duties to an alleged offence of gross negligence manslaughter. **11.45**

HSWA 1974 ss 2 and 3 require an employer to ensure, so far as is reasonably practicable, the safety of his employees at work and those others who may be affected by the conduct of his business.[22] These duties are not part of the civil law of negligence, nor is their breach actionable, as HSWA 1974 s 47(1)(a) of the Act excludes civil liability. Whether any breach of a statutory duty involving criminal liability **11.46**

---

[17] [2003] 1 Cr App R 22 para 37.
[18] ibid para 39.
[19] [2003] 1 Cr App R 22.
[20] [2005] 1 Cr App R 29.
[21] ibid para 23.
[22] See paras 6.01–6.42.

also gives rise to a civil cause of action is a question of construction[23] and the Act provides that a breach of health and safety regulations is actionable where it has caused damage unless the regulations in question provide otherwise.[24]

11.47 The common law duties owed by an employer to an employee or to someone affected by the conduct of his undertaking require a lower standard than that imposed by the statutory general health and safety duties, and by the regulations created pursuant to HSWA 1974. A number of health and safety regulations impose absolute liability, creating duties to ensure safety.

11.48 Statutory health and safety duties cannot be delegated and vicarious liability play no part;[25] at common law, liability in negligence may involve determination of an issue of vicarious liability. While such health and safety duties are non-delegable, the class of persons who owe such duties is narrowly, and to some extents arbitrarily, defined: the HSWA 1974 s 3 duty to ensure the safety of non-employees affected by a business is cast in identical terms upon employers (HSWA 1974 s 3(1))[26] and the self-employed (HSWA 1974 s 3(2)),[27] yet many self-employed persons are also employers, and some companies with extensive undertakings employ nobody, thus being neither an employer nor self-employed.

11.49 Thus an action for breach of a statutory duty remains different to and distinct from an action in negligence.

11.50 The judgment in *R v Singh (Gurphal)*[28] highlights the issue of the relevance of absolute statutory duties to an allegation of gross negligence manslaughter. The facts of that case concerned the death through carbon monoxide poisoning of a tenant of the appellant's father. The appellant accepted that he had helped his father in the running of tenanted properties and had acted as maintenance man. His father, the landlord, went away to India in early February 1996 and left the son in charge. The appellant had practical experience but no qualifications. The appellant accepted that ten days before the death of the tenant, he had been warned that the gas fire was faulty; he had looked at the fire for five minutes and had declared it safe.

11.51 A number of the statutory duties imposed on a landlord in respect of gas appliances are absolute non-delegable duties under the Gas Safety (Installation and Use) Regulations 1998 SI 1998/2451.[29] The same regulations impose absolute

---

[23] The general rule was stated by Lord Diplock in *Lonrho Ltd v Shell Petroleum Co Ltd (No 2)* [1982] AC 173, 185.
[24] s 47(2) Health and Safety at Work etc. Act 1974.
[25] See paras 6.75–6.79.
[26] See paras 6.24–6.28.
[27] See paras 6.29–6.32.
[28] [1999] Crim LR 582 (CA), but see transcript: (9803983 Z4), 19 February 1999.
[29] See App M and paras 12.220–12.224.

C. Manslaughter

duties on those carrying out 'work' to gas appliances, which is defined to include 'maintenance'.

11.52 During the course of his appeal, the appellant submitted that the judge in his summing-up had not made clear precisely what the duty was which the law imposed upon the appellant. He pointed out that the prosecution skeleton argument asserted, 'that Gurphal Singh had a positive duty *to ensure* the safety of the tenants which he breached by failing to take action to remedy the dangerous state of the fire'.

11.53 The appellant submitted that the judge never clarified in his own mind, or to the jury, what the nature of that duty was. Was it an absolute duty to ensure that the fire could be operated safely, as was imposed upon a landlord under the Gas Safety (Installation and Use) Regulations, or was it merely a duty to take care that the fire could be operated safely? The Court of Appeal accepted that the point was important, because:

> it was up to the jury to determine whether or no the duty had been broken. The same set of conclusions as to facts by the jury might in certain circumstances result in an acquittal if the duty was merely to take care whereas it might result in a conviction if the duty was an absolute one. If the jury were under the impression that the duty was an absolute one to ensure that the fire could be operated safely then it was clear that this duty had been broken and it would be relatively easy for the jury to conclude that the failure to carry out this duty involved a very high degree of negligence. If, by contrast, the duty was merely to take care that the fire could be operated safely then the jury would have to have been satisfied of a greater degree of culpability in the appellant.

11.54 The Court of Appeal held that the judge had only used the language of carelessness and negligence and that it was clear from the summing-up that there was no risk that the jury were not sure that the appellant had failed to take reasonable care and that this failure was grossly negligent.

11.55 The Court of Appeal held that whilst the appellant had:

> no contractual relationship with the tenant, that in itself is no bar to the imposition of a duty of care by the law of negligence. The son was not merely the representative of the landlord. His position was such that by reason of proximity and foreseeability of serious injury and death a duty of care arose. A landlord in the position of the father in this case is in any event now under a statutory duty to take care.

11.56 Whilst the judgment highlighted the potential difference between a jury's consideration of a failure to comply with an absolute statutory duty and a breach of a duty to take reasonable care, the case did not resolve whether a breach of an absolute statutory duty could or should form part of the relevant breach of duty for gross negligence manslaughter. Unfortunately, nowhere did the Court address what, if any, relevance had the various statutory health and safety duties arguably owed by the defendant to the issues that the jury had to determine.

The Caparo *test*

11.57 In September 2004, Mr Justice Mackay, dismissing counts of manslaughter against directors of a number of companies in the prosecution that followed the Hatfield train crash,[30] reviewed the conflicting authorities on the issue of how and by whom the existence of a duty of care is determined in individual gross negligence manslaughter. He identified how the offence was founded upon the 'ordinary principles of negligence' and, as such, the question of the existence of a duty care must be resolved in accordance with the classic tripartite test set out in the leading authority, namely *Caparo Industries plc v Dickman*,[31] which recognized explicitly how the ingredients of this tripartite test 'are not susceptible of any such precise definition as would be necessary to give them utility as practical tests'.[32]

11.58 The civil courts, applying the 'ordinary principles of negligence', have stated that such findings can and do involve uncertainty and a choice between competing issues. In *McFarlane v Tayside Health Board* [2000] 2 AC 59, Lord Steyn identified the essential determinant of duty of care questions in negligence as, 'a mosaic in which the principles of corrective and distributive justice are interwoven. And in situations of uncertainty and difficulty a choice sometimes has to be made between the two approaches.'[33]

11.59 Thus a range of different tests for determining whether a duty of care is owed have been developed. 'Proximity' and 'foreseeability of harm' are two parts of the tripartite test classically defined in *Caparo Industries plc v Dickman* [1990] 2 AC 605 (HL), with the third part of the test being the resolution of the nebulous issue, 'Is it fair, just and reasonable to impose a duty of care?'

11.60 In *R v Railtrack Plc & others*,[34] Mr Justice Mackay commented how he, 'found great difficulty in seeing how a jury could be helped to find its way through the tripartite *Caparo* test and in particular how the direction should be framed, if the exercise is as that case tells us an incremental one building on established examples of other duty situations'.[35] As a result, the prosecution in that case submitted that the question of the existence of a duty of care was a matter of law which he should resolve and thereafter direct the jury upon.

11.61 The judgment of the Court of Appeal in *Willoughby* [2005] 1 Cr App R 29 was given just months later, on 1 December 2004. The judgment of Rose LJ in that

---

[30] *R v Railtrack plc & others* (Central Criminal Court 1 September 2004).
[31] [1990] 1 All ER 568.
[32] ibid at 574.
[33] At 83A–C.
[34] *R v Railtrack plc & others* (Central Criminal Court 1 September 2004).
[35] Mackay J, *R v Railtrack plc & others* (Central Criminal Court 1 September 2004) transcript p 19.

C. Manslaughter

case was that whether a duty of care existed is normally a matter for the jury, but it was in this context that he referred to statutory duties, adding:

> We add that there may be exceptional cases, for example where a duty of care obviously exists, such as that arising between doctor and patient, or where Parliament has imposed a particular type of statutory duty, in which the judge can properly direct the jury that a duty exists.[36]

Statutory duties and the duty of care

11.62 It is submitted that the proper approach is that the existence of a statutory duty owed by a defendant will be relevant to the issue of 'proximity' of the defendant to the alleged victim: where the victim is in a class of persons either to whom the statutory duty can be said to be owed by the defendant, or such a class of persons likely to be affected by how the defendant performs that statutory duty, then the defendant can be said to have been in a proximate relationship to the victim; foreseeability of harm and the reasonableness/justice of imposing a duty of care are other relevant considerations; where such a duty of care is found to have been owed, the standard imposed by the law of negligence will be to take reasonable care; what 'reasonable care' amounts to may also be informed by what was required of the defendant by the statutory duty.

Company directors and statutory duties owed by the company as employer

11.63 The issue of the relevance of statutory health and safety duties to individual gross negligence manslaughter becomes most acute in the context of directors of employing companies. As set out above,[37] the general duties imposed upon employers under HSWA 1974 are non-delegable and personal to the employer. A director of a company does not owe any such duty as a result of holding such an office of a company that is an employer.

11.64 This same issue arises when considering the liability of a company director under HSWA 1974 s 37[38] for having caused, to some extent, through neglect an offence to have been committed by the company. In *Huckerby v Elliot* [1970] 1 All ER 189 (DC) the Court of Appeal was concerned with a prosecution in which the appellant, a director of the company, was charged with an offence against the Finance Act 1966 s 305(3), in that the company's offence was attributable to her neglect. Lord Parker CJ stated that a director of a company is not under a general duty to exercise a degree of control over the company's affairs nor to acquaint himself with all the details of the running of the company. Nor is a director under a duty to supervise the running of the company or his co-directors. Lord Parker

---

[36] [2005] 1 Cr App R 29 para 23.
[37] In detail at paras 6.75–6.79.
[38] See paras 5.46–5.82.

CJ quoted the judgment of Romer J in *Re City Equitable Fire Insurance Co Ltd* [1925] Ch 497 at 428–430, where it was held that amongst other things it is perfectly proper for a director to leave matters to another director or to an official of the company, and that he is under no obligation to test the accuracy of anything that he is told by such a person, or even to make certain that he is complying with the law.

11.65 As with HSWA 1974 s 37, the duty owed by a director of a company in respect of that company's compliance with its personal, non-delegable, health and safety duties requires proof of more than a mere failure to see that the law is observed, and requires the identification of a duty resting upon the director to fulfil a specific act or responsibility for the company's compliance, and failure so to do.

Corporate manslaughter

11.66 The Corporate Manslaughter and Homicide Act 2007 will, except in relation to deaths in custody, come into force on 6 April 2008. Section 20 of the Act abolishes the common law offence of manslaughter by gross negligence in its application to corporations and other organizations to which the new offence created by s 1 of the Act applies. By s 27 of the Act, liability under the common law will remain in respect of any offence committed wholly or partly before the commencement of the Act and such an offence includes one where any of the conduct or events alleged to constitute the offence occurred before that commencement date. Under the common law, a company is a person capable of being proceeded against and convicted of gross negligence manslaughter. However, such prosecutions have had a chequered history and few have succeeded, those being in relation to small companies with an easily identifiable 'directing mind' whose own acts or omissions have been a direct and substantial cause of death.

The common law

11.67 In the failed prosecution for manslaughter brought against a train operating company following the Southall rail disaster (the company pleaded guilty to breaches of HSWA 1974), the Court of Appeal was asked to rule upon whether a company could be convicted of the crime of manslaughter by gross negligence in the absence of evidence establishing the guilt of an identified human individual for the same crime *(Att-Gen's Reference (No 2 of 1999)* [2000] 2 Cr App R 207). The Court's judgment was that unless an identified individual's conduct, characterizable as gross criminal negligence, could be attributed to the company, the company was not liable for manslaughter.

11.68 In such a case the focus had to be upon the 'directing minds' of the company—its directors and senior managers. If any of them, individually, could be demonstrated to have been 'grossly negligent', the company itself could be convicted of manslaughter. However, in such circumstances the individual 'grossly negligent'

C. Manslaughter

director or directors themselves would additionally face prosecution for manslaughter. The company's liability could not arise through the aggregation of directors' negligence but only through such individual gross negligence.

The Corporate Manslaughter and Homicide Act 2007

11.69 Public demand for reform of the law relating to corporate manslaughter stemmed from the series of (mostly transport-related) public disasters, many of which were followed by failed prosecutions for corporate manslaughter. The slow pace of reform was due to the difficulties of reaching a consensus upon the best means for reform of the law of manslaughter and the simultaneous commitment to remove Crown immunity for such liability.

11.70 The draft Corporate Manslaughter Bill was published by the Home Office for consultation on 23 March 2005. On 20 December 2005 the Home Affairs and Work and Pensions Committees of the House of Commons published a three-volume joint report[39] considering the Bill. A wide panel of witnesses and various interest groups gave oral and written evidence. The select committee drew a number of conclusions which departed from those of the Government, in a number of important respects. For the most part, the Government rejected the report's recommendations, in a response published on 8 March 2006.[40]

11.71 After protracted 'ping-pong' proceedings between the two Houses over amendments to the Bill to extend its operation to deaths in custody, it finally passed through Parliament on 23 July 2007, a matter of days before an extended deadline for Royal Assent. The commencement date of the Act is proposed for 6 April 2008. Section 2(1) of the Act, which extends the offence to cover deaths in custody, can occur only through the affirmative resolution procedure.

Potential impact of the new corporate manslaughter offence

11.72 The Regulatory Impact Assessment[41] conducted by the Home Office anticipates that approximately five prosecutions of the new corporate offence will be launched per year. Currently, under the Work Related Death Protocol, all work-related deaths should be considered for potential investigation as suspected gross negligence manslaughter. The Regulatory Impact Assessment provides:

> Some investigations may need to be fuller in the future than is currently the case, although cases that are most resource intensive, involving major public incidents, are

---

[39] Home Affairs and Work and Pensions Committees: Draft Corporate Manslaughter Bill, First Joint Report of Session 2005–06 (HC 540).
[40] The Government Reply to the First Joint Report from The Home Affairs and Work and Pensions Committees Session 2005–06 (HC 540) (Cm 6755).
[41] Corporate Manslaughter: A Regulatory Impact Assessment of the Government's Draft Bill (Home Office, March 2005).

Chapter 11: Work-Related Deaths

already subject to full investigation. . . . The proposals do not in themselves affect the operation of the current protocol for liaison between enforcement bodies involved in investigating and prosecuting work-related deaths.[42]

11.73 The Assessment considers that, in terms of investigations passed by the police to the CPS for a decision on prosecution, 'a reasonable estimate overall to be an increase of 15 extra referrals a year, representing some 5% of reported work-related deaths'.[43]

11.74 Currently, where such an investigation is launched, the resources deployed by the police, in terms of investigative manpower, are far greater than that which the HSE is able to deploy. Whether or not the impact assessment proves accurate in its estimation of about five prosecutions for the new offence per year, corporations potentially involved in a work-related death must be more likely to find themselves the subject of far more intensive investigation than was previously the case, once the new offence is in force

The provisions of the Act

11.75 The Act makes provision for a new offence of corporate manslaughter (to be called corporate homicide in Scotland), punishable by way of unlimited fine, and for this to apply to the following organizations:

(a) a corporation;
(b) a Government department other public body listed in Sch 1 to the Act;
(c) a police force;
(d) a partnership, or a trade union or employers' association, that is an employer.

11.76 Section 1 defines the offence, which is committed if the way in which an organization's activities are managed or organized causes a person's death and amounts to a gross breach of a relevant duty of care owed by the organization to the deceased. The effect of ss 2 to 7 is to identify the range of activities covered by the new offence, and to specify certain functions performed by public authorities in relation to which the offence will not apply. Section 8 outlines factors for the jury to consider when assessing an organization's culpability. Section 9 makes provision for remedial orders to be made on conviction and section 10 gives a sentencing court power to make a 'publicity order', requiring a convicted organization to publicize its conviction in a specified manner.

11.77 The Act provides that no secondary liability can arise in respect of an offence and does not provide any form of individual liability for a corporate offence: the current

---

[42] Corporate Manslaughter: A Regulatory Impact Assessment of the Government's Draft Bill (Home Office, March 2005) at para 39.
[43] ibid.

C. Manslaughter

common law of individual gross negligence manslaughter will remain the only potential liability for company directors and other individuals. Section 19 of the Act provides that where, in the same proceedings, there is a charge of corporate manslaughter and a health and safety offence alleged against the same defendant arising out of some or all of the same circumstances, then the jury may, if the interests of justice so require, be invited to return a verdict on each charge.

11.78 Sections 11 to 13 deal with the application of the offence to the Crown, the armed forces, and police forces, where a number of provisions are required to reflect the particular status of Crown bodies and police forces. Section 14 provides that a partnership is to be treated as owing whatever duties of care it would owe if it were a body corporate, that proceedings against a partnership are to be brought in the name of the partnership, not its members, and that any fine imposed following conviction is to be paid out of partnership funds. Sections 15 and 16 make further supplemental provision to ensure that rules of procedure, evidence, and sentencing apply to Crown bodies and police forces and to set out where liability will fall following machinery of Government changes or other cases where functions are transferred.

11.79 Section 17 requires the consent of the Director of Public Prosecutions to commence proceedings in England and Wales or Northern Ireland. Further sections cover general points including the procedure for amending the list of Government departments and other bodies in Sch 1, extent, and jurisdiction.

11.80 The Schedules to the Act set out the Government departments and other similar bodies to which the offence will apply and a number of minor and consequential amendments.

'Relevant duty of care'

11.81 In respect of the new offence, the organization must owe a 'relevant duty of care' to the victim. The relevant duties of care are set out in s 2. Section 2(1) requires the duty of care to arise out of certain specific functions or activities performed by the organization. The effect is that the offence will only apply where an organization owes a duty of care:

(a) to its employees or to other persons working for the organization, thus including the employer's duty to provide a safe system of work for its employees and the common law duty of care to those whose work it is able to control or direct, even though they are not formally employed by it;
(b) as occupier of premises, which is defined to include land;
(c) when the organization is supplying goods or services, and thus this includes duties owed by organizations to their customers, duties owed by transport providers to their passengers and by retailers for the safety of their products. This will also include the supply of services by the public sector, for example, NHS bodies providing medical treatment;

(d) when constructing or maintaining buildings, infrastructure or vehicles etc, or when using plant or vehicles etc. In many circumstances such duties will be owed, in any event, by an organization in relation to the supply of a service or because it is operating commercially. This provision is intended to ensure that public sector bodies are included when carrying out such activities in all circumstances;

(e) when carrying out other activities on a commercial basis. This provision is intended to ensure that activities that are not the supply of goods and services but which are still performed by companies and others commercially, such as farming or mining, are covered by the offence;

(f) by section 2(1)(d), (a subsection not in force that can only be brought into force by an order affirmed by Parliamentary resolution), a duty owed to a person in any of many various forms of custody particularized in s 1(2), for whose safety the organization is responsible.

11.82 The Explanatory Notes[44] to the Bill state:

> Many functions that are peculiarly an aspect of government are not covered by the offence because they will not fall within any of the categories of duty of care in this clause. In particular, the offence will not extend to circumstances where public bodies perform activities for the benefit of the community at large but without supplying services to particular individuals. This includes wider policy-making activities on the part of central government, such as setting regulatory standards and issuing guidance to public bodies on the exercise of their functions.

11.83 The extension of the offence to cover deaths in custody proved to be the most controversial aspect of the Act in its passage through Parliament, with the compromise being reached that such extension will only be brought into force at a later date by resolution of Parliament affirming such an order.

11.84 Most radically, and in stark contrast to the common law position that will still pertain for individual gross negligence manslaughter, s 2(5) provides that the existence of a duty of care in a particular case is a matter of law for the judge to decide. The section further provides that: 'The judge must make any findings of fact necessary to decide that question', but does not further specify any particular procedure in this respect.

Senior management

11.85 For liability to arise, the organization must be in breach of the duty of care as a result of the way in which the activities of the organization were managed or organized. This test is not linked to a particular level of management but considers how an activity was managed within the organization as a whole. However, s 1(3) stipulates that an organization cannot be convicted of the offence unless a substantial

---

[44] HL Bill 19 EN 06-07 (6 December 2006).

element of the breach lies in the way the senior management of the organization managed or organized its activities. This provision has been the subject of challenge and criticism, being seen as reintroducing something that approaches the identification doctrine.

Causation

11.86 The way in which the organizations activities were managed or organized must have caused the victim's death and the ordinary principles of causation in the criminal law will apply to determine this question, ie any management failure need not have been the sole cause of death, rather a substantial, operating, cause.

Gross breach

11.87 Finally, the management failure must amount to a gross breach of the duty of care. Section 1(4)(b) sets out the test for whether a particular breach is 'gross'. In summary, the test is whether the conduct that constitutes the breach falls far below what could reasonably have been expected. However, s 8 sets out a number of factors for a jury to take into account when considering this issue. The section provides:

(2) The jury must consider whether the evidence shows that the organization failed to comply with any health and safety legislation that relates to the alleged breach, and if so-
  (a) how serious that failure was;
  (b) how much of a risk of death it posed.
(3) The jury may also-
  (a) consider the extent to which the evidence shows that there were attitudes. Policies, systems or accepted practices within the organization that were likely to have encouraged any such failure as is mentioned in subsection (2), or to have produced tolerance of it;
  (b) have regard to any health and safety guidance that relates to the alleged breach.

11.88 These factors are not intended to be exhaustive and s 8(4) provides that the jury is also to take account of any other relevant matters.

D. Coroner's Inquests and Work-Related Deaths

Proposed reform

11.89 The coroner's inquest system is set to be radically reformed in the future. Arguably, such reform is long overdue, and undoubtedly the very process of debate, analysis, and proposal for reform, which has extended over a number of years, has added to the strain that the system currently suffers from.

11.90 The Select Committee on Constitutional Affairs published a report[45] in August 2006. The Committee's report examined the system of death certification and investigation in England and Wales, following two earlier reports (the report of the Fundamental Review of Death Certification and Coroner Services (the Luce Review)[46] and the third report of the Shipman Inquiry;[47] and following the Government's draft Bill on reform of the coroner system,[48] which was published in June 2006.

11.91 In turn, the Government responded, both to the Select Committee's report, in November 2006,[49] and thereafter to public consultation on its draft Bill, in February 2007.[50] It promised a partial rethink and legislation, as soon as parliamentary time would allow. Bearing in mind the degree of controversy surrounding the proposals and comparing the length of time that the Corporate Manslaughter and Homicide Bill has taken to progress from a similar stage, it may still be a matter of years rather than months before a new system is in place.

11.92 The draft Bill contained provisions that provided powers to coroners but no change in the funding for investigations, with local authorities still to be financially responsible for coroners. The Bill proposed that coroners should have wide powers to require, by written notice, witness statements and/or any documents relating to the relevant investigation. It was proposed that a coroner would decide if it was in the public interest to require material to be provided to him, where the coroner does not have the funds to investigate for him or herself.

The provisions of the draft Bill

11.93 The Bill proposed a new system of appointing coroners that would involve the termination of all existing appointments, the appointment of a Senior Coroner for each coroner area and the appointment by local authorities of coroners, to be under 70 and legally qualified (where not previously having been appointed).

11.94 The proposed coroner's investigation powers included a power of entry for a Senior Coroner, and, coupled with this, a proposed power of search and seizure.

11.95 The inquest proposals provided that a coroner must sit without a jury unless the Senior Coroner has reason to suspect a death in custody or a death caused

---

[45] Constitutional Affairs Select Committee. Reform of the coroners' system and death certification: eighth report of session 2005–06 (HC 902).
[46] Cmnd 5831 June 2003.
[47] Cmnd 5854 July 2003.
[48] Cmnd 6849.
[49] Cmnd 6943 2006.
[50] Department for Constitutional Affairs: Draft Coroners Bill consultation ((R)6849/07).

D. Coroner's Inquests and Work-Related Deaths

by the police, or the Senior Coroner thinks there is sufficient reason for doing so, thus removing automatic jury inquests from work-related deaths.

In its most recent response, the Government gave an indication of the matters in the draft Bill that were potentially to be the subject of amendment:[51]   **11.96**

> The issues on which we intend to have further discussions before the Bill is introduced in Parliament are:
> - reporting restrictions;
> - sharing of information between coroners and other authorities;
> - disclosure of material to bereaved people;
> - improved support for families at inquests;
> - new coroner areas;
> - improved working relationships between coroners and registrars;
> - post-mortems; and
> - medical support to the coroner service.

The Government claimed that five key reforms would be delivered by enactment of its published Bill:[52]   **11.97**

(a) bereaved people will be able to contribute more to coroner's investigations, and there will be a new appeals system if they are unhappy about decisions taken;
(b) national leadership will be introduced through a new Chief Coroner and a Coronial Advisory Council;
(c) coroner posts will be full-time, and current boundaries will be reshaped to ensure a fair distribution of work and good links with relevant agencies;
(d) coroners will have new powers to obtain information and to summon witnesses, which will ensure better investigations and inquests; and
(e) coroners will have better medical support and advice at both local and national level.

Concerning the 'Use of juries and Deaths at Work', the Government stated:[53]   **11.98**

> Mixed views were received on this issue. Many coroners supported the Bill's proposal to have discretion to summon a jury in any case where the public interest demands. However, a number of trade unions, including the Trades Union Congress (TUC) itself, objected to the removal of juries from the mandatory category as they felt that it would take away the necessary transparency and accountability from the process. Varying views were received on the proposal to reduce the number of jurors. We intend to retain the statutory criteria, under the current law, for jury inquests,

---

[51] Department for Constitutional Affairs: Draft Coroners Bill consultation ((R)6849/07) para 9.
[52] ibid para 3.
[53] ibid para 7.

although we believe that a maximum of 9 jurors is sufficient for the purposes of a coroner's inquest.

The current system

11.99 The current legislation governing the role of the coroner is mostly comprised by the Coroners Act 1988 and the Coroners Rules 1984 SI 1984/552.

General principles

11.100 By the Coroners Act 1988 s 8(1) a coroner is required to hold an inquest where he has reasonable cause to suspect that a deceased:

(a) has died a violent or unnatural death; or
(b) has died a sudden death of which the cause is unknown; or
(c) has died in prison or in such other circumstances as to require an inquest under any other Act.

11.101 A violent death is one that is caused by any sort of traumatic event, accidental or deliberate, self-induced or otherwise. This involves no requirement that the violence was an intended act, nor is there a limit of time in which a resultant death must occur.

11.102 An unnatural death has been suggested to be one which was 'wholly or partly caused, or accelerated, by any act, intervention or omission other than a properly executed measure intended to prolong life'.[54]

Jury inquests

11.103 Whilst the vast majority of inquests are heard by the coroner sitting alone, inquests examining a work-related death will be held before a jury because of the terms of the Coroners Act 1988 s 8(3)(c), which requires that a coroner must summon a jury where he has reason to suspect:

> that the death was caused by an accident, poisoning or disease notice of which is required to be given under any Act to a government department, to any inspector or other officer of a government department or to an inspector appointed under section 19 of the Health and Safety at Work etc. Act 1974.

11.104 The terms of the Reporting of Injuries, Diseases and Dangerous Occurrences Regulations 1995 SI 1995/3163 (RIDDOR)[55] now require the reporting of all work-related accidents to the HSE and ensure that such a death falls within the terms of the Coroners Act 1988 s 8(3)(c)(1).[56]

---

[54] Herbert H. Pilling, *Medicine, Science & the Law*, April 1967, quoted in C Dorries, *Coroners Courts, a Guide to Law and Practice* (Blackstone Press, 1999).
[55] See paras 2.05–2.24.
[56] Unless death arises from an industrial disease but occurs after the employee has ceased to be employed.

D. Coroner's Inquests and Work-Related Deaths

11.105 The Coroners Act 1988 s 8(3)(d) further requires a coroner to summon a jury where a death has occurred in circumstances the continuance or possible recurrence of which is prejudicial to the health or safety of the public or any section of the public. By the Coroners Rules 1984 SI 1984/552 r 43, the coroner has the power to report the circumstances of any case to an appropriate authority, such as the HSE, with a view to ensuring that similar fatalities do not occur in the future.

The scope of an inquest

11.106 An inquest is a fact-finding exercise and not a method of apportioning guilt. It has been said that the procedures and rules of evidence that are suitable for one, are unsuitable for the other.[57]

11.107 Its purpose is a limited one and the coroner is not permitted to allow an inquest to stray beyond those narrow limits. Interested parties to an inquest are only permitted to ask questions of witnesses that are relevant to the statutory purpose of the inquest. It has also been said that it must never be forgotten that in an inquest there are no parties, there is no indictment, there is no prosecution, there is no defence, and there is no trial; there is simply an attempt to establish facts through an inquisitorial process.

11.108 However, recent cases concerned with inquests following deaths in custody have stressed the importance of the European Convention on the Protection of Human Rights and Fundamental Freedoms 1950 Art 2, where there is any potential involvement of the state in the death. The requirement is an enhanced form of inquest, known, by the House of Lords' decision that established the basis and ambit of such inquests, as a *Middleton*-type inquest.

11.109 Article 2 provides that 'Everyone's right to life shall be protected by law', imposing on states substantive obligations not to take life without justification and also to establish a framework of laws, precautions, procedures, and means of enforcement which, to the greatest extent reasonably practicable, protect life. Where an act or omission of the state may have been the cause of a person's death, there must be independent, thorough, and prompt investigation, which must be capable of leading to the identification and, where appropriate, the punishment of those responsible. There must be a sufficient element of public scrutiny of the investigation or its results to ensure accountability, and the next-of-kin must be involved to an extent necessary to safeguard their legitimate interests.[58] States must act of their own motion and cannot leave it to the next-of-kin to initiate a complaint or take proceedings.

---

[57] See *R v HM Coroner for North Humberside ex p Jamieson* (1993) 158 JP 1.
[58] See *R (on the application of Middleton) v West Somerset Coroner* [2004] UKHL 10, [2004] 2 All ER 465.

11.110 In *R (on the application of Middleton) v West Somerset Coroner* [2004] UKHL 10, [2004] 2 All ER 465, the issue was raised whether the current coroners' regime, under the Coroners Act 1988 and the Coroners Rules 1984,[59] met the requirements of a properly conducted official investigation into a death required by Art 2. Three issues were considered and answered in the single opinion of the Judicial Committee delivered by Bingham LJ.

11.111 The first was what, if anything, the Convention required (by way of verdict, judgment, findings, or recommendations) of a properly conducted official investigation into a death involving or possibly involving, a violation of Art 2, ie those deaths that may have been caused by an act or omission of the state. It was held that an inquest, being the means by which the state ordinarily discharged its procedural obligation to investigate under Art 2 of the Convention,[60] ought ordinarily to culminate in an expression, however brief, of the jury's conclusion on the disputed factual issues at the heart of the case.

11.112 The second issue was whether the regime for holding inquests met those requirements of the Convention: it was held that cases in which the inquest verdict did not express the jury's factual conclusion on the events leading up to the death, did not meet the requirements of the Convention.[61]

11.113 The third issue was whether the current regime governing the conduct of inquests should be revised, and if so, how. It was held that the only change needed to the current regime was to interpret the word 'how', in the phrase 'how ... the deceased came by his death'[62] found in the Coroners Act 1988 s 11(5)(b)(ii) and the Coroners Rules 1984[63] r 36(1)(b) as meaning not simply 'by what means' but 'by what means and in what circumstances'. Further, it was held that a change of approach was not required in all cases, where the traditional short form verdict would be quite satisfactory.

11.114 Where the terms of Art 2 did call for a change of approach, it was for the coroner, in the exercise of his discretion, to decide how best to elicit the jury's conclusion on the central issue or issues. The opinion of the Judicial Committee was that this could be done by inviting an expanded form of verdict, with a narrative form of verdict, in which the jury's factual conclusions were briefly summarized, or by inviting the jury's answers to factual questions put by the coroner.

---

[59] SI 1984/552.
[60] See also *Jordan v UK* (2001) 11 EHRC 1 and *Keenan v UK* (2001) 10 EHRC 319.
[61] See also *R v North Humberside and Scunthorpe Coroner ex p Jamieson* [1994] 3 All ER 972.
[62] [2004] UKHL 10, para 35.
[63] SI 1984/552.

D. Coroner's Inquests and Work-Related Deaths

The opinion set out how, if the coroner invited a narrative verdict or answers to questions, he might find it helpful to direct the jury with reference to some of the following matters: **11.115**

(a) where and when the death took place;
(b) the cause or causes of such death;
(c) the defects in the system which contributed to the death;
(d) any other factors relevant to the circumstances of the death.

Following the ruling in *Middleton*,[64] the requirements imposed upon the inquest system to ensure compliance with the terms of Art 2 in such situations have influenced, and continue to influence, the conduct by coroners of inquests generally, irrespective of whether any issue in relation to the right to life arises. **11.116**

Matters to be ascertained at an inquest

Rule 36 of the Coroners Rules 1984[65] provides: **11.117**

(1) The proceedings and evidence at an inquest shall be directed solely to ascertaining the following matters, namely—
   (a) who the deceased was;
   (b) how, when and where the deceased came by his death;
   (c) the particulars for the time being required by the Registration Acts to be registered concerning the death.
(2) Neither the coroner nor the jury shall express any opinion on any other matters.

Rule 42 of the Coroners Rules 1984[66] provides: **11.118**

No verdict shall be framed in such a way as to appear to determine any question of—

(a) criminal liability on the part of a named person, or civil liability;
(b) civil liability.

The obvious conflict between the duty to inquire into acts and omissions causing the death and the restrictions imposed by r 42 has been the subject of judicial guidance. In *R v Surrey Coroner ex parte Campbell* [1982] QB 661 it was said that: **11.119**

Such conflict as may in any given circumstances appear to arise between [r 42] and the duty to inquire 'how' must be resolved in favour of the statutory duty to inquire, whatever the consequences of this may be.

In *R v HM Coroner for North Humberside and Scunthorpe ex parte Jamieson* (1994) 3 All ER 972, Lord Bingham (then Master of the Rolls) held that whilst it may be accepted that in case of conflict the statutory duty to ascertain how the deceased **11.120**

---

[64] *R (on the application of Middleton) v West Somerset Coroner* [2004] UKHL 10, [2004] 2 All ER 465.
[65] SI 1984/552.
[66] ibid.

came by his death must prevail over the prohibition in the Coroners Rules 1984 r 42, nevertheless:

> the scope for conflict is small. Rule 42 applies, and applies only, to the verdict. Plainly the coroner and the jury may explore facts bearing on criminal and civil liability. But the verdict may not appear to determine any question of criminal liability on the part of a named person nor any question of civil liability.

11.121 In *R (on the application of Middleton) v West Somerset Coroner* [2004] UKHL 10, [2004] 2 All ER 465,[67] the Judicial Committee's opinion was that the rules did not preclude conclusions of fact as opposed to expressions of opinion. In respect of the requirements of ECHR Art 2, it was held that however an inquest jury's factual conclusion was conveyed, r 42 of the Coroners Rules 1984[68] should not be infringed.

Persons whose conduct may be called into question

11.122 Rule 24 of the Coroners Rules 1984 requires a coroner to give notice of the inquest to any person 'whose conduct is likely to be called into question' at the inquest.

Entitlement to examine witnesses

11.123 Rule 20 of the Coroners Rules 1984 provides:

> (1) Without prejudice to any enactment with regard to the examination of witnesses at an inquest, any person who satisfies the coroner that he is within paragraph (2) shall be entitled to examine any witness at an inquest either in person or by [an authorized advocate as defined by s 119(1) of the Courts and Legal Services Act 1990]:
> Provided that—
> (a) the chief officer of police, unless interested otherwise than in that capacity, shall only be entitled to examine a witness by [such an advocate];
> (b) the coroner shall disallow any question which in his opinion is not relevant or is otherwise not a proper question.
> (2) Each of the following persons shall have the rights conferred by paragraph—
> (a) a parent, child, spouse and any personal representative of the deceased;
> (b) any beneficiary under a policy of insurance issued on the life of the deceased;
> (c) the insurer who issued such a policy of insurance;
> (d) any person whose act or omission or that of his agent or servant may in the opinion of the coroner have caused, or contributed to, the death of the deceased;
> (e) any person appointed by a trade union to which the deceased at the time of his death belonged, if the death of the deceased may have been caused by an injury received in the course of his employment or by an industrial disease;

---

[67] See para 9.109.
[68] SI 1984/552.

D. Coroner's Inquests and Work-Related Deaths

    (f) an inspector appointed by, or a representative of, an enforcing authority, or any person appointed by a government department to attend the inquest;
    (g) the chief officer of police;
    (h) any other person who, in the opinion of the coroner, is a properly interested person.

11.124 The term in r 20(2)(h), 'properly interested person', is not further defined in the Coroners Rules 1984.[69] In relation to work-related deaths, r 20(2)(d) and (h) allow sufficient scope to ensure that an employer of the deceased or any employer whose conduct of his undertaking may have contributed to the death will fall within the terms of the rule.

Witness statements and evidence at an inquest

11.125 An inquest involving the hearing of witnesses before a jury is likely to occur only after an investigation into the death has been made by coroner's officers and police officers and/or an initial investigation by a health and safety inspector. It is for the coroner to decide which witnesses are to be called so as to ensure that at the inquest the relevant facts are 'fully, fairly and fearlessly investigated'.[70]

11.126 The purpose of the police and HSE providing copies of material obtained during an investigation concerning the death is to assist the coroner in carrying out his statutory functions and to avoid the necessity for duplication of effort by the coroner. In *Peach v Metropolitan Police Commissioner* (1986) 2 All ER 129, Fox LJ said:

> As a matter of sensible public administration it seems essential that the coroner should have the material obtained by the police so that he can decide which witnesses to call and investigate the matter generally. He could, of course, conduct his own enquiry *de novo*. But that would, in any event, be wasteful of public resources.[71]

The privilege against self-incrimination

11.127 Rule 22 of the Coroners Rules 1984 provides:

    (1) No witness at an inquest shall be obliged to answer any question tending to incriminate himself.
    (2) Where it appears to the coroner that a witness has been asked such a question, the coroner shall inform the witness that he may refuse to answer.

11.128 The rule reproduces the common law privilege against self-incrimination. It does not represent a right not to give evidence or prevent a person from being compelled to be a witness but is a protection in relation to specific questions.

---

[69] SI 1984/552.
[70] See *R v HM Coroner for North Humberside and Scunthorpe ex p Jamieson* (1994) 3 All ER 972.
[71] (1986) 2 All ER 129 at 138.

Disclosure and documentary evidence

11.129 By Coroners Rules 1984 r 57 a coroner must supply to any person who is, in the opinion of the coroner, a properly interested person a copy of any report of a post-mortem examination or special examination or of any notes of evidence or of any document put in evidence at an inquest.

11.130 Rule 37 provides:

(1) Subject to the provisions of paragraphs (2) to (4), the coroner may admit at an inquest documentary evidence relevant to the purposes of the inquest from any living person which in his opinion is unlikely to be disputed, unless a person who in the opinion of the coroner is within Rule 20(2) objects to the documentary evidence being admitted.

(2) Documentary evidence so objected to may be admitted if in the opinion of the coroner the maker of the document is unable to give oral evidence within a reasonable period.

(3) Subject to paragraph (4), before admitting such documentary evidence the coroner shall at the beginning of the inquest announce publicly—
  (a) that the documentary evidence may be admitted, and
  (b) (i) the full name of the maker of the document to be admitted in evidence, and
      (ii) a brief account of such document, and
  (c) that any person who in the opinion of the coroner is within Rule 20(2) may object to the admission of any such documentary evidence, and
  (d) that any person who in the opinion of the coroner is within Rule 20(2) is entitled to see a copy of any such documentary evidence if he so wishes.

(4) If during the course of an inquest it appears that there is available at the inquest documentary evidence which in the opinion of the coroner is relevant to the purposes of the inquest but the maker of the document is not present and in the opinion of the coroner the content of the documentary evidence is unlikely to be disputed, the coroner shall at the earliest opportunity during the course of the inquest comply with the provisions of paragraph (3).

(5) A coroner may admit as evidence at an inquest any document made by a deceased person if he is of the opinion that the contents of the document are relevant to the purposes of the inquest.

(6) Any documentary evidence admitted under this Rule shall, unless the coroner otherwise directs, be read aloud at the inquest.

11.131 Coroners habitually give advance notice and supply an advance copy of any witness statement which it is proposed to use as documentary evidence and seek the consent of interested persons prior to the start of the inquest in order to prevent disruption and adjournment of an inquest through a late objection pursuant to Coroners Rules 1984 r 37(3).

11.132 Statements from witnesses from whom the coroner intends to hear evidence in person at the inquest are not required to be disclosed to parties, and the statements are considered to have been provided for the coroner's own use. This is an area of controversy, where the issue of disclosure has been influenced by the

D. Coroner's Inquests and Work-Related Deaths

interpretation given to European Convention on the Protection of Human Rights and Fundamental Freedoms 1950 Art 2.[72] Increasingly, disclosure of such witness statements is being provided by coroners to interested persons in advance of inquests. In general, coroners have adopted the practice of providing a list of prospective witnesses to interested persons, together with copies of the statements. At the same time the coroner invites the interested persons to indicate which statements can be used as documentary evidence pursuant to Coroners Rules 1984[73] r 37.

Coroner's disclosure and work-related deaths

11.133 Paragraph 10.2 of the Work Related Death Protocol provides that:

> Where the CPS is prosecuting, and HSE, the local authority or other enforcing authority has submitted documents or a report to the coroner about a work-related death, the CPS and the police shall also be given a copy. Similarly, where an enforcing authority is prosecuting, and the police or CPS has submitted documents or a report to the coroner about a work-related death, the enforcing authority shall also be given a copy. In all cases, documents or reports may not be disclosed to any party without the consent of the party that originally submitted them.

11.134 Both police and health and safety inspector's reports include the officers' conclusions, opinions, and recommendations concerning prosecution. These are prepared in the course of the investigation and attract both legal privilege and public interest immunity from disclosure. Where a separate report has been drafted for a coroner, it is often supplied expressly on the basis that it will not be disclosed to any party, because to do so may prejudice any future criminal proceedings. In such circumstances, it would not be open to the coroner, nor could an interested party compel him, to disclose such reports to interested persons.

11.135 In any subsequent criminal proceedings, such a report that has been prepared for and supplied to a coroner, arguably, will fall within a class of documentation for which a public interest immunity from disclosure can be claimed.

11.136 Copies of statements and documentary exhibits obtained during the investigation are usually attached to officers' reports to a coroner. This is to enable him to carry out his functions and to prevent unnecessary duplication of effort. Coroners enquiring into work-related deaths often request all statements obtained by the police or HSE during an investigation. In work-related death cases where an investigation has established that no manslaughter charge is proposed, then the inquest will almost always occur prior to any decision to prosecute by the HSE.

---

[72] See paras 11.107–11.109.
[73] SI 1984/552.

11.137 Where a coroner requests all obtained statements from the HSE, this tends to be complied with, but an assurance is sought from the coroner that disclosure will not be made to interested parties without the express consent of the HSE. In this way, the coroner can decide which statements are relevant to the inquest and then revert to the HSE to discuss disclosure. Clearly, the HSE may be concerned that further disclosure could prejudice future prosecution and competing public interests can exist.[74] Difficulty can arise where statements from witnesses contain material relevant to a potential future prosecution that is beyond and irrelevant to evidence to be given at an inquest.

11.138 In such cases there may clearly be an overriding public interest in non-disclosure of such a statement, or parts of it, to interested persons prior to an inquest. In this respect, there are a number of authorities that establish that material obtained during the course of an investigation is confidential and should only be disclosed in furtherance of the purposes for which it was obtained.[75]

11.139 Disclosure of statements and material obtained by an inspector using his HSWA 1974 s 20 powers is restricted by the terms of HSWA 1974 s 28. However, the exceptions to the rule against disclosure contained in HSWA 1974 ss 28(3)(e) and 28(7) make lawful the disclosure of material obtained pursuant to HSWA 1974 s 20 'for the purposes of any legal proceedings' which must include inquests before coroners.

HSWA 1974 s 20(2)(j) statements and inquests

11.140 Compulsory statements obtained from witnesses by inspectors using their power under HSWA 1974 s 20(2)(j) may be supplied to a coroner. As such, the statement may become documentary evidence pursuant to Coroners Rules 1984[76] r 37 or the witness may be compelled to give evidence at an inquest.

11.141 HSWA 1974 s 20(7)1 provides that no answer obtained from a person by an inspector using his power under HSWA 1974 s 20(2)(j) 'shall be admissible in evidence against that person'. Evidence is admitted at an inquest for the purposes of assisting the coroner and jury in fulfilling their statutory functions and is not admitted 'against' any person. It follows that whilst such a person could invoke his right against self-incrimination in response to any question should he give evidence in accordance with a s 20(2)(j) statement, or confirm the truth of its contents, then such evidence itself could be admissible, subject to the discretion under

---

[74] See *Frankson v Home Office; Johns v Home Office* [2003] EWCA Civ 655 for a consideration of competing public interests involving disclosure of confidential material.
[75] See *Hellewell v Chief Constable of Derbyshire* [1995] 4 All ER 473; *Bunn v British Broadcasting Corporation* [1998] 3 All ER 552; *Marcel v Metropolitan Police Commissioner* [1992] 1 All ER 72; and *Taylor v SFO* [1998] 4 All ER 801.
[76] SI 1984/552.

D. Coroner's Inquests and Work-Related Deaths

the Police and Criminal Evidence Act 1984 s 78, in future proceedings against him. The reading of such a HSWA 1974 s 20(2)(j) statement, or its admission as documentary evidence at an inquest, would not affect the operation of HSWA 1974 s 20(7)[77] (the prohibition on use in evidence of such a statement in proceedings against the maker or his spouse), nor provide a source of evidence that could be admitted against the maker of the statement in subsequent proceedings.

Verdicts

11.142 There are a number of different verdicts that can be returned by juries at inquests consideration of which are beyond the scope of this work. Two verdicts are particularly relevant in inquests involving work-related deaths.

11.143 A verdict of accidental death is most often a source of confusion. It has been defined as 'the unexpected result of a deliberate act'[78] but would equally apply to the unexpected result of a non-deliberate act. The verdict encompasses genuine mischance and deaths which are wholly the result of another's carelessness. The verdict is neutral in terms of liability and does not affect civil remedies. It requires proof on the balance of probabilities.

11.144 Unlawful killing is a verdict returnable only if the jury finds the necessary facts proven beyond reasonable doubt. The jury is prohibited from returning a verdict that accuses a named person of criminal liability but not from stating that the deceased was unlawfully killed. In work-related death cases, such a verdict arises where the jury has been satisfied that the death has arisen as a result of 'gross' negligence by a person owing a duty of care to the deceased. The returning of such a verdict in a work-related death case will result in a review by the CPS concerning manslaughter.

Coroners Act 1988 s 16 and adjourned inquests

11.145 Section 16 of the Coroners Act 1988 provides:

(1) If on an inquest into a death the coroner before the conclusion of the inquest—
  (a) is informed by the [designated officer for] a magistrates' court under section 17(1) below that some person has been charged before a magistrates' court with—
    (i) the murder, manslaughter or infanticide of the deceased;
    (ii) an offence under [section 1[, 2B][, 3ZB] or 3A of] the Road Traffic Act 1988 (dangerous driving[, careless driving][, unlicensed, disqualified or uninsured drivers][or careless driving when under the influence of drink or drugs)] committed by causing the death of the deceased; . . .
    (iii) an offence under section 2(1) of the Suicide Act 1961 consisting of aiding, abetting, counselling or procuring the suicide of the deceased; or

---

[77] See paras 2.132–2.146.
[78] *Jervis on Coroners* (11th edn, para 13–24).

(iv) an offence under section 5 of the Domestic Violence, Crime and Victims Act 2004 (causing or allowing the death of a child or vulnerable adult); or]

(b) is informed by the Director of Public Prosecutions that some person has been charged before examining justices with [sent for trial for] an offence (whether or not involving the death of a person other than the deceased) alleged to have been committed in circumstances connected with the death of the deceased, not being an offence within paragraph (a) above, and is requested by the Director to adjourn the inquest,

then, subject to subsection (2) below, the coroner shall, in the absence of reason to the contrary, adjourn the inquest until after the conclusion of the relevant criminal proceedings and, if a jury has been summoned, may, if he thinks fit, discharge them.

(2) The coroner—

(a) need not adjourn the inquest in a case within subsection (1)(a) above if, before he has done so, the Director of Public Prosecutions notifies him that adjournment is unnecessary; and

(b) may in any case resume the adjourned inquest before the conclusion of the relevant criminal proceedings if notified by the Director that it is open to him to do so.

(3) After the conclusion of the relevant criminal proceedings, or on being notified under paragraph (b) of subsection (2) above before their conclusion, the coroner may, subject to the following provisions of this section, resume the adjourned inquest if in his opinion there is sufficient cause to do so.

11.146 The effect of this provision is that whenever the police conduct an investigation into a work-related death pursuant to the protocol,[79] ie where there is the potential for a manslaughter prosecution, then a file will have been submitted to the CPS, the coroner will be notified and the inquest will be formally opened and adjourned pending the outcome of the police investigation and CPS prosecution.

11.147 No adjourned inquest before a jury with witnesses will be resumed until either a decision is made not to prosecute for manslaughter or such proceedings are completed. The Coroners Act 1988 s 16(7) provides that where a coroner resumes an inquest which has been adjourned in compliance with subs (1) then the finding of the inquest as to the cause of death must not be inconsistent with the outcome of the relevant criminal proceedings.

11.148 A decision not to prosecute for manslaughter may be arrived at only after what may turn out to be a lengthy investigation. Paragraph 10.3 of the protocol provides:

Where the CPS have reviewed the case and decided not to prosecute, HSE, the local authority or other enforcing authority will await the result of the coroner's inquest before charging any health and safety offences, unless to await the result of the

---

[79] The Work Related Death Protocol; see paras 11.08–11.87.

D. Coroner's Inquests and Work-Related Deaths

coroner's inquest would prejudice the case. Where, following an inquest . . . it is necessary for the CPS to review, or re-review the case, HSE, local authority or other enforcing authority will wait until the review by the CPS has been completed before instigating its own proceedings.

11.149 The need to await the outcome of the inquest before instituting proceedings for a health and safety offence was highlighted by the decision in *R v Beedie*[80] but appears to be a requirement beyond good practice. The starting point is the decision of Devlin J in *Re Beresford* (1952) 36 Cr App R 1. In the course of giving judgment in that case, Devlin J said:

> It is common-place that in all cases of sudden death the law requires that an inquest should be held and, as a general rule, the inquest should be concluded before any other step is taken. There is one exception to that rule, provided by statute and that is when a charge of murder or manslaughter is preferred. In any other case the question of criminal responsibility for the death should be decided at the inquest before any lesser charge arising out of the same circumstances is tried. It is, I believe, a general practice, and it is certainly the proper practice, that the coroner should hold and complete the inquest unless he is informed that a charge of murder or manslaughter has been preferred. Conversely, the police should not prefer any lesser charge, or, if for some reason such a charge is preferred, it should not be heard and determined, until after the inquest is concluded. If this practice is followed, the possibility of two tribunals arriving at incongruous results will be lessened.[81]

11.150 That decision in *Beresford* predated the amendment to the law introduced by s 16 of the Coroners Act 1988. Section 16 expressly relates to cases before examining justices and the Crown Court, but does not apply to summary-only matters. In *Smith v Director of Public Prosecutions and Morris* [2000] RTR 36, a case concerning a summary-only matter of careless driving, Lord Justice Rose stated:

> In my judgment, there is no doubt that *Beresford* is still good law in relation to cases outwith s 16 of the 1988 Act. There is no doubt that, as a matter of practice, for the reasons explained by Devlin J, justices ought not to be proceed with a summary-only case . . . until the inquest has been held.

11.151 Similar dicta concerning the desirability of awaiting a coroner's inquest before a final decision upon criminal proceedings emerge from the Divisional Court decision in *R (on the application of Stanley) v Inner North London Coroner* [2003] EWHC 1180 (Admin).

11.152 The net effect can often be an extended period of delay prior to the commencement of proceedings for health and safety offences in cases involving work-related deaths. A number of such health and safety prosecutions have followed a re-review by the CPS after an inquest has returned an unlawful killing verdict.

---

[80] [1997] 2 Cr App R 167.
[81] (1952) 36 Cr App R 1 at p 2.

12

PRINCIPAL HEALTH AND SAFETY REGULATIONS

A. Health and Safety Regulations	12.01		The nature of the duty	12.57
Introduction	12.01	D.	**Manual Handling**	
B. **The Management of Health and Safety at Work Regulations 1999**	12.09		**Operations Regulations 1992**	12.60
			Introduction	12.60
Risk assessment	12.12		Relationship between Manual Handling Operations Regulations 1992 and other provisions	12.62
Particular requirements for risk assessments under other regulations	12.16		The primary duties under Manual Handling Operations Regulations 1992	12.63
What does a risk assessment entail?	12.20			
'Suitable and sufficient'	12.28			
Relevance of the size of the undertaking in question	12.29		What is a manual handling operation?	12.70
Model or generic risk assessments	12.30	E.	**Work Equipment: Provision and Use of Work Equipment Regulations 1998**	12.77
Identifying who might be affected by the activities in question	12.31			
Young people	12.34		Introduction	12.77
Pregnant women	12.35		Suitability of work equipment	12.80
Principles of prevention	12.36		Risk assessment and the duty under Provision and Use of Work Equipment Regulations 1998 reg 4	12.84
Continuing review and surveillance	12.37			
Information and training	12.41			
Procedures for evacuation in the event of imminent danger	12.45		Ergonomics	12.87
			Maintenance of work equipment	12.88
The operation and effect of the Management of Health and Safety at Work Regulations 1999 reg 21	12.46		Inspection of work equipment	12.97
			Keeping records of inspections	12.100
			Extent and frequency of the inspections	12.103
C. **The Workplace (Health, Safety and Welfare) Regulations 1992**	12.47		Competence of those who carry out inspections	12.106
Introduction	12.47		Information and instructions	12.108
Who owes the duty?	12.49		Training	12.110
The extent of the duty	12.50		Dangerous parts of machinery	12.112
To whom is the duty owed?	12.52			
What is a workplace?	12.53		Dangerous part	12.115

Protection against specified hazards	12.116	
F. **Substances Hazardous to Health: the COSHH Regulations 2002**	12.120	
The Control of Substances Hazardous to Health Regulations	12.120	
'Substance hazardous to health'	12.122	
Who owes duties under the Regulations?	12.123	
The duties	12.124	
G. **Construction**	12.139	
Introduction	12.139	
Work at Height Regulations 2005	12.140	
Construction (Design and Management) Regulations 2007	12.146	
Contractors	12.187	
H. **Asbestos**	12.192	
Introduction	12.192	
Duty-holders in premises	12.194	
Duties on employers	12.199	
I. **Gas: Gas Safety (Installation and Use) Regulations 1998**	12.211	
Gas Safety (Installation and Use) Regulations 1998	12.211	
Competent person	12.215	
Work	12.216	
Gas Safety (Installation and Use) Regulations 1998 reg 26	12.219	
Landlords	12.220	
J. **Electricity: Electricity at Work Regulations 1989**	12.225	
Systems, work activities, and protective equipment	12.228	
Protective equipment under the Electricity at Work Regulations 1989 reg 4(4)	12.233	
Electricity at Work Regulations 1989 regs 5–12	12.234	

A. Health and Safety Regulations

Introduction

12.01 A significant development in health and safety regulation came with the introduction in 1992 of six different sets of regulations, known as the 'six pack', which sought to give effect to the Directives that had emanated from the European Commission following the Commission's Third Programme on health and safety at work. Since 1992 almost all health and safety regulations enacted in the UK have been passed to implement EC Directives, and the original 'six pack' has, in part, been replaced by fresh regulations.

12.02 The regulations were introduced in part under the Health and Safety at Work etc. Act 1974 (HSWA 1974) s 15, and consequently provisions of that Act are relevant to the interpretation of the regulations. See Chapter 5 for a consideration of many of the terms used in HSWA 1974 ss 2–7 which are similarly employed by the regulations. See paragraphs 1.07–1.09 for a consideration of criminal liability for breach of regulations and paragraphs 1.45–1.59 for the relevance of European Directives to interpretation.

12.03 The HSE's publication, 'Health and Safety Regulation—a short guide' (HSC13) describes how HSWA 1974 and the duties in the Management of Health and Safety at Work Regulations 1999:

> are goal-setting . . . and leave employers freedom to decide how to control risks which they identify. Guidance and Approved Codes of Practice give advice. But some risks

A. Health and Safety Regulations

are so great, or the proper control measures so costly, that it would not be appropriate to leave employers discretion in deciding what to do about them. Regulations identify these risks and set out specific action that must be taken. Often these requirements are absolute—to do something without qualification by whether it is reasonably practicable.

Thus, regulations fall into two categories: those with general application across all workplaces, and those addressed to specific hazards arising in particular industries or workplaces. Generally, the former tend to be drafted in 'goal-setting' terms, setting out what ought to be done, and not how it should be done; while the latter are often *prescriptive*, that is providing mandatory minimum requirements in relation to a specific matter. **12.04**

By HSWA 1974 s 16 the Health and Safety Commission (HSC) is empowered to approve codes of practice with the consent of the appropriate Secretary of State. Such Approved Codes of Practice (ACOPs) give advice on how to comply with the law. While the codes represent good practice and have a legal status, they do not themselves impose legally enforceable duties. HSWA 1974 s 17 provides that while a failure on the part of any person to observe any provision of an approved code of practice shall not of itself render him liable to any civil or criminal proceedings, failure to comply with the provisions may be taken by a court in criminal proceedings as evidence of a failure to comply with the requirements of HSWA 1974 or of regulations to which the Code relates, unless it can be shown by the defendant that those requirements were complied with in some equally effective way. **12.05**

Beneath the layers of Regulations and Approved Codes of Practice comes guidance. The Health and Safety Executive (HSE) publishes guidance[1] on a range of subjects, its primary purpose being to assist in the understanding of and compliance with health and safety law. It is not compulsory to follow such guidance; but the HSE expresses the view that if duty-holders do follow guidance they will normally be doing enough to comply with the law. **12.06**

In respect of all of the principal health and safety regulations considered below, the HSE publishes books in a form which sets out each individual regulation, followed by a numbered paragraph marked in bold with the terms of any provision of an Approved Code of Practice for that particular regulation and thereafter, sequentially numbered, guidance relevant to that particular regulation and Code provision. This guidance tends to expand or explain the provision of the Code. **12.07**

In the remainder of this chapter, where reference is made to the terms of any Approved Code of Practice and guidance, this is described as a reference to the **12.08**

---

[1] See paras 1.26–1.27 for HSE publications.

Approved Code of Practice for short, and those terms that appear only in guidance are not further distinguished from the Code.

B. The Management of Health and Safety at Work Regulations 1999

12.09 The most important element of the 'six-pack' is the Management of Health and Safety at Work Regulations 1999 SI 1999/3242 (reproduced at Appendix E), which replaced the like-named 1992 regulations. They are of wide application, applying to all areas of activity except for ships.

12.10 As the HSE publication, 'Health and Safety Regulation—a short guide' (HSC13), describes:

> The Act[2] sets out the general duties which employers have towards employees and members of the public, and employees have to themselves and to each other...
> The Management of Health and Safety at Work Regulations 1999 SI 1999/3242 generally make more explicit what employers are required to do to manage health and safety under the Health and Safety at Work Act. Like the Act, they apply to every work activity.

12.11 Breach of the the Management of Health and Safety at Work Regulations 1999 SI 1999/3242 now gives rise to civil liability in the majority of cases.[3] The Regulations largely implement the Framework Directive which, as the name suggests, is considered to be the overall framework of EU legislation concerning health and safety. The Framework Directive provided for the creation of a number of individual Directives, known as the 'daughter Directives', dealing with specific areas of regulation. (See Chapter 1.)

Risk assessment

12.12 The major provision of the Management of Health and Safety at Work Regulations 1999 SI 1999/3242 is the imposition of a duty upon employers and the self-employed to make a suitable and sufficient assessment of the risks faced by employees at work, and by other persons, arising out of the conduct of the employer's undertaking. The process of risk assessment is considered to be the cornerstone of good health and safety management. Failure to comply with the duty to carry out a suitable and sufficient risk assessment is one of the most commonly prosecuted health and safety offences.

---

[2] Health and Safety at Work etc. Act 1974 (HSWA 1974).
[3] See the Management of Health and Safety at Work and Fire Precautions (Workplace) (Amendment) Regulations 2003 SI 2003/2457.

B. The Management of Health and Safety at Work Regulations 1999

Regulation 3 provides: **12.13**

(1) Every employer shall make a suitable and sufficient assessment of—
   (a) the risks to the health and safety of his employees to which they are exposed while they are at work; and
   (b) the risks to the health and safety of persons not in his employment arising out of or in connection with the conduct by him of his undertaking,

for the purpose of identifying the measures he needs to take to comply with the requirements and prohibitions imposed upon him by or under the relevant statutory provisions.

(2) Every self-employed person shall make a suitable and sufficient assessment of—
   (a) the risks to his own health and safety to which he is exposed while he is at work; and
   (b) the risks to the health and safety of persons not in his employment arising out of or in connection with the conduct by him of his undertaking,

for the purpose of identifying the measures he needs to take to comply with the requirements and prohibitions imposed upon him by or under the relevant statutory provisions.

The duty to make a risk assessment is a strict one. The 'suitable and sufficient' element applies only to the nature and quality of the assessment, not the duty to conduct one in the first place. For example, where, as commonly occurs, employees are in the habit of carrying out a particular task in an unsafe manner, without their employer's knowledge, it is unlikely that this lack of knowledge will provide the employer with any defence to a charge of breaching reg 3(1). **12.14**

By reg 3(6), where the employer employs five or more people, he must record the significant findings of the risk assessment and any group of his employees it identifies as being especially at risk. **12.15**

Particular requirements for risk assessments under other regulations

Other regulations also contain requirements for risk assessment specific to the hazards and risks they cover. For example, the Manual Handling Operations Regulations 1992 SI 1992/2793 reg 4(1)(b)(i) requires that an employer who is unable to avoid his employees carrying out some degree of lifting, must make a suitable assessment of the activity. **12.16**

The Personal Protective Equipment at Work Regulations 1992 SI 1992/2966 reg 6 requires an assessment to be made which includes both risks to health and safety which have not been avoided by other means, and a definition of the characteristics which personal protective equipment must have in order to be effective against those risks, and a comparison of the equipment available which has the characteristics referred to. **12.17**

Under the Control of Substances Hazardous to Health Regulations 2002 SI 2002/2677 reg 6, an employer may not carry out any work which exposes his **12.18** employees to substances hazardous to health unless he has first made a suitable and sufficient assessment of the risks created by that work.

12.19 An assessment made for the purpose of other regulations will partly cover the obligation to make assessments under the Management of Health and Safety at Work Regulations 1999, and will be helpful in identifying the areas it should address. Where an employer has already carried out an assessment under other regulations, they need not repeat those assessments as long as they remain valid; but a check should be made to ensure that it covers all significant risks.

What does a risk assessment entail?

12.20 Some limited guidance about the nature of risk assessment has been given in the context of particular civil cases,[4] but there has been no authoritative guidance in a criminal context. Practical guidance as to what a risk assessment should entail is provided by the Approved Code of Practice and Guidance[5] that accompanies the Regulations. By s 17 HSWA 1974 the Code is admissible in criminal proceedings. The general purpose of an assessment, the Code maintains, is to identify the measures that need to be taken to manage, reduce, or eliminate any actual or potential hazards in the workplace so as to comply with any relevant statutory health and safety duties.

12.21 A 'hazard' is defined widely as anything with the potential to cause harm, and may include articles, substances, plant or machines, methods of work, the working environment, and other aspects of work organization. A 'risk' is the likelihood of potential harm from that hazard being realized, the potential severity of that harm, and the number of people who may be affected by it. The assessment should include the risks of psychological as well as physical injuries.[6]

12.22 The diversity of work situations to which the Management of Health and Safety at Work Regulations 1999 SI 1999/3242 apply makes it impossible to provide fixed rules as to exactly how a risk assessment should be carried out. But the Approved Code of Practice and Guidance does set out the general principles to be followed.[7] A risk assessment should:

(a) ensure that significant risks and hazards are addressed;
(b) ensure that all aspects of the work activity are reviewed, including routine and non-routine activities, and those that are not under the immediate supervision

---

[4] See for instance *Bailey v Command Security Services Ltd* [2001] All ER 352 (QB).
[5] Management of Health and Safety at Work (2000) Approved Code of Practice L21. See also *5 steps to risk assessment* INDG1639(rev) (HSE Books, 1998).
[6] See Management of Health and Safety at Work (2000) Approved Code of Practice L21 paras 9–12.
[7] ibid para 18.

of the employer, such as employees working off-site as contractors, employees working from home, employees who visit members of the public at home, and mobile workers;

(c) take account of the management of incidents, such as interruptions to the work activity, which frequently cause accidents; and consider what procedures should be followed to mitigate the effects of the incident;

(d) adopt a systematic and structured approach to identifying hazards, whether one risk assessment covers the whole activity, or the assessment is divided up. For example, it may make sense to look at things in groups, such as machinery, transport, and substances; or to divide the work site on a geographical basis;

(e) take account of the way in which work is organized and the effects this can have on health;

(f) take account of risks to the public;

(g) take account of the need to cover fire risks.

12.23 Once any hazards have been identified, the risks arising from them need to be evaluated. Where measures are already in place to control certain risks, the effectiveness of those measures needs to be considered when assessing the extent of risk which remains. In doing so regard should be had to the reality of what actually happens in the workplace as opposed simply to what should take place in theory.[8] Accordingly, employees and safety representatives should be observed and consulted.

12.24 The Code acknowledges the importance of taking human factors into account when assessing risk. For example, it provides that there is a higher likelihood of human error between 2 am and 5 am, when physiology dictates that the human body should be asleep. The risks will also be influenced by how well-trained people are, whether they have had sufficient rest before starting a shift, and whether they have taken alcohol or drugs.

12.25 The Code suggests that in some cases, employers may find it efficient to make a rough assessment first, in order to eliminate from consideration those risks that require no further action and to identify areas where a fuller assessment is needed, if appropriate, using more sophisticated techniques.

12.26 Some difficulty arises in situations where the nature of the work changes on a frequent basis or where the workplace itself changes or develops, such as a building site. In such cases, an assessment of the broad range of foreseeable risks should be carried out at the outset, and further detailed planning and employee training provided when other less common risks arise.

---

[8] Management of Health and Safety at Work (2000) Approved Code of Practice L21 para 22.

12.27 Where two or more employers do share a workplace, they must cooperate with each other, take all reasonable steps to coordinate their efforts in complying with the Regulations, and keep each other informed about the risks arising from their work.[9]

'Suitable and sufficient'

12.28 The duty is to carry out a suitable and sufficient assessment. 'Suitable and sufficient' is not defined in the Management of Health and Safety at Work Regulations 1999 SI 1999/3242, but the Approved Code of Practice and Guidance provides some assistance.[10] The assessment must be suitable and sufficient to identify measures necessary to comply with the duties under the statutory provisions—ie health and safety legislation. The level of detail in a risk assessment should be proportionate to the risk: thus insignificant risks, once identified, can usually be ignored, along with risks arising from everyday activities associated with life in general. The assessment need include only what an employer or self-employed person could reasonably be expected to know: they are not expected to anticipate risks which are not foreseeable.[11] In general, the risk assessment process needs to be practical and take account of the views of employees and their safety representatives who may have practical knowledge to contribute. Where employees of different employers work in the same workplace, their respective employers may have to cooperate to produce an overall risk assessment.

Relevance of the size of the undertaking in question

12.29 In the case of small businesses posing few or simple hazards, the assessment can, the Code suggests, be a very straightforward procedure based on common sense and reference to appropriate guidance. Appropriate sources of information might include the trade press, supplier manuals and manufacturers' instructions, as well as the relevant legislation. In most cases, however, the assessment will need to be slightly more sophisticated and may call for specialist advice or analytical techniques from a suitably qualified person. Clearly, the most developed and detailed assessments will be reserved for the largest and most hazardous workplaces, particularly where there are complex or novel processes involved. For example, in the case of certain manufacturing sites that use or store bulk hazardous substances, large-scale mineral extraction, or nuclear plant, such techniques as quantified risk assessment are likely to be called for.

---

[9] Management of Health and Safety at Work Regulations 1999 reg 11(1)(a)–(c).
[10] Management of Health and Safety at Work (2000) Approved Code of Practice, L21 paras 13, 14.
[11] In this regard see *R v HTM Ltd* [2006] EWCA Crim 1156.

B. The Management of Health and Safety at Work Regulations 1999

Model or generic risk assessments

12.30 It is acceptable, the Code maintains,[12] for employers who control a number of similar workplaces containing similar activities to produce a 'model' risk assessment reflecting the core hazards and risks associated with these activities. Model assessments may also be developed by trade associations, employers' bodies, or other organizations. Such model assessments may be applied by employers or managers in each workplace, but only if they satisfy themselves that the model assessment is appropriate to their type of work and take any steps necessary to adapt it to their work situation, including any extension necessary to cover hazards and risks not referred to in the 'model'.

Identifying who might be affected by the activities in question

12.31 An important part of the process is to identify exactly who might be harmed by the hazard; whether it be office staff, night cleaners, maintenance staff, security guards, visitors, or members of the public. Groups of workers who might be particularly at risk, such as young people or expectant mothers, should also be pin-pointed. In addition, any risk assessment should include consideration of the risk of stress.

12.32 The assessment must also address any risks faced by members of the public who are not employees. For example, the risk assessment produced by a railway company will, inter alia, need to consider the hazards and risks which arise from the operation and maintenance of rail vehicles and train services, and which might adversely affect workers, passengers, or any person, such as a level-crossing user, who could foreseeably be affected.

12.33 An identification of the period of time for which the assessment is likely to remain valid should also be included so that the relevant manager is in a position to recognize when short-term control measures need to be reviewed and modified, and to put in place medium- and long-term controls where these are necessary.

Young people

12.34 An employer who employs or will employ a person under the age of 18 is required by the regulation to take particular account of various specified features.[13] These include general characteristics of the young—'inexperience, lack of awareness of risk, and immaturity'—and features of the working environment, or methods of working, which may have a particular affect upon them. The risk assessment may

---

[12] Management of Health and Safety at Work (2000) Approved Code of Practice L21 para 17.
[13] Management of Health and Safety at Work Regulations 1999 SI 1999/3242 reg 3(5).

necessitate prohibiting the young person from certain work activities, except under close control. Parents of school-age children (eg those on work experience programmes) must be provided with information about risks and control measures. The HSE booklet, *Young People at Work* (HSG165), provides guidance on the Health and Safety (Young Persons) Regulations 1997 SI 1997/135.

Pregnant women

12.35 An employer who employs a woman of child-bearing age, and believes that the nature of the work she undertakes might involve a risk to the health and safety of a new or expectant mother, or her baby, must also take those risks into account when conducting a risk assessment under Management of Health and Safety at Work Regulations 1999 SI 1999/3242 reg 3(1).[14] He may also be required to take special measures tailored to an individual employee when she has notified him that she is pregnant, or has given birth in the previous six months.[15]

Principles of prevention

12.36 The Management of Health and Safety at Work Regulations 1999 SI 1999/3242 reg 4 requires an employer to implement any preventative and protective measures on the basis of the principles of prevention which are set out in Art 6(2) of the Directive. These are a defined set of principles which are also set out in Sch 1 to the Regulations and contained in the Code. In broad terms, these principles require the employer to avoid a risk altogether if that is possible, and when that is not possible, to evaluate the remaining risks and combat them at source. Importantly, they also require that the work is adapted to the requirements of the individual, and that technological developments are employed where possible. Risk prevention measures should be implemented as part of coherent policy and priority should be given to measures which give collective rather than individual protection.

Continuing review and surveillance

12.37 The Management of Health and Safety at Work Regulations 1999 SI 1999/3242 continue to define a number of other duties which are intended to ensure that the contents of the assessment remain valid, and that suitable control measures are in place at all times.

12.38 Every employer must put into effect appropriate arrangements for the effective planning, organization, control, monitoring, and review of the preventative and

---

[14] Management of Health and Safety at Work Regulations 1999 SI 1999/3242 reg 16(1).
[15] ibid reg 18.

B. The Management of Health and Safety at Work Regulations 1999

protective measures identified as result of the risk assessment.[16] The arrangements must be recorded in writing where more than five persons are employed.[17]

Appropriate health surveillance must be provided where risks to health and safety have been identified by the assessment.[18] **12.39**

Every employer has to appoint one or more competent persons to assist in undertaking the health and safety measures required. A competent person will have sufficient training and experience or knowledge to enable him properly to assist.[19] **12.40**

Information and training

The Regulations require that the employer must provide his employees and anyone else working in the undertaking (eg sub-contractors' staff, temporary workers, and agency workers) with information on the risks identified by the assessment and the preventative measures adopted.[20] The same information should be provided to any other employer who shares the workplace. This provision is consistent with the general principle that employers owe a duty to third parties as well as their direct employees. It also reflects the fact that the conventional master/servant relationship is no longer the norm, with looser agency-type relationships becoming more common. **12.41**

The employer must ensure that his employees are provided with adequate health and safety training when they are recruited, or in the event that the nature of their work or responsibilities changes over time.[21] The training must be repeated periodically. He must take an employee's capabilities into account when entrusting him with a task. **12.42**

Temporary and agency workers are further specifically provided for in the Management of Health and Safety at Work Regulations 1999 SI 1999/3242. Where temporary workers are provided by an agency, the necessary health and safety information must be provided to the agency for passing to the temporary staff.[22] **12.43**

Employees are under a corresponding duty to act in accordance with the training given, and to inform their employers of work situations which represent a serious and imminent danger or any deficiency in the arrangements made for health and safety.[23] **12.44**

---

[16] Management of Health and Safety at Work Regulations 1999 SI 1999/3242 reg 5(1).
[17] ibid reg 5(2).
[18] ibid reg 6.
[19] ibid reg 7(5).
[20] ibid reg 10.
[21] ibid reg 13(2).
[22] ibid reg 15.
[23] ibid reg 14.

Procedures for evacuation in the event of imminent danger

12.45 By the Management of Health and Safety at Work Regulations 1999 SI 1999/3242 reg 8, the employer must establish procedures to deal with serious and imminent danger such as fire; to appoint competent[24] staff to take charge of evacuating buildings in emergencies; to take measures to prevent untrained staff from having access to places where particular hazards exist; and to permit employees to take action to save themselves. These procedures must be communicated to all employees.

The operation and effect of the Management of Health and Safety at Work Regulations 1999 reg 21

12.46 In many ways, the Management of Health and Safety at Work Regulations 1999 SI 1999/3242 reg 21 sits uneasily in the overall framework of health and safety legislation. It was assumed that the intended purpose of the regulation was to reverse the effect of the decision in *R v Nelson Group Services Ltd*, although this was not in fact the case.[25] On one interpretation, reg 21 may be said to make an employer strictly liable for the acts of his employees: the Approved Code of Practice and Guidance provides[26] that those who enforce the legislation will take account of the fact that the employer has taken reasonable steps to ensure his employees' competence. The scope and effect of reg 21 was considered—in many ways unsatisfactorily—in *R v HTM Ltd* [2006] EWCA Crim 1156. The interpretation and effect of the regulation upon the general duties under HSWA 1974 is considered in respect of health and safety regulations at paragraphs 6.102–6.115.

C. The Workplace (Health, Safety and Welfare) Regulations 1992

Introduction

12.47 The Workplace (Health, Safety and Welfare) Regulations 1992 SI 1992/3004 (reproduced at Appendix F) impact on virtually every employer and place of work. They expand on the general duty under HSWA 1974 s 2 to ensure, so far as is reasonably practicable, the health, safety, and welfare of their employees at work; and the duty under HSWA 1974 s 4, owed by employers to non-employees who may use

---

[24] A person is to be regarded as 'competent' where he has sufficient training and experience or knowledge and other qualities to enable him to properly implement the evacuation procedures.
[25] For which see paras 6.91–6.100.
[26] Management of Health and Safety at Work (2000) Approved Code of Practice, L21 para 102.

C. The Workplace (Health, Safety and Welfare) Regulations 1992

their premises. The Regulations are to be read in the light of the extensive guidance contained within the Workplace Health, Safety and Welfare Approved Code of Practice and Guidance.[27]

12.48 The Workplace (Health, Safety and Welfare) Regulations 1992 SI 1992/3004 contain a wide range of provisions for maintaining the workplace to the standards demanded by the EC Workplace Directive (89/654/EEC) and set out specific requirements relating to certain equipment, devices, and systems; ventilation; temperature; lighting; cleanliness; washing, changing, and sanitary facilities; room dimensions and space; doors and windows; workstations and seating; escalators and moving walkways; and the condition of floors and traffic routes. They also contain a requirement that employees should be safeguarded against being hit by falling objects, or from falling themselves a distance likely to cause injury, or into a tank, pit, or structure containing a dangerous substance.

Who owes the duty?

12.49 The duty to comply with the regulations is imposed on employers in control of a workplace (modification, extension, or conversion) as well as on persons who have, to any extent, control of a workplace.[28] No duty is placed upon the self-employed in respect of their own or their partner's workplace.[29] In the normal course of events, the Crown is not exempted.

The extent of the duty

12.50 The duty under the Workplace (Health, Safety and Welfare) Regulations 1992 SI 1992/3004 reg 4(2) is expressly limited to 'matters within that person's control'. In each case the question of whether a person had control is a matter of fact; and there is nothing to prevent more than one party having control at a given time. For the application of this principle, see *Bailey v Command Security Services Ltd* [2001] All ER 352 (QB), where a security guard patrolling a warehouse fell down an unguarded lift shaft. It was held that the while the injured party's employer had the power to report such matters, this was not sufficient to establish control; rather, control was in the hands of the warehouse owner, who had the power to alter the warehouse so as to make it comply with the Regulations.

12.51 That case may be contrasted with *King v RCO Support Services Ltd* [2001] ICR 608 (CA), where it was found that a bus company which had contracted out the

---

[27] Workplace (Health, Safety and Welfare) Regulations 1992. Approved Code of Practice L24.
[28] Reg 4(2).
[29] Reg 4(4).

gritting of its garage was not in control for the purposes of the Regulations at a time when the employee of the gritting contractor fell and injured himself.

To whom is the duty owed?

12.52 The duty under the Workplace (Health, Safety and Welfare) Regulations 1992 SI 1992/3004 reg 5 is for the protection of workers and is only owed to those who are at work in the workplace—whether or not they are employed by the person in control. It is not owed to a member of the public who may simply be visiting.[30]

What is a workplace?

12.53 The Workplace (Health, Safety and Welfare) Regulations 1992 SI 1992/3004 are also significant in that they provide a definition of the term 'workplace' which is adopted by many other health and safety regulations. Regulation 2(1) defines workplace as 'any premises or part of premises which are not domestic premises and are made available to any person as a place of work'. It includes:

> any place within the premises to which such person has access while at work; and any room, lobby, corridor, staircase, road or other place used as a means of access to or egress from that place of work; or where facilities are provided for use in connection with the place of work other than a public road.

However, a modification, extension, or conversion of any of the above is excluded until it has been completed.

12.54 What is and what is not a 'workplace' has been addressed in a number of civil cases. In *Lewis v Avidan* [2005] EWCA Civ 670 May LJ held that a 'workplace' could not be said to include a concealed water pipe which burst and caused water to be on the floor:[31] the floor—not the pipe—being the workplace.

12.55 The term 'work' itself is defined in HSWA 1974 s 52 as including work as an employee or self-employed person. It provides: 'an employee is at work throughout the time when he is in the course of his employment, but not otherwise; and a self-employed person is at work throughout such time as he devotes to work as a self-employed person.' In certain cases specific regulations define what is and what is not work: for example the (Health and Safety (Training for Employment) Regulations 1990, set out the circumstances in which a person undergoing training may also be 'at work'.

---

[30] See *Donaldson v Hays Distribution Services Ltd* 2005 SLT 733; *Ricketts v Torbay Council* [2003] EWCA Civ 613 (CA).
[31] Although he did hold that it was capable of being 'equipment'.

C. The Workplace (Health, Safety and Welfare) Regulations 1992

Having set out what exactly a workplace is, Workplace (Health, Safety and Welfare) Regulations 1992 SI 1992/3004 reg 3(1) go on to exclude certain types of workplaces from the requirements of the Regulations; namely a workplace on a ship, a workplace where the only activities being undertaken are building operations or works of engineering construction, and a workplace below ground at a mine (reg 3). In relation to certain other types of workplace, including in or on an aircraft and certain other locomotives, or in a forest or on agricultural land, only some of Workplace (Health, Safety and Welfare) Regulations 1992 SI 1992/3004 apply (reg 3(3)–(5)). **12.56**

The nature of the duty

The principal duty is contained in the Workplace (Health, Safety and Welfare) Regulations 1992 SI 1992/3004 reg 5(1): 'The workplace and the equipment, devices and systems to which the regulation applies shall be maintained (including cleaned as appropriate) in an efficient state, in efficient working order and in good repair.' **12.57**

Regulation 5(1) has been held to create a strict duty and is not dependent upon, for instance, whether the cause of the failure of equipment could be ascertained;[32] or whether the hazard was created by an independent contractor.[33] The expressions 'maintain' and 'in an efficient state', 'in efficient working order' and 'in good repair' have been the subject of extensive and detailed judicial consideration. For example, in *Lewis v Avidan Ltd* [2005] EWCA Civ 670, the fact that a floor may have become flooded by a burst pipe did not mean that it was not 'maintained in an efficient state': it was held that the word 'maintained' imported the concept of doing something to the floor itself. In *Coates v Jaguar Cars* [2004] EWCA Civ 337 it was held that the absence of a handrail from a flight of stairs did not imply that there had been a failure to maintain, as the regulation was concerned with what was provided, not what ought to have been provided. 'Efficient' is to be considered in the context of safety, and not productivity. **12.58**

Regulation 12(1) of the Workplace (Health, Safety and Welfare) Regulations 1992 SI 1992/3004 imposes a number of duties in relation to workplace floors and traffic routes. The HSE has proposed a route map for workplace transport as part of a consultation process: it can be found at <www.hse.gov.uk/consult/condocs/routemap.htm>. **12.59**

---

[32] *Galashiels Gas Co Ltd v O'Donnell* [1949] AC 275 (HL) and *McLaughlin v East and Midlothian NHS Trust*, 2000 Rep LR 87 (OH (Scot)).
[33] *Malcolm v Metropolitan Police Commissioner* [1999] CLY 1494.

D. Manual Handling Operations Regulations 1992

Introduction

12.60 More than a quarter of the accidents reported each year to the enforcing authorities are associated with the transporting or supporting of loads by hand or bodily force. The Manual Handling Operations Regulations 1992 SI 1992/2793 (reproduced at Appendix G) came into force on 1 January 1993[34] and implement European Directive 90/269/EEC, which provides that the employer shall take appropriate measures 'in order to avoid the need for manual handling of loads by employers'. The Regulations aim to reduce the number of sprains and strains caused in the workplace by poor lifting techniques. Breach of the Manual Handling Operations Regulations 1992 SI 1992/2793 will give rise to civil liability where damage occurs. They apply whenever an employee is engaged in a manual handling operation involving a risk of injury.

12.61 Although primarily directed at ensuring employers have regard for the physical safety of their employees, reg 2(2) makes the self-employed responsible for their own safety during manual handling. The Manual Handling Operations Regulations 1992 SI 1992/2793 also have relevance to those who manufacture items to be used in the workplace. HSWA 1974 requires designers and manufacturers to ensure the safety so far as is reasonably practicable of any article for use at work and to provide adequate information about the conditions necessary to ensure use without risk to health. To this end, the Guidance suggests that information is provided on the article or package as to its weight and, where appropriate, its centre of gravity.

Relationship between Manual Handling Operations Regulations 1992 and other provisions

12.62 Manual Handling Operations Regulations 1992 SI 1992/2793 supplement the general duties placed on employers and others by HSWA 1974 and the broad requirements of the Management of Health and Safety at Work Regulations 1999 SI 1999/3242. As the Guidance to the Regulations confirms, they should not be considered in isolation. Particular regard should be given to reg 3(1) of the Management of Health and Safety at Work Regulations 1999 SI 1999/3242, which requires employers to make a suitable and sufficient assessment of the risks to the health and safety of their employees while at work. Where this general assessment indicates the possibility of risks to employees from the manual handling

---

[34] They were amended by the Health and Safety (Miscellaneous Amendments) Regulations 2002.

D. Manual Handling Operations Regulations 1992

of loads, the requirements of the Manual Handling Operations Regulations 1992 SI 1992/2793 should be followed. Appendix 1 and the accompanying Guidance to the Regulations provide detailed advice on the carrying out of such an assessment.

The primary duties under Manual Handling Operations Regulations 1992

12.63 The Manual Handling Operations Regulations 1992 SI 1992/2793 adopt a clear hierarchical approach to tackling the risk of injury through manual handling operations. The primary duty is contained within reg 4(1)(a): it obliges employers to avoid the need for employees to undertake manual handling operations which involve a risk of their being injured so far as is reasonably practicable. Guidance issued to accompany the Regulations provides that this may involve either redesigning the task to avoid moving the load, or automation, or mechanization.

12.64 As regards the burden of proof, it is likely to be for the prosecution to prove that the activity in question was a manual handling activity which was likely to involve a risk of injury. Once the prosecution has proved this, it will be for the defendant to prove that it was not reasonably practicable to avoid the need for manual handling. A failure to adduce any evidence on the point is likely to mean that a defendant will fail to discharge the burden of proof placed on him, although in civil contexts the courts have shown themselves to be willing to infer—even in the absence of any evidence—that manual handling operations could not have been avoided.

12.65 What amounts to a 'risk' in the context of reg 4 has been the subject of much judicial debate. The pronouncement of Hale LJ in *Koonjul v Thameslink Healthcare Services Ltd* [2000] PIQR 123 (CA) that 'There must be a real risk, a foreseeable possibility of injury; certainly nothing approaching a probability' has been cited and approved in a number of subsequent cases.

12.66 Moving down the hierarchy of measures, by the Manual Handling Operations Regulations 1992 SI 1992/2793 reg 4(1)(b)(i), where it is not reasonably practicable for the employer to take steps to avoid the need for his employees to undertake manual handling operations, a suitable and sufficient assessment of the operation should be carried out. Schedule 1 to the Regulations provides detailed guidance on the factors to be taken into account.

12.67 Further to this, according to reg 4(1)(b)(ii), steps should then be taken to reduce the risk of injury to the lowest level reasonably practicable, and appropriate steps should be taken to provide any employee who may be undertaking such work with precise information about the load.

12.68 In *Swain v Denso Martin Ltd* [2000] ICR 1079 the Court of Appeal held that the Manual Handling Operations Regulations 1992 SI 1992/2793 reg 4(1)(b) imposed three separate obligations, each of which had to be complied with. They unsurprisingly rejected a submission that an employer who had failed to carry out a risk assessment in accordance with sub-para (i) was not therefore under a duty to comply with the remaining two sub-paragraphs.

12.69 The Regulations make it plain that the need to carry out a risk assessment is not a one-off requirement. Under the Manual Handling Operations Regulations 1992 SI 1992/2793 reg 4(2), the assessment must be kept up to date. It should be reviewed if new information comes to light, or if there has been a change in the manual handling operations which, in either case, could materially have affected the conclusion reached previously. The assessment should also be reviewed if a reportable injury occurs and corrected or modified where necessary.

What is a manual handling operation?

12.70 The question of what acts or processes should be classified as manual handling operations has been considered in a number of civil cases.

12.71 In *O'Neil v DSG Retail Ltd* [2002] EWCA Civ 1139, it was held that an employer was obliged to consider the specific task in the context of where it was to be performed and the particular employee by which it was to be performed.

12.72 In *Koonjul v Thameslink Healthcare Services* [2000] PIQR, [2000] CLY 2983, K was employed as care assistant in a small residential home. She sustained a back injury when she leant down to pull a low wooden bed from against the wall. It was held that though nothing more than a real risk was necessary to bring the task within the scope of Manual Handling Operations Regulations 1992 SI 1992/2793 reg 4, it was also necessary to consider the background of the incident. The appellant was experienced and had received prior training in lifting and bending techniques and it would be impracticable to provide guidance in the performance of innumerable everyday domestic tasks.

12.73 In *Postle v Norfolk and Norwich NHS Healthcare Trust* [2000] CLY 2970, it was held that whereas the Manual Handling Operations Regulations 1992 SI 1992/2793 applied, there had been no breach in the circumstances of the case, as a suitable and sufficient risk assessment would not have recognized the risks of injury to one nurse from the pulling of a trolley by another; nor would it have been necessary to give instructions to pull the trolley forward, or tell the other nurses that she was about to move.

12.74 In *Swain v Denso Marston Ltd* [2000] ICR 1079, S was injured when stripping down a newly delivered conveyor. It was argued that as S had been involved in the assessment process, the liability to provide guidance under the Manual Handling Operations Regulations 1992 SI 1992/2793 reg 4(1)(b)(iii) did not apply. However, it was held that the Manual Handling Operations Regulations

1992 SI 1992/2793 did not impose three discrete duties but should be read conjunctively; a detailed assessment might have involved obtaining manufacturers' manuals and specifications to highlight potential problems, and that if there had been insufficient information, then caution should have resulted in a presumption that the roller might be unduly heavy; and that this should have been communicated to S.

12.75 In *King v RCO Support Services Ltd* [2001] ICR 608, an employee who was engaged in gritting a yard slipped on an icy area. In upholding a wide construction of the Manual Handling Operations Regulations 1992 SI 1992/2793 reg 2, it was held that this did constitute a manual handling operation. However in *Gissing v Walkers Smith Snack Foods Ltd* [1999] CLY 3983 it was held that the Manual Handling Operations Regulations 1992 SI 1992/2793 did not apply to the repetitive action of packing crisp bags.

12.76 In *Wells v West Hertfordshire HA* (QBD 5 April 2000) it was held that the requirement to make a risk assessment under the Manual Handling Operations Regulations 1992 SI 1992/2793 reg 4(1) was applicable to the work of a midwife.

E. Work Equipment: Provision and Use of Work Equipment Regulations 1998

Introduction

12.77 The Provision and Use of Work Equipment Regulations 1998 SI 1998/2306 SI 1998/2306 (reproduced in Appendix H) are intended to implement the Work Equipment Directive 89/655 'concerning the minimum safety and health requirements for the use of work equipment by workers at work' and the Amending Directive to the Use of Work Equipment Directive 95/63/EC. The 1998 Regulations replaced the Provision and Use of Work Equipment Regulations 1992 and came into force on 5 November 1998. Their purpose, according to the Approved Code of Practice[35] is to 'ensure that work equipment should not result in health and safety risks regardless of its age, condition or origin.'

12.78 The Regulations impose the principal obligations on employers in relation to equipment used in the workplace, and include requirements relating to the suitability, maintenance, and inspection of work equipment. Whereas previous regulations covered specific types of equipment, the Provision and Use of Work Equipment Regulations 1998 SI 1998/2306 are all-embracing and apply a wide

---

[35] Provision and Use of Work Equipment Regulations 1998 SI 1998/2306. Approved Code of Practice and Guidance L22.

definition of 'equipment', including mobile equipment and lifting equipment, and impose general obligations regarding its safety.

12.79 The Provision and Use of Work Equipment Regulations 1998 SI 1998/2306, like the other regulations in the 'six-pack', cannot be considered in isolation of other statutory duties under health and safety legislation, in particular the general duties of HSWA 1974 ss 2–7. As the Approved Code of Practice and Guidance[36] acknowledges, there is some overlap between them and other regulations which are intended to cover more specific situations.

Suitability of work equipment

12.80 Regulation 4 of the Provision and Use of Work Equipment Regulations 1998 SI 1998/2306 provides:

(1) Every employer shall ensure that work equipment is so constructed or adapted as to be suitable for the purpose for which it is to be used or provided.
(2) In selecting work equipment, every employer shall have regard to the working conditions and to the risks to the health and safety of persons that exist in the premises or undertaking in which that work equipment is to be used and any additional risks posed by the use of that work equipment.
(3) Every employer shall ensure that work equipment is used only for operations for which, and under conditions for which, it is suitable.

12.81 Regulation 4(1) of the Provision and Use of Work Equipment Regulations 1998 SI 1998/2306 replaces the more limited provisions of s 17 of the Factories Act 1961 and requires an employer to ensure that work equipment is so constructed or adapted as to be suitable for the purpose for which it is used or provided. The regulation places a heavy burden on employers, and is not limited by any criterion of reasonable practicability. 'Suitable' is defined by the regulation as meaning in all respects in which it is reasonably foreseeable that it will affect the health or safety of 'any person'.[37] Although the Regulations undoubtedly impose a high standard on employers, the courts have viewed cases on a 'qualitative' basis and looked at the wider balance of risk, rather than finding a breach in which any risk could be shown. For example, in *Yorkshire Traction Co Ltd v Searby* [2003] EWCA Civ 1856, the Court found that whereas the absence of a screen separating a bus driver from his passengers did create a foreseeable risk of assault on the particular route, it was also entitled to take into account the fact that the very presence of the screen created its own risks. They reiterated that the regulation did not require an employer to ensure complete protection from all foreseeable risks.

---

[36] Provision and Use of Work Equipment Regulations 1998 SI 1998/2306. Approved Code of Practice and Guidance L22.
[37] Provision and Use of Work Equipment Regulations 1998 SI 1998/2306 reg 4(4).

E. Work Equipment: Provision and Use of Work Equipment Regulations 1998

12.82 Regulation 4(2) of the Provision and Use of Work Equipment Regulations 1998 SI 1998/2306 concerns the place where equipment is to be used. It requires an employer, in selecting work equipment, to have regard to the working conditions and to the risks to the health and safety of persons that exist in the premises or undertaking in which that work equipment is to be used, and any additional risk posed by the use of that work equipment. The guidance in the Approved Code of Practice and Guidance suggests, by way of illustration of the point, that electrical equipment is not suitable for use in wet conditions unless it is designed for the purpose or suitably protected. It also addresses the need to ensure that adequate ventilation is available when the equipment being used produces exhaust gases.

12.83 Regulation 4(3) of the Provision and Use of Work Equipment Regulations 1998 SI 1998/2306 concerns the purpose for which the equipment will be used. It provides that work equipment be used only for suitable operations and under suitable conditions. For example, scissors or other cutting tools may be used in preference to knives with unprotected blades, so as to reduce the possibility of injury. It is important to note that it is not enough that the equipment is suitable for the primary use to which it is put; reg 4(3) demands that it is suitable for all operations in which it is involved.

Risk assessment and the duty under Provision and Use of Work Equipment Regulations 1998 reg 4

12.84 The Approved Code of Practice and Guidance gives some guidance on the approach an employer should take when seeking to comply with his duties under the regulation. The Approved Code of Practice and Guidance provides: 'Because of the general risk assessment requirements in the Management Regulations, there is no specific regulation requiring a risk assessment in Provision and Use of Work Equipment Regulations.'[38]

12.85 It suggests that most employers should be capable of making the risk assessment appropriate to Provision and Use of Work Equipment Regulations 1998 SI 1998/2306 reg 4 themselves, using expertise within their own organization to identify the measures that need to be taken in relation to their work equipment. However, where the equipment or hazards are particularly complex, the assistance of an external health and safety adviser may be required.

12.86 The Approved Code of Practice and Guidance also points out that many employers will be familiar with the measures that need to be taken to comply with wider legal requirements and these should generally be adequate for the purpose of

---

[38] Provision and Use of Work Equipment Regulations 1998 SI 1998/2306. Approved Code of Practice and Guidance L22 para 24.

complying with the Provision and Use of Work Equipment Regulations 1998 SI 1998/2306. Where this is not so, general guidance or guidance specific to a particular industry or piece of equipment should be helpful, providing that it is appropriate. Where such guidance is inappropriate, or does not exist, the key principles that apply to all risk assessments should be used in relation to any piece of equipment: namely, the severity of any injury or ill-health likely to result from any hazard present; the likelihood of that happening and the numbers of people exposed to the risk.

Ergonomics

12.87 In addition, the Approved Code of Practice specifically provides that the employer should consider the matter of ergonomics when carrying out any assessment.[39] This involves taking account of the size and shape of the human body and ensuring that the design of the equipment is compatible with human dimensions. Such matters as operating positions, working heights, and reach distances should be adapted to accommodate the intended operator. The user should not be placed under undue strain when operating the equipment; nor should he be expected to exert undue force or stretch or reach in carrying out any task. This is particularly so where highly repetitive tasks, such as working on a supermarket checkout, are involved.

Maintenance of work equipment

12.88 Regulation 5 of the Provision and Use of Work Equipment Regulations 1998 SI 1998/2306 provides:

(1) Every employer shall ensure that work equipment is maintained in an efficient state, in efficient working order and in good repair.
(2) Every employer shall ensure that where any machinery has a maintenance log, the log is kept up to date.

12.89 The Provision and Use of Work Equipment Regulations 1998 SI 1998/2306 also build on a general duty under HSWA 1974 to maintain work equipment so that it is safe and impose an obligation on the employer to keep his equipment in good condition. Under reg 5(1), it is an offence to fail to ensure that it is maintained 'in an efficient state, in efficient working order and in good repair'. In this context, the word 'efficient' relates to the condition of the equipment in regard to health and safety matters, and not to productivity. The regulation is not concerned with the maintenance process itself, but rather the end result of that process. The liability imposed by the regulation is strict, so that an offence will

---

[39] Provision and Use of Work Equipment Regulations 1998 SI 1998/2306. Approved Code of Practice and Guidance L22 para 100.

E. Work Equipment: Provision and Use of Work Equipment Regulations 1998

be committed whenever injury is caused by a faulty machine, regardless of how conscientious the employer was in his system of maintenance, and regardless of whether the fault was a latent one.

12.90 The wide scope of the Provision and Use of Work Equipment Regulations 1998 SI 1998/2306 reg 5(1) was explored in *Stark v Post Office* [2000] ICR 1013, a case that concerned an accident caused by the mechanical failure of a postman's bicycle. Although it was accepted by the Court that even a thorough examination would not have revealed the fault, it was nevertheless held by the Court that reg 5(1) imposes a strict obligation to ensure that the equipment is at all times properly maintained; an obligation that had been breached. The question of foreseeability was irrelevant in this context. The Court confirmed that the regulation sets out a result to be achieved, rather than a means of achieving it. That the reasonableness of the system of inspection and maintenance is irrelevant was confirmed in *Cadger v Vauxhall Motors* (Cty Ct, 28 March 2000).

12.91 Although there is no requirement to keep a maintenance log, where one does exist, reg 5(2) of the Provision and Use of Work Equipment Regulations 1998 SI 1998/2306 requires the employer to keep it up to date. However, the Approved Code of Practice and Guidance recommends that a detailed log is kept in relation to all high-risk equipment to provide information for future maintenance planning and to inform any relevant person of previous action taken.[40]

12.92 Guidance provided in the Approved Code of Practice and Guidance explains the significance of the regulation in terms of maintaining equipment so that its performance does not deteriorate to the extent that it puts people at risk. It points out that some parts of equipment, such as guards, ventilation fittings, emergency shutdown systems, and pressure relief devices, have to be maintained to do their job properly. The need to maintain other parts, however, may not be so obvious. For example, failure to lubricate bearings or replace clogged filters might lead to danger because of seized parts or overheating. Further, some maintenance routines affect both the way the equipment works and its safety. Checking and replacing worn or damaged friction linings in the clutch on a guillotine will ensure it operates correctly, but could also prevent the driving mechanism jamming, thereby reducing the risk of uncontrolled and potentially dangerous strokes.

12.93 The Approved Code of Practice and Guidance recognizes that, given the diversity of equipment employed in the workplace, it would be impossible to lay down specific guidelines as to the frequency with which maintenance checks should be

---

[40] Provision and Use of Work Equipment Regulations 1998 SI 1998/2306. Approved Code of Practice and Guidance L22 para 129.

carried out, although it is better to err on the side of caution. A fault that affects production is normally apparent within a short time. However, a fault in a safety-critical system could remain undetected unless appropriate safety checks are included in maintenance activities. Common sense dictates that the extent and complexity of maintenance will range from the simple to the sophisticated depending on the complexity of the equipment. Maintenance procedures should be carried out in accordance with any manufacturer's recommendations relating to the equipment. However, additional measures may be necessary where the conditions of use are particularly arduous.

12.94 The Approved Code of Practice and Guidance[41] advises that the frequency at which maintenance activities are carried out should also take into account the:

(a) intensity of use—frequency and maximum working limits;
(b) operating environment—for example, whether marine or outdoors;
(c) variety of operations—whether the equipment is performing the same task all the time or changes;
(d) risk to health and safety from malfunction or failure.

12.95 The Approved Code of Practice and Guidance[42] identifies three different maintenance management techniques, referring to them as 'the planned preventive', 'the condition-based', and 'the breakdown', with each offering different benefits. Planned preventive maintenance involves replacing parts and consumables or making necessary adjustments at preset intervals so that risks do not occur as a result of the deterioration or failure of the equipment. Condition-based maintenance involves monitoring the condition of safety-critical parts and carrying out maintenance whenever necessary to avoid hazards which could otherwise occur. Breakdown maintenance involves carrying out maintenance only after faults or failures have occurred. This approach is obviously appropriate only if the failure does not prevent an immediate risk and can be corrected before risk occurs; for example, through effective fault reporting and maintenance schemes. Where safety-critical parts could fail and cause the equipment, guards, or other protection devices to fail and lead to immediate or hidden potential risks, a formal system of planned preventive or condition-based maintenance is likely to be needed.

12.96 Where the equipment concerned is hired from a hire company rather than owned by the employer, the Approved Code of Practice and Guidance stresses the importance of establishing which party will carry out safety-related maintenance. This is

---

[41] Provision and Use of Work Equipment Regulations 1998 SI 1998/2306. Approved Code of Practice and Guidance L22 para 121.
[42] ibid para 125.

E. Work Equipment: Provision and Use of Work Equipment Regulations 1998

all the more important for equipment that is hired on a long-term basis where the terms of the agreement are set out or recorded in writing.

Inspection of work equipment

12.97 Under the Provision and Use of Work Equipment Regulations 1998 SI 1998/2306 reg 6, an employer who fails to inspect his work equipment in certain prescribed circumstances commits an offence. The circumstances are: where it is placed for the first time in a certain position, or assembled at a new location; where the safety of the work equipment depends upon the installation conditions; where the equipment is exposed to conditions liable to cause deterioration giving rise to dangerous situations; and each time that exceptional circumstances occur which are liable to jeopardize the safety of the work equipment.

12.98 According to the Approved Code of Practice,[43] the purpose of an inspection is to identify whether the equipment can be operated, adjusted, and maintained safely, and to detect and remedy any deterioration, such as everyday wear and tear, before it results in unacceptable risks.

12.99 The Approved Code of Practice and Guidance identifies certain equipment types of equipment which may present significant risks as a result of deterioration, and which may therefore need inspection under the Provision and Use of Work Equipment Regulations 1998 SI 1998/2306 reg 6(2). They include most fairground equipment, machines where there is a need to approach the danger zone during normal operation, such as horizontal injection moulding machines, paper-cutting guillotines, die-casting machines, and shell-moulding machines; complex automated equipment, and integrated production lines.

Keeping records of inspections

12.100 Regulation 6(3) of the Provision and Use of Work Equipment Regulations 1998 SI 1998/2306 requires the employer to keep a record of such inspections. Such a record may be handwritten or stored electronically. Although there are no legal requirements as to what the record should contain, the Approved Code of Practice and Guidance advises that it should usually include information on the type and model of equipment, any identification mark or number that it has, its normal location, the date that the inspection was carried out, who carried it out, any faults or action taken, to whom the faults have been reported, and the date when repairs or other necessary action was carried out.

---

[43] Provision and Use of Work Equipment Regulations 1998 SI 1998/2306. Approved Code of Practice and Guidance L22 para 140.

12.101 Regulation 6(4) of the Provision and Use of Work Equipment Regulations 1998 SI 1998/2306 requires an employer to ensure that no equipment can either leave his undertaking, or if obtained from another person, be used in his undertaking, without physical evidence of the last inspection having been carried out pursuant to this regulation. According to the Approved Code of Practice and Guidance, this physical evidence should be appropriate to the type of work equipment being inspected. For large items of equipment for which inspection is necessary, the evidence can be in the form of a copy of the record of the last inspection carried out. For smaller items, a tagging, colour-coding, or labelling system can be used. The purpose of the physical evidence is to enable a user to check whether an inspection has been carried out and whether or not it is current, and also to determine the results of that inspection by being able to link back from the physical evidence to the records.

12.102 Exceptional circumstances which may result in the need for inspection include major modifications, refurbishment or repair work; known or suspected serious damage; and substantial change in the nature of use, for example, from an extended period of inactivity.

Extent and frequency of the inspections

12.103 The extent of the inspection required will depend upon the type of equipment and where and how it is used, but should include, where appropriate, visual checks, functional checks, and testing. Accordingly, an inspection may vary from a simple visual external inspection to a detailed comprehensive inspection that may include some dismantling or testing. But an inspection should always include those safety-related parts which are necessary for the safe operation of equipment; for example, overload warning devices and limit switches. Where an inspection is necessary under other legislation, no further inspection will be necessary under the Provision and Use of Work Equipment Regulations unless it fails to fully cover all of the significant health and safety risks which are likely to arise from the use of the equipment. The Approved Code of Practice and Guidance[44] also advises that, as part of an inspection, a functional or other test may be necessary to check that the safety-related parts, such as interlocks, protection devices, and controls, are working as they should be and that the work equipment and relevant parts are structurally sound.

12.104 The frequency of inspections required will depend on how quickly the work equipment or parts of it are likely to deteriorate and therefore to give rise to a significant risk. In turn, this will depend on the type of equipment, how it is used, and the

---

[44] Provision and Use of Work Equipment Regulations 1998 SI 1998/2306. Approved Code of Practice and Guidance L22 para 141.

E. Work Equipment: Provision and Use of Work Equipment Regulations 1998

conditions to which it is exposed. Clearly, the same equipment will deteriorate more quickly that it otherwise would if, for example, it is used in a harsh outdoor environment rather than an indoor warehouse. The Approved Code of Practice suggests that in order to arrive at appropriate inspection intervals and procedures, equipment should be subject to periodic reviews so that intervals can be extended where history shows that deterioration is negligible, or shortened where substantial amounts of deterioration are detected at each inspection.

12.105 No inspection is required in the case of equipment whose failure or fault cannot lead to significant risk, or if safety is guaranteed through appropriate maintenance regimes. Accordingly, equipment that is unlikely to need an inspection includes office furniture, hand tools, non-powered machinery, and also certain powered machinery, such as a reciprocating fixed-blade metal-cutting saw.

Competence of those who carry out inspections

12.106 The Approved Code of Practice[45] also provides that the persons who determine the nature of the inspections required and who carry out inspections are competent to do so, in that they have the necessary knowledge and experience to decide what the inspection should include, how it should be done, and when it should be carried out. They should also be able to determine whether any tests are needed during the inspection to see if the equipment is working safely or is structurally sound. Experienced in-house employees such as a departmental manager or supervisor may be able to do this.

12.107 The person who actually carries out the inspection need not necessarily be the same person who determines the nature of the inspections. It is most likely that the actual inspection will be done by an in-house employee who is skilled enough to know what to look at, what to look for, and what to do. Where necessary, the employer should provide that person with appropriate information, instruction, and training. The level of competence required will vary according to the type of equipment, and where and how it is used. Where the necessary level of knowledge is not available in-house, it will be necessary to call in an outsider.

Information and instructions

12.108 Regulation 8(1) of the Provision and Use of Work Equipment Regulations 1998 SI 1998/2306 provides: 'Every employer shall ensure that all persons who use work equipment have available to them adequate health and safety information and, where appropriate, written instructions pertaining to the use of work equipment.'

---

[45] Provision and Use of Work Equipment Regulations 1998 SI 1998/2306. Approved Code of Practice and Guidance L22, paras 148–152.

12.109 The Approved Code of Practice and Guidance provides that this regulation builds on the wider duty under s 2 of the HSWA to provide employees with information and instructions to ensure their health and safety so far as is reasonably practicable. Employers should enable workers to have easy access to the information and be able to understand it.

Training

12.110 Regulation 9(1) of the Provision and Use of Work Equipment Regulations 1998 SI 1998/2306 provides: 'Every employer shall ensure that all persons who use work equipment have received adequate training for purposes of health and safety, including training in the methods which may be adopted when using the work equipment, any risks which such use may entail and precautions to be taken.'

12.111 The Guidance to the Approved Code of Practice provides that other health and safety legislation, including HSWA 1974 and reg 11 of Management of Health and Safety at Work Regulations 1999 SI 1999/3242, contain general requirements in relation to training, and that reg 9 of the Provision and Use of Work Equipment Regulations 1998 SI 1998/2306 is concerned with the specifics of what that training should include.

Dangerous parts of machinery

12.112 Regulation 11(1) of the Provision and Use of Work Equipment Regulations 1998 SI 1998/2306 provides: 'Every employer shall ensure that measures are taken . . . which are effective—(a) to prevent access to any dangerous part of machinery.'

12.113 Accidental contact with dangerously exposed parts of machinery is one of the most common causes of injury within the industrial sector. There is undoubtedly a correlation between the frequency of this type of injury and the number of prosecutions brought under this regulation.

12.114 Every employer must ensure that he takes measures that are effective to prevent access to any dangerous part of machinery, or to stop the movement of any dangerous part before a person enters the danger zone. What measures are effective will generally be established by carrying out a risk assessment under the Management of Health and Safety at Work Regulations 1999 SI 1999/3242 reg 3(1). The regulation prescribes a hierarchy of measures, in descending order of effectiveness, which must be taken in order to comply with the duty. It is only if compliance with the topmost measure is not practicable that a person is entitled to descend to the next. The hierarchy is:

(a) the provision of fixed guards to enclose every dangerous part;
(b) the provision of other guards or protection devices;
(c) the provision of jigs, holders, push sticks.

Dangerous part

12.115 The definition of 'dangerous part' has been considered in a number of cases, mainly under the repealed Factories Act 1961. The burden of authority was to the effect that a part is to be considered dangerous if it might be a reasonably foreseeable cause of injury to a person who was acting in a reasonable manner in circumstances which might be reasonably expected to occur. It is to be noted, however, that the Directive from which the regulation derives makes no reference to 'foreseeability', but refers simply to preventing access to danger zones, and halting dangerous parts.

Protection against specified hazards

12.116 Regulation 12 of the The Provision and Use of Work Equipment Regulations 1998 SI 1998/2306 provides that: 'Every employer shall take measures to ensure that the exposure of a person using work equipment to any risk to his health or safety from any hazard specified in paragraph (3) is either prevented or where that is not reasonably practicable, adequately controlled.'

12.117 The hazards referred to include:

(a) any article or substance falling or being ejected from work equipment;
(b) rupture or disintegration of parts of work equipment;
(c) work equipment catching fire or overheating.

12.118 The regulation is one of general application, but does not apply when the situation is one which is specifically covered by other regulations, such as the Control of Substances Hazardous to Health Regulations 2002 SI 2002/2677 and the Control of Asbestos Regulations 2006 (see below).[46]

12.119 The Guidance to the Approved Code of Practice provides that the hazards associated with the above should be identified through a risk assessment carried out under the Management of Health and Safety at Work Regulations 1999 SI 1999/3242 reg 3(1). Equipment should be designed in such a way as to avoid any of the above consequences occurring where that is possible.

F. Substances Hazardous to Health: the COSHH Regulations 2002

The Control of Substances Hazardous to Health Regulations

12.120 The principal requirements relating to hazardous substances in the workplace are those contained in the Control of Substances Hazardous to Health Regulations

---

[46] Provision and Use of Work Equipment Regulations 1998 SI 1998/2306 reg 12(5).

SI 2002 SI 2002/2677, known as the COSHH Regulations. The regulations were first introduced in 1988 and have been subject to many changes and amendments since. The 2002 Regulations which are currently in force have been significantly amended several times since enactment. The most significant recent amendments have resulted in the introduction of 'Workplace Exposure Limits' (WELs) to replace both 'Maximum Exposure Limits' (MELs) and 'Occupational Exposure Standards' (OESs).

12.121 The Approved Code of Practice and Guidance is currently in its fifth edition,[47] and both the provisions contained within the extensive schedules to the Regulations and the Workplace Exposure Limits are regularly amended and updated. For this reason, the current COSHH Regulations 2002 SI 2002/2677, but not the Schedules, are reproduced at Appendix I.

'Substance hazardous to health'

12.122 Regulation 2(1) of the the COSHH Regulations 2002 SI 2002/2677 includes a definition of 'substance hazardous to health', which extends to:

(a) substances designated as very toxic, toxic, corrosive, harmful, or irritant under product labelling legislation, namely the Chemicals (Hazard Information and Packaging for Supply) Regulations 2002 SI 2002/1689;

(b) substances for which the Health and Safety Commission has approved a Workplace Exposure Limit;

(c) biological agents (micro-organisms, cell cultures, or human endoparasites);

(d) other dust of any kind, when present at a concentration in air equal or greater to 10 mg/m^3 (inhalable dust) or 4 mg/m^3 (respirable dust) as a time weighted average over an eight-hour period; and

(e) any other substance creating a risk to health because of its chemical or toxicological properties and the way it is used or is present at the workplace.

Who owes duties under the Regulations?

12.123 Regulation 8 of the COSHH Regulations 2002 SI 2002/2677 places duties on employers in relation to their employees, and also on employees themselves. The self-employed have a duty equivalent to both an employer and employee. Regulation 3(1) of the COSHH Regulations 2002 SI 2002/2677 extends the employer's duties, 'so far as is reasonably practicable', to 'any other person, whether at work or not, who may be affected by the work carried on'.

---

[47] The Control of Substances Hazardous to health Regulations 2002 (as amended). Approved Code of Practice and Guidance L5 (5th edn).

The duties

12.124 Regulation 4 of the COSHH Regulations 2002 SI 2002/2677 provides prohibitions relating to certain substances, including the specified prohibition of the manufacture, use, importation, and supply of various substances particularized in COSHH Regulations 2002 SI 2002/2677 Sch 2.

Risk assessment

12.125 Regulation 6(1) of the COSHH Regulations 2002 SI 2002/2677 requires that, 'an employer shall not carry out work liable to expose any employees to any substance hazardous to health' unless he has both made an assessment of health risks created by the work to identify steps necessary to comply with the COSHH Regulations 2002 SI 2002/2677, and implemented those steps.

12.126 When carrying out assessments, reg 3(1) of the COSHH Regulations 2002 SI 2002/2677 requires employees to take account, so far as is reasonably practicable, of others who may be affected by the work, such as visitors, contractors, customers, passers-by, and members of the emergency services.

12.127 Regulation 6(2) of the COSHH Regulations 2002 SI 2002/2677 provides many specific factors which must be considered in the assessment:

(a) the hazardous properties of the substance;
(b) information on health effects provided by the supplier, including information contained in any relevant safety data sheet;
(c) the level, type, and duration of exposure;
(d) the circumstances of the work, including the amount of the substance involved;
(e) activities, such as maintenance, where there is the potential for a high level of exposure;
(f) any relevant occupational workplace exposure limit or similar occupational exposure limit;
(g) the effect of preventive and control measures which have been or will be taken in accordance with COSHH Regulations 2002 SI 2002/2677 reg 7;
(h) the results of relevant health surveillance;
(i) the results of monitoring of exposure in accordance with COSHH Regulations 2002 SI 2002/2677 reg 10;
(j) in circumstances where the work will involve exposure to more than one substance hazardous to health, the risk presented by exposure to such substances in combination;
(k) the approved classification of any biological agent; and
(l) such additional information as the employer may need in order to complete the risk assessment.

12.128 Regulation 6(3) of the COSHH Regulations 2002 SI 2002/2677 requires that such assessment must be the subject of periodic review and be reviewed in specified circumstances.

The duty to prevent or control exposure

12.129 Regulation 7(1) of the COSHH Regulations 2002 SI 2002/2677 imposes an absolute duty on an employer to ensure that the exposure of his employees to substances hazardous to health is either prevented, or where this is not reasonably practicable, adequately controlled. The nature of this duty was authoritatively considered by the Court of Appeal in *Dugmore v Swansea NHS Trust* [2003] 1 All ER 333, considered at paragraphs 6.118–6.126. The 'hierarchical' approach required to be taken by the duty contained in this first paragraph of reg 7 is set out in the following paragraphs of COSHH Regulations 2002 SI 2002/2677 reg 7:

> (2) In complying with his duty of prevention under paragraph (1), substitution shall by preference be undertaken, whereby the employer shall avoid, so far as is reasonably practicable, the use of a substance hazardous to health at the workplace by replacing it with a substance or process which, under the conditions of its use, either eliminates or reduces the risk to the health of the employees.
> (3) Where it is not reasonably practicable to prevent exposure to a substance hazardous to health, the employer shall comply with his duty of control under paragraph (1) by applying protection measures appropriate to the activity and consistent with the risk assessment, including in order of priority—
> (a) the design and use of appropriate work processes, systems and engineering controls and the provision and use of suitable work equipment and materials;
> (b) the control of exposure at source, including adequate ventilation systems and appropriate organizational measures; and
> (c) where adequate control of exposure cannot be achieved by other means, the provision of suitable personal protective equipment in addition to the measures required by sub-paragraphs (a) and (b).

12.130 Regulation 7(4) of the COSHH Regulations 2002 SI 2002/2677 sets out various matters which must be included in the control measures required by para (3). These include arrangements for the safe handling, storage, and transport of hazardous substances, and of waste containing such substances, at the workplace; the adoption of suitable maintenance procedures; the control of the working environment, including appropriate general ventilation; appropriate hygiene measures, including adequate washing facilities; and reducing, to the minimum required for the work concerned, the number of employees subject to exposure, the level and duration of exposure, and the quantity of substances hazardous to health present at the workplace.

12.131 The COSHH Regulations 2002 SI 2002/2677 regs 7(7) (as amended) and 7(11) provide a definition of what is meant by the term 'adequate control' in respect of risks from hazardous substances. In relation to certain specified substances this provides a defined limit through 'Workplace Exposure Limits', in respect of

F. Substances Hazardous to Health: the COSHH Regulations 2002

specified others it sets a level of 'as low as is reasonably practicable', but otherwise provides generally that 'adequate' means adequate having regard only to the nature of the substance and the nature and degree of exposure to substances hazardous to health, and 'adequately' shall be construed accordingly. The details of 'Workplace Exposure Limits' are published regularly by the HSE, including on the HSE website via the COSHH homepage <http://www.hse.gov.uk/coshh/>.

12.132 Regulation 7(5) and (6) of the COSHH Regulations 2002 SI 2002/2677 set out a number of specific measures which must be applied where it is not reasonably practicable to prevent exposure to carcinogens, mutagens, and biological agents.

Other duties

12.133 By COSHH Regulations 2002 SI 2002/2677 reg 8, employees are required to make 'full and proper use of any control measure, other thing or facility provided in accordance with these Regulations', and employers must 'take all reasonable steps' to ensure that the same are properly used or applied.

12.134 By COSHH Regulations 2002 SI 2002/2677 reg 9, every employer who provides control measures to meet the requirements of reg 7, is under a duty to ensure that it is maintained in an efficient state. Regulation 10 provides that where a risk assessment has identified the need for monitoring, such monitoring of exposure of employees shall take place at regular intervals and whenever a change occurs which may affect that exposure.

12.135 Regulation 11 of the COSHH Regulations 2002 SI 2002/2677 contains detailed provisions concerning health surveillance, which is defined[48] as 'assessment of the state of health of an employee', and is to be treated as 'being appropriate' where the employee is exposed to certain substances or processes listed in Sch 6 or where an identifiable disease may be related to or occur from exposure to any substance during work and there are valid techniques for detecting indications of the disease.

12.136 Regulation 12(1) of the COSHH Regulations 2002 SI 2002/2677 provides that every employer who undertakes work which is liable to expose an employee to a hazardous substance must provide such an employee with suitable and sufficient information, instruction, and training. Regulation 12(2) provides detail of what shall be included in such information, training, and instruction.

12.137 The arrangements to deal with accidents, incidents, and emergencies related to hazardous substances that an employer must ensure are in place is set out in COSHH Regulations 2002 SI 2002/2677 reg 13. It provides that emergency procedures must be established where significant risks exist, and information on emergency arrangements provided.

---

[48] In reg 2(1).

Exemptions

12.138 Regulation 5 of the COSHH Regulations 2002 SI 2002/2677 provides exceptions from the application of regs 6–13, where more specific regulations are in place. These more specific regulations relate to respirable dust in coal mines, lead, asbestos, the radioactive, explosive, or flammable properties of substances, substances at high or low temperatures or high pressure, and substances administered in the course of medical treatment.

G. Construction

Introduction

12.139 According to the HSE website, some 2.2 million people work in Britain's construction industry, making it the country's largest employment sector. Statistically, it is also one of the most dangerous. Over 2,800 people have died while engaged in construction work in the last twenty-five years, and countless others have suffered injury or ill-health. The major reason for this rate of injury and death is undoubtedly the high level of risk inherent in construction and demolition work generally, but risks increase as a result of the presence of numerous different trades or firms on a given site, and the difficulties involved in managing and coordinating their activities.

Work at Height Regulations 2005

12.140 The Work at Height Regulations 2005 SI 2005/735 replaced those parts of the Construction (Health Safety and Welfare) Regulations 1996 SI 1996/1592 which regulated the conduct of work at height. The Regulations are reproduced at Appendix K. They impose duties on employers in respect of their employees, and on the self-employed in respect of themselves. Importantly, they also impose duties on all persons who exercise control over work at height, whatever their status.[49]

12.141 Regulations 4 and 5 of the Work at Height Regulations 2005 SI 2005/735 impose mandatory duties on every employer to ensure that work is both properly planned and appropriately supervised, and that no person engages in any activity in relation to work, including organization, planning, and supervision, unless he is competent to do so. Overall the employer must ensure that the work is carried out in a manner which is safe so far as is reasonably practicable.[50]

12.142 Whereas the relevant portions of the Construction (Health Safety and Welfare) Regulations 1996 SI 1996/1592 stipulated particular control measures when work was carried out at heights above two metres, this threshold is abandoned in

---

[49] Work at Height Regulations 2005 SI 2005/735 reg 3(2), (3).
[50] ibid reg 4.

G. Construction

the new regulations. Instead, they require the employer to ensure that work is not carried out at height, where this is reasonably practicable;[51] but where it is not, he must take suitable and sufficient measures to prevent, so far as is reasonably practicable, any person falling a distance liable to cause personal injury.[52]

12.143 The Regulations then set out a detailed hierarchy of measures for compliance with this duty. First, the employer should ensure that the work is carried out from an existing place of work;[53] where this is not reasonably practicable, he should provide sufficient work equipment to prevent, so far as is reasonably practicable, a fall occurring.[54] Where these measures do not eliminate the risk of a fall occurring, the employer must so far as is reasonably practicable, provide sufficient work equipment to minimize the distance and consequences of the fall.[55] Where it is not reasonably practicable to minimize the distance of a fall, he must minimize the consequences. Additionally, he must provide the training and instruction, or take other suitable and sufficient measures to prevent, so far as is reasonably practicable, any person falling a distance liable to cause personal injury.

12.144 Regulation 14 of the Work at Height Regulations 2005 SI 2005/735 is the only one of the regulations which imposes duties on 'workers', as opposed to employers, the self-employed or those in control of work. It provides that every person working under the control of another person must report to him any activity or defect relating to work at height which he knows is likely to endanger the safety of himself or another person. Every person must also use any work equipment or safety device provided to him for work at height by his employer, or by a person under whose control he works, in accordance with his training in the use of the work equipment or device concerned which has been received by him; and the instructions respecting that use which have been provided to him by that employer or person in compliance with the requirements and prohibitions imposed upon that employer or person by or under the relevant statutory provisions.[56]

12.145 The remainder of the Work at Height Regulations 2005 SI 2005/735 relate to the specifications of equipment used for working at height,[57] the process for the inspection of this equipment,[58] work on fragile surfaces,[59] falling objects,[60] and danger areas.[61]

---

[51] Work at Height Regulations 2005 SI 2005/735 reg 6(2).
[52] ibid reg 6(3).
[53] ibid reg 6(4)(a)(i).
[54] ibid reg 6(4)(b).
[55] ibid reg 6(5).
[56] ibid reg 14(2).
[57] ibid regs 7 and 8.
[58] ibid reg 12.
[59] ibid reg 9.
[60] ibid reg 10.
[61] ibid reg 11.

Construction (Design and Management) Regulations 2007

12.146 The Construction (Design and Management) Regulations 2007 SI 2007/320 came into force on 6 April 2007 and are reproduced at Appendix J. They repealed the Construction (Design and Management) Regulations 1994 SI 1994/3140 and the remaining Construction (Health Safety and Welfare) Regulations 1996 SI 1996/1592, and brought the provisions together in a single document. The new Regulations are supported by an Approved Code of Practice and Guidance which was available three months before the Regulations themselves come into force.

12.147 Since the Construction (Design and Management) Regulations 1994 SI 1994/3140 were introduced in 1995, they had been the subject of criticism. While it was considered that the general principles were sound, it was widely felt that they were overly complex and failed to achieve their initial objectives—namely, to create a new approach to managing health and safety in construction projects by providing clear guidance to all parties. Despite the revision of the Approved Code of Practice in 2001, it was felt that rather than reducing and controlling risk, they encouraged an overly bureaucratic response on the part of many duty-holders. A further criticism was that paper was created for its own sake; or, more cynically, in order to protect duty-holders in the event of litigation. Particular concerns were expressed about the role of the client and the planning supervisor. Of course, the planning supervisor was a creation of the European Directive, and was based on a role that already existed in many continental construction projects. There was historically no directly equivalent role in UK construction, and it is arguable that the some of the ineffectiveness of the role stems from the fact that it did not sit easily with traditional methods for the organization of construction work in the UK.

12.148 Concerns such as these prompted the HSE to publish a discussion document in 2002 entitled 'Revitalising Health and Safety in Construction', which invited all interested parties to express their views on the future of safety management in the industry. In the light of the concerns expressed during this process, in 2003 the HSE proposed a revision of the regulations. In March 2005 the HSC published a consultation document which contained draft new regulations and an Approved Code of Practice, again inviting comments on the proposed revisions. Between then and 2007, there was a further period of consultation with interested parties which resulted in the production of the 2007 Regulations and Approved Code of Practice; these can be found on the HSE website at <http://www.hse.gov.uk/aboutus/hsc/meetings/2006/171006/c54Ann5.pdf>.

12.149 According to the HSE, the intention of the new Construction (Design and Management) Regulations 2007 SI 2007/320 and Approved Code of Practice[62]

---

[62] Managing health and safety in construction, Construction (Design and Management) Regulations 2007, Approved Code of Practice L144.

G. Construction

is to help the construction industry by simplifying and clarifying the law and raising health and safety standards through improved planning and management of construction projects. Though prescriptive, they are also intended to be flexible enough to cater for the wide range of contractual arrangements to be found in the construction industry.

The principal changes are the redefinition of the role of the client, the creation of a new role—the Construction Design Management Coordinator—to replace the Planning Supervisor, and the bringing together of all construction-specific requirements into one set of regulations. As was the case with the 1994 Regulations, domestic clients are exempted from the duties, even though there was no provision for this in the Directive. The specific practical requirements in the Construction (Health Safety and Welfare) Regulations 1996 SI 1996/1592, are reproduced largely unchanged. **12.150**

The Construction (Design and Management) Regulations 2007 SI 2007/320 are divided into five parts. Part 1 contains an interpretation section; Part 2 contains general management duties applicable to all construction projects; Part 3 contains provisions which apply only to notifiable projects;[63] and Part 4 contains detailed provisions previously found in the Construction (Health Safety and Welfare) Regulations 1996 SI 1996/1592. The regulations in Part 4 are of general application and apply to all contractors working within construction projects, and all those who control the way in which work is carried out. **12.151**

A number of the definitions that were in the 1994 Regulations have been changed. These include: 'client', 'construction phase plan', 'construction work', 'contractor', 'design' and 'designer', 'place of work', and 'structure'. **12.152**

General provisions for health and safety

Part 2 of the new Construction (Design and Management) Regulations 2007 SI 2007/320 contains general provisions for the management of health and safety which apply to all construction projects, whether notifiable or not. Regulation 7 is broad in scope, requiring all involved in the design or planning of construction work, and all those who are involved in the construction work itself, to take account of the principles of prevention contained in the Framework Directive, Management of Health and Safety at Work Regulations 1999 SI 1999/3242 Sch 1, and Construction (Design and Management) Regulations 2007 SI 2007/320 app 7. There was no such general provision for health and safety management in the 1994 Regulations. **12.153**

---

[63] A notifiable project is defined as before, ie one which lasts more than 30 days or involves over 500 man-hours of construction work, but the application provision relating to less than five workers on site has been removed. Domestic projects no longer need to be notified.

Competence

12.154 The issue of competence was the cause of much discussion during the consultation phase for the new regulations. The concern was to achieve a meaningful and demonstrable level of competence—particularly with regard to coordinators and contractors—without excluding those with more extensive practical experience than paper qualifications. The regulations provide that duty-holders cannot arrange for, or instruct anyone to carry out or manage design or construction work unless that person is competent, or is being supervised by someone who is. Neither can a duty-holder accept a construction (design and management) appointment or engagement unless they are themselves competent to carry it out. The assessment and demonstration of competence is now simplified, with new core criteria and specific Approved Code of Practice material on individual and corporate competence.

12.155 Chapter 6 of the Approved Code of Practice and Guidance[64] and appendices 4, 5, and 6 contain what is intended to be user-friendly and practical guidance on assessing competence. It suggests that assessments of competence should focus on the needs of the particular project and be proportionate to the risks, size, and complexity of the work. Any person who is assessing competence must make reasonable enquiries to check competence. To be competent, an organization or individual must have sufficient knowledge of the specific tasks to be undertaken and the risks which the work will entail; and sufficient experience and ability to carry out their duties in relation to the project. They must also recognize their limitations and take appropriate action in order to prevent harm to those carrying out construction work, or those affected by the work.

12.156 A two-stage process for assessing competence is advocated; this assessment should be carried out using a set of 'core criteria' in order to encourage consistency. With smaller projects, such as those falling below the notification threshold, companies need only be asked to provide the minimum paperwork necessary to show that they meet each element set out in the core criteria. For larger projects, or those where the risks are greater, a more in-depth assessment will be needed, but the assessment should not go beyond the elements set out in the core criteria.

Cooperation and coordination

12.157 The Construction (Design and Management) Regulations 2007 SI 2007/320 also impose general duties of cooperation and coordination on everyone involved in a project, and contain a specific requirement to implement any preventive and protective measures on the basis of the principles specified in the Management Regulations.

---

[64] Managing health and safety in construction, Construction (Design and Management) Regulations 2007, Approved Code of Practice L144.

G. Construction

Within the definitions section of the Approved Code of Practice and Guidance (L144) it is suggested that for low-risk projects, a low-key approach will suffice; whereas for higher risk projects a more rigorous approach will be necessary. The Approved Code of Practice and Guidance suggests that in the case of high-risk projects the parties should have regard to the approach mandated for those involved in notifiable projects.

Clients: changed and increased duties

12.158 The Construction (Design and Management) Regulations 2007 SI 2007/320 place increased duties on clients with the intention of making them more accountable, consistent with their position of influence in a construction project. However, during the consultation phase, concerns were expressed as to whether the enhanced duties placed too heavy a burden on small and medium-sized enterprises (the 'one-off' client), as opposed to the full-time client who might be expected to have a developed infrastructure in place for meeting his obligations. For example, concerns were expressed about how the inexperienced client would be able to assess whether the coordinator was doing a proper job, and how he might be expected to assess the work of construction professionals generally. In the light of these concerns, an increased period of consultation was embarked on, and independent reports commissioned.[65] This led to specific material being introduced into the Approved Code of Practice and Guidance which is aimed at making it clear to the inexperienced client how he should discharge his duty; and how he should assess competency. An insight into how this may be reflected in enforcement action is to be found in the HSE discussion document, where it is stated: 'If enforcement action is necessary, each will be judged according to their level of construction experience and competence (respectively), whether their actions were reasonable according to the facts and circumstances of the case.'[66]

12.159 The primary duty imposed on the client is to take reasonable steps to ensure that suitable management arrangements are maintained and reviewed throughout the project.[67] These management arrangements should be suitable to ensure that the construction work can be carried out so far as is reasonably practicable without risk to the health and safety of any person, that welfare arrangements are in place before work commences,[68] and that any structure designed for use as a workplace complies with the Workplace Regulations 1992.

---

[65] See, for example, 'Small business and one-off/occasional clients' responsibilities' by Tim Kind.
[66] See <http://www.hse.gov.uk/aboutus/hsc/iacs/coniac/200706/m220061.pdf>.
[67] Construction (Design and Management) Regulations 2007 SI 2007/320 reg 9(2).
[68] ibid reg 9(1).

12.160 A more detailed description of what these arrangements should include is contained in chapter 1 of the Approved Code of Practice and Guidance.[69] This suggests that the management arrangements detailed in reg 9 will include arrangements to achieve clarity as to the roles, functions, and responsibilities of members of the project team; those with duties under the regulations have sufficient time and resource to comply with their duties; there is good communication, coordination, and cooperation between members of the project team (eg between designers and contractors); designers are able to confirm that their designs (and any design changes) have taken account of the requirements of reg 11 (designers' duties), and that the different design elements will work together in a way which does not create risks to the health and safety of those constructing, using, or maintaining the structure; that the contractor is provided with the pre-construction information; contractors are able to confirm that health and safety standards on site will be controlled and monitored; and welfare facilities will be provided by the contractor from the start of the construction phase through to handover and completion.

12.161 Whereas reg 9 of the Construction (Design and Management) Regulations 2007 SI 2007/320 contains the duties exclusive to clients, by application of the general duties in regs 4–7, clients must also ensure that designers, contractors, and others are competent, adequately resourced, and appointed sufficiently early. They must also allow sufficient time for each stage of the project, from concept onwards. They must cooperate with others concerned in the project as is necessary to allow other duty-holders to comply with their own duties, and coordinate their own work with others involved with the project in order to ensure the safety of those carrying out the construction work, and others who may be affected by it.

12.162 The client also has a duty to ensure that every person designing the structure, and every contractor who has been or may be appointed by the client, is promptly provided with certain prescribed pre-construction information so as to ensure so far as is reasonably practicable the health and safety of persons engaged in construction work or liable to be affected by it.[70]

12.163 Chapter 1 of the Approved Code of Practice and Guidance provides some practical guidance as to how the client—of whatever size—should discharge the duties imposed by reg 9. It provides that although the Regulations make clients accountable for the impact their approach has on the health and safety of the project, most clients, particularly those who only occasionally commission construction work,

---

[69] Managing health and safety in construction, Construction (Design and Management) Regulations 2007, Approved Code of Practice L144.
[70] Construction (Design and Management) Regulations 2007 SI 2007/320 reg 10(1).

G. Construction

will not be experts in the construction process, and for this reason are not required to take an active role in managing the work. Equally, they are not expected to develop substantial expertise in construction health and safety, unless this is central to their business. In general terms, while clients must ensure that various things are done, they are not normally expected to do them themselves.

12.164 The extent to which clients are expected to police the arrangements is also considered in the Approved Code of Practice and Guidance. It suggests that having made these initial checks before work begins, clients should make periodic checks through the life of the project to make sure the arrangements which have been made are properly implemented and updated as the project progresses. It goes on to say that, when deciding whether management arrangements are suitable and maintained throughout the project, clients will need to make a judgment, taking account of the nature of the project and the risks that the work will entail. In terms of potential future enforcement action, the Approved Code of Practice and Guidance provides that if this judgment is reasonable and clearly based on the evidence requested and provided, clients will not be criticized if the arrangements subsequently prove to be inadequate; or if the company that has made the arrangements fails to implement them properly without the client's knowledge.

12.165 For non-notifiable projects, the Approved Code of Practice and Guidance suggests that only simple checks will be needed: for example, checking that there is adequate protection for the client's workers and/or members of the public, checking to make sure that adequate welfare facilities have been provided by the contractor, checking that there is good cooperation and communication between designers and contractors, and checking that the arrangements which the contractor agreed in order to control key risks on site have been implemented.

Notifiable projects

12.166 If the project is notifiable, further duties apply. In that case, the client must appoint a construction design and management coordinator[71] as soon as possible after design work or the preparation for construction has begun. He must also appoint a principal contractor to plan and manage the construction work.[72] If he fails to appoint either a coordinator or a principal contractor, he will be taken to be performing both of their roles.[73] The client also has the responsibility promptly to provide the coordinator with design and pre-construction information concerning health and safety[74] and all the information required to keep the health

---

[71] Construction (Design and Management) Regulations 2007 SI 2007/320 reg 14(1).
[72] ibid reg 14(2).
[73] ibid reg 14(4).
[74] ibid reg 15.

and safety plan up to date.[75] He must also make sure that the construction phase does not start unless the principal contractor has completed a construction phase plan.[76]

12.167 The Approved Code of Practice and Guidance provides that getting the right people for these roles, and making early appointments, is particularly important for clients with little construction or health and safety expertise, as they will need to rely on the advice given by the construction (design and management) coordinator on matters relating to the competence of those whom they intend to appoint, and the adequacy of the management arrangements made by appointees.

12.168 The Approved Code of Practice and Guidance also sets out a helpful summary of what clients are not required or expected to do. They are not expected to plan or manage construction projects themselves, or to specify how work must be done, eg requiring a structure to be demolished by hand. Indeed, they should not do so unless they have the expertise to assess the various options and risks involved. They are not expected to provide welfare facilities for those carrying out construction work, although they should cooperate with the contractor to assist with his arrangements; they are not expected to check designs to make sure that reg 11[77] has been complied with; nor are they expected to visit the site to supervise or check construction work standards. They are not required to employ third party assurance advisors to monitor health and safety standards on site although the Approved Code of Practice and Guidance notes that there may be benefits to the client in doing so; equally they are not expected to subscribe to third party competence assessment schemes, though again it is noted that there may be benefits from doing so.

The coordinator

12.169 The role of the planning supervisor was widely perceived to have been ineffective and has been abolished by the new regulations. The very title—*planning supervisor*—was considered to be a misnomer, in that he neither planned, nor supervised. He has now been replaced by the coordinator, who has more extensive powers and duties. Arguably, the principal difference between the coordinator and the planning supervisor is that the coordinator has a number of positive obligations with regards to the client—for instance, he must advise and assist the client in the discharge of his duties; whereas under the Construction (Design and Management) Regulations 1994 SI 1994/314, he was merely required to

---

[75] Construction (Design and Management) Regulations 2007 SI 2007/320 reg 17.
[76] ibid reg 16.
[77] See below for Construction (Design and Management) Regulations 2007 SI 2007/320 reg 11.

G. Construction

be in a position to give adequate advice.[78] He has been referred to as the 'power behind the throne',[79] and the intention of the Regulations is to match his empowerment with an increased accountability.

12.170 It remains to be seen what approach the HSE will take towards enforcement proceedings against the coordinator, although it seems likely that his increased accountability will make prosecution more common than was the case with the planning supervisor. The HSE paper presenting the draft regulations and Approved Code of Practice in July 2006 stated: 'If enforcement action is necessary, each [ie the coordinator and client] will be judged according to their level of construction experience and competence (respectively), whether their actions were reasonable according to the facts and circumstances of the case.'

12.171 The coordinator's duties are contained within regs 20 and 21 and are detailed in the Approved Code of Practice and Guidance.[80] He need only be appointed for notifiable projects, but should be appointed as soon as practicable after the commencement of initial design work or other preparation for construction work.[81] His major role is to act as the client's friend—ie to support and advise him in discharging his duties. He must give the client suitable and sufficient advice on undertaking the measures he needs to take to comply with his duties under the Regulations—in particular, he must advise the client on his key duty, ie whether suitable health and safety arrangements have been made, and are maintained.[82] This involves his giving the client suitable and sufficient advice in respect of the selection of competent designers and contractors. He should advise the client if surveys need to be commissioned to fill significant gaps.[83]

12.172 Further, he must coordinate design work, planning, and other preparation for construction where it is relevant to health and safety, and take all reasonable steps to ensure that designers comply with their duties under the regulations—which are onerous. He must take all reasonable steps to identify and collect the pre-construction information, and promptly provide it in a convenient form to every person designing the structure, and every contractor, including the principal contractor.[84] During the construction phase he must take all reasonable steps

---

[78] Construction (Design and Management) Regulations 1994 SI 1994/3140 reg 14.
[79] Health and Safety Commission Construction Industry Advisory Committee (CONIAC) Proposed Construction (Design and Management) Regulations 2007: Clearance of Draft Regulatory Package.
[80] Managing health and safety in construction, Construction (Design and Management) Regulations 2007, Approved Code of Practice L144, paras 84 *et seq*.
[81] Construction (Design and Management) Regulations 2007 SI 2007/320 reg 14(1).
[82] ibid reg 20(1).
[83] Approved Code of Practice para 93.
[84] Construction (Design and Management) Regulations 2007 SI 2007/320 reg 20(2).

to ensure cooperation between designers and the principal contractor in relation to any design or change to a design.

12.173 As far as project documentation is concerned, he must advise the client on the suitability of the initial construction phase plan and the arrangements made to ensure that welfare facilities are on site from the start. He must produce or update a relevant, user-friendly, health and safety file which is suitable to be passed on for future use at the end of the construction phase.

12.174 The Approved Code of Practice and Guidance also makes it clear what the coordinator does not have to do.[85] He need not approve the appointment of designers, principal contractors, or contractors, although he will normally advise clients about competence and resources; he need not approve or check designs, but should be satisfied that the design process addresses the need to eliminate and control risks; he need not approve the principal contractor's construction phase health and safety plan, or be able to advise clients on its adequacy at the start of construction; he need not supervise the principal contractor's implementation of the construction phase health and safety plan; nor must he supervise or monitor construction work, as this is the responsibility of the principal contractor.

Designers

12.175 The Construction (Design and Management) Regulations 2007 SI 2007/320 purport to remove the confusion which may have existed under the 1994 Regulations, and to strengthen existing duties. These changes are reflected to some extent in the changed definitions of 'design' and 'designer' in the new regulations. Designers must now eliminate hazards and reduce remaining risks, so far as is reasonably practicable. The obligation to eliminate 'hazards' is new to the Construction (Design and Management) Regulations 2007 SI 2007/320, with the 1994 Regulations referring only to the need to combat risks. This extra requirement is, apparently, an application of the principles of prevention to the role of the designer. They must also now ensure that any workplace they design complies with relevant sections of the Workplace (Health Safety and Welfare) Regulations 1992 SI 1998/3004—a duty which was previously the client's.

12.176 The designer's primary duty for all projects, whether notifiable or not, is to avoid foreseeable risks to the health and safety of any persons doing construction work, or those liable to be affected by it, so far as is reasonably practicable.[86] To do this, the designer should eliminate hazards which may give rise to risks; and reduce risks from any remaining hazards.[87] He should make sure that he is

---

[85] Approved Code of Practice para 108.
[86] Construction (Design and Management) Regulations 2007 SI 2007/320 reg 11(3).
[87] ibid reg 11(4).

G. Construction

competent and adequately resourced to address the health and safety issues likely to be involved in the design; and check that clients are aware of their duties.[88] He must also take all reasonable steps to provide with his design sufficient information about aspects of the design of the structure or its construction or maintenance as will adequately assist clients, other designers, and contractors to comply with their duties under these Regulations.

12.177 For notifiable projects, designers are prohibited from doing anything more than initial design work before the coordinator has been appointed.[89] They must also ensure that the client has appointed a Construction (Design and Management) coordinator and notified the HSE. The designer must take all reasonable steps to provide with his design sufficient information about aspects of the design of the structure or its construction or maintenance as to adequately assist the Construction (Design and Management) coordinator to comply with his duties under these Regulations, including his duties in relation to the health and safety file.[90]

12.178 The Approved Code of Practice and Guidance provides a detailed list of who may be designers[91] and goes on to consider their duties in more detail, providing practical examples of how each aspect may be discharged. It suggests that the designer need not advert to all risks, only to the significant ones, or those which are unusual and unlikely to be obvious to a competent contractor or designer, and which are likely to be difficult to manage effectively. By way of example, it suggests that providing generic risk information about the prevention of falls is pointless, because competent contractors will already know what needs to be done; but if the design gives rise to a specific and unusual fall risk which may not be obvious to contractors, designers should provide information about this risk.[92]

12.179 The Approved Code of Practice and Guidance also sets out what designers do not have to do: they do not have to take into account or provide information about unforeseeable hazards and risks; design for possible future uses of structures that cannot reasonably be anticipated from their design brief; specify construction methods, except where the design assumes or requires a particular construction or erection sequence, or where a competent contractor might need such information; exercise any health and safety management function over contractors or others; or worry about trivial risks.

---

[88] Construction (Design and Management) Regulations 2007 SI 2007/320 regs 4 and 5.
[89] ibid reg 18.
[90] ibid.
[91] Approved Code of Practice para 116.
[92] ibid para 131.

The principal contractor

12.180 A principal contractor need only be appointed where the project is notifiable, but the client must ensure that he is appointed as soon as is reasonably practicable after he has sufficient information to make an informed selection,[93] and that a principal contractor remains in post until the end of the project. Although there are no substantial changes to duties, the new regulations set out his duties in a more straightforward and direct way than had previously been the case. His primary duty is to 'plan, manage and monitor' the construction phase.[94] In the 1994 Regulations there was no such direct expression of his overall responsibility,[95] and while they undoubtedly placed him at the heart of health and safety management in the construction phase, his responsibilities were somewhat diffuse and largely defined by reference to the provisions of the health and safety plan. The consultation process identified this as something which had the potential to encourage a slavish adherence to project documentation, at the expense of effective planning and risk management.

12.181 In addition to his primary duty to plan, manage, and monitor the construction phase, reg 22 of the Construction (Design and Management) Regulations 2007 SI 2007/320 sets out no less than twelve separate duties which the principal contractor must comply with. It is interesting to note that while in the case of other duty-holders the Approved Code of Practice and Guidance contains a long list of the things that the Regulations do not require him to do, for the principal contractor only one matter is listed: 'Undertake detailed supervision of contractors' work.'

12.182 The principal contractor must:

(a) liaise with the construction (design and management) coordinator during the construction phase in relation to any design or change to a design;
(b) ensure that sufficient welfare facilities are provided throughout the construction phase;
(c) draw up rules (the 'site rules') which are appropriate to the construction site and the activities on it where these are necessary for health and safety;
(d) give reasonable directions to any contractor so far as is necessary to enable him to comply with his own duties;
(e) ensure that every contractor is informed of the minimum amount of time which will be allowed to him for planning and preparation before he begins construction work;

---

[93] Construction (Design and Management) Regulations 2007 SI 2007/320 reg 14(2).
[94] Reg 22(1)(a).
[95] The Approved Code of Practice to the 1994 Regulations identified the need for the principal contractor to monitor the work of contactors.

G. Construction

(f) consult a contractor before finalizing a part of the construction phase plan relevant to the work to be performed by him;

(g) ensure that every contractor is given access to such part of the construction phase plan as is relevant to the work to be performed by him before he begins construction work and in sufficient time to enable him to prepare properly;

(h) ensure that every contractor is given, before he begins work, and in sufficient time for proper preparation, such further information as he needs to comply punctually with the duty to provide welfare facilities and to carry out his work without risk, so far as is reasonably practicable, to the health and safety of any person;

(i) identify to each contractor the information relating to the contractor's activity which is likely to be required by the coordinator for inclusion in the health and safety file for future reference and make sure that it is given to him;

(j) ensure that the particulars required to be in the notice given to the HSE are prominently displayed; and

(k) take reasonable steps to prevent unauthorized access to the construction site.

12.183 He must also take all reasonable steps to ensure that every worker is provided with a site induction,[96] and that he has received the appropriate training from the contractor who controls his work.[97] As far as the types of matters to be covered in the inductions are concerned, the Approved Code of Practice and Guidance lists twenty-two separate factors. This provision is to be read as requiring the principal contractor to make reasonable checks to ensure that contractor's workers are competent and properly trained. It is not, as is made plain in the Approved Code of Practice,[98] to be read as requiring the principal contractor to train everyone on site.

12.184 The Approved Code of Practice and Guidance gives some guidance as to how the competence of a particular individual should be assessed.[99] The process is the same two-stage test recommended for duty-holders assessing the competence of contractors. In broad terms, an assessment should be made of the person's knowledge of the task in hand; and then an assessment of the individual's experience and track record, to establish that they are capable of doing the work, and conscious of their own limitations. In general terms, stage one will involve looking at the individual's qualifications and training records; stage two will concentrate on his past experience in the type of work at hand.

---

[96] Construction (Design and Management) Regulations 2007 SI 2007/320 reg 22(2)(a).
[97] ibid reg 22(2)(b).
[98] Managing health and safety in construction, Construction (Design and Management) Regulations 2007, Approved Code of Practice L144 para 173.
[99] Managing health and safety in construction, Construction (Design and Management) Regulations 2007, Approved Code of Practice L144 para 213 *et seq.*

12.185 The Approved Code of Practice and Guidance provides some practical guidance about assessing basic safety awareness; it makes reference to the Construction Industry Training Board's Construction Skills touch-screen test and equivalent schemes, which are designed specifically to test this basic knowledge and understanding. It suggests that passing the touch-screen test or equivalent schemes is one way of demonstrating competence, and that all those who work on or regularly visit sites (including individuals from client, designer, or construction (design and management) coordinator organizations) should be able to demonstrate that they have achieved at least this level of understanding before starting work on site.

12.186 The principal contractor also has important duties with regard to the construction phase plan. He must take all reasonable steps to ensure that the construction phase plan identifies the risks to health and safety arising from the construction work and includes suitable and sufficient measures to address such risks, including any site rules.[100] All of his efforts concerning the plan should be directed towards ensuring, as far as reasonably practicable, health and safety on site and elsewhere. In summary, he must prepare a construction phase plan before work starts, which should contain such information and arrangements as are necessary to enable the work to start safely. He should also review and revise the plan from time to time to ensure it is effective; and arrange for it to be implemented.[101] Chapter 4 (para 160) of the Approved Code of Practice and Guidance suggests that the plan should not be a repository for detailed generic risk assessments, records of how decisions were reached, or detailed method statements; but it may, for example, set out when such documents will need to be prepared. It should be well focused, clear and easy for contractors and others to understand, emphasizing key points and avoiding irrelevant material. It is crucial that all relevant parties are involved and cooperate in the development and implementation of the plan as work progresses.

Contractors

12.187 Anyone who directly employs or engages construction workers or controls or manages construction work is a contractor for the purposes of the Construction (Design and Management) Regulations 2007 SI 2007/320. The duties on contractors apply whether the workers are employees, self-employed, or agency workers. A contractor's primary duty for all projects—whether notifiable or not—is to plan, manage, and monitor the construction work which he carries out or controls so as to ensure that it creates no risks to health and safety, so far as

---

[100] Construction (Design and Management) Regulations 2007 SI 2007/320 reg 23(2).
[101] ibid reg 23(1).

G. Construction

is reasonably practicable.[102] He must also ensure so far as reasonably practicable that the principles of prevention are observed throughout the project in respect of any person under his control.[103]

Aside from these overriding duties, he must also ensure that any contractor he engages is informed of the minimum amount of time he will be allowed for planning and preparation before he begins work.[104] He must also provide every worker under his control with any information and training he needs to do the work safely and without risk to health. This should include suitable site induction, where not provided by the principal contractor; information on the risks to their health and safety identified by his own risk assessments, or by others' risk assessments when their work may affect his own workers; and the detail of the risk control measures identified. **12.188**

A contractor has additional duties where the project is notifiable. With a notifiable project, no contractor may carry out construction work unless he has been provided with the names of the construction (design and management) coordinator and principal contractor, and has been given detailed access to the parts of the construction phase plan relevant his work.[105] The contractor has a number of complex and largely reciprocal duties in connection with the principal contractor: he must make available to him all information in his possession or control, including risk assessments, which might affect the health or safety of any person,[106] or which the principal contractor may need to give to the coordinator, or which might justify a review of the construction phase plan.[107] **12.189**

A contractor must also promptly identify to the principal contractor any contractor he appoints or engages,[108] comply with any direction given by him in connection with the performance of his own duties, and provide him with any information that he himself is required to report under Reporting of Injuries, Diseases and Dangerous Occurrences Regulations 1995 SI 1995/3163.[109] **12.190**

In overall terms, every contractor must also take all reasonable steps to ensure that work is carried out in accordance with the construction phase plan, and notify the principal contractor of any significant finding which requires the construction phase plan to be changed. **12.191**

---

[102] Construction (Design and Management) Regulations 2007 SI 2007/320 reg 13(2).
[103] ibid reg 13(7).
[104] ibid reg 13(3).
[105] ibid reg 19(1).
[106] ibid reg 19(2)(a)(i).
[107] ibid reg 19(2)(a)(ii).
[108] ibid reg 19(2)(b).
[109] ibid reg 19(2)(d).

H. Asbestos

Introduction

12.192 Asbestos-related diseases continue to account for the highest number of work-related deaths in the UK. Some 3,000 people die annually as a result of past exposure. This figure is expected to continue to rise for the next ten years.

12.193 The Control of Asbestos Regulations 2006 SI 2006/2739 came into force on 13 November 2006 and are reproduced at Appendix L. They consolidate in a single instrument the three previous sets of Regulations: they cover the prohibition of asbestos, the control of asbestos at work and asbestos licensing. They are divided into four parts, with Parts 2 and 3 containing the major provisions. There is also a new Approved Code of Practice and Guidance which provides more detailed guidance on the scope and application of the regulations.

Duty-holders in premises

12.194 While the vast majority of the regulations in Part 2 place duties on employers, Control of Asbestos Regulations 2006 SI 2006/2739 reg 4 imposes obligations on 'duty-holders'. A 'duty-holder' is defined as the person who has the obligation to maintain or repair non-domestic premises by virtue of a contract or tenancy;[110] or in the absence of any such agreement, any person who has control of premises. Where there is more than one such duty-holder, the relative contribution to be made by each person in complying with the requirements of this regulation will be determined by the nature and extent of the maintenance and repair obligation owed by that person.

12.195 The duty-holder must ensure that a suitable and sufficient assessment is carried out as to whether asbestos is or is liable to be present in the premises.[111] When making the assessment, he must consider the condition of any asbestos and the age of the premises. The findings of the assessment should be recorded,[112] and he must ensure that the assessment is reviewed when there has been a significant change in the premises.

12.196 If the assessment shows that asbestos is or is liable to be present in any part of the premises, the duty-holder must then ensure that a determination of the risk from that asbestos is made, in the form of a written plan which details the measures

---

[110] Control of Asbestos Regulations 2006 SI 2006/2739 reg 4(1)(a).
[111] ibid reg 4(3).
[112] ibid reg 4(7).

which are to be taken for managing the risk, in accordance with set criteria.[113] This plan should also be reviewed when circumstances warrant it.

Thus the extent of the duty will depend on the nature of any agreement. The HSE provides the following example by way of guidance: where a building is occupied by a single leaseholder, the agreement might nevertheless be for either the owner or leaseholder to take on the duty for the whole building, or it might be to share the duty. In a multi-occupied building, the agreement might be that the owner takes on the full duty for the whole building, or again it might be that the duty is shared—for example, the owner takes responsibility for the common parts while the leaseholders take responsibility for the parts they occupy. Sometimes, there might be an agreement to pass the responsibilities to a managing agent. **12.197**

In some cases, there may be no tenancy agreement or contract. Or, if there is, it may not specify who has responsibility for the maintenance or repair of non-domestic premises. In these cases, or where the premises are unoccupied, the duty is placed on whoever has control of the premises, or part of the premises. Often this will be the owner. **12.198**

Duties on employers

An employer must not undertake work in demolition, maintenance, or any other work which exposes or is liable to expose his employees to asbestos, in respect of any premises unless either he has carried out a suitable and sufficient assessment as to whether asbestos (what type of asbestos, contained in what material and in what condition) is present or is liable to be present in those premises, or if there is doubt as to whether asbestos is present in those premises he assumes that asbestos is present, and that it is not chrysotile alone, and observes the applicable provisions of the Regulations.[114] **12.199**

An employer must not carry out work which is liable to expose his employees to asbestos unless he has carried out an assessment about the presence, nature, and type of asbestos, or whether it is liable to be there,[115] and taken the steps necessary to meet the requirements of the Regulations. The Regulations set out detailed criteria regarding the assessment. It should be recorded in writing and should identify the type of asbestos, determine the nature and degree of exposure, consider the effects of control measures, and set out the steps to be taken to prevent exposure **12.200**

---

[113] Control of Asbestos Regulations 2006 SI 2006/2739 reg 4(8).
[114] ibid reg 5.
[115] ibid reg 6.

Chapter 12: Principal Health and Safety Regulations

or reduce it to the lowest level reasonably practicable. It should be reviewed regularly, and whenever there is any significant change to the work being done.[116]

12.201 An employer must not undertake any work with asbestos unless he has prepared a suitable written plan of work detailing how the work is to be carried out.[117] The plan should be retained at the premises and should include a number of details: the nature and probable duration of the work, the location of the place where the work is to be carried out, the methods to be applied where the work involves the handling of asbestos or materials containing asbestos, the characteristics of the equipment to be used for protection and decontamination of those carrying out the work, and of other persons on or near the worksite. The employer must also ensure, so far as is reasonably practicable, that the work to which the plan of work relates is carried out in accordance with that plan and any subsequent written changes to it.

12.202 The employer must give detailed instruction and training to each of his employees working with asbestos, or to any person supervising.[118] The regulation contains an extensive list of the matters which employers must ensure employees are made aware of; and the Approved Code of Practice and Guidance contains a detailed account of the persons who should be given asbestos awareness training, and what it should consist of.[119]

12.203 The primary duties of employers are contained in reg 11. The overriding duty is to prevent the exposure of his employees to asbestos, so far as is reasonably practicable.[120] Where it is not reasonably practicable to prevent exposure, he must take the measures necessary to reduce the exposure to asbestos to the lowest level reasonably practicable, by measures other than the use of respiratory protective equipment,[121] and ensure that the number of his employees who are exposed to asbestos at any one time is as low as is reasonably practicable. The Approved Code of Practice and Guidance expresses the nature of the duty by saying that work which disturbs asbestos should only be carried out when there is no reasonably practicable method of doing the work, or the alternative method creates a more significant risk.[122]

---

[116] Control of Asbestos Regulations 2006 SI 2006/2739 reg 6.
[117] ibid reg 7.
[118] ibid reg 10.
[119] Control of Asbestos Regulations 2006, Approved Code of Practice and guidance L143, paras 124–154.
[120] Control of Asbestos Regulations 2006 SI 2006/2739 reg 11(1)(a).
[121] ibid reg 11(b)(1).
[122] Control of Asbestos Regulations 2006, Approved Code of Practice and guidance L143 para 155.

H. Asbestos

12.204 Where it is not reasonably practicable for the employer to prevent his employees being exposed, he must implement control measures in the following order of priority:[123] the design and use of appropriate work processes, systems and engineering controls, and the provision and use of suitable work equipment and materials in order to avoid or minimize the release of asbestos; and the control of exposure at source, including adequate ventilation systems and appropriate organizational measures. In addition to these measures the employer must, so far as is reasonably practicable, provide the employees concerned with suitable respiratory protective equipment.[124]

12.205 Where it is not reasonably practicable to reduce the exposure of an employee to asbestos to below the control limit by measures other than respiratory equipment, then, in addition to taking those measures, the employer must provide that employee with suitable respiratory protective equipment in order to reduce the concentration of asbestos in the air inhaled by the employee to a concentration which is below the control limit, and as low as is reasonably practicable.[125]

12.206 The employer must ensure that no employee is exposed to asbestos in a concentration in the air inhaled by that worker which exceeds the control limit;[126] if the control limit is exceeded, he must immediately inform any employees concerned, and their representatives, and ensure that work does not continue in the affected area until adequate measures have been taken to reduce employees' exposure to asbestos to below the control limit.[127] As soon as is reasonably practicable he must identify the reasons for the control limit being exceeded and take the appropriate measures to prevent it being exceeded again; and check the effectiveness of the measures taken by carrying out immediate air monitoring.

12.207 Regulation 12(1) of the Control of Asbestos Regulations 2006 SI 2006/2739 is the only regulation which imposes a duty on employees. Not only must every employer who provides any control measure take all reasonable steps to ensure that it is properly used, or applied as the case may be, but every employee must make full and proper use of any control measure, other thing or facility provided pursuant to the regulations. Where relevant, the employee must take all reasonable steps to ensure that the control measure is returned after use to any accommodation provided for it, and if he discovers a defect, report it forthwith to his employer.

12.208 Regulation 13 of the Control of Asbestos Regulations 2006 SI 2006/2739 mirrors the terms of reg 5 of the Provision and Use of Work Equipment Regulations 1998

---

[123] Control of Asbestos Regulations 2006 SI 2006/2739 reg 11(2).
[124] ibid reg 11(2)(b).
[125] ibid reg 11(3).
[126] ibid reg 11(5)(a).
[127] ibid reg 11(5)(b).

SI 1998/2306, in that every employer who provides any control measure to meet the requirements of the Regulations must ensure that in the case of plant and equipment, including engineering controls and personal protective equipment, the equipment is maintained in an efficient state, in efficient working order, in good repair, and in a clean condition; and that in the case of provision of systems of work and supervision and of any other measure, it is reviewed at suitable intervals and revised if necessary.

12.209 Every employer must ensure that thorough examinations and tests of that equipment are carried out at suitable intervals by a competent person. Every employer must also keep a suitable record of the examinations and tests carried out and of repairs carried out as a result of those examinations and tests, and that record or a suitable summary thereof is kept available for at least five years from the date on which it was made.[128]

12.210 The remaining regulations contain provisions relating to the provision and cleaning of protective clothing, arrangements to deal with accidents, incidents and emergencies, the duty to prevent or reduce the spread of asbestos, the cleanliness of premises and plant, designated areas, air monitoring, standards for air testing and site clearance certification, standards for analysis, health records, and medical surveillance, washing and changing facilities, and storage, distribution, and labelling of raw asbestos and asbestos waste.

I. Gas: Gas Safety (Installation and Use) Regulations 1998

Gas Safety (Installation and Use) Regulations 1998

12.211 There are numerous health and safety provisions relating to commercial gas supply,[129] but the principal regulations concerned with gas systems and appliances are the Gas Safety (Installation and Use) Regulations 1998 SI 1998/2451, which are reproduced at Appendix M. These Regulations provide detailed requirements in relation to the safe installation, maintenance, and use of gas systems, including gas fittings, appliances, and flues. They apply to both domestic and commercial premises, such as offices, shops, and public buildings, although factories and mines are expressly excluded.

12.212 The Gas Safety (Installation and Use) Regulations 1998 SI 1998/2451 generally apply to any 'gas' as defined in the Gas Act 1986, apart from any gas comprising

---

[128] Control of Asbestos Regulations 2006 SI 2006/2739 reg 13(3).
[129] Gas Acts 1986 and 1995, Gas Safety (Management) Regulations 1996 SI 1996/551 and Pipelines Safety Regulations 1996 SI 1996/825.

wholly or mainly of hydrogen when used in non-domestic premises, when these products are 'used' by means of a gas appliance. Such appliances must be designed for use by a gas-consumer for heating, lighting, and cooking. However, a gas appliance does not include a portable or mobile appliance supplied with gas from a cylinder, except when such an appliance is under the control of an employer or self-employed person at a place of work. Similarly, the Regulations do not cover gas appliances in domestic premises where a tenant is entitled to remove such appliances from the premises.

12.213 Enforced by the HSE and, in certain limited circumstances, by local authorities, the Regulations place a heavy responsibility on a wide range of people, including those installing, servicing, maintaining, or repairing gas appliances and other gas fittings; as well as suppliers, users of gas, and particularly upon landlords. Most of the duties impose absolute requirements and thus create strict liability.

12.214 The effect of the Gas Safety (Installation and Use) Regulations 1998 SI 1998/2451 reg 34 is to provide that generally it is an offence for the occupier, owner, or other responsible person to permit a gas appliance to be used if at any time he knows, or has reason to suspect, that the appliance is unsafe or that it cannot be used without constituting a danger to any person.

Competent person

12.215 One of the chief requirements imposed by the Gas Safety (Installation and Use) Regulations 1998 SI 1998/2451 is that any work on a gas fitting must be carried out by a competent person. Those employing gas-fitting operatives, together with other specified persons, such as building contractors who control such work, must therefore ensure that operatives have the required competence for the particular work being done. They are also required to ensure that employees and non-employees carrying out such work on their behalf are members of a class of persons approved by the HSE. At present, this means that they should be registered with the Council for Registered Gas Installers (CORGI).

Work

12.216 The Gas Safety (Installation and Use) Regulations 1998 SI 1998/2451 include a wide variety of provisions relating to all aspects of gas safety, including inspection, maintenance, and preventative measures to be taken to avoid gas leaks, steps to be taken should such a leak occur, the installation of gas meters, valves, and flues, the location of gas pipework and equipment, and the provision of safety notices, colour-coding, and instructions.

12.217 Work, in relation to a gas fitting, is defined in the Gas Safety (Installation and Use) Regulations 1998 SI 1998/2451 as including any of the following activities carried out by any person, whether an employee or not:

(a) installing or reconnecting the fitting;
(b) maintaining, servicing, permanently adjusting, disconnecting, repairing, altering, or renewing the fitting, or purging it of air or gas;
(c) where the fitting is not readily movable, changing its position; and
(d) removing the fitting.

12.218 A general duty is imposed by the Gas Safety (Installation and Use) Regulations 1998 SI 1998/2451 reg 6(1) on all persons in connection with work associated with gas fittings to prevent a release of gas unless steps are taken which ensure the safety of any person. Regulation 6(2)–(6) provides that the following precautions must be observed when carrying out work activities:

(a) a gas fitting must not be left unattended unless every complete gasway has been sealed with an appropriate fitting;
(b) a disconnected gas fitting must be sealed at every outlet;
(c) smoking or the use of any source of ignition is prohibited near exposed gasways;
(d) it is prohibited to use any source of ignition when searching for escapes of gas;
(e) any work in relation to a gas fitting which might affect the tightness of the installation must be tested immediately for gas tightness.

Gas Safety (Installation and Use) Regulations 1998 reg 26

12.219 Particularly striking are the broad but absolute nature of the duties imposed by reg 26 of the Gas Safety (Installation and Use) Regulations 1998 SI 1998/2451, namely that no person shall install a gas appliance unless it can be used without constituting a danger to any person, that a gas appliance is not connected to the gas supply unless it can be used safely, that it complies with other relevant safety requirements, and that any second-hand appliance is in a safe condition for further use. Further, any work on an appliance is required to maintain safety standards, the appliance must be examined after such work has been completed, and any defect must be notified to the owner/user.

Landlords

12.220 Regulation 36 of the Gas Safety (Installation and Use) Regulations 1998 SI 1998/2451 is also noteworthy in that it places a heavy burden on landlords to ensure safe maintenance of gas appliances, flues, and pipework installed in premises under their control; it has led to several successful prosecutions in recent years. To ensure the safety of their tenants, landlords are required to ensure that annual safety checks are carried out on gas appliances and flues and that a record is kept and issued or displayed to tenants.

12.221 The duty imposed by reg 36 of the Gas Safety (Installation and Use) Regulations 1998 SI 1998/2451 on landlords of premises occupied under a lease or a licence includes a duty to ensure that relevant gas fittings and associated flues are maintained in a safe condition. A relevant gas fitting includes any gas appliance other than an appliance which the tenant is entitled to remove from the premises.

12.222 The landlord is responsible for ensuring that each gas appliance and flue is checked by a competent person at least every twelve months. The Regulations require that such safety checks must include the effectiveness of any flue, the supply of combustion air, the operating pressure and/or heat input of the appliance, and the operation of the appliance to ensure its safe functioning. Nothing done or agreed to be done by a tenant can be considered as discharging the duties of a landlord in respect of maintenance, except in relation to access to the appliance for the purposes of carrying out the maintenance or check.

Landlord's inspection certificate

12.223 The landlord is under a duty to retain a record of each inspection of any gas appliance or flue for a period of two years from the date of the inspection. The record must include:

(a) the date on which the appliance or flue is checked;
(b) the address of the premises at which the appliance or flue is installed;
(c) the name and address of the landlord of the premises at which the appliance or flue is installed;
(d) a description of, and the location of, each appliance and flue that has been checked, together with details of defects and any remedial action taken;
(e) the person carrying out the inspection must certify that he has checked:
  (i) the effectiveness of any flue;
  (ii) the supply of combustion air;
  (iii) the operating pressure and/or heat input of the appliance;
  (iv) the operation of the appliance to ensure its safe functioning.

12.224 Within twenty-eight days of the date of the check, the landlord must ensure that a copy of the record is given to each tenant of the premises to which the record relates. Before any new tenant moves in, the landlord must provide any such tenant with a copy of the record.

J. Electricity: Electricity at Work Regulations 1989

12.225 The purpose of the Electricity at Work Regulations 1989 SI 1989/635, which are reproduced at Appendix N, is to require precautions to be taken against the risk of death or personal injury from electricity in work activities. They came into force on 1 April 1990. The regulations impose duties on employers, the self-employed, and employees when they are working on or near systems, electrical equipment and conductors, and in respect of matters which are in their control.[130]

12.226 Many of the duties imposed by the Electricity at Work Regulations 1989 SI 1989/635 are not subject to the qualification or defence of reasonable practicability.[131] However, a person charged with breaching one of these duties will have a defence if he can show that he took all reasonable steps and exercised all due diligence to avoid the commission of an offence. As is the case with other statutory defences or exceptions, it will be for the defendant to prove on the balance of probabilities that the defence is made out.

12.227 The most wide-ranging and commonly encountered regulations in terms of enforcement action are as considered below.

Systems, work activities, and protective equipment

12.228 Regulation 4 of the Electricity at Work Regulations 1989 SI 1989/635 provides:

(1) All systems shall at all times be of such construction as to prevent, so far as is reasonably practicable, danger.
(2) As may be necessary to prevent danger, all systems shall be maintained so as to prevent, so far as is reasonably practicable, such danger.
(3) Every work activity, including operation, use and maintenance of a system and work near a system, shall be carried out in such a manner as not to give rise, so far as is reasonably practicable to danger.
(4) Any equipment provided under these regulations for the purpose of protecting persons at work on or near electrical equipment shall be suitable for the use for which it is provided, be maintained in a condition suitable for that use, and be properly used.

12.229 'System' is defined as 'all the constituent parts of a system, eg conductors and electrical equipment in it, and is not a reference solely to the functional circuit as a whole. It follows that something required of a system is required both of the system as a whole and of the equipment and conductors in it'; danger is defined as 'risk of injury'.[132]

'Construction'

12.230 The guidance to the Electricity at Work Regulations 1989 SI 1989/635 provides that the term 'construction' has a wide application and covers the physical condition and arrangement of the components in a system throughout its life. Deciding whether the construction of an electrical system is suitable requires

---

130 Electricity at Work Regulations 1989 SI 1989/635 reg 3(1).
131 ibid regs 4(4), 5, 8, 9, 10, 11, 12, 13, 14, 15, 16, and 25.
132 ibid reg 2(1).

consideration to be given to all the likely or reasonably foreseeable conditions of its use, and the guidance sets out a number of factors which should be taken into account.

'Maintenance'

12.231 The guidance provides that the requirement for maintenance under reg 4(2) of the Electricity at Work Regulations 1989 SI 1989/635 is concerned with the maintenance that is required to ensure that the system is safe, rather than the activity of doing the maintenance in a safe manner, which is dealt with in reg 4(3). Suitable maintenance regimes should be determined in accordance with the provisions of regs 12–16 of the Electricity at Work Regulations 1989 SI 1989/635.

12.232 An effective programme for maintenance to prevent danger will involve regular inspection of the system in question. Records of the maintenance which has taken place should be kept, preferably throughout the working life of an electrical system.

Protective equipment under the Electricity at Work Regulations 1989 reg 4(4)

12.233 The duty to ensure that equipment which is used to protect those who work on or near electrical equipment is not qualified by any requirement of reasonable practicability, as is the case with the duties under the Electricity at Work Regulations 1989 SI 1989/635 reg 4(1)–(3), although the defence under reg 29 is available.

Electricity at Work Regulations 1989 regs 5–12

12.234 These concern:

(a) Strength and capability of electrical equipment.
(b) Adverse or hazardous environments.
(c) Insulation, protection, and placing of conductors.
(d) Earthing or other suitable precautions.
(e) Integrity of referenced conductors.
(f) Connections.
(g) Means for protecting from excess of current.

12.235 The defence in reg 29 of the Electricity at Work Regulations 1989 SI 1989/635 is available in respect of all save regs 6 and 7 (adverse or hazardous environments and insulation, protection and placing of conductors), where the duty is qualified by reasonable practicability.

12.236 The duties imposed by regs 13 and 14 of the Electricity at Work Regulations 1989 SI 1989/635 are of general application, and may be considered to be two of the principal duties under the Electricity at Work Regulations 1989 SI 1989/635. Regulation 13 provides:

> Adequate precautions shall be taken to prevent electrical equipment, which has been made dead, in order to prevent danger while work is carried out on or near that equipment, from becoming electrically charged during that work if danger may thereby arise.

12.237 Regulation 14 provides:

> No person shall be engaged in any work on or so near any live conductor (other than one suitably covered with insulating material so as to prevent danger) that danger may arise unless—
>
> (a) it is unreasonable in all the circumstances for it to be dead; and
>
> (b) it is reasonable in all the circumstances for him to be at work or near it while it is live; and
>
> (c) suitable precautions (including where necessary the provision of suitable protective equipment) are taken to prevent injury.

12.238 The defence under reg 29 is available under these provisions. In acknowledgement of the danger of working on live conductors, the scheme of reg 14 of the Electricity at Work Regulations 1989 SI 1989/635 is to prohibit live working unless all of the conditions in (a)–(c) are satisfied. The guidance to the Electricity at Work Regulations 1989 SI 1989/635 provides some examples of when it might be appropriate to work live. It provides that it may be unreasonable having regard to 'all the relevant factors' for work to be carried out on equipment which has been made dead. The 'relevant factors' are listed as:

(a) when it is not practicable to carry out the work with the conductors dead, ie for the purposes of testing;

(b) the creation of other hazards by making the conductors dead;

(c) the need to comply with other statutory requirements;

(d) the level of risk involved in working live and the effectiveness of the precautions available set against economic need to perform that work.

12.239 For instance, when it is necessary to do maintenance work on a busy section of electrified railway track and it would be disproportionately disruptive and expensive to require that the live conductors are isolated for the period of the work.

Appendices

A. Relevant Extracts from the Health and Safety at Work etc. Act 1974 — 407

B. Enforcement Concordat; Enforcement Policy Statement (HSC15); Work Related Death Protocol — 455

C. The Health and Safety (Enforcing Authority) Regulations 1998 (as amended) — 475

D. The Employment Tribunals (Constitution and Rules of Procedure) Regulations 2004 — 483

E. Health and Safety: The Management of Health and Safety at Work Regulations 1999 (as amended) — 525

F. Health and Safety: The Workplace (Health, Safety and Welfare) Regulations 1992 (as amended) — 537

G. Health and Safety: The Manual Handling Operations Regulations 1992 (as amended) — 547

H. Health and Safety: The Provision and Use of Work Equipment Regulations 1998 (as amended) — 551

I. Health and Safety: The Control of Substances Hazardous to Health Regulations 2002 (as amended) — 567

J. Health and Safety: The Construction (Design and Management) Regulations 2007 — 583

K. Health and Safety: The Work at Height Regulations 2005 (as amended) — 605

L. Health and Safety: The Control of Asbestos Regulations 2006 — 619

M. Health and Safety: The Gas Safety (Installation and Use) Regulations 1998 (as amended)   639

N. Health and Safety: The Electricity at Work Regulations 1989 (as amended)   659

APPENDIX A

Relevant Extracts from the Health and Safety at Work etc. Act 1974

1974 Chapter 37

An Act to make further provision for securing the health, safety and welfare of persons at work, for protecting others against risks to health or safety in connection with the activities of persons at work, for controlling the keeping and use and preventing the unlawful acquisition, possession and use of dangerous substances, and for controlling certain emissions into the atmosphere; to make further provision with respect to the employment medical advisory service; to amend the law relating to building regulations, and the Building (Scotland) Act 1959; and for connected purposes

[31st July 1974]

BE IT ENACTED by the Queen's most Excellent Majesty, by and with the advice and consent of the Lords Spiritual and Temporal, and Commons, in this present Parliament assembled, and by the authority of the same, as follows:–

Part I

Health, Safety and Welfare in Connection with Work, and Control of Dangerous Substances and Certain Emissions into the Atmosphere

Preliminary

1 Preliminary

(1) The provisions of this Part shall have effect with a view to—
   (a) securing the health, safety and welfare of persons at work;
   (b) protecting persons other than persons at work against risks to health or safety arising out of or in connection with the activities of persons at work;
   (c) controlling the keeping and use of explosive or highly flammable or otherwise dangerous substances, and generally preventing the unlawful acquisition, possession and use of such substances; . . .
   (d) [. . .][1]
(2) The provisions of this Part relating to the making of health and safety regulations. [. . .][2] and the preparation and approval of codes of practice shall in particular have effect with a view to enabling the enactments specified in the third column of Schedule 1 and the regulations, orders and other instruments in force under those enactments to be progressively replaced by a system of regulations and approved codes of practice operating in combination with the other provisions of this Part and designed to maintain or improve the standards of health, safety and welfare established by or under those enactments.

---

[1] Repealed by Environmental Protection Act 1990.
[2] Repealed by Employment Protection Act 1975.

(3) For the purposes of this Part risks arising out of or in connection with the activities of persons at work shall be treated as including risks attributable to the manner of conducting an undertaking, the plant or substances used for the purposes of an undertaking and the condition of premises so used or any part of them.

(4) References in this Part to the general purposes of this Part are references to the purposes mentioned in subsection (1) above.

General duties

2 General duties of employers to their employees

(1) It shall be the duty of every employer to ensure, so far as is reasonably practicable, the health, safety and welfare at work of all his employees.

(2) Without prejudice to the generality of an employer's duty under the preceding subsection, the matters to which that duty extends include in particular—
   (a) the provision and maintenance of plant and systems of work that are, so far as is reasonably practicable, safe and without risks to health;
   (b) arrangements for ensuring, so far as is reasonably practicable, safety and absence of risks to health in connection with the use, handling, storage and transport of articles and substances;
   (c) the provision of such information, instruction, training and supervision as is necessary to ensure, so far as is reasonably practicable, the health and safety at work of his employees;
   (d) so far as is reasonably practicable as regards any place of work under the employer's control, the maintenance of it in a condition that is safe and without risks to health and the provision and maintenance of means of access to and egress from it that are safe and without such risks;
   (e) the provision and maintenance of a working environment for his employees that is, so far as is reasonably practicable, safe, without risks to health, and adequate as regards facilities and arrangements for their welfare at work.

(3) Except in such cases as may be prescribed, it shall be the duty of every employer to prepare and as often as may be appropriate revise a written statement of his general policy with respect to the health and safety at work of his employees and the organisation and arrangements for the time being in force for carrying out that policy, and to bring the statement and any revision of it to the notice of all his employees.

(4) Regulations made by the Secretary of State may provide for the appointment in prescribed cases by recognised trade unions (within the meaning of the regulations) of safety representatives from amongst the employees, and those representatives shall represent the employees in consultations with the employers under subsection (6) below and shall have such other functions as may be prescribed.

(5) [. . .][3]

(6) It shall be the duty of every employer to consult any such representatives with a view to the making and maintenance of arrangements which will enable him and his employees to co-operate effectively in promoting and developing measures to ensure the health and safety at work of the employees, and in checking the effectiveness of such measures.

(7) In such cases as may be prescribed it shall be the duty of every employer, if requested to do so by the safety representatives mentioned in [subsection (4)][4] above, to establish, in accordance with regulations made by the Secretary of State, a safety committee having the function of keeping under review the measures taken to ensure the health and safety at work of his employees and such other functions as may be prescribed.

---

[3] Repealed by the Employment Protection Act 1975.
[4] Amended by the Employment Protection Act 1975.

Relevant Extracts from the Health and Safety at Work etc. Act 1974

3 General duties of employers and self-employed to persons other than their employees

(1) It shall be the duty of every employer to conduct his undertaking in such a way as to ensure, so far as is reasonably practicable, that persons not in his employment who may be affected thereby are not thereby exposed to risks to their health or safety.

(2) It shall be the duty of every self-employed person to conduct his undertaking in such a way as to ensure, so far as is reasonably practicable, that he and other persons (not being his employees) who may be affected thereby are not thereby exposed to risks to their health or safety.

(3) In such cases as may be prescribed, it shall be the duty of every employer and every self-employed person, in the prescribed circumstances and in the prescribed manner, to give to persons (not being his employees) who may be affected by the way in which he conducts his undertaking the prescribed information about such aspects of the way in which he conducts his undertaking as might affect their health or safety.

4 General duties of persons concerned with premises to persons other than their employees

(1) This section has effect for imposing on persons duties in relation to those who—
   (a) are not their employees; but
   (b) use non-domestic premises made available to them as a place of work or as a place where they may use plant or substances provided for their use there,
   and applies to premises so made available and other non-domestic premises used in connection with them.

(2) It shall be the duty of each person who has, to any extent, control of premises to which this section applies or of the means of access thereto or egress therefrom or of any plant or substance in such premises to take such measures as it is reasonable for a person in his position to take to ensure, so far as is reasonably practicable, that the premises, all means of access thereto or egress therefrom available for use by persons using the premises, and any plant or substance in the premises or, as the case may be, provided for use there, is or are safe and without risks to health.

(3) Where a person has, by virtue of any contract or tenancy, an obligation of any extent in relation to—
   (a) the maintenance or repair of any premises to which this section applies or any means of access thereto or egress therefrom; or
   (b) the safety of or the absence of risks to health arising from plant or substances in any such premises;
   that person shall be treated, for the purposes of subsection (2) above, as being a person who has control of the matters to which his obligation extends.

(4) Any reference in this section to a person having control of any premises or matter is a reference to a person having control of the premises or matter in connection with the carrying on by him of a trade, business or other undertaking (whether for profit or not).

5 [...]⁵

6 General duties of manufacturers etc as regards articles and substances for use at work

[(1) It shall be the duty of any person who designs, manufactures, imports or supplies any article for use at work or any article of fairground equipment—
   (a) to ensure, so far as is reasonably practicable, that the article is so designed and constructed that it will be safe and without risks to health at all times when it is being set, used, cleaned or maintained by a person at work;
   (b) to carry out or arrange for the carrying out of such testing and examination as may be necessary for the performance of the duty imposed on him by the preceding paragraph;

---

[5] Repealed by the Environmental Protection Act 1990.

(c) to take such steps as are necessary to secure that persons supplied by that person with the article are provided with adequate information about the use for which the article is designed or has been tested and about any conditions necessary to ensure that it will be safe and without risks to health at all such times as are mentioned in paragraph (a) above and when it is being dismantled or disposed of; and

(d) to take such steps as are necessary to secure, so far as is reasonably practicable, that persons so supplied are provided with all such revisions of information provided to them by virtue of the preceding paragraph as are necessary by reason of its becoming known that anything gives rise to a serious risk to health or safety.

(1A) It shall be the duty of any person who designs, manufactures, imports or supplies any article of fairground equipment—

(a) to ensure, so far as is reasonably practicable, that the article is so designed and constructed that it will be safe and without risks to health at all times when it is being used for or in connection with the entertainment of members of the public;

(b) to carry out or arrange for the carrying out of such testing and examination as may be necessary for the performance of the duty imposed on him by the preceding paragraph;

(c) to take such steps as are necessary to secure that persons supplied by that person with the article are provided with adequate information about the use for which the article is designed or has been tested and about any conditions necessary to ensure that it will be safe and without risks to health at all times when it is being used for or in connection with the entertainment of members of the public; and

(d) to take such steps as are necessary to secure, so far as is reasonably practicable, that persons so supplied are provided with all such revisions of information provided to them by virtue of the preceding paragraph as are necessary by reason of its becoming known that anything gives rise to a serious risk to health or safety.][6]

(2) It shall be the duty of any person who undertakes the design or manufacture of any article for use at work [or of any article of fairground equipment] to carry out or arrange for the carrying out of any necessary research with a view to the discovery and, so far as is reasonably practicable, the elimination or minimisation of any risks to health or safety to which the design or article may give rise.

(3) It shall be the duty of any person who erects or installs any article for use at work in any premises where that article is to be used by persons at work [or who erects or installs any article of fairground equipment] to ensure, so far as is reasonably practicable, that nothing about the way in which [the article is erected or installed makes it unsafe or a risk to health at any such time as is mentioned in paragraph (a) of subsection (1) or, as the case may be, in paragraph (a) of subsection (1) or (1A) above].

[(4) It shall be the duty of any person who manufactures, imports or supplies any substance—

(a) to ensure, so far as is reasonably practicable, that the substance will be safe and without risks to health at all times when it is being used, handled, processed, stored or transported by a person at work or in premises to which section 4 above applies;

(b) to carry out or arrange for the carrying out of such testing and examination as may be necessary for the performance of the duty imposed on him by the preceding paragraph;

(c) to take such steps as are necessary to secure that persons supplied by that person with the substance are provided with adequate information about any risks to health or safety to which the inherent properties of the substance may give rise, about the results of any relevant tests which have been carried out on or in connection with the substance and about any conditions necessary to ensure that the substance will be safe and without risks to health at

---

[6] Amended by the Consumer Protection Act 1987.

all such times as are mentioned in paragraph (a) above and when the substance is being disposed of; and

(d) to take such steps as are necessary to secure, so far as is reasonably practicable, that persons so supplied are provided with all such revisions of information provided to them by virtue of the preceding paragraph as are necessary by reason of its becoming known that anything gives rise to a serious risk to health or safety.]⁷

(5) It shall be the duty of any person who undertakes the manufacture of any [substance] to carry out or arrange for the carrying out of any necessary research with a view to the discovery and, so far as is reasonably practicable, the elimination or minimisation of any risks to health or safety to which the substance may give rise [at all such times as are mentioned in paragraph (a) of subsection (4) above].⁸

(6) Nothing in the preceding provisions of this section shall be taken to require a person to repeat any testing, examination or research which has been carried out otherwise than by him or at his instance, in so far as it is reasonable for him to rely on the results thereof for the purposes of those provisions.

(7) Any duty imposed on any person by any of the preceding provisions of this section shall extend only to things done in the course of a trade, business or other undertaking carried on by him (whether for profit or not) and to matters within his control.

(8) Where a person designs, manufactures, imports or supplies an article [for use at work or an article of fairground equipment and does so for or to another]⁹ on the basis of a written undertaking by that other to take specified steps sufficient to ensure, so far as is reasonably practicable, that the article will be safe and without risks to health [at all such times as are mentioned in paragraph (a) of subsection (1) or, as the case may be, in paragraph (a) of subsection (1) or (1A) above], the undertaking shall have the effect of relieving the first-mentioned person from the duty imposed [by virtue of that paragraph]¹⁰ to such extent as is reasonable having regard to the terms of the undertaking.

[(8A) Nothing in subsection (7) or (8) above shall relieve any person who imports any article or substance from any duty in respect of anything which—

(a) in the case of an article designed outside the United Kingdom, was done by and in the course of any trade, profession or other undertaking carried on by, or was within the control of, the person who designed the article; or

(b) in the case of an article or substance manufactured outside the United Kingdom, was done by and in the course of any trade, profession or other undertaking carried on by, or was within the control of, the person who manufactured the article or substance.]¹¹

(9) Where a person ("the ostensible supplier") supplies any [article or substance]¹² to another ("the customer") under a hire-purchase agreement, conditional sale agreement or credit-sale agreement, and the ostensible supplier—

(a) carries on the business of financing the acquisition of goods by others by means of such agreements; and

(b) in the course of that business acquired his interest in the article or substance supplied to the customer as a means of financing its acquisition by the customer from a third person ("the effective supplier"),

the effective supplier and not the ostensible supplier shall be treated for the purposes of this section as supplying the article or substance to the customer, and any duty imposed by the

---

[7] Amended by the Consumer Protection Act 1987.
[8] ibid.
[9] ibid.
[10] ibid.
[11] Inserted by the Consumer Protection Act 1987.
[12] Amended by the Consumer Protection Act 1987.

preceding provisions of this section on suppliers shall accordingly fall on the effective supplier and not on the ostensible supplier.

[(10) For the purposes of this section an absence of safety or a risk to health shall be disregarded in so far as the case in or in relation to which it would arise is shown to be one the occurrence of which could not reasonably be foreseen; and in determining whether any duty imposed by virtue of paragraph (a) of subsection (1), (1A) or (4) above has been performed regard shall be had to any relevant information or advice which has been provided to any person by the person by whom the article has been designed, manufactured, imported or supplied or, as the case may be, by the person by whom the substance has been manufactured, imported or supplied.][13]

7 General duties of employees at work

It shall be the duty of every employee while at work—
(a) to take reasonable care for the health and safety of himself and of other persons who may be affected by his acts or omissions at work; and
(b) as regards any duty or requirement imposed on his employer or any other person by or under any of the relevant statutory provisions, to co-operate with him so far as is necessary to enable that duty or requirement to be performed or complied with.

8 Duty not to interfere with or misuse things provided pursuant to certain provisions

No person shall intentionally or recklessly interfere with or misuse anything provided in the interests of health, safety or welfare in pursuance of any of the relevant statutory provisions.

9 Duty not to charge employees for things done or provided pursuant to certain specific requirements

No employer shall levy or permit to be levied on any employee of his any charge in respect of anything done or provided in pursuance of any specific requirement of the relevant statutory provisions.

The Health and Safety Commission and the Health and Safety Executive

10 Establishment of the Commission and the Executive

(1) There shall be two bodies corporate to be called the Health and Safety Commission and the Health and Safety Executive which shall be constituted in accordance with the following provisions of this section.
(2) The Health and Safety Commission (hereafter in this Act referred to as "the Commission") shall consist of a chairman appointed by the Secretary of State and not less than six nor more than nine other members appointed by the Secretary of State in accordance with subsection (3) below.
(3) Before appointing the members of the Commission (other than the chairman) the Secretary of State shall—
  (a) as to three of them, consult such organisations representing employers as he considers appropriate;
  (b) as to three others, consult such organisations representing employees as he considers appropriate; and
  (c) as to any other members he may appoint, consult such organisations representing local authorities and such other organisations, including professional bodies, the activities of whose members are concerned with matters relating to any of the general purposes of this Part, as he considers appropriate.
(4) The Secretary of State may appoint one of the members to be deputy chairman of the Commission.

---

[13] Amended by the Consumer Protection Act 1987.

(5) The Health and Safety Executive (hereafter in this Act referred to as "the Executive") shall consist of three persons of whom one shall be appointed by the Commission with the approval of the Secretary of State to be the director of the Executive and the others shall be appointed by the Commission with the like approval after consultation with the said director.

(6) The provisions of Schedule 2 shall have effect with respect to the Commission and the Executive.

(7) The functions of the Commission and of the Executive, and of their officers and servants, shall be performed on behalf of the Crown.

[(8) For the purposes of any civil proceedings arising out of those functions, the Crown Proceedings Act 1947 and the Crown Suits (Scotland) Act 1857 shall apply to the Commission and the Executive as if they were government departments within the meaning of the said Act of 1947 or, as the case may be, public departments within the meaning of the said Act of 1857.][14]

11 General functions of the Commission and the Executive

(1) In addition to the other functions conferred on the Commission by virtue of this Act, but [subject to subsections (2A) and (3)][15] below, it shall be the general duty of the Commission to do such things and make such arrangements as it considers appropriate for the general purposes of this Part [. . .][16]

(2) It shall be the duty of the Commission [. . .][17]—
   (a) to assist and encourage persons concerned with matters relevant to any of the general purposes of this Part to further those purposes;
   (b) to make such arrangements as it considers appropriate for the carrying out of research, the publication of the results of research and the provision of training and information in connection with those purposes, and to encourage research and the provision of training and information in that connection by others;
   (c) to make such arrangements as it considers appropriate for securing that government departments, employers, employees, organisations representing employers and employees respectively, and other persons concerned with matters relevant to any of those purposes are provided with an information and advisory service and are kept informed of, and adequately advised on, such matters;
   (d) to submit from time to time to the authority having power to make regulations under any of the relevant statutory provisions such proposals as the Commission considers appropriate for the making of regulations under that power.

[(2A) In subsections (1) and (2) above—
   (a) references to the general purposes of this Part do not include references to the railway safety purposes; and
   (b) the reference to a power to make regulations under the relevant statutory provisions does not include a reference to any power so far as it is exercisable for the railway safety purposes.][18]

(3) It shall be the duty of the Commission—
   (a) to submit to the Secretary of State from time to time particulars of what it proposes to do for the purpose of performing its functions; and
   (b) subject to the following paragraph, to ensure that its activities are in accordance with proposals approved by the Secretary of State; and
   (c) to give effect to any directions given to it by the Secretary of State.

---

[14] Inserted by the Employment Protection Act 1975.
[15] Amended by the Railways Act 2005.
[16] Repealed by the Employment Protection Act 1975.
[17] ibid.
[18] Inserted by the Railways Act 2005.

(4) In addition to any other functions conferred on the Executive by virtue of this Part, it shall be the duty of the Executive—
   (a) to exercise on behalf of the Commission such of the Commission's functions as the Commission directs it to exercise; and
   (b) to give effect to any directions given to it by the Commission otherwise than in pursuance of paragraph (a) above;
   but, except for the purpose of giving effect to directions given to the Commission by the Secretary of State, the Commission shall not give to the Executive any directions as to the enforcement of any of the relevant statutory provisions in a particular case.
(5) Without prejudice to subsection (2) above, it shall be the duty of the Executive, if so requested by a Minister of the Crown—
   (a) to provide him with information about the activities of the Executive in connection with any matter with which he is concerned; and
   (b) to provide him with advice on any matter with which he is concerned on which relevant expert advice is obtainable from any of the officers or servants of the Executive but which is not relevant to any of the general purposes of this Part.
(6) The Commission and the Executive shall, subject to any directions given to it in pursuance of this Part, have power to do anything (except borrow money) which is calculated to facilitate, or is conducive or incidental to, the performance of any function of the Commission or, as the case may be, the Executive (including a function conferred on it by virtue of this subsection).

12 Control of the Commission by the Secretary of State

The Secretary of State may—
(a) approve, with or without modifications, any proposals submitted to him in pursuance of section 11(3)(a);
(b) give to the Commission at any time such directions as he thinks fit with respect to its functions (including directions modifying its functions, but not directions conferring on it functions other than any of which it was deprived by previous directions given by virtue of this paragraph), and any directions which it appears to him requisite or expedient to give in the interests of the safety of the State.

13 Other powers of the Commission

(1) The Commission shall have power—
   (a) to make agreements with any government department or other person for that department or person to perform on behalf of the Commission or the Executive (with or without payment) any of the functions of the Commission or, as the case may be, of the Executive;
   (b) subject to subsection (2) below, to make agreements with any Minister of the Crown, government department or other public authority for the Commission to perform on behalf of that Minister, department or authority (with or without payment) functions exercisable by the Minister, department or authority (including, in the case of a Minister, functions not conferred by an enactment), being functions which in the opinion of the Secretary of State can appropriately be performed by the Commission in connection with any of the Commission's functions;
   (c) to provide (with or without payment) services or facilities required otherwise than for the general purposes of this Part in so far as they are required by any government department or other public authority in connection with the exercise by that department or authority of any of its functions;
   (d) to appoint persons or committees of persons to provide the Commission with advice in connection with any of its functions and (without prejudice to the generality of the following paragraph) to pay to persons so appointed such remuneration as the Secretary of State may with the approval of the Minister for the Civil Service determine;

(e) in connection with any of the functions of the Commission, to pay to any person such travelling and subsistence allowances and such compensation for loss of remunerative time as the Secretary of State may with the approval of the Minister for the Civil Service determine;

(f) to carry out or arrange for or make payments in respect of research into any matter connected with any of the Commission's functions, and to disseminate or arrange for or make payments in respect of the dissemination of information derived from such research;

(g) to include, in any arrangements made by the Commission for the provision of facilities or services by it or on its behalf, provision for the making of payments to the Commission or any person acting on its behalf by other parties to the arrangements and by persons who use those facilities or services.

(2) Nothing in subsection (1)(b) shall authorise the Commission to perform any function of a Minister, department or authority which consists of a power to make regulations or other instruments of a legislative character.

14 Power of the Commission to direct investigations and inquiries

(1) This section applies to the following matters, that is to say any accident, occurrence, situation or other matter whatsoever which the Commission thinks it necessary or expedient to investigate for any of the general purposes of this Part or with a view to the making of regulations for those purposes; and for the purposes of this subsection

[(a) those general purposes shall be treated as not including the railway safety purposes; but

(b) it is otherwise][19]

immaterial whether the Executive is or is not responsible for securing the enforcement of such (if any) of the relevant statutory provisions as relate to the matter in question.

(2) The Commission may at any time—
(a) direct the Executive or authorise any other person to investigate and make a special report on any matter to which this section applies; or
(b) with the consent of the Secretary of State direct an inquiry to be held into any such matter;

[...][20]

(3) Any inquiry held by virtue of subsection (2)(b) above shall be held in accordance with regulations made for the purposes of this subsection by the Secretary of State, and shall be held in public except where or to the extent that the regulations provide otherwise.

(4) Regulations made for the purposes of subsection (3) above may in particular include provision—
(a) conferring on the person holding any such inquiry, and any person assisting him in the inquiry, powers of entry and inspection;
(b) conferring on any such person powers of summoning witnesses to give evidence or produce documents and power to take evidence on oath and administer oaths or require the making of declarations;
(c) requiring any such inquiry to be held otherwise than in public where or to the extent that a Minister of the Crown so directs.

(5) In the case of a special report made by virtue of subsection (2)(a) above or a report made by the person holding an inquiry held by virtue of subsection (2)(b) above, the Commission may cause the report, or so much of it as the Commission thinks fit, to be made public at such time and in such manner as the Commission thinks fit.

(6) The Commission—
(a) in the case of an investigation and special report made by virtue of subsection (2)(a) above (otherwise than by an officer or servant of the Executive), may pay to the person making it

---

[19] Amended by the Railways Act 2005.
[20] Repealed by the Employment Protection Act 1975.

such remuneration and expenses as the Secretary of State may, with the approval of the Minister for the Civil Service, determine;
(b) in the case of an inquiry held by virtue of subsection (2)(b) above, may pay to the person holding it and to any assessor appointed to assist him such remuneration and expenses, and to persons attending the inquiry as witnesses such expenses, as the Secretary of State may, with the like approval, determine; and
(c) may, to such extent as the Secretary of State may determine, defray the other costs, if any, of any such investigation and special report or inquiry.
(7) Where an inquiry is directed to be held by virtue of subsection (2)(b) above into any matter to which this section applies arising in Scotland, being a matter which causes the death of any person, no inquiry with regard to that death shall, unless the Lord Advocate otherwise directs, be held in pursuance of the [Fatal Accidents and Sudden Deaths Inquiry (Scotland) Act 1976].[21]

Health and safety regulations and approved codes of practice

15 Health and safety regulations

[(1) Subject to the provisions of section 50, the Secretary of State [. . .][22] shall have power to make regulations under this section for any of the general purposes of this Part (and regulations so made are in this Part referred to as "health and safety regulations").][23]
(2) Without prejudice to the generality of the preceding subsection, health and safety regulations may for any of the general purposes of this Part make provision for any of the purposes mentioned in Schedule 3.
(3) Health and safety regulations—
  (a) may repeal or modify any of the existing statutory provisions;
  (b) may exclude or modify in relation to any specified class of case any of the provisions of sections 2 to 9 or any of the existing statutory provisions;
  (c) may make a specified authority or class of authorities responsible, to such extent as may be specified, for the enforcement of any of the relevant statutory provisions.
(4) Health and safety regulations—
  (a) may impose requirements by reference to the approval of the Commission or any other specified body or person;
  (b) may provide for references in the regulations to any specified document to operate as references to that document as revised or re-issued from time to time.
(5) Health and safety regulations—
  (a) may provide (either unconditionally or subject to conditions, and with or without limit of time) for exemptions from any requirement or prohibition imposed by or under any of the relevant statutory provisions;
  (b) may enable exemptions from any requirement or prohibition imposed by or under any of the relevant statutory provisions to be granted (either unconditionally or subject to conditions, and with or without limit of time) by any specified person or by any person authorised in that behalf by a specified authority.
(6) Health and safety regulations—
  (a) may specify the persons or classes of persons who, in the event of a contravention of a requirement or prohibition imposed by or under the regulations, are to be guilty of an offence, whether in addition to or to the exclusion of other persons or classes of persons;
  (b) may provide for any specified defence to be available in proceedings for any offence under the relevant statutory provisions either generally or in specified circumstances;

---

[21] Amended by the Fatal Accidents and Sudden Deaths Inquiry (Scotland) Act 1976.
[22] Repealed by SI 2002/794.
[23] Amended by the Employment Protection Act 1975.

(c) may exclude proceedings on indictment in relation to offences consisting of a contravention of a requirement or prohibition imposed by or under any of the existing statutory provisions, sections 2 to 9 or health and safety regulations;

(d) may restrict the punishments [(other than the maximum fine on conviction on indictment)] which can be imposed in respect of any such offence as is mentioned in paragraph (c) above;

[(e) in the case of regulations made for any purpose mentioned in section 1(1) of the Offshore Safety Act 1992, may provide that any offence consisting of a contravention of the regulations, or of any requirement or prohibition imposed by or under them, shall be punishable on conviction on indictment by imprisonment for a term not exceeding two years, or a fine, or both.]

(7) Without prejudice to section 35, health and safety regulations may make provision for enabling offences under any of the relevant statutory provisions to be treated as having been committed at any specified place for the purpose of bringing any such offence within the field of responsibility of any enforcing authority or conferring jurisdiction on any court to entertain proceedings for any such offence.

(8) Health and safety regulations may take the form of regulations applying to particular circumstances only or to a particular case only (for example, regulations applying to particular premises only).

(9) If an Order in Council is made under section 84(3) providing that this section shall apply to or in relation to persons, premises or work outside Great Britain then, notwithstanding the Order, health and safety regulations shall not apply to or in relation to aircraft in flight, vessels, hovercraft or offshore installations outside Great Britain or persons at work outside Great Britain in connection with submarine cables or submarine pipelines except in so far as the regulations expressly so provide.

(10) In this section "specified" means specified in health and safety regulations.

16 Approval of codes of practice by the Commission

(1) For the purpose of providing practical guidance with respect to the requirements of any provision of [any of the enactments or instruments mentioned in subsection (1A) below],[24] the Commission may, subject to the following subsection [. . .][25]—

(a) approve and issue such codes of practice (whether prepared by it or not) as in its opinion are suitable for that purpose;

(b) approve such codes of practice issued or proposed to be issued otherwise than by the Commission as in its opinion are suitable for that purpose.

[(1A) Those enactments and instruments are—

(a) sections 2 to 7 above;

(b) health and safety regulations, except so far as they make provision exclusively in relation to transport systems falling within paragraph 1(3) of Schedule 3 to the Railways Act 2005; and

(c) the existing statutory provisions that are not such provisions by virtue of section 117(4) of the Railways Act 1993.][26]

(2) The Commission shall not approve a code of practice under subsection (1) above without the consent of the Secretary of State, and shall, before seeking his consent, consult—

(a) any government department or other body that appears to the Commission to be appropriate (and, in particular, in the case of a code relating to electro-magnetic radiations, [the Health Protection Agency][27]); and

---

[24] Amended by the Railways Act 2005.
[25] Repealed by the Employment Protection Act 1975.
[26] Inserted by the Railways Act 2005.
[27] Amended by the Health Protection Agency Act 2004.

(b) such government departments and other bodies, if any, as in relation to any matter dealt with in the code, the Commission is required to consult under this section by virtue of directions given to it by the Secretary of State.

(3) Where a code of practice is approved by the Commission under subsection (1) above, the Commission shall issue a notice in writing—
   (a) identifying the code in question and stating the date on which its approval by the Commission is to take effect; and
   (b) specifying for which of the provisions mentioned in subsection (1) above the code is approved.

(4) The Commission may—
   (a) from time to time revise the whole or any part of any code of practice prepared by it in pursuance of this section;
   (b) approve any revision or proposed revision of the whole or any part of any code of practice for the time being approved under this section;

and the provisions of subsections (2) and (3) above shall, with the necessary modifications, apply in relation to the approval of any revision under this subsection as they apply in relation to the approval of a code of practice under subsection (1) above.

(5) The Commission may at any time with the consent of the Secretary of State withdraw its approval from any code of practice approved under this section, but before seeking his consent shall consult the same government departments and other bodies as it would be required to consult under subsection (2) above if it were proposing to approve the code.

(6) Where under the preceding subsection the Commission withdraws its approval from a code of practice approved under this section, the Commission shall issue a notice in writing identifying the code in question and stating the date on which its approval of it is to cease to have effect.

(7) References in this part to an approved code of practice are references to that code as it has effect for the time being by virtue of any revision of the whole or any part of it approved under this section.

(8) The power of the Commission under subsection (1)(b) above to approve a code of practice issued or proposed to be issued otherwise than by the Commission shall include power to approve a part of such a code of practice; and accordingly in this Part "code of practice" may be read as including a part of such a code of practice.

17 Use of approved codes of practice in criminal proceedings

(1) A failure on the part of any person to observe any provision of an approved code of practice shall not of itself render him liable to any civil or criminal proceedings; but where in any criminal proceedings a party is alleged to have committed an offence by reason of a contravention of any requirement or prohibition imposed by or under any such provision as is mentioned in section 16(1) being a provision for which there was an approved code of practice at the time of the alleged contravention, the following subsection shall have effect with respect to that code in relation to those proceedings.

(2) Any provision of the code of practice which appears to the court to be relevant to the requirement or prohibition alleged to have been contravened shall be admissible in evidence in the proceedings; and if it is proved that there was at any material time a failure to observe any provision of the code which appears to the court to be relevant to any matter which it is necessary for the prosecution to prove in order to establish a contravention of that requirement or prohibition, that matter shall be taken as proved unless the court is satisfied that the requirement or prohibition was in respect of that matter complied with otherwise than by way of observance of that provision of the code.

(3) In any criminal proceedings—
   (a) a document purporting to be a notice issued by the Commission under section 16 shall be taken to be such a notice unless the contrary is proved; and

(b) a code of practice which appears to the court to be the subject of such a notice shall be taken to be the subject of that notice unless the contrary is proved.

Enforcement

18 Authorities responsible for enforcement of the relevant statutory provisions

(1) It shall be the duty of the Executive to make adequate arrangements for the enforcement of the relevant statutory provisions except to the extent that some other authority or class of authorities is by any of those provisions or by regulations under subsection (2) below made responsible for their enforcement.
(2) The Secretary of State may by regulations—
   (a) make local authorities responsible for the enforcement of the relevant statutory provisions to such extent as may be prescribed;
   (b) make provision for enabling responsibility for enforcing any of the relevant statutory provisions to be, to such extent as may be determined under the regulations—
      (i) transferred from the Executive to local authorities or from local authorities to the Executive; or
      (ii) assigned to the Executive or to local authorities for the purpose of removing any uncertainty as to what are by virtue of this subsection their respective responsibilities for the enforcement of those provisions;
and any regulations made in pursuance of paragraph (b) above shall include provision for securing that any transfer or assignment effected under the regulations is brought to the notice of persons affected by it.
(3) Any provision made by regulations under the preceding subsection shall have effect subject to any provision made by health and safety regulations [. . .][28] in pursuance of section 15(3)(c).
(4) It shall be the duty of every local authority—
   (a) to make adequate arrangements for the enforcement within their area of the relevant statutory provisions to the extent that they are by any of those provisions or by regulations under subsection (2) above made responsible for their enforcement; and
   (b) to perform the duty imposed on them by the preceding paragraph and any other functions conferred on them by any of the relevant statutory provisions in accordance with such guidance as the Commission may give them.
(5) Where any authority other than [. . .][29] the Executive or a local authority is by any of the relevant statutory provisions [. . .][30] made responsible for the enforcement of any of those provisions to any extent, it shall be the duty of that authority—
   (a) to make adequate arrangements for the enforcement of those provisions to that extent; and
   (b) [except where that authority is the Office of Rail Regulation,][31] to perform the duty imposed on the authority by the preceding paragraph and any other functions conferred on the authority by any of the relevant statutory provisions in accordance with such guidance as the Commission may give to the authority.
(6) Nothing in the provisions of this Act or of any regulations made thereunder charging any person in Scotland with the enforcement of any of the relevant statutory provisions shall be construed as authorising that person to institute proceedings for any offence.
(7) In this Part—
   (a) "enforcing authority" means the Executive or any other authority which is by any of the relevant statutory provisions or by regulations under subsection (2) above made responsible for the enforcement of any of those provisions to any extent; and

---

[28] Repealed by the Employment Protection Act 1975.
[29] ibid.
[30] Repealed by the Railways Act 2005.
[31] Amended by the Railways Act 2005.

(b) any reference to an enforcing authority's field of responsibility is a reference to the field over which that authority's responsibility for the enforcement of those provisions extends for the time being;

but where by virtue of paragraph (a) of section 13(1) the performance of any function of the Commission or the Executive is delegated to a government department or person, references to the Commission or the Executive (or to an enforcing authority where that authority is the Executive) in any provision of this Part which relates to that function shall, so far as may be necessary to give effect to any agreement under that paragraph, be construed as references to that department or person; and accordingly any reference to the field of responsibility of an enforcing authority shall be construed as a reference to the field over which that department or person for the time being performs such a function.

19 Appointment of inspectors

(1) Every enforcing authority may appoint as inspectors (under whatever title it may from time to time determine) such persons having suitable qualifications as it thinks necessary for carrying into effect the relevant statutory provisions within its field of responsibility, and may terminate any appointment made under this section.
(2) Every appointment of a person as an inspector under this section shall be made by an instrument in writing specifying which of the powers conferred on inspectors by the relevant statutory provisions are to be exercisable by the person appointed; and an inspector shall in right of his appointment under this section—
    (a) be entitled to exercise only such of those powers as are so specified; and
    (b) be entitled to exercise the powers so specified only within the field of responsibility of the authority which appointed him.
(3) So much of an inspector's instrument of appointment as specifies the powers which he is entitled to exercise may be varied by the enforcing authority which appointed him.
(4) An inspector shall, if so required when exercising or seeking to exercise any power conferred on him by any of the relevant statutory provisions, produce his instrument of appointment or a duly authenticated copy thereof.

20 Powers of inspectors

(1) Subject to the provisions of section 19 and this section, an inspector may, for the purpose of carrying into effect any of the relevant statutory provisions within the field of responsibility of the enforcing authority which appointed him, exercise the powers set out in subsection (2) below.
(2) The powers of an inspector referred to in the preceding subsection are the following, namely—
    (a) at any reasonable time (or, in a situation which in his opinion is or may be dangerous, at any time) to enter any premises which he has reason to believe it is necessary for him to enter for the purpose mentioned in subsection (1) above;
    (b) to take with him a constable if he has reasonable cause to apprehend any serious obstruction in the execution of his duty;
    (c) without prejudice to the preceding paragraph, on entering any premises by virtue of paragraph (a) above to take with him—
        (i) any other person duly authorised by his (the inspector's) enforcing authority; and
        (ii) any equipment or materials required for any purpose for which the power of entry is being exercised;
    (d) to make such examination and investigation as may in any circumstances be necessary for the purpose mentioned in subsection (1) above;
    (e) as regards any premises which he has power to enter, to direct that those premises or any part of them, or anything therein, shall be left undisturbed (whether generally or in particular respects) for so long as is reasonably necessary for the purpose of any examination or investigation under paragraph (d) above;

(f) to take such measurements and photographs and make such recordings as he considers necessary for the purpose of any examination or investigation under paragraph (d) above;

(g) to take samples of any articles or substances found in any premises which he has power to enter, and of the atmosphere in or in the vicinity of any such premises;

(h) in the case of any article or substance found in any premises which he has power to enter, being an article or substance which appears to him to have caused or to be likely to cause danger to health or safety, to cause it to be dismantled or subjected to any process or test (but not so as to damage or destroy it unless this is in the circumstances necessary for the purpose mentioned in subsection (1) above);

(i) in the case of any such article or substance as is mentioned in the preceding paragraph, to take possession of it and detain it for so long as is necessary for all or any of the following purposes, namely—

  (i) to examine it and do to it anything which he has power to do under that paragraph;
  (ii) to ensure that it is not tampered with before his examination of it is completed;
  (iii) to ensure that it is available for use as evidence in any proceedings for an offence under any of the relevant statutory provisions or any proceedings relating to a notice under section 21 or 22;

(j) to require any person whom he has reasonable cause to believe to be able to give any information relevant to any examination or investigation under paragraph (d) above to answer (in the absence of persons other than a person nominated by him to be present and any persons whom the inspector may allow to be present) such questions as the inspector thinks fit to ask and to sign a declaration of the truth of his answers;

(k) to require the production of, inspect, and take copies of or of any entry in—

  (i) any books or documents which by virtue of any of the relevant statutory provisions are required to be kept; and
  (ii) any other books or documents which it is necessary for him to see for the purposes of any examination or investigation under paragraph (d) above;

(l) to require any person to afford him such facilities and assistance with respect to any matters or things within that person's control or in relation to which that person has responsibilities as are necessary to enable the inspector to exercise any of the powers conferred on him by this section;

(m) any other power which is necessary for the purpose mentioned in subsection (1) above.

(3) The Secretary of State may by regulations make provision as to the procedure to be followed in connection with the taking of samples under subsection (2)(g) above (including provision as to the way in which samples that have been so taken are to be dealt with).

(4) Where an inspector proposes to exercise the power conferred by subsection (2)(h) above in the case of an article or substance found in any premises, he shall, if so requested by a person who at the time is present in and has responsibilities in relation to those premises, cause anything which is to be done by virtue of that power to be done in the presence of that person unless the inspector considers that its being done in that person's presence would be prejudicial to the safety of the State.

(5) Before exercising the power conferred by subsection (2)(h) above in the case of any article or substance, an inspector shall consult such persons as appear to him appropriate for the purpose of ascertaining what dangers, if any, there may be in doing anything which he proposes to do under that power.

(6) Where under the power conferred by subsection (2)(i) above an inspector takes possession of any article or substance found in any premises, he shall leave there, either with a responsible person or, if that is impracticable, fixed in a conspicuous position, a notice giving particulars of that article or substance sufficient to identify it and stating that he has taken possession of it under that power; and before taking possession of any such substance under that power an inspector shall, if it is practicable for him to do so, take a sample thereof and give to a responsible person at the premises a portion of the sample marked in a manner sufficient to identify it.

(7) No answer given by a person in pursuance of a requirement imposed under subsection (2)(j) above shall be admissible in evidence against that person or the [spouse or civil partner][32] of that person in any proceedings.

(8) Nothing in this section shall be taken to compel the production by any person of a document of which he would on grounds of legal professional privilege be entitled to withhold production on an order for discovery in an action in the High Court or, as the case may be, on an order for the production of documents in an action in the Court of Session.

21 Improvement notices

If an inspector is of the opinion that a person—

(a) is contravening one or more of the relevant statutory provisions; or
(b) has contravened one or more of those provisions in circumstances that make it likely that the contravention will continue or be repeated,

he may serve on him a notice (in this Part referred to as "an improvement notice") stating that he is of that opinion, specifying the provision or provisions as to which he is of that opinion, giving particulars of the reasons why he is of that opinion, and requiring that person to remedy the contravention or, as the case may be, the matters occasioning it within such period (ending not earlier than the period within which an appeal against the notice can be brought under section 24) as may be specified in the notice.

22 Prohibition notices

(1) This section applies to any activities which are being or are [likely][33] to be carried on by or under the control of any person, being activities to or in relation to which any of the relevant statutory provisions apply or will, if the activities are so carried on, apply.

(2) If as regards any activities to which this section applies an inspector is of the opinion that, as carried on or [likely][34] to be carried on by or under the control of the person in question, the activities involve or, as the case may be, will involve a risk of serious personal injury, the inspector may serve on that person a notice (in this Part referred to as "a prohibition notice").

(3) A prohibition notice shall—

(a) state that the inspector is of the said opinion;
(b) specify the matters which in his opinion give or, as the case may be, will give rise to the said risk;
(c) where in his opinion any of those matters involves or, as the case may be, will involve a contravention of any of the relevant statutory provisions, state that he is of that opinion, specify the provision or provisions as to which he is of that opinion, and give particulars of the reasons why he is of that opinion; and
(d) direct that the activities to which the notice relates shall not be carried on by or under the control of the person on whom the notice is served unless the matters specified in the notice in pursuance of paragraph (b) above and any associated contraventions of provisions so specified in pursuance of paragraph (c) above have been remedied.

[(4) A direction contained in a prohibition notice in pursuance of subsection (3)(d) above shall take effect—

(a) at the end of the period specified in the notice; or
(b) if the notice so declares, immediately.][35]

---

[32] Amended by the Civil Partnership Act 2004.
[33] Amended by the Consumer Protection Act 1987.
[34] ibid.
[35] ibid.

23 Provisions supplementary to ss 21 and 22

(1) In this section "a notice" means an improvement notice or a prohibition notice.
(2) A notice may (but need not) include directions as to the measures to be taken to remedy any contravention or matter to which the notice relates; and any such directions—
   (a) may be framed to any extent by reference to any approved code of practice; and
   (b) may be framed so as to afford the person on whom the notice is served a choice between different ways of remedying the contravention or matter.
(3) Where any of the relevant statutory provisions applies to a building or any matter connected with a building and an inspector proposes to serve an improvement notice relating to a contravention of that provision in connection with that building or matter, the notice shall not direct any measures to be taken to remedy the contravention of that provision which are more onerous than those necessary to secure conformity with the requirements of any building regulations for the time being in force to which that building or matter would be required to conform if the relevant building were being newly erected unless the provision in question imposes specific requirements more onerous than the requirements of any such building regulations to which the building or matter would be required to conform as aforesaid.

In this subsection "the relevant building", in the case of a building, means that building, and, in the case of a matter connected with a building, means the building with which the matter is connected.

(4) Before an inspector serves in connection with any premises used or about to be used as a place of work a notice requiring or likely to lead to the taking of measures affecting the means of escape in case of fire with which the premises are or ought to be provided, he shall consult the [fire and rescue authority].[36]

In this subsection "[fire and rescue authority]"[37] *has the meaning assigned by section 43(1) of the Fire Precautions Act 1971* [, in relation to premises, means—
   (a) where the Regulatory Reform (Fire Safety) Order 2005 applies to the premises, the enforcing authority within the meaning given by article 25 of that Order;
   (b) in any other case, the fire and rescue authority under the Fire and Rescue Services Act 2004 for the area where the premises are (or are to be) situated].[38]

(5) Where an improvement notice or a prohibition notice which is not to take immediate effect has been served—
   (a) the notice may be withdrawn by an inspector at any time before the end of the period specified therein in pursuance of section 21 or section 22(4) as the case may be; and
   (b) the period so specified may be extended or further extended by an inspector at any time when an appeal against the notice is not pending.
(6) In the application of this section to Scotland—
   (a) in subsection (3) for the words from "with the requirements" to "aforesaid" there shall be substituted the words—
      "(a) to any provisions of the building standards regulations to which that building or matter would be required to conform if the relevant building were being newly erected; or
      (b) where the sheriff, on an appeal to him under section 16 of the Building (Scotland) Act 1959—
         (i) against an order under section 10 of that Act requiring the execution of operations necessary to make the building or matter conform to the building standards regulations, or

---

[36] Amended by Fire and Rescue Services Act 2004.
[37] ibid.
[38] Amended by SI 2005/1541.

(ii) against an order under section 11 of that Act requiring the building or matter to conform to a provision of such regulations,

has varied the order, to any provisions of the building standards regulations referred to in paragraph (a) above as affected by the order as so varied,

unless the relevant statutory provision imposes specific requirements more onerous than the requirements of any provisions of building standards regulations as aforesaid or, as the case may be, than the requirements of the order as varied by the sheriff.";

(b) after subsection (5) there shall be inserted the following subsection—

"(5A) In subsection (3) above ".building standards regulations"' has the same meaning as in section 3 of the Building (Scotland) Act 1959."

24 Appeal against improvement or prohibition notice

(1) In this section "a notice" means an improvement notice or a prohibition notice.
(2) A person on whom a notice is served may within such period from the date of its service as may be prescribed appeal to an [employment tribunal];[39] and on such an appeal the tribunal may either cancel or affirm the notice and, if it affirms it, may do so either in its original form or with such modifications as the tribunal may in the circumstances think fit.
(3) Where an appeal under this section is brought against a notice within the period allowed under the preceding subsection, then—
   (a) in the case of an improvement notice, the bringing of the appeal shall have the effect of suspending the operation of the notice until the appeal is finally disposed of or, if the appeal is withdrawn, until the withdrawal of the appeal;
   (b) in the case of a prohibition notice, the bringing of the appeal shall have the like effect if, but only if, on the application of the appellant the tribunal so directs (and then only from the giving of the direction).
(4) One or more assessors may be appointed for the purposes of any proceedings brought before an [employment tribunal][40] under this section.

25 Power to deal with cause of imminent danger

(1) Where, in the case of any article or substance found by him in any premises which he has power to enter, an inspector has reasonable cause to believe that, in the circumstances in which he finds it, the article or substance is a cause of imminent danger of serious personal injury, he may seize it and cause it to be rendered harmless (whether by destruction or otherwise).
(2) Before there is rendered harmless under this section—
   (a) any article that forms part of a batch of similar articles; or
   (b) any substance,
the inspector shall, if it is practicable for him to do so, take a sample thereof and give to a responsible person at the premises where the article or substance was found by him a portion of the sample marked in a manner sufficient to identify it.

(3) As soon as may be after any article or substance has been seized and rendered harmless under this section, the inspector shall prepare and sign a written report giving particulars of the circumstances in which the article or substance was seized and so dealt with by him, and shall—
   (a) give a signed copy of the report to a responsible person at the premises where the article or substance was found by him; and
   (b) unless that person is the owner of the article or substance, also serve a signed copy of the report on the owner;

---

[39] Amended by the Employment Rights (Dispute Resolution) Act 1998.
[40] ibid.

and if, where paragraph (b) above applies, the inspector cannot after reasonable enquiry ascertain the name or address of the owner, the copy may be served on him by giving it to the person to whom a copy was given under the preceding paragraph.

[25A Power of customs officer to detain articles and substances][41]

[(1) A customs officer may, for the purpose of facilitating the exercise or performance by an enforcing authority or inspector of any of the powers or duties of the authority or inspector under any of the relevant statutory provisions, seize any imported article or imported substance and detain it for not more than two working days.

(2) Anything seized and detained under this section shall be dealt with during the period of its detention in such manner as the Commissioners of Customs and Excise may direct.

(3) In subsection (1) above the reference to two working days is a reference to a period of forty-eight hours calculated from the time when the goods in question are seized but disregarding so much of any period as falls on a Saturday or Sunday or on Christmas Day, Good Friday or a day which is a bank holiday under the Banking and Financial Dealings Act 1971 in the part of Great Britain where the goods are seized.]

26 Power of enforcing authorities to indemnify their inspectors

Where an action has been brought against an inspector in respect of an act done in the execution or purported execution of any of the relevant statutory provisions and the circumstances are such that he is not legally entitled to require the enforcing authority which appointed him to indemnify him, that authority may, nevertheless, indemnify him against the whole or part of any damages and costs or expenses which he may have been ordered to pay or may have incurred, if the authority is satisfied that he honestly believed that the act complained of was within his powers and that his duty as an inspector required or entitled him to do it.

Obtaining and disclosure of information

27 Obtaining of information by the Commission, the Executive, enforcing authorities etc

(1) For the purpose of obtaining—
  (a) any information which the Commission needs for the discharge of its functions; or
  (b) any information which an enforcing authority needs for the discharge of the authority's functions,
the Commission may, with the consent of the Secretary of State, serve on any person a notice requiring that person to furnish to the Commission or, as the case may be, to the enforcing authority in question such information about such matters as may be specified in the notice, and to do so in such form and manner and within such time as may be so specified.

In this subsection "consent" includes a general consent extending to cases of any stated description.

(2) Nothing in section 9 of the Statistics of Trade Act 1947 (which restricts the disclosure of information obtained under that Act) shall prevent or penalise—
  (a) the disclosure by a Minister of the Crown to the Commission or the Executive of information obtained under that Act about any undertaking within the meaning of that Act, being information consisting of the names and addresses of the persons carrying on the undertaking, the nature of the undertaking's activities, the numbers of persons of different descriptions who work in the undertaking, the addresses or places where activities of the undertaking are or were carried on, the nature of the activities carried on there, or the numbers of persons of different descriptions who work or worked in the undertaking there; [. . .][42]
  (b) [43]

---

[41] Inserted by the Consumer Protection Act 1987.
[42] Repealed by the Employment Act 1989.
[43] ibid.

(3) In the preceding subsection any reference to a Minister of the Crown, the Commission, [or the Executive][44] includes respectively a reference to an officer of his or of that body and also, in the case of a reference to the Commission, includes a reference to—
   (a) a person performing any functions of the Commission or the Executive on its behalf by virtue of section 13(1)(a);
   (b) an officer of a body which is so performing any such functions; and
   (c) an adviser appointed in pursuance of section 13(1)(d).
(4) A person to whom information is disclosed in pursuance of subsection (2) above shall not use the information for a purpose other than a purpose of the Commission or, as the case may be, of the Executive.

[27A Information communicated by Commissioners for Revenue and Customs]

[(1) If they think it appropriate to do so for the purpose of facilitating the exercise or performance by any person to whom subsection (2) below applies of any of that person's powers or duties under any of the relevant statutory provisions, [the Commissioners for Her Majesty's Revenue and Custom][45] may authorise the disclosure to that person of any information obtained [or held][46] for the purposes of the exercise [by Her Majesty's Revenue and Customs][47,48] of their functions in relation to imports.
(2) This subsection applies to an enforcing authority and to an inspector.
(3) A disclosure of information made to any person under subsection (1) above shall be made in such manner as may be directed by [the Commissioners for Her Majesty's Revenue and Customs] and may be made through such persons acting on behalf of that person as may be so directed.
(4) Information may be disclosed to a person under subsection (1) above whether or not the disclosure of the information has been requested by or on behalf of that person.][49]

28 Restrictions on disclosure of information

(1) In this and the two following subsections—
   (a) "relevant information" means information obtained by a person under section 27(1) or furnished to any person [under section 27A above[, by virtue of section 43A(6) below][50] or][51] in pursuance of a requirement imposed by any of the relevant statutory provisions; and
   (b) "the recipient", in relation to any relevant information, means the person by whom that information was so obtained or to whom that information was so furnished, as the case may be.
(2) Subject to the following subsection, no relevant information shall be disclosed without the consent of the person by whom it was furnished.
(3) The preceding subsection shall not apply to—
   (a) disclosure of information to the Commission, the Executive, [the Environment Agency, the Scottish Environment Protection Agency,][52] a government department or any enforcing authority;
   (b) without prejudice to paragraph (a) above, disclosure by the recipient of information to any person for the purpose of any function conferred on the recipient by or under any of the relevant statutory provisions;

---

[44] Amended by the Employment Act 1989.
[45] ibid.
[46] ibid.
[47] ibid.
[48] Amended by the Commissioners for Revenue and Customs Act 2005.
[49] Inserted by the Consumer Protection Act 1987.
[50] Amended by the Railways and Transport Safety Act 2003.
[51] Amended by the Consumer Protection Act 1987.
[52] Amended by the Environment Act 1995.

(c) without prejudice to paragraph (a) above, disclosure by the recipient of information to—
    (i) an officer of a local authority who is authorised by that authority to receive it,
    [(ii) an officer [...]⁵³ of a water undertaker, sewerage undertaker, water authority or water development board who is authorised by that [...] undertaker, authority or board to receive it,]⁵⁴
    (iii) [...]⁵⁵
    (iv) a constable authorised by a chief officer of police to receive it;
(d) disclosure by the recipient of information in a form calculated to prevent it from being identified as relating to a particular person or case;
(e) disclosure of information for the purposes of any legal proceedings or any investigation or inquiry held by virtue of section 14(2), or for the purposes of a report of any such proceedings or inquiry or of a special report made by virtue of section 14(2);
[(f) any other disclosure of information by the recipient, if—
    (i) the recipient is, or is acting on behalf of a person who is, a public authority for the purposes of the Freedom of Information Act 2000, and
    (ii) the information is not held by the authority on behalf of another person].⁵⁶

(4) In the preceding subsection any reference to the Commission, the Executive, [the Environment Agency, the Scottish Environment Protection Agency,]⁵⁷ a government department or an enforcing authority includes respectively a reference to an officer of that body or authority (including, in the case of an enforcing authority, any inspector appointed by it), and also, in the case of a reference to the Commission, includes a reference to—
    (a) a person performing any functions of the Commission or the Executive on its behalf by virtue of section 13(1)(a);
    (b) an officer of a body which is so performing any such functions; and
    (c) an adviser appointed in pursuance of section 13(1)(d).

(5) A person to whom information is disclosed in pursuance of [any of paragraphs (a) to (e) of]⁵⁸ subsection (3) above shall not use the information for a purpose other than—
    (a) in a case falling within paragraph (a) of that subsection, a purpose of the Commission or of the Executive or [of the Environment Agency or of the Scottish Environment Protection Agency or]⁵⁹ of the government department in question, or the purposes of the enforcing authority in question in connection with the relevant statutory provisions, as the case may be;
    (b) in the case of information given to an officer of a [body which is a local authority, [...],⁶⁰ a water undertaker, a sewerage undertaker, a water authority, a river purification board or a water development board, the purposes of the body]⁶¹ in connection with the relevant statutory provisions or any enactment whatsoever relating to public health, public safety or the protection of the environment;
    (c) in the case of information given to a constable, the purposes of the police in connection with the relevant statutory provisions or any enactment whatsoever relating to public health, public safety or the safety of the State.

---

⁵³ Repealed by the Environment Act 1995.
⁵⁴ Amended by the Water Act 1989.
⁵⁵ Repealed by the Environment Act 1995.
⁵⁶ Amended by SI 2004/3363.
⁵⁷ Amended by the Environment Act 1995.
⁵⁸ Amended by SI 2004/3363.
⁵⁹ Amended by the Environment Act 1995.
⁶⁰ Repealed by the Environment Act 1995.
⁶¹ Amended by the Water Act 1989.

[(6) References in subsections (3) and (5) above to a local authority include [...]⁶² a joint authority established by Part IV of the Local Government Act 1985 [and the London Fire and Emergency Planning Authority]⁶³.]⁶⁴

(7) A person shall not disclose any information obtained by him as a result of the exercise of any power conferred by section 14(4)(a) or 20 (including, in particular, any information with respect to any trade secret obtained by him in any premises entered by him by virtue of any such power) except—
  (a) for the purposes of his functions; or
  (b) for the purposes of any legal proceedings or any investigation or inquiry held by virtue of section 14(2) or for the purposes of a report of any such proceedings or inquiry or of a special report made by virtue of section 14(2); or
  (c) with the relevant consent.

In this subsection "the relevant consent" means, in the case of information furnished in pursuance of a requirement imposed under section 20, the consent of the person who furnished it, and, in any other case, the consent of a person having responsibilities in relation to the premises where the information was obtained.

(8) Notwithstanding anything in the preceding subsection an inspector shall, in circumstances in which it is necessary to do so for the purpose of assisting in keeping persons (or the representatives of persons) employed at any premises adequately informed about matters affecting their health, safety and welfare, give to such persons or their representatives the following descriptions of information, that is to say—
  (a) factual information obtained by him as mentioned in that subsection which relates to those premises or anything which was or is therein or was or is being done therein; and
  (b) information with respect to any action which he has taken or proposes to take in or in connection with those premises in the performance of his functions;

and, where an inspector does as aforesaid, he shall give the like information to the employer of the first-mentioned persons.

[(9) Notwithstanding anything in subsection (7) above, a person who has obtained such information as is referred to in that subsection may furnish to a person who appears to him to be likely to be a party to any civil proceedings arising out of any accident, occurrence, situation or other matter, a written statement of relevant facts observed by him in the course of exercising any of the powers referred to in that subsection.]⁶⁵

[(9A) Subsection (7) above does not apply if—
  (a) the person who has obtained any such information as is referred to in that subsection is, or is acting on behalf of a person who is, a public authority for the purposes of the Freedom of Information Act 2000, and
  (b) the information is not held by the authority on behalf of another person.]⁶⁶

[(10) The Broads Authority and every National Park authority shall be deemed to be local authorities for the purposes of this section.]⁶⁷

29 [...]⁶⁸

30 [...]⁶⁹

---

⁶² Repealed by the Education Reform Act 1988.
⁶³ Amended by the Greater London Authority Act 1999.
⁶⁴ Inserted by the Local Government Act 1985.
⁶⁵ Inserted by the Employment Protection Act 1975.
⁶⁶ Inserted by SI 2004/3363.
⁶⁷ Inserted by the Norfolk and Suffolk Broads Act 1988.
⁶⁸ Repealed by the Employment Protection Act 1975.
⁶⁹ ibid.

Relevant Extracts from the Health and Safety at Work etc. Act 1974

31 [...][70]

32 [...][71]

Provisions as to offences

33 Offences

(1) It is an offence for a person—
   (a) to fail to discharge a duty to which he is subject by virtue of sections 2 to 7;
   (b) to contravene section 8 or 9;
   (c) to contravene any health and safety regulations [...][72] or any requirement or prohibition imposed under any such regulations (including any requirement or prohibition to which he is subject by virtue of the terms of or any condition or restriction attached to any licence, approval, exemption or other authority issued, given or granted under the regulations);
   (d) to contravene any requirement imposed by or under regulations under section 14 or intentionally to obstruct any person in the exercise of his powers under that section;
   (e) to contravene any requirement imposed by an inspector under section 20 or 25;
   (f) to prevent or attempt to prevent any other person from appearing before an inspector or from answering any question to which an inspector may by virtue of section 20(2) require an answer;
   (g) to contravene any requirement or prohibition imposed by an improvement notice or a prohibition notice (including any such notice as modified on appeal);
   (h) intentionally to obstruct an inspector in the exercise or performance of his powers or duties [or to obstruct a customs officer in the exercise of his powers under section 25A];[73]
   (i) to contravene any requirement imposed by a notice under section 27(1);
   (j) to use or disclose any information in contravention of section 27(4) or 28;
   (k) to make a statement which he knows to be false or recklessly to make a statement which is false where the statement is made—
      (i) in purported compliance with a requirement to furnish any information imposed by or under any of the relevant statutory provisions; or
      (ii) for the purpose of obtaining the issue of a document under any of the relevant statutory provisions to himself or another person;
   (l) intentionally to make a false entry in any register, book, notice or other document required by or under any of the relevant statutory provisions to be kept, served or given or, with intent to deceive, to make use of any such entry which he knows to be false;
   (m) with intent to deceive, to [...][74] use a document issued or authorised to be issued under any of the relevant statutory provisions or required for any purpose thereunder or to make or have in his possession a document so closely resembling any such document as to be calculated to deceive;
   (n) falsely to pretend to be an inspector;
   (o) to fail to comply with an order made by a court under section 42.
[(1A) Subject to any provision made by virtue of section 15(6)(d), a person guilty of an offence under subsection (1)(a) above consisting of failing to discharge a duty to which he is subject by virtue of sections 2 to 6 shall be liable—
   (a) on summary conviction, to a fine not exceeding £20,000;
   (b) on conviction on indictment, to a fine.][75]

---

[70] Repealed by the Employment Protection Act 1975.
[71] ibid.
[72] ibid.
[73] Amended by the Consumer Protection Act 1987.
[74] Repealed by the Forgery and Counterfeiting Act 1981.
[75] Inserted by the Offshore Safety Act 1992.

(2) A person guilty of an offence under paragraph (d), (f), (h) or (n) of subsection (1) above, or of an offence under paragraph (e) of that subsection consisting of contravening a requirement imposed by an inspector under section 20, shall be liable on summary conviction to a fine not exceeding [level 5 on the standard scale].[76]

[(2A) A person guilty of an offence under subsection (1)(g) or (o) above shall be liable—
  (a) on summary conviction, to imprisonment for a term not exceeding six months, or a fine not exceeding £20,000, or both;
  (b) on conviction on indictment, to imprisonment for a term not exceeding two years, or a fine, or both.][77]

(3) Subject to any provision made by virtue of section 15(6)(d) [or (e)][78] or by virtue of paragraph 2(2) of Schedule 3, a person guilty of [an offence under subsection (1) above not falling within subsection (1A), (2) or (2A) above],[79] or of an offence under any of the existing statutory provisions, being an offence for which no other penalty is specified, shall be liable—
  (a) on summary conviction, to a fine not exceeding [the prescribed sum];[80]
  (b) on conviction on indictment—
    (i) if the offence is one to which this sub-paragraph applies, to imprisonment for a term not exceeding two years, or a fine, or both;
    (ii) if the offence is not one to which the preceding sub-paragraph applies, to a fine.

(4) Subsection (3)(b)(i) above applies to the following offences—
  (a) an offence consisting of contravening any of the relevant statutory provisions by doing otherwise than under the authority of a licence issued by the Executive [. . .][81] something for the doing of which such a licence is necessary under the relevant statutory provisions;
  (b) an offence consisting of contravening a term of or a condition or restriction attached to any such licence as is mentioned in the preceding paragraph;
  (c) an offence consisting of acquiring or attempting to acquire, possessing or using an explosive article or substance (within the meaning of any of the relevant statutory provisions) in contravention of any of the relevant statutory provisions;
  (d) [. . .][82]
  (e) an offence under subsection (1)(j) above.

(5) [. . .][83]

(6) [. . .][84]

34 Extension of time for bringing summary proceedings

(1) Where—
  (a) a special report on any matter to which section 14 of this Act applies is made by virtue of subsection (2)(a) of that section; or
  (b) a report is made by the person holding an inquiry into any such matter by virtue of subsection (2)(b) of that section; or
  (c) a coroner's inquest is held touching the death of any person whose death may have been caused by an accident which happened while he was at work or by a disease which he contracted or probably contracted at work or by any accident, act or omission which occurred in connection with the work of any person whatsoever; or

---

[76] Amended by the Criminal Law Act 1977 and the Criminal Justice Act 1982.
[77] Inserted by the Offshore Safety Act 1992.
[78] Amended by the Offshore Safety Act 1992.
[79] ibid.
[80] Amended by the Magistrates' Courts Act 1980.
[81] Repealed by the Employment Protection Act 1975.
[82] Repealed by the Offshore Safety Act 1992.
[83] ibid.
[84] Repealed by the Forgery and Counterfeiting Act 1981.

(d) a public inquiry into any death that may have been so caused is held under the Fatal Accidents Inquiry (Scotland) Act 1895 or the Fatal Accidents and Sudden Deaths Inquiry (Scotland) Act 1906;

and it appears from the report or, in a case falling within paragraph (c) or (d) above, from the proceedings at the inquest or inquiry, that any of the relevant statutory provisions was contravened at a time which is material in relation to the subject-matter of the report, inquest or inquiry, summary proceedings against any person liable to be proceeded against in respect of the contravention may be commenced at any time within three months of the making of the report or, in a case falling within paragraph (c) or (d) above, within three months of the conclusion of the inquest or inquiry.

(2) Where an offence under any of the relevant statutory provisions is committed by reason of a failure to do something at or within a time fixed by or under any of those provisions, the offence shall be deemed to continue until that thing is done.

(3) Summary proceedings for an offence to which this subsection applies may be commenced at any time within six months from the date on which there comes to the knowledge of a responsible enforcing authority evidence sufficient in the opinion of that authority to justify a prosecution for that offence; and for the purposes of this subsection—

(a) a certificate of an enforcing authority stating that such evidence came to its knowledge on a specified date shall be conclusive evidence of that fact; and

(b) a document purporting to be such a certificate and to be signed by or on behalf of the enforcing authority in question shall be presumed to be such a certificate unless the contrary is proved.

(4) The preceding subsection applies to any offence under any of the relevant statutory provisions which a person commits by virtue of any provision or requirement to which he is subject as the designer, manufacturer, importer or supplier of any thing; and in that subsection "responsible enforcing authority" means an enforcing authority within whose field of responsibility the offence in question lies, whether by virtue of section 35 or otherwise.

(5) In the application of subsection (3) above to Scotland—

(a) for the words from "there comes" to "that offence" there shall be substituted the words "evidence, sufficient in the opinion of the enforcing authority to justify a report to the Lord Advocate with a view to consideration of the question of prosecution, comes to the knowledge of the authority";

(b) at the end of paragraph (b) there shall be added the words "and

(c) [section 331(3) of the Criminal Procedure (Scotland) Act 1975][85] (date of commencement of proceedings) shall have effect as it has effect for the purposes of that section."

[(6) In the application of subsection (4) above to Scotland, after the words "applies to" there shall be inserted the words "any offence under section 33(1)(c) above where the health and safety regulations concerned were made for the general purpose mentioned in section 18(1) of the Gas Act 1986 and".][86]

35 Venue

An offence under any of the relevant statutory provisions committed in connection with any plant or substance may, if necessary for the purpose of bringing the offence within the field of responsibility of any enforcing authority or conferring jurisdiction on any court to entertain proceedings for the offence, be treated as having been committed at the place where that plant or substance is for the time being.

---

[85] Amended by the Criminal Procedure (Scotland) Act 1975.
[86] Inserted by the Gas Act 1986. Amendment only applicable to Scotland.

36 Offences due to fault of other person

(1) Where the commission by any person of an offence under any of the relevant statutory provisions is due to the act or default of some other person, that other person shall be guilty of the offence, and a person may be charged with and convicted of the offence by virtue of this subsection whether or not proceedings are taken against the first-mentioned person.

(2) Where there would be or have been the commission of an offence under section 33 by the Crown but for the circumstance that that section does not bind the Crown, and that fact is due to the act or default of a person other than the Crown, that person shall be guilty of the offence which, but for that circumstance, the Crown would be committing or would have committed, and may be charged with and convicted of that offence accordingly.

(3) The preceding provisions of this section are subject to any provision made by virtue of section 15(6).

37 Offences by bodies corporate

(1) Where an offence under any of the relevant statutory provisions committed by a body corporate is proved to have been committed with the consent or connivance of, or to have been attributable to any neglect on the part of, any director, manager, secretary or other similar officer of the body corporate or a person who was purporting to act in any such capacity, he as well as the body corporate shall be guilty of that offence and shall be liable to be proceeded against and punished accordingly.

(2) Where the affairs of a body corporate are managed by its members, the preceding subsection shall apply in relation to the acts and defaults of a member in connection with his functions of management as if he were a director of the body corporate.

38 Restriction on institution of proceedings in England and Wales

Proceedings for an offence under any of the relevant statutory provisions shall not, in England and Wales, be instituted except by an inspector or [the Environment Agency or][87] by or with the consent of the Director of Public Prosecutions.

39 Prosecutions by inspectors

(1) An inspector, if authorised in that behalf by the enforcing authority which appointed him, may, although not of counsel or a solicitor, prosecute before a magistrates' court proceedings for an offence under any of the relevant statutory provisions.

(2) This section shall not apply to Scotland.

40 Onus of proving limits of what is practicable etc

In any proceedings for an offence under any of the relevant statutory provisions consisting of a failure to comply with a duty or requirement to do something so far as is practicable or so far as is reasonably practicable, or to use the best means to do something, it shall be for the accused to prove (as the case may be) that it was not practicable or not reasonably practicable to do more than was in fact done to satisfy the duty or requirement, or that there was no better practicable means than was in fact used to satisfy the duty or requirement.

41 Evidence

(1) Where an entry is required by any of the relevant statutory provisions to be made in any register or other record, the entry, if made, shall, as against the person by or on whose behalf it was made, be admissible as evidence or in Scotland sufficient evidence of the facts stated therein.

---

[87] Amended by the Environment Act 1995.

(2) Where an entry which is so required to be so made with respect to the observance of any of the relevant statutory provisions has not been made, that fact shall be admissible as evidence or in Scotland sufficient evidence that that provision has not been observed.

42 Power of court to order cause of offence to be remedied or, in certain cases, forfeiture

(1) Where a person is convicted of an offence under any of the relevant statutory provisions in respect of any matters which appear to the court to be matters which it is in his power to remedy, the court may, in addition to or instead of imposing any punishment, order him, within such time as may be fixed by the order, to take such steps as may be specified in the order for remedying the said matters.
(2) The time fixed by an order under subsection (1) above may be extended or further extended by order of the court on an application made before the end of that time as originally fixed or as extended under this subsection, as the case may be.
(3) Where a person is ordered under subsection (1) above to remedy any matters, that person shall not be liable under any of the relevant statutory provisions in respect of those matters in so far as they continue during the time fixed by the order or any further time allowed under subsection (2) above.
(4) Subject to the following subsection, the court by or before which a person is convicted of an offence such as is mentioned in section 33(4)(c) in respect of any such explosive article or substance as is there mentioned may order the article or substance in question to be forfeited and either destroyed or dealt with in such other manner as the court may order.
(5) The court shall not order anything to be forfeited under the preceding subsection where a person claiming to be the owner of or otherwise interested in it applies to be heard by the court, unless an opportunity has been given to him to show cause why the order should not be made.

Financial provisions

43 Financial provisions

(1) It shall be the duty of the Secretary of State to pay to the Commission such sums as are approved by the Treasury and as he considers appropriate for the purpose of enabling the Commission to perform its functions; and it shall be the duty of the Commission to pay to the Executive such sums as the Commission considers appropriate for the purpose of enabling the Executive to perform its functions.
(2) Regulations may provide for such fees as may be fixed by or determined under the regulations to be payable for or in connection with the performance by or on behalf of any authority to which this subsection applies of any function conferred on that authority by or under any of the relevant statutory provisions.
(3) Subsection (2) above applies to the following authorities, namely the Commission, the Executive, the Secretary of State, [. . .][88] every enforcing authority, and any other person on whom any function is conferred by or under any of the relevant statutory provisions.
(4) Regulations under this section may specify the person by whom any fee payable under the regulations is to be paid; but no such fee shall be made payable by a person in any of the following capacities, namely an employee, a person seeking employment, a person training for employment, and a person seeking training for employment.
(5) Without prejudice to section 82 (3), regulations under this section may fix or provide for the determination of different fees in relation to different functions, or in relation to the same function in different circumstances.

---

[88] Repealed by the Employment Protection Act 1975.

[(6) The power to make regulations under this section shall be exercisable by the Secretary of State [. . .][89][90]
(8) In subsection (4) above the references to a person training for employment and a person seeking training for employment shall include respectively a person attending an industrial rehabilitation course provided by virtue of the Employment and Training Act 1973 and a person seeking to attend such a course.
(9) For the purposes of this section the performance by an inspector of his functions shall be treated as the performance by the enforcing authority which appointed him of functions conferred on that authority by or under any of the relevant statutory provisions.

[43A Railway safety levy][91]

[(1) The Secretary of State may make regulations requiring persons who provide railway services to pay railway safety levy.
(2) Railway safety levy shall be applied only for the purpose of meeting expenses incurred—
    [(a) in respect of activities undertaken by the Office of Rail Regulation under or by virtue of this Act or Schedule 3 to the Railways Act 2005; or
    (b) in respect of activities in relation to a transport system falling within paragraph 1(3) of that Schedule that are undertaken by that Office under or by virtue of any other enactment].[92]
(3) The railway safety levy shall not be used to meet—
    (a) an expense in respect of which a fee is payable under regulations made under section 43, or
    (b) an expense in respect of a matter specified by the regulations for the purpose of this paragraph.
(4) Where an expense is incurred partly in respect of activity within subsection (2)(a) or (b) and partly in respect of other activity, the railway safety levy may be used to meet a part of that expense which is reasonably referable to activity within subsection (2)(a) or (b).
(5) Regulations under subsection (1) may, in particular, determine or enable the [Office of Rail Regulation][93] to determine—
    (a) the total amount of the railway safety levy to be imposed in respect of a specified period;
    (b) the persons by whom the levy is to be paid;
    (c) the criteria for assessing the proportion of the levy to be paid by a particular person (which may, in particular, refer to the size of a person's income or provide for an amount to be reduced or waived in specified circumstances);
    (d) the periods in respect of which the levy is to be paid;
    (e) the manner in which the levy is to be paid;
    (f) the person to whom the levy is to be paid;
    (g) when the levy is to be paid.
(6) Regulations under subsection (1) may, in particular, enable the [Office of Rail Regulation][94]—
    (a) to require a person who provides railway services to supply information for the purposes of the consideration of a matter specified in subsection (5);
    (b) where information requested is not supplied, to make assumptions;
    (c) to revise a determination of a matter specified in subsection (5) (whether before, during or after the period to which it relates);
    (d) to make refunds.
(7) Regulations by virtue of subsection (6)(a) may, in particular, make provision—
    (a) about the manner and timing of the supply of information;
    (b) about certification of the accuracy of information supplied;

---

[89] Repealed by SI 2002/794.
[90] Amended by the Employment Protection Act 1975.
[91] Inserted by the Railways and Transport Safety Act 2003.
[92] Amended by the Railways Act 2005.
[93] ibid.
[94] ibid.

(c) creating a criminal offence in connection with the supply of inaccurate or misleading information (but not an offence punishable with imprisonment).

(8) Regulations under subsection (1) may enable payment to be enforced by civil proceeding.

(9) For the purposes of this section a person provides railway services if he manages or controls, or participates in managing or controlling, a transport system [falling within paragraph 1(3) of Schedule 3 to the Railways Act 2005][95].[96]

Miscellaneous and supplementary

44 Appeals in connection with licensing provisions in the relevant statutory provisions

(1) Any person who is aggrieved by a decision of an authority having power to issue licences (other than [. . .][97] nuclear site licences) under any of the relevant statutory provisions—
   (a) refusing to issue him a licence, to renew a licence held by him, or to transfer to him a licence held by another;
   (b) issuing him a licence on or subject to any term, condition or restriction whereby he is aggrieved;
   (c) varying or refusing to vary any term, condition or restriction on or subject to which a licence is held by him; or
   (d) revoking a licence held by him,

may appeal to the Secretary of State.

(2) The Secretary of State may, in such cases as he considers it appropriate to do so, having regard to the nature of the questions which appear to him to arise, direct that an appeal under this section shall be determined on his behalf by a person appointed by him for that purpose.

(3) Before the determination of an appeal the Secretary of State shall ask the appellant and the authority against whose decision the appeal is brought whether they wish to appear and be heard on the appeal and—
   (a) the appeal may be determined without a hearing of the parties if both of them express a wish not to appear and be heard as aforesaid;
   (b) the Secretary of State shall, if either of the parties expresses a wish to appear and be heard, afford to both of them an opportunity of so doing.

(4) The Tribunals and Inquiries Act [1992][98] shall apply to a hearing held by a person appointed in pursuance of subsection (2) above to determine an appeal as it applies to a statutory inquiry held by the Secretary of State, but as if in [section 10(1)][99] of that Act (statement of reasons for decisions) the reference to any decision taken by the Secretary of State included a reference to a decision taken on his behalf by that person.

(5) A person who determines an appeal under this section on behalf of the Secretary of State and the Secretary of State, if he determines such an appeal, may give such directions as he considers appropriate to give effect to his determination.

(6) The Secretary of State may pay to any person appointed to hear or determine an appeal under this section on his behalf such remuneration and allowances as the Secretary of State may with the approval of the Minister for the Civil Service determine.

(7) In this section—
   (a) "licence" means a licence under any of the relevant statutory provisions other than [a nuclear site licence];[100]

---

[95] Amended by the Railways Act 2005.
[96] Inserted by the Railways and Transport Safety Act 2003.
[97] Repealed by Employment Protection Act 1975.
[98] Amended by Tribunals and Inquiries Act 1992.
[99] ibid.
[100] Amended by the Employment Protection Act 1975.

(b) "nuclear site licence" means a licence to use a site for the purpose of installing or operating a nuclear installation within the meaning of the following subsection.

(8) For the purposes of the preceding subsection "nuclear installation" means—
   (a) a nuclear reactor (other than such a reactor comprised in a means of transport, whether by land, water or air); or
   (b) any other installation of such class or description as may be prescribed for the purposes of this paragraph or section 1(1)(b) of the Nuclear Installations Act 1965, being an installation designed or adapted for—
      (i) the production or use of atomic energy, or
      (ii) the carrying out of any process which is preparatory or ancillary to the production or use of atomic energy and which involves or is capable of causing the emission of ionising radiations; or
      (iii) the storage, processing or disposal of nuclear fuel or of bulk quantities of other radioactive matter, being matter which has been produced or irradiated in the course of the production or use of nuclear fuel;

and in this subsection—

"atomic energy" has the meaning assigned by the Atomic Energy Act 1946;

"nuclear reactor" means any plant (including any machinery, equipment or appliance whether affixed to land or not) designed or adapted for the production of atomic energy by a fission process in which a controlled chain reaction can be maintained without an additional source of neutrons.

45 Default powers

(1) Where, in the case of a local authority who are an enforcing authority, the Commission is of the opinion that an investigation should be made as to whether that local authority have failed to perform any of their enforcement functions, the Commission may make a report to the Secretary of State.

(2) The Secretary of State may, after considering a report submitted to him under the preceding subsection, cause a local inquiry to be held; and the provisions of subsections (2) to (5) of section 250 of the Local Government Act 1972 as to local inquiries shall, without prejudice to the generality of subsection (1) of that section, apply to a local inquiry so held as they apply to a local inquiry held in pursuance of that section.

(3) If the Secretary of State is satisfied, after having caused a local inquiry to be held into the matter, that a local authority have failed to perform any of their enforcement functions, he may make an order declaring the authority to be in default.

(4) An order made by virtue of the preceding subsection which declares an authority to be in default may, for the purpose of remedying the default, direct the authority (hereafter in this section referred to as "the defaulting authority") to perform such of their enforcement functions as are specified in the order in such manner as may be so specified and may specify the time or times within which those functions are to be performed by the authority.

(5) If the defaulting authority fail to comply with any direction contained in such an order the Secretary of State may, instead of enforcing the order by mandamus, make an order transferring to the Executive such of the enforcement functions of the defaulting authority as he thinks fit.

(6) Where any enforcement functions of the defaulting authority are transferred in pursuance of the preceding subsection, the amount of any expenses which the Executive certifies were incurred by it in performing those functions shall on demand be paid to it by the defaulting authority.

(7) Any expenses which in pursuance of the preceding subsection are required to be paid by the defaulting authority in respect of any enforcement functions transferred in pursuance of this section shall be defrayed by the authority in the like manner, and shall be debited to the like

account, as if the enforcement functions had not been transferred and the expenses had been incurred by the authority in performing them.
(8) Where the defaulting authority are required to defray any such expenses the authority shall have the like powers for the purpose of raising the money for defraying those expenses as they would have had for the purpose of raising money required for defraying expenses incurred for the purpose of the enforcement functions in question.
(9) An order transferring any enforcement functions of the defaulting authority in pursuance of subsection (5) above may provide for the transfer to the Executive of such of the rights, liabilities and obligations of the authority as the Secretary of State considers appropriate; and where such an order is revoked the Secretary of State may, by the revoking order or a subsequent order, make such provision as he considers appropriate with respect to any rights, liabilities and obligations held by the Executive for the purposes of the transferred enforcement functions.
(10) The Secretary of State may by order vary or revoke any order previously made by him in pursuance of this section.
(11) In this section "enforcement functions", in relation to a local authority, means the functions of the authority as an enforcing authority.
(12) In the application of this section to Scotland—
  (a) in subsection (2) for the words " subsections (2) to (5) of section 250 of the Local Government Act 1972" there shall be substituted the words " subsections (2) to (8) of section 210 of the Local Government (Scotland) Act 1973", except that before 16th May 1975 for the said words there shall be substituted the words " subsections (2) to (9) of section 355 of the Local Government (Scotland) Act 1947";
  (b) in subsection (5) the words "instead of enforcing the order by mandamus" shall be omitted.

46 Service of notices

(1) Any notice required or authorised by any of the relevant statutory provisions to be served on or given to an inspector may be served or given by delivering it to him or by leaving it at, or sending it by post to, his office.
(2) Any such notice required or authorised to be served on or given to a person other than an inspector may be served or given by delivering it to him, or by leaving it at his proper address, or by sending it by post to him at that address.
(3) Any such notice may—
  (a) in the case of a body corporate, be served on or given to the secretary or clerk of that body;
  (b) in the case of a partnership, be served on or given to a partner or a person having the control or management of the partnership business or, in Scotland, the firm.
(4) For the purposes of this section and of section 26 of the Interpretation Act 1889 (service of documents by post) in its application to this section, the proper address of any person on or to whom any such notice is to be served or given shall be his last known address, except that—
  (a) in the case of a body corporate or their secretary or clerk, it shall be the address of the registered or principal office of that body;
  (b) in the case of a partnership or a person having the control or the management of the partnership business, it shall be the principal office of the partnership;
and for the purposes of this subsection the principal office of a company registered outside the United Kingdom or of a partnership carrying on business outside the United Kingdom shall be their principal office within the United Kingdom.
(5) If the person to be served with or given any such notice has specified an address within the United Kingdom other than his proper address within the meaning of subsection (4) above as the one at which he or someone on his behalf will accept notices of the same description as that notice, that address shall also be treated for the purposes of this section and section 26 of the Interpretation Act 1889 as his proper address.

(6) Without prejudice to any other provision of this section, any such notice required or authorised to be served on or given to the owner or occupier of any premises (whether a body corporate or not) may be served or given by sending it by post to him at those premises, or by addressing it by name to the person on or to whom it is to be served or given and delivering it to some responsible person who is or appears to be resident or employed in the premises.

(7) If the name or the address of any owner or occupier of premises on or to whom any such notice as aforesaid is to be served or given cannot after reasonable inquiry be ascertained, the notice may be served or given by addressing it to the person on or to whom it is to be served or given by the description of "owner" or "occupier" of the premises (describing them) to which the notice relates, and by delivering it to some responsible person who is or appears to be resident or employed in the premises, or, if there is no such person to whom it can be delivered, by affixing it or a copy of it to some conspicuous part of the premises.

(8) The preceding provisions of this section shall apply to the sending or giving of a document as they apply to the giving of a notice.

47 Civil liability

(1) Nothing in this Part shall be construed—
   (a) as conferring a right of action in any civil proceedings in respect of any failure to comply with any duty imposed by sections 2 to 7 or any contravention of section 8; or
   (b) as affecting the extent (if any) to which breach of a duty imposed by any of the existing statutory provisions is actionable; or
   (c) as affecting the operation of section 12 of the Nuclear Installations Act 1965 (right to compensation by virtue of certain provisions of that Act).

(2) Breach of a duty imposed by health and safety regulations [. . .][101] shall, so far as it causes damage, be actionable except in so far as the regulations provide otherwise.

(3) No provision made by virtue of section 15(6)(b) shall afford a defence in any civil proceedings, whether brought by virtue of subsection (2) above or not; but as regards any duty imposed as mentioned in subsection (2) above health and safety regulations [. . .][102] may provide for any defence specified in the regulations to be available in any action for breach of that duty.

(4) Subsections (1)(a) and (2) above are without prejudice to any right of action which exists apart from the provisions of this Act, and subsection (3) above is without prejudice to any defence which may be available apart from the provisions of the regulations there mentioned.

(5) Any term of an agreement which purports to exclude or restrict the operation of subsection (2) above, or any liability arising by virtue of that subsection, shall be void, except in so far as health and safety regulations [. . .][103] provide otherwise.

(6) In this section "damage" includes the death of, or injury to, any person (including any disease and any impairment of a person's physical or mental condition).

48 Application to Crown

(1) Subject to the provisions of this section, the provisions of this Part, except sections 21 to 25 and 33 to 42, and of regulations made under this Part shall bind the Crown.

(2) Although they do not bind the Crown, sections 33 to 42 shall apply to persons in the public service of the Crown as they apply to other persons.

(3) For the purposes of this Part and regulations made thereunder persons in the service of the Crown shall be treated as employees of the Crown whether or not they would be so treated apart from this subsection.

(4) Without prejudice to section 15(5), the Secretary of State may, to the extent that it appears to him requisite or expedient to do so in the interests of the safety of the State or the safe custody

---

[101] Repealed by the Employment Protection Act 1975.
[102] ibid.
[103] ibid.

Relevant Extracts from the Health and Safety at Work etc. Act 1974

of persons lawfully detained, by order exempt the Crown either generally or in particular respects from all or any of the provisions of this Part which would, by virtue of subsection (1) above, bind the Crown.

(5) The power to make orders under this section shall be exercisable by statutory instrument, and any such order may be varied or revoked by a subsequent order.

(6) Nothing in this section shall authorise proceedings to be brought against Her Majesty in her private capacity, and this subsection shall be construed as if section 38(3) of the Crown Proceedings Act 1947 (interpretation of references in that Act to Her Majesty in her private capacity) were contained in this Act.

49 Adaptation of enactments to metric units or appropriate metric units

(1) [Regulations made under this subsection may amend][104]—
   (a) any of the relevant statutory provisions; or
   (b) any provision of an enactment which relates to any matter relevant to any of the general purposes of this Part but is not among the relevant statutory provisions; or
   (c) any provision of an instrument made or having effect under any such enactment as is mentioned in the preceding paragraph,

by substituting an amount or quantity expressed in metric units for an amount or quantity not so expressed or by substituting an amount or quantity expressed in metric units of a description specified in the regulations for an amount or quantity expressed in metric units of a different description.

(2) The amendments shall be such as to preserve the effect of the provisions mentioned except to such extent as in the opinion of the [authority making the regulations][105] is necessary to obtain amounts expressed in convenient and suitable terms.

(3) Regulations made [...][106] under this subsection may, in the case of a provision which falls within any of paragraphs (a) to (c) of subsection (1) above and contains words which refer to units other than metric units, repeal those words [if the authority making the regulations][107] is of the opinion that those words could be omitted without altering the effect of that provision.

[(4) The power to make regulations under this section shall be exercisable by the Secretary of State [...][108]][109]

50 Regulations under the relevant statutory provisions

[(1) Where any power to make regulations under any of the relevant statutory provisions is exercisable by the Secretary of State, the Minister of Agriculture, Fisheries and Food or both of them acting jointly that power may be exercised either so as to give effect (with or without modifications) to proposals submitted by the Commission under section 11(2)(d) or independently of any such proposals; but the authority who is to exercise the power shall not exercise it independently of proposals from the Commission unless he has consulted the Commission and such other bodies as appear to him to be appropriate.][110]

[(1A) Subsection (1) does not apply to the exercise of a power to make regulations so far as it is exercised—
   (a) for giving effect (with or without modifications) to proposals submitted by the Office of Rail Regulation under paragraph 2(5) of Schedule 3 to the Railways Act 2005; or
   (b) otherwise for or in connection with the railway safety purposes.][111]

---

[104] Amended by the Employment Protection Act 1975.
[105] ibid.
[106] Repealed by the Employment Protection Act 1975.
[107] Amended by the Employment Protection Act 1975.
[108] Repealed by SI 2002/794.
[109] Amended by the Employment Protection Act 1975.
[110] Inserted by the Employment Protection Act 1975.
[111] Inserted by the Railways Act 2005.

(2) Where the [authority who is to exercise any such power as is mentioned in subsection (1) above proposes to exercise that power][112] so as to give effect to any such proposals as are there mentioned with modifications, he shall, before making the regulations, consult the Commission.

(3) Where the Commission proposes to submit [under section 11(2)(d)][113] any such proposals as are mentioned in subsection (1) above except proposals for the making of regulations under section 43(2), it shall, before so submitting them, consult—

(a) any government department or other body that appears to the Commission to be appropriate (and, in particular, in the case of proposals for the making of regulations under section 18(2), any body representing local authorities that so appears, and, in the case of proposals for the making of regulations relating to electro-magnetic radiations, [the Health Protection Agency][114]);

(b) such government departments and other bodies, if any, as, in relation to any matter dealt with in the proposals, the Commission is required to consult under this subsection by virtue of directions given to it by the Secretary of State.

(4) [...][115]

(5) [...][116]

51 Exclusion of application to domestic employment

Nothing in this Part shall apply in relation to a person by reason only that he employs another, or is himself employed, as a domestic servant in a private household.

[51A Application of Part to police][117]

[(1) For the purposes of this Part, a person who, otherwise than under a contract of employment, holds the office of constable or an appointment as police cadet shall be treated as an employee of the relevant officer.

(2) In this section "the relevant officer"—

(a) in relation to a member of a police force or a special constable or police cadet appointed for a police area, means the chief officer of police,

[(b) in relation to a member of a police force seconded to the Serious Organised Crime Agency to serve as a member of its staff, means that Agency, and][118]

(c) in relation to any other person holding the office of constable or an appointment as police cadet, means the person who has the direction and control of the body of constables or cadets in question.

[(2A) For the purposes of this Part the relevant officer, as defined by subsection (2)(a) or (c) above, shall be treated as a corporation sole.

(2B) Where, in a case in which the relevant officer, as so defined, is guilty of an offence by virtue of this section, it is proved—

(a) that the officer-holder personally consented to the commission of the offence,

(b) that he personally connived in its commission, or

(c) that the commission of the offence was attributable to personal neglect on his part,

the office-holder (as well as the corporation sole) shall be guilty of the offence and shall be liable to be proceeded against and punished accordingly.

(2C) In subsection (2B) above "the office-holder", in relation to the relevant officer, means an individual who, at the time of the consent, connivance or neglect—

---

[112] Amended by the Employment Protection Act 1975.
[113] ibid.
[114] Amended by the Health Protection Agency Act 2004.
[115] Repealed by the Employment Protection Act 1975.
[116] ibid.
[117] Inserted by the Police (Health and Safety) Act 1997.
[118] Amended by the Serious Organised Crime and Police Act 2005.

(a) held the office or other position mentioned in subsection (2) above as the office or position of that officer; or
(b) was for the time being responsible for exercising and performing the powers and duties of that office or position.

(2D) The provisions mentioned in subsection (2E) below (which impose the same liability for unlawful conduct of constables on persons having their direction or control as would arise if the constables were employees of those persons) do not apply to any liability by virtue of this Part.

(2E) Those provisions are—
   (a) section 39 of the Police (Scotland) Act 1967;
   (b) section 88(1) of the Police Act 1996;
   (c) section 97(9) of that Act;
   (d) paragraph 7(1) of Schedule 8 to the Police Act 1997;
   (e) paragraph 14(1) of Schedule 3 to the Criminal Justice and Police Act 2001;
   (f) section 28 of the Serious Organised Crime and Police Act 2005.

(2F) In the application of this section to Scotland—
   (a) subsection (2A) shall have effect as if for the words "corporation sole" there were substituted "distinct juristic person (that is to say, as a juristic person distinct from the individual who for the time being is the office-holder)";
   (b) subsection (2B) shall have effect as if for the words "corporation sole" there were substituted "juristic person"; and
   (c) subsection (2C) shall have effect as if for the words "subsection (2B)" there were substituted "subsections (2A) and (2B)".][119]

(3) For the purposes of regulations under section 2(4) above—
   (a) the Police Federation for England and Wales shall be treated as a recognised trade union recognised by each chief officer of police in England and Wales,
   (b) the Police Federation for Scotland shall be treated as a recognised trade union recognised by each chief officer of police in Scotland, and
   (c) any body recognised by the Secretary of State for the purposes of section 64 of the Police Act 1996 shall be treated as a recognised trade union recognised by each *chief officer of police* [*police authority*] in England, Wales and Scotland.

(4) Regulations under section 2(4) above may provide, in relation to persons falling within subsection (2)(b) or (c) above, that a body specified in the regulations is to be treated as a recognised trade union recognised by such person as may be specified.][120]

52 Meaning of work and at work

(1) For the purposes of this Part—
   (a) "work" means work as an employee or as a self-employed person;
   (b) an employee is at work throughout the time when he is in the course of his employment, but not otherwise;
   [(bb) a person holding the office of constable is at work throughout the time when he is on duty, but not otherwise; and][121]
   (c) a self-employed person is at work throughout such time as he devotes to work as a self-employed person;

and, subject to the following subsection, the expressions "work" and "at work", in whatever context, shall be construed accordingly.

(2) Regulations made under this subsection may—
   (a) extend the meaning of "work" and "at work" for the purposes of this Part; and

---

[119] Inserted by the Serious Organised Crime and Police Act 2005.
[120] Inserted by the Police (Health and Safety) Act 1997.
[121] Amended by the Police (Health and Safety) Act 1997.

(b) in that connection provide for any of the relevant statutory provisions to have effect subject to such adaptations as may be specified in the regulations.

[(3) The power to make regulations under subsection (2) above shall be exercisable by the Secretary of State [. . .][122].][123]

53 General interpretation of Part I

(1) In this Part, unless the context otherwise requires—

[. . .][124]

"article for use at work" means—

(a) any plant designed for use or operation (whether exclusively or not) by persons at work, and

(b) any article designed for use as a component in any such plant;

["article of fairground equipment" means any fairground equipment or any article designed for use as a component in any such equipment.][125]

"code of practice" (without prejudice to section 16(8)) includes a standard, a specification and any other documentary form of practical guidance;

"the Commission" has the meaning assigned by section 10(2);

"conditional sale agreement" means an agreement for the sale of goods under which the purchase price or part of it is payable by instalments, and the property in the goods is to remain in the seller (notwithstanding that the buyer is to be in possession of the goods) until such conditions as to the payment of instalments or otherwise as may be specified in the agreement are fulfilled;

"contract of employment" means a contract of employment or apprenticeship (whether express or implied and, if express, whether oral or in writing);

"credit-sale agreement" means an agreement for the sale of goods, under which the purchase price or part of it is payable by instalments, but which is not a conditional sale agreement;

["customs officer" means an officer within the meaning of the Customs and Excise Management Act 1979;][126]

"domestic premises" means premises occupied as a private dwelling (including any garden, yard, garage, outhouse or other appurtenance of such premises which is not used in common by the occupants of more than one such dwelling), and "non-domestic premises" shall be construed accordingly;

"employee" means an individual who works under a contract of employment [or is treated by section 51A as being an employee],[127] and related expressions shall be construed accordingly;

"enforcing authority" has the meaning assigned by section 18(7);

"the Executive" has the meaning assigned by section 10(5);

"the existing statutory provisions" means the following provisions while and to the extent that they remain in force, namely the provisions of the Acts mentioned in Schedule 1 which are specified in the third column of that Schedule and of the regulations, orders or other instruments of a legislative character made or having effect under any provision so specified;

[. . .][128]

["fairground equipment" means any fairground ride, any similar plant which is designed to be in motion for entertainment purposes with members of the public on or inside it or any plant which is designed to be used by members of the public for entertainment purposes

---

[122] Repealed by SI 2002/794.
[123] Inserted by the Employment Protection Act 1975.
[124] Repealed by the Employment Protection Act 1975.
[125] Amended by the Consumer Protection Act 1987.
[126] ibid.
[127] Amended by the Police (Health and Security) Act 1997.
[128] Repealed by the Employment Protection Act 1975.

either as a slide or for bouncing upon, and in this definition the reference to plant which is designed to be in motion with members of the public on or inside it includes a reference to swings, dodgems and other plant which is designed to be in motion wholly or partly under the control of, or to be put in motion by, a member of the public;][129]

"the general purposes of this Part" has the meaning assigned by section 1;

"health and safety regulations" has the meaning assigned by section 15(1);

"hire-purchase agreement" means an agreement other than a conditional sale agreement, under which—
- (a) goods are bailed or (in Scotland) hired in return for periodical payments by the person to whom they are bailed or hired; and
- (b) the property in the goods will pass to that person if the terms of the agreement are complied with and one or more of the following occurs:
    - (i) the exercise of an option to purchase by that person;
    - (ii) the doing of any other specified act by any party to the agreement;
    - (iii) the happening of any other event;

and "hire-purchase" shall be construed accordingly;

"improvement notice" means a notice under section 21;

"inspector" means an inspector appointed under section 19;

[...][130]

"local authority" means—
- (a) in relation to England [...],[131] a county council, [...],[132] a district council, a London borough council, the Common Council of the City of London, the Sub-Treasurer of the Inner Temple or the Under-Treasurer of the Middle Temple,
- [(aa) in relation to Wales, a county council or a county borough council,][133]
- (b) in relation to Scotland, a [council constituted under section 2 of the Local Government etc (Scotland) Act 1994];[134]

["micro-organism" includes any microscopic biological entity which is capable of replication;][135]

"offshore installation" means any installation which is intended for underwater exploitation of mineral resources or exploration with a view to such exploitation;

"personal injury" includes any disease and any impairment of a person's physical or mental condition;

"plant" includes any machinery, equipment or appliance;

"premises" includes any place and, in particular, includes—
- (a) any vehicle, vessel, aircraft or hovercraft,
- (b) any installation on land (including the foreshore and other land intermittently covered by water), any offshore installation, and any other installation (whether floating, or resting on the seabed or the subsoil thereof, or resting on other land covered with water or the subsoil thereof), and
- (c) any tent or movable structure;

"prescribed" means prescribed by regulations made by the Secretary of State;

"prohibition notice" means a notice under section 22;

---

[129] Amended by the Consumer Protection 1987.
[130] Repealed by the Employment Protection Act 1975.
[131] Repealed by the Local Government (Wales) Act 1994.
[132] Repealed by the Local Government Act 1985.
[133] Amended by the Local Government (Wales) Act 1994.
[134] Amended by the Local Government etc (Scotland) Act 1994.
[135] Amended by the Consumer Protection Act 1987.

["railway safety purposes" has the same meaning as in Schedule 3 to the Railways Act 2005;][136]
[. . .][137]
"the relevant statutory provisions" means—
   (a) the provisions of this Part and of any health and safety regulations [. . .];[138] and
   (b) the existing statutory provisions;
"self-employed person" means an individual who works for gain or reward otherwise than under a contract of employment, whether or not he himself employs others;
"substance" means any natural or artificial substance [(including micro-organisms)],[139] whether in solid or liquid form or in the form of a gas or vapour;
[. . .][140]
"supply", where the reference is to supplying articles or substances, means supplying them by way of sale, lease, hire or hire-purchase, whether as principal or agent for another.

(2) [. . .][141]

(3) [. . .][142]

(4) [. . .][143]

(5) [. . .][144]

(6) [. . .][145]

54 Application of Part I to Isles of Scilly

This Part, in its application to the Isles of Scilly, shall apply as if those Isles were a local government area and the Council of those Isles were a local authority.

[Part II not reproduced; Part III Repealed by the Building Act 1984, s 133(2), Sch 7.]

Part IV

Miscellaneous and General

77 [. . .][146]

78 [Amendment of Fire Precautions Act 1971][147]

79 [. . .][148]

80 General power to repeal or modify Acts and instruments

(1) Regulations made under this subsection may repeal or modify any provision to which this subsection applies if it appears to the authority making the regulations that the repeal or, as the case may be, the modification of that provision is expedient in consequence of or in connection with any provision made by or under Part I.

---

[136] Amended by the Railways Act 2005.
[137] Repealed by the Employment Protection Act 1975.
[138] ibid.
[139] Amended by the Consumer Protection Act 1987.
[140] Repealed by the Consumer Protection Act 1987.
[141] Repealed by the Employment Protection Act 1975.
[142] ibid.
[143] ibid.
[144] ibid.
[145] ibid.
[146] Repealed by the Health Protection Agency Act 2004.
[147] Repealed by SI 2005/1541.
[148] Repealed by the Companies Consolidation (Consequential Provisions) Act 1985.

(2) Subsection (1) above applies to any provision, not being among the relevant statutory provisions, which—
   (a) is contained in this Act or in any other Act passed before or in the same Session as this Act; or
   (b) is contained in any regulations, order or other instrument of a legislative character which was made under an Act before the passing of this Act; or
   (c) applies, excludes or for any other purpose refers to any of the relevant statutory provisions and is contained in any Act not falling within paragraph (a) above or in any regulations, order or other instrument of a legislative character which is made under an Act but does not fall within paragraph (b) above.
[(2A) Subsection (1) above shall apply to provisions in [the Employment Rights Act 1996 or the Trade Union and Labour Relations (Consolidation) Act 1992 which derive from provisions of the Employment Protection (Consolidation) Act 1978 which re-enacted] provisions previously contained in the Redundancy Payments Act 1965, the Contracts of Employment Act 1972 and the Trade Union and Labour Relations Act 1974 as it applies to provisions contained in Acts passed before or in the same Session as this Act.][149]
(3) Without prejudice to the generality of subsection (1) above, the modifications which may be made by regulations thereunder include modifications relating to the enforcement of provisions to which this section applies (including the appointment of persons for the purpose of such enforcement, and the powers of persons so appointed).
[(4) The power to make regulations under subsection (1) above shall be exercisable by the Secretary of State [. . .];[150] but the authority who is to exercise the power shall, before exercising it, consult such bodies as appear to him to be appropriate.
(5) In this section "the relevant statutory provisions" has the same meaning as in Part I.][151]

81 Expenses and receipts

There shall be paid out of money provided by Parliament—
   (a) any expenses incurred by a Minister of the Crown or government department for the purposes of this Act; and
   (b) any increase attributable to the provisions of this Act in the sums payable under any other Act out of money so provided;

and any sums received by a Minister of the Crown or government department by virtue of this Act shall be paid into the Consolidated Fund.

82 General provisions as to interpretation and regulations

(1) In this Act—
   (a) "Act" includes a provisional order confirmed by an Act;
   (b) "contravention" includes failure to comply, and "contravene" has a corresponding meaning;
   (c) "modifications" includes additions, omissions and amendments, and related expressions shall be construed accordingly;
   (d) any reference to a Part, section or Schedule not otherwise identified is a reference to that Part or section of, or Schedule to, this Act.
(2) Except in so far as the context otherwise requires, any reference in this Act to an enactment is a reference to it as amended, and includes a reference to it as applied, by or under any other enactment, including this Act.

---

[149] Inserted by the Employment Protection (Consolidation) Act 1978.
[150] Repealed by SI 2002/794.
[151] Inserted by the Employment Protection Act 1975.

(3) Any power conferred by Part I or II or this Part to make regulations—
  (a) includes power to make different provision by the regulations for different circumstances or cases and to include in the regulations such incidental, supplemental and transitional provisions as the authority making the regulations considers appropriate in connection with the regulations; and
  (b) shall be exercisable by statutory instrument, which [(unless subsection (4) applies)][152] shall be subject to annulment in pursuance of a resolution of either House of Parliament.
[(4) The first regulations under section 43A(1) shall not be made unless a draft has been laid before and approved by resolution of each House of Parliament.][153]

83 [...][154]

84 Extent, and application of Act

(1) This Act, except—
  (a) Part I and this Part so far as may be necessary to enable regulations under section 15 [...][155] to be made and operate for the purpose mentioned in paragraph 2 of Schedule 3; and
  (b) paragraph [...][156] 3 of Schedule 9,
does not extend to Northern Ireland.

(2) Part III, except section 75 and Schedule 7, does not extend to Scotland.

(3) Her Majesty may by Order in Council provide that the provisions of Parts I and II and this Part shall, to such extent and for such purposes as may be specified in the Order, apply (with or without modification) to or in relation to persons, premises, work, articles, substances and other matters (of whatever kind) outside Great Britain as those provisions apply within Great Britain or within a part of Great Britain so specified.

For the purposes of this subsection "premises", "work" and "substance" have the same meaning as they have for the purposes of Part I.

(4) An Order in Council under subsection (3) above—
  (a) may make different provision for different circumstances or cases;
  (b) may (notwithstanding that this may affect individuals or bodies corporate outside the United Kingdom) provide for any of the provisions mentioned in that subsection, as applied by such an Order, to apply to individuals whether or not they are British subjects and to bodies corporate whether or not they are incorporated under the law of any part of the United Kingdom;
  (c) may make provision for conferring jurisdiction on any court or class of courts specified in the Order with respect to offences under Part I committed outside Great Britain or with respect to causes of action arising by virtue of section 47 (2) in respect of acts or omissions taking place outside Great Britain, and for the determination, in accordance with the law in force in such part of Great Britain as may be specified in the Order, of questions arising out of such acts or omissions;
  (d) may exclude from the operation of section 3 of the Territorial Waters Jurisdiction Act 1878 (consents required for prosecutions) proceedings for offences under any provision of Part I committed outside Great Britain;
  (e) may be varied or revoked by a subsequent Order in Council under this section;
and any such Order shall be subject to annulment in pursuance of a resolution of either House of Parliament.

---

[152] Amended by the Railways and Transport Safety Act 2003.
[153] Inserted by the Railways and Transport Safety Act 2003.
[154] Repealed by the Statute Law (Repeals) Act 1993.
[155] Repealed by the Employment Protection Act 1975.
[156] Repealed by the House of Commons Disqualification Act 1975.

(5) [...]¹⁵⁷

(6) Any jurisdiction conferred on any court under this section shall be without prejudice to any jurisdiction exercisable apart from this section by that or any other court.

85 Short title and commencement

(1) This Act may be cited as the Health and Safety at Work etc. Act 1974.

(2) This Act shall come into operation on such day as the Secretary of State may by order made by statutory instrument appoint, and different days may be appointed under this subsection for different purposes.

(3) An order under this section may contain such transitional provisions and savings as appear to the Secretary of State to be necessary or expedient in connection with the provisions thereby brought into force, including such adaptations of those provisions or any provision of this Act then in force as appear to him to be necessary or expedient in consequence of the partial operation of this Act (whether before or after the day appointed by the order).

SCHEDULE 1

EXISTING ENACTMENTS WHICH ARE RELEVANT STATUTORY PROVISIONS

Sections 1, 53

Chapter. Short title: Provisions which are relevant statutory provisions

1875 c 17. The Explosives Act 1875: The whole Act except sections 30 to 32, 80 and 116 to 121.

1882 c 22. The Boiler Explosions Act 1882: The whole Act.

1890 c 35. The Boiler Explosions Act 1890: The whole Act.

1906 c 14. The Alkali, &c Works Regulation Act 1906: The whole Act.

1909 c 43. The Revenue Act 1909: Section 11.

[...]¹⁵⁸

1920 c 65. The Employment of Women, Young Persons and Children Act 1920: The whole Act.

1922 c 35. The Celluloid and Cinematograph Film Act 1922: The whole Act.

[...]¹⁵⁹

1926 c 43. The Public Health (Smoke Abatement) Act 1926: The whole Act.

1928 c 32. The Petroleum (Consolidation) Act 1928: The whole Act.

1936 c 22. The Hours of Employment (Conventions) Act 1936: The whole Act except section 5.

1936 c 27. The Petroleum (Transfer of Licences) Act 1936: The whole Act.

1937 c 45. The Hydrogen Cyanide (Fumigation) Act 1937: The whole Act.

1945 c 19. The Ministry of Fuel and Power Act 1945: Section 1(1) so far as it relates to maintaining and improving the safety, health and welfare of persons employed in or about mines and quarries in Great Britain.

1946 c 59. The Coal Industry Nationalisation Act 1946: Section 42(1) and (2).

1948 c 37. The Radioactive Substances Act 1948: Section 5(1)(a).

1951 c 21. The Alkali, &c Works Regulation (Scotland) Act 1951: The whole Act.

---

[157] Subs (5): repealed by the Offshore Safety Act 1992.
[158] Repealed by SI 2005/228.
[159] Repealed by SI 2005/1082.

Appendix A

[. . .]¹⁶⁰

1952 c 60. The Agriculture (Poisonous Substances) Act 1952: The whole Act.

[. . .]¹⁶¹

[. . .]¹⁶²

1954 c 70. The Mines and Quarries Act 1954: The whole Act except section 151.

1956 c 49. The Agriculture (Safety, Health and Welfare Provisions) Act 1956: The whole Act.

1961 c 34. The Factories Act 1961: The whole Act except section 135.

1961 c 64. The Public Health Act 1961: Section 73.

1962 c 58. The Pipe-lines Act 1962: Sections 20 to 26, 33, 34 and 42, Schedule 5.

1963 c 41. The Offices, Shops and Railway Premises Act 1963: The whole Act.

1965 c 57. The Nuclear Installations Act 1965: Sections 1, 3 to 6, 22 and [24A],¹⁶³ Schedule 2.

1969 c 10. The Mines and Quarries (Tips) Act 1969: Sections 1 to 10.

1971 c 20. The Mines Management Act 1971: The whole Act.

1972 c 28. The Employment Medical Advisory Service Act 1972: The whole Act except sections 1 and 6 and Schedule 1.

Schedule 2

Additional Provisions Relating to Constitution etc, of the Commission and Executive

Section 10

Tenure of office

1 Subject to paragraphs 2 to 4 below, a person shall hold and vacate office as a member or as chairman or deputy chairman in accordance with the terms of the instrument appointing him to that office.

2 A person may at any time resign his office as a member or as chairman or deputy chairman by giving the Secretary of State a notice in writing signed by that person and stating that he resigns that office.

3(1) If a member becomes or ceases to be the chairman or deputy chairman, the Secretary of State may vary the terms of the instrument appointing him to be a member so as to alter the date on which he is to vacate office as a member.

(2) If the chairman or deputy chairman ceases to be a member he shall cease to be chairman or deputy chairman, as the case may be.

4(1) If the Secretary of State is satisfied that a member—
   (a) has been absent from meetings of the Commission for a period longer than six consecutive months without the permission of the Commission; or
   (b) has become bankrupt or made an arrangement with his creditors; or
   (c) is incapacitated by physical or mental illness; or
   (d) is otherwise unable or unfit to discharge the functions of a member,

the Secretary of State may declare his office as a member to be vacant and shall notify the declaration in such manner as the Secretary of State thinks fit; and thereupon the office shall become vacant.

---

[160] Repealed by SI 2005/1082.
[161] ibid.
[162] Amended by the Sex Discrimination Act 1975. Amendment repealed by the Sex Discrimination Act 1986.
[163] Amended by the Atomic Energy Act 1989.

(2) In the application of the preceding sub-paragraph to Scotland for the references in paragraph (b) to a member's having become bankrupt and to a member's having made an arrangement with his creditors there shall be substituted respectively references to sequestration of a member's estate having been awarded and to a member's having made a trust deed for behoof of his creditors or a composition contract.

Remuneration etc of members

5   The Commission may pay to each member such remuneration and allowances as the Secretary of State may determine.
6   The Commission may pay or make provision for paying, to or in respect of any member, such sums by way of pension, superannuation allowances and gratuities as the Secretary of State may determine.
7   Where a person ceases to be a member otherwise than on the expiry of his term of office and it appears to the Secretary of State that there are special circumstances which make it right for him to receive compensation, the Commission may make to him a payment of such amount as the Secretary of State may determine.

Proceedings

8   The quorum of the Commission and the arrangements relating to meetings of the Commission shall be such as the Commission may determine.
9   The validity of any proceedings of the Commission shall not be affected by any vacancy among the members or by any defect in the appointment of a member.

Staff

10   It shall be the duty of the Executive to provide for the Commission such officers and servants as are requisite for the proper discharge of the Commission's functions; and any reference in this Act to an officer or servant of the Commission is a reference to an officer or servant provided for the Commission in pursuance of this paragraph.
11   The Executive may appoint such officers and servants as it may determine with the consent of the Secretary of State as to numbers and terms and conditions of service.
12   The Commission shall pay to the Minister for the Civil Service, at such times in each accounting year as may be determined by that Minister subject to any directions of the Treasury, sums of such amounts as he may so determine for the purposes of this paragraph as being equivalent to the increase during that year of such liabilities of his as are attributable to the provision of pensions, allowances or gratuities to or in respect of persons who are or have been in the service of the Executive in so far as that increase results from the service of those persons during that accounting year and to the expense to be incurred in administering those pensions, allowances or gratuities.

Performance of functions

13   The Commission may authorise any member of the Commission or any officer or servant of the Commission or of the Executive to perform on behalf of the Commission such of the Commission's functions (including the function conferred on the Commission by this paragraph) as are specified in the authorisation.

Accounts and reports

14(1)   It shall be the duty of the Commission—
  (a) to keep proper accounts and proper records in relation to the accounts;
  (b) to prepare in respect of each accounting year a statement of accounts in such form as the Secretary of State may direct with the approval of the Treasury; and
  (c) to send copies of the statement to the Secretary of State and the Comptroller and Auditor General before the end of the month of November next following the accounting year to which the statement relates.

(2) The Comptroller and Auditor General shall examine, certify and report on each statement received by him in pursuance of this Schedule and shall lay copies of each statement and of his report before each House of Parliament.

15 It shall be the duty of the Commission to make to the Secretary of State, as soon as possible after the end of each accounting year, a report on the performance of its functions during that year; and the Secretary of State shall lay before each House of Parliament a copy of each report made to him in pursuance of this paragraph.

Supplemental

16 The Secretary of State shall not make a determination or give his consent in pursuance of paragraph 5, 6, 7 or 11 of this Schedule except with the approval of the Minister for the Civil Service.

17 The fixing of the common seal of the Commission shall be authenticated by the signature of the secretary of the Commission or some other person authorised by the Commission to act for that purpose.

18 A document purporting to be duly executed under the seal of the Commission shall be received in evidence and shall, unless the contrary is proved, be deemed to be so executed.

19 In the preceding provisions of this Schedule—
   (a) "accounting year" means the period of twelve months ending with 31st March in any year except that the first accounting year of the Commission shall, if the Secretary of State so directs, be such period shorter or longer than twelve months (but not longer than two years) as is specified in the direction; and
   (b) "the chairman", "a deputy chairman" and "a member" mean respectively the chairman, a deputy chairman and a member of the Commission.

20(1) The preceding provisions of this Schedule (except paragraphs 10 to 12 and 15) shall have effect in relation to the Executive as if—
   (a) for any reference to the Commission there were substituted a reference to the Executive;
   (b) for any reference to the Secretary of State in paragraphs 2 to 4 and 19 and the first such reference in paragraph 7 there were substituted a reference to the Commission;
   (c) for any reference to the Secretary of State in paragraphs 5 to 7 (except the first such reference in paragraph 7) there were substituted a reference to the Commission acting with the consent of the Secretary of State;
   (d) for any reference to the chairman there were substituted a reference to the director and any reference to the deputy chairman were omitted;
   (e) in paragraph 14(1)(c) for the words from "Secretary" to "following" there were substituted the words "Commission by such date as the Commission may direct after the end of".

(2) It shall be the duty of the Commission to include in or send with the copies of the statement sent by it as required by paragraph 14(1)(c) of this Schedule copies of the statement sent to it by the Executive in pursuance of the said paragraph 14(1)(c) as adapted by the preceding sub-paragraph.

(3) The terms of an instrument appointing a person to be a member of the Executive shall be such as the Commission may determine with the approval of the Secretary of State and the Minister for the Civil Service.

Schedule 3

Subject Matter of Health and Safety Regulations

Section 15

1(1) Regulating or prohibiting—
   (a) the manufacture, supply or use of any plant;
   (b) the manufacture, supply, keeping or use of any substance;
   (c) the carrying on of any process or the carrying out of any operation.

Relevant Extracts from the Health and Safety at Work etc. Act 1974

(2) Imposing requirements with respect to the design, construction, guarding, siting, installation, commissioning, examination, repair, maintenance, alteration, adjustment, dismantling, testing or inspection of any plant.

(3) Imposing requirements with respect to the marking of any plant or of any articles used or designed for use as components in any plant, and in that connection regulating or restricting the use of specified markings.

(4) Imposing requirements with respect to the testing, labelling or examination of any substance.

(5) Imposing requirements with respect to the carrying out of research in connection with any activity mentioned in sub-paragraphs (1) to (4) above.

2(1) Prohibiting the importation into the United Kingdom or the landing or unloading there of articles or substances of any specified description, whether absolutely or unless conditions imposed by or under the regulations are complied with.

(2) Specifying, in a case where an act or omission in relation to such an importation, landing or unloading as is mentioned in the preceding sub-paragraph constitutes an offence under a provision of this Act and of [the Customs and Excise Acts 1979],[164] the Act under which the offence is to be punished.

3(1) Prohibiting or regulating the transport of articles or substances of any specified description.

(2) Imposing requirements with respect to the manner and means of transporting articles or substances of any specified description, including requirements with respect to the construction, testing and marking of containers and means of transport and the packaging and labelling of articles or substances in connection with their transport.

4(1) Prohibiting the carrying on of any specified activity or the doing of any specified thing except under the authority and in accordance with the terms and conditions of a licence, or except with the consent or approval of specified authority.

(2) Providing for the grant, renewal, variation, transfer and revocation of licences (including the variation and revocation of conditions attached to licences).

5 Requiring any person, premises or thing to be registered in any specified circumstances or as a condition of the carrying on of any specified activity or the doing of any specified thing.

6(1) Requiring, in specified circumstances, the appointment (whether in a specified capacity or not) of persons (or persons with specified qualifications or experience, or both) to perform specified functions, and imposing duties or conferring powers on persons appointed (whether in pursuance of the regulations or not) to perform specified functions.

(2) Restricting the performance of specified functions to persons possessing specified qualifications or experience.

7 Regulating or prohibiting the employment in specified circumstances of all persons or any class of persons.

8(1) Requiring the making of arrangements for securing the health of persons at work or other persons, including arrangements for medical examinations and health surveys.

(2) Requiring the making of arrangements for monitoring the atmospheric or other conditions in which persons work.

9 Imposing requirements with respect to any matter affecting the conditions in which persons work, including in particular such matters as the structural condition and stability of premises, the means of access to and egress from premises, cleanliness, temperature, lighting, ventilation, overcrowding, noise, vibrations, ionising and other radiations, dust and fumes.

10 Securing the provision of specified welfare facilities for persons at work, including in particular such things as an adequate water supply, sanitary conveniences, washing and bathing facilities, ambulance and first-aid arrangements, cloakroom accommodation, sitting facilities and refreshment facilities.

---

[164] Amended by the Customs and Excise Management Act 1979.

11 Imposing requirements with respect to the provision and use in specified circumstances of protective clothing or equipment, including affording protection against the weather.

12 Requiring in specified circumstances the taking of specified precautions in connection with the risk of fire.

13(1) Prohibiting or imposing requirements in connection with the emission into the atmosphere of any specified gas, smoke or dust or any other specified substance whatsoever.

(2) Prohibiting or imposing requirements in connection with the emission of noise, vibrations or any ionising or other radiations.

(3) Imposing requirements with respect to the monitoring of any such emission as is mentioned in the preceding sub-paragraphs.

14 Imposing requirements with respect to the instruction, training and supervision of persons at work.

15(1) Requiring, in specified circumstances, specified matters to be notified in a specified manner to specified persons.

(2) Empowering inspectors in specified circumstances to require persons to submit written particulars of measures proposed to be taken to achieve compliance with any of the relevant statutory provisions.

16 Imposing requirements with respect to the keeping and preservation of records and other documents, including plans and maps.

17 Imposing requirements with respect to the management of animals.

18 The following purposes as regards premises of any specified description where persons work, namely—
   (a) requiring precautions to be taken against dangers to which the premises or persons therein are or may be exposed by reason of conditions (including natural conditions) existing in the vicinity;
   (b) securing that persons in the premises leave them in specified circumstances.

19 Conferring, in specified circumstances involving a risk of fire or explosion, power to search a person or any article which a person has with him for the purpose of ascertaining whether he has in his possession any article of a specified kind likely in those circumstances to cause a fire or explosion, and power to seize and dispose of any article of that kind found on such a search.

20 Restricting, prohibiting or requiring the doing of any specified thing where any accident or other occurrence of a specified kind has occurred.

21 As regards cases of any specified class, being a class such that the variety in the circumstances of particular cases within it calls for the making of special provision for particular cases, any of the following purposes, namely—
   (a) conferring on employers or other persons power to make rules or give directions with respect to matters affecting health or safety;
   (b) requiring employers or other persons to make rules with respect to any such matters;
   (c) empowering specified persons to require employers or other persons either to make rules with respect to any such matters or to modify any such rules previously made by virtue of this paragraph; and
   (d) making admissible in evidence without further proof, in such circumstances and subject to such conditions as may be specified, documents which purport to be copies of rules or rules of any specified class made under this paragraph.

22 Conferring on any local or public authority power to make byelaws with respect to any specified matter, specifying the authority or person by whom any byelaws made in the exercise of that power need to be confirmed, and generally providing for the procedure to be followed in connection with the making of any such byelaws.

Relevant Extracts from the Health and Safety at Work etc. Act 1974

Interpretation

23(1) In this Schedule "specified" means specified in health and safety regulations.

(2) It is hereby declared that the mention in this Schedule of a purpose that falls within any more general purpose mentioned therein is without prejudice to the generality of the more general purpose.

Schedule 4

[...]¹⁶⁵

Schedule 5

[...]¹⁶⁶

Schedule 6

[...]¹⁶⁷

Schedule 7

[...]¹⁶⁸

Schedule 8

[...]¹⁶⁹

Schedule 9

[...]¹⁷⁰

Schedule 10

[...]¹⁷¹

---

[165] Repealed by the Employment Protection Act 1975.
[166] Repealed by the Building Act 1984.
[167] ibid.
[168] Repealed by the Building (Scotland) Act 2003.
[169] Repealed, in relation to England and Wales, by SI 2005/1541.
[170] Repealed by the Statute Law (Repeals) Act 1993.
[171] ibid.

APPENDIX B

Enforcement Concordat;
Enforcement Policy Statement (HSC15);
Work Related Death Protocol

The Principles of Good Enforcement: Policy and Procedures

This document sets out what business and others being regulated can expect from enforcement officers. It commits us to good enforcement policies and procedures. It may be supplemented by additional statements of enforcement policy.

The primary function of central and local government enforcement work is to protect the public, the environment and groups such as consumers and workers. At the same time, carrying out enforcement functions in an equitable, practical and consistent manner helps to promote a thriving national and local economy. We are committed to these aims and to maintaining a fair and safe trading environment.

The effectiveness of legislation in protecting consumers or sectors in society depends crucially on the compliance of those regulated. We recognise that most businesses want to comply with the law. We will, therefore, take care to help business and others meet their legal obligations without unnecessary expense, while taking firm action, including prosecution where appropriate, against those who flout the law or act irresponsibly. All citizens will reap the benefits of this policy through better information, choice and safety.

We have therefore adopted the central and local government Concordat on Good Enforcement. Included in the term 'enforcement' are advisory visits and assisting with compliance as well as licensing and formal enforcement action. By adopting the concordat we commit ourselves to the following policies and procedures, which contribute to best value, and will provide information to show that we are observing them.

Principles of Good Enforcement: Policy

Standards

In consultation with business and other relevant interested parties, including technical experts where appropriate, we will draw up clear standards setting out the level of service and performance the public and business people can expect to receive. We will publish these standards and our annual performance against them. The standards will be made available to businesses and others who are regulated.

Openness

We will provide information and advice in plain language on the rules that we apply and will disseminate this as widely as possible. We will be open about how we set about our work, including any charges that we set, consulting business, voluntary organisations, charities, consumers and workforce representatives. We will discuss general issues, specific compliance failures or problems with anyone experiencing difficulties.

Helpfulness

We believe that prevention is better than cure and that our role therefore involves actively working with business, especially small and medium sized businesses, to advise on and assist with compliance. We will provide a courteous and efficient service and our staff will identify themselves by name. We will provide a contact point and telephone number for further dealings with us and we will encourage business to seek advice/information from us. Applications for approval of establishments, licences, registrations, etc, will be dealt with efficiently and promptly. We will ensure that, wherever practicable, our enforcement services are effectively co-ordinated to minimise unnecessary overlaps and time delays.

Complaints about service

We will provide well publicised, effective and timely complaints procedures easily accessible to business, the public, employees and consumer groups. In cases where disputes cannot be resolved, any right of complaint or appeal will be explained, with details of the process and the likely time-scales involved.

Proportionality

We will minimise the costs of compliance for business by ensuring that any action we require is proportionate to the risks. As far as the law allows, we will take account of the circumstances of the case and the attitude of the operator when considering action. We will take particular care to work with small businesses and voluntary and community organisations so that they can meet their legal obligations without unnecessary expense, where practicable.

Consistency

We will carry out our duties in a fair, equitable and consistent manner. While inspectors are expected to exercise judgement in individual cases, we will have arrangements in place to promote consistency, including effective arrangements for liaison with other authorities and enforcement bodies through schemes such as those operated by the Local Authorities Co-Ordinating Body on Food and Trading Standards (LACOTS) and the Local Authority National Type Approval Confederation (LANTAC).

Principles of Good Enforcement: Procedures

Advice from an officer will be put clearly and simply and will be confirmed in writing, on request, explaining why any remedial work is necessary and over what time-scale, and making sure that legal requirements are clearly distinguished from best practice advice.

Before formal enforcement action is taken, officers will provide an opportunity to discuss the circumstances of the case and, if possible, resolve points of difference, unless immediate action is required (for example, in the interests of health and safety or environmental protection or to prevent evidence being destroyed).

Where immediate action is considered necessary, an explanation of why such action was required will be given at the time and confirmed in writing in most cases within 5 working days and, in all cases, within 10 working days.

Where there are rights of appeal against formal action, advice on the appeal mechanism will be clearly set out in writing at the time the action is taken (whenever possible this advice will be issued with the enforcement notice).

CABINET OFFICE
MARCH 1998

Enforcement Concordat; Enforcement Policy Statement (HSC15)

ENFORCEMENT POLICY STATEMENT – HEALTH AND SAFETY COMMISSION

Introduction

The Health and Safety Commission's (HSC) aims are to protect the health, safety and welfare of people at work, and to safeguard others, mainly members of the public, who may be exposed to risks from the way work is carried out.

HSC's statutory functions include proposing new or updated laws and standards, conducting research, and providing information and advice. HSC is advised and assisted by the Health and Safety Executive (HSE) which has statutory responsibilities to make adequate arrangements for the enforcement of health and safety law in relation to specified work activities. Local authorities also enforce health and safety law in workplaces allocated to them—including offices, shops, retail and wholesale distribution centres, leisure, hotel and catering premises.

This Enforcement Policy Statement sets out the general principles and approach which HSC expects the health and safety enforcing authorities (mainly HSE and local authorities) to follow. All local authority and HSE staff who take enforcement decisions are required to follow HSC's Enforcement Policy Statement. In general, those staff will be inspectors, so this policy refers to inspectors for simplicity.

The appropriate use of enforcement powers, including prosecution, is important, both to secure compliance with the law and to ensure that those who have duties under it may be held to account for failures to safeguard health, safety and welfare.

In allocating resources, enforcing authorities should have regard to the principles set out below, the objectives published in HSC's and the HSE/Local Authority Enforcement Liaison Committee's (HELA) strategic plans, and the need to maintain a balance between enforcement and other activities, including inspection.

The Health and Safety Commission's Policy Statement on Enforcement

The following is the full text of the statement:

The purpose and method of enforcement

1. The ultimate purpose of the enforcing authorities is to ensure that duty holders manage and control risks effectively, thus preventing harm. The term 'enforcement' has a wide meaning and applies to all dealings between enforcing authorities and those on whom the law places duties (employers, the self-employed, employees and others).

2. The purpose of enforcement is to:
   - ensure that duty holders take action to deal immediately with serious risks;
   - promote and achieve sustained compliance with the law;
   - ensure that duty holders who breach health and safety requirements, and directors or managers who fail in their responsibilities, may be held to account, which may include bringing alleged offenders before the courts in England and Wales, or recommending prosecution in Scotland, in the circumstances set out later in this policy.

   Enforcement is distinct from civil claims for compensation and is not undertaken in all circumstances where civil claims may be pursued, nor to assist such claims.

3. The enforcing authorities have a range of tools at their disposal in seeking to secure compliance with the law and to ensure a proportionate response to criminal offences. Inspectors may offer duty holders information, and advice, both face to face and in writing. This may include warning a duty holder that in the opinion of the inspector, they are failing to comply with the law. Where appropriate, inspectors may also serve improvement and prohibition

notices, withdraw approvals, vary licence conditions or exemptions, issue formal cautions[1] (England and Wales only), and they may prosecute (or report to the Procurator Fiscal with a view to prosecution in Scotland).

4 Giving information and advice, issuing improvement or prohibition notices, and withdrawal or variation of licences or other authorisations are the main means which inspectors use to achieve the broad aim of dealing with serious risks, securing compliance with health and safety law and preventing harm. A prohibition notice stops work in order to prevent serious personal injury. Information on improvement and prohibition notices should be made publicly available.

5 Every improvement notice contains a statement that in the opinion of an inspector an offence has been committed. Improvement and prohibition notices, and written advice, may be used in court proceedings.

6 Formal cautions and prosecution are important ways to bring duty holders to account for alleged breaches of the law. Where it is appropriate to do so in accordance with this policy, enforcing authorities should use one of these measures in addition to issuing an improvement or prohibition notice.

7 Investigating the circumstances encountered during inspections or following incidents or complaints is essential before taking any enforcement action. In deciding what resources to devote to these investigations, enforcing authorities should have regard to the principles of enforcement set out in this statement and the objectives published in HSC and HELA strategic plans. In particular, in allocating resources, enforcing authorities must strike a balance between investigations and mainly preventive activity.

8 Sometimes the law is prescriptive—spelling out in detail what must be done. However, much of modern health and safety law is goal setting, setting out what must be achieved, but not how it must be done. Advice on how to achieve the goals is often set out in Approved Codes of Practice (ACOPs). These give practical advice on compliance and have a special legal status. If someone is prosecuted for a breach of health and safety law and did not follow the relevant provisions of an ACOP, then the onus is on them to show that they complied with the law in another way. Advice is also contained in other HSC, HSE and HELA guidance material describing good practice. Following this guidance is not compulsory, but doing so is normally enough to comply with the law. Neither ACOPs nor guidance material are in terms which necessarily fit every case. In considering whether the law has been complied with, inspectors will need to take relevant ACOPs and guidance into account, using sensible judgement about the extent of the risks and the effort that has been applied to counter them. More is said about these matters in this statement.

9 HSC expects enforcing authorities to use discretion in deciding when to investigate or what enforcement action may be appropriate. Enforcing authorities should set down in writing the decision-making process which inspectors will follow when deciding on enforcement action, and make this publicly available. HSC expects that such judgements will be made in accordance with the following principles. These are in accordance with the *Enforcement Concordat* agreed between the Cabinet, Home and Scottish (now the Scottish Executive) Offices and local authority associations.

The principles of enforcement

10 HSC believes in firm but fair enforcement of health and safety law. This should be informed by the principles of *proportionality* in applying the law and securing compliance; *consistency* of

---

[1] A formal caution is a statement by an inspector, that is accepted in writing by the duty holder, that the duty holder has committed an offence for which there is a realistic prospect of conviction. A formal caution may only be used where a prosecution could be properly brought. 'Formal cautions' are entirely distinct from a caution given under the Police and Criminal Evidence Act by an inspector before questioning a suspect about an alleged offence. Enforcing authorities should take account of current Home Office guidelines when considering whether to offer a formal caution.

approach; *targeting* of enforcement action; *transparency* about how the regulator operates and what those regulated may expect; and *accountability* for the regulator's actions. These principles should apply both to enforcement in particular cases and to the health and safety enforcing authorities' management of enforcement activities as a whole.

Proportionality

11 Proportionality means relating enforcement action to the risks.[2] Those whom the law protects and those on whom it places duties (duty holders) expect that action taken by enforcing authorities to achieve compliance or bring duty holders to account for non-compliance should be proportionate to any risks to health and safety, or to the seriousness of any breach, which includes any actual or potential harm arising from a breach of the law.

12 In practice, applying the principle of proportionality means that enforcing authorities should take particular account of how far the duty holder has fallen short of what the law requires and the extent of the risks to people arising from the breach.

13 Some health and safety duties are specific and absolute. Others require action so far as is reasonably practicable. Enforcing authorities should apply the principle of proportionality in relation to both kinds of duty.

14 Deciding what is reasonably practicable to control risks involves the exercise of judgement. Where duty holders must control risks so far as is reasonably practicable, enforcing authorities considering protective measures taken by duty holders must take account of the degree of risk on the one hand, and on the other the sacrifice, whether in money, time or trouble, involved in the measures necessary to avert the risk. Unless it can be shown that there is gross disproportion between these factors and that the risk is insignificant in relation to the cost, the duty holder must take measures and incur costs to reduce the risk.

15 The authorities will expect relevant good practice to be followed. Where relevant good practice in particular cases is not clearly established, health and safety law effectively requires duty holders to establish explicitly the significance of the risks to determine what action needs to be taken. Ultimately, the courts determine what is reasonably practicable in particular cases.

16 Some irreducible risks may be so serious that they cannot be permitted irrespective of the consequences.

Targeting

17 Targeting means making sure that contacts are targeted primarily on those whose activities give rise to the most serious risks or where the hazards are least well controlled; and that action is focused on the duty holders who are responsible for the risk and who are best placed to control it—whether employers, manufacturers, suppliers, or others.

18 HSC expects enforcing authorities to have systems for deciding which inspections, investigations or other regulatory contacts should take priority according to the nature and extent of risks posed by a duty holder's operations. The duty holder's management competence is important, because a relatively low hazard site poorly managed can entail greater risk to workers or the public than a higher hazard site where proper and adequate risk control measures are in place. Certain very high hazard sites will receive regular inspections so that enforcing authorities can give public assurance that such risks are properly controlled.

19 Any enforcement action will be directed against duty holders responsible for a breach. This may be employers in relation to workers or others exposed to risks; the self-employed; owners of premises; suppliers of equipment; designers or clients of projects; or employees themselves. Where several duty holders have responsibilities, enforcing authorities may take action against more than one when it is appropriate to do so in accordance with this policy.

---

[2] In this policy, 'risk' (where the term is used alone) is defined broadly to include a source of possible harm, the likelihood of that harm occurring, and the severity of any harm.

20 When inspectors issue improvement or prohibition notices; withdraw approvals; vary licence conditions or exemptions; issue formal cautions; or prosecute; enforcing authorities should ensure that a senior officer of the duty holder concerned, at board level, is also notified.

Consistency

21 Consistency of approach does not mean uniformity. It means taking a similar approach in similar circumstances to achieve similar ends.

22 Duty holders managing similar risks expect a consistent approach from enforcing authorities in the advice tendered; the use of enforcement notices, approvals etc; decisions on whether to prosecute; and in the response to incidents.

23 HSC recognises that in practice consistency is not a simple matter. HSE and local authority inspectors are faced with many variables including the degree of risk, the attitude and competence of management, any history of incidents or breaches involving the duty holder, previous enforcement action, and the seriousness of any breach, which includes any potential or actual harm arising from a breach of the law. Decisions on enforcement action are discretionary, involving judgement by the enforcer. All enforcing authorities should have arrangements in place to promote consistency in the exercise of discretion, including effective arrangements for liaison with other enforcing authorities.

Transparency

24 Transparency means helping duty holders to understand what is expected of them and what they should expect from the enforcing authorities. It also means making clear to duty holders not only what they have to do but, where this is relevant, what they don't. That means distinguishing between statutory requirements and advice or guidance about what is desirable but not compulsory.

25 Transparency also involves the enforcing authorities in having arrangements for keeping employees, their representatives, and victims or their families informed. These arrangements must have regard to legal constraints and requirements.

26 This statement sets out the general policy framework within which enforcing authorities should operate. Duty holders, employees, their representatives and others also need to know what to expect when an inspector calls and what rights of complaint are open to them. All enforcing authority inspectors are required to issue the HSC leaflet *What to expect when a health and safety inspector calls* to those they visit. This explains what employers and employees and their representatives can expect when a health and safety inspector calls at a workplace. In particular:
- when inspectors offer duty holders information, or advice, face to face or in writing, including any warning, inspectors will tell the duty holder what to do to comply with the law, and explain why. Inspectors will, if asked, write to confirm any advice, and to distinguish legal requirements from best practice advice;
- in the case of improvement notices the inspector will discuss the notice and, if possible, resolve points of difference before serving it. The notice will say what needs to be done, why, and by when, and that in the inspector's opinion a breach of the law has been committed;
- in the case of a prohibition notice the notice will explain why the prohibition is necessary.

In addition, in response to *Service First* HSE has issued two publications, *The Health and Safety Executive: Working with employers* and *The Health and Safety Executive and you*, which reflect the principles of the *Enforcement Concordat*.

Accountability

27 Regulators are accountable to the public for their actions. This means that enforcing authorities must have policies and standards (such as the four enforcement principles above) against which they can be judged, and an effective and easily accessible mechanism for dealing with comments and handling complaints.

28 HSE's procedures for dealing with comments and handling complaints are set out in the publications referred to in paragraph 26. In particular, they:
- describe a complaints procedure in the case of decisions by officials, or if procedures have not been followed; and
- explain about the right of appeal to an Employment Tribunal in the case of statutory notices.
29 Local authorities have their own complaints procedures—details are available from individual authorities.

Investigation

30 As with prosecution, HSC expects enforcing authorities to use discretion in deciding whether incidents, cases of ill health, or complaints should be investigated. Indicative targets related to levels of investigation by HSE are normally specified in HSC's Strategic Plan, which is approved by the Government. HSC's priorities are also reflected in the HELA Strategy which is used by local authorities to target their activities and resources, via their Departmental Service Plans.

31 Investigations are undertaken in order to determine:
- causes;
- whether action has been taken or needs to be taken to prevent a recurrence and to secure compliance with the law;
- lessons to be learnt and to influence the law and guidance;
- what response is appropriate to a breach of the law.

To maintain a proportionate response, most resources available for investigation of incidents will be devoted to the more serious circumstances. HSC's Strategic Plan recognises that is neither possible nor necessary for the purposes of the Act to investigate all issues of noncompliance with the law which are uncovered in the course of preventive inspection, or in the investigation of reported events.

32 The enforcing authorities should carry out a site investigation of a reportable work-related death, unless it is an instance of adult trespass or apparent suicide on the railway[3] or there are other specific reasons for not doing so, in which case those reasons should be recorded.

33 In selecting which complaints or reports of injury or occupational ill health to investigate and in deciding the level of resources to be used, the enforcing authorities should take account of the following factors:
- the severity and scale of potential or actual harm;
- the seriousness of any potential breach of the law;
- knowledge of the duty holder's past health and safety performance;
- the enforcement priorities;
- the practicality of achieving results;
- the wider relevance of the event, including serious public concern.

Prosecution

England and Wales

34 In England and Wales the decision to proceed with a court case rests with the enforcing authorities. Enforcing authorities must use discretion in deciding whether to bring a prosecution.

35 In England and Wales the decision whether to prosecute should take account of the evidential test and the relevant public interest factors set down by the Director of Public Prosecutions in the Code for Crown Prosecutors. No prosecution may go ahead unless the prosecutor finds there is sufficient evidence to provide a realistic prospect of conviction, and decides that prosecution would be in the public interest.

---

[3] Where the police will always investigate and advise HSE if railway operational matters are at issue.

36 While the primary purpose of the enforcing authorities is to ensure that duty holders manage and control risks effectively, thus preventing harm, prosecution is an essential part of enforcement. HSC expects that where in the course of an investigation an enforcing authority has collected sufficient evidence to provide a realistic prospect of conviction and has decided, in accordance with this policy and taking account of the Code for Crown Prosecutors, that it is in the public interest to prosecute, then that prosecution should go ahead. Where the circumstances warrant it and the evidence to support a case is available, enforcing authorities may prosecute without prior warning or recourse to alternative sanctions.

Scotland

37 In Scotland the Procurator Fiscal decides whether to bring a prosecution. This may be on the basis of a recommendation by an enforcing authority; although the Procurator Fiscal may investigate the circumstances and institute proceedings independently of an enforcing authority. Enforcing authorities must use discretion in deciding whether to report to the Procurator Fiscal with a view to prosecution. The Crown Office and the Procurator Fiscal Service endorse this Statement by HSC, and acknowledge that action on reports of offences submitted to them by the enforcing authorities should reflect the approach set out here.

38 In Scotland, before prosecutions can be instituted, the Procurator Fiscal will need to be satisfied that there is sufficient evidence and that prosecution is in the public interest. In Scotland therefore the decision as to proceedings is one for the prosecutor rather than the enforcing authority whose views will, however, be taken into account.

39 Subject to the above, HSC expects that, in the public interest, enforcing authorities should normally prosecute, or recommend prosecution, where, following an investigation or other regulatory contact, one or more of the following circumstances apply. Where:
- death was a result of a breach of the legislation;[4]
- the gravity of an alleged offence, taken together with the seriousness of any actual or potential harm, or the general record and approach of the offender warrants it;
- there has been reckless disregard of health and safety requirements;
- there have been repeated breaches which give rise to significant risk, or persistent and significant poor compliance;
- work has been carried out without or in serious non-compliance with an appropriate licence or safety case;
- a duty holder's standard of managing health and safety is found to be far below what is required by health and safety law and to be giving rise to significant risk;
- there has been a failure to comply with an improvement or prohibition notice; or there has been a repetition of a breach that was subject to a formal caution;
- false information has been supplied wilfully, or there has been an intent to deceive, in relation to a matter which gives rise to significant risk;
- inspectors have been intentionally obstructed in the lawful course of their duties.

Where inspectors are assaulted, enforcing authorities will seek police assistance, with a view to seeking the prosecution of offenders.

40 HSC also expects that, in the public interest, enforcing authorities will consider prosecution, or consider recommending prosecution, where following an investigation or other regulatory contact, one or more of the following circumstances apply:
- it is appropriate in the circumstances as a way to draw general attention to the need for compliance with the law and the maintenance of standards required by law, and conviction may deter others from similar failures to comply with the law;

---

[4] Health and safety sentencing guidelines regard death resulting from a criminal act as an aggravating feature of the offence. If there is sufficient evidence, HSC considers that normally such cases should be brought before the court. However, there will be occasions where the public interest does not require a prosecution, depending on the nature of the breach and the surrounding circumstances of the death.

- a breach which gives rise to significant risk has continued despite relevant warnings from employees, or their representatives, or from others affected by a work activity.

Prosecution of individuals

41 Subject to the above, enforcing authorities should identify and prosecute or recommend prosecution of individuals if they consider that a prosecution is warranted. In particular, they should consider the management chain and the role played by individual directors and managers, and should take action against them where the inspection or investigation reveals that the offence was committed with their consent or connivance or to have been attributable to neglect on their part and where it would be appropriate to do so in accordance with this policy. Where appropriate, enforcing authorities should seek disqualification of directors under the Company Directors Disqualification Act 1986.

Publicity

42 Enforcing authorities in England and Wales should make arrangements for the publication annually of the names of all the companies and individuals who have been convicted in the previous 12 months of breaking health and safety law. They should also have arrangements for making publicly available information on these convictions and on improvement and prohibition notices which they have issued.

43 Enforcing authorities in England and Wales should also consider in all cases drawing media attention to factual information about charges which have been laid before the courts, but great care must be taken to avoid any publicity which could prejudice a fair trial. They should also consider publicising any conviction which could serve to draw attention to the need to comply with health and safety requirements, or deter anyone tempted to disregard their duties under health and safety law. In Scotland, decisions in relation to publicity of prosecutions are a matter for the Crown Office.

Action by the courts

44 Health and safety law gives the courts considerable scope to punish offenders and to deter others, including imprisonment for some offences. Unlimited fines may be imposed by higher courts. HSC will continue to seek to raise the courts' awareness of the gravity of health and safety offences and of the full extent of their sentencing powers, while recognising that it is for the courts to decide whether or not someone is guilty and what penalty if any to impose on conviction. A list of the sanctions presently available to the courts is attached to this statement.

45 In England and Wales, the enforcing authorities should, when appropriate, draw to the court's attention all the factors which are relevant to the court's decision as to what sentence is appropriate on conviction. The Court of Appeal has given guidance on some of the factors which should inform the courts in health and safety cases (*R* v *F Howe and Son (Engineers) Ltd* [1999] 2 All ER, and subsequent judgments). HSC notes that the Lord Chancellor has said that someone injured by a breach of health and safety legislation is no less a victim than someone who is assaulted.

Representations to the courts

46 In cases of sufficient seriousness, and when given the opportunity, the enforcing authorities in England and Wales should consider indicating to the magistrates that the offence is so serious that they may send it to be heard or sentenced in the higher court where higher penalties can be imposed. In considering what representations to make, enforcing authorities should have regard to Court of Appeal guidance: the Court of Appeal has said 'In our judgment magistrates should always think carefully before accepting jurisdiction in health and safety at work cases, where it is arguable that the fine may exceed the limit of their jurisdiction or where death or serious injury has resulted from the offence'.

47 In Scotland it would fall to the Procurator Fiscal to draw the court's attention to the seriousness of any offence.

Appendix B

Death at Work

48 Where there has been a breach of the law leading to a work-related death, enforcing authorities need to consider whether the circumstances of the case might justify a charge of manslaughter (culpable homicide in Scotland).

49 In England and Wales, to ensure decisions on investigation and prosecution are closely co-ordinated following a work-related death, HSE, the Association of Chief Police Officers (ACPO) and the Crown Prosecution Service (CPS) have jointly agreed and published *Work-related deaths: A protocol for liaison*. The Local Government Association has agreed that local authorities should take account of the protocol when responding to work-related deaths.

50 The police are responsible for deciding whether to pursue a manslaughter investigation and whether to refer a case to the CPS to consider possible manslaughter charges. The enforcing authorities are responsible for investigating possible health and safety offences. If in the course of their health and safety investigation, the enforcing authorities find evidence suggesting manslaughter, they should pass it on to the police. If the police or the CPS decide not to pursue a manslaughter case, the enforcing authorities will normally bring a health and safety prosecution in accordance with this policy.

51 In Scotland, responsibility for investigating sudden or suspicious deaths rests with the Procurator Fiscal. Unless a prosecution takes place in the same circumstances, the Procurator Fiscal is required to hold a Fatal Accident Inquiry into the circumstances of a death resulting from a work related[5] accident. An Inquiry may also be held where it appears to be in the public interest on the ground that the death was sudden, suspicious or unexplained, or has occurred in circumstances such as to give rise to serious public concern.

Crown bodies

52 Crown bodies must comply with health and safety requirements, but they are not subject to statutory enforcement, including prosecution. The Cabinet Office has established non-statutory arrangements for enforcing health and safety requirements in Crown bodies. These arrangements allow HSE to issue non-statutory improvement and prohibition notices, and for the censure of Crown bodies in circumstances where, but for Crown immunity, prosecution would have been justified. In deciding when to investigate or what form of enforcement action to take, HSE should follow as far as possible the same approach as for non-Crown bodies, in accordance with this enforcement policy.

Penalties for Health and Safety Offences[6]

The Health and Safety at Work etc. Act 1974 (the HSW Act), section 33 (as amended) sets out the offences and maximum penalties under health and safety legislation.

Failing to comply with an improvement or prohibition notice, or a court remedy order (issued under the HSW Act sections 21, 22 and 42 respectively):
Lower court maximum   £20 000 and/or 6 months' imprisonment
Higher court maximum   Unlimited fine and/or 2 years' imprisonment

Breach of sections 2-6 of the HSW Act, which set out the general duties of employers, self-employed persons, manufacturers and suppliers to safeguard the health and safety of workers and members of the public who may be affected by work activities:
Lower court maximum   £20 000
Higher court maximum   Unlimited fine

Other breaches of the HSW Act, and breaches of 'relevant statutory provisions' under the Act, which include all health and safety regulations. These impose both general and more specific requirements,

---

[5] In this case, an accident in the course of employment, if the deceased was an employee, or while engaged in their occupation, if an employer or self-employed person.

[6] As at January 2002. These penalties can change from time to time.

such as requirements to carry out a suitable and sufficient risk assessment or to provide suitable personal protective equipment:

Lower court maximum   £5000
Higher court maximum   Unlimited fine

Contravening licence requirements or provisions relating to explosives. Licensing requirements apply to nuclear installations, asbestos removal, and storage and manufacture of explosives. All entail serious hazards which must be rigorously controlled.

Lower court maximum   £5000
Higher court maximum   Unlimited fine and/or 2 years' imprisonment

On conviction of directors for indictable offences in connection with the management of a company (all of the above, by virtue of the HSW Act sections 36 and 37), the courts may also make a disqualification order (Company Directors Disqualification Act 1986, sections 1 and 2). The courts have exercised this power following health and safety convictions. Health and safety inspectors draw this power to the court's attention whenever appropriate.

Lower court maximum   5 years' disqualification
Higher court maximum   15 years' disqualification

Further information

Further copies of this leaflet are available from HSE Books or the HSE website.

HSE has prepared a Quality Statement which explains how it responds to the Commission's Enforcement Policy, and to the Cabinet Office Enforcement Concordat on good enforcement practice. The HSE Statement forms part of the HSE Quality Framework, which sets out how HSE will meet its aim of being a quality organisation. A copy of the HSE Statement *(Quality Statement aim 2, to secure compliance)* is available from HSE Information Centres (phone HSE's Infoline for details of your nearest Information Centre—see below for details).

More information about the way health and safety legislation is enforced and about health and safety legislation generally can be found in these free leaflets:

Successful health and safety management HSG65 (Second edition) HSE Books 1997 ISBN 0 7176 1276 7

The Health and Safety Executive working with employers: 2000 edition Leaflet HSE38 HSE Books 2000 (single copy free)

The Health and Safety Executive and you: 2000 edition Leaflet HSE37 HSE Books 2000 (single copy free)

What to expect when a health and safety inspector calls: A brief guide for businesses, employees and their representatives Leaflet HSC14 HSE Books 1998 (single copy free)

Work-related deaths: A protocol for liaison Booklet MISC491 HSE Books 2003

Local authorities may produce their own further information on enforcing health and safety. You can find your local authority's address and telephone number in your local telephone directory.

HSE priced and free publications are available by mail order from

HSE Books, PO Box 1999, Sudbury, Suffolk CO10 2WA Tel: 01787 881165 Fax: 01787 313995 Website: <http://www.hsebooks.co.uk>.

(HSE priced publications are also available from bookshops and free leaflets can be downloaded from HSE's website: <http://www.hse.gov.uk>.)

For information about health and safety ring HSE's Infoline Tel: 08701 545500 Fax: 02920 859260 e-mail: hseinformationservices@natbrit.com or write to HSE Information Services, Caerphilly Business Park, Caerphilly CF83 3GG.

© Crown copyright

First published 01/02. Please acknowledge the source as HSE.

Appendix B

WORK-RELATED DEATHS

A protocol for liaison

Foreword

The first version of this protocol was introduced in 1998, and was signed by the police, the Health and Safety Executive and the Crown Prosecution Service. This is the second version, revised to extend our partnership to include local authorities through their representative body, the Local Government Association, and the British Transport Police, and to emphasise further the importance of working together to investigate thoroughly, and to prosecute appropriately, those responsible for work-related deaths in England and Wales.

We were pleased by the level of response to the public consultation exercises, which produced some excellent and extremely helpful suggestions. We are acutely conscious of the strength of public feeling about workplace fatalities, and how these tragic incidents devastate people's lives.

All five signatory organisations recognise the need for investigating authorities to talk to each other and to share information and best practice. We appreciate that people want to be confident that we are doing all that we can to co-ordinate our efforts and to co-operate with each other in the best interests of public safety and of those affected by work-related deaths.

We endorse this revised protocol. We believe that it provides an enhanced framework for liaison, and that its introduction will help ensure that all five signatory organisations work in partnership to deliver the high standard of professionalism that the public requires and deserves.

T J Stoddart,
Assistant Chief Constable,
Association of Chief Police Officers

Ian Johnston,
Chief Constable,
British Transport Police

Sir David Calvert-Smith,
Director of Public Prosecutions,
Crown Prosecution Service Health and Safety Executive

Timothy Walker,
Director General,
Health and Safety Executive

Sir Brian Briscoe,
Chief Executive,
The Local Government Association

Introduction

This protocol has been agreed between the Health and Safety Executive (HSE), the Association of Chief Police Officers (ACPO), the British Transport Police (BTP), the Local Government Association and the Crown Prosecution Service (CPS). It sets out the principles for effective liaison between the parties in relation to work-related deaths in England and Wales, and is available to the public. In particular, it deals with incidents where, following a death, evidence indicates that a serious criminal offence other than a health and safety offence may have been committed. The protocol addresses issues concerning general liaison and is not intended to cover the operational practices of the signatory organisations.

HSE, local authorities, the police and the CPS have different roles and responsibilities in relation to a work-related death.

Enforcement Concordat; Enforcement Policy Statement (HSC15)

HSE and local authorities are responsible, under section 18 of the Health and Safety at Work etc. Act 1974 (HSW Act), for making adequate arrangements for the enforcement of health and safety legislation with a view to securing the health, safety and welfare of workers and protecting others, principally the public. Each has specific areas of responsibility, further details of which are set out in Annex A of this protocol.

Please note that this protocol does not take into account co-operation and liaison with the Rail Accident Investigation Branch (RAIB), which, at the time this protocol was written, had not come into being. When the RAIB is introduced, it is envisaged that a separate protocol, or protocols, governing liaison between the RAIB and the signatory organisations will be developed.

There are other enforcing authorities that have responsibility for enforcing health and safety, or other similar legislation. Some of these are listed in Annex A. The Civil Aviation Authority and the Maritime and Coastguard Agency are not signatories to this protocol, but each has agreed to abide by the protocol's principles.

At present, only the police can investigate serious criminal offences (other than health and safety offences) such as manslaughter, and only the CPS can decide whether such a case will proceed. The police will also have an interest in establishing the circumstances surrounding a work-related death in order to assist the coroner's inquest.

Health and safety offences are usually prosecuted by HSE, the local authority or other enforcing authority in accordance with current enforcement policy. The CPS may also prosecute health and safety offences, but usually does so only when prosecuting other serious criminal offences, such as manslaughter, arising out of the same circumstances.

When making a decision whether to prosecute, the CPS, HSE, the local authority or other enforcing authority will review the evidence according to the Code for Crown Prosecutors to decide if there is a realistic prospect of conviction and, if so, whether a prosecution is needed in the public interest. Local authorities that have signed up to the Enforcement Concordat must follow the principles and procedures within it.

The underlying principles of this protocol are as follows:

- an appropriate decision concerning prosecution will be made based on a sound investigation of the circumstances surrounding work-related deaths;
- the police will conduct an investigation where there is an indication of the commission of a serious criminal offence (other than a health and safety offence), and HSE, the local authority or other enforcing authority will investigate health and safety offences. There will usually be a joint investigation, but on the rare occasions where this would not be appropriate, there will still be liaison and co-operation between the investigating parties;
- the decision to prosecute will be co-ordinated, and made without undue delay;
- the bereaved and witnesses will be kept suitably informed; and
- the parties to the protocol will maintain effective mechanisms for liaison.

In What Circumstances Will This Protocol Apply?

For the purposes of this protocol, a work-related death is a fatality resulting from an incident arising out of, or in connection with, work. The principles set out in this protocol also apply to cases where the victim suffers injuries in such an incident that are so serious that there is a clear indication, according to medical opinion, of a strong likelihood of death.

There will be cases in which it is difficult to determine whether a death is work-related within the application of this protocol; for example, those arising out of some road traffic incidents, or in prisons or health care institutions, or following a gas leak. Each fatality must be considered individually, on its particular facts, according to organisational internal guidance, and a decision made as to whether it should be classed as a work-related death. In determining the question, the enforcing authorities will hold discussions and agree upon a conclusion without delay.

Appendix B

1 Statement of Intent

1.1 In the early stages of an investigation, whether any serious criminal offence has been committed is not always apparent. The parties to the protocol are committed to ensuring that any investigation into a work-related death is thorough and appropriate, and agree to work closely together in order to achieve this. Decisions in relation to who will lead the investigation, and the direction it will take, should be timely, informed by the best available evidence and technical expertise, and should take account of the wider public interest. Should there be any issue as to who is to be involved in investigating any work-related death, then the parties will work together to reach a conclusion.

2 Initial Action

2.1 A police officer attending an incident involving a work-related death should arrange, according to the officer's own force procedures governing unexplained deaths, to:
- identify, secure, preserve and take control of the scene, and any other relevant place;
- supervise and record all activity;
- inform a senior supervisory officer;
- enquire whether the employer or other responsible person in control of the premises or activity has informed HSE, the local authority or other investigating or enforcing authority; and
- contact and discuss the incident with HSE, the local authority or other enforcing authority, and agree arrangements for controlling the scene, for considering access to others, and for other local handling procedures to ensure the safety of the public.

2.2 A police officer of supervisory rank should attend the scene and any other relevant place to assess the situation, review actions taken to date and assume responsibility for the investigation. Should any other investigating or enforcing authority have staff in attendance before the police arrive, it should ensure that the police have been called, and preserve the scene in accordance with the initial actions (above) until the police get there.

3 Management of the Investigation

3.1 Investigations should always be managed professionally, with communications between the signatory organisations continually maintained. Investigations should generally be jointly conducted, with one of the parties taking the lead, or primacy, as appropriate. An investigation may also require liaison with any other enforcing authority that may have an interest, and may include liaison with the CPS.

3.2 Throughout the period of the investigation, the police and HSE, the local authority or other enforcing authority should keep the progress of the investigation under review. Milestones should be agreed and monitored, and policy and key decisions recorded.

3.3 The police, HSE, the local authority or other enforcing authority should agree upon:
- how resources are to be specifically used;
- how evidence is to be disclosed between the parties;
- how the interviewing of witnesses, the instruction of experts and the forensic examination of exhibits is to be co-ordinated;
- how, and to what extent, corporate or organisational failures should be investigated;
- a strategy for keeping the bereaved, witnesses, and other interested parties such as the coroner, informed of developments in the investigation; and
- a media strategy to take account of the sensitivities of the bereaved and those involved in the incident, and to encourage consistency of approach in reporting.

3.4 In certain large-scale investigations it may be beneficial to form a strategic liaison group to ensure effective inter-organisational communication, and to share relevant information and experiences.

4 Decision Making

4.1 Where the investigation gives rise to a suspicion that a serious criminal offence (other than a health and safety offence) may have caused the death, the police will assume primacy for the investigation and will work in partnership with HSE, the local authority or other enforcing authority.

4.2 Where it becomes apparent during the investigation that there is insufficient evidence that a serious criminal offence (other than a health and safety offence) caused the death, the investigation should, by agreement, be taken over by HSE, the local authority or other enforcing authority. Both parties should record such a decision in writing.

4.3 Where HSE, the local authority or other enforcing authority is investigating the death, and new information is discovered which may assist the police in considering whether a serious criminal offence (other than a health and safety offence) has been committed, then the enforcing authority will pass that new information to the police. An enforcing authority inspector may do this, but it may also be from the enforcing authority's solicitors via the CPS. The police should then consider whether to resume primacy for the investigation. The decision and reasons should be recorded in writing.

4.4 There will also be rare occasions where as a result of the coroner's inquest, judicial review or other legal proceedings, further consideration of the evidence and surrounding facts may need to be made. Where this takes place the police, the enforcing authority with primacy for the investigation and the CPS will work in partnership to ensure an early decision. There may also be a need for further investigation.

5 Disclosure of Material

5.1 Disclosure must always follow the established law and procedure.

5.2 Where there has been an investigation, any material obtained should be shared, subject to any legal restrictions, between the police, HSE, the local authority or other enforcing authority and the CPS. Special handling procedures may be necessary in certain cases. The organisation responsible for retaining the exhibits, documents and other relevant material should also be agreed upon.

6 Special Inquiries

6.1 In the case of some incidents, particularly those involving multiple fatalities, the Health and Safety Commission may, with the consent of the Secretary of State, direct that a public inquiry be held. Alternatively, the Commission may authorise HSE, or any other person, to investigate and produce a special report.

6.2 In such circumstances, the police will provide any necessary support and evidence to the person appointed to conduct the public inquiry, or to the special investigation, subject to the relevant regulations.

6.3 Complex legal issues may arise when there are parallel public inquiries and criminal investigations or prosecutions. The signatories will aim to keep inquiry chairs informed of the progress of the investigation.

6.4 Sometimes the report of a public inquiry may be delayed to await the conclusion of criminal proceedings, and on other occasions, there may be no such delay because of strong public interest in publishing the report and the recommendations of a public inquiry quickly. In either event, the signatories to the protocol will work together to ensure that the decision to prosecute is made as expeditiously as possible and any criminal proceedings commenced without delay.

7 Advice Prior to Charge

7.1 Early liaison by the police, HSE, the local authority or other enforcing authority with the CPS, is to be encouraged in the best interests of the investigation and prosecution process as a whole. There is no need to wait until a file is ready to be submitted before the police open discussions with the CPS. The police are encouraged, at any stage following a work-related death, to consult the CPS for advice, not only about the nature of any charges, but also as to the legal and evidential issues surrounding the investigation, including advice about expert evidence.

7.2 The police should seek the advice of the CPS before charging an individual with any serious criminal offence (other than a health and safety offence) arising out of a work-related death.

7.3 The police must consult the CPS Casework Directorate for advice when there is any consideration of charging a company or corporation with any serious criminal offence (other than a health and safety offence).

8 The Decision to Prosecute

8.1 The decision to prosecute any serious criminal offence (other than a health and safety offence) arising out of the death will be taken by the CPS according to *The code for crown prosecutors*. Such an offence may be prosecuted either with or without related health and safety offences. The decision will be made following discussion with the police, and, where appropriate, HSE the local authority or another enforcing authority.

8.2 There should be no undue delay in reaching the prosecution decision. If there is a delay, then the CPS will notify the police and the enforcing authority and explain the reasons for the delay, and will keep them informed of the progress of the decision making.

8.3 The CPS should always take into account the consequences for the bereaved of the decision whether or not to prosecute, and of any views expressed by them.

8.4 When the CPS has made its decision, it must be communicated to the police, HSE, the local authority or other enforcing authority as soon as practicable, so that HSE, the local authority or other enforcing authority can decide as expeditiously as possible whether to prosecute for health and safety offences if the CPS is not doing so.

8.5 Where HSE, the local authority or other enforcing authority has primacy for the investigation, then the decision whether to prosecute for health and safety offences will be taken without undue delay. The relevant enforcing authority should then inform the police of its decision.

8.6 No prosecution decision will be made public until the bereaved, the Coroner's Office and any potential defendants have been notified according to the previously agreed strategy.

8.7 The public announcement of the decision will be made according to the agreed media strategy.

8.8 Where there is to be no CPS prosecution, the announcement of the CPS's decision shall include the fact that the decision of HSE, the local authority or other enforcing authority will be made after the inquest (subject to paragraph 10.3, below). It is CPS policy to set out its reasons in writing and send them to the bereaved, and to offer to meet them to discuss the reasons for reaching the decision.

9 The Prosecution

9.1 Where the CPS and HSE, the local authority or another enforcing authority seek to prosecute offences arising out of the same incident, the prosecution(s) shall be initiated and managed jointly.

9.2 There should be an early conference attended by the CPS, the police and HSE, the local authority or other enforcing authority to consider the management of the proceedings. In particular, the following issues should be discussed, agreed and recorded:
- who will take lead responsibility for the prosecution;
- the nature and the wording of the charges (including, where appropriate, consideration of any alternative charges and acceptable pleas);
- arrangements for the retention and disclosure of material;
- a case management timetable;
- arrangements for keeping the bereaved and witnesses informed;
- the announcement of the decision;

- arrangements for maintaining contact during the prosecution, and an agreement as to a mechanism for consulting, should an issue arise which results in the discontinuance of the proceedings or no evidence being offered;
- an agreement as to any specific instructions to the prosecuting advocate; and
- any other case management issues.

9.3 Where the CPS is prosecuting, and there is no prosecution by HSE, the local authority or other enforcing authority, but an enforcing authority wishes to retain an interest in the case, the CPS will keep that enforcing authority informed of the progress and outcome of the case.

10 HM Coroner

10.1 The police or the CPS will notify the coroner when a serious criminal offence arising out of a work-related death (other than a health and safety offence) has been charged. The coroner may then adjourn the inquest until the end of the criminal prosecution. The Director of Public Prosecutions may also ask the coroner to adjourn the inquest where there are certain proceedings before a magistrates' court that are related to a death.

10.2 Where the CPS is prosecuting, and HSE, the local authority or other enforcing authority has submitted documents or a report to the coroner about a work-related death, the CPS and the police shall also be given a copy. Similarly, where an enforcing authority is prosecuting, and the police or CPS has submitted documents or a report to the coroner about a work-related death, the enforcing authority shall also be given a copy. In all cases, documents or reports may not be disclosed to any party without the consent of the party that originally submitted them.

10.3 Where the CPS has reviewed the case and decided not to prosecute, HSE, the local authority or other enforcing authority will await the result of the coroner's inquest before charging any health and safety offences, unless to wait would prejudice the case. Where, following an inquest, public inquiry, judicial review or other legal proceedings, it is necessary for the CPS to review or re-review the case, HSE, the local authority or other enforcing authority will wait until the review by the CPS has been completed before instigating or continuing its own proceedings.

11 National Liaison

11.1 The National Liaison Committee comprises representatives from the police, BTP, the CPS, HSE and the Local Government Association. It will meet at least twice a year to review the operation of the protocol and consider the need for changes to the arrangements.

12 Local Liaison

12.1 The Regional Liaison Committees comprise representatives from the signatories, nominated at local levels. These committees will meet on a regular basis to discuss issues of mutual interest and concern, and in particular, the operation of the protocol from a local standpoint, to monitor the protocol's effectiveness, and to communicate any issues to the National Liaison Committee.

12.2 The Regional Liaison Committees will be responsible for ensuring that there is an identified and effective line of local communication between the five organisations.

Annex A: A General Guide to the Enforcement of the Health and Safety at Work etc. Act 1974 (HSW ACT) and Related Legislation

Enforcement of the HSW Act and the related legislation is generally shared between HSE and local authorities. A general guide to the allocation of the main activity is detailed below. For more detailed guidance on allocation of specific activities or premises refer to HSE's website: <http://www.hse.gov.uk/lau/lacs/23-15.htm>.

The Health and Safety Executive

HSE is responsible for enforcing work-related health and safety legislation in:
- factories and other manufacturing premises, including motor vehicle repair
- chemical plants and refineries
- construction
- railways, tram and underground systems
- mines, quarries and landfill sites
- farms, agriculture and forestry
- hospitals, including nursing homes
- local government, including their offices and facilities run by them
- schools, colleges and universities
- domestic gas installation, maintenance or repair
- utilities, including power generation, water, and waste
- fairgrounds (travelling or fixed)
- airports (except terminal buildings, car parks and office buildings)
- police and fire authorities; Crown bodies, including the Ministry of Defence
- prisons
- docks
- nuclear installations
- offshore gas and oil installations and associated activities, including pipe-laying barges, and diving support vessels
- onshore major hazards, including pipelines, gas transmission and distribution
- transport of dangerous goods by road and rail
- manufacture, transport, handling and security of explosives
- common parts of domestic premises.

Local authorities

In England and Wales, local authorities enforce the HSW Act in respect of certain non-domestic premises, including:
- shops and retailing, including market stalls, coin-operated launderettes, and mobile vendors
- most office-based activities
- some wholesale and retail warehouses
- hotels, guest houses, hostels, caravan and camping sites, restaurants, public houses and other licensed premises
- leisure and entertainment, including night clubs, cinemas, social clubs, circuses, sports facilities, health clubs, gyms, riding schools, racecourses, pleasure boat hire, motor-racing circuits, museums, theatres, art galleries and exhibition centres
- places of worship and undertakers
- animal care, including zoos, livery stables and kennels
- therapeutic and beauty services, including massage, saunas, solariums, tattooing, skin and body piercing, and hairdressing
- residential care homes
- privately run pre-school child care, eg nurseries.

There are other authorities and agencies with responsibilities for the investigation and enforcement of the HSW Act and other similar legislation. These include, for example:
- the Fire Authorities
- the Maritime & Coastguard Agency (on board ships)
- the Care Standards Commission (health care premises)
- the Environment Agency

Enforcement Concordat; Enforcement Policy Statement (HSC15)

- the Civil Aviation Authority
- Trading Standards
- the Department of Trade and Industry (DTI)
- the Marine Accident Investigation Branch.

Contacting HSE out of hours

HSE is not an emergency service. It has produced guidance for police and other emergency service control rooms describing how to contact HSE inspectors out of hours.

Contacting local authorities out of hours

There will be local arrangements in place for contacting the authorised health and safety inspectors within local authorities. Contact can usually be made through the local town hall or council offices during office hours and on an emergency number out of office hours.

Acknowledgements

The National Liaison Committee is extremely grateful to those people and organisations that responded to the public consultation exercises. Their contributions were invaluable to the revision process, and, as a result, many of the ideas and proposals have now been included in this protocol.

Further Reading

More information can be found in these free publications:

Advice and information for bereaved families (England and Wales) MISC446a 2002 (available from HSE inspectors, **not** HSE Books)

Advice and information for bereaved families (Scotland) MISC200 2000 (available from HSE inspectors, **not** HSE Books)

Enforcement policy statement HSC15 HSE Books 2002

Health and safety in local authority enforced sectors. Section 18: HSC guidance to local authorities MISC488 HSE Books 2002

The code for crown prosecutors Crown Prosecution Service, Communications Branch, 50 Ludgate Hill, London EC4M 7EX, Tel: 020 7796 8442, Website: <http://www.cps.gov.uk/>.

Further Information

HSE priced and free publications are available by mail order from HSE Books, PO Box 1999, Sudbury, Suffolk CO10 2WA. Tel: 01787 881165 Fax: 01787 313995 Website: <http://www.hsebooks.co.uk> (HSE priced publications are also available from bookshops and free leaflets can be downloaded from HSE's website: <http://www.hse.gov.uk>).

For information about health and safety ring HSE's Infoline Tel: 08701 545500, Fax: 02920 859260 e-mail: hseinformationservices@natbrit.com or write to HSE Information Services, Caerphilly Business Park, Caerphilly BF83 3GG.

CPS publications are available from CPS Communications Branch, 50 Ludgate Hill, London, EC4M 7EX; Tel: 020 7796 8442, Website: <http://www.cps.gov.uk>.

ACPO website: <http://www.acpo.police.uk/>.

The information is current at 02/03.

APPENDIX C

The Health and Safety (Enforcing Authority) Regulations 1998 (as amended)

1998 No. 494

Made	*2nd March 1998*
Laid before Parliament	*6th March 1998*
Coming into force	*1st April 1998*

The Secretary of State, in exercise of the powers conferred on him by sections 15(1) and (3)(a) and (c), 18(2) and 82(3)(a) of the Health and Safety at Work etc. Act 1974 ("the 1974 Act") and of all other powers enabling him in that behalf and for the purpose of giving effect without modifications to proposals submitted to him by the Health and Safety Commission under section 11(2)(d) of the 1974 Act after the carrying out by the said Commission of consultations in accordance with section 50(3) of that Act, hereby makes the following Regulations:

1 Citation and commencement

These Regulations may be cited as the Health and Safety (Enforcing Authority) Regulations 1998 and shall come into force on 1st April 1998.

2 Interpretation

(1) In these Regulations, unless the context otherwise requires—
"the 1974 Act" means the Health and Safety at Work etc. Act 1974;
"agricultural activities"—
  (a) includes horticulture, fruit growing, seed growing, dairy farming, livestock breeding and keeping, including the management of livestock up to the point of slaughter or export from Great Britain, forestry, the use of land as grazing land, market gardens and nursery grounds and the preparation of land for agricultural use;
  (b) does not include such activities at a garden centre or other shop,
and for this purpose "livestock breeding and keeping" does not include activities the main purpose of which is entertainment;
["bus" means a motor vehicle which is designed or adapted to travel along roads and to carry more than eight passengers but which is not a tramcar;][1]
"the Commission" means the Health and Safety Commission;
"common parts" means those parts of premises used in common by, or for providing common services to or common facilities for, the occupiers of the premises;
"construction work" and "contractor" have the meanings assigned to them by regulation 2(1) of the Construction (Design and Management) Regulations 1994;
"dock premises" has the meaning assigned to it by regulation 2(1) of the Docks Regulations 1988;
"electricity system" does not include the consumer's installation within the meaning of regulation 3(1) of the Electricity Supply Regulations 1988;
"the Executive" means the Health and Safety Executive;

---

[1] Amended by SI 2006/557.

"fairground" means such part of premises as is for the time being used wholly or mainly for the operation of any fairground equipment, other than a coin-operated ride, non-powered children's playground equipment, swimming pool slide, go-kart, or plant designed to be used by members of the public for entertainment purposes for bouncing upon;

"gas" has the meaning assigned to it by section 48 of the Gas Act 1986;

"gas fitting" has the meaning assigned to it by section 48 of the Gas Act 1986;

"gas system" does not include a portable or mobile appliance supplied with gas from a cylinder, or the cylinder, pipes and other fittings used for supplying gas to that appliance;

["guided bus system" means a system of transport, used wholly or mainly for the carriage of passengers, that employs buses which for some or all of the time when they are in operation—
  (a) travel along roads; and
  (b) are guided (whether while on the road or at other times) by means of—
    (i) apparatus, a structure or other device which is fixed and not part of the bus; or
    (ii) a guidance system which is automatic;

"guided transport" means a system of transport, used wholly or mainly for the carriage of passengers, employing vehicles which for some or all of the time when they are in operation are guided by means of—
  (a) rails, beams, slots, guides or other apparatus, structures or devices which are fixed and not part of the vehicle; or
  (b) a guidance system which is automatic;
and for this purpose "vehicle" includes a mobile traction unit;][2]

"ionising radiation" has the meaning assigned to it by regulation 2(1) of [the Ionising Radiations Regulations 1999 [SI 1999/3232]];[3]

"livestock" means any creature kept for the production of food, wool, skins or fur or for the purpose of any agricultural activity;

"local authority" means—
  (a) in relation to England, a county council so far as they are the council for an area for which there are no district councils, a district council, a London borough council, the Common Council of the City of London, the Sub-Treasurer of the Inner Temple, the Under-Treasurer of the Middle Temple or the Council of the Isles of Scilly;
  (b) in relation to Scotland, the council for a local government area; and
  (c) in relation to Wales, a county council or a county borough council;

"mine" has the meaning assigned to it by section 180 of the Mines and Quarries Act 1954 [but, notwithstanding subsection (5) of that section, does not include any railway serving the mine unless and to the extent that the railway is located within the curtilage of the mine];[4]

"office activities" includes any activity for the purposes of administration, clerical work, handling money, telephone and telegraph operating and the production of computer software by the use of computers; and for this purpose "clerical work" includes writing, book-keeping, sorting papers, filing, typing, duplicating, machine calculating, drawing and the editorial preparation of matter for publication except where that preparation is on the premises where newspapers, magazines, periodicals or books are printed;

"pleasure craft" has the meaning assigned to it by regulation 2(1) of the Docks Regulations 1988;

"preparation dangerous for supply" has the meaning assigned to it by regulation 2(1) of the Chemicals (Hazard Information and Packaging for Supply) Regulations 1994;

[...][5]

---

[2] Amended by SI 2006/557.
[3] Amended by SI 1999/3232.
[4] Amended by SI 2006/557.
[5] Repealed by SI 2006/557.

The Health and Safety (Enforcing Authority) Regulations 1998 (as amended)

["quarry" has the meaning assigned to it by regulation 3 of the Quarries Regulations 1999;][6]

["railway" means any system of transport the operation of which is specified in regulation 3(2) of the Health and Safety (Enforcing Authority for Railways and Other Guided Transport Systems) Regulations 2006;][7]

["road"—
  (a) in England and Wales, means any length of highway or of any other road to which the public has access, and includes bridges over which a road passes; and
  (b) in Scotland, has the same meaning as in the Roads (Scotland) Act 1984;][8]

"substance dangerous for supply" has the meaning assigned to it by regulation 2(1) of the Chemicals (Hazard Information and Packaging for Supply) Regulations 1994;

["trolley vehicle system" means a system of transport by vehicles constructed or adapted for use on roads without rails under electric power transmitted to them by overhead wires (whether or not there is in addition a source of power on board the vehicles);][9]

"veterinary surgery" has the meaning assigned to it by section 27 of the Veterinary Surgeons Act 1966;

"work" in relation to a gas fitting has the meaning assigned to it by regulation 2(1) of the Gas Safety (Installation and Use) Regulations 1994;

"zoo" has the meaning assigned to it by section 1(2) of the Zoo Licensing Act 1981.

(2) In these Regulations (except regulation 4(7)), unless the context otherwise requires, any reference to the enforcing authority for premises or parts of premises is a reference to the enforcing authority for the relevant statutory provisions in relation to those premises or parts, as the case may be, and to any activity carried on in them.

(3) In these Regulations, unless the context otherwise requires, any reference to—
  (a) a numbered regulation or Schedule is a reference to the regulation of or Schedule to these Regulations so numbered; and
  (b) a numbered paragraph is a reference to the paragraph so numbered in the regulation or Schedule in which that reference appears.

3 Local authorities to be enforcing authorities in certain cases

(1) Where the main activity carried on in non-domestic premises is specified in Schedule 1, the local authority for the area in which those premises are situated shall be the enforcing authority for them, and the Executive shall be the enforcing authority in any other case including the common parts of domestic premises.

(2) Where such premises are occupied by more than one occupier each part separately occupied shall be regarded as being separate premises for the purposes of paragraph (1).

(3) While a vehicle is parked in connection with the sale from it of food, drink or other articles the vehicle together with its pitch shall be regarded as separate premises for the purposes of paragraph (1).

(4) Where paragraph (2) applies, the local authority shall be the enforcing authority for the common parts, except that—
  (a) if the Executive is the enforcing authority for—
    (i) all other parts of the premises, the Executive shall be the enforcing authority for the common parts;
    (ii) any other part of the premises and the occupier of that part has any obligations under the relevant statutory provisions for any matters appertaining to the common parts, the Executive shall be the enforcing authority for those provisions in respect of such matters;

---

[6] Amended by SI 1999/2024.
[7] Amended by SI 2006/557.
[8] ibid.
[9] ibid.

(b) in the case of land within the perimeter of an airport the Executive shall be the enforcing authority for the common parts—
  (i) which are not within a building; or
  (ii) to which passengers are admitted but other members of the public are not admitted.
(5) Paragraph (2) shall not apply to—
  (a) the tunnel system within the meaning it would have in section 1(7) of the Channel Tunnel Act 1987 if the words "to be" did not appear;
  (b) an offshore installation within the meaning of regulation 3 of the Offshore Installations and Pipeline Works (Management and Administration) Regulations 1995;
  (c) a building or construction site, that is to say, premises where the only activities being undertaken are construction work and activities for the purposes of or in connection with such work;
  (d) the campus of a university, polytechnic, college, school or similar educational establishment;
  (e) a hospital;
  and the Executive shall be the enforcing authority for the whole of any such premises.
(6) Notwithstanding paragraph (4), the [Office of Rail Regulation][10] shall be the enforcing authority for the common parts in a railway station or terminal or in a goods yard which is served by a railway.
(7) This regulation shall have effect subject to regulations 4, 5 and 6.

4 Exceptions

(1) The Executive shall be the enforcing authority for—
  (a) the enforcement of any of the relevant statutory provisions against a body specified in paragraph (3) or the officers or servants of such a body;
  (b) any part of premises occupied by such a body.
(2) Where premises are mainly occupied by a body specified in paragraph (3) and are partly occupied by another person for the purpose of providing services at the premises for that body, the Executive shall be the enforcing authority for the part of the premises occupied by that other person.
(3) The bodies referred to in paragraphs (1) and (2) are—
  (a) a county council;
  (b) any other local authority as defined in regulation 2;
  (c) a parish council in England or a community council in Wales or Scotland;
  (d) a police authority or the Receiver for the Metropolitan Police District;
  (e) *a [relevant authority (as defined in section 6 of the Fire (Scotland) Act 2005 (asp 5))]*;[11]
  [(e) a fire and rescue authority under the Fire and Rescue Services Act 2004;][12]
  (f) a headquarters or an organisation designated for the purposes of the International Headquarters and Defence Organisation Act 1964; or a service authority of a visiting force within the meaning of section 12 of the Visiting Forces Act 1952;
  (g) the United Kingdom Atomic Energy Authority;
  (h) the Crown, but regulation 3 shall apply to any part of premises occupied by the Executive and to any activity carried on there.
(4) The Executive shall be the enforcing authority for—
  (a) section 6 of the 1974 Act;
  (b) the other relevant statutory provisions in respect of any activity specified in Schedule 2 (whether or not it is the main activity carried on in premises).

---

[10] Amended by SI 2006/557.
[11] Amended by SI 2005/2060.
[12] Amended by SI 2005/1541.

The Health and Safety (Enforcing Authority) Regulations 1998 (as amended)

(5) Regulation 3 and the preceding provisions of this regulation shall have effect subject to any provisions made for enforcement responsibility by other regulations made under the 1974 Act or by any of the existing statutory provisions.

(6) The preceding provisions of this regulation shall have effect subject to regulations 5 and 6.

[(7) Notwithstanding the preceding provisions of this regulation and subject to paragraphs (8) to (10), a licensing authority shall be the enforcing authority for the Manufacture and Storage of Explosives Regulations 2005 ("the 2005 Regulations") and section 23 of the Explosives Act 1875—

   (a) for a site in relation to which it has granted a person a licence for the manufacture or storage of explosives at that site under regulation 13 of the 2005 Regulations or registered a person in respect of such storage at that site under regulation 11 of those Regulations;

   (b) where, in relation to a deemed licence or deemed registration, it would have been the licensing authority by virtue of paragraph 1 of Schedule 1 to the 2005 Regulations if an application for a licence or registration had been made under those Regulations; and

   (c) where, in any other case than those referred to in sub-paragraphs (a) and (b)—

      (i) it would be the licensing authority by virtue of paragraph 1 of Schedule 1 to those Regulations if an application for a licence or registration is, or should have been, made under those Regulations, or

      (ii) it would have been the licensing authority had not the requirements of regulation 9(1) or, as the case may be, 10(1) been disapplied by virtue of any of the provisions of, respectively, regulation 9(2) or 10(2) applying in the case concerned.

(8) The Executive shall be the enforcing authority for the 2005 Regulations in respect of the manufacture of ammonium nitrate blasting intermediate.

(9) A licensing authority which is a local authority shall be the enforcing authority for regulation 25 of the 2005 Regulations in the area of that local authority.

(10) The enforcing authority for regulation 6 of the 2005 Regulations where a person disposes of explosives or decontaminates explosive-contaminated items at a place other than at a site in relation to which a person has a licence to manufacture or store explosives under regulation 13 of the 2005 Regulations or is registered in respect of such storage under regulation 11 of those Regulations, shall be—

   (a) where the disposal or decontamination is carried out by, or on behalf of, a person who holds a licence granted by the Executive under those Regulations in a case in which the assent of the local authority was required under regulation 13(3) of those Regulations before the licence was granted, the Executive;

   (b) subject to sub-paragraph (a), where the local authority is by virtue of the Health and Safety (Enforcing Authority) Regulations 1998 the enforcing authority for the premises or part of premises at which the disposal or decontamination is carried out, the local authority; or

   (c) in any other case, the Executive.

(11) For the purposes of paragraphs (7) to (10), "ammonium nitrate blasting intermediate", "disposes", "licence", "licensing authority", "registered" and "site" have the same meanings as they are given by regulation 2(1) of the 2005 Regulations and "deemed licence" and "deemed registration" have the same meanings as they are given by regulation 27(19) of those Regulations.][13]

---

[13] Amended by SI 2005/1082.

5 Arrangements enabling responsibility for enforcement to be transferred

(1) The responsibility for enforcing any of the relevant statutory provisions in respect of any particular premises, part of premises, or any activity carried on there may be transferred from the Executive to the local authority or from the local authority to the Executive.

(2) A transfer may be made only by agreement between the enforcing authority which has the current responsibility and the authority to which it proposed to transfer it, or by the Commission.

(3) Where a transfer has been made, the authority to which responsibility has been transferred shall cause notice of the transfer to be given to persons affected by it, and where a transfer has been made by the Commission the Commission shall cause notice of it to be given to both enforcing authorities concerned.

(4) The preceding provisions of this regulation shall not apply to any part of premises occupied by the Crown or to any activity carried on there but responsibility for enforcing any of the relevant statutory provisions in respect of office activities and the premises used for them may be transferred by an agreement between the Executive, the local authority concerned and the Government Department or other public body concerned.

6 Arrangements enabling responsibility for enforcement to be assigned in cases of uncertainty

(1) The responsibility for enforcing any of the relevant statutory provisions in respect of any particular premises, part of premises or any activity carried on there may be assigned to the Executive or to the local authority; and an assignment under this paragraph may be made only by the Executive and the local authority jointly and only where they agree—
   (a) that there is uncertainty in the particular case as to what are their respective responsibilities by virtue of regulations made under section 18(2) of the 1974 Act; and
   (b) which authority is more appropriate to be responsible for enforcement in that case;
   and where such an assignment is made the authority to which responsibility has been assigned shall cause notice of assignment to be given to persons affected by it.

(2) For the purpose of removing uncertainty in any particular case as to what are their respective responsibilities by virtue of regulations made under section 18(2) of the 1974 Act either the Executive or the local authority may apply to the Commission and where the Commission considers that there is uncertainty it shall, after considering the circumstances and any views which may have been expressed to them by either enforcing authority or by persons affected, assign responsibility to whichever authority it considers appropriate; and where such an assignment is made the Commission shall cause notice of the assignment to be given to both enforcing authorities concerned and to persons affected by it.

7 Revocation of instruments

The instruments specified in column 1 of Schedule 3 are revoked to the extent specified in column 3 of that Schedule.

Signed by authority of the Secretary of State

Angela Eagle

Parliamentary Under Secretary of State,

Department of the Environment, Transport and the Regions

2nd March 1998

The Health and Safety (Enforcing Authority) Regulations 1998 (as amended)

SCHEDULE 1

Main Activities which Determine Whether Local Authorities will be Enforcing Authorities

Regulation 3(1)

1. The sale of goods, or the storage of goods for retail or wholesale distribution, except—
   (a) at container depots where the main activity is the storage of goods in the course of transit to or from dock premises, an airport or a railway;
   (b) where the main activity is the sale or storage for wholesale distribution of any substance or preparation dangerous for supply;
   (c) where the main activity is the sale or storage of water or sewage or their by-products or natural or town gas;
   and for the purposes of this paragraph where the main activity carried on in premises is the sale and fitting of motor car tyres, exhausts, windscreens or sunroofs the main activity shall be deemed to be the sale of goods.
2. The display or demonstration of goods at an exhibition for the purposes of offer or advertisement for sale.
3. Office activities.
4. Catering services.
5. The provision of permanent or temporary residential accommodation including the provision of a site for caravans or campers.
6. Consumer services provided in a shop except dry cleaning or radio and television repairs, and in this paragraph "consumer services" means services of a type ordinarily supplied to persons who receive them otherwise than in the course of a trade, business or other undertaking carried on by them (whether for profit or not).
7. Cleaning (wet or dry) in coin operated units in launderettes and similar premises.
8. The use of a bath, sauna or solarium, massaging, hair transplanting, skin piercing, manicuring or other cosmetic services and therapeutic treatments, except where they are carried out under the supervision or control of a registered medical practitioner, a dentist registered under the Dentists Act 1984, a physiotherapist, an osteopath or a chiropractor.
9. The practice or presentation of the arts, sports, games, entertainment or other cultural or recreational activities except where the main activity is the exhibition of a cave to the public.
10. The hiring out of pleasure craft for use on inland waters.
11. The care, treatment, accommodation or exhibition of animals, birds or other creatures, except where the main activity is horse breeding or horse training at a stable, or is an agricultural activity or veterinary surgery.
12. The activities of an undertaker, except where the main activity is embalming or the making of coffins.
13. Church worship or religious meetings.
14. The provision of car parking facilities within the perimeter of an airport.
15. The provision of child care, or playgroup or nursery facilities.

SCHEDULE 2

Activities in Respect of which the Health and Safety Executive is the Enforcing Authority

Regulation 4(4)(b)

1. Any activity in a mine or quarry other than a quarry in respect of which notice of abandonment has been given under [regulation 45(1) of the Quarries Regulations 1999].[14]
2. Any activity in a fairground.

---

[14] Amended by SI 1999/2024.

3 Any activity in premises occupied by a radio, television or film undertaking in which the activity of broadcasting, recording or filming is carried on, and the activity of broadcasting, recording or filming wherever carried on, and for this purpose "film" includes video.

4 The following activities carried on at any premises by persons who do not normally work in the premises—
   (a) construction work if—
      (i) regulation 7(1) of the Construction (Design and Management) Regulations 1994 (which requires projects which include or are intended to include construction work to be notified to the Executive) applies to the project which includes the work; or
      (ii) the whole or part of the work contracted to be undertaken by the contractor at the premises is to the external fabric or other external part of a building or structure; or
      (iii) it is carried out in a physically segregated area of the premises, the activities normally carried out in that area have been suspended for the purpose of enabling the construction work to be carried out, the contractor has authority to exclude from that area persons who are not attending in connection with the carrying out of the work and the work is not the maintenance of insulation on pipes, boilers or other parts of heating or water systems or its removal from them;
   (b) the installation, maintenance or repair of any gas system, or any work in relation to a gas fitting;
   (c) the installation, maintenance or repair of electricity systems;
   (d) work with ionising radiations except work in one or more of the categories set out in [Schedule 1 of the Ionising Radiations Regulations 1999 [SI 1999/3232]].[15]

[4A The reference in paragraph 4(a)(iii) to a physically segregated area does not include an area segregated only in order to prevent the escape of asbestos; and in this paragraph "asbestos" has the meaning assigned to it by regulation 2(1) of the Control of Asbestos at Work Regulations 2002.][16]

5 The use of ionising radiations for medical exposure (within the meaning of regulation 2(1) of [the Ionising Radiations Regulations 1999 [SI 1999/3232]][17]).

6 Any activity in premises occupied by a radiography undertaking in which there is carried on any work with ionising radiations.

7 Agricultural activities, and any activity at an agricultural show which involves the handling of livestock or the working of agricultural equipment.

8 Any activity on board a sea-going ship.

9 Any activity in relation to a ski slope, ski lift, ski tow or cable car.

10 Fish, maggot and game breeding except in a zoo.

11 Any activity in relation to a pipeline within the meaning of regulation 3 of the Pipelines Safety Regulations 1996.

[12 The operation of—
   (a) a guided bus system; or
   (b) any other system of guided transport, other than a railway, that employs vehicles which for some or all of the time when they are in operation travel along roads.][18]

[13 The operation of a trolley vehicle system.][19]

---

[15] Amended by SI 1999/3232.
[16] Inserted by SI 2002/2675.
[17] Amended by SI 1999/3232.
[18] Inserted by SI 2006/557.
[19] ibid.

APPENDIX D

The Employment Tribunals (Constitution and Rules of Procedure) Regulations 2004

2004 No. 1861

Made   *19th July 2004*
Laid before Parliament   *20th July 2004*
Coming into force   *1st October 2004*

The Secretary of State, in exercise of the powers conferred on her by section 24(2) of the Health and Safety at Work etc. Act 1974, sections 1(1), 4(6), and (6A), 7(1), (3), (3ZA), (3A), and (5), 7A(1) and (2), 9(1), (2) and (4), 10(2), (5), (6) and (7), 10A(1), 11(1), 12(2), 13, 13A(1) and (2), 19 and 41(4) of the Employment Tribunals Act 1996 and paragraph 36 of Schedule 8 to the Government of Wales Act 1998, and paragraph 37 of Schedule 6 to the Scotland Act 1998, and after consultation with the Council of Tribunals, in accordance with section 8(1) of the Tribunals and Inquiries Act 1992, hereby makes the following Regulations:

Citation, commencement and revocation

1.—[(1) These Regulations may be cited as the Employment Tribunals (Constitution and Rules of Procedure) Regulations 2004 and the Rules of Procedure contained in Schedules 1, 2, 3, 4 and 5 to these Regulations may be referred to, respectively, as—
   (a) the Employment Tribunals Rules of Procedure [2004];[1]
   (b) the Employment Tribunals (National Security) Rules of Procedure [2004];[2]
   (c) the Employment Tribunals (Levy Appeals) Rules of Procedure [2004];[3]
   (d) the Employment Tribunals (Health and Safety—Appeals against Improvement and Prohibition Notices) Rules of Procedure [2004]; and
   (e) the Employment Tribunals (Non-Discrimination Notices Appeals) Rules of Procedure][4] [2004].[5]
(2) These Regulations shall come into force on 1 October 2004.
(3) Subject to the savings in regulation 20, the Employment Tribunals (Constitution and Rules of Procedure) Regulations 2001 and the Employment Tribunals (Constitution and Rules of Procedure) (Scotland) Regulations 2001 are revoked.

Interpretation

2. [1][6] [In these Regulations and in Schedules 1, 2, 3, 4, 5 and 6][7]—
   "ACAS" means the Advisory, Conciliation and Arbitration Service referred to in section 247 of TULR(C)A;

---

[1] Amended by SI 2005/1865.
[2] ibid.
[3] ibid.
[4] Amended by SI 2004/2351.
[5] Amended by SI 2005/1865.
[6] Amended by SI 2006/680.
[7] Amended by SI 2004/2351.

"appointing office holder" means, in England and Wales, the Lord Chancellor, and in Scotland, the Lord President;

"chairman" means the President or a member of the panel of chairmen appointed in accordance with regulation 8(3)(a), or, for the purposes of national security proceedings, a member of the panel referred to in regulation 10 selected in accordance with regulation 11(a), and in relation to particular proceedings it means the chairman to whom the proceedings have been referred by the President, Vice President or a Regional Chairman;

"compromise agreement" means an agreement to refrain from continuing proceedings where the agreement meets the conditions in section 203(3) of the Employment Rights Act;

"constructive dismissal" has the meaning set out in section 95(1)(c) of the Employment Rights Act;

"Disability Discrimination Act" means the Disability Discrimination Act 1995;

"electronic communication" has the meaning given to it by section 15(1) of the Electronic Communications Act 2000;

"Employment Act" means the Employment Act 2002;

"Employment Rights Act" means the Employment Rights Act 1996;

"Employment Tribunals Act" means the Employment Tribunals Act 1996;

"Employment Tribunal Office" means any office which has been established for any area in either England & Wales or Scotland specified by the President and which carries out administrative functions in support of functions being carried out by a tribunal or chairman, and in relation to particular proceedings it is the office notified to the parties in accordance with rule 61(3) of Schedule 1;

"enactment" includes an enactment comprised in, or in an instrument made under, an Act of the Scottish Parliament;

"Equal Pay Act" means the Equal Pay Act 1970;

"excluded person" means, in relation to any proceedings, a person who has been excluded from all or part of the proceedings by virtue of—
  (a) a direction of a Minister of the Crown under rule 54(1)(b) or (c) of Schedule 1, or
  (b) an order of the tribunal under rule 54(2)(a) read with 54(1)(b) or(c) of Schedule 1;

"hearing" means a case management discussion, pre-hearing review, review hearing or Hearing (as those terms are defined in Schedule 1) or a sitting of a chairman or a tribunal duly constituted for the purpose of receiving evidence, hearing addresses and witnesses or doing anything lawful to enable the chairman or tribunal to reach a decision on any question;

"legally represented" has the meaning set out in rule 38(5) of Schedule 1;

"Lord President" means the Lord President of the Court of Session;

"misconceived" includes having no reasonable prospect of success;

"national security proceedings" means proceedings in relation to which a direction is given under rule 54(1) of Schedule 1, or an order is made under rule 54(2) of that Schedule;

"old (England & Wales) regulations" means the Employment Tribunals (Constitution and Rules of Procedure) [. . .][8] Regulations 2001;

"old (Scotland) regulations" means the Employment Tribunals (Constitution and Rules of Procedure) [(Scotland)][9] Regulations 2001;

"panel of chairmen" means a panel referred to in regulation 8(3)(a);

"President" means, in England and Wales, the person [appointed by the Lord Chancellor or nominated by the Lord Chief Justice][10] to discharge for the time being the functions of the President of Employment Tribunals (England and Wales), and, in Scotland, the person

---

[8] Repealed by SI 2005/1865.
[9] Amended by SI 2005/1865.
[10] Amended by SI 2006/680.

appointed or nominated by the Lord President to discharge for the time being the functions of the President of Employment Tribunals (Scotland);

"Race Relations Act" means the Race Relations Act 1976;

"Regional Chairman" means a member of the panel of chairmen who has been appointed to the position of Regional Chairman in accordance with regulation 6 or who has been nominated to discharge the functions of a Regional Chairman in accordance with regulation 6;

"Register" means the Register of judgments and written reasons kept in accordance with regulation 17;

"Secretary" means a person for the time being appointed to act as the Secretary of employment tribunals either in England and Wales or in Scotland;

"Sex Discrimination Act" means the Sex Discrimination Act 1975;

"special advocate" means a person appointed in accordance with rule 8 of Schedule 2;

"tribunal" means an employment tribunal established in accordance with regulation 5, and in relation to any proceedings means the tribunal to which the proceedings have been referred by the President, Vice President or a Regional Chairman;

"TULR(C)A" means the Trade Union and Labour Relations (Consolidation) Act 1992;

"Vice President" means a person who has been appointed to the position of Vice President in accordance with regulation 7 or who has been nominated to discharge the functions of the Vice President in accordance with that regulation;

"writing" includes writing delivered by means of electronic communication.

Overriding objective

3.—[(1) The overriding objective of these regulations and the rules in Schedules 1, 2, 3, 4, 5 and 6 is to enable tribunals and chairmen to deal with cases justly.][11]

(2) Dealing with a case justly includes, so far as practicable—
 (a) ensuring that the parties are on an equal footing;
 (b) dealing with the case in ways which are proportionate to the complexity or importance of the issues;
 (c) ensuring that it is dealt with expeditiously and fairly; and
 (d) saving expense.

[(3) A tribunal or chairman shall seek to give effect to the overriding objective when it or he:—
 (a) exercises any power given to it or him by these regulations or the rules in Schedules 1, 2, 3, 4, 5 and 6; or
 (b) interprets these regulations or any rule in Schedules 1, 2, 3, 4, 5 and 6.][12]

(4) The parties shall assist the tribunal or the chairman to further the overriding objective.

President of Employment Tribunals

4.—(1) There shall be a President of Employment Tribunals (England and Wales), responsible for the administration of justice by tribunals and chairmen in England and Wales, who shall be appointed by the Lord Chancellor and shall be a person described in paragraph (3).

(2) There shall be a President of Employment Tribunals (Scotland), responsible for the administration of justice by tribunals and chairmen in Scotland, who shall be appointed by the Lord President and shall be a person described in paragraph (3).

(3) A President shall be a person:—
 (a) having a seven year general qualification within the meaning of section 71 of the Courts and Legal Services Act 1990;
 (b) being an advocate or solicitor admitted in Scotland of at least seven years standing; or

---

[11] Amended by SI 2004/2351.
[12] ibid.

(c) being a member of the Bar of Northern Ireland or solicitor of the Supreme Court of Northern Ireland of at least seven years standing.

(4) A President may resign his office by notice in writing to the appointing office holder.

(5) If the appointing office holder is satisfied that the President is incapacitated by infirmity of mind or body from discharging the duties of his office, or the President is adjudged to be bankrupt or makes a composition or arrangement with his creditors, the appointing office holder may revoke his appointment.

[(5A) Where the Lord Chancellor is the appointing office holder, he may revoke an appointment in accordance with paragraph (5) only with the concurrence of the Lord Chief Justice.][13]

(6) The functions of President under these Regulations may, if he is for any reason unable to act or during any vacancy in his office, be discharged by a person nominated for that purpose by the appointing office holder [where that is the Lord President, or, where the appointing office holder is the Lord Chancellor, by the Lord Chief Justice after consulting the Lord Chancellor].[14]

[(7) The Lord Chief Justice may nominate a judicial office holder (as defined in section 109(4) of the Constitutional Reform Act 2005) to exercise his functions under this regulation.][15]

Establishment of employment tribunals

5.—(1) Each President shall, in relation to that part of Great Britain for which he has responsibility, from time to time determine the number of tribunals to be established for the purposes of determining proceedings.

(2) The President, a Regional Chairman or the Vice President shall determine, in relation to the area specified in relation to him, at what times and in what places in that area tribunals and chairmen shall sit.

Regional Chairmen

6.—(1) The Lord Chancellor may from time to time appoint Regional Chairmen from the panel of full-time chairmen and each Regional Chairman shall be responsible to the President (England and Wales) for the administration of justice by tribunals and chairmen in the area specified by the President (England and Wales) in relation to him.

(2) The President (England and Wales) or the Regional Chairman for an area may from time to time nominate a member of the panel of full time chairmen to discharge for the time being the functions of the Regional Chairman for that area.

Vice President

7.—(1) The Lord President may from time to time appoint a Vice President from the panel of full time chairmen and the Vice President shall be responsible to the President (Scotland) for the administration of justice by tribunals and chairmen in Scotland.

(2) The President (Scotland) or the Vice President may from time to time nominate a member of the panel of full time chairmen to discharge for the time being the functions of the Vice President.

Panels of members of tribunals—general

8.—(1) There shall be three panels of members of Employment Tribunals (England and Wales), as set out in paragraph (3).

(2) There shall be three panels of members of Employment Tribunals (Scotland), as set out in paragraph (3).

(3) The panels referred to in paragraphs (1) and (2) are—

---

[13] Inserted by SI 2006/680.
[14] ibid.
[15] ibid.

The Employment Tribunals (Constitution and Rules of Procedure) Regulations 2004

(a) a panel of full-time and part-time chairmen appointed by the appointing office holder consisting of persons—
  (i) having a seven year general qualification within the meaning of section 71 of the Courts and Legal Services Act 1990;
  (ii) being an advocate or solicitor admitted in Scotland of at least seven years standing; or
  (iii) being a member of the Bar of Northern Ireland or solicitor of the Supreme Court of Northern Ireland of at least seven years standing;
(b) a panel of persons appointed by the Secretary of State after consultation with such organisations or associations of organisations representative of employees as she sees fit; and
(c) a panel of persons appointed by the Secretary of State after consultation with such organisations or associations of organisations representative of employers as she sees fit.

(4) Members of the panels constituted under these Regulations shall hold and vacate office under the terms of the instrument under which they are appointed but may resign their office by notice in writing, in the case of a member of the panel of chairmen, to the appointing office holder and, in any other case, to the Secretary of State; and any such member who ceases to hold office shall be eligible for reappointment.

(5) The President may establish further specialist panels of chairmen and persons referred to in paragraphs (3)(b) and (c) and may select persons from such specialist panels in order to deal with proceedings in which particular specialist knowledge would be beneficial.

Composition of tribunals—general

9.—(1) For each hearing, the President, Vice President or the Regional Chairman shall select a chairman, who shall, subject to regulation 11, be a member of the panel of chairmen, and the President, Vice President or the Regional Chairman may select himself.

(2) In any proceedings which are to be determined by a tribunal comprising a chairman and two other members, the President, Regional Chairman or Vice President shall, subject to regulation 11, select one of those other members from the panel of persons appointed by the Secretary of State under regulation 8(3)(b) and the other from the panel of persons appointed under regulation 8(3)(c).

(3) In any proceedings which are to be determined by a tribunal whose composition is described in paragraph (2) or, as the case may be, regulation 11(b), those proceedings may, with the consent of the parties, be heard and determined in the absence of any one member other than the chairman.

(4) The President, Vice President, or a Regional Chairman may at any time select from the appropriate panel another person in substitution for the chairman or other member of the tribunal previously selected to hear any proceedings before a tribunal or chairman.

Panels of members of tribunals—national security proceedings

10. In relation to national security proceedings, the President shall:—
(a) select a panel of persons from the panel of chairmen to act as chairmen in such cases; and
(b) select—
  (i) a panel of persons from the panel referred to in regulation 8(3)(b) as persons suitable to act as members in such cases; and
  (ii) a panel of persons from the panel referred to in regulation 8(3)(c) as persons suitable to act as members in such cases.

Composition of tribunals—national security proceedings

11. In relation to national security proceedings:—
(a) the President, the Regional Chairman or the Vice President shall select a chairman, who shall be a member of the panel selected in accordance with regulation 10(a), and the President, Regional Chairman or Vice President may select himself; and

(b) in any such proceedings which are to be determined by a tribunal comprising a chairman and two other members, the President, Regional Chairman or Vice President shall select one of those other members from the panel selected in accordance with regulation 10(b)(i) and the other from the panel selected in accordance with regulation 10(b)(ii).

Modification of section 4 of the Employment Tribunals Act (national security proceedings)

12.—(1) For the purposes of national security proceedings section 4 of the Employment Tribunals Act shall be modified as follows.

(2) In section 4(1)(a), for the words "in accordance with regulations made under section 1(1)" substitute the words "in accordance with regulation 11(a) of the Employment Tribunals (Constitution and Rules of Procedure) Regulations 2004".

(3) In section 4(1)(b), for the words "in accordance with regulations so made" substitute the words "in accordance with regulation 11(b) of those Regulations".

(4) In section 4(5), for the words "in accordance with Regulations made under section 1(1)" substitute the words "in accordance with regulation 10(a) of the Employment Tribunals (Constitution and Rules of Procedure) Regulations 2004".

Practice directions

13.—(1) The President may make practice directions about the procedure of employment tribunals in the area for which he is responsible, including practice directions about the exercise by tribunals or chairmen of powers under these Regulations or the Schedules to them.

(2) The power of the President to make practice directions under paragraph (1) includes power—
  (a) to vary or revoke practice directions;
  (b) to make different provision for different cases or different areas, including different provision for specific types of proceedings.

(3) The President shall publish a practice direction made under paragraph (1), and any revocation or variation of it, in such manner as he considers appropriate for bringing it to the attention of the persons to whom it is addressed.

Power to prescribe

14.—(1) The Secretary of State may prescribe—
  (a) one or more versions of a form, one of which shall be used by all claimants for the purpose of commencing proceedings in an employment tribunal ("claim form") except any claim or proceedings listed in paragraph (3);
  (b) one or more versions of a form, one of which shall be used by all respondents to a claim for the purpose of responding to a claim before an employment tribunal ("response form") except respondents to a claim or proceedings listed in paragraph (3); and
  (c) that the provision of certain information and answering of certain questions in a claim form or in a response form is mandatory in all proceedings save those listed in paragraph (3).

(2) The Secretary of State shall publish the forms and matters prescribed pursuant to paragraph (1) in such manner as she considers appropriate in order to bring them to the attention of potential claimants, respondents and their advisers.

(3) The proceedings referred to in paragraph (1) are:—
  (a) those referred to an employment tribunal by a court;
  (b) proceedings to which any of Schedules 3 to 5 apply; or
  (c) proceedings brought under any of the following enactments—
    (i) sections 19, 20 or 22 of the National Minimum Wage Act 1998;
    (ii) section 11 of the Employment Rights Act where the proceedings are brought by the employer.

Calculation of time limits

15.—[(1) Any period of time for doing any act required or permitted to be done under any of the rules in Schedules 1, 2, 3, 4, 5 and 6, or under any decision, order or judgment of a tribunal or a chairman, shall be calculated in accordance with paragraphs (2) to (6).][16]

(2) Where any act must or may be done within a certain number of days of or from an event, the date of that event shall not be included in the calculation. For example, a respondent is sent a copy of a claim on 1st October. He must present a response to the Employment Tribunal Office within 28 days of the date on which he was sent the copy. The last day for presentation of the response is 29th October.

(3) Where any act must or may be done not less than a certain number of days before or after an event, the date of that event shall not be included in the calculation. For example, if a party wishes to submit representations in writing for consideration by a tribunal at a hearing, he must submit them not less than 7 days before the hearing. If the hearing is fixed for 8th October, the representations must be submitted no later than 1st October.

(4) Where the tribunal or a chairman gives any decision, order or judgment which imposes a time limit for doing any act, the last date for compliance shall, wherever practicable, be expressed as a calendar date.

(5) In rule 14(4) of Schedule 1 the requirement to send the notice of hearing to the parties not less than 14 days before the date fixed for the hearing shall not be construed as a requirement for service of the notice to have been effected not less than 14 days before the hearing date, but as a requirement for the notice to have been placed in the post not less than 14 days before that date. For example, a hearing is fixed for 15th October. The last day on which the notice may be placed in the post is 1st October.

(6) Where any act must or may have been done within a certain number of days of a document being sent to a person by the Secretary, the date when the document was sent shall, unless the contrary is proved, be regarded as the date on the letter from the Secretary which accompanied the document. For example, a respondent must present his response to a claim to the Employment Tribunal Office [within 28 days on the date of which][17] he was sent a copy of the claim. If the letter from the Secretary sending him a copy of the claim is dated 1st October, the last day for presentation of the response is 29th October.

Application of Schedules 1–5 to proceedings

16.—(1) [Subject to paragraphs (2), (3) and (4)],[18] the rules in Schedule 1 shall apply in relation to all proceedings before an employment tribunal except where separate rules of procedure made under the provisions of any enactment are applicable.

(2) In proceedings to which the rules in Schedule 1 apply and in which any power conferred on the Minister, the tribunal or a chairman by rule 54 (national security proceedings) of Schedule 1 is exercised, Schedule 1 shall be modified in accordance with Schedule 2.

(3) The rules in Schedules 3, 4 and 5 shall apply to modify the rules in Schedule 1 in relation to proceedings which consist, respectively, in:—
  (a) an appeal by a person assessed to levy imposed under a levy order made under section 12 of the Industrial Training Act 1982;
  (b) (b) an appeal against an improvement or prohibition notice under section 24 of the Health and Safety at Work etc. Act 1974; and

---

[16] Amended by SI 2004/2351.
[17] ibid.
[18] ibid.

(c) (c) an appeal against a non-discrimination notice under section 68 of the Sex Discrimination Act, section 59 of the Race Relations Act or paragraph 10 of Schedule 3 to the Disability Rights Commission Act 1999.

[(4) In proceedings which involve an equal value claim (as defined in rule 2 of schedule 6), schedule 1 shall be modified in accordance with schedule 6.][19]

Register

17.—(1) The Secretary shall maintain a Register which shall be open to the inspection of any person without charge at all reasonable hours.

(2) The Register shall contain a copy of all judgments and any written reasons issued by any tribunal or chairman which are required to be entered in the Register in accordance with the rules in Schedules 1 to 5.

(3) The Register, or any part of it, may be kept by means of a computer.

Proof of decisions of tribunals

18. The production in any proceedings in any court of a document purporting to be certified by the Secretary to be a true copy of an entry of a judgment in the Register shall, unless the contrary is proved, be sufficient evidence of the document and of the facts stated therein.

Jurisdiction of tribunals in Scotland and in England & Wales

19.—(1) An employment tribunal in England or Wales shall only have jurisdiction to deal with proceedings (referred to as "English and Welsh proceedings") where—
  (a) the respondent or one of the respondents resides or carries on business in England and Wales;
  (b) had the remedy been by way of action in the county court, the cause of action would have arisen wholly or partly in England and Wales;
  (c) the proceedings are to determine a question which has been referred to the tribunal by a court in England and Wales; or
  (d) in the case of proceedings to which Schedule 3, 4 or 5 applies, the proceedings relate to matters arising in England and Wales.

(2) An employment tribunal in Scotland shall only have jurisdiction to deal with proceedings (referred to as "Scottish proceedings") where—
  (a) the respondent or one of the respondents resides or carries on business in Scotland;
  (b) the proceedings relate to a contract of employment the place of execution or performance of which is in Scotland;
  (c) the proceedings are to determine a question which has been referred to the tribunal by a sheriff in Scotland; or
  (d) in the case of proceedings to which Schedule 3, 4 or 5 applies, the proceedings relate to matters arising in Scotland.

Transitional provisions

20.—(1) [These Regulations and Schedules 1 to 6][20] to them shall apply in relation to all proceedings to which they relate where those proceedings were commenced on or after 1 October 2004.

(2) These Regulations and Schedules 1 and 2 to them (with the exception of rules 1 to 3 and 38 to 48 of Schedule 1) shall apply to proceedings:—
  (a) which were commenced prior to 1 October 2004; and
  (b) to which Schedule 1 to either the old (England & Wales) regulations or the old (Scotland) regulations applied;

---

[19] Inserted by SI 2004/2351.
[20] Amended by SI 2004/2351.

The Employment Tribunals (Constitution and Rules of Procedure) Regulations 2004

provided that a copy of the originating application was not sent to the respondent prior to 1 October 2004.

(3) In relation to the proceedings described in paragraph (2), the following provisions of Schedule 1 to the old (England & Wales) regulations or the old (Scotland) regulations (as the case may be) shall continue to apply—
  (a) rule 1 (originating application);
  (b) rule 2 (action upon receipt of originating application) with the exception of paragraphs (2), (4) and (5) of that rule; and
  (c) rule 14 (costs).

(4) In relation to proceedings described in paragraph (2) but where a copy of the originating application was sent to the respondent prior to 1 October 2004, Schedules 1 and 2 to these Regulations shall apply with the exception of rules 1 to 9, 21 to 24, 33 and 38 to 48 of Schedule 1 and rules 2, 3 and 4 of Schedule 2.

(5) In relation to proceedings described in paragraph (4), the following provisions of the old (England & Wales) regulations or the old (Scotland) regulations (as the case may be) shall continue to apply:—
  (a) in Schedule 1—
    (i) rule 1 (originating application);
    (ii) rule 2 (action upon receipt of originating application) with the exception of paragraphs (2), (4) and (5) of that rule;
    (iii) rule 3 (appearance by respondent);
    (iv) rule 8 (national security);
    (v) rule 14 (costs); and
  (b) rule 1 of Schedule 2.

(6) In relation to proceedings commenced prior to 1 October 2004 and to which Schedule 4, 5 or 6 to the old (England & Wales) regulations or the old (Scotland) regulations (as the case may be) applied, the provisions of those schedules shall continue to apply to such proceedings.

[(7) In relation to proceedings:—
  (i) which were commenced prior to 1 October 2004;
  (ii) to which Schedule 3 to either the old (England & Wales) regulations or the old (Scotland) regulations applied; and
  (iii) in which the tribunal has not, prior to 1 October 2004, required a member of the panel of independent experts to prepare a report under section 2A(1)(b) of the Equal Pay Act;
  these Regulations and rules 1 to 13 of Schedule 6, with the exception of rule 4(3)(a), shall apply.

(8) In relation to proceedings:—
  (i) which were commenced prior to 1 October 2004;
  (ii) to which Schedule 3 to either the old (England & Wales) regulations or the old (Scotland) regulations applied; and
  (iii) in which the tribunal has, prior to 1 October 2004, required a member of the panel of independent experts to prepare a report under section 2A(1)(b) of the Equal Pay Act;
  Schedule 3 to either the old (England & Wales) regulations or the old (Scotland) regulations (as the case may be) shall continue to apply.

(9) In relation to proceedings described in paragraph (8), the following rules of Schedule 6 shall also apply and shall take precedence over any conflicting provision in Schedule 3 to either the old (England & Wales) regulations or the old (Scotland) regulations, namely rules 3, 11(2), 11(4), 12, 13(1) and 13(3).

(10) Rule 14 of Schedule 6 shall apply to all proceedings to which, in accordance with this regulation, rule 10 of Schedule 2 applies.]²¹

Gerry Sutcliffe

Parliamentary Under Secretary of State for Employment Relations, Competition and Consumers, Department of Trade and Industry

19th July 2004

SCHEDULE 1

Regulation 16

THE EMPLOYMENT TRIBUNALS RULES OF PROCEDURE
How to Bring a Claim

Starting a claim

1.—(1) A claim shall be brought before an employment tribunal by the claimant presenting to an Employment Tribunal Office the details of the claim in writing. Those details must include all the relevant required information (subject to paragraph (5) of this rule and to rule 53 (Employment Agencies Act 1973)).

(2) The claim may only be presented to an Employment Tribunal Office in England and Wales if it relates to English and Welsh proceedings (defined in regulation 19(1)). The claim may only be presented to an Employment Tribunal Office in Scotland if it relates to Scottish proceedings (defined in regulation 19(2)).

(3) Unless it is a claim in proceedings described in regulation 14(3), a claim which is presented on or after [1st October 2005]²² must be presented on a claim form which has been prescribed by the Secretary of State in accordance with regulation 14.

(4) Subject to paragraph (5) and to rule 53, the required information in relation to the claim is—
  (a) each claimant's name;
  (b) each claimant's address;
  (c) the name of each person against whom the claim is made ("the respondent");
  (d) each respondent's address;
  (e) details of the claim;
  (f) whether or not the claimant is or was an employee of the respondent;
  (g) whether or not the claim includes a complaint that the respondent has dismissed the claimant or has contemplated doing so;
  (h) whether or not the claimant has raised the subject matter of the claim with the respondent in writing at least 28 days prior to presenting the claim to an Employment Tribunal Office;
  (i) if the claimant has not done as described in (h), why he has not done so.

(5) In the following circumstances the required information identified below is not required to be provided in relation to that claim—
  (a) if the claimant is not or was not an employee of the respondent, the information in paragraphs (4)(g) to (i) is not required;
  (b) if the claimant was an employee of the respondent and the claim consists only of a complaint that the respondent has dismissed the claimant or has contemplated doing so, the information in paragraphs (4)(h) and (i) is not required;
  (c) if the claimant was an employee of the respondent and the claim does not relate to the claimant being dismissed or a contemplated dismissal by the respondent, and the claimant

---

[21] Inserted by SI 2004/2351.
[22] Amended by SI 2005/435.

has raised the subject matter of the claim with the respondent as described in paragraph (4)(h), the information in paragraph (4)(i) is not required.

(6) References in this rule to being dismissed or a dismissal by the respondent do not include references to constructive dismissal.

(7) Two or more claimants may present their claims in the same document if their claims arise out of the same set of facts.

(8) When section 32 of the Employment Act applies to the claim or part of one and a chairman considers in accordance with subsection (6) of section 32 that there has been a breach of subsections (2) to (4) of that section, neither a chairman nor a tribunal shall consider the substance of the claim (or the relevant part of it) until such time as those subsections have been complied with in relation to the claim or the relevant part of it.

Acceptance of Claim Procedure

What the tribunal does after receiving the claim

2.—(1) On receiving the claim the Secretary shall consider whether the claim or part of it should be accepted in accordance with rule 3. If a claim or part of one is not accepted the tribunal shall not proceed to deal with any part which has not been accepted (unless it is accepted at a later date). If no part of a claim is accepted the claim shall not be copied to the respondent.

(2) If the Secretary accepts the claim or part of it, he shall—
  (a) send a copy of the claim to each respondent and record in writing the date on which it was sent;
  (b) inform the parties in writing of the case number of the claim (which must from then on be referred to in all correspondence relating to the claim) and the address to which notices and other communications to the Employment Tribunal Office must be sent;
  (c) inform the respondent in writing about how to present a response to the claim, the time limit for doing so, what may happen if a response is not entered within the time limit and that the respondent has a right to receive a copy of any judgment disposing of the claim;
  (d) when any enactment relevant to the claim provides for conciliation, notify the parties that the services of a conciliation officer are available to them;
  (e) when rule 22 (fixed period for conciliation) applies, notify the parties of the date on which the conciliation officer's duty to conciliate ends and that after that date the services of a conciliation officer shall be available to them only in limited circumstances; and
  (f) if only part of the claim has been accepted, inform the claimant and any respondent which parts of the claim have not been accepted and that the tribunal shall not proceed to deal with those parts unless they are accepted at a later date.

When the claim will not be accepted by the Secretary

3.—(1) When a claim is required by rule 1(3) to be presented using a prescribed form, but the prescribed form has not been used, the Secretary shall not accept the claim and shall return it to the claimant with an explanation of why the claim has been rejected and provide a prescribed claim form.

(2) The Secretary shall not accept the claim (or a relevant part of one) if it is clear to him that one or more of the following circumstances applies—
  (a) the claim does not include all the relevant required information;
  (b) the tribunal does not have power to consider the claim (or that relevant part of it); or
  (c) section 32 of the Employment Act (complaints about grievances) applies to the claim or part of it and the claim has been presented to the tribunal in breach of subsections (2) to (4) of section 32.

(3) If the Secretary decides not to accept a claim or part of one for any of the reasons in paragraph (2), he shall refer the claim together with a statement of his reasons for not accepting it to a

chairman. The chairman shall decide in accordance with the criteria in paragraph (2) whether the claim or part of it should be accepted and allowed to proceed.

(4) If the chairman decides that the claim or part of one should be accepted he shall inform the Secretary in writing and the Secretary shall accept the relevant part of the claim and then proceed to deal with it in accordance with rule 2(2).

(5) If the chairman decides that the claim or part of it should not be accepted he shall record his decision together with the reasons for it in writing in a document signed by him. The Secretary shall as soon as is reasonably practicable inform the claimant of that decision and the reasons for it in writing together with information on how that decision may be reviewed or appealed.

(6) Where a claim or part of one has been presented to the tribunal in breach of subsections (2) to (4) of section 32 of the Employment Act, the Secretary shall notify the claimant of the time limit which applies to the claim or the part of it concerned and shall inform the claimant of the consequences of not complying with section 32 of that Act.

(7) Except for the purposes of paragraph (6) and (8) or any appeal to the Employment Appeal Tribunal, where a chairman has decided that a claim or part of one should not be accepted such a claim (or the relevant part of it) is to be treated as if it had not been received by the Secretary on that occasion.

(8) Any decision by a chairman not to accept a claim or part of one may be reviewed in accordance with rules 34 to 36. If the result of such review is that any parts of the claim should have been accepted, then paragraph (7) shall not apply to the relevant parts of that claim and the Secretary shall then accept such parts and proceed to deal with it as described in rule 2(2).

(9) A decision to accept or not to accept a claim or part of one shall not bind any future tribunal or chairman where any of the issues listed in paragraph (2) fall to be determined later in the proceedings.

(10) Except in rule 34 (review of other judgments and decisions), all references to a claim in the remainder of these rules are to be read as references to only the part of the claim which has been accepted.

Response

Responding to the claim

4.—(1) If the respondent wishes to respond to the claim made against him he must present his response to the Employment Tribunal Office within 28 days of the date on which he was sent a copy of the claim. The response must include all the relevant required information. The time limit for the respondent to present his response may be extended in accordance with paragraph (4).

(2) Unless it is a response in proceedings described in regulation 14(3), any response presented on or after [1st October 2005][23] must be on a response form prescribed by the Secretary of State pursuant to regulation 14.

(3) The required information in relation to the response is—
   (a) the respondent's full name;
   (b) the respondent's address;
   (c) whether or not the respondent wishes to resist the claim in whole or in part; and
   (d) if the respondent wishes to so resist, on what grounds.

(4) The respondent may apply under rule 11 for an extension of the time limit within which he is to present his response. The application must be presented to the Employment Tribunal Office within 28 days of the date on which the respondent was sent a copy of the claim (unless the

---

[23] Amended by SI 2005/435.

application is made under rule 33(1)) and must explain why the respondent cannot comply with the time limit. Subject to rule 33, the chairman shall only extend the time within which a response may be presented if he is satisfied that it is just and equitable to do so.
(5) A single document may include the response to more than one claim if the relief claimed arises out of the same set of facts, provided that in respect of each of the claims to which the single response relates—
   (a) the respondent intends to resist all the claims and the grounds for doing so are the same in relation to each claim; or
   (b) the respondent does not intend to resist any of the claims.
(6) A single document may include the response of more than one respondent to a single claim provided that—
   (a) each respondent intends to resist the claim and the grounds for doing so are the same for each respondent; or
   (b) none of the respondents intends to resist the claim.

ACCEPTANCE OF RESPONSE PROCEDURE

What the tribunal does after receiving the response

5.—(1) On receiving the response the Secretary shall consider whether the response should be accepted in accordance with rule 6. If the response is not accepted it shall be returned to the respondent and (subject to paragraphs (5) and (6) of rule 6) the claim shall be dealt with as if no response to the claim had been presented.
(2) If the Secretary accepts the response he shall send a copy of it to all other parties and record in writing the date on which he does so.

When the response will not be accepted by the Secretary

6.—(1) Where a response is required to be presented using a prescribed form by rule 4(2), but the prescribed form has not been used, the Secretary shall not accept the response and shall return it to the respondent with an explanation of why the response has been rejected and provide a prescribed response form.
(2) The Secretary shall not accept the response if it is clear to him that any of the following circumstances apply—
   (a) the response does not include all the required information (defined in rule 4(3));
   (b) the response has not been presented within the relevant time limit.
(3) If the Secretary decides not to accept a response for either of the reasons in paragraph (2), he shall refer the response together with a statement of his reasons for not accepting the response to a chairman. The chairman shall decide in accordance with the criteria in paragraph (2) whether the response should be accepted.
(4) If the chairman decides that the response should be accepted he shall inform the Secretary in writing and the Secretary shall accept the response and then deal with it in accordance with rule 5(2).
(5) If the chairman decides that the response should not be accepted he shall record his decision together with the reasons for it in writing in a document signed by him. The Secretary shall inform both the claimant and the respondent of that decision and the reasons for it. The Secretary shall also inform the respondent of the consequences for the respondent of that decision and how it may be reviewed or appealed.
(6) Any decision by a chairman not to accept a response may be reviewed in accordance with rules 34 to 36. If the result of such a review is that the response should have been accepted, then the Secretary shall accept the response and proceed to deal with the response as described in rule 5(2).

Counterclaims

7.—(1) When a respondent wishes to present a claim against the claimant ("a counterclaim") in accordance with article 4 of the Employment Tribunals Extension of Jurisdiction (England and Wales) Order 1994, or as the case may be, article 4 of the Employment Tribunals Extension of Jurisdiction (Scotland) Order 1994, he must present the details of his counterclaim to the Employment Tribunal Office in writing. Those details must include—
   (a) the respondent's name;
   (b) the respondent's address;
   (c) the name of each claimant whom the counterclaim is made against;
   (d) the claimant's address;
   (e) details of the counterclaim.
(2) A chairman may in relation to particular proceedings by order made under rule 10(1) establish the procedure which shall be followed by the respondent making the counterclaim and any claimant responding to the counterclaim.
(3) The President may by a practice direction made under regulation 13 make provision for the procedure which is to apply to counterclaims generally.

Consequences of a Response not Being Presented or Accepted

Default judgments

8.—(1) In any proceedings if the relevant time limit for presenting a response has passed, a chairman may, in the circumstances listed in paragraph (2), issue a default judgment to determine the claim without a hearing if he considers it appropriate to do so.
[(2) Those circumstances are when either—
   (a) no response in those proceedings has been presented to the Employment Tribunal Office within the relevant time limit; or
   (b)
   (c) a response has been so presented, but a decision has been made not to accept the response either by the Secretary under rule 6(1) or by a chairman under rule 6(3), and the Employment Tribunal Office has not received an application under rule 34 to have that decision reviewed; or
   (d) a response has been accepted in those proceedings, but the respondent has stated in the response that he does not intend to resist the claim][24]
and the claimant has not informed the Employment Tribunal Office in writing either that he does not wish a default judgment to be issued or that the claim has been settled.
(3) A default judgment may determine liability only or it may determine liability and remedy. If a default judgment determines remedy it shall be such remedy as it appears to the chairman that the claimant is entitled to on the basis of the information before him.
(4) Any default judgment issued by a chairman under this rule shall be recorded in writing and shall be signed by him. The Secretary shall send a copy of that judgment to the parties, to ACAS, and, if the proceedings were referred to the tribunal by a court, to that court. The Secretary shall also inform the parties of their right to have the default judgment reviewed under rule 33. The Secretary shall put a copy of the default judgment on the Register (subject to rule 49 (sexual offences and the Register)).
(5) The claimant or respondent may apply to have the default judgment reviewed in accordance with rule 33.

---

[24] Amended by SI 2004/2351.

(6) If the parties settle the proceedings (either by means of a compromise agreement (as defined in rule 23(2)) or through ACAS) before or on the date on which a default judgment in those proceedings is issued, the default judgment shall have no effect.

(7) When paragraph (6) applies, either party may apply under rule 33 to have the default judgment revoked.

Taking no further part in the proceedings

9. A respondent who has not presented a response to a claim or whose response has not been accepted shall not be entitled to take any part in the proceedings except to—
   (a) make an application under rule 33 (review of default judgments);
   (b) make an application under rule 35 (preliminary consideration of application for review) in respect of [rule 34(3)(a) and (b) or (e)];[25]
   (c) be called as a witness by another person; or
   (d) be sent a copy of a document or corrected entry in accordance with rule 8(4), 29(2) or 37;
and in these rules the word "party" or "respondent" includes a respondent only in relation to his entitlement to take such a part in the proceedings, and in relation to any such part which he takes.

CASE MANAGEMENT

General power to manage proceedings

10.—(1) Subject to the following rules, the chairman may at any time either on the application of a party or on his own initiative make an order in relation to any matter which appears to him to be appropriate. Such orders may be any of those listed in paragraph (2) or such other orders as he thinks fit. Subject to the following rules, orders may be issued as a result of a chairman considering the papers before him in the absence of the parties, or at a hearing (see regulation 2 for the definition of "hearing").

(2) Examples of orders which may be made under paragraph (1) are orders—
   (a) as to the manner in which the proceedings are to be conducted, including any time limit to be observed;
   (b) that a party provide additional information;
   (c) requiring the attendance of any person in Great Britain either to give evidence or to produce documents or information;
   (d) requiring any person in Great Britain to disclose documents or information to a party to allow a party to inspect such material as might be ordered by a County Court (or in Scotland, by a sheriff);
   (e) extending any time limit, whether or not expired (subject to rules 4(4), 11(2), 25(5), 30(5), 33(1), 35(1), 38(7) and 42(5) of this Schedule, and to rule 3(4) of Schedule 2);
   (f) requiring the provision of written answers to questions put by the tribunal or chairman;
   (g) that, subject to rule 22(8), a short conciliation period be extended into a standard conciliation period;
   (h) staying (in Scotland, sisting) the whole or part of any proceedings;
   (i) that part of the proceedings be dealt with separately;
   (j) that different claims be considered together;
   (k) that any person who the chairman or tribunal considers may be liable for the remedy claimed should be made a respondent in the proceedings;
   (l) dismissing the claim against a respondent who is no longer directly interested in the claim;
   (m) postponing or adjourning any hearing;

---

[25] Amended by SI 2005/1865.

(n) varying or revoking other orders;
(o) giving notice to the parties of a pre-hearing review or the Hearing;
(p) giving notice under rule 19;
(q) giving leave to amend a claim or response;
(r) that any person who the chairman or tribunal considers has an interest in the outcome of the proceedings may be joined as a party to the proceedings;
(s) that a witness statement be prepared or exchanged; or
(t) as to the use of experts or interpreters in the proceedings.

(3) An order may specify the time at or within which and the place at which any act is required to be done. An order may also impose conditions and it shall inform the parties of the potential consequences of non-compliance set out in rule 13.

(4) When a requirement has been imposed under paragraph (1) the person subject to the requirement may make an application under rule 11 (applications in proceedings) for the order to be varied or revoked.

(5) An order described in [. . .][26] paragraph (2)(d) which requires a person other than a party to grant disclosure or inspection of material may be made only when the disclosure sought is necessary in order to dispose fairly of the claim or to save expense.

(6) Any order containing a requirement described in either sub-paragraph (2)(c) or (d) shall state that under section 7(4) of the Employment Tribunals Act, any person who without reasonable excuse fails to comply with the requirement shall be liable on summary conviction to a fine, and the document shall also state the amount of the maximum fine.

(7) An order as described in paragraph (2)(j) may be made only if all relevant parties have been given notice that such an order may be made and they have been given the opportunity to make oral or written representations as to why such an order should or should not be made.

(8) Any order made under this rule shall be recorded in writing and signed by the chairman and the Secretary shall inform all parties to the proceedings of any order made as soon as is reasonably practicable.

Applications in proceedings

11.—(1) At any stage of the proceedings a party may apply for an order to be issued, varied or revoked or for a case management discussion or pre-hearing review to be held.

(2) An application for an order must be made not less than 10 days before the date of the hearing at which it is to be considered (if any) unless it is not reasonably practicable to do so, or the chairman or tribunal considers it in the interests of justice that shorter notice be allowed. The application must (unless a chairman orders otherwise) be in writing to the Employment Tribunal Office and include the case number for the proceedings and the reasons for the request. If the application is for a case management discussion or a pre-hearing review to be held, it must identify any orders sought.

(3) An application for an order must include an explanation of how the order would assist the tribunal or chairman in dealing with the proceedings efficiently and fairly.

(4) When a party is legally represented in relation to the application (except where the application is for a witness order described in rule 10(2)(c) only), that party or his representative must, at the same time as the application is sent to the Employment Tribunal Office, provide all other parties with the following information in writing—
  (a) details of the application and the reasons why it is sought;
  (b) notification that any objection to the application must be sent to the Employment Tribunal Office within 7 days of receiving the application, or before the date of the hearing (whichever date is the earlier);

---

[26] Repealed by SI 2005/1865.

(c) that any objection to the application must be copied to both the Employment Tribunal Office and all other parties;

and the party or his representative must confirm in writing to the Employment Tribunal Office that this rule has been complied with.

(5) Where a party is not legally represented in relation to the application, the Secretary shall inform all other parties of the matters listed in paragraphs (4)(a) to (c).

(6) A chairman may refuse a party's application and if he does so the Secretary shall inform the parties in writing of such refusal unless the application is refused at a hearing.

Chairman acting on his own initiative

12.—(1) Subject to paragraph (2) and to rules 10(7) and 18(7), a chairman may make an order on his own initiative with or without hearing the parties or giving them an opportunity to make written or oral representations. He may also decide to hold a case management discussion or pre-hearing review on his own initiative.

(2) Where a chairman makes an order without giving the parties the opportunity to make representations—
   (a) the Secretary must send to the party affected by such order a copy of the order and a statement explaining the right to make an application under paragraph (2)(b); and
   (b) a party affected by the order may apply to have it varied or revoked.

(3) An application under paragraph (2)(b) must (subject to rule 10(2)(e)) be made before the time at which, or the expiry of the period within which, the order was to be complied with. Such an application must (unless a chairman orders otherwise) be made in writing to an Employment Tribunal Office and it must include the reasons for the application. Paragraphs (4) and (5) of rule 11 apply in relation to informing the other parties of the application.

Compliance with orders and practice directions

13.—(1) If a party does not comply with an order made under these rules, under rule 8 of Schedule 3, rule 7 of Schedule 4 or a practice direction, a chairman or tribunal—
   (a) may make an order in respect of costs or preparation time under rules 38 to 46; or
   (b) may (subject to paragraph (2) and rule 19) at a pre-hearing review or a Hearing make an order to strike out the whole or part of the claim or, as the case may be, the response and, where appropriate, order that a respondent be debarred from responding to the claim altogether.

(2) An order may also provide that unless the order is complied with, the claim or, as the case may be, the response shall be struck out on the date of non-compliance without further consideration of the proceedings or the need to give notice under rule 19 or hold a pre-hearing review or Hearing.

(3) Chairmen and tribunals shall comply with any practice directions issued under regulation 13.

Different Types of Hearing

Hearings—general

14.—(1) A chairman or a tribunal (depending on the relevant rule) may hold the following types of hearing—
   (a) a case management discussion under rule 17;
   (b) a pre-hearing review under rule 18;
   (c) a Hearing under rule 26; or
   (d) a review hearing under rule 33 or 36.

(2) So far as it appears appropriate to do so, the chairman or tribunal shall seek to avoid formality in his or its proceedings and shall not be bound by any enactment or rule of law relating to the admissibility of evidence in proceedings before the courts.

(3) The chairman or tribunal (as the case may be) shall make such enquiries of persons appearing before him or it and of witnesses as he or it considers appropriate and shall otherwise conduct

the hearing in such manner as he or it considers most appropriate for the clarification of the issues and generally for the just handling of the proceedings.

(4) Unless the parties agree to shorter notice, the Secretary shall send notice of any hearing (other than a case management discussion) to every party not less than 14 days before the date fixed for the hearing and shall inform them that they have the opportunity to submit written representations and to advance oral argument. The Secretary shall give the parties reasonable notice before a case management discussion is held.

(5) If a party wishes to submit written representations for consideration at a hearing (other than a case management discussion) he shall present them to the Employment Tribunal Office not less than 7 days before the hearing and shall at the same time send a copy to all other parties.

(6) The tribunal or chairman may, if it or he considers it appropriate, consider representations in writing which have been submitted otherwise than in accordance with paragraph (5).

Use of electronic communications

15.—(1) A hearing (other than those mentioned in sub-paragraphs (c) and (d) of rule 14(1)) may be conducted by use of electronic communications provided that the chairman or tribunal conducting the hearing considers it just and equitable to do so.

(2) Where a hearing is required by these rules to be held in public and it is to be conducted by use of electronic communications in accordance with this rule then, subject to rule 16, it must be held in a place to which the public has access and using equipment so that the public is able to hear all parties to the communication.

Hearings which may be held in private

16.—(1) A hearing or part of one may be conducted in private for the purpose of hearing from any person evidence or representations which in the opinion of the tribunal or chairman is likely to consist of information—
  (a) which he could not disclose without contravening a prohibition imposed by or by virtue of any enactment;
  (b) which has been communicated to him in confidence, or which he has otherwise obtained in consequence of the confidence placed in him by another person; or
  (c) the disclosure of which would, for reasons other than its effect on negotiations with respect to any of the matters mentioned in section 178(2) of TULR(C)A, cause substantial injury to any undertaking of his or any undertaking in which he works.

(2) Where a tribunal or chairman decides to hold a hearing or part of one in private, it or he shall give reasons for doing so. A member of the Council on Tribunals (in Scotland, a member of the Council on Tribunals or its Scottish Committee) shall be entitled to attend any Hearing or pre-hearing review taking place in private in his capacity as a member.

CASE MANAGEMENT DISCUSSIONS

Conduct of case management discussions

17.—(1) Case management discussions are interim hearings and may deal with matters of procedure and management of the proceedings and they [shall be held in private].[27] Case management discussions shall be conducted by a chairman.

(2) Any determination of a person's civil rights or obligations shall not be dealt with in a case management discussion. The matters listed in rule 10(2) are examples of matters which may be dealt with at case management discussions. Orders and judgments listed in rule 18(7) may not be made at a case management discussion.

---

[27] Amended by SI 2005/1865.

Pre-Hearing Reviews

Conduct of pre-hearing reviews

18.—(1) Pre-hearing reviews are interim hearings and shall be conducted by a chairman unless the circumstances in paragraph (3) are applicable. Subject to rule 16, they shall take place in public.

(2) At a pre-hearing review the chairman may carry out a preliminary consideration of the proceedings and he may—
   (a) determine any interim or preliminary matter relating to the proceedings;
   (b) issue any order in accordance with rule 10 or do anything else which may be done at a case management discussion;
   (c) order that a deposit be paid in accordance with rule 20 without hearing evidence;
   (d) consider any oral or written representations or evidence;
   (e) deal with an application for interim relief made under section 161 of TULR(C)A or section 128 of the Employment Rights Act.

(3) Pre-hearing reviews shall be conducted by a tribunal composed in accordance with section 4(1) and (2) of the Employment Tribunals Act if—
   (a) a party has made a request in writing not less than 10 days before the date on which the pre-hearing review is due to take place that the pre-hearing review be conducted by a tribunal instead of a chairman; and
   (b) a chairman considers that one or more substantive issues of fact are likely to be determined at the pre-hearing review, that it would be desirable for the pre-hearing review to be conducted by a tribunal and he has issued an order that the pre-hearing review be conducted by a tribunal.

(4) If an order is made under paragraph (3), any reference to a chairman in relation to a pre-hearing review shall be read as a reference to a tribunal.

(5) Notwithstanding the preliminary or interim nature of a pre-hearing review, at a pre-hearing review the chairman may give judgment on any preliminary issue of substance relating to the proceedings. Judgments or orders made at a pre-hearing review may result in the proceedings being struck out or dismissed or otherwise determined with the result that a Hearing is no longer necessary in those proceedings.

(6) Before a judgment or order listed in paragraph (7) is made, notice must be given in accordance with rule 19. The judgments or [orders][28] listed in paragraph (7) must be made at a pre-hearing review or a Hearing if one of the parties has so requested. If no such request has been made such judgments or [orders][29] may be made in the absence of the parties.

(7) Subject to paragraph (6), a chairman or tribunal may make a judgment or order—
   (a) as to the entitlement of any party to bring or contest particular proceedings;
   (b) striking out or amending all or part of any claim or response on the grounds that it is scandalous, or vexatious or has no reasonable prospect of success;
   (c) striking out any claim or response (or part of one) on the grounds that the manner in which the proceedings have been conducted by or on behalf of the claimant or the respondent (as the case may be) has been scandalous, unreasonable or vexatious;
   (d) striking out a claim which has not been actively pursued;
   (e) striking out a claim or response (or part of one) for non-compliance with an order or practice direction;
   (f) striking out a claim where the chairman or tribunal considers that it is no longer possible to have a fair Hearing in those proceedings;
   (g) making a restricted reporting order (subject to rule 50).

---

[28] Amended by SI 2005/1865.
[29] ibid.

(8) A claim or response or any part of one may be struck out under these rules only on the grounds stated in sub-paragraphs (7)(b) to (f).

(9) If at a pre-hearing review a requirement to pay a deposit under rule 20 has been considered, the chairman who conducted that pre-hearing review shall not be a member of the tribunal at the Hearing in relation to those proceedings.

Notice requirements

19.—(1) Before a chairman or a tribunal makes a judgment or order described in rule 18(7), except where the order is one described in rule 13(2) or it is a temporary restricted reporting order made in accordance with rule 50, the Secretary shall send notice to the party against whom it is proposed that the order or judgment should be made. The notice shall inform him of the order or judgment to be considered and give him the opportunity to give reasons why the order or judgment should not be made. This paragraph shall not be taken to require the Secretary to send such notice to that party if that party has been given an opportunity to give reasons orally to the chairman or the tribunal as to why the order should not be made.

(2) Where a notice required by paragraph (1) is sent in relation to an order to strike out a claim which has not been actively pursued, unless the contrary is proved, the notice shall be treated as if it were received by the addressee if it has been sent to the address specified in the claim as the address to which notices are to be sent (or to any subsequent replacement for that address which has been notified to the Employment Tribunal Office).

Payment of a Deposit

Requirement to pay a deposit in order to continue with proceedings

20.—(1) At a pre-hearing review if a chairman considers that the contentions put forward by any party in relation to a matter required to be determined by a tribunal have little reasonable prospect of success, the chairman may make an order against that party requiring the party to pay a deposit of an amount not exceeding £500 as a condition of being permitted to continue to take part in the proceedings relating to that matter.

(2) No order shall be made under this rule unless the chairman has taken reasonable steps to ascertain the ability of the party against whom it is proposed to make the order to comply with such an order, and has taken account of any information so ascertained in determining the amount of the deposit.

(3) An order made under this rule, and the chairman's grounds for making such an order, shall be recorded in a document signed by the chairman. A copy of that document shall be sent to each of the parties and shall be accompanied by a note explaining that if the party against whom the order is made persists in making those contentions relating to the matter to which the order relates, he may have an award of costs or preparation time made against him and could lose his deposit.

(4) If a party against whom an order has been made does not pay the amount specified in the order to the Secretary either—
  (a) within the period of 21 days of the day on which the document recording the making of the order is sent to him; or
  (b) within such further period, not exceeding 14 days, as the chairman may allow in the light of representations made by that party within the period of 21 days;
a chairman shall strike out the claim or response of that party or, as the case may be, the part of it to which the order relates.

(5) The deposit paid by a party under an order made under this rule shall be refunded to him in full except where rule 47 applies.

Conciliation

Documents to be sent to conciliators

21. In proceedings brought under the provisions of any enactment providing for conciliation, the Secretary shall send copies of all documents, orders, judgments, written reasons and notices to an ACAS conciliation officer except where the Secretary and ACAS have agreed otherwise.

Fixed period for conciliation

22.—(1) This rule and rules 23 and 24 apply to all proceedings before a tribunal which are brought under any enactment which provides for conciliation except national security proceedings and proceedings which include a claim made under one or more of the following enactments—
   (a) the Equal Pay Act, section 2(1);
   (b) the Sex Discrimination Act, Part II, section 63;
   (c) the Race Relations Act, Part II, section 54;
   (d) the Disability Discrimination Act, Part II, section 17A or 25(8);
   (e) the Employment Equality (Sexual Orientation) Regulations 2003;
   (f) the Employment Equality (Religion or Belief) Regulations 2003; [...][30]
   (g) Employment Rights Act, sections 47B, 103A and 105(6A)[; and
   (h) the Employment Equality (Age) Regulations 2006].[31]

(2) In all proceedings to which this rule applies there shall be a conciliation period to give a time limited opportunity for the parties to reach an ACAS conciliated settlement (the "conciliation period"). In proceedings in which there is more than one respondent there shall be a conciliation period in relation to each respondent.

(3) In any proceedings to which this rule applies a Hearing shall not take place during a conciliation period and where the time and place of a Hearing has been fixed to take place during a conciliation period, such Hearing shall be postponed until after the end of any conciliation period. The fixing of the time and place for the Hearing may take place during a conciliation period. Pre-hearing reviews and case management discussions may take place during a conciliation period.

(4) In relation to each respondent the conciliation period commences on [the day following][32] the date on which the Secretary sends a copy of the claim to that respondent. The duration of the conciliation period shall be determined in accordance with the following paragraphs and rule 23.

(5) In any proceedings which consist of claims under any of the following enactments (but no other enactments) the conciliation period is seven weeks (the "short conciliation period")—
   (a) Employment Tribunals Act, section 3 (breach of contract);
   (b) the following provisions of the Employment Rights Act—
      (i) sections 13 to 27 (failure to pay wages or an unauthorised deduction of wages);
      (ii) section 28 (right to a guarantee payment);
      (iii) section 50 (right to time off for public duties);
      (iv) section 52 (right to time off to look for work or arrange training);
      (v) section 53 (right to remuneration for time off under section 52);
      (vi) section 55 (right to time off for ante-natal care);
      (vii) section 56 (right to remuneration for time off under section 55);
      (viii) section 64 (failure to pay remuneration whilst suspended for medical reasons);
      (ix) section 68 (right to remuneration whilst suspended on maternity grounds);
      (x) sections 163 or 164 (failure to pay a redundancy payment);

---

[30] Repealed by SI 2007/825.
[31] Amended by SI 2007/825.
[32] Amended by SI 2005/1865.

(c) the following provisions of TULR(C)A—
   (i) section 68 (right not to suffer deduction of unauthorised subscriptions);
   (ii) section 168 (time off for carrying out trade union duties);
   (iii) section 169 (payment for time off under section 168);
   (iv) section 170 (time off for trade union activities);
   (v) section 192 (failure to pay remuneration under a protective award);
(d) [regulation 15(10) of the Transfer of Undertakings (Protection of Employment) Regulations 2006][33] (failure to pay compensation following failure to inform or consult).
[(e) regulations 13, 14(2) or 16(1) of the Working Time Regulations 1998c (right to paid annual leave.)][34]

(6) In all other proceedings to which this rule applies the conciliation period is thirteen weeks (the "standard conciliation period").

(7) In proceedings to which the standard conciliation period applies, that period shall be extended by a period of a further two weeks if ACAS notifies the Secretary in writing that all of the following circumstances apply before the expiry of the standard conciliation period—
   (a) all parties to the proceedings agree to the extension of any relevant conciliation period;
   (b) a proposal for settling the proceedings has been made by a party and is under consideration by the other parties to the proceedings; and
   (c) ACAS considers it probable that the proceedings will be settled during the further extended conciliation period.

(8) A short conciliation period in any proceedings may, if that period has not already ended, be extended into a standard conciliation period if a chairman considers on the basis of the complexity of the proceedings that a standard conciliation period would be more appropriate. Where a chairman makes an order extending the conciliation period in such circumstances, the Secretary shall inform the parties to the proceedings and ACAS in writing as soon as is reasonably practicable.

Early termination of conciliation period

23.—(1) Should one of the following circumstances arise during a conciliation period (be it short or standard) which relates to a particular respondent (referred to in this rule as the relevant respondent), that conciliation period shall terminate early on the relevant date specified (and if more than one circumstance or date listed below is applicable to any conciliation period, that conciliation period shall terminate on the earliest of those dates)—
   (a) where a default judgment is issued against the relevant respondent which determines both liability and remedy, the date on which the default judgment is signed;
   (b) where a default judgment is issued against the relevant respondent which determines liability only, the date which is 14 days after the date on which the default judgment is signed;
   (c) where either the claim or the response entered by the relevant respondent is struck out, the date on which the judgment to strike out is signed;
   (d) where the claim is withdrawn, the date of receipt by the Employment Tribunal Office of the notice of withdrawal;
   (e) where the claimant or the relevant respondent has informed ACAS in writing that they do not wish to proceed with attempting to conciliate in relation to those proceedings, the date on which ACAS sends notice of such circumstances to the parties and to the Employment Tribunal Office;

---

[33] Amended by SI 2006/2405.
[34] Amended by SI 2005/1865.

(f) where the claimant and the relevant respondent have reached a settlement by way of a compromise agreement (including a compromise agreement to refer proceedings to arbitration), the date on which the Employment Tribunal Office receives notice from both of those parties to that effect;

(g) where the claimant and the relevant respondent have reached a settlement through a conciliation officer (including a settlement to refer the proceedings to arbitration), the date of the settlement;

(h) where no response presented by the relevant respondent has been accepted in the proceedings and no default judgment has been issued against that respondent, the date which is 14 days after the expiry of the time limit for presenting the response to the Secretary.

(2) Where a chairman or tribunal makes an order which re-establishes the relevant respondent's right to respond to the claim (for example, revoking a default judgment) and when that order is made, the conciliation period in relation to that respondent has terminated early under paragraph (1) or has otherwise expired, the chairman or tribunal may order that a further conciliation period shall apply in relation to that respondent if they consider it appropriate to do so.

(3) When an order is made under paragraph (2), the further conciliation period commences on the date of that order and the duration of that period shall be determined in accordance with paragraphs (5) to (8) of rule 22 and paragraph (1) of this rule as if the earlier conciliation period in relation to that respondent had not taken place.

Effect of staying or sisting proceedings on the conciliation period

24. Where during a conciliation period an order is made to stay (or in Scotland, sist) the proceedings, that order has the effect of suspending any conciliation period in those proceedings. Any unexpired portion of a conciliation period takes effect from the date on which the stay comes to an end (or in Scotland, the sist is recalled) and continues for the duration of the unexpired portion of that conciliation period or two weeks (whichever is the greater).

WITHDRAWAL OF PROCEEDINGS

Right to withdraw proceedings

25.—(1) A claimant may withdraw all or part of his claim at any time—this may be done either orally at a hearing or in writing in accordance with paragraph (2).

(2) To withdraw a claim or part of one in writing the claimant must inform the Employment Tribunal Office of the claim or the parts of it which are to be withdrawn. Where there is more than one respondent the notification must specify against which respondents the claim is being withdrawn.

(3) The Secretary shall inform all other parties of the withdrawal. Withdrawal takes effect on the date on which the Employment Tribunal Office (in the case of written notifications) or the tribunal (in the case of oral notification) receives notice of it and where the whole claim is withdrawn, subject to paragraph (4), proceedings are brought to an end against the relevant respondent on that date. Withdrawal does not affect proceedings as to costs, preparation time or wasted costs.

(4) Where a claim has been withdrawn, a respondent may make an application to have the proceedings against him dismissed. Such an application must be made by the respondent in writing to the Employment Tribunal Office within 28 days of the notice of the withdrawal being sent to the respondent. If the respondent's application is granted and the proceedings are dismissed those proceedings cannot be continued by the claimant (unless the decision to dismiss is successfully reviewed or appealed).

(5) The time limit in paragraph (4) may be extended by a chairman if he considers it just and equitable to do so.

The Hearing

Hearings

26.—(1) A Hearing is held for the purpose of determining outstanding procedural or substantive issues or disposing of the proceedings. In any proceedings there may be more than one Hearing and there may be different categories of Hearing, such as a Hearing on liability, remedies, costs (in Scotland, expenses) or preparation time.

(2) Any Hearing of a claim shall be heard by a tribunal composed in accordance with section 4(1) and (2) of the Employment Tribunals Act.

(3) Any Hearing of a claim shall take place in public, subject to rule 16.

What happens at the Hearing

27.—(1) The President, Vice President or a Regional Chairman shall fix the date, time and place of the Hearing and the Secretary shall send to each party a notice of the Hearing together with information and guidance as to procedure at the Hearing.

(2) Subject to rule 14(3), at the Hearing a party shall be entitled to give evidence, to call witnesses, to question witnesses and to address the tribunal.

(3) The tribunal shall require parties and witnesses who attend the Hearing to give their evidence on oath or affirmation.

(4) The tribunal may exclude from the Hearing any person who is to appear as a witness in the proceedings until such time as they give evidence if it considers it in the interests of justice to do so.

(5) If a party fails to attend or to be represented (for the purpose of conducting the party's case at the Hearing) at the time and place fixed for the Hearing, the tribunal may dismiss or dispose of the proceedings in the absence of that party or may adjourn the Hearing to a later date.

(6) If the tribunal wishes to dismiss or dispose of proceedings in the circumstances described in paragraph (5), it shall first consider any information in its possession which has been made available to it by the parties.

(7) At a Hearing a tribunal may exercise any powers which may be exercised by a chairman under these rules.

Orders, Judgments and Reasons

Orders and judgments

28.—(1) Chairmen or tribunals may issue the following—
  (a) a "judgment", which is a final determination of the proceedings or of a particular issue in those proceedings; it may include an award of compensation, a declaration or recommendation and it may also include orders for costs, preparation time or wasted costs;
  (b) an "order", which may be issued in relation to interim matters and it will require a person to do or not to do something.

(2) If the parties agree in writing upon the terms of any order or judgment a chairman or tribunal may, if he or it thinks fit, make such order or judgment.

(3) At the end of a hearing the chairman (or, as the case may be, the tribunal) shall either issue any order or judgment orally or shall reserve the judgment or order to be given in writing at a later date.

(4) Where a tribunal is composed of three persons any order or judgment may be made or issued by a majority; and if a tribunal is composed of two persons only, the chairman has a second or casting vote.

Form and content of judgments

29.—(1) When judgment is reserved a written judgment shall be sent to the parties as soon as practicable. All judgments (whether issued orally or in writing) shall be recorded in writing and signed by the chairman.

The Employment Tribunals (Constitution and Rules of Procedure) Regulations 2004

(2) The Secretary shall provide a copy of the judgment to each of the parties and, where the proceedings were referred to the tribunal by a court, to that court. The Secretary shall include guidance to the parties on how the judgment may be reviewed or appealed.

(3) Where the judgment includes an award of compensation or a determination that one party is required to pay a sum to another (excluding an order for costs, expenses, allowances, preparation time or wasted costs), the document shall also contain a statement of the amount of compensation awarded, or of the sum required to be paid.

Reasons

30.—(1) A tribunal or chairman must give reasons (either oral or written) for any—
   (a) judgment; or
   (b) order, if a request for reasons is made before or at the hearing at which the order is made.
(2) Reasons may be given orally at the time of issuing the judgment or order or they may be reserved to be given in writing at a later date. If reasons are reserved, they shall be signed by the chairman and sent to the parties by the Secretary.
(3) [[Where oral reasons have been provided],[35] written reasons shall only be provided][36]—
   (a) in relation to judgments if requested by one of the parties within the time limit set out in paragraph (5); or
   (b) in relation to any judgment or order if requested by the Employment Appeal Tribunal at any time.
(4) When written reasons are provided, the Secretary shall send a copy of the reasons to all parties to the proceedings and record the date on which the reasons were sent. Written reasons shall be signed by the chairman.
(5) A request for written reasons for a judgment must be made by a party either orally at the hearing (if the judgment is issued at a hearing), or in writing within 14 days of the date on which the judgment was sent to the parties. This time limit may be extended by a chairman where he considers it just and equitable to do so.
(6) Written reasons for a judgment shall include the following information—
   (a) the issues which the tribunal or chairman has identified as being relevant to the claim;
   (b) if some identified issues were not determined, what those issues were and why they were not determined;
   (c) findings of fact relevant to the issues which have been determined;
   (d) a concise statement of the applicable law;
   (e) how the relevant findings of fact and applicable law have been applied in order to determine the issues; and
   (f) where the judgment includes an award of compensation or a determination that one party make a payment to the other, a table showing how the amount or sum has been calculated or a description of the manner in which it has been calculated.

Absence of chairman

31. Where it is not possible for a judgment, order or reasons to be signed by the chairman due to death, incapacity or absence—
   (a) if the chairman has dealt with the proceedings alone the document shall be signed by the Regional Chairman, Vice President or President when it is practicable for him to do so; and
   (b) if the proceedings have been dealt with by a tribunal composed of two or three persons, the document shall be signed by the other person or persons;
and any person who signs the document shall certify that the chairman is unable to sign.

---

[35] Amended by SI 2005/1865.
[36] Amended by SI 2004/2351.

The Register

32.—(1) Subject to rule 49, the Secretary shall enter a copy of the following documents in the Register—
   (a) any judgment (including any costs, expenses, preparation time or wasted costs order); and
   (b) any written reasons provided in accordance with rule 30 in relation to any judgment.
(2) Written reasons for judgments shall be omitted from the Register in any case in which evidence has been heard in private and the tribunal or chairman so orders. In such a case the Secretary shall send the reasons to each of the parties and where there are proceedings before a superior court relating to the judgment in question, he shall send the reasons to that court, together with a copy of the entry in the Register of the judgment to which the reasons relate.

POWER TO REVIEW JUDGMENTS AND DECISIONS

Review of default judgments

33.—(1) A party may apply to have a default judgment against or in favour of him reviewed. An application must be made in writing and presented to the Employment Tribunal Office within 14 days of the date on which the default judgment was sent to the parties. The 14 day time limit may be extended by a chairman if he considers that it is just and equitable to do so.
(2) The application must state the reasons why the default judgment should be varied or revoked. When it is the respondent applying to have the default judgment reviewed, the application must include with it the respondent's proposed response to the claim, an application for an extension of the time limit for presenting the response and an explanation of why rules 4(1) and (4) were not complied with.
(3) A review of a default judgment shall be conducted by a chairman in public. Notice of the Hearing and a copy of the application shall be sent by the Secretary to all other parties.
(4) The chairman may—
   (a) refuse the application for a review;
   (b) vary the default judgment;
   (c) revoke all or part of the default judgment;
   (d) confirm the default judgment;
and all parties to the proceedings shall be informed by the Secretary in writing of the chairman's judgment on the application.
(5) A default judgment must be revoked if the whole of the claim was satisfied before the judgment was issued or if rule 8(6) applies. A chairman may revoke or vary all or part of a default judgment if the respondent has a reasonable prospect of successfully responding to the claim or part of it.
(6) In considering the application for a review of a default judgment the chairman must have regard to whether there was good reason for the response not having been presented within the applicable time limit.
(7) If the chairman decides that the default judgment should be varied or revoked and that the respondent should be allowed to respond to the claim the Secretary shall accept the response and proceed in accordance with rule 5(2).

Review of other judgments and decisions

34.—(1) Parties may apply to have certain judgments and decisions made by a tribunal or a chairman reviewed under rules 34 to 36. Those judgments and decisions are—
   (a) a decision not to accept a claim, response or counterclaim;
   (b) a judgment (other than a default judgment but including an order for costs, expenses, preparation time or wasted costs); and
   (c) a decision made under rule 6(3) of Schedule 4;
and references to "decision" in rules 34 to 37 are references to the above judgments and decisions only. Other decisions or orders may not be reviewed under these rules.

(2) In relation to a decision not to accept a claim or response, only the party against whom the decision is made may apply to have the decision reviewed.
(3) Subject to paragraph (4), decisions may be reviewed on the following grounds only—
   (a) the decision was wrongly made as a result of an administrative error;
   (b) a party did not receive notice of the proceedings leading to the decision;
   (c) the decision was made in the absence of a party;
   (d) new evidence has become available since the conclusion of the hearing to which the decision relates, provided that its existence could not have been reasonably known of or foreseen at that time; or
   (e) the interests of justice require such a review.
(4) A decision not to accept a claim or response may only be reviewed on the grounds listed in paragraphs (3)(a) and (e).
(5) A tribunal or chairman may on its or his own initiative review a decision made by it or him on the grounds listed in paragraphs (3) or (4).

Preliminary consideration of application for review

35.—(1) An application under rule 34 to have a decision reviewed must be made to the Employment Tribunal Office within 14 days of the date on which the decision was sent to the parties. The 14 day time limit may be extended by a chairman if he considers that it is just and equitable to do so.
(2) The application must be in writing and must identify the grounds of the application in accordance with rule 34(3), but if the decision to be reviewed was made at a hearing, an application may be made orally at that hearing.
(3) The application to have a decision reviewed shall be considered (without the need to hold a hearing) by the chairman of the tribunal which made the decision or, if that is not practicable, by—
   (a) a Regional Chairman or the Vice President;
   (b) any chairman nominated by a Regional Chairman or the Vice President; or
   (c) the President;
and that person shall refuse the application if he considers that there are no grounds for the decision to be reviewed under rule 34(3) or there is no reasonable prospect of the decision being varied or revoked.
(4) If an application for a review is refused after such preliminary consideration the Secretary shall inform the party making the application in writing of the chairman's decision and his reasons for it. If the application for a review is not refused the decision shall be reviewed under rule 36.

The review

36.—(1) When a party has applied for a review and the application has not been refused after the preliminary consideration above, the decision shall be reviewed by the chairman or tribunal who made the original decision. If that is not practicable a different chairman or tribunal (as the case may be) shall be appointed by a Regional Chairman, the Vice President or the President.
(2) Where no application has been made by a party and the decision is being reviewed on the initiative of the tribunal or chairman, the review must be carried out by the same tribunal or chairman who made the original decision and—
   (a) a notice must be sent to each of the parties explaining in summary the grounds upon which it is proposed to review the decision and giving them an opportunity to give reasons why there should be no review; and
   (b) such notice must be sent before the expiry of 14 days from the date on which the original decision was sent to the parties.

(3) A tribunal or chairman who reviews a decision under paragraph (1) or (2) may confirm, vary or revoke the decision. If the decision is revoked, the tribunal or chairman must order the decision to be taken again. When an order is made that the original decision be taken again, if the original decision was taken by a chairman without a hearing, the new decision may be taken without hearing the parties and if the original decision was taken at a hearing, a new hearing must be held.

Correction of judgments, decisions or reasons

37.—(1) Clerical mistakes in any order, judgment, decision or reasons, or errors arising in those documents from an accidental slip or omission, may at any time be corrected by certificate by the chairman, Regional Chairman, Vice President or President.

(2) If a document is corrected by certificate under paragraph (1), or if a decision is revoked or varied under rules 33 or 36 or altered in any way by order of a superior court, the Secretary shall alter any entry in the Register which is so affected to conform with the certificate or order and send a copy of any entry so altered to each of the parties and, if the proceedings have been referred to the tribunal by a court, to that court.

(3) Where a document omitted from the Register under rules 32 or 49 is corrected by certificate under this rule, the Secretary shall send a copy of the corrected document to the parties; and where there are proceedings before any superior court relating to the decision or reasons in question, he shall send a copy to that court together with a copy of the entry in the Register of the decision, if it has been altered under this rule.

(4) In Scotland, the references in paragraphs (2) and (3) to superior courts shall be read as referring to appellate courts.

Costs Orders and Orders for Expenses

General power to make costs and expenses orders

38.—(1) Subject to paragraph (2) and in the circumstances listed in rules 39, 40 and 47 a tribunal or chairman may make an order ("a costs order") that—
  (a) a party ("the paying party") make a payment in respect of the costs incurred by another party ("the receiving party");
  (b) the paying party pay to the Secretary of State, in whole or in part, any allowances (other than allowances paid to members of tribunals) paid by the Secretary of State under section 5(2) or (3) of the Employment Tribunals Act to any person for the purposes of, or in connection with, that person's attendance at the tribunal.

(2) A costs order may be made under rules 39, 40 and 47 only where the receiving party has been legally represented at the Hearing or, in proceedings which are determined without a Hearing, if the receiving party is legally represented when the proceedings are determined. If the receiving party has not been so legally represented a tribunal [or chairman][37] may make a preparation time order (subject to rules 42 to 45). (See rule 46 on the restriction on making a costs order and a preparation time order in the same proceedings.)

(3) For the purposes of these rules "costs" shall mean fees, charges, disbursements or expenses incurred by or on behalf of a party, in relation to the proceedings. In Scotland all references to costs (except when used in the expression "wasted costs") or costs orders shall be read as references to expenses or orders for expenses.

(4) A costs order may be made against or in favour of a respondent who has not had a response accepted in the proceedings in relation to the conduct of any part which he has taken in the proceedings.

---

[37] Amended by SI 2005/1865.

The Employment Tribunals (Constitution and Rules of Procedure) Regulations 2004

(5) In these rules legally represented means having the assistance of a person (including where that person is the receiving party's employee) who—
  (a) has a general qualification within the meaning of section 71 of the Courts and Legal Services Act 1990;
  (b) is an advocate or solicitor in Scotland; or
  (c) is a member of the Bar of Northern Ireland or a solicitor of the Supreme Court of Northern Ireland.

(6) Any costs order made under rules 39, 40 or 47 shall be payable by the paying party and not his representative.

(7) A party may apply for a costs order to be made at any time during the proceedings. An application may be made at the end of a hearing, or in writing to the Employment Tribunal Office. An application for costs which is received by the Employment Tribunal Office later than 28 days from the issuing of the judgment determining the claim shall not be accepted or considered by a tribunal or chairman unless it or he considers that it is in the interests of justice to do so.

(8) In paragraph (7), the date of issuing of the judgment determining the claim shall be either—
  (a) the date of the Hearing if the judgment was issued orally; or
  (b) if the judgment was reserved, the date on which the written judgment was sent to the parties.

(9) No costs order shall be made unless the Secretary has sent notice to the party against whom the order may be made giving him the opportunity to give reasons why the order should not be made. This paragraph shall not be taken to require the Secretary to send notice to that party if the party has been given an opportunity to give reasons orally to the chairman or tribunal as to why the order should not be made.

(10) Where a tribunal or chairman makes a costs order it or he shall provide written reasons for doing so if a request for written reasons is made within 14 days of the date of the costs order. The Secretary shall send a copy of the written reasons to all parties to the proceedings.

When a costs or expenses order must be made

39.—(1) Subject to rule 38(2), a tribunal [or chairman][38] must make a costs order against a respondent where in proceedings for unfair dismissal a Hearing has been postponed or adjourned and—
  (a) the claimant has expressed a wish to be reinstated or re-engaged which has been communicated to the respondent not less than 7 days before the Hearing; and
  (b) the postponement or adjournment of that Hearing has been caused by the respondent's failure, without a special reason, to adduce reasonable evidence as to the availability of the job from which the claimant was dismissed, or of comparable or suitable employment.

(2) A costs order made under paragraph (1) shall relate to any costs incurred as a result of the postponement or adjournment of the Hearing.

When a costs or expenses order may be made

40.—(1) A tribunal or chairman may make a costs order when on the application of a party it has postponed the day or time fixed for or adjourned a Hearing or pre-hearing review. The costs order may be against or, as the case may require, in favour of that party as respects any costs incurred or any allowances paid as a result of the postponement or adjournment.

(2) A tribunal or chairman shall consider making a costs order against a paying party where, in the opinion of the tribunal or chairman (as the case may be), any of the circumstances in paragraph (3) apply. Having so considered, the tribunal or chairman may make a costs order against the paying party if it or he considers it appropriate to do so.

---

[38] Amended by SI 2005/1865.

(3) The circumstances referred to in paragraph (2) are where the paying party has in bringing the proceedings, or he or his representative has in conducting the proceedings, acted vexatiously, abusively, disruptively or otherwise unreasonably, or the bringing or conducting of the proceedings by the paying party has been misconceived.

(4) A tribunal or chairman may make a costs order against a party who has not complied with an order or practice direction.

The amount of a costs or expenses order

41.—(1) The amount of a costs order against the paying party shall be determined in any of the following ways—
   (a) the tribunal may specify the sum which the paying party must pay to the receiving party, provided that sum does not exceed £10,000;
   (b) the parties may agree on a sum to be paid by the paying party to the receiving party and if they do so the costs order shall be for the sum so agreed;
   (c) the tribunal may order the paying party to pay the receiving party the whole or a specified part of the costs of the receiving party with the amount to be paid being determined by way of detailed assessment in a County Court in accordance with the Civil Procedure Rules 1998 or, in Scotland, as taxed according to such part of the table of fees prescribed for proceedings in the sheriff court as shall be directed by the order.

(2) The tribunal or chairman may have regard to the paying party's ability to pay when considering whether it or he shall make a costs order or how much that order should be.

(3) For the avoidance of doubt, the amount of a costs order made under paragraphs (1)(b) or (c) may exceed £10,000.

Preparation Time Orders

General power to make preparation time orders

42.—(1) Subject to paragraph (2) and in the circumstances described in rules 43, 44 and 47 a tribunal or chairman may make an order ("a preparation time order") that a party ("the paying party") make a payment in respect of the preparation time of another party ("the receiving party").

(2) A preparation time order may be made under rules 43, 44 or 47 only where the receiving party has not been legally represented at a Hearing or, in proceedings which are determined without a Hearing, if the receiving party has not been legally represented when the proceedings are determined. (See: rules 38 to 41 on when a costs order may be made; rule 38(5) for the definition of legally represented; and rule 46 on the restriction on making a costs order and a preparation time order in the same proceedings.)

(3) For the purposes of these rules preparation time shall mean time spent by—
   (a) the receiving party or his employees carrying out preparatory work directly relating to the proceedings; and
   (b) the receiving party's legal or other advisers relating to the conduct of the proceedings;
up to but not including time spent at any Hearing.

(4) A preparation time order may be made against a respondent who has not had a response accepted in the proceedings in relation to the conduct of any part which he has taken in the proceedings.

(5) A party may apply to the tribunal for a preparation time order to be made at any time during the proceedings. An application may be made at the end of a hearing or in writing to the Secretary. An application for preparation time which is received by the Employment Tribunal Office later than 28 days from the issuing of the judgment determining the claim shall not be accepted or considered by a tribunal or chairman unless they consider that it is in the interests of justice to do so.

(6) In paragraph (5) the date of issuing of the judgment determining the claim shall be either—
   (a) the date of the Hearing if the judgment was issued orally; or,
   (b) if the judgment was reserved, the date on which the written judgment was sent to the parties.
(7) No preparation time order shall be made unless the Secretary has sent notice to the party against whom the order may be made giving him the opportunity to give reasons why the order should not be made. This paragraph shall not be taken to require the Secretary to send notice to that party if the party has been given an opportunity to give reasons orally to the chairman or tribunal as to why the order should not be made.
(8) Where a tribunal or chairman makes a preparation time order it or he shall provide written reasons for doing so if a request for written reasons is made within 14 days of the date of the preparation time order. The Secretary shall send a copy of the written reasons to all parties to the proceedings.

When a preparation time order must be made

43.—(1) Subject to rule 42(2), a tribunal [or chairman]³⁹ must make a preparation time order against a respondent where in proceedings for unfair dismissal a Hearing has been postponed or adjourned and—
   (a) the claimant has expressed a wish to be reinstated or re-engaged which has been communicated to the respondent not less than 7 days before the Hearing; and
   (b) the postponement or adjournment of that Hearing has been caused by the respondent's failure, without a special reason, to adduce reasonable evidence as to the availability of the job from which the claimant was dismissed, or of comparable or suitable employment.
(2) A preparation time order made under paragraph (1) shall relate to any preparation time spent as a result of the postponement or adjournment of the Hearing.

When a preparation time order may be made

44.—(1) A tribunal or chairman may make a preparation time order when on the application of a party it has postponed the day or time fixed for or adjourned a Hearing or a pre-hearing review. The preparation time order may be against or, as the case may require, in favour of that party as respects any preparation time spent as a result of the postponement or adjournment.
(2) A tribunal or chairman shall consider making a preparation time order against a party (the paying party) where, in the opinion of the tribunal or the chairman (as the case may be), any of the circumstances in paragraph (3) apply. Having so considered the tribunal or chairman may make a preparation time order against that party if it considers it appropriate to do so.
(3) The circumstances described in paragraph (2) are where the paying party has in bringing the proceedings, or he or his representative has in conducting the proceedings, acted vexatiously, abusively, disruptively or otherwise unreasonably, or the bringing or conducting of the proceedings by the paying party has been misconceived.
(4) A tribunal or chairman may make a preparation time order against a party who has not complied with an order or practice direction.

Calculation of a preparation time order

45.—(1) In order to calculate the amount of preparation time the tribunal or chairman shall make an assessment of the number of hours spent on preparation time on the basis of—
   (a) information on time spent provided by the receiving party; and
   (b) the tribunal or chairman's own assessment of what it or he considers to be a reasonable and proportionate amount of time to spend on such preparatory work and with reference to,

---

[39] Amended by SI 2005/1865.

for example, matters such as the complexity of the proceedings, the number of witnesses and documentation required.

(2) Once the tribunal or chairman has assessed the number of hours spent on preparation time in accordance with paragraph (1), it or he shall calculate the amount of the award to be paid to the receiving party by applying an hourly rate of £25.00 to that figure (or such other figure calculated in accordance with paragraph (4)). No preparation time order made under these rules may exceed the sum of £10,000.

(3) The tribunal or chairman may have regard to the paying party's ability to pay when considering whether it or he shall make a preparation time order or how much that order should be.

(4) For the year commencing on 6th April 2006, the hourly rate of £25 shall be increased by the sum of £1.00 and for each subsequent year commencing on 6 April, the hourly rate for the previous year shall also be increased by the sum of £1.00.

Restriction on making costs or expenses orders and preparation time orders

46.—(1) A tribunal or chairman may not make a preparation time order and a costs order in favour of the same party in the same proceedings. However where a preparation time order is made in favour of a party in proceedings, the tribunal or chairman may make a costs order in favour of another party or in favour of the Secretary of State under rule 38(1)(b) in the same proceedings.

(2) If a tribunal or a chairman wishes to make either a costs order or a preparation time order in proceedings, before the claim has been determined, it or he may make an order that either costs or preparation time be awarded to the receiving party. In such circumstances a tribunal or chairman may decide whether the award should be for costs or preparation time after the proceedings have been determined.

Costs, expenses or preparation time orders when a deposit has been taken

47.—(1) When—
 (a) a party has been ordered under rule 20 to pay a deposit as a condition of being permitted to continue to participate in proceedings relating to a matter;
 (b) in respect of that matter, the tribunal or chairman has found against that party in its or his judgment; and
 (c) no award of costs or preparation time has been made against that party arising out of the proceedings on the matter;

the tribunal or chairman shall consider whether to make a costs or preparation time order against that party on the ground that he conducted the proceedings relating to the matter unreasonably in persisting in having the matter determined; but the tribunal or chairman shall not make a costs or preparation time order on that ground unless it has considered the document recording the order under rule 20 and is of the opinion that the grounds which caused the tribunal or chairman to find against the party in its judgment were substantially the same as the grounds recorded in that document for considering that the contentions of the party had little reasonable prospect of success.

(2) When a costs or preparation time order is made against a party who has had an order under rule 20 made against him (whether the award arises out of the proceedings relating to the matter in respect of which the order was made or out of proceedings relating to any other matter considered with that matter), his deposit shall be paid in part or full settlement of the costs or preparation time order—
 (a) when an order is made in favour of one party, to that party; and
 (b) when orders are made in favour of more than one party, to all of them or any one or more of them as the tribunal or chairman thinks fit, and if to all or more than one, in such proportions as the tribunal or chairman considers appropriate;

and if the amount of the deposit exceeds the amount of the costs or preparation time order, the balance shall be refunded to the party who paid it.

Wasted Costs Orders against Representatives

Personal liability of representatives for costs

48.—(1) A tribunal or chairman may make a wasted costs order against a party's representative.

(2) In a wasted costs order the tribunal or chairman may—

(a) disallow, or order the representative of a party to meet the whole or part of any wasted costs of any party, including an order that the representative repay to his client any costs which have already been paid; and

(b) order the representative to pay to the Secretary of State, in whole or in part, any allowances (other than allowances paid to members of tribunals) paid by the Secretary of State under section 5(2) or (3) of the Employment Tribunals Act to any person for the purposes of, or in connection with, that person's attendance at the tribunal by reason of the representative's conduct of the proceedings.

(3) "Wasted costs" means any costs incurred by a party—

(a) as a result of any improper, unreasonable or negligent act or omission on the part of any representative; or

(b) which, in the light of any such act or omission occurring after they were incurred, the tribunal considers it unreasonable to expect that party to pay.

(4) In this rule "representative" means a party's legal or other representative or any employee of such representative, but it does not include a representative who is not acting in pursuit of profit with regard to those proceedings. A person is considered to be acting in pursuit of profit if he is acting on a conditional fee arrangement.

(5) A wasted costs order may be made in favour of a party whether or not that party is legally represented and such an order may also be made in favour of a representative's own client. A wasted costs order may not be made against a representative where that representative is an employee of a party.

(6) Before making a wasted costs order, the tribunal or chairman shall give the representative a reasonable opportunity to make oral or written representations as to reasons why such an order should not be made. [The tribunal or chairman may also have regard to the representative's ability to pay][40] when considering whether it shall make a wasted costs order or how much that order should be.

(7) When a tribunal or chairman makes a wasted costs order, it must specify in the order the amount to be disallowed or paid.

(8) The Secretary shall inform the representative's client in writing—

(a) of any proceedings under this rule; or

(b) of any order made under this rule against the party's representative.

(9) Where a tribunal or chairman makes a wasted costs order it or he shall provide written reasons for doing so if a request is made for written reasons within 14 days of the date of the wasted costs order. This 14 day time limit may not be extended under rule 10. The Secretary shall send a copy of the written reasons to all parties to the proceedings.

Powers in Relation to Specific Types of Proceedings

Sexual offences and the Register

49. In any proceedings appearing to involve allegations of the commission of a sexual offence the tribunal, the chairman or the Secretary shall omit from the Register, or delete from the Register or any judgment, document or record of the proceedings, which is available to the public, any identifying matter which is likely to lead members of the public to identify any person affected by or making such an allegation.

---

[40] Amended by SI 2004/2351.

Restricted reporting orders

50.—(1) A restricted reporting order may be made in the following types of proceedings—
  (a) any case which involves allegations of sexual misconduct;
  (b) a complaint under section 17A or 25(8) of the Disability Discrimination Act in which evidence of a personal nature is likely to be heard by the tribunal or a chairman.
(2) A party (or where a complaint is made under the Disability Discrimination Act, the complainant) may apply for a restricted reporting order (either temporary or full) in writing to the Employment Tribunal Office, or orally at a hearing, or the tribunal or chairman may make the order on its or his own initiative without any application having been made.
(3) A chairman or tribunal may make a temporary restricted reporting order without holding a hearing or sending a copy of the application to other parties.
(4) Where a temporary restricted reporting order has been made the Secretary shall inform all parties to the proceedings in writing as soon as possible of—
  (a) the fact that the order has been made; and
  (b) their right to apply to have the temporary restricted reporting order revoked or converted into a full restricted reporting order within 14 days of the temporary order having been made.
(5) If no application under paragraph (4)(b) is made within the 14 days, the temporary restricted reporting order shall lapse and cease to have any effect on the fifteenth day after the order was made. If such an application is made the temporary restricted reporting order shall continue to have effect until the pre-hearing review or Hearing at which the application is considered.
(6) All parties must be given an opportunity to advance oral argument at a pre-hearing review or a Hearing before a tribunal or chairman decides whether or not to make a full restricted reporting order (whether or not there was previously a temporary restricted reporting order in the proceedings).
(7) Any person may make an application to the chairman or tribunal to have a right to make representations before a full restricted reporting order is made. The chairman or tribunal shall allow such representations to be made where he or it considers that the applicant has a legitimate interest in whether or not the order is made.
(8) Where a tribunal or chairman makes a restricted reporting order—
  (a) it shall specify in the order the persons who may not be identified;
  (b) a full order shall remain in force until both liability and remedy have been determined in the proceedings unless it is revoked earlier; and
  (c) the Secretary shall ensure that a notice of the fact that a restricted reporting order has been made in relation to those proceedings is displayed on the notice board of the employment tribunal with any list of the proceedings taking place before the employment tribunal, and on the door of the room in which the proceedings affected by the order are taking place.
(9) Where a restricted reporting order has been made under this rule and that complaint is being dealt with together with any other proceedings, the tribunal or chairman may order that the restricted reporting order applies also in relation to those other proceedings or a part of them.
(10) A tribunal or chairman may revoke a restricted reporting order at any time.
(11) For the purposes of this rule liability and remedy are determined in the proceedings on the date recorded as being the date on which the judgment disposing of the claim was sent to the parties, and references to a restricted reporting order include references to both a temporary and a full restricted reporting order.

Proceedings involving the National Insurance Fund

51. The Secretary of State shall be entitled to appear as if she were a party and be heard at any hearing in relation to proceedings which may involve a payment out of the National Insurance Fund, and in that event she shall be treated for the purposes of these rules as if she were a party.

Collective agreements

52. Where a claim includes a complaint under section 6(4A) of the Sex Discrimination Act 1986 relating to a term of a collective agreement, the following persons, whether or not identified in the claim, shall be regarded as the persons against whom a remedy is claimed and shall be treated as respondents for the purposes of these rules, that is to say—
    (a) the claimant's employer (or prospective employer); and
    (b) every organisation of employers and organisation of workers, and every association of or representative of such organisations, which, if the terms were to be varied voluntarily, would be likely, in the opinion of a chairman, to negotiate the variation;

provided that such an organisation or association shall not be treated as a respondent if the chairman, having made such enquiries of the claimant and such other enquiries as he thinks fit, is of the opinion that it is not reasonably practicable to identify the organisation or association.

Employment Agencies Act 1973

53. In relation to any claim in respect of an application under section 3C of the Employment Agencies Act 1973 for the variation or revocation of a prohibition order, the Secretary of State shall be treated as the respondent in such proceedings for the purposes of these rules. In relation to such an application the claim does not need to include the name and address of the persons against whom the claim is being made.

National security proceedings

54.—(1) A Minister of the Crown (whether or not he is a party to the proceedings) may, if he considers it expedient in the interests of national security, direct a tribunal or chairman by notice to the Secretary to—
    (a) conduct proceedings in private for all or part of particular Crown employment proceedings;
    (b) exclude the claimant from all or part of particular Crown employment proceedings;
    (c) exclude the claimant's representative from all or part of particular Crown employment proceedings;
    (d) take steps to conceal the identity of a particular witness in particular Crown employment proceedings.
(2) A tribunal or chairman may, if it or he considers it expedient in the interests of national security, by order—
    (a) do [in relation to particular proceedings before it][41] anything which can be required by direction to be done [in relation to particular Crown employment proceedings][42] under paragraph (1);
    (b) order any person to whom any document (including any judgment or record of the proceedings) has been provided for the purposes of the proceedings not to disclose any such document or the content thereof—
        (i) to any excluded person;
        (ii) in any case in which a direction has been given under paragraph (1)(a) or an order has been made under paragraph (2)(a) read with paragraph (1)(a), to any person excluded from all or part of the proceedings by virtue of such direction or order; or
        (iii) in any case in which a Minister of the Crown has informed the Secretary in accordance with paragraph (3) that he wishes to address the tribunal or chairman with a view to an order being made under paragraph (2)(a) read with paragraph (1)(b) or (c), to any person who may be excluded from all or part of the proceedings by virtue of such an order, if an order is made, at any time before the tribunal or chairman decides whether or not to make such an order;

---

[41] Amended by SI 2005/1865.
[42] ibid.

(c) take steps to keep secret all or part of the reasons for its judgment.

The tribunal or chairman (as the case may be) shall keep under review any order it or he has made under this paragraph.

(3) In any proceedings in which a Minister of the Crown considers that it would be appropriate for a tribunal or chairman to make an order as referred to in paragraph (2), he shall (whether or not he is a party to the proceedings) be entitled to appear before and to address the tribunal or chairman thereon. The Minister shall inform the Secretary by notice that he wishes to address the tribunal or chairman and the Secretary shall copy the notice to the parties.

(4) When exercising its or his functions, a tribunal or chairman shall ensure that information is not disclosed contrary to the interests of national security.

Dismissals in connection with industrial action

55.—(1) In relation to a complaint under section 111 of the Employment Rights Act 1996 (unfair dismissal: complaints to employment tribunal) that a dismissal is unfair by virtue of section 238A of TULR(C)A (participation in official industrial action) a tribunal or chairman may adjourn the proceedings where civil proceedings have been brought until such time as interim proceedings arising out of the civil proceedings have been concluded.

(2) In this rule—
  (a) "civil proceedings" means legal proceedings brought by any person against another person in which it is to be determined whether an act of that other person, which induced the claimant to commit an act, or each of a series of acts, is by virtue of section 219 of TULR(C)A not actionable in tort or in delict; and
  (b) the interim proceedings shall not be regarded as having concluded until all rights of appeal have been exhausted or the time for presenting any appeal in the course of the interim proceedings has expired.

Devolution issues

56.—(1) In any proceedings in which a devolution issue within the definition of the term in paragraph 1 of Schedule 6 to the Scotland Act 1998 arises, the Secretary shall as soon as reasonably practicable by notice inform the Advocate General for Scotland and the Lord Advocate thereof (unless they are a party to the proceedings) and shall at the same time—
  (a) send a copy of the notice to the parties to the proceedings; and
  (b) send the Advocate General for Scotland and the Lord Advocate a copy of the claim and the response.

(2) In any proceedings in which a devolution issue within the definition of the term in paragraph 1 of Schedule 8 to the Government of Wales Act 1998 arises, the Secretary shall as soon as reasonably practicable by notice inform the Attorney General and the National Assembly for Wales thereof (unless they are a party to the proceedings) and shall at the same time—
  (a) send a copy of the notice to the parties to the proceedings; and
  (b) send the Attorney General and the National Assembly for Wales a copy of the claim and the response.

(3) A person to whom notice is given in pursuance of paragraph (1) or (2) may within 14 days of receiving it, by notice to the Secretary, take part as a party in the proceedings, so far as they relate to the devolution issue. The Secretary shall send a copy of the notice to the other parties to the proceedings.

Transfer of proceedings between Scotland and England & Wales

57.—(1) The President (England and Wales) or a Regional Chairman may at any time, with the consent of the President (Scotland), order any proceedings in England and Wales to be transferred to an Employment Tribunal Office in Scotland if it appears to him that the proceedings could be (in accordance with regulation 19), and would more conveniently be, determined in an employment tribunal located in Scotland.

(2) The President (Scotland) or the Vice President may at any time, with the consent of the President (England and Wales), order any proceedings in Scotland to be transferred to an Employment Tribunal Office in England and Wales if it appears to him that the proceedings could be (in accordance with regulation 19), and would more conveniently be, determined in an employment tribunal located in England or Wales.

(3) An order under paragraph (1) or (2) may be made by the President, Vice President or Regional Chairman without any application having been made by a party. A party may apply for an order under paragraph (1) or (2) in accordance with rule 11.

(4) Where proceedings have been transferred under this rule, they shall be treated as if in all respects they had been presented to the Secretary by the claimant.

References to the European Court of Justice

58. Where a tribunal or chairman makes an order referring a question to the European Court of Justice for a preliminary ruling under Article 234 of the Treaty establishing the European Community, the Secretary shall send a copy of the order to the Registrar of that Court.

Transfer of proceedings from a court

59. Where proceedings are referred to a tribunal by a court, these rules shall apply to them as if the proceedings had been sent to the Secretary by the claimant.

General Provisions

Powers

60.—(1) Subject to the provisions of these rules and any practice directions, a tribunal or chairman may regulate its or his own procedure.

(2) At a Hearing, or a pre-hearing review held in accordance with rule 18(3), a tribunal may make any order which a chairman has power to make under these rules, subject to compliance with any relevant notice or other procedural requirements.

(3) Any function of the Secretary may be performed by a person acting with the authority of the Secretary.

Notices, etc

61.—(1) Any notice given or document sent under these rules shall (unless a chairman or tribunal orders otherwise) be in writing and may be given or sent—
  (a) by post;
  (b) by fax or other means of electronic communication; or
  (c) by personal delivery.

(2) Where a notice or document has been given or sent in accordance with paragraph (1), that notice or document shall, unless the contrary is proved, be taken to have been received by the party to whom it is addressed—
  (a) in the case of a notice or document given or sent by post, on the day on which the notice or document would be delivered in the ordinary course of post;
  (b) in the case of a notice or document transmitted by fax or other means of electronic communication, on the day on which the notice or document is transmitted;
  (c) in the case of a notice or document delivered in person, on the day on which the notice or document is delivered.

(3) All notices and documents required by these rules to be presented to the Secretary or an Employment Tribunal Office, other than a claim, shall be presented at the Employment Tribunal Office as notified by the Secretary to the parties.

(4) All notices and documents required or authorised by these rules to be sent or given to any person listed below may be sent to or delivered at—
  (a) in the case of a notice or document directed to the Secretary of State in proceedings to which she is not a party and which are brought under section 170 of the Employment

Rights Act, the offices of the Redundancy Payments Directorate of the Insolvency Service at PO Box 203, 21 Bloomsbury Street, London WC1B 3QW, or such other office as may be notified by the Secretary of State;

(b) in the case of any other notice or document directed to the Secretary of State in proceedings to which she is not a party (or in respect of which she is treated as a party for the purposes of these rules by rule 51), the offices of the Department of Trade and Industry (Employment Relations Directorate) at 1 Victoria Street, London, SW1H 0ET, or such other office as be notified by the Secretary of State;

(c) in the case of a notice or document directed to the Attorney General under rule 56, the Attorney General's Chambers, 9 Buckingham Gate, London, SW1E 7JP;

(d) in the case of a notice or document directed to the National Assembly for Wales under rule 56, the Counsel General to the National Assembly for Wales, Crown Buildings, Cathays Park, Cardiff, CF10 3NQ;

(e) in the case of a notice or document directed to the Advocate General for Scotland under rule 56, the Office of the Solicitor to the Advocate General for Scotland, Victoria Quay, Edinburgh, EH6 6QQ;

(f) in the case of a notice or document directed to the Lord Advocate under rule 56, the Legal Secretariat to the Lord Advocate, 25 Chambers Street, Edinburgh, EH1 1LA;

(g) in the case of a notice or document directed to a court, the office of the clerk of the court;

(h) in the case of a notice or document directed to a party—
  (i) the address specified in the claim or response to which notices and documents are to be sent, or in a notice under paragraph (5); or
  (ii) if no such address has been specified, or if a notice sent to such an address has been returned, to any other known address or place of business in the United Kingdom or, if the party is a corporate body, the body's registered or principal office in the United Kingdom, or, in any case, such address or place outside the United Kingdom as the President, Vice President or a Regional Chairman may allow;

(i) in the case of a notice or document directed to any person (other than a person specified in the foregoing provisions of this paragraph), his address or place of business in the United Kingdom or, if the person is a corporate body, the body's registered or principal office in the United Kingdom;

and a notice or document sent or given to the authorised representative of a party shall be taken to have been sent or given to that party.

(5) A party may at any time by notice to the Employment Tribunal Office and to the other party or parties (and, where appropriate, to the appropriate conciliation officer) change the address to which notices and documents are to be sent or transmitted.

(6) The President, Vice President or a Regional Chairman may order that there shall be substituted service in such manner as he may deem fit in any case he considers appropriate.

(7) In proceedings which may involve a payment out of the National Insurance Fund, the Secretary shall, where appropriate, send copies of all documents and notices to the Secretary of State whether or not she is a party.

(8) Copies of every document sent to the parties under rules 29, 30 or 32 shall be sent by the Secretary—
  (a) in the case of proceedings under the Equal Pay Act, the Sex Discrimination Act or the Sex Discrimination Act 1986, to the Equal Opportunities Commission;
  (b) in the case of proceedings under the Race Relations Act, to the Commission for Racial Equality; and
  (c) in the case of proceedings under the Disability Discrimination Act, to the Disability Rights Commission.

SCHEDULE 4

Regulation 16(3)(b)

THE EMPLOYMENT TRIBUNALS (HEALTH AND SAFETY—APPEALS AGAINST IMPROVEMENT AND PROHIBITION NOTICES) RULES OF PROCEDURE

For use only in proceedings in an appeal against an improvement or prohibition notice

Application of Schedule 1

1. Subject to rules 11 and 12 of this Schedule, Schedule 1 shall apply to appeals against an improvement notice or a prohibition notice. The rules in this Schedule modify the rules in Schedule 1 in relation to such appeals. If there is conflict between the rules contained in this Schedule and those in Schedule 1, the rules in this Schedule shall prevail.

Definitions

2. In this Schedule and in relation to proceedings to which this Schedule applies—
   "Health and Safety Act" means the Health and Safety at Work etc. Act 1974;
   "improvement notice" means a notice under section 21 of the Health and Safety Act;
   "inspector" means a person appointed under section 19(1) of the Health and Safety Act;
   "prohibition notice" means a notice under section 22 of the Health and Safety Act; and
   "respondent" means the inspector who issued the improvement notice or prohibition notice which is the subject of the appeal.

Notice of appeal

3. A person wishing to appeal an improvement notice or a prohibition notice (the appellant) shall do so by sending to the Employment Tribunal Office [. . .][43] a notice of appeal which must include the following—
   (a) the name and address of the appellant and, if different, an address to which he requires notices and documents relating to the appeal to be sent;
   (b) the date of the improvement notice or prohibition notice appealed against and the address of the premises or the place concerned;
   (c) the name and address of the respondent;
   (d) details of the requirements or directions which are being appealed; and
   (e) the grounds for the appeal.

Time limit for bringing appeal

4.—(1) Subject to paragraph (2), the notice of appeal must be sent to the Employment Tribunal Office within 21 days from the date of the service on the appellant of the notice appealed against.
(2) A tribunal may extend the time mentioned above where it is satisfied, on an application made in writing to the Secretary either before or after the expiration of that time, that it is or was not reasonably practicable for an appeal to be brought within that time.

Action on receipt of appeal

5. On receiving the notice of appeal the Secretary shall—
   (a) send a copy of the notice of appeal to the respondent; and
   (b) inform the parties in writing of the case number of the appeal (which must from then on be referred to in all correspondence relating to the appeal) and of the address to which notices and other communications to the Employment Tribunal Office shall be sent.

---

[43] Repealed by SI 2004/2351.

Application for a direction suspending the operation of a prohibition notice

6.—(1) When an appeal is brought against a prohibition notice, an application may be made by the appellant under section 24(3)(b) of the Health and Safety Act for a direction suspending the operation of the prohibition notice until the appeal is determined or withdrawn. The application must be presented to the Employment Tribunal Office in writing and shall include—
   (a) the case number of the appeal, or if there is no case number sufficient details to identify the appeal; and
   (b) the grounds on which the application is made.
(2) The Secretary shall send a copy of the application to the respondent as soon as practicable after it has been received and shall inform the respondent that he has the opportunity to submit representations in writing if he so wishes, within a specified time but not less than 7 days.
(3) The chairman shall consider the application and any representations submitted by the respondent, and may—
   (a) order that the application should not be determined separately from the full hearing of the appeal;
   (b) order that the operation of the prohibition notice be suspended until the appeal is determined or withdrawn;
   (c) dismiss the appellant's application; or
   (d) order that the application be determined at a Hearing (held in accordance with rule 26 of Schedule 1).
(4) The chairman must give reasons for any decision made under paragraph (3) or made following a Hearing ordered under paragraph (3)(d).
(5) A decision made under paragraph (3) or made following a Hearing ordered under paragraph (3)(d) shall be treated as a decision which may be reviewed upon the application of a party under rule 34 of Schedule 1.

General power to manage proceedings

7.—(1) The chairman may at any time on the application of a party, make an order in relation to any matter which appears to him to be appropriate. Such orders may be those listed in rule 10(2) of Schedule 1 (subject to rule 11 below) or such other orders as he thinks fit. Subject to the case management rules in Schedule 1, orders may be issued as a result of a chairman considering the papers before him in the absence of the parties, or at a hearing (see regulation 2 for the definition of "hearing").
(2) If the parties agree in writing upon the terms of any decision to be made by the tribunal or chairman, the chairman may, if he thinks fit, decide accordingly.

Appointment of an assessor

8. The President, Vice President or a Regional Chairman may, if he thinks fit, appoint in accordance with section 24(4) of the Health and Safety Act a person having special knowledge or experience in relation to the subject matter of the appeal to sit with the tribunal or chairman as an assessor.

Right to withdraw proceedings

9.—(1) An appellant may withdraw all or part of the appeal at any time. This may be done either orally at a hearing or in writing in accordance with paragraph (2).
(2) To withdraw an appeal or part of one in writing the appellant must inform the Employment Tribunal Office in writing of the appeal or the parts of it which are to be withdrawn.
(3) The Secretary shall inform all other parties of the withdrawal. Withdrawal takes effect on the date on which the Employment Tribunal Office (in the case of written notifications) or the tribunal or chairman receives notice of it and where the whole appeal is withdrawn proceedings are brought to an end against the respondent on that date and the tribunal or chairman shall dismiss the appeal.

Costs and expenses

10.—(1) A tribunal or chairman may make an order ("a costs order") that a party ("the paying party") make a payment in respect of the costs incurred by another party ("the receiving party").

(2) For the purposes of paragraph (1) "costs" shall mean fees, charges, disbursements [or expenses][44] incurred by or on behalf of a party in relation to the proceedings. In Scotland all references in this Schedule to costs or costs orders shall be read as references to expenses or orders for expenses.

(3) The amount of a costs order against the paying party can be determined in the following ways—
   (a) the tribunal may specify the sum which the party must pay to the receiving party, provided that sum does not exceed £10,000;
   (b) the parties may agree on a sum to be paid by the paying party to the receiving party and if they do so the costs order shall be for the sum so agreed;
   (c) the tribunal may order the paying party to pay the receiving party the whole or a specified part of the costs of the second party with the amount to be paid being determined by way of detailed assessment in a County Court in accordance with the Civil Procedure Rules or, in Scotland, as taxed according to such part of the table of fees prescribed for proceedings in the sheriff court as shall be directed by the order.

(4) The tribunal or chairman shall have regard to the paying party's ability to pay when considering whether it or he shall make a costs order or how much that order should be.

(5) For the avoidance of doubt, the amount of a costs order made under either paragraph (4)(b) or (c) may exceed £10,000.

Provisions of Schedule 1 which do not apply to appeals against improvement notices or prohibition notices

11. The following rules in Schedule 1 shall not apply in relation to appeals against improvement and prohibition notices: rules 1 to 9, 10(1), 10(2)(g), (i), (k), (l) and (r), 12, 13, 16(1)(c), 18(2)(c) and (e), 18(8), 20 to 25, 29(3), 33, 34(1)(a), 34(2), 38 to 47, 49 to 53, 55, and 61(4)(a), (7) and (8). All references in Schedule 1 to the rules listed in this rule shall have no effect in relation to an appeal against an improvement notice or a prohibition notice.

Modification of Schedule 1

12. Schedule 1 shall be further modified so that all references in Schedule 1 to a claim shall be read as references to a notice of appeal or to an appeal against an improvement notice or a prohibition notice, as the context may require, and all references to the claimant shall be read as references to the appellant in such an appeal.

---

[44] Amended by SI 2005/1865.

APPENDIX E

Health and Safety: The Management of Health and Safety at Work Regulations 1999 (as amended)

1999 No. 3242

Made	*3rd December 1999*
Laid before Parliament	*8th December 1999*
Coming into force	*29th December 1999*

The Secretary of State, being a Minister designated for the purposes of section 2(2) of the European Communities Act 1972 in relation to measures relating to employers' obligations in respect of the health and safety of workers and in relation to measures relating to the minimum health and safety requirements for the workplace that relate to fire safety and in exercise of the powers conferred on him by the said section 2 and by sections 15(1), (2), (3)(a), (5), and (9), 47(2), 52(2), and (3), 80(1) and 82(3)(a) of and paragraphs 6(1), 7, 8(1), 10, 14, 15, and 16 of Schedule 3 to, the Health and Safety at Work etc. Act 1974 ("the 1974 Act") and of all other powers enabling him in that behalf—

(a) for the purpose of giving effect without modifications to proposals submitted to him by the Health and Safety Commission under section 11(2)(d) of the 1974 Act after the carrying out by the Commission of consultations in accordance with section 50(3) of that Act; and
(b) it appearing to him that the modifications to the Regulations marked with an asterisk in Schedule 2 are expedient and that it also appearing to him not to be appropriate to consult bodies in respect of such modifications in accordance with section 80(4) of the 1974 Act,

hereby makes the following Regulations:

1 Citation, commencement and interpretation

(1) These Regulations may be cited as the Management of Health and Safety at Work Regulations 1999 and shall come into force on 29th December 1999.
(2) In these Regulations—
"the 1996 Act" means the Employment Rights Act 1996;
"the assessment" means, in the case of an employer or self-employed person, the assessment made or changed by him in accordance with regulation 3;
"child"—
   (a) as respects England and Wales, means a person who is not over compulsory school age, construed in accordance with section 8 of the Education Act 1996; and
   (b) as respects Scotland, means a person who is not over school age, construed in accordance with section 31 of the Education (Scotland) Act 1980;
"employment business" means a business (whether or not carried on with a view to profit and whether or not carried on in conjunction with any other business) which supplies persons (other than seafarers) who are employed in it to work for and under the control of other persons in any capacity;
"fixed-term contract of employment" means a contract of employment for a specific term which is fixed in advance or which can be ascertained in advance by reference to some relevant circumstance;

"given birth" means delivered a living child or, after twenty-four weeks of pregnancy, a stillborn child;

"new or expectant mother" means an employee who is pregnant; who has given birth within the previous six months; or who is breastfeeding;

"the preventive and protective measures" means the measures which have been identified by the employer or by the self-employed person in consequence of the assessment as the measures he needs to take to comply with the requirements and prohibitions imposed upon him by or under the relevant statutory provisions [...];[1]

"young person" means any person who has not attained the age of eighteen.

(3) Any reference in these Regulations to—
  (a) a numbered regulation or Schedule is a reference to the regulation or Schedule in these Regulations so numbered; or
  (b) a numbered paragraph is a reference to the paragraph so numbered in the regulation in which the reference appears.

[2 **Disapplication of these Regulations**]

[(1) These Regulations shall not apply to or in relation to the master or crew of a ship, or to the employer of such persons, in respect of the normal ship-board activities of a ship's crew which are carried out solely by the crew under the direction of the master.

(2) Regulations 3(4), (5), 10(2) and 19 shall not apply to occasional work or short-term work involving work regarded as not being harmful, damaging or dangerous to young people in a family undertaking.

(3) In this regulation—
"normal ship-board activities" include—
  (a) the construction, reconstruction or conversion of a ship outside, but not inside, Great Britain; and
  (b) the repair of a ship save repair when carried out in dry dock;
"ship" includes every description of vessel used in navigation, other than a ship belonging to Her Majesty which forms part of Her Majesty's Navy.][2]

3 Risk assessment

(1) Every employer shall make a suitable and sufficient assessment of—
  (a) the risks to the health and safety of his employees to which they are exposed whilst they are at work; and
  (b) the risks to the health and safety of persons not in his employment arising out of or in connection with the conduct by him of his undertaking,
  for the purpose of identifying the measures he needs to take to comply with the requirements and prohibitions imposed upon him by or under the relevant statutory provisions [...].[3]

(2) Every self-employed person shall make a suitable and sufficient assessment of—
  (a) the risks to his own health and safety to which he is exposed whilst he is at work; and
  (b) the risks to the health and safety of persons not in his employment arising out of or in connection with the conduct by him of his undertaking,
  for the purpose of identifying the measures he needs to take to comply with the requirements and prohibitions imposed upon him by or under the relevant statutory provisions.

---

[1] Repealed by SI 2005/1541.
[2] Inserted by SI 2003/2457.
[3] Repealed by SI 2005/1541.

(3) Any assessment such as is referred to in paragraph (1) or (2) shall be reviewed by the employer or self-employed person who made it if—
   (a) there is reason to suspect that it is no longer valid; or
   (b) there has been a significant change in the matters to which it relates;
   and where as a result of any such review changes to an assessment are required, the employer or self-employed person concerned shall make them.
(4) An employer shall not employ a young person unless he has, in relation to risks to the health and safety of young persons, made or reviewed an assessment in accordance with paragraphs (1) and (5).
(5) In making or reviewing the assessment, an employer who employs or is to employ a young person shall take particular account of—
   (a) the inexperience, lack of awareness of risks and immaturity of young persons;
   (b) the fitting-out and layout of the workplace and the workstation;
   (c) the nature, degree and duration of exposure to physical, biological and chemical agents;
   (d) the form, range, and use of work equipment and the way in which it is handled;
   (e) the organisation of processes and activities;
   (f) the extent of the health and safety training provided or to be provided to young persons; and
   (g) risks from agents, processes and work listed in the Annex to Council Directive 94/33/EC on the protection of young people at work.
(6) Where the employer employs five or more employees, he shall record—
   (a) the significant findings of the assessment; and
   (b) any group of his employees identified by it as being especially at risk.

4 Principles of prevention to be applied

Where an employer implements any preventive and protective measures he shall do so on the basis of the principles specified in Schedule 1 to these Regulations.

5 Health and safety arrangements

(1) Every employer shall make and give effect to such arrangements as are appropriate, having regard to the nature of his activities and the size of his undertaking, for the effective planning, organisation, control, monitoring and review of the preventive and protective measures.
(2) Where the employer employs five or more employees, he shall record the arrangements referred to in paragraph (1).

6 Health surveillance

Every employer shall ensure that his employees are provided with such health surveillance as is appropriate having regard to the risks to their health and safety which are identified by the assessment.

7 Health and safety assistance

(1) Every employer shall, subject to paragraphs (6) and (7), appoint one or more competent persons to assist him in undertaking the measures he needs to take to comply with the requirements and prohibitions imposed upon him by or under the relevant statutory provisions [. . .].[4]
(2) Where an employer appoints persons in accordance with paragraph (1), he shall make arrangements for ensuring adequate co-operation between them.
(3) The employer shall ensure that the number of persons appointed under paragraph (1), the time available for them to fulfil their functions and the means at their disposal are adequate having

---

[4] Repealed by SI 2005/1541.

regard to the size of his undertaking, the risks to which his employees are exposed and the distribution of those risks throughout the undertaking.

(4) The employer shall ensure that—
  (a) any person appointed by him in accordance with paragraph (1) who is not in his employment—
    (i) is informed of the factors known by him to affect, or suspected by him of affecting, the health and safety of any other person who may be affected by the conduct of his undertaking, and
    (ii) has access to the information referred to in regulation 10; and
  (b) any person appointed by him in accordance with paragraph (1) is given such information about any person working in his undertaking who is—
    (i) employed by him under a fixed-term contract of employment, or
    (ii) employed in an employment business,
  as is necessary to enable that person properly to carry out the function specified in that paragraph.

(5) A person shall be regarded as competent for the purposes of paragraphs (1) and (8) where he has sufficient training and experience or knowledge and other qualities to enable him properly to assist in undertaking the measures referred to in paragraph (1).

(6) Paragraph (1) shall not apply to a self-employed employer who is not in partnership with any other person where he has sufficient training and experience or knowledge and other qualities properly to undertake the measures referred to in that paragraph himself.

(7) Paragraph (1) shall not apply to individuals who are employers and who are together carrying on business in partnership where at least one of the individuals concerned has sufficient training and experience or knowledge and other qualities—
  (a) properly to undertake the measures he needs to take to comply with the requirements and prohibitions imposed upon him by or under the relevant statutory provisions; and
  (b) properly to assist his fellow partners in undertaking the measures they need to take to comply with the requirements and prohibitions imposed upon them by or under the relevant statutory provisions.

(8) Where there is a competent person in the employer's employment, that person shall be appointed for the purposes of paragraph (1) in preference to a competent person not in his employment.

8 Procedures for serious and imminent danger and for danger areas

(1) Every employer shall—
  (a) establish and where necessary give effect to appropriate procedures to be followed in the event of serious and imminent danger to persons at work in his undertaking;
  (b) nominate a sufficient number of competent persons to implement those procedures in so far as they relate to the evacuation from premises of persons at work in his undertaking; and
  (c) ensure that none of his employees has access to any area occupied by him to which it is necessary to restrict access on grounds of health and safety unless the employee concerned has received adequate health and safety instruction.

(2) Without prejudice to the generality of paragraph (1)(a), the procedures referred to in that sub-paragraph shall—
  (a) so far as is practicable, require any persons at work who are exposed to serious and imminent danger to be informed of the nature of the hazard and of the steps taken or to be taken to protect them from it;
  (b) enable the persons concerned (if necessary by taking appropriate steps in the absence of guidance or instruction and in the light of their knowledge and the technical means at their disposal) to stop work and immediately proceed to a place of safety in the event of their being exposed to serious, imminent and unavoidable danger; and

(c) save in exceptional cases for reasons duly substantiated (which cases and reasons shall be specified in those procedures), require the persons concerned to be prevented from resuming work in any situation where there is still a serious and imminent danger.

(3) A person shall be regarded as competent for the purposes of paragraph (1)(b) where he has sufficient training and experience or knowledge and other qualities to enable him properly to implement the evacuation procedures referred to in that sub-paragraph.

9 Contacts with external services

Every employer shall ensure that any necessary contacts with external services are arranged, particularly as regards first-aid, emergency medical care and rescue work.

10 Information for employees

(1) Every employer shall provide his employees with comprehensible and relevant information on—
   (a) the risks to their health and safety identified by the assessment;
   (b) the preventive and protective measures;
   (c) the procedures referred to in regulation 8(1)(a) [. . .];[5]
   (d) the identity of those persons nominated by him in accordance with regulation 8(1)(b) [. . .];[6] and
   (e) the risks notified to him in accordance with regulation 11(1)(c).
(2) Every employer shall, before employing a child, provide a parent of the child with comprehensible and relevant information on—
   (a) the risks to his health and safety identified by the assessment;
   (b) the preventive and protective measures; and
   (c) the risks notified to him in accordance with regulation 11(1)(c).
(3) The reference in paragraph (2) to a parent of the child includes—
   (a) in England and Wales, a person who has parental responsibility, within the meaning of section 3 of the Children Act 1989, for him; and
   (b) in Scotland, a person who has parental rights, within the meaning of section 8 of the Law Reform (Parent and Child) (Scotland) Act 1986 for him.

11 Co-operation and co-ordination

(1) Where two or more employers share a workplace (whether on a temporary or a permanent basis) each such employer shall—
   (a) co-operate with the other employers concerned so far as is necessary to enable them to comply with the requirements and prohibitions imposed upon them by or under the relevant statutory provisions [. . .];[7]
   (b) (taking into account the nature of his activities) take all reasonable steps to co-ordinate the measures he takes to comply with the requirements and prohibitions imposed upon him by or under the relevant statutory provisions [. . .][8] with the measures the other employers concerned are taking to comply with the requirements and prohibitions imposed upon them by that legislation; and
   (c) take all reasonable steps to inform the other employers concerned of the risks to their employees' health and safety arising out of or in connection with the conduct by him of his undertaking.

---

[5] Repealed by SI 2005/1541.
[6] ibid.
[7] ibid.
[8] ibid.

(2) Paragraph (1) [...]⁹ shall apply to employers sharing a workplace with self-employed persons and to self-employed persons sharing a workplace with other self-employed persons as it applies to employers sharing a workplace with other employers; and the references in that paragraph to employers and the reference in the said paragraph to their employees shall be construed accordingly.

12 Persons working in host employers' or self-employed persons' undertakings

(1) Every employer and every self-employed person shall ensure that the employer of any employees from an outside undertaking who are working in his undertaking is provided with comprehensible information on—
   (a) the risks to those employees' health and safety arising out of or in connection with the conduct by that first-mentioned employer or by that self-employed person of his undertaking; and
   (b) the measures taken by that first-mentioned employer or by that self-employed person in compliance with the requirements and prohibitions imposed upon him by or under the relevant statutory provisions [...]¹⁰ in so far as the said requirements and prohibitions relate to those employees.

(2) Paragraph (1) [...]¹¹ shall apply to a self-employed person who is working in the undertaking of an employer or a self-employed person as it applies to employees from an outside undertaking who are working therein; and the reference in that paragraph to the employer of any employees from an outside undertaking who are working in the undertaking of an employer or a self-employed person and the references in the said paragraph to employees from an outside undertaking who are working in the undertaking of an employer or a self-employed person shall be construed accordingly.

(3) Every employer shall ensure that any person working in his undertaking who is not his employee and every self-employed person (not being an employer) shall ensure that any person working in his undertaking is provided with appropriate instructions and comprehensible information regarding any risks to that person's health and safety which arise out of the conduct by that employer or self-employed person of his undertaking.

(4) Every employer shall—
   (a) ensure that the employer of any employees from an outside undertaking who are working in his undertaking is provided with sufficient information to enable that second-mentioned employer to identify any person nominated by that first mentioned employer in accordance with regulation 8(1)(b) to implement evacuation procedures as far as those employees are concerned; and
   (b) take all reasonable steps to ensure that any employees from an outside undertaking who are working in his undertaking receive sufficient information to enable them to identify any person nominated by him in accordance with regulation 8(1)(b) to implement evacuation procedures as far as they are concerned.

(5) Paragraph (4) shall apply to a self-employed person who is working in an employer's undertaking as it applies to employees from an outside undertaking who are working therein; and the reference in that paragraph to the employer of any employees from an outside undertaking who are working in an employer's undertaking and the references in the said paragraph to employees from an outside undertaking who are working in an employer's undertaking shall be construed accordingly.

---

⁹ Repealed by SI 2005/1541.
¹⁰ ibid.
¹¹ ibid.

13 Capabilities and training

(1) Every employer shall, in entrusting tasks to his employees, take into account their capabilities as regards health and safety.

(2) Every employer shall ensure that his employees are provided with adequate health and safety training—
   (a) on their being recruited into the employer's undertaking; and
   (b) on their being exposed to new or increased risks because of—
      (i) their being transferred or given a change of responsibilities within the employer's undertaking,
      (ii) the introduction of new work equipment into or a change respecting work equipment already in use within the employer's undertaking,
      (iii) the introduction of new technology into the employer's undertaking, or
      (iv) the introduction of a new system of work into or a change respecting a system of work already in use within the employer's undertaking.

(3) The training referred to in paragraph (2) shall—
   (a) be repeated periodically where appropriate;
   (b) be adapted to take account of any new or changed risks to the health and safety of the employees concerned; and
   (c) take place during working hours.

14 Employees' duties

(1) Every employee shall use any machinery, equipment, dangerous substance, transport equipment, means of production or safety device provided to him by his employer in accordance both with any training in the use of the equipment concerned which has been received by him and the instructions respecting that use which have been provided to him by the said employer in compliance with the requirements and prohibitions imposed upon that employer by or under the relevant statutory provisions.

(2) Every employee shall inform his employer or any other employee of that employer with specific responsibility for the health and safety of his fellow employees—
   (a) of any work situation which a person with the first-mentioned employee's training and instruction would reasonably consider represented a serious and immediate danger to health and safety; and
   (b) of any matter which a person with the first-mentioned employee's training and instruction would reasonably consider represented a shortcoming in the employer's protection arrangements for health and safety,

in so far as that situation or matter either affects the health and safety of that first mentioned employee or arises out of or in connection with his own activities at work, and has not previously been reported to his employer or to any other employee of that employer in accordance with this paragraph.

15 Temporary workers

(1) Every employer shall provide any person whom he has employed under a fixed-term contract of employment with comprehensible information on—
   (a) any special occupational qualifications or skills required to be held by that employee if he is to carry out his work safely; and
   (b) any health surveillance required to be provided to that employee by or under any of the relevant statutory provisions,

and shall provide the said information before the employee concerned commences his duties.

(2) Every employer and every self-employed person shall provide any person employed in an employment business who is to carry out work in his undertaking with comprehensible information on—
   (a) any special occupational qualifications or skills required to be held by that employee if he is to carry out his work safely; and

(b) health surveillance required to be provided to that employee by or under any of the relevant statutory provisions.

(3) Every employer and every self-employed person shall ensure that every person carrying on an employment business whose employees are to carry out work in his undertaking is provided with comprehensible information on—
   (a) any special occupational qualifications or skills required to be held by those employees if they are to carry out their work safely; and
   (b) the specific features of the jobs to be filled by those employees (in so far as those features are likely to affect their health and safety);

and the person carrying on the employment business concerned shall ensure that the information so provided is given to the said employees.

16 Risk assessment in respect of new or expectant mothers

(1) Where—
   (a) the persons working in an undertaking include women of child-bearing age; and
   (b) the work is of a kind which could involve risk, by reason of her condition, to the health and safety of a new or expectant mother, or to that of her baby, from any processes or working conditions, or physical, biological or chemical agents, including those specified in Annexes I and II of Council Directive 92/85/EEC on the introduction of measures to encourage improvements in the safety and health at work of pregnant workers and workers who have recently given birth or are breastfeeding,

the assessment required by regulation 3(1) shall also include an assessment of such risk.

(2) Where, in the case of an individual employee, the taking of any other action the employer is required to take under the relevant statutory provisions would not avoid the risk referred to in paragraph (1) the employer shall, if it is reasonable to do so, and would avoid such risks, alter her working conditions or hours of work.

(3) If it is not reasonable to alter the working conditions or hours of work, or if it would not avoid such risk, the employer shall, subject to section 67 of the 1996 Act suspend the employee from work for so long as is necessary to avoid such risk.

(4) In paragraphs (1) to (3) references to risk, in relation to risk from any infectious or contagious disease, are references to a level of risk at work which is in addition to the level to which a new or expectant mother may be expected to be exposed outside the workplace.

17 Certificate from registered medical practitioner in respect of new or expectant mothers

Where—
   (a) a new or expectant mother works at night; and
   (b) a certificate from a registered medical practitioner or a registered midwife shows that it is necessary for her health or safety that she should not be at work for any period of such work identified in the certificate,

the employer shall, subject to section 67 of the 1996 Act, suspend her from work for so long as is necessary for her health or safety.

18 Notification by new or expectant mothers

(1) Nothing in paragraph (2) or (3) of regulation 16 shall require the employer to take any action in relation to an employee until she has notified the employer in writing that she is pregnant, has given birth within the previous six months, or is breastfeeding.

(2) Nothing in paragraph (2) or (3) of regulation 16 or in regulation 17 shall require the employer to maintain action taken in relation to an employee—
   (a) in a case—
      (i) to which regulation 16(2) or (3) relates; and
      (ii) where the employee has notified her employer that she is pregnant, where she has failed, within a reasonable time of being requested to do so in writing by her employer,

to produce for the employer's inspection a certificate from a registered medical practitioner or a registered midwife showing that she is pregnant;
   (b) once the employer knows that she is no longer a new or expectant mother; or
   (c) if the employer cannot establish whether she remains a new or expectant mother.

19 Protection of young persons

(1) Every employer shall ensure that young persons employed by him are protected at work from any risks to their health or safety which are a consequence of their lack of experience, or absence of awareness of existing or potential risks or the fact that young persons have not yet fully matured.
(2) Subject to paragraph (3), no employer shall employ a young person for work—
   (a) which is beyond his physical or psychological capacity;
   (b) involving harmful exposure to agents which are toxic or carcinogenic, cause heritable genetic damage or harm to the unborn child or which in any other way chronically affect human health;
   (c) involving harmful exposure to radiation;
   (d) involving the risk of accidents which it may reasonably be assumed cannot be recognised or avoided by young persons owing to their insufficient attention to safety or lack of experience or training; or
   (e) in which there is a risk to health from—
      (i) extreme cold or heat;
      (ii) noise; or
      (iii) vibration,
   and in determining whether work will involve harm or risks for the purposes of this paragraph, regard shall be had to the results of the assessment.
(3) Nothing in paragraph (2) shall prevent the employment of a young person who is no longer a child for work—
   (a) where it is necessary for his training;
   (b) where the young person will be supervised by a competent person; and
   (c) where any risk will be reduced to the lowest level that is reasonably practicable.
(4) [. . .][12]

20 Exemption certificates

(1) The Secretary of State for Defence may, in the interests of national security, by a certificate in writing exempt—
   (a) any of the home forces, any visiting force or any headquarters from those requirements of these Regulations which impose obligations other than those in regulations 16–18 on employers; or
   (b) any member of the home forces, any member of a visiting force or any member of a headquarters from the requirements imposed by regulation 14;
   and any exemption such as is specified in sub-paragraph (a) or (b) of this paragraph may be granted subject to conditions and to a limit of time and may be revoked by the said Secretary of State by a further certificate in writing at any time.
(2) In this regulation—
   (a) "the home forces" has the same meaning as in section 12(1) of the Visiting Forces Act 1952;
   (b) "headquarters" means a headquarters for the time being specified in Schedule 2 to the Visiting Forces and International Headquarters (Application of Law) Order 1999;

---

[12] Repealed by SI 2003/2457.

(c) "member of a headquarters" has the same meaning as in paragraph 1(1) of the Schedule to the International Headquarters and Defence Organisations Act 1964; and
(d) "visiting force" has the same meaning as it does for the purposes of any provision of Part I of the Visiting Forces Act 1952.

21 Provisions as to liability

Nothing in the relevant statutory provisions shall operate so as to afford an employer a defence in any criminal proceedings for a contravention of those provisions by reason of any act or default of—
(a) an employee of his, or
(b) a person appointed by him under regulation 7.

[22 Restriction of civil liability for breach of statutory duty]

[(1) Breach of a duty imposed on an employer by these Regulations shall not confer a right of action in any civil proceedings insofar as that duty applies for the protection of a third party.
(2) Breach of a duty imposed on an employee by regulation 14 shall not confer a right of action in any civil proceedings insofar as that duty applies for the protection of a third party.
(3) In this regulation, "third party", in relation to the undertaking, means any person who may be affected by that undertaking other than the employer whose undertaking it is and persons in his employment.][13]

23 Extension outside Great Britain

(1) These Regulations shall, subject to regulation 2, apply to and in relation to the premises and activities outside Great Britain to which sections 1 to 59 and 80 to 82 of the Health and Safety at Work etc. Act 1974 apply by virtue of the Health and Safety at Work etc. Act 1974 (Application Outside Great Britain) Order 1995 as they apply within Great Britain.
(2) For the purposes of Part I of the 1974 Act, the meaning of "at work" shall be extended so that an employee or a self-employed person shall be treated as being at work throughout the time that he is present at the premises to and in relation to which these Regulations apply by virtue of paragraph (1); and, in that connection, these Regulations shall have effect subject to the extension effected by this paragraph.

24 Amendment of the Health and Safety (First-Aid) Regulations 1981

Regulation 6 of the Health and Safety (First-Aid) Regulations 1981 is hereby revoked.

25 Amendment of the Offshore Installations and Pipeline Works (First-Aid) Regulations 1989

(1) The Offshore Installations and Pipeline Works (First-Aid) Regulations 1989 shall be amended in accordance with the following provisions of this regulation.
(2) In regulation 7(1) for the words "from all or any of the requirements of these Regulations", there shall be substituted the words "from regulation 5(1)(b) and (c) and (2)(a) of these Regulations".
(3) After regulation 7(2) the following paragraph shall be added—
"(3) An exemption granted under paragraph (1) above from the requirements in regulation 5(2)(a) of these Regulations shall be subject to the condition that a person provided under regulation 5(1)(a) of these Regulations shall have undergone adequate training.".

26 Amendment of the Mines Miscellaneous Health and Safety Provisions Regulations 1995

(1) The Mines Miscellaneous Health and Safety Provisions Regulations 1995 shall be amended in accordance with the following provisions of this regulation.

---

[13] Inserted by SI 2006/438.

The Management of Health and Safety at Work Regulations 1999 (as amended)

(2) Paragraph (2)(b) of regulation 4 shall be deleted.
(3) After paragraph (4) of regulation 4 there shall be added the following paragraph—
"(5) In relation to fire, the health and safety document prepared pursuant to paragraph (1) shall—
  (a) include a fire protection plan detailing the likely sources of fire, and the precautions to be taken to protect against, to detect and combat the outbreak and spread of fire; and
  (b) in respect of every part of the mine other than any building on the surface of that mine—
    (i) include the designation of persons to implement the plan, ensuring that the number of such persons, their training and the equipment available to them is adequate, taking into account the size of, and the specific hazards involved in the mine concerned; and
    (ii) include the arrangements for any necessary contacts with external emergency services, particularly as regards rescue work and fire-fighting; and
    (iii) be adapted to the nature of the activities carried on at that mine, the size of the mine and take account of the persons other than employees who may be present.".

27 Amendment of the Construction (Health, Safety and Welfare) Regulations 1996

(1) The Construction (Health, Safety and Welfare) Regulations 1996 shall be amended in accordance with the following provisions of this regulation.
(2) Paragraph (2) of regulation 20 shall be deleted and the following substituted—
"(2) Without prejudice to the generality of paragraph (1), arrangements prepared pursuant to that paragraph shall—
  (a) have regard to those matters set out in paragraph (4) of regulation 19;
  (b) designate an adequate number of persons who will implement the arrangements; and
  (c) include any necessary contacts with external emergency services, particularly as regards rescue work and fire-fighting.".

28 [. . .][14]

29 Revocations and consequential amendments

(1) The Management of Health and Safety at Work Regulations 1992, the Management of Health and Safety at Work (Amendment) Regulations 1994, the Health and Safety (Young Persons) Regulations 1997 and Part III of the Fire Precautions (Workplace) Regulations 1997 are hereby revoked.
(2) The instruments specified in column 1 of Schedule 2 shall be amended in accordance with the corresponding provisions in column 3 of that Schedule.

30 Transitional provision

The substitution of provisions in these Regulations for provisions of the Management of Health and Safety at Work Regulations 1992 shall not affect the continuity of the law; and accordingly anything done under or for the purposes of such provision of the 1992 Regulations shall have effect as if done under or for the purposes of any corresponding provision of these Regulations.

---

[14] Repealed by SI 2005/1541.

Signed by authority of the Secretary of State

Whitty,

Parliamentary Under Secretary of State,

Department of the Environment, Transport and the Regions

3rd December 1999

Schedule 1
General Principles of Prevention

Regulation 4

(This Schedule specifies the general principles of prevention set out in Article 6(2) of Council Directive 89/391/EEC)

- (a) avoiding risks;
- (b) evaluating the risks which cannot be avoided;
- (c) combating the risks at source;
- (d) adapting the work to the individual, especially as regards the design of workplaces, the choice of work equipment and the choice of working and production methods, with a view, in particular, to alleviating monotonous work and work at a predetermined work-rate and to reducing their effect on health;
- (e) adapting to technical progress;
- (f) replacing the dangerous by the non-dangerous or the less dangerous;
- (g) developing a coherent overall prevention policy which covers technology, organisation of work, working conditions, social relationships and the influence of factors relating to the working environment;
- (h) giving collective protective measures priority over individual protective measures; and
- (i) giving appropriate instructions to employees.

APPENDIX F

Health and Safety: The Workplace (Health, Safety and Welfare) Regulations 1992 (as amended)

1992 No. 3004

Made *1st December 1992*

The Secretary of State, in exercise of the powers conferred on her by sections 15(1), (2), (3)(a) and (5)(b), and 82(3)(a) of, and paragraphs 1(2), 9 and 10 of Schedule 3 to, the Health and Safety at Work etc. Act 1974 ("the 1974 Act") and of all other powers enabling her in that behalf and for the purpose of giving effect without modifications to proposals submitted to her by the Health and Safety Commission under section 11(2)(d) of the 1974 Act after the carrying out by the said Commission of consultations in accordance with section 50(3) of that Act, hereby makes the following Regulations—

1 Citation and commencement

(1) These Regulations may be cited as the Workplace (Health, Safety and Welfare) Regulations 1992.

(2) Subject to paragraph (3), these Regulations shall come into force on 1st January 1993.

(3) Regulations 5 to 27 and the Schedules shall come into force on 1st January 1996 with respect to any workplace or part of a workplace which is not—
 (a) a new workplace; or
 (b) a modification, an extension or a conversion.

2 Interpretation

(1) In these Regulations, unless the context otherwise requires—

["disabled person" has the meaning given by section 1 of the Disability Discrimination Act 1995;][1]

["mine" means a mine within the meaning of the Mines and Quarries Act 1954;][2]

"new workplace" means a workplace used for the first time as a workplace after 31st December 1992;

"public road" means (in England and Wales) a highway maintainable at public expense within the meaning of section 329 of the Highways Act 1980 and (in Scotland) a public road within the meaning assigned to that term by section 151 of the Roads (Scotland) Act 1984;

["quarry" means a quarry within the meaning of the Quarries Regulations 1999;][3]

"traffic route" means a route for pedestrian traffic, vehicles or both and includes any stairs, staircase, fixed ladder, doorway, gateway, loading bay or ramp;

---

[1] Amended by SI 2002/2174.
[2] Amended by SI 1995/2036.
[3] Amended by SI 1999/2024.

"workplace" means, subject to paragraph (2), any premises or part of premises which are not domestic premises and are made available to any person as a place of work, and includes—
   (a) any place within the premises to which such person has access while at work; and
   (b) any room, lobby, corridor, staircase, road or other place used as a means of access to or egress from that place of work or where facilities are provided for use in connection with the place of work other than a public road;
[...]⁴

(2) Any reference in these Regulations, except in paragraph (1), to a modification, an extension or a conversion is a reference, as the case may be, to a modification, an extension or a conversion of a workplace started after 31st December 1992.

(3) Any requirement that anything done or provided in pursuance of these Regulations shall be suitable shall be construed to include a requirement that it is suitable for any person in respect of whom such thing is so done or provided.

(4) Any reference in these Regulations to—
   (a) a numbered regulation or Schedule is a reference to the regulation in or Schedule to these Regulations so numbered; and
   (b) a numbered paragraph is a reference to the paragraph so numbered in the regulation in which the reference appears.

3 Application of these Regulations

(1) These Regulations apply to every workplace but shall not apply to—
   (a) a workplace which is or is in or on a ship within the meaning assigned to that word by regulation 2(1) of the Docks Regulations 1988;
   [(b) a workplace where the only activity being undertaken is construction work within the meaning assigned to that phrase by regulation 2(1) of the Construction (Health, Safety and Welfare) Regulations 1996 (SI 1996/1592), except for any workplace from which the application of the said Regulations is excluded by regulation 3(2) of those Regulations;]⁵ [or
   (c) a workplace located below ground at a mine]⁶
   (d) [...]⁷

(2) In their application to temporary work sites, any requirement to ensure a workplace complies with any of regulations 20 to 25 shall have effect as a requirement to so ensure so far as is reasonably practicable.

(3) As respects any workplace which is or is in or on an aircraft, locomotive or rolling stock, trailer or semi-trailer used as a means of transport or a vehicle for which a licence is in force under the Vehicles (Excise) Act 1971 or a vehicle exempted from duty under that Act—
   (a) regulations 5 to 12 and 14 to 25 shall not apply to any such workplace; and
   (b) regulation 13 shall apply to any such workplace only when the aircraft, locomotive or rolling stock, trailer or semi-trailer or vehicle is stationary inside a workplace and, in the case of a vehicle for which a licence is in force under the Vehicles (Excise) Act 1971, is not on a public road.

(4) As respects any workplace which is in fields, woods or other land forming part of an agricultural or forestry undertaking but which is not inside a building and is situated away from the undertaking's main buildings—
   (a) regulations 5 to 19 and 23 to 25 shall not apply to any such workplace; and
   (b) any requirement to ensure that any such workplace complies with any of regulations 20 to 22 shall have effect as a requirement to so ensure so far as is reasonably practicable.

---

⁴ Repealed by SI 2002/2174.
⁵ Amended by SI 1996/1592.
⁶ Amended by SI 1995/2036.
⁷ Repealed by SI 1995/2036.

[(5) As respects any workplace which is at a quarry or above ground at a mine regulation 12 shall only apply to a floor or traffic route which is located inside a building.][8]

4 Requirements under these Regulations

(1) Every employer shall ensure that every workplace, modification, extension or conversion which is under his control and where any of his employees works complies with any requirement of these Regulations which—
   (a) applies to that workplace or, as the case may be, to the workplace which contains that modification, extension or conversion; and
   (b) is in force in respect of the workplace, modification, extension or conversion.
(2) Subject to paragraph (4), every person who has, to any extent, control of a workplace, modification, extension or conversion shall ensure that such workplace, modification, extension or conversion complies with any requirement of these Regulations which—
   (a) applies to that workplace or, as the case may be, to the workplace which contains that modification, extension or conversion;
   (b) is in force in respect of the workplace, modification, extension, or conversion; and
   (c) relates to matters within that person's control.
(3) Any reference in this regulation to a person having control of any workplace, modification, extension or conversion is a reference to a person having control of the workplace, modification, extension or conversion in connection with the carrying on by him of a trade, business or other undertaking (whether for profit or not).
(4) Paragraph (2) shall not impose any requirement upon a self-employed person in respect of his own work or the work of any partner of his in the undertaking.
(5) Every person who is deemed to be the occupier of a factory by virtue of section 175(5) of the Factories Act 1961 shall ensure that the premises which are so deemed to be a factory comply with these Regulations.

[4A Stability and solidity]

[Where a workplace is in a building, the building shall have a stability and solidity appropriate to the nature of the use of the workplace.][9]

5 Maintenance of workplace, and of equipment, devices and systems

(1) The workplace and the equipment, devices and systems to which this regulation applies shall be maintained (including cleaned as appropriate) in an efficient state, in efficient working order and in good repair.
(2) Where appropriate, the equipment, devices and systems to which this regulation applies shall be subject to a suitable system of maintenance.
(3) The equipment, devices and systems to which this regulation applies are—
   (a) equipment and devices a fault in which is liable to result in a failure to comply with any of these Regulations; [. . .][10]
   (b) mechanical ventilation systems provided pursuant to regulation 6 (whether or not they include equipment or devices within sub-paragraph (a) of this paragraph); [and][11]
   [(c) equipment and devices intended to prevent or reduce hazards].[12]

---

[8] Amended by SI 1995/2036.
[9] Inserted by SI 2002/2174.
[10] Repealed by SI 2002/2174.
[11] Amended by SI 2002/2174.
[12] ibid.

6 Ventilation[13]

(1) Effective and suitable provision shall be made to ensure that every enclosed workplace is ventilated by a sufficient quantity of fresh or purified air.
(2) Any plant used for the purpose of complying with paragraph (1) shall include an effective device to give visible or audible warning of any failure of the plant where necessary for reasons of health or safety.
(3) [...]

7 Temperature in indoor workplaces

(1) During working hours, the temperature in all workplaces inside buildings shall be reasonable.
[(1A) Without prejudice to the generality of paragraph (1)—
  (a) a workplace shall be adequately thermally insulated where it is necessary, having regard to the type of work carried out and the physical activity of the persons carrying out the work; and
  (b) excessive effects of sunlight on temperature shall be avoided.][14]
(2) A method of heating or cooling shall not be used which results in the escape into a workplace of fumes, gas or vapour of such character and to such extent that they are likely to be injurious or offensive to any person.
(3) A sufficient number of thermometers shall be provided to enable persons at work to determine the temperature in any workplace inside a building.

8 Lighting

(1) Every workplace shall have suitable and sufficient lighting.
(2) The lighting mentioned in paragraph (1) shall, so far as is reasonably practicable, be by natural light.
(3) Without prejudice to the generality of paragraph (1), suitable and sufficient emergency lighting shall be provided in any room in circumstances in which persons at work are specially exposed to danger in the event of failure of artificial lighting.

9 Cleanliness and waste materials

(1) Every workplace and the furniture, furnishings and fittings therein shall be kept sufficiently clean.
(2) The surfaces of the floors, walls and ceilings of all workplaces inside buildings shall be capable of being kept sufficiently clean.
(3) So far as is reasonably practicable, waste materials shall not be allowed to accumulate in a workplace except in suitable receptacles.

10 Room dimensions and space

(1) Every room where persons work shall have sufficient floor area, height and unoccupied space for purposes of health, safety and welfare.
(2) It shall be sufficient compliance with this regulation in a workplace which is not a new workplace, a modification, an extension or a conversion and which, immediately before this regulation came into force in respect of it, was subject to the provisions of the Factories Act 1961, if the workplace does not contravene the provisions of Part I of Schedule 1.

---

[13] Repealed by SI 2002/2174.
[14] Inserted by SI 2002/2174.

The Workplace (Health, Safety and Welfare) Regulations 1992 (as amended)

11 Workstations and seating

(1) Every workstation shall be so arranged that it is suitable both for any person at work in the workplace who is likely to work at that workstation and for any work of the undertaking which is likely to be done there.

(2) Without prejudice to the generality of paragraph (1), every workstation outdoors shall be so arranged that—
   (a) so far as is reasonably practicable, it provides protection from adverse weather;
   (b) it enables any person at the workstation to leave it swiftly or, as appropriate, to be assisted in the event of an emergency; and
   (c) it ensures that any person at the workstation is not likely to slip or fall.

(3) A suitable seat shall be provided for each person at work in the workplace whose work includes operations of a kind that the work (or a substantial part of it) can or must be done sitting.

(4) A seat shall not be suitable for the purpose of paragraph (3) unless—
   (a) it is suitable for the person for whom it is provided as well as for the operations to be performed; and
   (b) a suitable footrest is also provided where necessary.

12 Condition of floors and traffic routes

(1) Every floor in a workplace and the surface of every traffic route in a workplace shall be of a construction such that the floor or surface of the traffic route is suitable for the purpose for which it is used.

(2) Without prejudice to the generality of paragraph (1), the requirements in that paragraph shall include requirements that—
   (a) the floor, or surface of the traffic route, shall have no hole or slope, or be uneven or slippery so as, in each case, to expose any person to a risk to his health or safety; and
   (b) every such floor shall have effective means of drainage where necessary.

(3) So far as is reasonably practicable, every floor in a workplace and the surface of every traffic route in a workplace shall be kept free from obstructions and from any article or substance which may cause a person to slip, trip or fall.

(4) In considering whether for the purposes of paragraph (2)(a) a hole or slope exposes any person to a risk to his health or safety—
   (a) no account shall be taken of a hole where adequate measures have been taken to prevent a person falling; and
   (b) account shall be taken of any handrail provided in connection with any slope.

(5) Suitable and sufficient handrails and, if appropriate, guards shall be provided on all traffic routes which are staircases except in circumstances in which a handrail can not be provided without obstructing the traffic route.

13 Falls or falling objects

(1) [. . .][15]
(2) [. . .][16]
(3) [. . .][17]
(4) [. . .][18]
(5) So far as is practicable, every tank, pit or structure where there is a risk of a person in the workplace falling into a dangerous substance in the tank, pit or structure, shall be securely covered or fenced.

---

[15] Repealed by SI 2005/735.
[16] ibid.
[17] ibid.
[18] ibid.

(6) Every traffic route over, across or in an uncovered tank, pit or structure such as is mentioned in paragraph (5) shall be securely fenced.
(7) In this regulation, "dangerous substance" means—
   (a) any substance likely to scald or burn;
   (b) any poisonous substance;
   (c) any corrosive substance;
   (d) any fume, gas or vapour likely to overcome a person; or
   (e) any granular or free-flowing solid substance, or any viscous substance which, in any case, is of a nature or quantity which is likely to cause danger to any person.

14 Windows, and transparent or translucent doors, gates and walls

(1) Every window or other transparent or translucent surface in a wall or partition and every transparent or translucent surface in a door or gate shall, where necessary for reasons of health or safety—
   (a) be of safety material or be protected against breakage of the transparent or translucent material; and
   (b) be appropriately marked or incorporate features so as, in either case, to make it apparent.

15 Windows, skylights and ventilators

(1) No window, skylight or ventilator which is capable of being opened shall be likely to be opened, closed or adjusted in a manner which exposes any person performing such operation to a risk to his health or safety.
(2) No window, skylight or ventilator shall be in a position when open which is likely to expose any person in the workplace to a risk to his health or safety.

16 Ability to clean windows etc safely

(1) All windows and skylights in a workplace shall be of a design or be so constructed that they may be cleaned safely.
(2) In considering whether a window or skylight is of a design or so constructed as to comply with paragraph (1), account may be taken of equipment used in conjunction with the window or skylight or of devices fitted to the building.

17 Organisation etc of traffic routes

(1) Every workplace shall be organised in such a way that pedestrians and vehicles can circulate in a safe manner.
(2) Traffic routes in a workplace shall be suitable for the persons or vehicles using them, sufficient in number, in suitable positions and of sufficient size.
(3) Without prejudice to the generality of paragraph (2), traffic routes shall not satisfy the requirements of that paragraph unless suitable measures are taken to ensure that—
   (a) pedestrians or, as the case may be, vehicles may use a traffic route without causing danger to the health or safety of persons at work near it;
   (b) there is sufficient separation of any traffic route for vehicles from doors or gates or from traffic routes for pedestrians which lead onto it; and
   (c) where vehicles and pedestrians use the same traffic route, there is sufficient separation between them.
(4) All traffic routes shall be suitably indicated where necessary for reasons of health or safety.
(5) Paragraph (2) shall apply so far as is reasonably practicable, to a workplace which is not a new workplace, a modification, an extension or a conversion.

18 Doors and gates

(1) Doors and gates shall be suitably constructed (including being fitted with any necessary safety devices).

(2) Without prejudice to the generality of paragraph (1), doors and gates shall not comply with that paragraph unless—
   (a) any sliding door or gate has a device to prevent it coming off its track during use;
   (b) any upward opening door or gate has a device to prevent it falling back;
   (c) any powered door or gate has suitable and effective features to prevent it causing injury by trapping any person;
   (d) where necessary for reasons of health or safety, any powered door or gate can be operated manually unless it opens automatically if the power fails; and
   (e) any door or gate which is capable of opening by being pushed from either side is of such a construction as to provide, when closed, a clear view of the space close to both sides.

19 Escalators and moving walkways

Escalators and moving walkways shall—
   (a) function safely;
   (b) be equipped with any necessary safety devices;
   (c) be fitted with one or more emergency stop controls which are easily identifiable and readily accessible.

20 Sanitary conveniences

(1) Suitable and sufficient sanitary conveniences shall be provided at readily accessible places.
(2) Without prejudice to the generality of paragraph (1), sanitary conveniences shall not be suitable unless—
   (a) the rooms containing them are adequately ventilated and lit;
   (b) they and the rooms containing them are kept in a clean and orderly condition; and
   (c) separate rooms containing conveniences are provided for men and women except where and so far as each convenience is in a separate room the door of which is capable of being secured from inside.
(3) It shall be sufficient compliance with the requirement in paragraph (1) to provide sufficient sanitary conveniences in a workplace which is not a new workplace, a modification, an extension or a conversion and which, immediately before this regulation came into force in respect of it, was subject to the provisions of the Factories Act 1961, if sanitary conveniences are provided in accordance with the provisions of Part II of Schedule 1.

21 Washing facilities

(1) Suitable and sufficient washing facilities, including showers if required by the nature of the work or for health reasons, shall be provided at readily accessible places.
(2) Without prejudice to the generality of paragraph (1), washing facilities shall not be suitable unless—
   (a) they are provided in the immediate vicinity of every sanitary convenience, whether or not provided elsewhere as well;
   (b) they are provided in the vicinity of any changing rooms required by these Regulations, whether or not provided elsewhere as well;
   (c) they include a supply of clean hot and cold, or warm, water (which shall be running water so far as is practicable);
   (d) they include soap or other suitable means of cleaning;
   (e) they include towels or other suitable means of drying;
   (f) the rooms containing them are sufficiently ventilated and lit:
   (g) they and the rooms containing them are kept in a clean and orderly condition; and
   (h) separate facilities are provided for men and women, except where and so far as they are provided in a room the door of which is capable of being secured from inside and the facilities in each such room are intended to be used by only one person at a time.
(3) Paragraph (2)(h) shall not apply to facilities which are provided for washing hands, forearms and face only.

22 Drinking water

(1) An adequate supply of wholesome drinking water shall be provided for all persons at work in the workplace.
(2) Every supply of drinking water required by paragraph (1) shall—
    (a) be readily accessible at suitable places; and
    (b) be conspicuously marked by an appropriate sign where necessary for reasons of health or safety.
(3) Where a supply of drinking water is required by paragraph (1), there shall also be provided a sufficient number of suitable cups or other drinking vessels unless the supply of drinking water is in a jet from which persons can drink easily.

23 Accommodation for clothing

(1) Suitable and sufficient accommodation shall be provided—
    (a) for the clothing of any person at work which is not worn during working hours; and
    (b) for special clothing which is worn by any person at work but which is not taken home.
(2) Without prejudice to the generality of paragraph (1), the accommodation mentioned in that paragraph shall not be suitable unless—
    (a) where facilities to change clothing are required by regulation 24, it provides suitable security for the clothing mentioned in paragraph (1)(a);
    (b) where necessary to avoid risks to health or damage to the clothing, it includes separate accommodation for clothing worn at work and for other clothing;
    (c) so far as is reasonably practicable, it allows or includes facilities for drying clothing; and
    (d) it is in a suitable location.

24 Facilities for changing clothing

(1) Suitable and sufficient facilities shall be provided for any person at work in the workplace to change clothing in all cases where—
    (a) the person has to wear special clothing for the purpose of work; and
    (b) the person can not, for reasons of health or propriety, be expected to change in another room.
(2) Without prejudice to the generality of paragraph (1), the facilities mentioned in that paragraph shall not be suitable unless they include separate facilities for, or separate use of facilities by, men and women where necessary for reasons of propriety [and the facilities are easily accessible, of sufficient capacity and provided with seating].[19]

25 Facilities for rest and to eat meals

(1) Suitable and sufficient rest facilities shall be provided at readily accessible places.
(2) Rest facilities provided by virtue of paragraph (1) shall—
    (a) where necessary for reasons of health or safety include, in the case of a new workplace an extension or a conversion, rest facilities provided in one or more rest rooms, or, in other cases, in rest rooms or rest areas;
    (b) include suitable facilities to eat meals where food eaten in the workplace would otherwise be likely to become contaminated.
[(3) Rest rooms and rest areas shall—
    (a) include suitable arrangements to protect non-smokers from discomfort caused by tobacco smoke; and
    (b) be equipped with—
        (i) an adequate number of tables and adequate seating with backs for the number of persons at work likely to use them at any one time; and

---

[19] Amended by SI 2002/2174.

(ii) seating which is adequate for the number of disabled persons at work and suitable for them.]²⁰
(4) Suitable facilities shall be provided for any person at work who is a pregnant woman or nursing mother to rest.
(5) Suitable and sufficient facilities shall be provided for persons at work to eat meals where meals are regularly eaten in the workplace.

[25A Disabled persons]

[Where necessary, those parts of the workplace (including in particular doors, passageways, stairs, showers, washbasins, lavatories and workstations) used or occupied directly by disabled persons at work shall be organised to take account of such persons.]²¹

26 Exemption certificates

(1) The Secretary of State for Defence may, in the interests of national security, by a certificate in writing exempt any of the home forces, any visiting force or any headquarters from the requirements of these Regulations and any exemption may be granted subject to conditions and to a limit of time and may be revoked by the said Secretary of State by a further certificate in writing at any time.
(2) In this regulation—
    (a) "the home forces" has the same meaning as in section 12(1) of the Visiting Forces Act 1952;
    (b) "headquarters" has the same meaning as in article 3(2) of the Visiting Forces and International Headquarters (Application of Law) Order 1965;
    (c) "visiting force" has the same meaning as it does for the purposes of any provision of Part I of the Visiting Forces Act 1952.

27 Repeals, saving and revocations

(1) The enactments mentioned in column 2 of Part I of Schedule 2 are repealed to the extent specified in column 3 of that Part.
(2) Nothing in this regulation shall affect the operation of any provision of the Offices, Shops and Railway Premises Act 1963 as that provision has effect by virtue of section 90(4) of that Act.
(3) The instruments mentioned in column 1 of Part II of Schedule 2 are revoked to the extent specified in column 3 of that Part.

SCHEDULE 1

PROVISIONS APPLICABLE TO FACTORIES WHICH ARE NOT NEW WORKPLACES, MODIFICATIONS, EXTENSIONS OR CONVERSIONS

Regulations 10, 20

PART I

SPACE

1 No room in the workplace shall be so overcrowded as to cause risk to the health or safety of persons at work in it.
2 Without prejudice to the generality of paragraph 1, the number of persons employed at a time in any workroom shall not be such that the amount of cubic space allowed for each is less than 11 cubic metres.

---

[20] Inserted by SI 2002/2174.
[21] ibid.

3   In calculating for the purposes of this Part of this Schedule the amount of cubic space in any room no space more than 4.2 metres from the floor shall be taken into account and, where a room contains a gallery, the gallery shall be treated for the purposes of this Schedule as if it were partitioned off from the remainder of the room and formed a separate room.

Part II
Number of Sanitary Conveniences

4   In workplaces where females work, there shall be at least one suitable water closet for use by females only for every 25 females.
5   In workplaces where males work, there shall be at least one suitable water closet for use by males only for every 25 males.
6   In calculating the number of males or females who work in any workplace for the purposes of this Part of this Schedule, any number not itself divisible by 25 without fraction or remainder shall be treated as the next number higher than it which is so divisible

APPENDIX G

Health and Safety: The Manual Handling Operations Regulations 1992 (as amended)

1992 No. 2793

Made   *5th November 1992*
Laid before Parliament   *16th November 1992*
Coming into force   *1st January 1993*

The Secretary of State, in exercise of the powers conferred on her by sections 15(1), (2), (3)(a), (5)(a) and (9) and 80(1), (2)(a) and (4) of, and paragraphs 1(1)(a) and (c) and 8 of Schedule 3 to, the Health and Safety at Work etc. Act 1974 ("the 1974 Act") and of all other powers enabling her in that behalf and—

(a) for the purpose of giving effect without modifications to proposals submitted to her by the Health and Safety Commission under section 11(2)(d) of the 1974 Act after the carrying out by the said Commission of consultations in accordance with section 50(3) of that Act; and

(b) it appearing to her that the repeal of section 18(1)(f) of the Children and Young Persons Act 1933 and section 28(1)(f) of the Children and Young Persons (Scotland) Act 1937 except insofar as those provisions apply to such employment as is permitted under section 1(2) of the Employment of Women, Young Persons, and Children Act 1920 is expedient in consequence of the Regulations referred to below after the carrying out by her of consultations in accordance with section 80(4) of the 1974 Act,

hereby makes the following Regulations:

1 Citation and commencement

These Regulations may be cited as the Manual Handling Operations Regulations 1992 and shall come into force on 1st January 1993.

2 Interpretation

(1) In these Regulations, unless the context otherwise requires—
"injury" does not include injury caused by any toxic or corrosive substance which—
(a) has leaked or spilled from a load;
(b) is present on the surface of a load but has not leaked or spilled from it; or
(c) is a constituent part of a load;
and "injured" shall be construed accordingly;
"load" includes any person and any animal;
"manual handling operations" means any transporting or supporting of a load (including the lifting, putting down, pushing, pulling, carrying or moving thereof) by hand or by bodily force.

(2) Any duty imposed by these Regulations on an employer in respect of his employees shall also be imposed on a self-employed person in respect of himself.

Appendix G

3 Disapplication of Regulations

These Regulations shall not apply to or in relation to the master or crew of a sea-going ship or to the employer of such persons in respect of the normal ship-board activities of a ship's crew under the direction of the master.

4 Duties of employers

(1) Each employer shall—
- (a) so far as is reasonably practicable, avoid the need for his employees to undertake any manual handling operations at work which involve a risk of their being injured; or
- (b) where it is not reasonably practicable to avoid the need for his employees to undertake any manual handling operations at work which involve a risk of their being injured—
    - (i) make a suitable and sufficient assessment of all such manual handling operations to be undertaken by them, having regard to the factors which are specified in column 1 of Schedule 1 to these Regulations and considering the questions which are specified in the corresponding entry in column 2 of that Schedule,
    - (ii) take appropriate steps to reduce the risk of injury to those employees arising out of their undertaking any such manual handling operations to the lowest level reasonably practicable, and
    - (iii) take appropriate steps to provide any of those employees who are undertaking any such manual handling operations with general indications and, where it is reasonably practicable to do so, precise information on—
        - (aa) the weight of each load, and
        - (bb) the heaviest side of any load whose centre of gravity is not positioned centrally.

(2) Any assessment such as is referred to in paragraph (1)(b)(i) of this regulation shall be reviewed by the employer who made it if—
- (a) there is reason to suspect that it is no longer valid; or
- (b) there has been a significant change in the manual handling operations to which it relates;

and where as a result of any such review changes to an assessment are required, the relevant employer shall make them.

[(3) In determining for the purposes of this regulation whether manual handling operations at work involve a risk of injury and in determining the appropriate steps to reduce that risk regard shall be had in particular to—
- (a) the physical suitability of the employee to carry out the operations;
- (b) the clothing, footwear or other personal effects he is wearing;
- (c) his knowledge and training;
- (d) the results of any relevant risk assessment carried out pursuant to regulation 3 of the Management of Health and Safety at Work Regulations 1999;
- (e) whether the employee is within a group of employees identified by that assessment as being especially at risk; and
- (f) the results of any health surveillance provided pursuant to regulation 6 of the Management of Health and Safety Regulations 1999.][1]

5 Duty of employees

Each employee while at work shall make full and proper use of any system of work provided for his use by his employer in compliance with regulation 4(1)(b)(ii) of these Regulations.

---

[1] Inserted by SI 2002/2174.

6 Exemption certificates

(1) The Secretary of State for Defence may, in the interests of national security, by a certificate in writing exempt—
   (a) any of the home forces, any visiting force or any headquarters from any requirement imposed by regulation 4 of these Regulations; or
   (b) any member of the home forces, any member of a visiting force or any member of a headquarters from the requirement imposed by regulation 5 of these Regulations;
   and any exemption such as is specified in sub-paragraph (a) or (b) of this paragraph may be granted subject to conditions and to a limit of time and may be revoked by the said Secretary of State by a further certificate in writing at any time.
(2) In this regulation—
   (a) "the home forces" has the same meaning as in section 12(1) of the Visiting Forces Act 1952;
   (b) "headquarters" has the same meaning as in article 3(2) of the Visiting Forces and International Headquarters (Application of Law) Order 1965;
   (c) "member of a headquarters" has the same meaning as in paragraph 1(1) of the Schedule to the International Headquarters and Defence Organisations Act 1964; and
   (d) "visiting force" has the same meaning as it does for the purposes of any provision of Part I of the Visiting Forces Act 1952.

7 Extension outside Great Britain

These Regulations shall, subject to regulation 3 hereof, apply to and in relation to the premises and activities outside Great Britain to which sections 1 to 59 and 80 to 82 of the Health and Safety at Work etc. Act 1974 apply by virtue of the Health and Safety at Work etc. Act 1974 (Application Outside Great Britain) Order 1989 as they apply within Great Britain.

8 Repeals and revocations

(1) The enactments mentioned in column 1 of Part I of Schedule 2 to these Regulations are repealed to the extent specified in the corresponding entry in column 3 of that part.
(2) The Regulations mentioned in column 1 of Part II of Schedule 2 to these Regulations are revoked to the extent specified in the corresponding entry in column 3 of that part.

SCHEDULE 1

Regulation 4(1)(b)(i)

FACTORS TO WHICH THE EMPLOYER MUST HAVE REGARD AND QUESTIONS HE MUST CONSIDER WHEN MAKING AN ASSESSMENT OF MANUAL HANDLING OPERATIONS

Column 1 Factors	Column 2 Questions
1. The tasks	**Do they involve:** —holding or manipulating loads at distance from trunk? —unsatisfactory bodily movement or posture, especially: —twisting the trunk? —stooping? —reaching upwards? —excessive movement of loads, especially: —excessive lifting or lowering distances? —excessive carrying distances? —excessive pushing or pulling of loads? —risk of sudden movement of loads?

	—frequent or prolonged physical effort? —insufficient rest or recovery periods? —a rate of work imposed by a process?
2. The loads	**Are they:** —heavy? —bulky or unwieldy? —difficult to grasp? —unstable, or with contents likely to shift? —sharp, hot or otherwise potentially damaging?
3. The working environment	**Are there:** —space constraints preventing good posture? —uneven, slippery or unstable floors? —variations in level of floors or work surfaces? —extremes of temperature or humidity? —conditions causing ventilation problems or gusts of wind? —poor lighting conditions?
4. Individual capability	**Does the job:** —require unusual strength, height, etc? —create a hazard to those who might reasonably be considered to be —pregnant or to have a health problem? —require special information or training for its safe performance?
5. Other factors	**Is movement or posture hindered by personal protective equipment by clothing?**

APPENDIX H

Health and Safety: The Provision and Use of Work Equipment Regulations 1998 (as amended)

1998 No. 2306

Made   *15th September 1998*
Laid before Parliament   *25th September 1998*
Coming into force   *5th December 1998*

The Secretary of State, in the exercise of the powers conferred on him by sections 15(1), (2), (3)(a), (5) and (6)(a), 49 and 82(3)(a) of, and paragraphs 1(1), (2) and (3), 9, 14, 15(1) and 16 of Schedule 3 to, the Health and Safety at Work etc. Act 1974 ("the 1974 Act") and of all other powers enabling him in that behalf and for the purpose of giving effect without modifications to proposals submitted to him by the Health and Safety Commission under section 11(2)(d) of the 1974 Act, after the carrying out by the said Commission of consultations in accordance with section 50(3) of that Act, hereby makes the following Regulations:

Part I
Introduction

1 Citation and commencement

These Regulations may be cited as the Provision and Use of Work Equipment Regulations 1998 and shall come into force on 5th December 1998.

2 Interpretation

(1) In these Regulations, unless the context otherwise requires—
   "the 1974 Act" means the Health and Safety at Work etc. Act 1974;
   "employer" except in regulation 3(2) and (3) includes a person to whom the requirements imposed by these Regulations apply by virtue of regulation 3(3)(a) and (b);
   "essential requirements" means requirements described in regulation 10(1);
   "the Executive" means the Health and Safety Executive;
   "inspection" in relation to an inspection under paragraph (1) or (2) of regulation 6—
   (a) means such visual or more rigorous inspection by a competent person as is appropriate for the purpose described in the paragraph;
   (b) where it is appropriate to carry out testing for the purpose, includes testing the nature and extent of which are appropriate for the purpose;
   "power press" means a press or press brake for the working of metal by means of tools, or for die proving, which is power driven and which embodies a flywheel and clutch;
   "thorough examination" in relation to a thorough examination under paragraph (1), (2), (3) or (4) of regulation 32—
   (a) means a thorough examination by a competent person;
   (b) includes testing the nature and extent of which are appropriate for the purpose described in the paragraph;

"use" in relation to work equipment means any activity involving work equipment and includes starting, stopping, programming, setting, transporting, repairing, modifying, maintaining, servicing and cleaning;

"work equipment" means any machinery, appliance, apparatus, tool or installation for use at work (whether exclusively or not);

and related expressions shall be construed accordingly.

(2) Any reference in regulations 32 to 34 or Schedule 3 to a guard or protection device is a reference to a guard or protection device provided for the tools of a power press.

(3) Any reference in regulation 32 or 33 to a guard or protection device being on a power press shall, in the case of a guard or protection device designed to operate while adjacent to a power press, be construed as a reference to its being adjacent to it.

(4) Any reference in these Regulations to—
   (a) a numbered regulation or Schedule is a reference to the regulation or Schedule in these Regulations so numbered; and
   (b) a numbered paragraph is a reference to the paragraph so numbered in the regulation in which the reference appears.

3 Application

(1) These Regulations shall apply—
   (a) in Great Britain; and
   (b) outside Great Britain as sections 1 to 59 and 80 to 82 of the 1974 Act apply by virtue of the Health and Safety at Work etc. Act 1974 (Application outside Great Britain) Order 1995 ("the 1995 Order").

(2) The requirements imposed by these Regulations on an employer in respect of work equipment shall apply to such equipment provided for use or used by an employee of his at work.

(3) The requirements imposed by these Regulations on an employer shall also apply—
   (a) to a self-employed person, in respect of work equipment he uses at work;
   (b) subject to paragraph (5), to a person who has control to any extent of—
      (i) work equipment;
      (ii) a person at work who uses or supervises or manages the use of work equipment; or
      (iii) the way in which work equipment is used at work,
   and to the extent of his control.

(4) Any reference in paragraph (3)(b) to a person having control is a reference to a person having control in connection with the carrying on by him of a trade, business or other undertaking (whether for profit or not).

(5) The requirements imposed by these Regulations shall not apply to a person in respect of work equipment supplied by him by way of sale, agreement for sale or hire-purchase agreement.

(6) Subject to paragraphs (7) to (10), these Regulations shall not impose any obligation in relation to a ship's work equipment (whether that equipment is used on or off the ship).

(7) Where merchant shipping requirements are applicable to a ship's work equipment, paragraph (6) shall relieve the shore employer of his obligations under these Regulations in respect of that equipment only where he has taken all reasonable steps to satisfy himself that the merchant shipping requirements are being complied with in respect of that equipment.

(8) In a case where the merchant shipping requirements are not applicable to the ship's work equipment by reason only that for the time being there is no master, crew or watchman on the ship, those requirements shall nevertheless be treated for the purpose of paragraph (7) as if they were applicable.

(9) Where the ship's work equipment is used in a specified operation paragraph (6) shall not apply to regulations 7 to 9, 11 to 13, 20 to 22 and 30 (each as applied by regulation 3).

(10) Paragraph (6) does not apply to a ship's work equipment provided for use or used in an activity (whether carried on in or outside Great Britain) specified in the 1995 Order save that it does apply to—
   (a) the loading, unloading, fuelling or provisioning of the ship; or
   (b) the construction, reconstruction, finishing, refitting, repair, maintenance, cleaning or breaking up of the ship.
(11) In this regulation—
   "master" has the meaning assigned to it by section 313(1) of the Merchant Shipping Act 1995;
   "merchant shipping requirements" means the requirements of regulations 3 and 4 of the Merchant Shipping (Guarding of Machinery and Safety of Electrical Equipment) Regulations 1988 and regulations 5 to 10 of the Merchant Shipping (Hatches and Lifting Plant) Regulations 1988;
   "ship" has the meaning assigned to it by section 313(1) of the Merchant Shipping Act 1995 save that it does not include an offshore installation;
   "shore employer" means an employer of persons (other than the master and crew of any ship) who are engaged in a specified operation;
   "specified operation" means an operation in which the ship's work equipment is used—
      (a) by persons other than the master and crew; or
      (b) where persons other than the master and crew are liable to be exposed to a risk to their health or safety from its use.

Part II
General

4 Suitability of work equipment

(1) Every employer shall ensure that work equipment is so constructed or adapted as to be suitable for the purpose for which it is used or provided.
(2) In selecting work equipment, every employer shall have regard to the working conditions and to the risks to the health and safety of persons which exist in the premises or undertaking in which that work equipment is to be used and any additional risk posed by the use of that work equipment.
(3) Every employer shall ensure that work equipment is used only for operations for which, and under conditions for which, it is suitable.
[(4) In this regulation "suitable"—
   (a) subject to sub-paragraph (b), means suitable in any respect which it is reasonably foreseeable will affect the health or safety of any person;
   (b) in relation to—
      (i) an offensive weapon within the meaning of section 1(4) of the Prevention of Crime Act 1953 provided for use as self-defence or as deterrent equipment; and
      (ii) work equipment provided for use for arrest or restraint,
   by a person who holds the office of constable or an appointment as police cadet, means suitable in any respect which it is reasonably foreseeable will affect the health or safety of such person.][1]

---

[1] Inserted by SI 1999/860.

5 Maintenance

(1) Every employer shall ensure that work equipment is maintained in an efficient state, in efficient working order and in good repair.
(2) Every employer shall ensure that where any machinery has a maintenance log, the log is kept up to date.

6 Inspection

(1) Every employer shall ensure that, where the safety of work equipment depends on the installation conditions, it is inspected—
    (a) after installation and before being put into service for the first time; or
    (b) after assembly at a new site or in a new location,
    to ensure that it has been installed correctly and is safe to operate.
(2) Every employer shall ensure that work equipment exposed to conditions causing deterioration which is liable to result in dangerous situations is inspected—
    (a) at suitable intervals; and
    (b) each time that exceptional circumstances which are liable to jeopardise the safety of the work equipment have occurred,
    to ensure that health and safety conditions are maintained and that any deterioration can be detected and remedied in good time.
(3) Every employer shall ensure that the result of an inspection made under this regulation is recorded and kept until the next inspection under this regulation is recorded.
(4) Every employer shall ensure that no work equipment—
    (a) leaves his undertaking; or
    (b) if obtained from the undertaking of another person, is used in his undertaking,
    unless it is accompanied by physical evidence that the last inspection required to be carried out under this regulation has been carried out.
(5) This regulation does not apply to—
    (a) a power press to which regulations 32 to 35 apply;
    (b) a guard or protection device for the tools of such power press;
    (c) work equipment for lifting loads including persons;
    (d) winding apparatus to which the Mines (Shafts and Winding) Regulations 1993 apply;
    (e) work equipment required to be inspected by regulation 29 of the Construction (Health, Safety and Welfare) Regulations 1996;
    [(f) work equipment to which regulation 12 of the Work at Height Regulations 2005 applies].[2]

7 Specific risks

(1) Where the use of work equipment is likely to involve a specific risk to health or safety, every employer shall ensure that—
    (a) the use of that work equipment is restricted to those persons given the task of using it; and
    (b) repairs, modifications, maintenance or servicing of that work equipment is restricted to those persons who have been specifically designated to perform operations of that description (whether or not also authorised to perform other operations).
(2) The employer shall ensure that the persons designated for the purposes of sub-paragraph (b) of paragraph (1) have received adequate training related to any operations in respect of which they have been so designated.

---

[2] Amended by SI 2005/735.

8 Information and instructions

(1) Every employer shall ensure that all persons who use work equipment have available to them adequate health and safety information and, where appropriate, written instructions pertaining to the use of the work equipment.
(2) Every employer shall ensure that any of his employees who supervises or manages the use of work equipment has available to him adequate health and safety information and, where appropriate, written instructions pertaining to the use of the work equipment.
(3) Without prejudice to the generality of paragraphs (1) or (2), the information and instructions required by either of those paragraphs shall include information and, where appropriate, written instructions on—
   (a) the conditions in which and the methods by which the work equipment may be used;
   (b) foreseeable abnormal situations and the action to be taken if such a situation were to occur; and
   (c) any conclusions to be drawn from experience in using the work equipment.
(4) Information and instructions required by this regulation shall be readily comprehensible to those concerned.

9 Training

(1) Every employer shall ensure that all persons who use work equipment have received adequate training for purposes of health and safety, including training in the methods which may be adopted when using the work equipment, any risks which such use may entail and precautions to be taken.
(2) Every employer shall ensure that any of his employees who supervises or manages the use of work equipment has received adequate training for purposes of health and safety, including training in the methods which may be adopted when using the work equipment, any risks which such use may entail and precautions to be taken.

10 Conformity with Community requirements

[(1) Every employer shall ensure that an item of work equipment conforms at all times with any essential requirements, other than requirements which, at the time of its being first supplied or put into service in any place in which these Regulations apply, did not apply to work equipment of its type.
(2) In this regulation "essential requirements", in relation to an item of work equipment, means requirements relating to the design and construction of work equipment of its type in any of the instruments listed in Schedule 1 (being instruments which give effect to Community directives concerning the safety of products).]³
(3) This regulation applies to items of work equipment provided for use in the premises or undertaking of the employer for the first time after 31st December 1992.

11 Dangerous parts of machinery

(1) Every employer shall ensure that measures are taken in accordance with paragraph (2) which are effective—
   (a) to prevent access to any dangerous part of machinery or to any rotating stock-bar; or
   (b) to stop the movement of any dangerous part of machinery or rotating stock-bar before any part of a person enters a danger zone.
[(2) The measures required by paragraph (1) shall consist of—
   (a) the provision of fixed guards enclosing every dangerous part or rotating stock-bar where and to the extent that it is practicable to do so, but where or to the extent that it is not, then

---

³ Inserted by SI 2002/2174.

(b) the provision of other guards or protection devices where and to the extent that it is practicable to do so, but where or to the extent that it is not, then

(c) the provision of jigs, holders, push-sticks or similar protection appliances used in conjunction with the machinery where and to the extent that it is practicable to do so,

and the provision of such information, instruction, training and supervision as is necessary.][4]

(3) All guards and protection devices provided under sub-paragraphs (a) or (b) of paragraph (2) shall—

(a) be suitable for the purpose for which they are provided;
(b) be of good construction, sound material and adequate strength;
(c) be maintained in an efficient state, in efficient working order and in good repair;
(d) not give rise to any increased risk to health or safety;
(e) not be easily bypassed or disabled;
(f) be situated at sufficient distance from the danger zone;
(g) not unduly restrict the view of the operating cycle of the machinery, where such a view is necessary;
(h) be so constructed or adapted that they allow operations necessary to fit or replace parts and for maintenance work, restricting access so that it is allowed only to the area where the work is to be carried out and, if possible, without having to dismantle the guard or protection device.

(4) All protection appliances provided under sub-paragraph (c) of paragraph (2) shall comply with sub-paragraphs (a) to (d) and (g) of paragraph (3).

(5) In this regulation—

"danger zone" means any zone in or around machinery in which a person is exposed to a risk to health or safety from contact with a dangerous part of machinery or a rotating stock-bar;

"stock-bar" means any part of a stock-bar which projects beyond the head-stock of a lathe.

12 Protection against specified hazards

(1) Every employer shall take measures to ensure that the exposure of a person using work equipment to any risk to his health or safety from any hazard specified in paragraph (3) is either prevented, or, where that is not reasonably practicable, adequately controlled.

(2) The measures required by paragraph (1) shall—

(a) be measures other than the provision of personal protective equipment or of information, instruction, training and supervision, so far as is reasonably practicable; and
(b) include, where appropriate, measures to minimise the effects of the hazard as well as to reduce the likelihood of the hazard occurring.

(3) The hazards referred to in paragraph (1) are—

(a) any article or substance falling or being ejected from work equipment;
(b) rupture or disintegration of parts of work equipment;
(c) work equipment catching fire or overheating;
(d) the unintended or premature discharge of any article or of any gas, dust, liquid, vapour or other substance which, in each case, is produced, used or stored in the work equipment;
(e) the unintended or premature explosion of the work equipment or any article or substance produced, used or stored in it.

(4) For the purposes of this regulation "adequately" means adequately having regard only to the nature of the hazard and the nature and degree of exposure to the risk.

(5) This regulation shall not apply where any of the following Regulations apply in respect of any risk to a person's health or safety for which such Regulations require measures to be taken to prevent or control such risk, namely—

(a) the Ionising Radiations Regulations 1985;
(b) the Control of Asbestos at Work Regulations 1987;

---

[4] Inserted by SI 2002/2174.

(c) the Control of Substances Hazardous to Health Regulations 1994;
(d) [the Noise at Work Regulations 1989][5] [the Control of Noise at Work Regulations 2005];[6]
(e) the Construction (Head Protection) Regulations 1989;
(f) the Control of Lead at Work Regulations 1998;
[(g) the Control of Vibration at Work Regulations 2005].[7]

13 High or very low temperature

Every employer shall ensure that work equipment, parts of work equipment and any article or substance produced, used or stored in work equipment which, in each case, is at a high or very low temperature shall have protection where appropriate so as to prevent injury to any person by burn, scald or sear.

14 Controls for starting or making a significant change in operating conditions

(1) Every employer shall ensure that, where appropriate, work equipment is provided with one or more controls for the purposes of—
   (a) starting the work equipment (including re-starting after a stoppage for any reason); or
   (b) controlling any change in the speed, pressure or other operating conditions of the work equipment where such conditions after the change result in risk to health and safety which is greater than or of a different nature from such risks before the change.
(2) Subject to paragraph (3), every employer shall ensure that, where a control is required by paragraph (1), it shall not be possible to perform any operation mentioned in sub-paragraph (a) or (b) of that paragraph except by a deliberate action on such control.
(3) Paragraph (1) shall not apply to re-starting or changing operating conditions as a result of the normal operating cycle of an automatic device.

15 Stop controls

(1) Every employer shall ensure that, where appropriate, work equipment is provided with one or more readily accessible controls the operation of which will bring the work equipment to a safe condition in a safe manner.
(2) Any control required by paragraph (1) shall bring the work equipment to a complete stop where necessary for reasons of health and safety.
(3) Any control required by paragraph (1) shall, if necessary for reasons of health and safety, switch off all sources of energy after stopping the functioning of the work equipment.
(4) Any control required by paragraph (1) shall operate in priority to any control which starts or changes the operating conditions of the work equipment.

16 Emergency stop controls

(1) Every employer shall ensure that, where appropriate, work equipment is provided with one or more readily accessible emergency stop controls unless it is not necessary by reason of the nature of the hazards and the time taken for the work equipment to come to a complete stop as a result of the action of any control provided by virtue of regulation 15(1).
(2) Any control required by paragraph (1) shall operate in priority to any control required by regulation 15(1).

17 Controls

(1) Every employer shall ensure that all controls for work equipment are clearly visible and identifiable, including by appropriate marking where necessary.

---

[5] Repealed by SI 2005/1643 (date in force: 6 April 2006) except for the music and entertainment sectors (for these sectors the repeal will be effective from 6 April 2008).
[6] Amended by SI 2005/1643 (date in force: 6 April 2006) except for the music and entertainment sectors (for these sectors the amendment will be effective from 6 April 2008).
[7] Amended by SI 2005/1093.

(2) Except where necessary, the employer shall ensure that no control for work equipment is in a position where any person operating the control is exposed to a risk to his health or safety.

(3) Every employer shall ensure where appropriate—
  (a) that, so far as is reasonably practicable, the operator of any control is able to ensure from the position of that control that no person is in a place where he would be exposed to any risk to his health or safety as a result of the operation of that control, but where or to the extent that it is not reasonably practicable;
  (b) that, so far as is reasonably practicable, systems of work are effective to ensure that, when work equipment is about to start, no person is in a place where he would be exposed to a risk to his health or safety as a result of the work equipment starting, but where neither of these is reasonably practicable;
  (c) that an audible, visible or other suitable warning is given by virtue of regulation 24 whenever work equipment is about to start.

(4) Every employer shall take appropriate measures to ensure that any person who is in a place where he would be exposed to a risk to his health or safety as a result of the starting or stopping of work equipment has sufficient time and suitable means to avoid that risk.

18 Control systems

[(1) Every employer shall ensure, so far as is reasonably practicable, that all control systems of work equipment—
  (a) are safe; and
  (b) are chosen making due allowance for the failures, faults and constraints to be expected in the planned circumstances of use.][8]

(2) Without prejudice to the generality of paragraph (1), a control system shall not be safe unless—
  (a) its operation does not create any increased risk to health or safety;
  (b) it ensures, so far as is reasonably practicable, that any fault in or damage to any part of the control system or the loss of supply of any source of energy used by the work equipment cannot result in additional or increased risk to health or safety;
  (c) it does not impede the operation of any control required by regulation 15 or 16.

19 Isolation from sources of energy

(1) Every employer shall ensure that where appropriate work equipment is provided with suitable means to isolate it from all its sources of energy.

(2) Without prejudice to the generality of paragraph (1), the means mentioned in that paragraph shall not be suitable unless they are clearly identifiable and readily accessible.

(3) Every employer shall take appropriate measures to ensure that re-connection of any energy source to work equipment does not expose any person using the work equipment to any risk to his health or safety.

20 Stability

Every employer shall ensure that work equipment or any part of work equipment is stabilised by clamping or otherwise where necessary for purposes of health or safety.

21 Lighting

Every employer shall ensure that suitable and sufficient lighting, which takes account of the operations to be carried out, is provided at any place where a person uses work equipment.

---

[8] Inserted by SI 2002/2174.

22 Maintenance operations

Every employer shall take appropriate measures to ensure that work equipment is so constructed or adapted that, so far as is reasonably practicable, maintenance operations which involve a risk to health or safety can be carried out while the work equipment is shut down, or in other cases—
(a) maintenance operations can be carried out without exposing the person carrying them out to a risk to his health or safety; or
(b) appropriate measures can be taken for the protection of any person carrying out maintenance operations which involve a risk to his health or safety.

23 Markings

Every employer shall ensure that work equipment is marked in a clearly visible manner with any marking appropriate for reasons of health and safety.

24 Warnings

(1) Every employer shall ensure that work equipment incorporates any warnings or warning devices which are appropriate for reasons of health and safety.
(2) Without prejudice to the generality of paragraph (1), warnings given by warning devices on work equipment shall not be appropriate unless they are unambiguous, easily perceived and easily understood.

Part III
Mobile Work Equipment

25 Employees carried on mobile work equipment

Every employer shall ensure that no employee is carried by mobile work equipment unless—
(a) it is suitable for carrying persons; and
(b) it incorporates features for reducing to as low as is reasonably practicable risks to their safety, including risks from wheels or tracks.

26 Rolling over of mobile work equipment

(1) Every employer shall ensure that where there is a risk to an employee riding on mobile work equipment from its rolling over, it is minimised by—
  (a) stabilising the work equipment;
  (b) a structure which ensures that the work equipment does no more than fall on its side;
  (c) a structure giving sufficient clearance to anyone being carried if it overturns further than that; or
  (d) a device giving comparable protection.
(2) Where there is a risk of anyone being carried by mobile work equipment being crushed by its rolling over, the employer shall ensure that it has a suitable restraining system for him.
(3) This regulation shall not apply to a fork-lift truck having a structure described in sub-paragraph (b) or (c) of paragraph (1).
(4) Compliance with this regulation is not required where—
  (a) it would increase the overall risk to safety;
  (b) it would not be reasonably practicable to operate the mobile work equipment in consequence; or
  (c) in relation to an item of work equipment provided for use in the undertaking or establishment before 5th December 1998 it would not be reasonably practicable.

27 Overturning of fork-lift trucks

Every employer shall ensure that a fork-lift truck to which regulation 26(3) refers and which carries an employee is adapted or equipped to reduce to as low as is reasonably practicable the risk to safety from its overturning.

28 Self-propelled work equipment

Every employer shall ensure that, where self-propelled work equipment may, while in motion, involve risk to the safety of persons—
(a) it has facilities for preventing its being started by an unauthorised person;
(b) it has appropriate facilities for minimising the consequences of a collision where there is more than one item of rail-mounted work equipment in motion at the same time;
(c) it has a device for braking and stopping;
(d) where safety constraints so require, emergency facilities operated by readily accessible controls or automatic systems are available for braking and stopping the work equipment in the event of failure of the main facility;
(e) where the driver's direct field of vision is inadequate to ensure safety, there are adequate devices for improving his vision so far as is reasonably practicable;
(f) if provided for use at night or in dark places—
   (i) it is equipped with lighting appropriate to the work to be carried out; and
   (ii) is otherwise sufficiently safe for such use;
(g) if it, or anything carried or towed by it, constitutes a fire hazard and is liable to endanger employees, it carries appropriate fire-fighting equipment, unless such equipment is kept sufficiently close to it.

29 Remote-controlled self-propelled work equipment

Every employer shall ensure that where remote-controlled self-propelled work equipment involves a risk to safety while in motion—
(a) it stops automatically once it leaves its control range; and
(b) where the risk is of crushing or impact it incorporates features to guard against such risk unless other appropriate devices are able to do so.

30 Drive shafts

(1) Where the seizure of the drive shaft between mobile work equipment and its accessories or anything towed is likely to involve a risk to safety every employer shall—
   (a) ensure that the work equipment has a means of preventing such seizure; or
   (b) where such seizure cannot be avoided, take every possible measure to avoid an adverse effect on the safety of an employee.
(2) Every employer shall ensure that—
   (a) where mobile work equipment has a shaft for the transmission of energy between it and other mobile work equipment; and
   (b) the shaft could become soiled or damaged by contact with the ground while uncoupled,
   the work equipment has a system for safeguarding the shaft.

Part IV
Power Presses

31 Power presses to which Part IV does not apply

Regulations 32 to 35 shall not apply to a power press of a kind which is described in Schedule 2.

32 Thorough examination of power presses, guards and protection devices

(1) Every employer shall ensure that a power press is not put into service for the first time after installation, or after assembly at a new site or in a new location unless—
   (a) it has been thoroughly examined to ensure that it—
      (i) has been installed correctly; and
      (ii) would be safe to operate; and
   (b) any defect has been remedied.

The Provision and Use of Work Equipment Regulations 1998 (as amended)

(2) Every employer shall ensure that a guard, other than one to which paragraph (3) relates, or protection device is not put into service for the first time on a power press unless—
  (a) it has been thoroughly examined when in position on that power press to ensure that it is effective for its purpose; and
  (b) any defect has been remedied.
(3) Every employer shall ensure that that part of a closed tool which acts as a fixed guard is not used on a power press unless—
  (a) it has been thoroughly examined when in position on any power press in the premises to ensure that it is effective for its purpose; and
  (b) any defect has been remedied.
(4) For the purpose of ensuring that health and safety conditions are maintained, and that any deterioration can be detected and remedied in good time, every employer shall ensure that—
  (a) every power press is thoroughly examined, and its guards and protection devices are thoroughly examined when in position on that power press—
    (i) at least every 12 months, where it has fixed guards only; or
    (ii) at least every 6 months, in other cases; and
    (iii) each time that exceptional circumstances have occurred which are liable to jeopardise the safety of the power press or its guards or protection devices; and
  (b) any defect is remedied before the power press is used again.
(5) Where a power press, guard or protection device was before the coming into force of these Regulations required to be thoroughly examined by regulation 5(2) of the Power Presses Regulations 1965 the first thorough examination under paragraph (4) shall be made before the date by which a thorough examination would have been required by regulation 5(2) had it remained in force.
(6) Paragraph (4) shall not apply to that part of a closed tool which acts as a fixed guard.
(7) In this regulation "defect" means a defect notified under regulation 34 other than a defect which has not yet become a danger to persons.

33 Inspection of guards and protection devices

(1) Every employer shall ensure that a power press is not used after the setting, re-setting or adjustment of its tools, save in trying out its tools or save in die proving, unless—
  (a) its every guard and protection device has been inspected and tested while in position on the power press by a person appointed in writing by the employer who is—
    (i) competent; or
    (ii) undergoing training for that purpose and acting under the immediate supervision of a competent person,
   and who has signed a certificate which complies with paragraph (3); or
  (b) the guards and protection devices have not been altered or disturbed in the course of the adjustment of its tools.
(2) Every employer shall ensure that a power press is not used after the expiration of the fourth hour of a working period unless its every guard and protection device has been inspected and tested while in position on the power press by a person appointed in writing by the employer who is—
  (a) competent; or
  (b) undergoing training for that purpose and acting under the immediate supervision of a competent person,
   and who has signed a certificate which complies with paragraph (3).
(3) A certificate referred to in this regulation shall—
  (a) contain sufficient particulars to identify every guard and protection device inspected and tested and the power press on which it was positioned at the time of the inspection and test;
  (b) state the date and time of the inspection and test; and

(c) state that every guard and protection device on the power press is in position and effective for its purpose.

(4) In this regulation "working period", in relation to a power press, means—
   (a) the period in which the day's or night's work is done; or
   (b) in premises where a shift system is in operation, a shift.

34 Reports

(1) A person making a thorough examination for an employer under regulation 32 shall—
   (a) notify the employer forthwith of any defect in a power press or its guard or protection device which in his opinion is or could become a danger to persons;
   (b) as soon as is practicable make a report of the thorough examination to the employer in writing authenticated by him or on his behalf by signature or equally secure means and containing the information specified in Schedule 3; and
   (c) where there is in his opinion a defect in a power press or its guard or protection device which is or could become a danger to persons, send a copy of the report as soon as is practicable to the enforcing authority for the premises in which the power press is situated.

(2) A person making an inspection and test for an employer under regulation 33 shall forthwith notify the employer of any defect in a guard or protection device which in his opinion is or could become a danger to persons and the reason for his opinion.

35 Keeping of information

(1) Every employer shall ensure that the information in every report made pursuant to regulation 34(1) is kept available for inspection for 2 years after it is made.

(2) Every employer shall ensure that a certificate under regulation [33][9] is kept available for inspection—
   (a) at or near the power press to which it relates until superseded by a later certificate; and
   (b) after that, until 6 months have passed since it was signed.

Part V
Miscellaneous

36 Exemption for the armed forces

(1) The Secretary of State for Defence may, in the interests of national security, by a certificate in writing exempt any of the home forces, any visiting force or any headquarters from any requirement or prohibition imposed by these Regulations and any such exemption may be granted subject to conditions and to a limit of time and may be revoked by the said Secretary of State by a certificate in writing at any time.

(2) In this regulation—
   (a) "the home forces" has the same meaning as in section 12(1) of the Visiting Forces Act 1952;
   (b) "headquarters" has the same meaning as in article 3(2) of the Visiting Forces and International Headquarters (Application of Law) Order 1965;
   (c) "visiting force" has the same meaning as it does for the purposes of any provision of Part I of the Visiting Forces Act 1952.

37 Transitional provision

The requirements in regulations 25 to 30 shall not apply to work equipment provided for use in the undertaking or establishment before 5th December 1998 until 5th December 2002.

---

[9] Amended by SI 2002/2174.

38 Repeal of enactment

Section 19 of the Offices, Shops and Railway Premises Act 1963 is repealed.

39 Revocation of instruments

The instruments specified in column 1 of Schedule 4 are revoked to the extent specified in column 3 of that Schedule.

Signed by authority of the Secretary of State

Alan Meale

Parliamentary Under Secretary of State

Department of the Environment, Transport and the Regions

15th September 1998

SCHEDULE 1

Regulation 10

INSTRUMENTS WHICH GIVE EFFECT TO COMMUNITY DIRECTIVES CONCERNING THE SAFETY OF PRODUCTS

(1) Title	(2) Reference
The Construction Plant and Equipment (Harmonisation of Noise Emission Standards) Regulations 1985	SI 1985/1968, amended by SI 1989/1127
The Construction Plant and Equipment (Harmonisation of Noise Emission Standards) Regulations 1988	SI 1988/361, amended by SI 1992/488, 1995/2357
The Electro-medical Equipment (EEC Requirements) Regulations 1988	SI 1988/1586, amended by SI 1994/3017
The Low Voltage Electrical Equipment (Safety) Regulations 1989	SI 1989/728, amended by SI 1994/3260
The Construction Products Regulations 1991	SI 1991/1620, amended by SI 1994/3051
The Simple Pressure Vessels (Safety) Regulations 1991	SI 1991/2749, amended by SI 1994/3098
The Lawnmowers (Harmonisation of Noise Emission Standards) Regulations 1992	SI 1992/168
The Gas Appliances (Safety) Regulations 1992	SI 1992/711
The Electromagnetic Compatibility Regulations 1992	SI 1992/2372, amended by SI 1994/3080
The Supply of Machinery (Safety) Regulations 1992	SI 1992/3073, amended by SI 1994/2063 [SI 2005/831][10]

---

[10] Amended by SI 2005/831.

The Personal Protective Equipment (EC Directive) Regulations 1992	SI 1992/3139, amended by SI 1993/3074, 1994/2326, 1996/3039
The Active Implantable Medical Devices Regulations 1992	SI 1992/3146, amended by SI 1995/1671
The Medical Devices Regulations 1994	SI 1994/3017
The Electrical Equipment (Safety) Regulations 1994	SI 1994/3260
The Gas Appliances (Safety) Regulations 1995	SI 1995/1629
The Equipment and Protective Systems Intended for Use in Potentially Explosive Atmospheres Regulations 1996	[SI 1996/192, amended by SI 1998/81, 2001/3766, 2005/830][11]
The Lifts Regulations 1997	SI 1997/831[, amended by SI 2005/831][12]
[The Pressure Equipment Regulations 1999	SI 1999/2001][13]
[The Noise Emission in the Environment by Equipment for use Outdoors Regulations 2001	SI 2001/1701][14]
[The Cableway Installations Regulations 2004	SI 2004/129][15]

SCHEDULE 2

Regulation 31

POWER PRESSES TO WHICH REGULATIONS 32 TO 35 DO NOT APPLY

1. A power press for the working of hot metal.
2. A power press not capable of a stroke greater than 6 millimetres.
3. A guillotine.
4. A combination punching and shearing machine, turret punch press or similar machine for punching, shearing or cropping.
5. A machine, other than a press brake, for bending steel sections.
6. A straightening machine.
7. An upsetting machine.
8. A heading machine.
9. A riveting machine.
10. An eyeletting machine.
11. A press-stud attaching machine.
12. A zip fastener bottom stop attaching machine.
13. A stapling machine.
14. A wire stitching machine.
15. A power press for the compacting of metal powders.

---

[11] Amended by SI 2005/830.
[12] Amended by SI 2005/831.
[13] Inserted by SI 1999/2001.
[14] Inserted by SI 2001/1701.
[15] Inserted by SI 2004/129.

The Provision and Use of Work Equipment Regulations 1998 (as amended)

SCHEDULE 3

Regulation 34(1)(b)

INFORMATION TO BE CONTAINED IN A REPORT OF A THOROUGH EXAMINATION OF A POWER PRESS, GUARD OR PROTECTION DEVICE

1. The name of the employer for whom the thorough examination was made.
2. The address of the premises at which the thorough examination was made.
3. In relation to each item examined—
   (a) that it is a power press, interlocking guard, fixed guard or other type of guard or protection device;
   (b) where known its make, type and year of manufacture;
   (c) the identifying mark of—
      (i) the manufacture,
      (ii) the employer.
4. In relation to the first thorough examination of a power press after installation or after assembly at a new site or in a new location—
   (a) that it is such thorough examination;
   (b) either that it has been installed correctly and would be safe to operate or the respects in which it has not been installed correctly or would not be safe to operate;
   (c) identification of any part found to have a defect, and a description of the defect.
5. In relation to a thorough examination of a power press other than one to which paragraph 4 relates—
   (a) that it is such other thorough examination;
   (b) either that the power press would be safe to operate or the respects in which it would not be safe to operate;
   (c) identification of any part found to have a defect which is or could become a danger to persons, and a description of the defect.
6. In relation to a thorough examination of a guard or protection device—
   (a) either that it is effective for its purpose or the respects in which it is not effective for its purpose;
   (b) identification of any part found to have a defect which is or could become a danger to persons, and a description of the defect.
7. Any repair, renewal or alteration required to remedy a defect found to be a danger to persons.
8. In the case of a defect which is not yet but could become a danger to persons—
   (a) the time by which it could become such danger;
   (b) any repair, renewal or alteration required to remedy it.
9. Any other defect which requires remedy.
10. Any repair, renewal or alteration referred to in paragraph 7 which has already been effected.
11. The date on which any defect referred to in paragraph 8 was notified to the employer under regulation 34(1)(a).
12. The qualification and address of the person making the report; that he is self-employed or if employed, the name and address of his employer.
13. The date of the thorough examination.
14. The date of the report.
15. The name of the person making the report and where different the name of the person signing or otherwise authenticating it.

APPENDIX I

Health and Safety: The Control of Substances Hazardous to Health Regulations 2002 (as amended)

2002 No. 2677

Made   *24th October 2002*
Laid before Parliament   *31st October 2002*
Coming into force   *21st November 2002*

The Secretary of State being the Minister designated for the purpose of section 2(2) of the European Communities Act 1972 in relation to the abolition of restrictions on the import or export of goods, in the exercise of the powers conferred on him by the said section 2(2) and sections 15(1), (2), (3)(b), (4), (5)(b), (6)(b) and (9), 52(2) and (3) and 82(3)(a) of, and paragraphs 1(1) to (4), 2, 6(1), 8, 9, 11, 14, 15(1), 16 and 20 of Schedule 3 to, the Health and Safety at Work etc. Act 1974 ("the 1974 Act") and of all other powers enabling him in that behalf and for the purpose of giving effect without modifications to proposals submitted to him by the Health and Safety Commission under section 11(2)(d) of the 1974 Act after the carrying out by the said Commission of consultations in accordance with section 50(3) of that Act, hereby makes the following Regulations:

1 Citation and commencement

These Regulations may be cited as the Control of Substances Hazardous to Health Regulations 2002 and shall come into force on 21st November 2002.

2 Interpretation

(1) In these Regulations—

"the 1974 Act" means the Health and Safety at Work etc. Act 1974;

"the Agreement" means the Agreement on the European Economic Area signed at Oporto on 2nd May 1992 as adjusted by the Protocol signed at Brussels on 17th March 1993 and adopted as respects Great Britain by the European Economic Area Act 1993;

"appointed doctor" means a registered medical practitioner appointed for the time being in writing by the Executive for the purpose of these Regulations;

"approved" means approved for the time being in writing;

"approved classification" of a biological agent means the classification of that agent approved by the Health and Safety Commission;

"approved supply list" has the meaning assigned to it in regulation 2(1) of the CHIP Regulations;

"biological agent" means a micro-organism, cell culture, or human endoparasite, whether or not genetically modified, which may cause infection, allergy, toxicity or otherwise create a hazard to human health;

"carcinogen" means—

(a) a substance or preparation which if classified in accordance with the classification provided for by regulation 4 of the CHIP Regulations would be in the category of danger,

carcinogenic (category 1) or carcinogenic (category 2) whether or not the substance or preparation would be required to be classified under those Regulations; or
(b) a substance or preparation—
  (i) listed in Schedule 1, or
  (ii) arising from a process specified in Schedule 1 which is a substance hazardous to health;

"cell culture" means the in-vitro growth of cells derived from multicellular organisms;

"the CHIP Regulations" means the Chemicals (Hazard Information and Packaging for Supply) Regulations 2002;

"control measure" means a measure taken to reduce exposure to a substance hazardous to health (including the provision of systems of work and supervision, the cleaning of workplaces, premises, plant and equipment, the provision and use of engineering controls and personal protective equipment);

"employment medical adviser" means an employment medical adviser appointed under section 56 of the Health and Safety at Work etc. Act 1974;

"the Executive" means the Health and Safety Executive;

"fumigation" means an operation in which a substance is released into the atmosphere so as to form a gas to control or kill pests or other undesirable organisms and "fumigate" and "fumigant" shall be construed accordingly;

"Group", in relation to a biological agent, means one of the four hazard Groups specified in paragraph 2 of Schedule 3 to which that agent is assigned;

"hazard", in relation to a substance, means the intrinsic property of that substance which has the potential to cause harm to the health of a person, and "hazardous" shall be construed accordingly;

"health surveillance" means assessment of the state of health of an employee, as related to exposure to substances hazardous to health, and includes biological monitoring;

"inhalable dust" means airborne material which is capable of entering the nose and mouth during breathing, as defined by BS EN 481 1993;

[. . .][1]

"medical examination" includes any laboratory tests and X-rays that a relevant doctor may require;

"member State" means a State which is a Contracting Party to the Agreement;

"micro-organism" means a microbiological entity, cellular or non-cellular, which is capable of replication or of transferring genetic material;

"mine" has the meaning assigned to it by section 180 of the Mines and Quarries Act 1954;

["mutagen" means a substance or preparation which if classified in accordance with the classification provided for by regulation 4 of the Chemicals (Hazard Information and Packaging for Supply) Regulations 2002 would be in the category of danger, mutagenic (category 1) or mutagenic (category 2) whether or not the substance or preparation would be required to be classified under those Regulations;][2]

[. . .][3]

"personal protective equipment" means all equipment (including clothing) which is intended to be worn or held by a person at work and which protects that person against one or more risks to his health, and any addition or accessory designed to meet that objective;

"preparation" means a mixture or solution of two or more substances;

---

[1] Repealed by SI 2004/3386.
[2] Amended by SI 2003/978.
[3] Repealed by SI 2004/3386.

"public road" means (in England and Wales) a highway maintainable at the public expense within the meaning of section 329 of the Highways Act 1980 and (in Scotland) a public road within the meaning assigned to that term by section 151 of the Roads (Scotland) Act 1984;

"registered dentist" has the meaning assigned to it in section 53(1) of the Dentists Act 1984;

"relevant doctor" means an appointed doctor or an employment medical adviser;

"respirable dust" means airborne material which is capable of penetrating to the gas exchange region of the lung, as defined by BS EN 481 1993;

"risk", in relation to the exposure of an employee to a substance hazardous to health, means the likelihood that the potential for harm to the health of a person will be attained under the conditions of use and exposure and also the extent of that harm;

"the risk assessment" means the assessment of risk required by regulation 6(1)(a);

["risk phrase" has the meaning assigned to it in regulation 2(1) of the CHIP Regulations;][4]

"safety data sheet" means a safety data sheet within the meaning of regulation 5 of the CHIP Regulations;

"substance" means a natural or artificial substance whether in solid or liquid form or in the form of a gas or vapour (including micro-organisms);

"substance hazardous to health" means a substance (including a preparation)—

(a) which is listed in Part I of the approved supply list as dangerous for supply within the meaning of the CHIP Regulations and for which an indication of danger specified for the substance is very toxic, toxic, harmful, corrosive or irritant;

(b) for which the Health and Safety Commission has approved [a workplace exposure limit];

(c) which is a biological agent;

(d) which is dust of any kind, except dust which is a substance within paragraph (a) or (b) above, when present at a concentration in air equal to or greater than—

(i) 10 mg/m^3, as a time-weighted average over an 8-hour period, of inhalable dust, or

(ii) 4 mg/m^3, as a time-weighted average over an 8-hour period, of respirable dust;

(e) which, not being a substance falling within sub-paragraphs (a) to (d), because of its chemical or toxicological properties and the way it is used or is present at the workplace creates a risk to health;

"workplace" means any premises or part of premises used for or in connection with work, and includes—

(a) any place within the premises to which an employee has access while at work; and

(b) any room, lobby, corridor, staircase, road or other place—

(i) used as a means of access to or egress from that place of work, or

(ii) where facilities are provided for use in connection with that place of work,

other than a public road[;

"workplace exposure limit" for a substance hazardous to health means the exposure limit approved by the Health and Safety Commission for that substance in relation to the specified reference period when calculated by a method approved by the Health and Safety Commission, as contained in HSE publication "EH/40 Workplace Exposure Limits 2005" as updated from time to time].[5]

(2) In these Regulations, a reference to an employee being exposed to a substance hazardous to health is a reference to the exposure of that employee to a substance hazardous to health arising out of or in connection with work at the workplace.

---

[4] Amended by SI 2004/3386.
[5] ibid.

(3) Where a biological agent has an approved classification, any reference in these Regulations to a particular Group in relation to that agent shall be taken as a reference to the Group to which that agent has been assigned in that approved classification.

3 Duties under these Regulations

(1) Where a duty is placed by these Regulations on an employer in respect of his employees, he shall, so far as is reasonably practicable, be under a like duty in respect of any other person, whether at work or not, who may be affected by the work carried out by the employer except that the duties of the employer—
   (a) under regulation 11 (health surveillance) shall not extend to persons who are not his employees; and
   (b) under regulations 10, 12(1) and (2) and 13 (which relate respectively to monitoring, information and training and dealing with accidents) shall not extend to persons who are not his employees, unless those persons are on the premises where the work is being carried out.
(2) These Regulations shall apply to a self-employed person as they apply to an employer and an employee and as if that self-employed person were both an employer and an employee, except that regulations 10 and 11 shall not apply to a self-employed person.
[(3) These Regulations shall not apply to the master or crew of a ship or to the employer of such persons in respect of the normal shipboard activities of a ship's crew which—
   (a) are carried out solely by the crew under the direction of the master; and
   (b) are not liable to expose persons other than the master and crew to a risk to their health and safety,
and for the purposes of this paragraph "ship" includes every description of vessel used in navigation, other than a ship forming part of Her Majesty's Navy.][6]

4 Prohibitions relating to certain substances

(1) Those substances described in Column 1 of Schedule 2 are prohibited to the extent set out in the corresponding entry in Column 2 of that Schedule.
(2) The importation into the United Kingdom, other than from another member State, of the following substances and articles is prohibited, namely—
   (a) 2-naphthylamine, benzidine, 4-aminodiphenyl, 4-nitrodiphenyl, their salts and any substance containing any of those compounds in a total concentration equal to or greater than 0.1 per cent by mass;
   (b) matches made with white phosphorus,
   and a contravention of this paragraph shall be punishable under the Customs and Excise Management Act 1979 and not as a contravention of a health and safety regulation.
(3) A person shall not supply during the course of or for use at work a substance or article specified in paragraph (2).
(4) A person shall not supply during the course of or for use at work, benzene or a substance containing benzene unless its intended use is not prohibited by item 11 of Schedule 2.

5 Application of regulations 6 to 13

(1) Regulations 6 to 13 shall have effect with a view to protecting persons against a risk to their health, whether immediate or delayed, arising from exposure to substances hazardous to health except—
   (a) where and to the extent that the following Regulations apply, namely—
      (i) the Coal Mines (Respirable Dust) Regulations 1975,
      (ii) the Control of Lead at Work Regulations 2002,

---

[6] Inserted by SI 2004/3386.

(iii) the Control of Asbestos at Work Regulations 2002;
(b) where the substance is hazardous to health solely by virtue of its radioactive, explosive or flammable properties, or solely because it is at a high or low temperature or a high pressure;
(c) where the risk to health is a risk to the health of a person to whom the substance is administered in the course of his medical treatment.

(2) In paragraph (1)(c) "medical treatment" means medical or dental examination or treatment which is conducted by, or under the direction of a—
(a) registered medical practitioner;
(b) registered dentist; or
(c) other person who is an appropriate practitioner for the purposes of section 58 of the Medicines Act 1968,
and includes any such examination or treatment conducted for the purpose of research.

6 Assessment of the risk to health created by work involving substances hazardous to health

(1) An employer shall not carry out work which is liable to expose any employees to any substance hazardous to health unless he has—
(a) made a suitable and sufficient assessment of the risk created by that work to the health of those employees and of the steps that need to be taken to meet the requirements of these Regulations; and
(b) implemented the steps referred to in sub-paragraph (a).

(2) The risk assessment shall include consideration of—
(a) the hazardous properties of the substance;
(b) information on health effects provided by the supplier, including information contained in any relevant safety data sheet;
(c) the level, type and duration of exposure;
(d) the circumstances of the work, including the amount of the substance involved;
(e) activities, such as maintenance, where there is the potential for a high level of exposure;
(f) any relevant [workplace exposure limit][7] or similar occupational exposure limit;
(g) the effect of preventive and control measures which have been or will be taken in accordance with regulation 7;
(h) the results of relevant health surveillance;
(i) the results of monitoring of exposure in accordance with regulation 10;
(j) in circumstances where the work will involve exposure to more than one substance hazardous to health, the risk presented by exposure to such substances in combination;
(k) the approved classification of any biological agent; and
(l) such additional information as the employer may need in order to complete the risk assessment.

(3) The risk assessment shall be reviewed regularly and forthwith if—
(a) there is reason to suspect that the risk assessment is no longer valid;
(b) there has been a significant change in the work to which the risk assessment relates; or
(c) the results of any monitoring carried out in accordance with regulation 10 show it to be necessary,
and where, as a result of the review, changes to the risk assessment are required, those changes shall be made.

---

[7] Amended by SI 2004/3386.

(4) Where the employer employs 5 or more employees, he shall record—
   (a) the significant findings of the risk assessment as soon as is practicable after the risk assessment is made; and
   (b) the steps which he has taken to meet the requirements of regulation 7.

7 Prevention or control of exposure to substances hazardous to health

(1) Every employer shall ensure that the exposure of his employees to substances hazardous to health is either prevented or, where this is not reasonably practicable, adequately controlled.

(2) In complying with his duty of prevention under paragraph (1), substitution shall by preference be undertaken, whereby the employer shall avoid, so far as is reasonably practicable, the use of a substance hazardous to health at the workplace by replacing it with a substance or process which, under the conditions of its use, either eliminates or reduces the risk to the health of his employees.

(3) Where it is not reasonably practicable to prevent exposure to a substance hazardous to health, the employer shall comply with his duty of control under paragraph (1) by applying protection measures appropriate to the activity and consistent with the risk assessment, including, in order of priority—
   (a) the design and use of appropriate work processes, systems and engineering controls and the provision and use of suitable work equipment and materials;
   (b) the control of exposure at source, including adequate ventilation systems and appropriate organisational measures; and
   (c) where adequate control of exposure cannot be achieved by other means, the provision of suitable personal protective equipment in addition to the measures required by sub-paragraphs (a) and (b).

(4) The measures referred to in paragraph (3) shall include—
   (a) arrangements for the safe handling, storage and transport of substances hazardous to health, and of waste containing such substances, at the workplace;
   (b) the adoption of suitable maintenance procedures;
   (c) reducing, to the minimum required for the work concerned—
      (i) the number of employees subject to exposure,
      (ii) the level and duration of exposure, and
      (iii) the quantity of substances hazardous to health present at the workplace;
   (d) the control of the working environment, including appropriate general ventilation; and
   (e) appropriate hygiene measures including adequate washing facilities.

(5) Without prejudice to the generality of paragraph (1), where it is not reasonably practicable to prevent exposure to a carcinogen [or mutagen],[8] the employer shall apply the following measures in addition to those required by paragraph (3)—
   (a) totally enclosing the process and handling systems, unless this is not reasonably practicable;
   (b) the prohibition of eating, drinking and smoking in areas that may be contaminated by carcinogens [or mutagens];[9]
   (c) cleaning floors, walls and other surfaces at regular intervals and whenever necessary;
   (d) designating those areas and installations which may be contaminated by carcinogens [or mutagens][10] and using suitable and sufficient warning signs; and
   (e) storing, handling and disposing of carcinogens [or mutagens][11] safely, including using closed and clearly labelled containers.

---

[8] Amended by SI 2003/978.
[9] ibid.
[10] ibid.
[11] ibid.

(6) Without prejudice to the generality of paragraph (1), where it is not reasonably practicable to prevent exposure to a biological agent, the employer shall apply the following measures in addition to those required by paragraph (3)—
   (a) displaying suitable and sufficient warning signs, including the biohazard sign shown in Part IV of Schedule 3;
   (b) specifying appropriate decontamination and disinfection procedures;
   (c) instituting means for the safe collection, storage and disposal of contaminated waste, including the use of secure and identifiable containers, after suitable treatment where appropriate;
   (d) testing, where it is necessary and technically possible, for the presence, outside the primary physical confinement, of biological agents used at work;
   (e) specifying procedures for working with, and transporting at the workplace, a biological agent or material that may contain such an agent;
   (f) where appropriate, making available effective vaccines for those employees who are not already immune to the biological agent to which they are exposed or are liable to be exposed;
   (g) instituting hygiene measures compatible with the aim of preventing or reducing the accidental transfer or release of a biological agent from the workplace, including—
      (i) the provision of appropriate and adequate washing and toilet facilities, and
      (ii) where appropriate, the prohibition of eating, drinking, smoking and the application of cosmetics in working areas where there is a risk of contamination by biological agents; and
   (h) where there are human patients or animals which are, or are suspected of being, infected with a Group 3 or 4 biological agent, the employer shall select the most suitable control and containment measures from those listed in Part II of Schedule 3 with a view to controlling adequately the risk of infection.
[(7) Without prejudice to the generality of paragraph (1), where there is exposure to a substance hazardous to health, control of that exposure shall only be treated as adequate if—
   (a) the principles of good practice for the control of exposure to substances hazardous to health set out in Schedule 2A are applied;
   (b) any workplace exposure limit approved for that substance is not exceeded; and
   (c) for a substance—
      (i) which carries the risk phrase R45, R46 or R49, or for a substance or process which is listed in Schedule 1; or
      (ii) which carries the risk phrase R42 or R42/43, or which is listed in section C of HSE publication "Asthmagen? Critical assessments of the evidence for agents implicated in occupational asthma" as updated from time to time, or any other substance which the risk assessment has shown to be a potential cause of occupational asthma,
exposure is reduced to as low a level as is reasonably practicable.][12]
(8) [...][13]
(9) Personal protective equipment provided by an employer in accordance with this regulation shall be suitable for the purpose and shall—
   (a) comply with any provision in the Personal Protective Equipment Regulations 2002 which is applicable to that item of personal protective equipment; or
   (b) in the case of respiratory protective equipment, where no provision referred to in sub-paragraph (a) applies, be of a type approved or shall conform to a standard approved, in either case, by the Executive.

---

[12] Inserted by SI 2004/3386.
[13] Repealed by SI 2004/3386.

(10) Without prejudice to the provisions of this regulation, Schedule 3 shall have effect in relation to work with biological agents.

(11) In this regulation, "adequate" means adequate having regard only to the nature of the substance and the nature and degree of exposure to substances hazardous to health and "adequately" shall be construed accordingly.

8 Use of control measures etc

(1) Every employer who provides any control measure, other thing or facility in accordance with these Regulations shall take all reasonable steps to ensure that it is properly used or applied as the case may be.

(2) Every employee shall make full and proper use of any control measure, other thing or facility provided in accordance with these Regulations and, where relevant, shall—
   (a) take all reasonable steps to ensure it is returned after use to any accommodation provided for it; and
   (b) if he discovers a defect therein, report it forthwith to his employer.

9 Maintenance, examination and testing of control measures

[(1) Every employer who provides any control measure to meet the requirements of regulation 7 shall ensure that—
   (a) in the case of plant and equipment, including engineering controls and personal protective equipment, it is maintained in an efficient state, in efficient working order, in good repair and in a clean condition; and
   (b) in the case of the provision of systems of work and supervision and of any other measure, it is reviewed at suitable intervals and revised if necessary.][14]

(2) Where engineering controls are provided to meet the requirements of regulation 7, the employer shall ensure that thorough examination and testing of those controls is carried out—
   (a) in the case of local exhaust ventilation plant, at least once every 14 months, or for local exhaust ventilation plant used in conjunction with a process specified in Column 1 of Schedule 4, at not more than the interval specified in the corresponding entry in Column 2 of that Schedule; or
   (b) in any other case, at suitable intervals.

(3) Where respiratory protective equipment (other than disposable respiratory protective equipment) is provided to meet the requirements of regulation 7, the employer shall ensure that thorough examination and, where appropriate, testing of that equipment is carried out at suitable intervals.

(4) Every employer shall keep a suitable record of the examinations and tests carried out in accordance with paragraphs (2) and (3) and of repairs carried out as a result of those examinations and tests, and that record or a suitable summary thereof shall be kept available for at least 5 years from the date on which it was made.

(5) Every employer shall ensure that personal protective equipment, including protective clothing, is:
   (a) properly stored in a well-defined place;
   (b) checked at suitable intervals; and
   (c) when discovered to be defective, repaired or replaced before further use.

(6) Personal protective equipment which may be contaminated by a substance hazardous to health shall be removed on leaving the working area and kept apart from uncontaminated clothing and equipment.

---

[14] Inserted by SI 2004/3386.

(7) The employer shall ensure that the equipment referred to in paragraph (6) is subsequently decontaminated and cleaned or, if necessary, destroyed.

10 Monitoring exposure at the workplace

(1) Where the risk assessment indicates that—
   (a) it is requisite for ensuring the maintenance of adequate control of the exposure of employees to substances hazardous to health; or
   (b) it is otherwise requisite for protecting the health of employees,
   the employer shall ensure that the exposure of employees to substances hazardous to health is monitored in accordance with a suitable procedure.
(2) Paragraph (1) shall not apply where the employer is able to demonstrate by another method of evaluation that the requirements of regulation 7(1) have been complied with.
(3) The monitoring referred to in paragraph (1) shall take place—
   (a) at regular intervals; and
   (b) when any change occurs which may affect that exposure.
(4) Where a substance or process is specified in Column 1 of Schedule 5, monitoring shall be carried out at least at the frequency specified in the corresponding entry in Column 2 of that Schedule.
(5) The employer shall ensure that a suitable record of monitoring carried out for the purpose of this regulation is made and maintained and that that record or a suitable summary thereof is kept available—
   (a) where the record is representative of the personal exposures of identifiable employees, for at least 40 years; or
   (b) in any other case, for at least 5 years,
   from the date of the last entry made in it.
(6) Where an employee is required by regulation 11 to be under health surveillance, an individual record of any monitoring carried out in accordance with this regulation shall be made, maintained and kept in respect of that employee.
(7) The employer shall—
   (a) on reasonable notice being given, allow an employee access to his personal monitoring record;
   (b) provide the Executive with copies of such monitoring records as the Executive may require; and
   (c) if he ceases to trade, notify the Executive forthwith in writing and make available to the Executive all monitoring records kept by him.

11 Health surveillance

(1) Where it is appropriate for the protection of the health of his employees who are, or are liable to be, exposed to a substance hazardous to health, the employer shall ensure that such employees are under suitable health surveillance.
(2) Health surveillance shall be treated as being appropriate where—
   (a) the employee is exposed to one of the substances specified in Column 1 of Schedule 6 and is engaged in a process specified in Column 2 of that Schedule, and there is a reasonable likelihood that an identifiable disease or adverse health effect will result from that exposure; or
   (b) the exposure of the employee to a substance hazardous to health is such that—
      (i) an identifiable disease or adverse health effect may be related to the exposure,
      (ii) there is a reasonable likelihood that the disease or effect may occur under the particular conditions of his work, and
      (iii) there are valid techniques for detecting indications of the disease or effect,
   and the technique of investigation is of low risk to the employee.
(3) The employer shall ensure that a health record, containing particulars approved by the Executive, in respect of each of his employees to whom paragraph (1) applies, is made and maintained and

that that record or a copy thereof is kept available in a suitable form for at least 40 years from the date of the last entry made in it.
(4) The employer shall—
  (a) on reasonable notice being given, allow an employee access to his personal health record;
  (b) provide the Executive with copies of such health records as the Executive may require; and
  (c) if he ceases to trade, notify the Executive forthwith in writing and make available to the Executive all health records kept by him.
(5) If an employee is exposed to a substance specified in Schedule 6 and is engaged in a process specified therein, the health surveillance required under paragraph (1) shall include medical surveillance under the supervision of a relevant doctor at intervals of not more than 12 months or at such shorter intervals as the relevant doctor may require.
(6) Where an employee is subject to medical surveillance in accordance with paragraph (5) and a relevant doctor has certified by an entry in the health record of that employee that in his professional opinion that employee should not be engaged in work which exposes him to that substance or that he should only be so engaged under conditions specified in the record, the employer shall not permit the employee to be engaged in such work except in accordance with the conditions, if any, specified in the health record, unless that entry has been cancelled by a relevant doctor.
(7) Where an employee is subject to medical surveillance in accordance with paragraph (5) and a relevant doctor has certified by an entry in his health record that medical surveillance should be continued after his exposure to that substance has ceased, the employer shall ensure that the medical surveillance of that employee is continued in accordance with that entry while he is employed by the employer, unless that entry has been cancelled by a relevant doctor.
(8) An employee to whom this regulation applies shall, when required by his employer and at the cost of the employer, present himself during his working hours for such health surveillance procedures as may be required for the purposes of paragraph (1) and, in the case of an employee who is subject to medical surveillance in accordance with paragraph (5), shall furnish the relevant doctor with such information concerning his health as the relevant doctor may reasonably require.
(9) Where, as a result of health surveillance, an employee is found to have an identifiable disease or adverse health effect which is considered by a relevant doctor or other occupational health professional to be the result of exposure to a substance hazardous to health the employer of that employee shall—
  (a) ensure that a suitably qualified person informs the employee accordingly and provides the employee with information and advice regarding further health surveillance;
  (b) review the risk assessment;
  (c) review any measure taken to comply with regulation 7, taking into account any advice given by a relevant doctor, occupational health professional or by the Executive;
  (d) consider assigning the employee to alternative work where there is no risk of further exposure to that substance, taking into account any advice given by a relevant doctor or occupational health professional; and
  (e) provide for a review of the health of any other employee who has been similarly exposed, including a medical examination where such an examination is recommended by a relevant doctor, occupational health professional or by the Executive.
(10) Where, for the purpose of carrying out his functions under these Regulations, a relevant doctor requires to inspect any workplace or any record kept for the purposes of these Regulations, the employer shall permit him to do so.
(11) Where an employee or an employer is aggrieved by a decision recorded in the health record by a relevant doctor to suspend an employee from work which exposes him to a substance hazardous to health (or to impose conditions on such work), he may, by an application in writing to the Executive within 28 days of the date on which he was notified of the decision, apply for that

decision to be reviewed in accordance with a procedure approved for the purposes of this paragraph by the Health and Safety Commission, and the result of that review shall be notified to the employee and employer and entered in the health record in accordance with the approved procedure.

12 Information, instruction and training for persons who may be exposed to substances hazardous to health

(1) Every employer who undertakes work which is liable to expose an employee to a substance hazardous to health shall provide that employee with suitable and sufficient information, instruction and training.
(2) Without prejudice to the generality of paragraph (1), the information, instruction and training provided under that paragraph shall include—
  (a) details of the substances hazardous to health to which the employee is liable to be exposed including—
    (i) the names of those substances and the risk which they present to health,
    (ii) any relevant [workplace exposure limit][15] or similar occupational exposure limit,
    (iii) access to any relevant safety data sheet, and
    (iv) other legislative provisions which concern the hazardous properties of those substances;
  (b) the significant findings of the risk assessment;
  (c) the appropriate precautions and actions to be taken by the employee in order to safeguard himself and other employees at the workplace;
  (d) the results of any monitoring of exposure in accordance with regulation 10 and, in particular, in the case of a substance hazardous to health for which a [workplace exposure limit][16] has been approved, the employee or his representatives shall be informed forthwith, if the results of such monitoring show that the [workplace exposure limit][17] has been exceeded;
  (e) the collective results of any health surveillance undertaken in accordance with regulation 11 in a form calculated to prevent those results from being identified as relating to a particular person; and
  (f) where employees are working with a Group 4 biological agent or material that may contain such an agent, the provision of written instructions and, if appropriate, the display of notices which outline the procedures for handling such an agent or material.
(3) The information, instruction and training required by paragraph (1) shall be—
  (a) adapted to take account of significant changes in the type of work carried out or methods of work used by the employer; and
  (b) provided in a manner appropriate to the level, type and duration of exposure identified by the risk assessment.
(4) Every employer shall ensure that any person (whether or not his employee) who carries out work in connection with the employer's duties under these Regulations has suitable and sufficient information, instruction and training.
(5) Where containers and pipes for substances hazardous to health used at work are not marked in accordance with any relevant legislation listed in Schedule 7, the employer shall, without prejudice to any derogations provided for in that legislation, ensure that the contents of those containers and pipes, together with the nature of those contents and any associated hazards, are clearly identifiable.

---

[15] Amended by SI 2004/3386.
[16] ibid.
[17] ibid.

13 Arrangements to deal with accidents, incidents and emergencies

(1) Subject to paragraph (4) and without prejudice to the relevant provisions of the Management of Health and Safety at Work Regulations 1999, in order to protect the health of his employees from an accident, incident or emergency related to the presence of a substance hazardous to health at the workplace, the employer shall ensure that—
   (a) procedures, including the provision of appropriate first-aid facilities and relevant safety drills (which shall be tested at regular intervals), have been prepared which can be put into effect when such an event occurs;
   (b) information on emergency arrangements, including—
      (i) details of relevant work hazards and hazard identification arrangements, and
      (ii) specific hazards likely to arise at the time of an accident, incident or emergency, is available; and
   (c) suitable warning and other communication systems are established to enable an appropriate response, including remedial actions and rescue operations, to be made immediately when such an event occurs.

(2) The employer shall ensure that information on the procedures and systems required by paragraph (1)(a) and (c) and the information required by paragraph(1)(b) is—
   (a) made available to relevant accident and emergency services to enable those services, whether internal or external to the workplace, to prepare their own response procedures and precautionary measures; and
   (b) displayed at the workplace, if this is appropriate.

(3) Subject to paragraph (4), in the event of an accident, incident or emergency related to the presence of a substance hazardous to health at the workplace, the employer shall ensure that—
   (a) immediate steps are taken to—
      (i) mitigate the effects of the event,
      (ii) restore the situation to normal, and
      (iii) inform those of his employees who may be affected;
   (b) only those persons who are essential for the carrying out of repairs and other necessary work are permitted in the affected area and they are provided with—
      (i) appropriate personal protective equipment, and
      (ii) any necessary specialised safety equipment and plant,
   which shall be used until the situation is restored to normal; and
   (c) in the case of an incident or accident which has or may have resulted in the release of a biological agent which could cause severe human disease, as soon as practicable thereafter his employees or their representatives are informed of—
      (i) the causes of that incident or accident, and
      (ii) the measures taken or to be taken to rectify the situation.

(4) Paragraph (1) and, provided the substance hazardous to health is not a [carcinogen, mutagen or biological agent],[18] paragraph (3) shall not apply where—
   (a) the results of the risk assessment show that, because of the quantity of each substance hazardous to health present at the workplace, there is only a slight risk to the health of employees; and
   (b) the measures taken by the employer to comply with the duty under regulation 7(1) are sufficient to control that risk.

(5) An employee shall report forthwith, to his employer or to any other employee of that employer with specific responsibility for the health and safety of his fellow employees, any accident or incident which has or may have resulted in the release of a biological agent which could cause severe human disease.

---

[18] Amended by SI 2003/978.

14 Provisions relating to certain fumigations

(1) This regulation shall apply to fumigations in which the fumigant used or intended to be used is hydrogen cyanide, phosphine or methyl bromide, except that paragraph (2) shall not apply to fumigations using the fumigant specified in Column 1 of Schedule 8 when the nature of the fumigation is that specified in the corresponding entry in Column 2 of that Schedule.

(2) An employer shall not undertake fumigation to which this regulation applies unless he has—
   (a) notified the persons specified in Part I of Schedule 9 of his intention to undertake the fumigation; and
   (b) provided to those persons the information specified in Part II of that Schedule,
   at least 24 hours in advance, or such shorter time in advance as the persons required to be notified may agree.

(3) An employer who undertakes a fumigation to which this regulation applies shall ensure that, before the fumigant is released, suitable warning notices have been affixed at all points of reasonable access to the premises or to those parts of the premises in which the fumigation is to be carried out and that after the fumigation has been completed, and the premises are safe to enter, those warning notices are removed.

15 Exemption certificates

(1) Subject to paragraph (2) the Executive may, by a certificate in writing, exempt any person or class of persons or any substance or class of substances from all or any of the requirements or prohibitions imposed by regulations 4 (to the extent permitted by article 9 of Council Directive 98/24/EC), 8, 9, 11(8), (10) and (11) and 14 of these Regulations and any such exemption may be granted subject to conditions and to a limit of time and may be revoked by a certificate in writing at any time.

(2) The Executive shall not grant any such exemption unless having regard to the circumstances of the case and, in particular, to—
   (a) the conditions, if any, which it proposes to attach to the exemption; and
   (b) any requirements imposed by or under any enactments which apply to the case,
   it is satisfied that the health and safety of persons who are likely to be affected by the exemption will not be prejudiced in consequence of it.

16 Exemptions relating to the Ministry of Defence etc

(1) In this regulation—
   (a) "Her Majesty's Forces" means any of the naval, military or air forces of the Crown, whether raised inside or outside the United Kingdom and whether any such force is a regular, auxiliary or reserve force, and includes any civilian employed by those forces;
   (b) "visiting force" has the same meaning as it does for the purposes of any provision of Part I of the Visiting Forces Act 1952; and
   (c) "headquarters" means a headquarters for the time being specified in Schedule 2 to the Visiting Forces and International Headquarters (Application of Law) Order 1999.

(2) The Secretary of State for Defence may, in the interests of national security, by a certificate in writing exempt—
   (a) any of Her Majesty's Forces;
   (b) any visiting force;
   (c) members of a visiting force working in or attached to a headquarters; or
   (d) any person engaged in work involving substances hazardous to health, if that person is under the direct supervision of a representative of the Secretary of State for Defence,
   from all or any of the requirements or prohibitions imposed by these Regulations and any such exemption may be granted subject to conditions and to a limit of time and may be revoked at any time by a certificate in writing, except that, where any such exemption is granted, suitable

arrangements shall be made for the assessment of the health risk created by the work involving substances hazardous to health and for adequately controlling the exposure to those substances of persons to whom the exemption relates.

(3) Regulation 11(11) shall not apply in relation to—
   (a) any visiting force; or
   (b) members of a visiting force working in or attached to a headquarters.

[16A Modifications relating to the Office of Rail Regulation]

[(1) In so far as these Regulations apply to, or in connection with, any activities in relation to which the Office of Rail Regulation is made the enforcing authority by regulation 3(1) of the Health and Safety (Enforcing Authority for Railways and Other Guided Transport Systems) Regulations 2006, they shall have effect as if any reference to the Executive in the provisions specified in paragraph (2) were a reference to the Office of Rail Regulation.

(2) The provisions referred to in paragraph (1) are as follows—
   (a) regulation 10(7)(b) (monitoring exposure at the workplace);
   (b) regulation 11(4)(b) (health surveillance); and
   (c) regulation 18 (revocation and savings).][19]

17 Extension outside Great Britain

(1) Subject to paragraph (2), these Regulations shall apply to and in relation to any activity outside Great Britain to which sections 1 to 59 and 80 to 82 of the 1974 Act apply by virtue of the Health and Safety at Work etc. Act 1974 (Application outside Great Britain) Order 2001 as those provisions apply within Great Britain.

(2) These Regulations shall not extend to Northern Ireland except insofar as they relate to imports of substances and articles referred to in regulation 4(2) into the United Kingdom.

18 Revocation and savings

(1) The Control of Substances Hazardous to Health Regulations 1999 are revoked.

(2) Any record or register required to be kept under the Regulations revoked by paragraph (1) shall, notwithstanding that revocation, be kept in the same manner and for the same period as specified in those Regulations as if these Regulations had not been made, except that the Executive may approve the keeping of records at a place or in a form other than at the place where, or in the form in which, records were required to be kept under the Regulations so revoked.

19 Extension of meaning of "work"

For the purposes of Part I of the 1974 Act the meaning of "work" shall be extended to include any activity involving the consignment, storage or use of a Group 2, 3 or 4 biological agent and the meaning of "at work" shall be extended accordingly, and in that connection the references to employer in paragraphs 5 and 6 of Schedule 3 include references to any persons carrying out such an activity.

20 Modification of section 3(2) of the 1974 Act

Section 3(2) of the 1974 Act shall be modified in relation to an activity involving the consignment, storage or use of any of the biological agents referred to in regulation 19 so as to have effect as if the reference therein to a self-employed person is a reference to any person who is not an employer or an employee and the reference therein to his undertaking includes a reference to such an activity.

---

[19] Inserted by SI 2006/557.

21 Defence

Subject to regulation 21 of the Management of Health and Safety at Work Regulations 1999, in any proceedings for an offence consisting of a contravention of these Regulations it shall be a defence for any person to prove that he took all reasonable precautions and exercised all due diligence to avoid the commission of that offence.

Signed by authority of the Secretary of State

N Brown

Minister of State,

Department for Work and Pensions

24th October 2002

APPENDIX J

Health and Safety: The Construction (Design and Management) Regulations 2007

2007 No. 320

Made	*7th February 2007*
Laid before Parliament	*15th February 2007*
Coming into force	*6 April 2007*

The Secretary of State makes the following Regulations in the exercise of the powers conferred upon him by sections 15(1), (2), (3)(a) and (c), 5(a), (6)(a) and (b), (8) and (9), 47(2) and (3), 80(1) and (2) and 82(3)(a) of, and paragraphs 1(1) and (2), 6, 7, 8(1), 9 to 12, 14, 15(1), 16, 18, 20 and 21 of Schedule 3 to, the Health and Safety at Work etc. Act 1974("the 1974 Act").

In doing so he gives effect without modifications to proposals submitted to him by the Health and Safety Commission under section 11(2)(d) of the 1974 Act after the carrying out by the said Commission of consultations in accordance with section 50(3) of that Act, and it appearing expedient to him after consulting such bodies as appear to him to be appropriate in accordance with section 80(4) of that Act.

PART I
INTRODUCTION

Citation and commencement

1. These Regulations may be cited as the Construction (Design and Management) Regulations 2007 and shall come into force on 6th April 2007.

Interpretation

2. —(1) In these Regulations, unless the context otherwise requires—
   "business" means a trade, business or other undertaking (whether for profit or not);
   "client" means a person who in the course or furtherance of a business—
   (a) seeks or accepts the services of another which may be used in the carrying out of a project for him; or
   (b) carries out a project himself;
   "CDM co-ordinator" means the person appointed as the CDM co-ordinator under regulation 14(1);
   "construction site" includes any place where construction work is being carried out or to which the workers have access, but does not include a workplace within it which is set aside for purposes other than construction work;
   "construction phase" means the period of time starting when construction work in any project starts and ending when construction work in that project is completed;
   "construction phase plan" means a document recording the health and safety arrangements, site rules and any special measures for construction work;

"construction work" means the carrying out of any building, civil engineering or engineering construction work and includes—
   (a) the construction, alteration, conversion, fitting out, commissioning, renovation, repair, upkeep, redecoration or other maintenance (including cleaning which involves the use of water or an abrasive at high pressure or the use of corrosive or toxic substances), de-commissioning, demolition or dismantling of a structure;
   (b) the preparation for an intended structure, including site clearance, exploration, investigation (but not site survey) and excavation, and the clearance or preparation of the site or structure for use or occupation at its conclusion;
   (c) the assembly on site of prefabricated elements to form a structure or the disassembly on site of prefabricated elements which, immediately before such disassembly, formed a structure;
   (d) the removal of a structure or of any product or waste resulting from demolition or dismantling of a structure or from disassembly of prefabricated elements which immediately before such disassembly formed such a structure; and
   (e) the installation, commissioning, maintenance, repair or removal of mechanical, electrical, gas, compressed air, hydraulic, telecommunications, computer or similar services which are normally fixed within or to a structure,
but does not include the exploration for or extraction of mineral resources or activities preparatory thereto carried out at a place where such exploration or extraction is carried out;
"contractor" means any person (including a client, principal contractor or other person referred to in these Regulations) who, in the course or furtherance of a business, carries out or manages construction work;
"design" includes drawings, design details, specification and bill of quantities (including specification of articles or substances) relating to a structure, and calculations prepared for the purpose of a design;
"designer" means any person (including a client, contractor or other person referred to in these Regulations) who in the course or furtherance of a business—
   (a) prepares or modifies a design; or
   (b) arranges for or instructs any person under his control to do so,
relating to a structure or to a product or mechanical or electrical system intended for a particular structure, and a person is deemed to prepare a design where a design is prepared by a person under his control;
"excavation" includes any earthwork, trench, well, shaft, tunnel or underground working;
"the Executive" means the Health and Safety Executive;
"the general principles of prevention" means the general principles of prevention specified in Schedule 1 to the Management of Health and Safety at Work Regulations 1999;
"health and safety file"—
   (a) means the record referred to in regulation 20(2)(e); and
   (b) includes a health and safety file prepared under regulation 14(d) of the Construction (Design and Management) Regulations 1994;
"loading bay" means any facility for loading or unloading;
"place of work" means any place which is used by any person at work for the purposes of construction work or for the purposes of any activity arising out of or in connection with construction work;
"pre-construction information" means the information described in regulation 10 and, where the project is notifiable, regulation 15.
"principal contractor" means the person appointed as the principal contractor under regulation 14(2);
"project" means a project which includes or is intended to include construction work and includes all planning, design, management or other work involved in a project until the end of the construction phase;
"site rules" means the rules described in regulation 22(1)(d);

"structure" means—
  (a) any building, timber, masonry, metal or reinforced concrete structure, railway line or siding, tramway line, dock, harbour, inland navigation, tunnel, shaft, bridge, viaduct, waterworks, reservoir, pipe or pipe-line, cable, aqueduct, sewer, sewage works, gasholder, road, airfield, sea defence works, river works, drainage works, earthworks, lagoon, dam, wall, caisson, mast, tower, pylon, underground tank, earth retaining structure or structure designed to preserve or alter any natural feature, fixed plant and any structure similar to the foregoing; or
  (b) any formwork, falsework, scaffold or other structure designed or used to provide support or means of access during construction work,
and any reference to a structure includes a part of a structure.

"traffic route" means a route for pedestrian traffic or for vehicles and includes any doorway, gateway, loading bay or ramp;

"vehicle" includes any mobile work equipment;

"work equipment" means any machinery, appliance, apparatus, tool or installation for use at work (whether exclusively or not);

"workplace" means a workplace within the meaning of regulation 2(1) of the Workplace (Health, Safety and Welfare) Regulations 1992 other than a construction site; and

"writing" includes writing which is kept in electronic form and which can be printed.

(2) Any reference in these Regulations to a plan, rules, document, report or copy includes a plan, rules, document, report or copy which is kept in a form—
  (a) in which it is capable of being reproduced as a printed copy when required; and
  (b) which is secure from loss or unauthorised interference.

(3) For the purposes of these Regulations, a project is notifiable if the construction phase is likely to involve more than—
  (a) 30 days; or
  (b) 500 person days,
of construction work.

Application

3. —(1) These Regulations shall apply—
  (a) in Great Britain; and
  (b) outside Great Britain as sections 1 to 59 and 80 to 82 of the 1974 Act apply by virtue of article 8(1)(a) of the Health and Safety at Work etc. Act 1974 (Application outside Great Britain) Order 2001.
(2) Subject to the following paragraphs of this regulation, these Regulations shall apply to and in relation to construction work.
(3) The duties under Part 3 shall apply only where a project—
  (a) is notifiable; and
  (b) is carried out for or on behalf of, or by, a client.
(4) Part 4 shall apply only in relation to a construction site.
(5) Regulations 9(1)(b), 13(7), 22(1)(c), and Schedule 2 shall apply only in relation to persons at work who are carrying out construction work.

PART 2
GENERAL MANAGEMENT DUTIES APPLYING TO CONSTRUCTION PROJECTS

Competence

4. —(1) No person on whom these Regulations place a duty shall—
  (a) appoint or engage a CDM co-ordinator, designer, principal contractor or contractor unless he has taken reasonable steps to ensure that the person to be appointed or engaged is competent;

(b) accept such an appointment or engagement unless he is competent;
(c) arrange for or instruct a worker to carry out or manage design or construction work unless the worker is—
   (i) competent, or
   (ii) under the supervision of a competent person.

(2) Any reference in this regulation to a person being competent shall extend only to his being competent to—
   (a) perform any requirement; and
   (b) avoid contravening any prohibition,
imposed on him by or under any of the relevant statutory provisions.

Co-operation

5. —(1) Every person concerned in a project on whom a duty is placed by these Regulations, including paragraph (2), shall—
   (a) seek the co-operation of any other person concerned in any project involving construction work at the same or an adjoining site so far as is necessary to enable himself to perform any duty or function under these Regulations; and
   (b) co-operate with any other person concerned in any project involving construction work at the same or an adjoining site so far as is necessary to enable that person to perform any duty or function under these Regulations.

(2) Every person concerned in a project who is working under the control of another person shall report to that person anything which he is aware is likely to endanger the health or safety of himself or others.

Co-ordination

6. All persons concerned in a project on whom a duty is placed by these Regulations shall co-ordinate their activities with one another in a manner which ensures, so far as is reasonably practicable, the health and safety of persons—
   (a) carrying out the construction work; and
   (b) affected by the construction work.

General principles of prevention

7. —(1) Every person on whom a duty is placed by these Regulations in relation to the design, planning and preparation of a project shall take account of the general principles of prevention in the performance of those duties during all the stages of the project.

(2) Every person on whom a duty is placed by these Regulations in relation to the construction phase of a project shall ensure so far as is reasonably practicable that the general principles of prevention are applied in the carrying out of the construction work.

Election by clients

8. Where there is more than one client in relation to a project, if one or more of such clients elect in writing to be treated for the purposes of these Regulations as the only client or clients, no other client who has agreed in writing to such election shall be subject after such election and consent to any duty owed by a client under these Regulations save the duties in regulations 5(1)(b), 10(1), 15 and 17(1) insofar as those duties relate to information in his possession.

Client's duty in relation to arrangements for managing projects

9. —(1) Every client shall take reasonable steps to ensure that the arrangements made for managing the project (including the allocation of sufficient time and other resources) by persons with a duty under these Regulations (including the client himself) are suitable to ensure that—
   (a) the construction work can be carried out so far as is reasonably practicable without risk to the health and safety of any person;

(b) the requirements of Schedule 2 are complied with in respect of any person carrying out the construction work; and

(c) any structure designed for use as a workplace has been designed taking account of the provisions of the Workplace (Health, Safety and Welfare) Regulations 1992 which relate to the design of, and materials used in, the structure.

(2) The client shall take reasonable steps to ensure that the arrangements referred to in paragraph (1) are maintained and reviewed throughout the project.

Client's duty in relation to information

10. —(1) Every client shall ensure that
   (a) every person designing the structure; and
   (b) every contractor who has been or may be appointed by the client,

is promptly provided with pre-construction information in accordance with paragraph (2).

(2) The pre-construction information shall consist of all the information in the client's possession (or which is reasonably obtainable), including—
   (a) any information about or affecting the site or the construction work;
   (b) any information concerning the proposed use of the structure as a workplace;
   (c) the minimum amount of time before the construction phase which will be allowed to the contractors appointed by the client for planning and preparation for construction work; and
   (d) any information in any existing health and safety file,

which is relevant to the person to whom the client provides it for the purposes specified in paragraph (3).

(3) The purposes referred to in paragraph (2) are—
   (a) to ensure so far as is reasonably practicable the health and safety of persons—
      (i) engaged in the construction work,
      (ii) liable to be affected by the way in which it is carried out, and
      (iii) who will use the structure as a workplace; and
   (b) without prejudice to sub-paragraph (a), to assist the persons to whom information is provided under this regulation—
      (i) to perform their duties under these Regulations, and
      (ii) to determine the resources referred to in regulation 9(1) which they are to allocate for managing the project.

Duties of designers

11. —(1) No designer shall commence work in relation to a project unless any client for the project is aware of his duties under these Regulations.

(2) The duties in paragraphs (3) and (4) shall be performed so far as is reasonably practicable, taking due account of other relevant design considerations.

(3) Every designer shall in preparing or modifying a design which may be used in construction work in Great Britain avoid foreseeable risks to the health and safety of any person—
   (a) carrying out construction work;
   (b) liable to be affected by such construction work;
   (c) cleaning any window or any transparent or translucent wall, ceiling or roof in or on a structure;
   (d) maintaining the permanent fixtures and fittings of a structure; or
   (e) using a structure designed as a workplace.

(4) In discharging the duty in paragraph (3), the designer shall—
   (a) eliminate hazards which may give rise to risks; and
   (b) reduce risks from any remaining hazards,
   and in so doing shall give collective measures priority over individual measures.

(5) In designing any structure for use as a workplace the designer shall take account of the provisions of the Workplace (Health, Safety and Welfare) Regulations 1992 which relate to the design of, and materials used in, the structure.

(6) The designer shall take all reasonable steps to provide with his design sufficient information about aspects of the design of the structure or its construction or maintenance as will adequately assist—
   (a) clients;
   (b) other designers; and
   (c) contractors,
to comply with their duties under these Regulations.

Designs prepared or modified outside Great Britain

12. Where a design is prepared or modified outside Great Britain for use in construction work to which these Regulations apply—
    (a) the person who commissions it, if he is established within Great Britain; or
    (b) if that person is not so established, any client for the project,
shall ensure that regulation 11 is complied with.

Duties of contractors

13. —(1) No contractor shall carry out construction work in relation to a project unless any client for the project is aware of his duties under these Regulations.

(2) Every contractor shall plan, manage and monitor construction work carried out by him or under his control in a way which ensures that, so far as is reasonably practicable, it is carried out without risks to health and safety.

(3) Every contractor shall ensure that any contractor whom he appoints or engages in his turn in connection with a project is informed of the minimum amount of time which will be allowed to him for planning and preparation before he begins construction work.

(4) Every contractor shall provide every worker carrying out the construction work under his control with any information and training which he needs for the particular work to be carried out safely and without risk to health, including—
   (a) suitable site induction, where not provided by any principal contractor;
   (b) information on the risks to their health and safety—
      (i) identified by his risk assessment under regulation 3 of the Management of Health and Safety at Work Regulations 1999, or
      (ii) arising out of the conduct by another contractor of his undertaking and of which he is or ought reasonably to be aware;
   (c) the measures which have been identified by the contractor in consequence of the risk assessment as the measures he needs to take to comply with the requirements and prohibitions imposed upon him by or under the relevant statutory provisions;
   (d) any site rules;
   (e) the procedures to be followed in the event of serious and imminent danger to such workers; and
   (f) the identity of the persons nominated to implement those procedures.

(5) Without prejudice to paragraph (4), every contractor shall in the case of any of his employees provide those employees with any health and safety training which he is required to provide to them in respect of the construction work by virtue of regulation 13(2)(b) of the Management of Health and Safety at Work Regulations 1999.

(6) No contractor shall begin work on a construction site unless reasonable steps have been taken to prevent access by unauthorised persons to that site.

(7) Every contractor shall ensure, so far as is reasonably practicable, that the requirements of Schedule 2 are complied with throughout the construction phase in respect of any person at work who is under his control.

PART 3
ADDITIONAL DUTIES WHERE PROJECT IS NOTIFIABLE

Appointments by the client where a project is notifiable

14.—(1) Where a project is notifiable, the client shall appoint a person ("the CDM co-ordinator") to perform the duties specified in regulations 20 and 21 as soon as is practicable after initial design work or other preparation for construction work has begun.

(2) After appointing a CDM co-ordinator under paragraph (1), the client shall appoint a person ("the principal contractor") to perform the duties specified in regulations 22 to 24 as soon as is practicable after the client knows enough about the project to be able to select a suitable person for such appointment.

(3) The client shall ensure that appointments under paragraphs (1) and (2) are changed or renewed as necessary to ensure that there is at all times until the end of the construction phase a CDM co-ordinator and principal contractor.

(4) The client shall—
  (a) be deemed for the purposes of these Regulations, save paragraphs (1) and (2) and regulations 18(1) and 19(1)(a) to have been appointed as the CDM co-ordinator or principal contractor, or both, for any period for which no person (including himself) has been so appointed; and
  (b) accordingly be subject to the duties imposed by regulations 20 and 21 on a CDM co-ordinator or, as the case may be, the duties imposed by regulations 22 to 24 on a principal contractor, or both sets of duties.

(5) Any reference in this regulation to appointment is to appointment in writing.

Client's duty in relation to information where a project is notifiable

15. Where the project is notifiable, the client shall promptly provide the CDM co-ordinator with pre-construction information consisting of—
  (a) all the information described in regulation 10(2) to be provided to any person in pursuance of regulation 10(1);
  (b) any further information as described in regulation 10(2) in the client's possession (or which is reasonably obtainable) which is relevant to the CDM co-ordinator for the purposes specified in regulation 10(3), including the minimum amount of time before the construction phase which will be allowed to the principal contractor for planning and preparation for construction work.

The client's duty in relation to the start of the construction phase where a project is notifiable

16. Where the project is notifiable, the client shall ensure that the construction phase does not start unless—
  (a) the principal contractor has prepared a construction phase plan which complies with regulations 23(1)(a) and 23(2); and
  (b) he is satisfied that the requirements of regulation 22(1)(c) (provision of welfare facilities) will be complied with during the construction phase.

The client's duty in relation to the health and safety file

17.—(1) The client shall ensure that the CDM co-ordinator is provided with all the health and safety information in the client's possession (or which is reasonably obtainable) relating to the project which is likely to be needed for inclusion in the health and safety file, including information specified in regulation 4(9)(c) of the Control of Asbestos Regulations 2006.

(2) Where a single health and safety file relates to more than one project, site or structure, or where it includes other related information, the client shall ensure that the information relating to each site or structure can be easily identified.

(3) The client shall take reasonable steps to ensure that after the construction phase the information in the health and safety file—
   (a) is kept available for inspection by any person who may need it to comply with the relevant statutory provisions; and
   (b) is revised as often as may be appropriate to incorporate any relevant new information.

(4) It shall be sufficient compliance with paragraph (3)(a) by a client who disposes of his entire interest in the structure if he delivers the health and safety file to the person who acquires his interest in it and ensures that he is aware of the nature and purpose of the file.

Additional duties of designers

18. —(1) Where a project is notifiable, no designer shall commence work (other than initial design work) in relation to the project unless a CDM co-ordinator has been appointed for the project.

(2) The designer shall take all reasonable steps to provide with his design sufficient information about aspects of the design of the structure or its construction or maintenance as will adequately assist the CDM co-ordinator to comply with his duties under these Regulations, including his duties in relation to the health and safety file.

Additional duties of contractors

19. —(1) Where a project is notifiable, no contractor shall carry out construction work in relation to the project unless—
   (a) he has been provided with the names of the CDM co-ordinator and principal contractor;
   (b) he has been given access to such part of the construction phase plan as is relevant to the work to be performed by him, containing sufficient detail in relation to such work; and
   (c) notice of the project has been given to the Executive, or as the case may be the Office of Rail Regulation, under regulation 21.

(2) Every contractor shall—
   (a) promptly provide the principal contractor with any information (including any relevant part of any risk assessment in his possession or control) which—
      (i) might affect the health or safety of any person carrying out the construction work or of any person who may be affected by it,
      (ii) might justify a review of the construction phase plan, or
      (iii) has been identified for inclusion in the health and safety file in pursuance of regulation 22(1)(j);
   (b) promptly identify any contractor whom he appoints or engages in his turn in connection with the project to the principal contractor;
   (c) comply with—
      (i) any directions of the principal contractor given to him under regulation 22(1)(e), and
      (ii) any site rules;
   (d) promptly provide the principal contractor with the information in relation to any death, injury, condition or dangerous occurrence which the contractor is required to notify or report under the Reporting of Injuries, Diseases and Dangerous Occurrences Regulations 1995.

(3) Every contractor shall—
   (a) in complying with his duty under regulation 13(2) take all reasonable steps to ensure that the construction work is carried out in accordance with the construction phase plan;
   (b) take appropriate action to ensure health and safety where it is not possible to comply with the construction phase plan in any particular case; and

(c) notify the principal contractor of any significant finding which requires the construction phase plan to be altered or added to.

General duties of CDM co-ordinators

20. —(1) The CDM co-ordinator shall—
   (a) give suitable and sufficient advice and assistance to the client on undertaking the measures he needs to take to comply with these Regulations during the project (including, in particular, assisting the client in complying with regulations 9 and 16);
   (b) ensure that suitable arrangements are made and implemented for the co-ordination of health and safety measures during planning and preparation for the construction phase, including facilitating—
      (i) co-operation and co-ordination between persons concerned in the project in pursuance of regulations 5 and 6, and
      (ii) the application of the general principles of prevention in pursuance of regulation 7; and
   (c) liaise with the principal contractor regarding—
      (i) the contents of the health and safety file,
      (ii) the information which the principal contractor needs to prepare the construction phase plan, and
      (iii) any design development which may affect planning and management of the construction work.
(2) Without prejudice to paragraph (1) the CDM co-ordinator shall—
   (a) take all reasonable steps to identify and collect the pre-construction information;
   (b) promptly provide in a convenient form to—
      (i) every person designing the structure, and
      (ii) every contractor who has been or may be appointed by the client (including the principal contractor),
      such of the pre-construction information in his possession as is relevant to each;
   (c) take all reasonable steps to ensure that designers comply with their duties under regulations 11 and 18(2);
   (d) take all reasonable steps to ensure co-operation between designers and the principal contractor during the construction phase in relation to any design or change to a design;
   (e) prepare, where none exists, and otherwise review and update a record ("the health and safety file") containing information relating to the project which is likely to be needed during any subsequent construction work to ensure the health and safety of any person, including the information provided in pursuance of regulations 17(1), 18(2) and 22(1)(j); and
   (f) at the end of the construction phase, pass the health and safety file to the client.

Notification of project by the CDM co-ordinator

21. —(1) The CDM co-ordinator shall as soon as is practicable after his appointment ensure that notice is given to the Executive containing such of the particulars specified in Schedule 1 as are available.
(2) Where any particulars specified in Schedule 1 have not been notified under paragraph (1) because a principal contractor has not yet been appointed, notice of such particulars shall be given to the Executive as soon as is practicable after the appointment of the principal contractor, and in any event before the start of the construction work.
(3) Any notice under paragraph (1) or (2) shall be signed by or on behalf of the client or, if sent by electronic means, shall otherwise show that he has approved it.
(4) Insofar as the project includes construction work of a description for which the Office of Rail Regulation is made the enforcing authority by regulation 3(1) of the Health and Safety (Enforcing Authority for Railways and Other Guided Transport Systems) Regulations 2006,

paragraphs (1) and (2) shall have effect as if any reference to the Executive were a reference to the Office of Rail Regulation.

Duties of the principal contractor

22. —(1) The principal contractor for a project shall—
   (a) plan, manage and monitor the construction phase in a way which ensures that, so far as is reasonably practicable, it is carried out without risks to health or safety, including facilitating—
      (i) co-operation and co-ordination between persons concerned in the project in pursuance of regulations 5 and 6, and
      (ii) the application of the general principles of prevention in pursuance of regulation 7;
   (b) liaise with the CDM co-ordinator in performing his duties in regulation 20(2)(d) during the construction phase in relation to any design or change to a design;
   (c) ensure that welfare facilities sufficient to comply with the requirements of Schedule 2 are provided throughout the construction phase;
   (d) where necessary for health and safety, draw up rules which are appropriate to the construction site and the activities on it (referred to in these Regulations as "site rules");
   (e) give reasonable directions to any contractor so far as is necessary to enable the principal contractor to comply with his duties under these Regulations;
   (f) ensure that every contractor is informed of the minimum amount of time which will be allowed to him for planning and preparation before he begins construction work;
   (g) where necessary, consult a contractor before finalising such part of the construction phase plan as is relevant to the work to be performed by him;
   (h) ensure that every contractor is given, before he begins construction work and in sufficient time to enable him to prepare properly for that work, access to such part of the construction phase plan as is relevant to the work to be performed by him;
   (i) ensure that every contractor is given, before he begins construction work and in sufficient time to enable him to prepare properly for that work, such further information as he needs—
      (i) to comply punctually with the duty under regulation 13(7), and
      (ii) to carry out the work to be performed by him without risk, so far as is reasonably practicable, to the health and safety of any person;
   (j) identify to each contractor the information relating to the contractor's activity which is likely to be required by the CDM co-ordinator for inclusion in the health and safety file in pursuance of regulation 20(2)(e) and ensure that such information is promptly provided to the CDM co-ordinator;
   (k) ensure that the particulars required to be in the notice given under regulation 21 are displayed in a readable condition in a position where they can be read by any worker engaged in the construction work; and
   (l) take reasonable steps to prevent access by unauthorised persons to the construction site.
(2) The principal contractor shall take all reasonable steps to ensure that every worker carrying out the construction work is provided with—
   (a) a suitable site induction;
   (b) the information and training referred to in regulation 13(4) by a contractor on whom a duty is placed by that regulation; and
   (c) any further information and training which he needs for the particular work to be carried out without undue risk to health or safety.

The principal contractor's duty in relation to the construction phase plan

23. —(1) The principal contractor shall—
   (a) before the start of the construction phase, prepare a construction phase plan which is sufficient to ensure that the construction phase is planned, managed and monitored in a way

which enables the construction work to be started so far as is reasonably practicable without risk to health or safety, paying adequate regard to the information provided by the designer under regulations 11(6) and 18(2) and the pre-construction information provided under regulation 20(2)(b);
  (b) from time to time and as often as may be appropriate throughout the project update, review, revise and refine the construction phase plan so that it continues to be sufficient to ensure that the construction phase is planned, managed and monitored in a way which enables the construction work to be carried out so far as is reasonably practicable without risk to health or safety; and
  (c) arrange for the construction phase plan to be implemented in a way which will ensure so far as is reasonably practicable the health and safety of all persons carrying out the construction work and all persons who may be affected by the work.
(2) The principal contractor shall take all reasonable steps to ensure that the construction phase plan identifies the risks to health and safety arising from the construction work (including the risks specific to the particular type of construction work concerned) and includes suitable and sufficient measures to address such risks, including any site rules.

The principal contractor's duty in relation to co-operation and consultation with workers

24. The principal contractor shall—
  (a) make and maintain arrangements which will enable him and the workers engaged in the construction work to co-operate effectively in promoting and developing measures to ensure the health, safety and welfare of the workers and in checking the effectiveness of such measures;
  (b) consult those workers or their representatives in good time on matters connected with the project which may affect their health, safety or welfare, so far as they or their representatives are not so consulted on those matters by any employer of theirs;
  (c) ensure that such workers or their representatives can inspect and take copies of any information which the principal contractor has, or which these Regulations require to be provided to him, which relates to the planning and management of the project, or which otherwise may affect their health, safety or welfare at the site, except any information—
    (i) the disclosure of which would be against the interests of national security,
    (ii) which he could not disclose without contravening a prohibition imposed by or under an enactment,
    (iii) relating specifically to an individual, unless he has consented to its being disclosed,
    (iv) the disclosure of which would, for reasons other than its effect on health, safety or welfare at work, cause substantial injury to his undertaking or, where the information was supplied to him by some other person, to the undertaking of that other person, or
    (v) obtained by him for the purpose of bringing, prosecuting or defending any legal proceedings.

Part 4
Duties Relating to Health and Safety on Construction Sites

Application of Regulations 26 to 44

25.—(1) Every contractor carrying out construction work shall comply with the requirements of regulations 26 to 44 insofar as they affect him or any person carrying out construction work under his control or relate to matters within his control.
(2) Every person (other than a contractor carrying out construction work) who controls the way in which any construction work is carried out by a person at work shall comply with the requirements of regulations 26 to 44 insofar as they relate to matters which are within his control.

(3) Every person at work on construction work under the control of another person shall report to that person any defect which he is aware may endanger the health and safety of himself or another person.

(4) Paragraphs (1) and (2) shall not apply to regulation 33, which expressly says on whom the duties in that regulation are imposed.

Safe places of work

26. —(1) There shall, so far as is reasonably practicable, be suitable and sufficient safe access to and egress from every place of work and to and from every other place provided for the use of any person while at work, which access and egress shall be properly maintained.

(2) Every place of work shall, so far as is reasonably practicable, be made and kept safe for, and without risks to health to, any person at work there.

(3) Suitable and sufficient steps shall be taken to ensure, so far as is reasonably practicable, that no person uses access or egress, or gains access to any place, which does not comply with the requirements of paragraph (1) or (2) respectively.

(4) Every place of work shall, so far as is reasonably practicable, have sufficient working space and be so arranged that it is suitable for any person who is working or who is likely to work there, taking account of any necessary work equipment present.

Good order and site security

27. —(1) Every part of a construction site shall, so far as is reasonably practicable, be kept in good order and every part of a construction site which is used as a place of work shall be kept in a reasonable state of cleanliness.

(2) Where necessary in the interests of health and safety, a construction site shall, so far as is reasonably practicable and in accordance with the level of risk posed, either—
   (a) have its perimeter identified by suitable signs and be so arranged that its extent is readily identifiable; or
   (b) be fenced off,
   or both.

(3) No timber or other material with projecting nails (or similar sharp object) shall—
   (a) be used in any work; or
   (b) be allowed to remain in any place,
   if the nails (or similar sharp object) may be a source of danger to any person.

Stability of structures

28. —(1) All practicable steps shall be taken, where necessary to prevent danger to any person, to ensure that any new or existing structure or any part of such structure which may become unstable or in a temporary state of weakness or instability due to the carrying out of construction work does not collapse.

(2) Any buttress, temporary support or temporary structure must be of such design and so installed and maintained as to withstand any foreseeable loads which may be imposed on it, and must only be used for the purposes for which it is so designed, installed and maintained.

(3) No part of a structure shall be so loaded as to render it unsafe to any person.

Demolition or dismantling

29. —(1) The demolition or dismantling of a structure, or part of a structure, shall be planned and carried out in such a manner as to prevent danger or, where it is not practicable to prevent it, to reduce danger to as low a level as is reasonably practicable.

(2) The arrangements for carrying out such demolition or dismantling shall be recorded in writing before the demolition or dismantling work begins.

Explosives

30. —(1) So far as is reasonably practicable, explosives shall be stored, transported and used safely and securely.

(2) Without prejudice to paragraph (1), an explosive charge shall be used or fired only if suitable and sufficient steps have been taken to ensure that no person is exposed to risk of injury from the explosion or from projected or flying material caused thereby.

Excavations

31. —(1) All practicable steps shall be taken, where necessary to prevent danger to any person, including, where necessary, the provision of supports or battering, to ensure that—
  (a) any excavation or part of an excavation does not collapse;
  (b) no material from a side or roof of, or adjacent to, any excavation is dislodged or falls; and
  (c) no person is buried or trapped in an excavation by material which is dislodged or falls.

(2) Suitable and sufficient steps shall be taken to prevent any person, work equipment, or any accumulation of material from falling into any excavation

(3) Without prejudice to paragraphs (1) and (2), suitable and sufficient steps shall be taken, where necessary, to prevent any part of an excavation or ground adjacent to it from being overloaded by work equipment or material;

(4) Construction work shall not be carried out in an excavation where any supports or battering have been provided pursuant to paragraph (1) unless—
  (a) the excavation and any work equipment and materials which affect its safety, have been inspected by a competent person—
    (i) at the start of the shift in which the work is to be carried out,
    (ii) after any event likely to have affected the strength or stability of the excavation, and
    (iii) after any material unintentionally falls or is dislodged; and
  (b) the person who carried out the inspection is satisfied that the work can be carried out there safely.

(5) Where the person who carried out the inspection has under regulation 33(1)(a) informed the person on whose behalf the inspection was carried out of any matter about which he is not satisfied, work shall not be carried out in the excavation until the matters have been satisfactorily remedied.

Cofferdams and caissons

32. —(1) Every cofferdam or caisson shall be—
  (a) of suitable design and construction;
  (b) appropriately equipped so that workers can gain shelter or escape if water or materials enter it; and
  (c) properly maintained.

(2) A cofferdam or caisson shall be used to carry out construction work only if—
  (a) the cofferdam or caisson, and any work equipment and materials which affect its safety, have been inspected by a competent person—
    (i) at the start of the shift in which the work is to be carried out, and
    (ii) after any event likely to have affected the strength or stability of the cofferdam or caisson; and
  (b) the person who carried out the inspection is satisfied that the work can be safely carried out there.

(3) Where the person who carried out the inspection has under regulation 33(1)(a) informed the person on whose behalf the inspection was carried out of any matter about which he is not satisfied, work shall not be carried out in the cofferdam or caisson until the matters have been satisfactorily remedied.

Reports of inspections

33. —(1) Subject to paragraph (5), the person who carries out an inspection under regulation 31 or 32 shall, before the end of the shift within which the inspection is completed—
   (a) where he is not satisfied that the construction work can be carried out safely at the place inspected, inform the person for whom the inspection was carried out of any matters about which he is not satisfied; and
   (b) prepare a report which shall include the particulars set out in Schedule 3.
(2) A person who prepares a report under paragraph (1) shall, within 24 hours of completing the inspection to which the report relates, provide the report or a copy of it to the person on whose behalf the inspection was carried out.
(3) Where the person owing a duty under paragraph (1) or (2) is an employee or works under the control of another, his employer or, as the case may be, the person under whose control he works shall ensure that he performs the duty.
(4) The person on whose behalf the inspection was carried out shall—
   (a) keep the report or a copy of it available for inspection by an inspector appointed under section 19 of the Health and Safety at Work etc. Act 1974—
      (i) at the site of the place of work in respect of which the inspection was carried out until that work is completed, and
      (ii) after that for 3 months,
   and send to the inspector such extracts from or copies of it as the inspector may from time to time require.
(5) Nothing in this regulation shall require as regards an inspection carried out on a place of work for the purposes of regulations 31(4)(a)(i) and 32(2)(a)(i), the preparation of more than one report within a period of 7 days

Energy distribution installations

34. —(1) Where necessary to prevent danger, energy distribution installations shall be suitably located, checked and clearly indicated.
(2) Where there is a risk from electric power cables—
   (a) they shall be directed away from the area of risk; or
   (b) the power shall be isolated and, where necessary, earthed; or
   (c) if it is not reasonably practicable to comply with paragraph (a) or (b), suitable warning notices and—
      (i) barriers suitable for excluding work equipment which is not needed, or
      (ii) where vehicles need to pass beneath the cables, suspended protections, or
      (iii) in either case, measures providing an equivalent level of safety,
   shall be provided or (in the case of measures) taken.
(3) No construction work which is liable to create a risk to health or safety from an underground service, or from damage to or disturbance of it, shall be carried out unless suitable and sufficient steps (including any steps required by this regulation) have been taken to prevent such risk, so far as is reasonably practicable.

Prevention of drowning

35. —(1) Where in the course of construction work any person is liable to fall into water or other liquid with a risk of drowning, suitable and sufficient steps shall be taken—
   (a) to prevent, so far as is reasonably practicable, such person from so falling;
   (b) to minimise the risk of drowning in the event of such a fall; and
   (c) to ensure that suitable rescue equipment is provided, maintained and, when necessary, used so that such person may be promptly rescued in the event of such a fall.
(2) Suitable and sufficient steps shall be taken to ensure the safe transport of any person conveyed by water to or from any place of work.

(3) Any vessel used to convey any person by water to or from a place of work shall not be overcrowded or overloaded.

Traffic routes

36. —(1) Every construction site shall be organised in such a way that, so far as is reasonably practicable, pedestrians and vehicles can move safely and without risks to health.
(2) Traffic routes shall be suitable for the persons or vehicles using them, sufficient in number, in suitable positions and of sufficient size.
(3) A traffic route shall not satisfy sub-paragraph (2) unless suitable and sufficient steps are taken to ensure that—
    (a) pedestrians or vehicles may use it without causing danger to the health or safety of persons near it;
    (b) any door or gate for pedestrians which leads onto a traffic route is sufficiently separated from that traffic route to enable pedestrians to see any approaching vehicle or plant from a place of safety;
    (c) there is sufficient separation between vehicles and pedestrians to ensure safety or, where this is not reasonably practicable —
        (i) there are provided other means for the protection of pedestrians, and
        (ii) there are effective arrangements for warning any person liable to be crushed or trapped by any vehicle of its approach;
    (d) any loading bay has at least one exit point for the exclusive use of pedestrians; and
    (e) where it is unsafe for pedestrians to use a gate intended primarily for vehicles, one or more doors for pedestrians is provided in the immediate vicinity of the gate, is clearly marked and is kept free from obstruction.
(4) Every traffic route shall be—
    (a) indicated by suitable signs where necessary for reasons of health or safety;
    (b) regularly checked; and
    (c) properly maintained.
(5) No vehicle shall be driven on a traffic route unless, so far as is reasonably practicable, that traffic route is free from obstruction and permits sufficient clearance.

Vehicles

37. —(1) Suitable and sufficient steps shall be taken to prevent or control the unintended movement of any vehicle.
(2) Suitable and sufficient steps shall be taken to ensure that, where any person may be endangered by the movement of any vehicle, the person having effective control of the vehicle shall give warning to any person who is liable to be at risk from the movement of the vehicle.
(3) Any vehicle being used for the purposes of construction work shall when being driven, operated or towed—
    (a) be driven, operated or towed in such a manner as is safe in the circumstances; and
    (b) be loaded in such a way that it can be driven, operated or towed safely.
(4) No person shall ride or be required or permitted to ride on any vehicle being used for the purposes of construction work otherwise than in a safe place thereon provided for that purpose.
(5) No person shall remain or be required or permitted to remain on any vehicle during the loading or unloading of any loose material unless a safe place of work is provided and maintained for such person.
(6) Suitable and sufficient measures shall be taken so as to prevent any vehicle from falling into any excavation or pit, or into water, or overrunning the edge of any embankment or earthwork.

Prevention of risk from fire etc.

38. Suitable and sufficient steps shall be taken to prevent, so far as is reasonably practicable, the risk of injury to any person during the carrying out of construction work arising from—
  (a) fire or explosion;
  (b) flooding; or
  (c) any substance liable to cause asphyxiation.

Emergency procedures

39. —(1) Where necessary in the interests of the health and safety of any person on a construction site, there shall be prepared and, where necessary, implemented suitable and sufficient arrangements for dealing with any foreseeable emergency, which arrangements shall include procedures for any necessary evacuation of the site or any part thereof.
(2) In making arrangements under paragraph (1), account shall be taken of—
  (a) the type of work for which the construction site is being used;
  (b) the characteristics and size of the construction site and the number and location of places of work on that site;
  (c) the work equipment being used;
  (d) the number of persons likely to be present on the site at any one time; and
  (e) the physical and chemical properties of any substances or materials on or likely to be on the site.
(3) Where arrangements are prepared pursuant to paragraph (1), suitable and sufficient steps shall be taken to ensure that—
  (a) every person to whom the arrangements extend is familiar with those arrangements; and
  (b) the arrangements are tested by being put into effect at suitable intervals.

Emergency routes and exits

40. —(1) Where necessary in the interests of the health and safety of any person on a construction site, a sufficient number of suitable emergency routes and exits shall be provided to enable any person to reach a place of safety quickly in the event of danger.
(2) An emergency route or exit provided pursuant to paragraph (1) shall lead as directly as possible to an identified safe area.
(3) Any emergency route or exit provided in accordance with paragraph (1), and any traffic route giving access thereto, shall be kept clear and free from obstruction and, where necessary, provided with emergency lighting so that such emergency route or exit may be used at any time.
(4) In making provision under paragraph (1), account shall be taken of the matters in regulation 39(2).
(5) All emergency routes or exits shall be indicated by suitable signs.

Fire detection and fire-fighting

41. —(1) Where necessary in the interests of the health and safety of any person at work on a construction site there shall be provided suitable and sufficient—
  (a) fire-fighting equipment; and
  (b) fire detection and alarm systems,
which shall be suitably located.
(2) In making provision under paragraph (1), account shall be taken of the matters in regulation 39(2).
(3) Any fire-fighting equipment and any fire detection and alarm system provided under paragraph (1) shall be examined and tested at suitable intervals and properly maintained.
(4) Any fire-fighting equipment which is not designed to come into use automatically shall be easily accessible.
(5) Every person at work on a construction site shall, so far as is reasonably practicable, be instructed in the correct use of any fire-fighting equipment which it may be necessary for him to use.

(6) Where a work activity may give rise to a particular risk of fire, a person shall not carry out such work unless he is suitably instructed.

(7) Fire-fighting equipment shall be indicated by suitable signs.

Fresh air

42.—(1) Suitable and sufficient steps shall be taken to ensure, so far as is reasonably practicable, that every place of work or approach thereto has sufficient fresh or purified air to ensure that the place or approach is safe and without risks to health.

(2) Any plant used for the purpose of complying with paragraph (1) shall, where necessary for reasons of health or safety, include an effective device to give visible or audible warning of any failure of the plant.

Temperature and weather protection

43.—(1) Suitable and sufficient steps shall be taken to ensure, so far as is reasonably practicable, that during working hours the temperature at any place of work indoors is reasonable having regard to the purpose for which that place is used.

(2) Every place of work outdoors shall, where necessary to ensure the health and safety of persons at work there, be so arranged that, so far as is reasonably practicable and having regard to the purpose for which that place is used and any protective clothing or work equipment provided for the use of any person at work there, it provides protection from adverse weather.

Lighting

44.—(1) Every place of work and approach thereto and every traffic route shall be provided with suitable and sufficient lighting, which shall be, so far as is reasonably practicable, by natural light.

(2) The colour of any artificial lighting provided shall not adversely affect or change the perception of any sign or signal provided for the purposes of health and safety.

(3) Without prejudice to paragraph (1), suitable and sufficient secondary lighting shall be provided in any place where there would be a risk to the health or safety of any person in the event of failure of primary artificial lighting.

PART 5
GENERAL

Civil liability

45. Breach of a duty imposed by the preceding provisions of these Regulations, other than those imposed by regulations 9(1)(b), 13(6) and (7), 16, 22(1)(c) and (l), 25(1), (2) and (4), 26 to 44 and Schedule 2, shall not confer a right of action in any civil proceedings insofar as that duty applies for the protection of a person who is not an employee of the person on whom the duty is placed.

Enforcement in respect of fire

46.—(1) Subject to paragraphs (2) and (3)—
  (a) in England and Wales the enforcing authority within the meaning of article 25 of the Regulatory Reform (Fire Safety) Order 2005; or
  (b) in Scotland the enforcing authority within the meaning of section 61 of the Fire (Scotland) Act 2005,
shall be the enforcing authority in respect of a construction site which is contained within, or forms part of, premises which are occupied by persons other than those carrying out the construction work or any activity arising from such work as regards regulations 39 and 40, in so far as those regulations relate to fire, and regulation 41.

(2) In England and Wales paragraph (1) only applies in respect of premises to which the Regulatory Reform (Fire Safety) Order 2005 applies.

(3) In Scotland paragraph (1) only applies in respect of premises to which Part 3 of the Fire (Scotland) Act 2005 applies.

Transitional provisions

47.—(1) These Regulations shall apply in relation to a project which began before their coming into force, with the following modifications.

(2) Subject to paragraph (3), where the time specified in paragraph (1) or (2) of regulation 14 for the appointment of the CDM co-ordinator or the principal contractor occurred before the coming into force of these Regulations, the client shall appoint the CDM co-ordinator or, as the case may be, the principal contractor, as soon as is practicable.

(3) Where a client appoints any planning supervisor or principal contractor already appointed under regulation 6 of the Construction (Design and Management) Regulations 1994 (referred to in this regulation as "the 1994 Regulations") as the CDM co-ordinator or the principal contractor respectively pursuant to paragraph (2), regulation 4(1) shall have effect so that the client shall within twelve months of the coming into force of these Regulations take reasonable steps to ensure that any CDM co-ordinator or principal contractor so appointed is competent within the meaning of regulation 4(2).

(4) Any planning supervisor or principal contractor appointed under regulation 6 of the 1994 Regulations shall, in the absence of an express appointment by the client, be treated for the purposes of paragraph (2) as having been appointed as the CDM co-ordinator, or the principal contractor, respectively.

(5) Any person treated as having been appointed as the CDM co-ordinator or the principal contractor pursuant to paragraph (4) shall within twelve months of the coming into force of these Regulations take such steps as are necessary to ensure that he is competent within the meaning of regulation 4(2).

(6) Any agent appointed by a client under regulation 4 of the 1994 Regulations before the coming into force of these Regulations may, if requested by the client and if he himself consents, continue to act as the agent of that client and shall be subject to such requirements and prohibitions as are placed by these Regulations on that client, unless or until such time as such appointment is revoked by that client, or the project comes to an end, or five years elapse from the coming into force of these Regulations, whichever arises first.

(7) Where notice has been given under regulation 7 of the 1994 Regulations, the references in regulations 19(1)(c) and 22(1)(k) to notice under regulation 21 shall be construed as being to notice under that regulation.

Revocations and amendments

48.—(1) The revocations listed in Schedule 4 shall have effect.

(2) The amendments listed in Schedule 5 shall have effect.

Signed by authority of the Secretary of State for Work and Pensions.

Bill McKenzie

Parliamentary Under Secretary of State, Department for Work and Pensions

7th February 2007

Schedule 1

Regulation 21(1), (2) and (4)

Particulars to be Notified to the Executive (or Office of Rail Regulation)

1. Date of forwarding.
2. Exact address of the construction site.

3. The name of the local authority where the site is located.
4. A brief description of the project and the construction work which it includes.
5. Contact details of the client (name, address, telephone number and any e-mail address).
6. Contact details of the CDM co-ordinator (name, address, telephone number and any e-mail address).
7. Contact details of the principal contractor (name, address, telephone number and any e-mail address).
8. Date planned for the start of the construction phase.
9. The time allowed by the client to the principal contractor referred to in regulation 15(b) for planning and preparation for construction work.
10. Planned duration of the construction phase.
11. Estimated maximum number of people at work on the construction site.
12. Planned number of contractors on the construction site.
13. Name and address of any contractor already appointed.
14. Name and address of any designer already engaged.
15. A declaration signed by or on behalf of the client that he is aware of his duties under these Regulations.

Schedule 2

Regulations 9(1)(b), 13(7) and 22(1)(c)

Welfare Facilities

Sanitary conveniences

1. Suitable and sufficient sanitary conveniences shall be provided or made available at readily accessible places. So far as is reasonably practicable, rooms containing sanitary conveniences shall be adequately ventilated and lit.
2. So far as is reasonably practicable, sanitary conveniences and the rooms containing them shall be kept in a clean and orderly condition.
3. Separate rooms containing sanitary conveniences shall be provided for men and women, except where and so far as each convenience is in a separate room, the door of which is capable of being secured from the inside.

Washing facilities

4. Suitable and sufficient washing facilities, including showers if required by the nature of the work or for health reasons, shall so far as is reasonably practicable be provided or made available at readily accessible places.
5. Washing facilities shall be provided—
   (a) in the immediate vicinity of every sanitary convenience, whether or not provided elsewhere; and
   (b) in the vicinity of any changing rooms required by paragraph 14 whether or not provided elsewhere.
6. Washing facilities shall include—
   (a) a supply of clean hot and cold, or warm, water (which shall be running water so far as is reasonably practicable);
   (b) soap or other suitable means of cleaning; and
   (c) towels or other suitable means of drying.
7. Rooms containing washing facilities shall be sufficiently ventilated and lit.
8. Washing facilities and the rooms containing them shall be kept in a clean and orderly condition.
9. Subject to paragraph 10 below, separate washing facilities shall be provided for men and women, except where and so far as they are provided in a room the door of which is capable of being

secured from inside and the facilities in each such room are intended to be used by only one person at a time.
10. Paragraph 9 above shall not apply to facilities which are provided for washing hands, forearms and face only.

Drinking water

11. An adequate supply of wholesome drinking water shall be provided or made available at readily accessible and suitable places.
12. Every supply of drinking water shall be conspicuously marked by an appropriate sign where necessary for reasons of health and safety.
13. Where a supply of drinking water is provided, there shall also be provided a sufficient number of suitable cups or other drinking vessels unless the supply of drinking water is in a jet from which persons can drink easily.

Changing rooms and lockers

14. —(1) Suitable and sufficient changing rooms shall be provided or made available at readily accessible places if—
    (a) a worker has to wear special clothing for the purposes of his work; and
    (b) he cannot, for reasons of health or propriety, be expected to change elsewhere,
being separate rooms for, or separate use of rooms by, men and women where necessary for reasons of propriety.

(2) Changing rooms shall—
    (a) be provided with seating; and
    (b) include, where necessary, facilities to enable a person to dry any such special clothing and his own clothing and personal effects.

(3) Suitable and sufficient facilities shall, where necessary, be provided or made available at readily accessible places to enable persons to lock away—
    (a) any such special clothing which is not taken home;
    (b) their own clothing which is not worn during working hours; and
    (c) their personal effects.

Facilities for rest

15. —(1) Suitable and sufficient rest rooms or rest areas shall be provided or made available at readily accessible places.

(2) Rest rooms and rest areas shall—
    (a) include suitable arrangements to protect non-smokers from discomfort caused by tobacco smoke;
    (b) be equipped with an adequate number of tables and adequate seating with backs for the number of persons at work likely to use them at any one time;
    (c) where necessary, include suitable facilities for any person at work who is a pregnant woman or nursing mother to rest lying down;
    (d) include suitable arrangements to ensure that meals can be prepared and eaten;
    (e) include the means for boiling water; and
    (f) be maintained at an appropriate temperature.

Schedule 3

Regulation 33(1)(b)

Particulars to be Included in a Report of Inspection

1. Name and address of the person on whose behalf the inspection was carried out.
2. Location of the place of work inspected.

3. Description of the place of work or part of that place inspected (including any work equipment and materials).
4. Date and time of the inspection.
5. Details of any matter identified that could give rise to a risk to the health or safety of any person.
6. Details of any action taken as a result of any matter identified in paragraph 5 above.
7. Details of any further action considered necessary.
8. Name and position of the person making the report.

APPENDIX K

Health and Safety: The Work at Height Regulations 2005 (as amended)

2005 No. 735

Made	16th March 2005
Laid before Parliament	16th March 2005
Coming into force	6th April 2005

The Secretary of State, in the exercise of the powers conferred on him by sections 15(1), (2), (3)(a), (5)(b), (6)(a) and 82(3)(a) of, and paragraphs 1(1), (2) and (3), 9, 11, 14, 15(1) and 16 of Schedule 3 to, the Health and Safety at Work etc. Act 1974 ("the 1974 Act") and for the purpose of giving effect without modifications to proposals submitted to him by the Health and Safety Commission under section 11(2)(d) of the 1974 Act, after the carrying out by the said Commission of consultations in accordance with section 50(3) of that Act, hereby makes the following Regulations:

1 Citation and commencement

These Regulations may be cited as the Work at Height Regulations 2005 and shall come into force on 6th April 2005.

2 Interpretation

(1) In these Regulations, unless the context otherwise requires—

"the 1974 Act" means the Health and Safety at Work etc. Act 1974;

"access" and "egress" include ascent and descent;

"construction work" has the meaning assigned to it by regulation 2(1) of the [Construction (Design and Management) Regulations 2007];[1]

"fragile surface" means a surface which would be liable to fail if any reasonably foreseeable loading were to be applied to it;

"ladder" includes a fixed ladder and a stepladder;

"line" includes rope, chain or webbing;

"the Management Regulations" means the Management of Health and Safety at Work Regulations 1999;

"personal fall protection system" means—

(a) a fall prevention, work restraint, work positioning, fall arrest or rescue system, other than a system in which the only safeguards are collective safeguards; or

(b) rope access and positioning techniques;

"suitable" means suitable in any respect which it is reasonably foreseeable will affect the safety of any person;

"work at height" means—

(a) work in any place, including a place at or below ground level;

---

[1] Amended by SI 2007/320 date in force: 6 April 2007.

(b) obtaining access to or egress from such place while at work, except by a staircase in a permanent workplace,

where, if measures required by these Regulations were not taken, a person could fall a distance liable to cause personal injury;

"work equipment" means any machinery, appliance, apparatus, tool or installation for use at work (whether exclusively or not) and includes anything to which regulation 8 and Schedules 2 to 6 apply;

"working platform"—
- (a) means any platform used as a place of work or as a means of access to or egress from a place of work;
- (b) includes any scaffold, suspended scaffold, cradle, mobile platform, trestle, gangway, gantry and stairway which is so used.

(2) Any reference in these Regulations to the keeping of a report or copy of a report or plan shall include reference to its being kept in a form—
- (a) in which it is capable of being reproduced as a printed copy when required;
- (b) which is secure from loss or unauthorised interference.

3 Application

(1) These Regulations shall apply—
- (a) in Great Britain; and
- (b) outside Great Britain as sections 1 to 59 and 80 to 82 of the 1974 Act apply by virtue of the Health and Safety at Work etc. Act 1974 (Application outside Great Britain) Order 2001.

(2) The requirements imposed by these Regulations on an employer shall apply in relation to work—
- (a) by an employee of his; or
- (b) by any other person under his control, to the extent of his control.

(3) The requirements imposed by these Regulations on an employer shall also apply to—
- (a) a self-employed person, in relation to work—
    - (i) by him; or
    - (ii) by a person under his control, to the extent of his control; and
- (b) to any person other than a self-employed person, in relation to work by a person under his control, to the extent of his control.

(4) Regulations 4 to 16 of these Regulations shall not apply to or in relation to—
- (a) the master and crew of a ship, or to the employer of such persons, in respect of the normal ship-board activities of a ship's crew which—
    - (i) are carried out solely by the crew under the direction of the master; and
    - (ii) are not liable to expose persons at work other than the master and crew to a risk to their safety;
- (b) a place specified in regulation 7(6) of the Docks Regulations 1988 where persons are engaged in dock operations; [or][2]
- (c) a place specified in regulation 5(3) of the Loading and Unloading of Fishing Vessels Regulations 1988 where persons are engaged in fish loading processes; [. . .][3]
- (d) [. . .][4]

(5) Regulation 11 of these Regulations shall not apply to an installation while regulation 12 of the Offshore Installations and Wells (Design and Construction, etc) Regulations 1996 apply to it.

(6) In this regulation—

---

[2] Amended by SI 2007/114.
[3] Repealed by SI 2007/114.
[4] ibid.

(a) [...]⁵
(b) [...]⁶
(c) "ship" includes every description of vessel used in navigation, other than a ship which forms part of Her Majesty's Navy.

4 Organisation and planning

(1) Every employer shall ensure that work at height is—
   (a) properly planned;
   (b) appropriately supervised; and
   (c) carried out in a manner which is so far as is reasonably practicable safe,
and that its planning includes the selection of work equipment in accordance with regulation 7.

(2) Reference in paragraph (1) to planning of work includes planning for emergencies and rescue.

(3) Every employer shall ensure that work at height is carried out only when the weather conditions do not jeopardise the health or safety of persons involved in the work.

(4) Paragraph (3) shall not apply where members of the police, fire, ambulance or other emergency services are acting in an emergency.

5 Competence

Every employer shall ensure that no person engages in any activity, including organisation, planning and supervision, in relation to work at height or work equipment for use in such work unless he is competent to do so or, if being trained, is being supervised by a competent person.

6 Avoidance of risks from work at height

(1) In identifying the measures required by this regulation, every employer shall take account of a risk assessment under regulation 3 of the Management Regulations.

(2) Every employer shall ensure that work is not carried out at height where it is reasonably practicable to carry out the work safely otherwise than at height.

(3) Where work is carried out at height, every employer shall take suitable and sufficient measures to prevent, so far as is reasonably practicable, any person falling a distance liable to cause personal injury.

(4) The measures required by paragraph (3) shall include—
   (a) his ensuring that the work is carried out—
      (i) from an existing place of work; or
      (ii) (in the case of obtaining access or egress) using an existing means,
      which complies with Schedule 1, where it is reasonably practicable to carry it out safely and under appropriate ergonomic conditions; and
   (b) where it is not reasonably practicable for the work to be carried out in accordance with sub-paragraph (a), his providing sufficient work equipment for preventing, so far as is reasonably practicable, a fall occurring.

(5) Where the measures taken under paragraph (4) do not eliminate the risk of a fall occurring, every employer shall—
   (a) so far as is reasonably practicable, provide sufficient work equipment to minimise—
      (i) the distance and consequences; or
      (ii) where it is not reasonably practicable to minimise the distance, the consequences, of a fall; and

---

⁵ Repealed by SI 2007/114.
⁶ ibid.

(b) without prejudice to the generality of paragraph (3), provide such additional training and instruction or take other additional suitable and sufficient measures to prevent, so far as is reasonably practicable, any person falling a distance liable to cause personal injury.

7 Selection of work equipment for work at height

(1) Every employer, in selecting work equipment for use in work at height, shall—
 (a) give collective protection measures priority over personal protection measures; and
 (b) take account of—
  (i) the working conditions and the risks to the safety of persons at the place where the work equipment is to be used;
  (ii) in the case of work equipment for access and egress, the distance to be negotiated;
  (iii) the distance and consequences of a potential fall;
  (iv) the duration and frequency of use;
  (v) the need for easy and timely evacuation and rescue in an emergency;
  (vi) any additional risk posed by the use, installation or removal of that work equipment or by evacuation and rescue from it; and
  (vii) the other provisions of these Regulations.
(2) An employer shall select work equipment for work at height which—
 (a) has characteristics including dimensions which—
  (i) are appropriate to the nature of the work to be performed and the foreseeable loadings; and
  (ii) allow passage without risk; and
 (b) is in other respects the most suitable work equipment, having regard in particular to the purposes specified in regulation 6.

8 Requirements for particular work equipment

Every employer shall ensure that, in the case of—
 (a) a guard-rail, toe-board, barrier or similar collective means of protection, Schedule 2 is complied with;
 (b) a working platform—
  (i) Part 1 of Schedule 3 is complied with; and
  (ii) where scaffolding is provided, Part 2 of Schedule 3 is also complied with;
 (c) a net, airbag or other collective safeguard for arresting falls which is not part of a personal fall protection system, Schedule 4 is complied with;
 (d) a personal fall protection system, Part 1 of Schedule 5 and—
  (i) in the case of a work positioning system, Part 2 of Schedule 5;
  (ii) in the case of rope access and positioning techniques, Part 3 of Schedule 5;
  (iii) in the case of a fall arrest system, Part 4 of Schedule 5;
  (iv) in the case of a work restraint system, Part 5 of Schedule 5,
  are complied with; and
 (e) a ladder, Schedule 6 is complied with.

9 Fragile surfaces

(1) Every employer shall ensure that no person at work passes across or near, or works on, from or near, a fragile surface where it is reasonably practicable to carry out work safely and under appropriate ergonomic conditions without his doing so.
(2) Where it is not reasonably practicable to carry out work safely and under appropriate ergonomic conditions without passing across or near, or working on, from or near, a fragile surface, every employer shall—
 (a) ensure, so far as is reasonably practicable, that suitable and sufficient platforms, coverings, guard rails or similar means of support or protection are provided and used so that any foreseeable loading is supported by such supports or borne by such protection;

(b) where a risk of a person at work falling remains despite the measures taken under the preceding provisions of this regulation, take suitable and sufficient measures to minimise the distances and consequences of his fall.

(3) Where any person at work may pass across or near, or work on, from or near, a fragile surface, every employer shall ensure that—
   (a) prominent warning notices are so far as is reasonably practicable affixed at the approach to the place where the fragile surface is situated; or
   (b) where that is not reasonably practicable, such persons are made aware of it by other means.

(4) Paragraph (3) shall not apply where members of the police, fire, ambulance or other emergency services are acting in an emergency.

10 Falling objects

(1) Every employer shall, where necessary to prevent injury to any person, take suitable and sufficient steps to prevent, so far as is reasonably practicable, the fall of any material or object.

(2) Where it is not reasonably practicable to comply with the requirements of paragraph (1), every employer shall take suitable and sufficient steps to prevent any person being struck by any falling material or object which is liable to cause personal injury.

(3) Every employer shall ensure that no material or object is thrown or tipped from height in circumstances where it is liable to cause injury to any person.

(4) Every employer shall ensure that materials and objects are stored in such a way as to prevent risk to any person arising from the collapse, overturning or unintended movement of such materials or objects.

11 Danger areas

Without prejudice to the preceding requirements of these Regulations, every employer shall ensure that—
   (a) where a workplace contains an area in which, owing to the nature of the work, there is a risk of any person at work—
      (i) falling a distance; or
      (ii) being struck by a falling object,
      which is liable to cause personal injury, the workplace is so far as is reasonably practicable equipped with devices preventing unauthorised persons from entering such area; and
   (b) such area is clearly indicated.

12 Inspection of work equipment

(1) This regulation applies only to work equipment to which regulation 8 and Schedules 2 to 6 apply.

(2) Every employer shall ensure that, where the safety of work equipment depends on how it is installed or assembled, it is not used after installation or assembly in any position unless it has been inspected in that position.

(3) Every employer shall ensure that work equipment exposed to conditions causing deterioration which is liable to result in dangerous situations is inspected—
   (a) at suitable intervals; and
   (b) each time that exceptional circumstances which are liable to jeopardise the safety of the work equipment have occurred,
to ensure that health and safety conditions are maintained and that any deterioration can be detected and remedied in good time.

(4) Without prejudice to paragraph (2), every employer shall ensure that a working platform—
   (a) used for construction work; and
   (b) from which a person could fall 2 metres or more,

is not used in any position unless it has been inspected in that position or, in the case of a mobile working platform, inspected on the site, within the previous 7 days.

(5) Every employer shall ensure that no work equipment, other than lifting equipment to which the requirement in regulation 9(4) of the Lifting Operations and Lifting Equipment Regulations 1998 ("LOLER") applies—
   (a) leaves his undertaking; or
   (b) if obtained from the undertaking of another person, is used in his undertaking,

unless it is accompanied by physical evidence that the last inspection required to be carried out under this regulation has been carried out.

(6) Every employer shall ensure that the result of an inspection under this regulation is recorded and, subject to paragraph (8), kept until the next inspection under this regulation is recorded.

(7) A person carrying out an inspection of work equipment to which paragraph (4) applies shall—
   (a) before the end of the working period within which the inspection is completed, prepare a report containing the particulars set out in Schedule 7; and
   (b) within 24 hours of completing the inspection, provide the report or a copy thereof to the person on whose behalf the inspection was carried out.

(8) An employer receiving a report or copy under paragraph (7) shall keep the report or a copy thereof—
   (a) at the site where the inspection was carried out until the construction work is completed; and
   (b) thereafter at an office of his for 3 months.

(9) Where a thorough examination has been made of lifting equipment under regulation 9 of LOLER—
   (a) it shall for the purposes of this regulation, other than paragraphs (7) and (8), be treated as an inspection of the lifting equipment; and
   (b) the making under regulation 10 of LOLER of a report of such examination shall for the purposes of paragraph (6) of this regulation be treated as the recording of the inspection.

(10) In this regulation "inspection", subject to paragraph (9)—
   (a) means such visual or more rigorous inspection by a competent person as is appropriate for safety purposes;
   (b) includes any testing appropriate for those purposes,

and "inspected" shall be construed accordingly.

13 Inspection of places of work at height

Every employer shall so far as is reasonably practicable ensure that the surface and every parapet, permanent rail or other such fall protection measure of every place of work at height are checked on each occasion before the place is used.

14 Duties of persons at work

(1) Every person shall, where working under the control of another person, report to that person any activity or defect relating to work at height which he knows is likely to endanger the safety of himself or another person.

(2) Every person shall use any work equipment or safety device provided to him for work at height by his employer, or by a person under whose control he works, in accordance with—
   (a) any training in the use of the work equipment or device concerned which have been received by him; and
   (b) the instructions respecting that use which have been provided to him by that employer or person in compliance with the requirements and prohibitions imposed upon that employer or person by or under the relevant statutory provisions.

[14A Special provision in relation to caving and climbing]

[(1) Paragraph (2) applies in relation to the application of these Regulations to work concerning the provision of instruction or leadership to one or more persons in connection with their engagement in caving or climbing by way of sport, recreation, team building or similar activities.

(2) Where this paragraph applies, an employer, self-employed person or other person shall be taken to have complied with the caving and climbing requirements, if, by alternative means to any requirement of those requirements, he maintains in relation to a person at such work as is referred to in paragraph (1) a level of safety equivalent to that required by those requirements.

(3) For the purposes of paragraph (2), in determining whether an equivalent level of safety is maintained, regard shall be had to—
   (a) the nature of the activity;
   (b) any publicly available and generally accepted procedures for the activity; and
   (c) any other relevant circumstances.

(4) In this regulation—
   (a) "caving" includes the exploration of parts of mines which are no longer worked;
   (b) "climbing" means climbing, traversing, abseiling or scrambling over natural terrain or man-made structures; and
   (c) "the caving and climbing requirements" means regulation 8(d)(ii), so far as it relates to paragraph 1 in Part 3 of Schedule 5, and that paragraph.][7]

15 Exemption by the Health and Safety Executive

(1) Subject to paragraph (2), the Health and Safety Executive ("the Executive") may, by a certificate in writing, exempt—
   (a) any person or class of persons;
   (b) any premises or class of premises;
   (c) any work equipment; or
   (d) any work activity,

from the requirements imposed by paragraph 3(a) and (c) of Schedule 2, and any such exemption may be granted subject to conditions and to a limit of time and may be revoked at any time by a certificate in writing.

(2) The Executive shall not grant any such exemption unless, having regard to the circumstances of the case and in particular to—
   (a) the conditions, if any, which it proposes to attach to the exemption; and
   (b) any other requirements imposed by or under any enactments which apply to the case,

it is satisfied that the health and safety of persons who are likely to be affected by the exemption will not be prejudiced in consequence of it.

16 Exemption for the armed forces

(1) Subject to paragraph (2), the Secretary of State for Defence may, in the interests of national security, by a certificate in writing exempt any person or class of persons from any requirement or prohibition imposed by these Regulations in respect of activities carried out in the interests of national security, and any such exemption may be granted subject to conditions and may be revoked by the Secretary of State by a certificate in writing at any time.

(2) The Secretary of State shall not grant any such exemption unless he is satisfied that the health and safety of the employees concerned are ensured as far as possible in the light of the objectives of these Regulations.

---

[7] Inserted by SI 2007/114.

17 Amendment of the Provision and Use of Work Equipment Regulations 1998

There shall be added to regulation 6(5) of the Provision and Use of Work Equipment Regulations 1998 the following sub-paragraph—

"(f) work equipment to which regulation 12 of the Work at Height Regulations 2005 applies".

18 Repeal of section 24 of the Factories Act 1961

Section 24 of the Factories Act 1961 is repealed.

19 Revocation of instruments

The instruments specified in column 1 of Schedule 8 are revoked to the extent specified in column 3 of that Schedule.

Signed by authority of the Secretary of State

Jane Kennedy

Minister of State,

Department for Work and Pensions

16th March 2005

SCHEDULE 1

Regulation 6(4)(a)

REQUIREMENTS FOR EXISTING PLACES OF WORK AND MEANS OF ACCESS OR EGRESS AT HEIGHT

Every existing place of work or means of access or egress at height shall—
- (a) be stable and of sufficient strength and rigidity for the purpose for which it is intended to be or is being used;
- (b) where applicable, rest on a stable, sufficiently strong surface;
- (c) be of sufficient dimensions to permit the safe passage of persons and the safe use of any plant or materials required to be used and to provide a safe working area having regard to the work to be carried out there;
- (d) possess suitable and sufficient means for preventing a fall;
- (e) possess a surface which has no gap—
    - (i) through which a person could fall;
    - (ii) through which any material or object could fall and injure a person; or
    - (iii) giving rise to other risk of injury to any person, unless measures have been taken to protect persons against such risk;
- (f) be so constructed and used, and maintained in such condition, as to prevent, so far as is reasonably practicable—
    - (i) the risk of slipping or tripping; or
    - (ii) any person being caught between it and any adjacent structure;
- (g) where it has moving parts, be prevented by appropriate devices from moving inadvertently during work at height.

Schedule 2

Regulation 8(a)

Requirements for Guard-rails, Toe-boards, Barriers and Similar Collective means of Protection

1. Unless the context otherwise requires, any reference in this Schedule to means of protection is to a guard-rail, toe-board, barrier or similar collective means of protection.
2. Means of protection shall—
   (a) be of sufficient dimensions, of sufficient strength and rigidity for the purposes for which they are being used, and otherwise suitable;
   (b) be so placed, secured and used as to ensure, so far as is reasonably practicable, that they do not become accidentally displaced; and
   (c) be so placed as to prevent, so far as is practicable, the fall of any person, or of any material or object, from any place of work.
3. In relation to work at height involved in construction work—
   (a) the top guard-rail or other similar means of protection shall be at least 950 millimetres or, in the case of such means of protection already fixed at the coming into force of these Regulations, at least 910 millimetres above the edge from which any person is liable to fall;
   (b) toe-boards shall be suitable and sufficient to prevent the fall of any person, or any material or object, from any place of work; and
   (c) any intermediate guard-rail or similar means of protection shall be positioned so that any gap between it and other means of protection does not exceed 470 millimetres.
4. Any structure or part of a structure which supports means of protection or to which means of protection are attached shall be of sufficient strength and suitable for the purpose of such support or attachment.
5. —(1) Subject to sub-paragraph (2), there shall not be a lateral opening in means of protection save at a point of access to a ladder or stairway where an opening is necessary.
   (2) Means of protection shall be removed only for the time and to the extent necessary to gain access or egress or for the performance of a particular task and shall be replaced as soon as practicable.
   (3) The task shall not be performed while means of protection are removed unless effective compensatory safety measures are in place.

Schedule 3

Regulation 8(b)

Requirements for Working Platforms
Part 1
Requirements for All Working Platforms

Interpretation

1. In this Schedule, "supporting structure" means any structure used for the purpose of supporting a working platform and includes any plant used for that purpose.

Condition of surfaces

2. Any surface upon which any supporting structure rests shall be stable, of sufficient strength and of suitable composition safely to support the supporting structure, the working platform and any loading intended to be placed on the working platform.

Stability of supporting structure

3. Any supporting structure shall—
   (a) be suitable and of sufficient strength and rigidity for the purpose for which it is being used;
   (b) in the case of a wheeled structure, be prevented by appropriate devices from moving inadvertently during work at height;
   (c) in other cases, be prevented from slipping by secure attachment to the bearing surface or to another structure, provision of an effective anti-slip device or by other means of equivalent effectiveness;
   (d) be stable while being erected, used and dismantled; and
   (e) when altered or modified, be so altered or modified as to ensure that it remains stable.

Stability of working platforms

4. A working platform shall—
   (a) be suitable and of sufficient strength and rigidity for the purpose or purposes for which it is intended to be used or is being used;
   (b) be so erected and used as to ensure that its components do not become accidentally displaced so as to endanger any person;
   (c) when altered or modified, be so altered or modified as to ensure that it remains stable; and
   (d) be dismantled in such a way as to prevent accidental displacement.

Safety on working platforms

5. A working platform shall—
   (a) be of sufficient dimensions to permit the safe passage of persons and the safe use of any plant or materials required to be used and to provide a safe working area having regard to the work being carried out there;
   (b) possess a suitable surface and, in particular, be so constructed that the surface of the working platform has no gap—
      (i) through which a person could fall;
      (ii) through which any material or object could fall and injure a person; or
      (iii) giving rise to other risk of injury to any person, unless measures have been taken to protect persons against such risk; and
   (c) be so erected and used, and maintained in such condition, as to prevent, so far as is reasonably practicable—
      (i) the risk of slipping or tripping; or
      (ii) any person being caught between the working platform and any adjacent structure.

Loading

6. A working platform and any supporting structure shall not be loaded so as to give rise to a risk of collapse or to any deformation which could affect its safe use.

Part 2
Additional Requirements for Scaffolding

Additional requirements for scaffolding

7. Strength and stability calculations for scaffolding shall be carried out unless—
   (a) a note of the calculations, covering the structural arrangements contemplated, is available; or
   (b) it is assembled in conformity with a generally recognised standard configuration.
8. Depending on the complexity of the scaffolding selected, an assembly, use and dismantling plan shall be drawn up by a competent person. This may be in the form of a standard plan, supplemented by items relating to specific details of the scaffolding in question.

9. A copy of the plan, including any instructions it may contain, shall be kept available for the use of persons concerned in the assembly, use, dismantling or alteration of scaffolding until it has been dismantled.
10. The dimensions, form and layout of scaffolding decks shall be appropriate to the nature of the work to be performed and suitable for the loads to be carried and permit work and passage in safety.
11. While a scaffold is not available for use, including during its assembly, dismantling or alteration, it shall be marked with general warning signs in accordance with the Health and Safety (Safety Signs and Signals) Regulations 1996 and be suitably delineated by physical means preventing access to the danger zone.
12. Scaffolding may be assembled, dismantled or significantly altered only under the supervision of a competent person and by persons who have received appropriate and specific training in the operations envisaged which addresses specific risks which the operations may entail and precautions to be taken, and more particularly in—
    (a) understanding of the plan for the assembly, dismantling or alteration of the scaffolding concerned;
    (b) safety during the assembly, dismantling or alteration of the scaffolding concerned;
    (c) measures to prevent the risk of persons, materials or objects falling;
    (d) safety measures in the event of changing weather conditions which could adversely affect the safety of the scaffolding concerned;
    (e) permissible loadings;
    (f) any other risks which the assembly, dismantling or alteration of the scaffolding may entail.

Schedule 4

Regulation 8(c)

Requirements for Collective Safeguards for Arresting Falls

1. Any reference in this Schedule to a safeguard is to a collective safeguard for arresting falls.
2. A safeguard shall be used only if—
    (a) a risk assessment has demonstrated that the work activity can so far as is reasonably practicable be performed safely while using it and without affecting its effectiveness;
    (b) the use of other, safer work equipment is not reasonably practicable; and
    (c) a sufficient number of available persons have received adequate training specific to the safeguard, including rescue procedures.
3. A safeguard shall be suitable and of sufficient strength to arrest safely the fall of any person who is liable to fall.
4. A safeguard shall—
    (a) in the case of a safeguard which is designed to be attached, be securely attached to all the required anchors, and the anchors and the means of attachment thereto shall be suitable and of sufficient strength and stability for the purpose of safely supporting the foreseeable loading in arresting any fall and during any subsequent rescue;
    (b) in the case of an airbag, landing mat or similar safeguard, be stable; and
    (c) in the case of a safeguard which distorts in arresting a fall, afford sufficient clearance.
5. Suitable and sufficient steps shall be taken to ensure, so far as practicable, that in the event of a fall by any person the safeguard does not itself cause injury to that person.

Schedule 5

Regulation 8(d)

Requirements for Personal Fall Protection Systems
Part 1
Requirements for all Personal Fall Protection Systems

1. A personal fall protection system shall be used only if—
   (a) a risk assessment has demonstrated that—
      (i) the work can so far as is reasonably practicable be performed safely while using that system; and
      (ii) the use of other, safer work equipment is not reasonably practicable; and
   (b) the user and a sufficient number of available persons have received adequate training specific to the operations envisaged, including rescue procedures.
2. A personal fall protection system shall—
   (a) be suitable and of sufficient strength for the purposes for which it is being used having regard to the work being carried out and any foreseeable loading;
   (b) where necessary, fit the user;
   (c) be correctly fitted;
   (d) be designed to minimise injury to the user and, where necessary, be adjusted to prevent the user falling or slipping from it, should a fall occur; and
   (e) be so designed, installed and used as to prevent unplanned or uncontrolled movement of the user.
3. A personal fall protection system designed for use with an anchor shall be securely attached to at least one anchor, and each anchor and the means of attachment thereto shall be suitable and of sufficient strength and stability for the purpose of supporting any foreseeable loading.
4. Suitable and sufficient steps shall be taken to prevent any person falling or slipping from a personal fall protection system.

Part 2
Additional Requirements for Work Positioning Systems

A work positioning system shall be used only if either—
   (a) the system includes a suitable backup system for preventing or arresting a fall; and
   (b) where the system includes a line as a backup system, the user is connected to it; or
   (c) where it is not reasonably practicable to comply with sub-paragraph (a), all practicable measures are taken to ensure that the work positioning system does not fail.

Part 3
Additional Requirements for Rope Access and Positioning Techniques

1. [Except as provided in paragraph 3,][8] a rope access or positioning technique shall be used only if—
   (a) [. . .],[9] it involves a system comprising at least two separately anchored lines, of which one ("the working line") is used as a means of access, egress and support and the other is the safety line;

---

[8] Amended by SI 2007/114.
[9] Repealed by SI 2007/114.

(b) the user is provided with a suitable harness and is connected by it to the working line and the safety line;

(c) the working line is equipped with safe means of ascent and descent and has a self-locking system to prevent the user falling should he lose control of his movements; and

(d) the safety line is equipped with a mobile fall protection system which is connected to and travels with the user of the system.

2. Taking the risk assessment into account and depending in particular on the duration of the job and the ergonomic constraints, provision must be made for a seat with appropriate accessories.

3. The system may comprise a single rope where—

(a) a risk assessment has demonstrated that the use of a second line would entail higher risk to persons; and

(b) appropriate measures have been taken to ensure safety.

Part 4
Additional Requirements for Fall Arrest Systems

1. A fall arrest system shall incorporate a suitable means of absorbing energy and limiting the forces applied to the user's body.

2. A fall arrest system shall not be used in a manner—

(a) which involves the risk of a line being cut;

(b) where its safe use requires a clear zone (allowing for any pendulum effect), which does not afford such zone; or

(c) which otherwise inhibits its performance or renders its use unsafe.

Part 5
Additional Requirements for Work Restraint Systems

A work restraint system shall—

(a) be so designed that, if used correctly, it prevents the user from getting into a position in which a fall can occur; and

(b) be used correctly.

Schedule 6

Regulation 8(e)

Requirements for Ladders

1. Every employer shall ensure that a ladder is used for work at height only if a risk assessment under regulation 3 of the Management Regulations has demonstrated that the use of more suitable work equipment is not justified because of the low risk and -

(a) the short duration of use; or

(b) existing features on site which he cannot alter.

2. Any surface upon which a ladder rests shall be stable, firm, of sufficient strength and of suitable composition safely to support the ladder so that its rungs or steps remain horizontal, and any loading intended to be placed on it.

3. A ladder shall be so positioned as to ensure its stability during use.

4. A suspended ladder shall be attached in a secure manner and so that, with the exception of a flexible ladder, it cannot be displaced and swinging is prevented.

5. A portable ladder shall be prevented from slipping during use by—

(a) securing the stiles at or near their upper or lower ends;

(b) an effective anti-slip or other effective stability device; or

(c) any other arrangement of equivalent effectiveness.

6. A ladder used for access shall be long enough to protrude sufficiently above the place of landing to which it provides access, unless other measures have been taken to ensure a firm handhold.
7. No interlocking or extension ladder shall be used unless its sections are prevented from moving relative to each other while in use.
8. A mobile ladder shall be prevented from moving before it is stepped on.
9. Where a ladder or run of ladders rises a vertical distance of 9 metres or more above its base, there shall, where reasonably practicable, be provided at suitable intervals sufficient safe landing areas or rest platforms.
10. Every ladder shall be used in such a way that—
    (a) a secure handhold and secure support are always available to the user; and
    (b) the user can maintain a safe handhold when carrying a load unless, in the case of a step ladder, the maintenance of a handhold is not practicable when a load is carried, and a risk assessment under regulation 3 of the Management Regulations has demonstrated that the use of a stepladder is justified because of—
        (i) the low risk; and
        (ii) the short duration of use.

Schedule 7

Regulation 12(7)

Particulars to be Included in a Report of Inspection

1. The name and address of the person for whom the inspection was carried out.
2. The location of the work equipment inspected.
3. A description of the work equipment inspected.
4. The date and time of the inspection.
5. Details of any matter identified that could give rise to a risk to the health or safety of any person.
6. Details of any action taken as a result of any matter identified in paragraph 5.
7. Details of any further action considered necessary.
8. The name and position of the person making the report.

APPENDIX L

Health and Safety: The Control of Asbestos Regulations 2006

2006 No. 2739

Made   *12th October 2006*
Laid before Parliament   *20th October 2006*
Coming into force in accordance with regulation 1

The Secretary of State makes the following Regulations in the exercise of the powers conferred upon him by sections 15(1), (2), (3), (4), (5), (6)(b), (9), 18(2), 80(1) and 82(3) of, and paragraphs 1(1) to (4), 2, 3(2), 4, 6, 8 to 11, 13(1) and (3), 14, 15(1), 16 and 20 of Schedule 3 to, the Health and Safety at Work etc. Act 1974 ("the 1974 Act") and section 2(2) of the European Communities Act 1972 ("the 1972 Act").

In doing so he gives effect without modifications to proposals submitted to him by the Health and Safety Commission under section 11(2)(d) of the 1974 Act after the carrying out by the Commission of consultations in accordance with section 50(3) of that Act, and it appearing expedient to him after consulting such bodies as appear to him to be appropriate in accordance with section 80(4) of that Act.

He is a Minister designated for the purpose of section 2(2) of the 1972 Act in relation to the regulation and control of classification, packaging and labelling of dangerous substances and preparations, and persistent organic pollutants, dangerous substances, preparations and chemicals.

Part I
Preliminary

1. Citation and Commencement

These Regulations may be cited as the Control of Asbestos Regulations 2006 and shall come into force on 13th November 2006, except regulation 20(4) which shall come into force on 6th April 2007.

2. Interpretation

(1) In these Regulations—

"adequate" means adequate having regard only to the nature and degree of exposure to asbestos, and "adequately" shall be construed accordingly;

"appointed doctor" means a registered medical practitioner appointed for the time being in writing by the Executive for the purpose of these Regulations;

"approved" means approved for the time being in writing by the Health and Safety Commission or the Executive as the case may be;

"asbestos" means the following fibrous silicates—

    (a) asbestos actinolite, CAS No 77536-66-4(*);

    (b) asbestos grunerite (amosite), CAS No 12172-73-5(*);

    (c) asbestos anthophyllite, CAS No 77536-67-5(*);

(d) chrysotile, CAS No 12001-29-5;
(e) crocidolite, CAS NO 12001-28-4(*); and
(f) asbestos tremolite, CAS No 77536-68-6(*),

and references to "CAS" followed by a numerical sequence are references to CAS Registry Numbers assigned to chemicals by the Chemical Abstracts Service, a division of the American Chemical Society;

"the control limit" means a concentration of asbestos in the atmosphere when measured in accordance with the 1997 WHO recommended method, or by a method giving equivalent results to that method approved by the Health and Safety Commission, of 0.1 fibres per cubic centimetre of air averaged over a continuous period of 4 hours;

"control measure" means a measure taken to prevent or reduce exposure to asbestos (including the provision of systems of work and supervision, the cleaning of workplaces, premises, plant and equipment, and the provision and use of engineering controls and personal protective equipment);

"emergency services" include—
(a) police, fire, rescue and ambulance services;
(b) Her Majesty's Coastguard;

"employment medical adviser" means an employment medical adviser appointed under section 56 of the Health and Safety at Work etc. Act 1974;

"enforcing authority" means the Executive, local authority or Office of Rail Regulation, determined in accordance with the provisions of the Health and Safety (Enforcing Authority) Regulations 1998 and the provisions of the Health and Safety (Enforcing Authority for Railways and Other Guided Transport Systems) Regulations 2006;

"the Executive" means the Health and Safety Executive;

"ISO 17020" means European Standard EN ISO/IEC 17020, "General criteria for the operation of various types of bodies performing inspection" as revised or reissued from time to time and accepted by the Comité Européen de Normalisation Electrotechnique (CEN/CENELEC);

"ISO 17025" means European Standard EN ISO/IEC 17025, "General requirements for the competence of testing and calibration laboratories" as revised or reissued from time to time and accepted by the Comité Européen de Normalisation Electrotechnique (CEN/CENELEC);

"medical examination" includes any laboratory tests and X-rays that a relevant doctor may require;

"personal protective equipment" means all equipment (including clothing) which is intended to be worn or held by a person at work and which protects that person against one or more risks to his health, and any addition or accessory designed to meet that objective;

"relevant doctor" means an appointed doctor or an employment medical adviser;

"risk assessment" means the assessment of risk required by regulation 6(1)(a);

"the 1997 WHO recommended method" means the publication "Determination of airborne fibre concentrations. A recommended method, by phase-contrast optical microscopy (membrane filter method)", WHO (World Health Organisation), Geneva 1997.

(2) For the purposes of these Regulations, except in accordance with regulation 11(3) and (5), in determining whether an employee is exposed to asbestos or whether the extent of such exposure exceeds the control limit, no account shall be taken of respiratory protective equipment which, for the time being, is being worn by that employee.

(3) A reference to work with asbestos in these Regulations shall include—
(a) work which consists of the removal, repair or disturbance of asbestos or materials containing asbestos;
(b) work which is ancillary to such work; and
(c) supervision of such work and such ancillary work.

3. Application of these Regulations

(1) These Regulations shall apply to a self-employed person as they apply to an employer and an employee and as if that self-employed person were both an employer and an employee.

(2) Subject to paragraph (3), regulations 8 (licensing), 9 (notification of work with asbestos), 15(1) (arrangements to deal with accidents, incidents and emergencies), 18(1)(a) (asbestos areas) and 22 (health records and medical surveillance) shall not apply where—
  (a) the exposure of employees to asbestos is sporadic and of low intensity;
  (b) it is clear from the risk assessment that the exposure of any employee to asbestos will not exceed the control limit; and
  (c) the work involves—
    (i) short, non-continuous maintenance activities,
    (ii) removal of materials in which the asbestos fibres are firmly linked in a matrix,
    (iii) encapsulation or sealing of asbestos-containing materials which are in good condition, or
    (iv) air monitoring and control, and the collection and analysis of samples to ascertain whether a specific material contains asbestos.

(3) No exposure to asbestos will be sporadic and of low intensity within the meaning of paragraph (2)(a) if the concentration of asbestos in the atmosphere when measured in accordance with the 1997 WHO recommended method or by a method giving equivalent results to that method approved by the Health and Safety Commission exceeds or is liable to exceed the concentration approved in relation to a specified reference period for the purposes of this paragraph by the Health and Safety Commission.

(4) Where a duty is placed by these Regulations on an employer in respect of his employees, he shall, so far as is reasonably practicable, be under a like duty in respect of any other person, whether at work or not, who may be affected by the work activity carried out by the employer except that the duties of the employer—
  (a) under regulation 10 (information, instruction and training) shall not extend to persons who are not his employees unless those persons are on the premises where the work is being carried out; and
  (b) under regulation 22 (health records and medical surveillance) shall not extend to persons who are not his employees.

(5) Regulation 17, insofar as it requires an employer to ensure that premises are thoroughly cleaned, shall not apply—
  (a) in England and Wales, to a fire and rescue authority within the meaning of section 1 of the Fire and Rescue Services Act 2004, or in Scotland to a relevant authority within the meaning of section 6 of the Fire (Scotland) Act 2005, in respect of premises attended by its employees for the purpose of fighting a fire or in an emergency; or
  (b) to the employer of persons who attend a ship in dock premises for the purpose of fighting a fire or in an emergency, in respect of any ship so attended,
  and for the purposes of this paragraph "ship" includes all vessels and hovercraft which operate on water or land and water, and "dock premises" means a dock, wharf, quay, jetty or other place at which ships load or unload goods or embark or disembark passengers, together with neighbouring land or water which is used or occupied, or intended to be used or occupied, for those or incidental activities, and any part of a ship when used for those or incidental activities.

(6) These Regulations shall not apply to the master or crew of a ship or to the employer of such persons in respect of the normal shipboard activities of a ship's crew which are carried out solely by the crew under the direction of the master, and for the purposes of this paragraph "ship" includes every description of vessel used in navigation, other than a ship forming part of Her Majesty's Navy.

Part 2
General Requirements

4. Duty to manage asbestos in non-domestic premises

(1) In this regulation "the dutyholder" means—
   (a) every person who has, by virtue of a contract or tenancy, an obligation of any extent in relation to the maintenance or repair of non-domestic premises or any means of access thereto or egress therefrom; or
   (b) in relation to any part of non-domestic premises where there is no such contract or tenancy, every person who has, to any extent, control of that part of those non-domestic premises or any means of access thereto or egress therefrom,

   and where there is more than one such dutyholder, the relative contribution to be made by each such person in complying with the requirements of this regulation will be determined by the nature and extent of the maintenance and repair obligation owed by that person.

(2) Every person shall cooperate with the dutyholder so far as is necessary to enable the dutyholder to comply with his duties under this regulation.

(3) In order to enable him to manage the risk from asbestos in non-domestic premises, the dutyholder shall ensure that a suitable and sufficient assessment is carried out as to whether asbestos is or is liable to be present in the premises.

(4) In making the assessment—
   (a) such steps as are reasonable in the circumstances shall be taken; and
   (b) the condition of any asbestos which is, or has been assumed to be, present in the premises shall be considered.

(5) Without prejudice to the generality of paragraph (4), the dutyholder shall ensure that—
   (a) account is taken of building plans or other relevant information and of the age of the premises; and
   (b) an inspection is made of those parts of the premises which are reasonably accessible.

(6) The dutyholder shall ensure that the assessment is reviewed forthwith if—
   (a) there is reason to suspect that the assessment is no longer valid; or
   (b) there has been a significant change in the premises to which the assessment relates.

(7) The dutyholder shall ensure that the conclusions of the assessment and every review are recorded.

(8) Where the assessment shows that asbestos is or is liable to be present in any part of the premises the dutyholder shall ensure that—
   (a) a determination of the risk from that asbestos is made;
   (b) a written plan identifying those parts of the premises concerned is prepared; and
   (c) the measures which are to be taken for managing the risk are specified in the written plan.

(9) The measures to be specified in the plan for managing the risk shall include adequate measures for—
   (a) monitoring the condition of any asbestos or any substance containing or suspected of containing asbestos;
   (b) ensuring any asbestos or any such substance is properly maintained or where necessary safely removed; and
   (c) ensuring that information about the location and condition of any asbestos or any such substance is—
      (i) provided to every person liable to disturb it, and
      (ii) made available to the emergency services.

(10) The dutyholder shall ensure that—
   (a) the plan is reviewed and revised at regular intervals, and forthwith if—
      (i) there is reason to suspect that the plan is no longer valid, or
      (ii) there has been a significant change in the premises to which the plan relates;

(b) the measures specified in the plan are implemented; and
(c) the measures taken to implement the plan are recorded.

(11) In this regulation, a reference to—
(a) "the assessment" is a reference to the assessment required by paragraph (3);
(b) "the premises" is a reference to the non-domestic premises referred to in paragraph (1); and
(c) "the plan" is a reference to the plan required by paragraph (8).

5. Identification of the presence of asbestos

An employer shall not undertake work in demolition, maintenance, or any other work which exposes or is liable to expose his employees to asbestos in respect of any premises unless either—
(a) he has carried out a suitable and sufficient assessment as to whether asbestos, what type of asbestos, contained in what material and in what condition is present or is liable to be present in those premises; or
(b) if there is doubt as to whether asbestos is present in those premises he—
  (i) assumes that asbestos is present, and that it is not chrysotile alone, and
  (ii) observes the applicable provisions of these Regulations.

6. Assessment of work which exposes employees to asbestos

(1) An employer shall not carry out work which is liable to expose his employees to asbestos unless he has—
(a) made a suitable and sufficient assessment of the risk created by that exposure to the health of those employees and of the steps that need to be taken to meet the requirements of these Regulations;
(b) recorded the significant findings of that risk assessment as soon as is practicable after the risk assessment is made; and
(c) implemented the steps referred to in sub-paragraph (a).

(2) Without prejudice to the generality of paragraph (1), the risk assessment shall—
(a) subject to regulation 5, identify the type of asbestos to which employees are liable to be exposed;
(b) determine the nature and degree of exposure which may occur in the course of the work;
(c) consider the effects of control measures which have been or will be taken in accordance with regulation 11;
(d) consider the results of monitoring of exposure in accordance with regulation 19;
(e) set out the steps to be taken to prevent that exposure or reduce it to the lowest level reasonably practicable;
(f) consider the results of any medical surveillance that is relevant; and
(g) include such additional information as the employer may need in order to complete the risk assessment.

(3) The risk assessment shall be reviewed regularly, and forthwith if—
(a) there is reason to suspect that the existing risk assessment is no longer valid;
(b) there is a significant change in the work to which the risk assessment relates; or
(c) the results of any monitoring carried out pursuant to regulation 19 show it to be necessary,
and where, as a result of the review, changes to the risk assessment are required, those changes shall be made and, where they relate to the significant findings of the risk assessment or are themselves significant, recorded.

(4) Where, in accordance with the requirement in paragraph (2)(b), the risk assessment has determined that the exposure of his employees to asbestos may exceed the control limit, the employer shall keep a copy of the significant findings of the risk assessment at those premises at which, and for such time as, the work to which that risk assessment relates is being carried out.

7. Plans of work

(1) An employer shall not undertake any work with asbestos unless he has prepared a suitable written plan of work detailing how that work is to be carried out.
(2) The employer shall keep a copy of the plan of work at those premises at which the work to which the plan relates is being carried out for such time as that work continues.
(3) In cases of final demolition or major refurbishment of premises, the plan of work shall, so far as is reasonably practicable, and unless it would cause a greater risk to employees than if the asbestos had been left in place, specify that asbestos shall be removed before any other major works begin.
(4) The plan of work shall include in particular details of—
   (a) the nature and probable duration of the work;
   (b) the location of the place where the work is to be carried out;
   (c) the methods to be applied where the work involves the handling of asbestos or materials containing asbestos;
   (d) the characteristics of the equipment to be used for—
      (i) protection and decontamination of those carrying out the work, and
      (ii) protection of other persons on or near the worksite;
   (e) the measures which the employer intends to take in order to comply with the requirements of regulation 11; and
   (f) the measures which the employer intends to take in order to comply with the requirements of regulation 17.
(5) The employer shall ensure, so far as is reasonably practicable, that the work to which the plan of work relates is carried out in accordance with that plan and any subsequent written changes to it.

8. Licensing of work with asbestos

(1) Subject to regulation 3(2), an employer shall not undertake any work with asbestos unless he holds a licence granted under paragraph (2) of this regulation.
(2) The Executive may grant a licence for work with asbestos if it considers it appropriate to do so and—
   (a) the person who wishes the licence to be granted to him has made application for it on a form approved for the purposes of this regulation by the Executive; and
   (b) the application was made at least 28 days before the date from which the licence is to run, or such shorter period as the Executive may allow.
(3) A licence under this regulation—
   (a) shall come into operation on the date specified in the licence, and shall be valid for any period up to a maximum of three years that the Executive may specify in it; and
   (b) may be granted subject to such conditions as the Executive may consider appropriate.
(4) The Executive may vary the terms of a licence under this regulation if it considers it appropriate to do so and in particular may—
   (a) add further conditions and vary or omit existing ones; and
   (b) reduce the period for which the licence is valid or extend that period up to a maximum of three years from the date on which the licence first came into operation.
(5) The Executive may revoke a licence under this regulation if it considers it appropriate to do so.
(6) The holder of a licence under this regulation shall return the licence to the Executive—
   (a) when required by the Executive for any amendment; or
   (b) following its revocation.

9. Notification of work with asbestos

(1) Subject to regulation 3(2), an employer shall not undertake any work with asbestos unless he has notified the appropriate office of the enforcing authority in writing of the particulars specified in Schedule 1 at least 14 days before commencing that work or such shorter time before as the enforcing authority may agree.

(2) Where an employer has notified work in accordance with paragraph (1) and there is a material change in that work which might affect the particulars so notified (including the cessation of the work), the employer shall forthwith notify the appropriate office of the enforcing authority in writing of that change.

10. Information, instruction and training

(1) Every employer shall ensure that adequate information, instruction and training is given to those of his employees—
   (a) who are or who are liable to be exposed to asbestos, or who supervise such employees, so that they are aware of—
      (i) the properties of asbestos and its effects on health, including its interaction with smoking,
      (ii) the types of products or materials likely to contain asbestos,
      (iii) the operations which could result in asbestos exposure and the importance of preventive controls to minimise exposure,
      (iv) safe work practices, control measures, and protective equipment,
      (v) the purpose, choice, limitations, proper use and maintenance of respiratory protective equipment,
      (vi) emergency procedures,
      (vii) hygiene requirements,
      (viii) decontamination procedures,
      (ix) waste handling procedures,
      (x) medical examination requirements, and
      (xi) the control limit and the need for air monitoring,
      in order to safeguard themselves and other employees; and
   (b) who carry out work in connection with the employer's duties under these Regulations, so that they can carry out that work effectively.
(2) The information, instruction and training required by paragraph (1) shall be—
   (a) given at regular intervals;
   (b) adapted to take account of significant changes in the type of work carried out or methods of work used by the employer; and
   (c) provided in a manner appropriate to the nature and degree of exposure identified by the risk assessment, and so that the employees are aware of—
      (i) the significant findings of the risk assessment, and
      (ii) the results of any air monitoring carried out with an explanation of the findings.

11. Prevention or reduction of exposure to asbestos

(1) Every employer shall—
   (a) prevent the exposure of his employees to asbestos so far as is reasonably practicable;
   (b) where it is not reasonably practicable to prevent such exposure—
      (i) take the measures necessary to reduce the exposure of his employees to asbestos to the lowest level reasonably practicable by measures other than the use of respiratory protective equipment, and
      (ii) ensure that the number of his employees who are exposed to asbestos at any one time is as low as is reasonably practicable.
(2) Where it is not reasonably practicable for the employer to prevent the exposure of his employees to asbestos in accordance with paragraph (1)(a), the measures referred to in paragraph (1)(b)(i) shall include, in order of priority—
   (a) the design and use of appropriate work processes, systems and engineering controls and the provision and use of suitable work equipment and materials in order to avoid or minimise the release of asbestos; and

(b) the control of exposure at source, including adequate ventilation systems and appropriate organisational measures,

and the employer shall so far as is reasonably practicable provide the employees concerned with suitable respiratory protective equipment in addition to the measures required by sub-paragraphs (a) and (b).

(3) Where it is not reasonably practicable to reduce the exposure of an employee to asbestos to below the control limit by the measures referred to in paragraph (1)(b)(i), then, in addition to taking those measures, the employer shall provide that employee with suitable respiratory protective equipment which will reduce the concentration of asbestos in the air inhaled by the employee (after taking account of the effect of that respiratory protective equipment) to a concentration which is—

(a) below the control limit; and

(b) is as low as is reasonably practicable.

(4) Personal protective equipment provided by an employer in accordance with this regulation or with regulation 14(1) shall be suitable for its purpose and shall—

(a) comply with any provision of the Personal Protective Equipment Regulations 2002 which is applicable to that item of personal protective equipment; or

(b) in the case of respiratory protective equipment, where no provision referred to in sub-paragraph (a) applies, be of a type approved or shall conform to a standard approved, in either case, by the Executive.

(5) The employer shall—

(a) ensure that no employee is exposed to asbestos in a concentration in the air inhaled by that worker which exceeds the control limit; or

(b) if the control limit is exceeded—

(i) forthwith inform any employees concerned and their representatives and ensure that work does not continue in the affected area until adequate measures have been taken to reduce employees' exposure to asbestos to below the control limit,

(ii) as soon as is reasonably practicable identify the reasons for the control limit being exceeded and take the appropriate measures to prevent it being exceeded again, and

(iii) check the effectiveness of the measures taken pursuant to sub-paragraph (ii) by carrying out immediate air monitoring.

12. **Use of control measures etc.**

(1) Every employer who provides any control measure, other thing or facility pursuant to these Regulations shall take all reasonable steps to ensure that it is properly used or applied as the case may be.

(2) Every employee shall make full and proper use of any control measure, other thing or facility provided pursuant to these Regulations and, where relevant, shall—

(a) take all reasonable steps to ensure that it is returned after use to any accommodation provided for it; and

(b) if he discovers a defect therein report it forthwith to his employer.

13. **Maintenance of control measures etc.**

(1) Every employer who provides any control measure to meet the requirements of these Regulations shall ensure that—

(a) in the case of plant and equipment, including engineering controls and personal protective equipment, it is maintained in an efficient state, in efficient working order, in good repair and in a clean condition; and

(b) in the case of provision of systems of work and supervision and of any other measure, it is reviewed at suitable intervals and revised if necessary.

(2) Where exhaust ventilation equipment or respiratory protective equipment (except disposable respiratory protective equipment) is provided to meet the requirements of these Regulations, the employer shall ensure that thorough examinations and tests of that equipment are carried out at suitable intervals by a competent person.

(3) Every employer shall keep a suitable record of the examinations and tests carried out in accordance with paragraph (2) and of repairs carried out as a result of those examinations and tests, and that record or a suitable summary thereof shall be kept available for at least 5 years from the date on which it was made.

14. Provision and cleaning of protective clothing

(1) Every employer shall provide adequate and suitable protective clothing for such of his employees as are exposed or are liable to be exposed to asbestos, unless no significant quantity of asbestos is liable to be deposited on the clothes of the employee while he is at work.

(2) The employer shall ensure that protective clothing provided in pursuance of paragraph (1) is either disposed of as asbestos waste or adequately cleaned at suitable intervals.

(3) The cleaning required by paragraph (2) shall be carried out either on the premises where the exposure to asbestos has occurred, where those premises are suitably equipped for such cleaning, or in a suitably equipped laundry.

(4) The employer shall ensure that protective clothing which has been used and is to be removed from the premises referred to in paragraph (3) (whether for cleaning, further use or disposal) is packed, before being removed, in a suitable receptacle which shall be labelled in accordance with the provisions of Schedule 2 as if it were a product containing asbestos or, in the case of protective clothing intended for disposal as waste, in accordance with regulation 24(3).

(5) Where, as a result of the failure or improper use of the protective clothing provided in pursuance of paragraph (1), a significant quantity of asbestos is deposited on the personal clothing of an employee, then for the purposes of paragraphs (2), (3) and (4) that personal clothing shall be treated as if it were protective clothing.

15. Arrangements to deal with accidents, incidents and emergencies

(1) Subject to regulation 3(2) and to paragraph (3) of this regulation, and without prejudice to the relevant provisions of the Management of Health and Safety at Work Regulations 1999, in order to protect the health of his employees from an accident, incident or emergency related to the use of asbestos in a work process or to the removal or repair of asbestos-containing materials at the workplace, the employer shall ensure that—
   (a) procedures, including the provision of relevant safety drills (which shall be tested at regular intervals), have been prepared which can be put into effect when such an event occurs;
   (b) information on emergency arrangements, including—
      (i) details of relevant work hazards and hazard identification arrangements, and
      (ii) specific hazards likely to arise at the time of an accident, incident or emergency, is available; and
   (c) suitable warning and other communication systems are established to enable an appropriate response, including remedial actions and rescue operations, to be made immediately when such an event occurs.

(2) The employer shall ensure that information on the procedure and systems required by paragraph (1)(a) and (c) and the information required by paragraph (1)(b) is—
   (a) made available to the relevant accident and emergency services to enable those services, whether internal or external to the workplace, to prepare their own response procedures and precautionary measures; and
   (b) displayed at the workplace, if this is appropriate.

(3) Paragraph (1) shall not apply where—
   (a) the results of the risk assessment show that, because of the quantity of asbestos present at the workplace, there is only a slight risk to the health of employees; and

(b) the measures taken by the employer to comply with the duty under regulation 11(1) are sufficient to control that risk.

(4) In the event of an accident, incident or emergency related to the unplanned release of asbestos at the workplace, the employer shall ensure that—
   (a) immediate steps are taken to—
      (i) mitigate the effects of the event,
      (ii) restore the situation to normal, and
      (iii) inform any person who may be affected; and
   (b) only those persons who are responsible for the carrying out of repairs and other necessary work are permitted in the affected area and they are provided with—
      (i) appropriate respiratory protective equipment and protective clothing, and
      (ii) any necessary specialised safety equipment and plant,
      which shall be used until the situation is restored to normal.

16. Duty to prevent or reduce the spread of asbestos

Every employer shall prevent or, where this is not reasonably practicable, reduce to the lowest level reasonably practicable the spread of asbestos from any place where work under his control is carried out.

17. Cleanliness of premises and plant

Every employer who undertakes work which exposes or is liable to expose his employees to asbestos shall ensure that—
   (a) the premises, or those parts of the premises where that work is carried out, and the plant used in connection with that work are kept in a clean state; and
   (b) where such work has been completed, the premises, or those parts of the premises where the work was carried out, are thoroughly cleaned.

18. Designated Areas

(1) Every employer shall ensure that any area in which work under his control is carried out is designated as—
   (a) an asbestos area, subject to regulation 3(2), where any employee would be liable to be exposed to asbestos in that area;
   (b) a respirator zone where the concentration of asbestos fibres in the air in that area would exceed or would be liable to exceed the control limit.

(2) Asbestos areas and respirator zones shall be clearly and separately demarcated and identified by notices indicating—
   (a) that the area is an asbestos area or a respirator zone or both, as the case may be; and
   (b) in the case of a respirator zone, that the exposure of an employee who enters it is liable to exceed the control limit and that respiratory protective equipment must be worn.

(3) The employer shall not permit any employee, other than an employee who by reason of his work is required to be in an area designated as an asbestos area or a respirator zone, to enter or remain in any such area and only employees who are so permitted shall enter or remain in any such area.

(4) Every employer shall ensure that only competent employees shall—
   (a) enter a respirator zone; and
   (b) supervise any employees who enter a respirator zone,
and for the purposes of this paragraph a competent employee means an employee who has received adequate information, instruction and training.

(5) Every employer shall ensure that—
   (a) his employees do not eat, drink or smoke in an area designated as an asbestos area or a respirator zone; and
   (b) arrangements are made for such employees to eat or drink in some other place.

19. Air Monitoring

(1) Subject to paragraph (2), every employer shall monitor the exposure of his employees to asbestos by measurement of asbestos fibres present in the air—
   (a) at regular intervals; and
   (b) when a change occurs which may affect that exposure.
(2) Paragraph (1) shall not apply where—
   (a) the exposure of an employee is not liable to exceed the control limit; or
   (b) the employer is able to demonstrate by another method of evaluation that the requirements of regulation 11(1) and (5) have been complied with.
(3) The employer shall keep a suitable record of—
   (a) monitoring carried out in accordance with paragraph (1); or
   (b) where he decides that monitoring is not required because paragraph 2(b) applies, the reason for that decision.
(4) The record required by paragraph (3), or a suitable summary thereof, shall be kept—
   (a) in a case where exposure is such that a health record is required to be kept under regulation 22 for at least 40 years; or
   (b) in any other case, for at least 5 years,
   from the date of the last entry made in it.
(5) In relation to the record required by paragraph (3), the employer shall—
   (a) on reasonable notice being given, allow an employee access to his personal monitoring record;
   (b) provide the Executive with copies of such monitoring records as the Executive may require; and
   (c) if he ceases to trade, notify the Executive forthwith in writing and make available to the Executive all monitoring records kept by him.

20. Standards for air testing and site clearance certification

(1) In paragraph (4) "site clearance certificate for reoccupation" means a certificate issued to confirm that premises or parts of premises where work with asbestos has been carried out have been thoroughly cleaned upon completion of that work in accordance with regulation 17(b).
(2) Every employer who carries out any measurement of the concentration of asbestos fibres present in the air shall ensure that he meets criteria equivalent to those set out in the paragraphs of ISO 17025 which cover organisation, quality systems, control of records, personnel, accommodation and environmental conditions, test and calibration methods, method validation, equipment, handling of test and calibration items, and reporting results.
(3) Every employer who requests a person to carry out any measurement of the concentration of asbestos fibres present in the air shall ensure that that person is accredited by an appropriate body as competent to perform work in compliance with ISO 17025.
(4) Every employer who requests a person to assess whether premises or parts of premises where work with asbestos has been carried out have been thoroughly cleaned upon completion of that work and are suitable for reoccupation such that a site clearance certificate for reoccupation can be issued shall ensure that that person is accredited by an appropriate body as competent to perform work in compliance with the paragraphs of ISO 17020 and ISO 17025 which cover organisation, quality systems, control of records, personnel, accommodation and environmental conditions, test and calibration methods, method validation, equipment, handling of test and calibration items, and reporting results.
(5) Paragraphs (2) and (3) shall not apply to work carried out in a laboratory for the purposes only of research.

21. Standards for analysis

(1) Every employer who analyses a sample of any material to determine whether it contains asbestos shall ensure that he meets criteria equivalent to those set out in the paragraphs of ISO 17025 which cover organisation, quality systems, control of records, personnel, accommodation and environmental conditions, test and calibration methods, method validation, equipment, handling of test and calibration items, and reporting results.

(2) Every employer who requests a person to analyse a sample of any material taken to determine whether it contains asbestos shall ensure that that person is accredited by an appropriate body as competent to perform work in compliance with ISO 17025.

(3) Paragraphs (1) and (2) shall not apply to work carried out in a laboratory for the purposes only of research.

22. Health records and medical surveillance

(1) Subject to regulation 3(2), every employer shall ensure that—
  (a) a health record, containing particulars approved by the Executive, relating to each of his employees who is exposed to asbestos is maintained; and
  (b) that record or a copy thereof is kept available in a suitable form for at least 40 years from the date of the last entry made in it.

(2) Subject to regulation 3(2), every employer shall ensure that each of his employees who is exposed to asbestos is under adequate medical surveillance by a relevant doctor.

(3) The medical surveillance required by paragraph (2) shall include—
  (a) a medical examination not more than 2 years before the beginning of such exposure; and
  (b) periodic medical examinations at intervals of not more than 2 years or such shorter time as the relevant doctor may require while such exposure continues, and each such medical examination shall include a specific examination of the chest.

(4) Where an employee has been examined in accordance with paragraph (3), the relevant doctor shall issue a certificate to the employer and employee stating—
  (a) that the employee has been so examined; and
  (b) the date of the examination,

and the employer shall keep that certificate or a copy thereof for at least 4 years from the date on which it was issued.

(5) An employee to whom this regulation applies shall, when required by his employer and at the cost of the employer, present himself during his working hours for such examination and tests as may be required for the purposes of paragraph (3) and shall furnish the relevant doctor with such information concerning his health as the relevant doctor may reasonably require.

(6) Where, for the purpose of carrying out his functions under these Regulations, a relevant doctor requires to inspect any record kept for the purposes of these Regulations, the employer shall permit him to do so.

(7) Where medical surveillance is carried out on the premises of the employer, the employer shall ensure that suitable facilities are made available for the purpose.

(8) The employer shall—
  (a) on reasonable notice being given, allow an employee access to his personal health record;
  (b) provide the Executive with copies of such personal health records as the Executive may require; and
  (c) if he ceases to trade, notify the Executive forthwith in writing and make available to the Executive all personal health records kept by him.

(9) Where, as a result of medical surveillance, an employee is found to have an identifiable disease or adverse health effect which is considered by a relevant doctor to be the result of exposure to asbestos at work the employer of that employee shall—
  (a) ensure that a suitable person informs the employee accordingly and provides the employee with information and advice regarding further medical surveillance;

(b) review the risk assessment;
(c) review any measure taken to comply with regulation 11 taking into account any advice given by a relevant doctor or by the Executive;
(d) consider assigning the employee to alternative work where there is no risk of further exposure to asbestos, taking into account any advice given by a relevant doctor; and
(e) provide for a review of the health of every other employee who has been similarly exposed, including a medical examination (which shall include a specific examination of the chest) where such an examination is recommended by a relevant doctor or by the Executive.

23. Washing and changing facilities

(1) Every employer shall ensure that, for any of his employees who is exposed or liable to be exposed to asbestos, there be provided—
 (a) adequate washing and changing facilities;
 (b) where he is required to provide protective clothing, adequate facilities for the storage of—
  (i) that protective clothing, and
  (ii) personal clothing not worn during working hours; and
 (c) where he is required to provide respiratory protective equipment, adequate facilities for the storage of that equipment.
(2) The facilities provided under paragraph (1) for the storage of—
 (a) personal protective clothing;
 (b) personal clothing not worn during working hours; and
 (c) respiratory protective equipment,
shall be separate from each other.

24. Storage, distribution and labelling of raw asbestos and asbestos waste

(1) Every employer who undertakes work with asbestos shall ensure that raw asbestos or waste which contains asbestos is not—
 (a) stored;
 (b) received into or despatched from any place of work; or
 (c) distributed within any place of work, except in a totally enclosed distribution system, unless it is in a sealed receptacle or, where more appropriate, sealed wrapping, clearly marked in accordance with paragraphs (2) and (3) showing that it contains asbestos.
(2) Raw asbestos shall be labelled in accordance with the provisions of Schedule 2.
(3) Waste containing asbestos shall be labelled—
 (a) where the Carriage of Dangerous Goods and Use of Transportable Pressure Equipment Regulations 2004 apply, in accordance with those Regulations; and
 (b) in any other case in accordance with the provisions of Schedule 2.

PART 3
Prohibitions and Related Provisions

25. Interpretation of prohibitions

(1) In this Part —
 "asbestos cement" means a material which is predominantly a mixture of cement and chrysotile and which when in a dry state absorbs less than 30% water by weight;
 "asbestos spraying" means the application by spraying of any material containing asbestos to form a continuous surface coating;
 "extraction of asbestos" means the extraction by mining or otherwise of asbestos as the primary product of such extraction, but shall not include extraction which produces asbestos as a by-product of the primary activity of extraction;

"supply" means supply by way of sale, lease, hire, hire-purchase, loan, gift or exchange for a consideration other than money, whether (in all cases) as principal or as agent for another; and

"use" in relation to asbestos or any product to which asbestos has intentionally been added means—
   (a) putting asbestos or any product to which asbestos has intentionally been added to use for the first time; or
   (b) putting asbestos or any product to which asbestos has intentionally been added which has been in use before to a new use.

(2) Any prohibition imposed on any person by this Part shall apply only to acts done in the course of a trade, business or other undertaking (whether for profit or not) carried on by him.

(3) Any prohibition imposed by this Part on the importation into the United Kingdom, or on the supply or use of asbestos shall not apply to the importation, supply or use of asbestos solely for the purposes of research, development or analysis.

(4) Where in this Part it is stated that asbestos has intentionally been added to a product or is intentionally added, it will be presumed where—
   (a) asbestos is present in any product; and
   (b) asbestos is not a naturally occurring impurity of that product, or of any component or constituent thereof,

that the asbestos has intentionally been added or is intentionally added, as the case may be, subject to evidence to the contrary being adduced in any proceedings.

26. Prohibitions of exposure to asbestos

(1) No person shall undertake asbestos spraying or working procedures that involve using low-density (less than $1g/cm^3$) insulating or soundproofing materials which contain asbestos.

(2) Every employer shall ensure that no employees are exposed to asbestos during the extraction of asbestos.

(3) Every employer shall ensure that no employees are exposed to asbestos during the manufacture of asbestos products or of products containing intentionally added asbestos.

(4) In the case of chrysotile only, the prohibition in paragraph (3) is subject to the exception in paragraph 2 of Schedule 3.

27. Prohibition of the importation of asbestos

(1) Subject to paragraph (2), the importation into the United Kingdom of asbestos or of any product to which asbestos has intentionally been added is prohibited and any contravention of this paragraph shall be punishable under the Customs and Excise Management Act 1979 and not as a contravention of a health and safety regulation.

(2) In the case of chrysotile only, the prohibition in paragraph (1) is subject to the exceptions in paragraphs 1, 2 and 3 of Schedule 3.

28. Prohibition of the supply of asbestos

(1) Subject to paragraphs (2) and (3), no person shall supply, other than solely for the purpose of disposal, asbestos or any product to which asbestos has intentionally been added.

(2) In the case of chrysotile only, the prohibition in paragraph (1) shall not apply where the asbestos or the product was in use before 24th November 1999, except in the case of a product to which asbestos has intentionally been added of which the supply was prohibited by regulation 7 of the Asbestos (Prohibitions) Regulations 1992 as in force immediately before 24th November 1999.

(3) In the case of chrysotile only, the prohibition in paragraph (1) is subject to the exceptions in paragraphs 1 and 2 of Schedule 3.

29. Prohibition of the use of asbestos

(1) Subject to paragraphs (2) to (6), no person shall use, except in the course of any activity in connection with its disposal, asbestos or any product to which asbestos has intentionally been added.
(2) In the case of products containing crocidolite or asbestos grunerite (amosite), the prohibition in paragraph (1) shall not apply where the product was in use before 1st January 1986.
(3) In the case of products containing any other form of asbestos than crocidolite or asbestos grunerite (amosite), but excepting chrysotile, the prohibition in paragraph (1) shall not apply where the product was in use before 1st January 1993.
(4) In the case of chrysotile only, the prohibition in paragraph (1) shall not apply where the asbestos or product was in use before 24th November 1999, except in the case of a product containing chrysotile of which the supply was prohibited by regulation 7 of the Asbestos (Prohibitions) Regulations 1992 as in force immediately before 24th November 1999.
(5) Notwithstanding paragraph (4), no person shall use, except in the course of any activity in connection with its disposal—
    (a) asbestos cement;
    (b) any board, panel or tile, all or part of which has been painted with paint containing chrysotile; or
    (c) any board, panel or tile, all or part of which has been covered in a textured finishing plaster used for decorative purposes and containing chrysotile,
unless it is installed in or forms part of any premises or plant and, before 24th November 1999, it was installed in or formed part of those same premises or plant.
(6) In the case of chrysotile only, the prohibition in paragraph (1) is subject to the exceptions in paragraphs 1 and 2 of Schedule 3.

30. Labelling of products containing asbestos

(1) Subject to paragraph (2), a person shall not supply under an exception in Schedule 3 or an exemption granted pursuant to regulation 32 or regulation 33 a product which contains asbestos unless that product is labelled in accordance with the provisions of Schedule 2.
(2) Where a component of a product contains asbestos, it shall be sufficient compliance with this regulation if that component is labelled in accordance with the provisions of Schedule 2 except that where the size of that component makes it impossible for a label to be fixed to it neither that component nor the product need be labelled.

31. Additional provisions in the case of exceptions and exemptions

(1) Where under an exception in Schedule 3 or an exemption granted pursuant to regulation 32 or regulation 33 asbestos is used in a work process or is produced by a work process, the employer shall ensure that the quantity of asbestos and materials containing asbestos at the premises where the work is carried out is reduced to as low a level as is reasonably practicable.
(2) Subject to paragraph (3), where under an exception in Schedule 3 or an exemption granted pursuant to regulation 32 or regulation 33 a manufacturing process which gives rise to asbestos dust is carried out in a building, the employer shall ensure that any part of the building in which the process is carried out is—
    (a) so designed and constructed as to facilitate cleaning; and
    (b) is equipped with an adequate and suitable vacuum cleaning system which shall, where reasonably practicable, be a fixed system.
(3) Paragraph 2(a) shall not apply to a building in which, prior to 1st March 1988, there was carried out a process to which either—
    (a) as then in force, regulation 13 of the Asbestos Regulations 1969 applied and the process was carried out in compliance with that regulation; or
    (b) that regulation did not apply.

Part 4
Miscellaneous

32. Exemption certificates

(1) Subject to paragraph (4), the Executive may, by a certificate in writing, exempt any person or class of persons or any product containing asbestos or class of such products from all or any of the requirements or prohibitions imposed by regulations 4, 8, 12, 13, 21, 22(5) to (7) and 27 and any such exemption may be granted subject to conditions and to a limit of time and may be varied or revoked by a further certificate in writing at any time.

(2) Subject to paragraph (4) and to the provisions of Council Directive 76/769/EEC on the marketing and use of certain dangerous substances and preparations, the Executive may, by a certificate in writing, exempt any person or class of persons or any product containing asbestos or class of such products from the prohibitions imposed by regulations 28(1) and 29(1) and any such exemption may be granted subject to conditions and to a limit of time and may be varied or revoked by a further certificate in writing at any time.

(3) Subject to paragraph (4), the Executive may exempt emergency services from all or any of the requirements or prohibitions imposed by regulations 7 and 9 and any such exemption may be granted subject to conditions and to a limit of time and may be varied or revoked by a further certificate in writing at any time.

(4) The Executive shall not grant any exemption under paragraph (1), (2) or (3) unless having regard to the circumstances of the case and in particular to—
  (a) the conditions, if any, which it proposes to attach to the exemption; and
  (b) any other requirements imposed by or under any enactments which apply to the case,
  it is satisfied that the health or safety of persons who are likely to be affected by the exemption will not be prejudiced in consequence of it.

33. Exemptions relating to the Ministry of Defence

The Secretary of State for Defence may, in the interests of national security, exempt any person or class of persons from all or any of the prohibitions imposed by Part 3 of these Regulations by a certificate in writing, and any such exemption may be granted subject to conditions and to a limit of time and may be varied or revoked by a further certificate in writing at any time.

34. Extension outside Great Britain

These Regulations shall apply to any work outside Great Britain to which sections 1 to 59 and 80 to 82 of the Health and Safety at Work etc. Act 1974 apply by virtue of the Health and Safety at Work etc. Act 1974 (Application Outside Great Britain) Order 2001 as they apply to work in Great Britain.

35. Existing licences and exemption certificates

(1) An existing licence issued by the Executive under regulation 4(1) of the Asbestos (Licensing) Regulations 1983 shall—
  (a) continue to have effect as if it had been granted under regulation 8(2) of these Regulations;
  (b) be of the duration and subject to the conditions specified in it as if that duration and those conditions had been specified under regulation 8(3); and
  (c) be liable to variation and revocation under regulation 8(4) and (5),
  and any requirement in such a licence concerning notification and any exception to such a requirement shall have effect as a requirement for notification under regulation 9 of and as an exception to such a requirement under regulation 3(2) of these Regulations.

(2) An existing exemption granted by the Executive under regulation 7(1) of the Asbestos (Licensing) Regulations 1983, regulation 8(1) of the Asbestos (Prohibitions) Regulations 1992, or regulation 25(1) of the Control of Asbestos at Work Regulations 2002 shall continue to have effect

and be subject to any limitation of time or any conditions specified in it and liable to revocation as if it had been granted under regulation 32(1), (2) or (3) of these Regulations.

(3) An existing exemption granted by the Secretary of State for Defence under regulation 8(3) of the Asbestos (Prohibitions) Regulations 1992 shall continue to have effect and be subject to any limitation of time or any conditions specified in it and liable to revocation as if it had been granted under regulation 33 of these Regulations.

36. Revocations, amendments and savings

(1) The revocations listed in Schedule 4 shall have effect.
(2) The amendments listed in Schedule 5 shall have effect.
(3) Any record or register required to be kept under any Regulations revoked either by paragraph (1) or by regulation 27(1) of the Control of Asbestos at Work Regulations 2002 shall, notwithstanding that revocation, be kept in the same manner and for the same period as specified in those Regulations as if these Regulations had not been made, except that the Executive may approve the keeping of records at a place or in a form other than at the place where, or in the form which, records were required to be kept under the Regulations so revoked.

37. Defence

Subject to regulation 21 of the Management of Health and Safety at Work Regulations 1999, in any proceedings for an offence consisting of a contravention of Part 2 of these Regulations it shall be a defence for any person to prove that he took all reasonable precautions and exercised all due diligence to avoid the commission of that offence.

Signed by authority of the Secretary of State for Work and Pensions.

Philip Hunt

Parliamentary Under-Secretary of State, Department for Work and Pensions
12th October 2006

Schedule 1

Regulation 9(1)

Particulars to be Included in a Notification

The following particulars are to be included in a notification made in accordance with regulation 9(1), namely—
   (a) the name and address of the notifier and the address and telephone number of his usual place of business;
   (b) a brief description of—
      (i) the location of the work site,
      (ii) the types of asbestos to be used or handled (classified in accordance with regulation 2),
      (iii) the maximum quantity of asbestos of each type to be held at any one time at the premises at which the work is to take place,
      (iv) the activities and processes involved,
      (v) the number of workers involved, and
      (vi) the measures taken to limit the exposure of employees to asbestos, and
   (c) the date of the commencement of the work and its expected duration.

Schedule 2

Regulations 14(4), 24(2) and (3) and 30(1) and (2)

The Labelling of Raw Asbestos, Asbestos Waste and Products Containing Asbestos

1. —(1) Subject to sub-paragraphs (2) and (3) of this paragraph, the label to be used on—
  (a) raw asbestos (together with the labelling required under the Chemicals (Hazard Information and Packaging for Supply) Regulations 2002 and the Carriage of Dangerous Goods and Use of Transportable Pressure Equipment Regulations 2004;
  (b) asbestos waste, when required to be so labelled by regulation 24(3); and
  (c) products containing asbestos, including used protective clothing to which regulation 14(2) applies,
shall be in the form and in the colours of the following diagram and shall comply with the specifications set out in paragraphs 2 and 3.

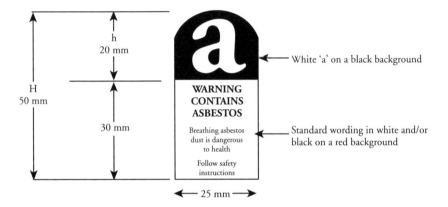

(2) In the case of a product containing crocidolite, the words "contains asbestos" shown in the diagram shall be replaced by the words "contains crocidolite/blue asbestos".

(3) Where the label is printed directly onto a product, a single colour contrasting with the background colour may be used.

2. The dimensions in millimetres of the label referred to in paragraph 1(1) shall be those shown on the diagram in that paragraph, except that larger measurements may be used, but in that case the dimension indicated as h on the diagram shall be 40% of the dimension indicated as H.

3. The label shall be clearly and indelibly printed so that the words in the lower half of the label can be easily read, and those words shall be printed in black or white.

4.—(1) Where a product containing asbestos may undergo processing or finishing it shall bear a label containing safety instructions appropriate to the particular product and in particular the following instructions—

"operate if possible out of doors in a well-ventilated place";

"preferably use hand tools or low speed tools equipped, if necessary, with an appropriate dust extraction facility. If high speed tools are used, they should always be so equipped";

"if possible, dampen before cutting or drilling"; and

"dampen dust, place it in a properly closed receptacle and dispose of it safely".

(2) Additional safety information given on a label shall not detract from or contradict the safety information given in accordance with sub-paragraph (1).

5.—(1) Labelling of packaged and unpackaged products containing asbestos in accordance with the foregoing paragraphs shall be effected by means of—
  (a) an adhesive label firmly affixed to the product or its packaging, as the case may be;
  (b) a tie-on label firmly attached to the product or its packaging, as the case may be; or
  (c) direct printing onto the product or its packaging, as the case may be.
(2) Where, in the case of an unpackaged product containing asbestos, it is not reasonably practicable to comply with the provisions of sub-paragraph (1) the label shall be printed on a suitable sheet accompanying the product.
(3) Labelling of raw asbestos and asbestos waste shall be effected in accordance with sub-paragraph (1)(a) or (c).
(4) For the purposes of this Schedule but subject to sub-paragraph (5), a product supplied in loose plastic or other similar wrapping (including plastic and paper bags) but no other packaging, shall be treated as being supplied in a package whether the product is placed in such wrapping at the time of its supply or was already so wrapped previously.
(5) No wrapping in which a product is placed at the time of its supply shall be regarded as packaging if any product contained in it is labelled in accordance with the requirements of this Schedule or any other packaging in which that product is contained is so labelled.

Schedule 3

Regulations 26(4), 27(2), 28(3) and 29(6)

Exceptions to the Prohibitions on the Importation, Supply and use of Chrysotile

1. Regulations 27(1), 28(1) and 29(1) shall not apply to brake linings within the meaning of the Road Vehicles (Brake Linings Safety) Regulations 1999.

2. Where it is not practicable for an employer to substitute for chrysotile a substance which, under the conditions of its use, does not create a risk to the health of his employees or creates a lesser risk than that created by chrysotile, regulations 26(3), 27(1), 28(1) and 29(1) shall not apply to—
  (a) Diaphragms for use in electrolytic cells in existing electrolysis plants for chlor-alkali manufacture;
  (b) chrysotile, or products to which chrysotile has intentionally been added, required solely for the manufacture of the products described in sub-paragraph (a).

3. Regulation 27(1) shall not apply to receptacles used for the storage of acetylene gas under pressure and in use before 24th November 1999.

APPENDIX M

Health and Safety: The Gas Safety (Installation and Use) Regulations 1998 (as amended)

1998 No. 2451

Made	*3rd October 1998*
Laid before Parliament	*9th October 1998*
Coming into force	*31st October 1998*

The Secretary of State, in exercise of the powers conferred on him by sections 15(1), (2), (4)(a), (5), (6)(b) and 82(3)(a) of, and paragraphs 1(1), (2) and (3), 4(1), 12, 15(1) and 16 of Schedule 3 to, the Health and Safety at Work etc. Act 1974 ("the 1974 Act") and of all other powers enabling him in that behalf and for the purpose of giving effect without modifications to proposals submitted to him by the Health and Safety Commission under section 11(2)(d) of the 1974 Act after the carrying out by the said Commission of consultations in accordance with section 50(3) of that Act, hereby makes the following Regulations—

PART A
GENERAL

1 Citation and commencement

These Regulations may be cited as the Gas Safety (Installation and Use) Regulations 1998 and shall come into force on 31st October 1998.

2 General interpretation and application

(1) In these Regulations, unless the context otherwise requires—

"appropriate fitting" means a fitting which—
  (a) has been designed for the purpose of effecting a gas tight seal in a pipe or other gasway;
  (b) achieves that purpose when fitted; and
  (c) is secure, so far as is reasonably practicable, against unauthorised opening or removal;

"distribution main" means any main through which a transporter is for the time being distributing gas and which is not being used only for the purpose of conveying gas in bulk;

"emergency control" means a valve for shutting off the supply of gas in an emergency, being a valve intended for use by a consumer of gas;

"flue" means a passage for conveying the products of combustion from a gas appliance to the external air and includes any part of the passage in a gas appliance duct which serves the purpose of a flue;

"gas" means any substance which is or (if it were in a gaseous state) would be gas within the meaning of the Gas Act 1986 except that it does not include gas consisting wholly or mainly of hydrogen when used in non-domestic premises;

"gas appliance" means an appliance designed for use by a consumer of gas for heating, lighting, cooking or other purposes for which gas can be used but it does not include a portable or mobile appliance supplied with gas from a cylinder, or the cylinder, pipes and other fittings used for supplying gas to that appliance, save that, for the purposes of regulations 3, 35 and 36 of these Regulations, it does include a portable or mobile space heater supplied with gas from a cylinder, and the cylinder, pipes and other fittings used for supplying gas to that heater;

"gas fittings" means gas pipework, valves (other than emergency controls), regulators and meters, and fittings, apparatus and appliances designed for use by consumers of gas for heating, lighting, cooking or other purposes for which gas can be used (other than the purpose of an industrial process carried out on industrial premises), but it does not mean—
(a) any part of a service pipe;
(b) any part of a distribution main or other pipe upstream of the service pipe;
(c) a gas storage vessel; or
(d) a gas cylinder or cartridge designed to be disposed of when empty;

"gas storage vessel" means a storage container designed to be filled or re-filled with gas at the place where it is connected for use or a re-fillable cylinder designed to store gas, and includes the vapour valve; but it does not include a cylinder or cartridge designed to be disposed of when empty;

"gas water heater" includes a gas fired central heating boiler;

"installation pipework" means any pipework for conveying gas for a particular consumer and any associated valve or other gas fitting including any pipework used to connect a gas appliance to other installation pipework and any shut off device at the inlet to the appliance, but it does not mean—
(a) a service pipe;
(b) a pipe comprised in a gas appliance;
(c) any valve attached to a storage container or cylinder; or
(d) service pipework;

"meter by pass" means any pipe and other gas fittings used in connection with it through which gas can be conveyed from a service pipe or service pipework to installation pipework without passing through the meter;

"primary meter" means the meter nearest to and downstream of a service pipe or service pipework for ascertaining the quantity of gas supplied through that pipe or pipework by a supplier;

"re-fillable cylinder" means a cylinder which is filled other than at the place where it is connected for use;

"the responsible person", in relation to any premises, means the occupier of the premises or, where there is no occupier or the occupier is away, the owner of the premises or any person with authority for the time being to take appropriate action in relation to any gas fitting therein;

"room-sealed appliance" means an appliance whose combustion system is sealed from the room in which the appliance is located and which obtains air for combustion from a ventilated uninhabited space within the premises or directly from the open air outside the premises and which vents the products of combustion directly to open air outside the premises;

"service pipe" means a pipe for distributing gas to premises from a distribution main, being any pipe between the distribution main and the outlet of the first emergency control downstream from the distribution main;

"service pipework" means a pipe for supplying gas to premises from a gas storage vessel, being any pipe between the gas storage vessel and the outlet of the emergency control;

"service valve" means a valve (other than an emergency control) for controlling a supply of gas, being a valve—
(a) incorporated in a service pipe; and
(b) intended for use by a transporter of gas; and
(c) not situated inside a building;

The Gas Safety (Installation and Use) Regulations 1998 (as amended)

"supplier" in relation to gas means—
- (a) a person who supplies gas to any premises through a primary meter; or
- (b) a person who provides a supply of gas to a consumer by means of the filling or re-filling of a storage container designed to be filled or re-filled with gas at the place where it is connected for use whether or not such container is or remains the property of the supplier; or
- (c) a person who provides gas in re-fillable cylinders for use by a consumer whether or not such cylinders are filled or re-filled directly by that person and whether or not such cylinders are or remain the property of that person, but a retailer shall not be deemed to be a supplier when he sells a brand of gas other than his own;

"transporter" in relation to gas means a person who conveys gas through a distribution main;

"work" in relation to a gas fitting includes any of the following activities carried out by any person, whether an employee or not, that is to say—
- (a) installing or re-connecting the fitting;
- (b) maintaining, servicing, permanently adjusting, disconnecting, repairing, altering or renewing the fitting or purging it of air or gas;
- (c) where the fitting is not readily movable, changing its position; and
- (d) removing the fitting;

but the expression does not include the connection or disconnection of a bayonet fitting or other self-sealing connector.

(2) For the purposes of these Regulations—
- (a) any reference to installing a gas fitting includes a reference to converting any pipe, fitting, meter, apparatus or appliance to gas use; and
- (b) a person to whom gas is supplied and who provides that gas for use in a flat or part of premises let by him shall not in so doing be deemed to be supplying gas.

(3) Subject to paragraphs (4) and (5) below, these Regulations shall apply to or in relation to gas fittings used in connection with—
- (a) gas which has been conveyed to premises through a distribution main; or
- (b) gas conveyed from a gas storage vessel.

(4) Save for regulations 37, 38 and 41 and subject to regulation 3(8), these Regulations shall not apply in relation to the supply of gas to, or anything done in respect of a gas fitting at, the following premises, that is to say—

[(a)
- (i) a mine within the meaning of the Mines and Quarries Act 1954 or any place deemed to form part of a mine for the purposes of that Act, or
- (ii) a quarry within the meaning of the Quarries Regulations 1999 or any place deemed to form part of a quarry for the purposes of those Regulations;][1]
- (b) a factory within the meaning of the Factories Act 1961 or any place to which any provisions of the said Act apply by virtue of sections 123 to 126 of that Act;
- (c) agricultural premises, being agricultural land, including land being or forming part of a market garden, and any building thereon which is used in connection with agricultural operations;
- (d) temporary installations used in connection with any construction work within the meaning assigned to that phrase by regulation 2(1) of the Construction (Design and Management) Regulations 1994;
- (e) premises used for the testing of gas fittings; or
- (f) premises used for the treatment of sewage,

---

[1] Amended by SI 1999/2024.

but they shall apply in relation to such premises or part thereof used for domestic or residential purposes or as sleeping accommodation.

(5) Nothing in these Regulations shall apply in relation to the supply of gas to, or anything done in respect of a gas fitting on—
   (a) a self-propelled vehicle except when such a vehicle is—
      (i) hired out in the course of a business; or
      (ii) made available to members of the public in the course of a business carried on from that vehicle;
   (b) a sea-going ship;
   (c) a vessel not requiring a national or international load line certificate except when such vessel is—
      (i) hired out in the course of a business;
      (ii) made available to members of the public in the course of a business carried out from that vessel; or
      (iii) used primarily for domestic or residential purposes;
   (d) a hovercraft; or
   (e) a caravan used for touring otherwise than when hired out in the course of a business.

(6) Nothing in these Regulations shall apply in relation to—
   (a) the supply of gas to the propulsion system of any vehicle or to any gas fitting forming part of such propulsion system;
   (b) the supply of gas to, or anything done in respect of, a bunsen burner used in an educational establishment; or
   (c) work in relation to a control device on a gas appliance if—
      (i) the device is intended primarily for use by a consumer of gas; and
      (ii) the work does not involve breaking into a gasway.

(7) These Regulations shall not apply in relation to a gas fitting used for the purpose of training gas fitting operatives in a college or other training establishment, except that paragraphs (1) to (5) and (7) of regulation 3 shall apply to work in relation to a gas fitting carried out by a person providing such training.

(8) These Regulations shall not apply in relation to a gas fitting used for the purpose of assessing the competence of a gas fitting operative at an assessment centre where such assessment is carried out for the purposes of a nationally accredited certification scheme, except that regulation 3(1) and (2) shall apply to work in relation to a gas fitting carried out by a person carrying out such assessment.

Part B
Gas Fittings—General Provisions

3 Qualification and supervision

(1) No person shall carry out any work in relation to a gas fitting or gas storage vessel unless he is competent to do so.

(2) The employer of any person carrying out such work for that employer, every other employer and self-employed person who has control to any extent of such work and every employer and self-employed person who has required such work to be carried out at any place of work under his control shall ensure that paragraph (1) above is complied with in relation to such work.

(3) Without prejudice to the generality of paragraphs (1) and (2) above and subject to paragraph (4) below, no employer shall allow any of his employees to carry out any work in relation to a gas fitting or service pipework and no self-employed person shall carry out any such work, unless the employer or self-employed person, as the case may be, is a member of a class of persons approved for the time being by the Health and Safety Executive for the purposes of this paragraph.

The Gas Safety (Installation and Use) Regulations 1998 (as amended)

(4) The requirements of paragraph (3) above shall not apply in respect of—
   (a) the replacement of a hose or regulator on a portable or mobile space heater; or
   (b) the replacement of a hose connecting a re-fillable cylinder to installation pipework.
(5) An approval given pursuant to paragraph (3) above (and any withdrawal of such approval) shall be in writing and notice of it shall be given to such persons and in such manner as the Health and Safety Executive considers appropriate.
(6) The employer of any person carrying out any work in relation to a gas fitting or gas storage vessel in the course of his employment shall ensure that such of the following provisions of these Regulations as impose duties upon that person and are for the time being in force are complied with by that person.
(7) No person shall falsely pretend to be a member of a class of persons required to be approved under paragraph (3) above.
(8) Notwithstanding sub-paragraph (b) of regulation 2(4), when a person is carrying out work in premises referred to in that sub-paragraph in relation to a gas fitting in a vehicle, vessel or caravan—
   (a) paragraphs (1), (2) and (6) of this regulation shall be complied with as respects thereto; and
   (b) he shall ensure, so far as is reasonably practicable, that the installation of the gas fittings and flues will not contravene the provisions of these Regulations when the gas fittings are connected to a gas supply,
except that this paragraph shall not apply where the person has reasonable grounds for believing that the vehicle, vessel or caravan will be first used for a purpose which when so used will exclude it from the application of these Regulations by virtue of sub-paragraphs (a), (c) or (e) of regulation 2(5).

4 Duty on employer

Where an employer or a self-employed person requires any work in relation to a gas fitting to be carried out at any place of work under his control or where an employer or self-employed person has control to any extent of work in relation to a gas fitting, he shall take reasonable steps to ensure that the person undertaking that work is, or is employed by, a member of a class of persons approved by the Health and Safety Executive under regulation 3(3) above.

5 Materials and workmanship

(1) No person shall install a gas fitting unless every part of it is of good construction and sound material, of adequate strength and size to secure safety and of a type appropriate for the gas with which it is to be used.
(2) Without prejudice to the generality of paragraph (1) above, no person shall install in a building any pipe or pipe fitting for use in the supply of gas which is—
   (a) made of lead or lead alloy; or
   (b) made of a non-metallic substance unless it is—
      (i) a pipe connected to a readily movable gas appliance designed for use without a flue; or
      (ii) a pipe entering the building and that part of it within the building is placed inside a metallic sheath which is so constructed and installed as to prevent, so far as is reasonably practicable, the escape of gas into the building if the pipe should fail.
(3) No person shall carry out any work in relation to a gas fitting or gas storage vessel otherwise than in accordance with appropriate standards and in such a way as to prevent danger to any person.

6 General safety precautions

(1) No person shall carry out any work in relation to a gas fitting in such a manner that gas could be released unless steps are taken to prevent the gas so released constituting a danger to any person.
(2) No person carrying out work in relation to a gas fitting shall leave the fitting unattended unless every incomplete gasway has been sealed with the appropriate fitting or the gas fitting is otherwise safe.

(3) Any person who disconnects a gas fitting shall, with the appropriate fitting, seal off every outlet of every pipe to which it was connected.

(4) No person carrying out work in relation to a gas fitting which involves exposing gasways which contain or have contained flammable gas shall smoke or use any source of ignition in such a manner as may lead to the risk of fire or explosion.

(5) No person searching for an escape of gas shall use any source of ignition.

(6) Where a person carries out any work in relation to a gas fitting which might affect the gas tightness of the gas installation he shall immediately thereafter test the installation for gas tightness at least as far as the nearest valves upstream and downstream in the installation.

(7) No person shall install a gas storage vessel unless the site where it is to be installed is such as to ensure that the gas storage vessel can be used, filled or refilled without causing a danger to any person.

(8) No person shall install in a cellar or basement—
   (a) a gas storage vessel; or
   (b) an appliance fuelled by liquefied petroleum gas which has an automatic ignition device or a pilot light.

(9) No person shall intentionally or recklessly interfere with a gas storage vessel or otherwise do anything which might affect a gas storage vessel so that the subsequent use of that vessel might cause a danger to any person.

(10) No person shall store or keep gas consisting wholly or mainly of methane on domestic premises, and, for the purpose of this paragraph, such gas from time to time present in pipes or in the fuel tank of any vehicle propelled by gas shall be deemed not to be so stored or kept.

7 Protection against damage

(1) Any person installing a gas fitting shall ensure that it is properly supported and so placed or protected as to avoid any undue risk of damage to the fitting.

(2) No person shall install a gas fitting if he has reason to suspect that foreign matter may block or otherwise interfere with the safe operation of the fitting unless he has fitted to the gas inlet of, and any airway in, the fitting a suitable filter or other suitable protection.

(3) No person shall install a gas fitting in a position where it is likely to be exposed to any substance which may corrode gas fittings unless the fitting is constructed of materials which are inherently resistant to being so corroded or it is suitably protected against being so corroded.

8 Existing gas fittings

(1) No person shall make any alteration to any premises in which a gas fitting or gas storage vessel is fitted if that alteration would adversely affect the safety of the fitting or vessel in such a manner that, if the fitting or the vessel had been installed after the alteration, there would have been a contravention of, or failure to comply with, these Regulations.

(2) No person shall do anything which would affect a gas fitting or any flue or means of ventilation used in connection with the fitting in such a manner that the subsequent use of the fitting might constitute a danger to any person, except that this paragraph does not apply to an alteration to premises.

(3) In relation to any place of work under his control, an employer or a self-employed person shall ensure, so far as is reasonably practicable, that the provisions of paragraphs (1) and (2) above are complied with.

9 Emergency controls

(1) No person shall for the first time enable gas to be supplied for use in any premises unless there is provided an appropriately sited emergency control to which there is adequate access.

(2) Any person installing an emergency control shall ensure that—
   (a) any key, lever or hand-wheel of the control is securely attached to the operating spindle of the control;

(b) any such key or lever is attached so that—
   (i) the key or lever is parallel to the axis of the pipe in which the control is installed when the control is in the open position; and
   (ii) where the key or lever is not attached so as to move only horizontally, gas cannot pass beyond the control when the key or lever has been moved as far as possible downwards;
(c) either the means of operating the key or lever is clearly and permanently marked or a notice in permanent form is prominently displayed near such means so as to indicate when the control is open and when the control is shut; and
(d) any hand-wheel indicates the direction of opening or closing of the control.

(3) Where a person installs an emergency control which is not adjacent to a primary meter, he shall immediately thereafter prominently display on or near the means of operating the control a suitably worded notice in permanent form indicating the procedure to be followed in the event of an escape of gas.

(4) Where any person first supplies gas to premises where an emergency control is installed, he shall ensure that the notice required by paragraph (3) above remains suitably worded or shall, where necessary, forthwith amend or replace that notice so as to give effect to the provisions of that paragraph.

(5) This regulation shall not apply where gas is supplied in a refillable cylinder except where two or more cylinders are connected by means of an automatic change-over device.

10 Maintaining electrical continuity

In any case where it is necessary to prevent danger, no person shall carry out work in relation to a gas fitting without using a suitable bond to maintain electrical continuity until the work is completed and permanent electrical continuity has been restored.

Part C
Meters and Regulators

11 Interpretation of Part C

In this Part—

"meter box" means a receptacle or compartment designed and constructed to contain a meter with its associated fittings;

"meter compound" means an area or room designed and constructed to contain one or more meters with their associated fittings;

"secondary meter" means a meter, other than a primary meter, for ascertaining the quantity of gas provided by a person for use by another person.

12 Meters—general provisions

(1) No person shall install a meter in any premises unless the site where it is to be installed is such as to ensure so far as is reasonably practicable that the means of escape from those premises in the event of fire is not adversely affected.

(2) No person shall install a meter in any premises unless it is of sound construction adequate to ensure so far as is reasonably practicable that in the event of fire gas is not able to escape in hazardous quantities, save that this paragraph shall not apply to any meter installed in non-domestic premises to which gas is supplied through a readily accessible service valve.

(3) No person shall install a meter unless the installation is so placed as to ensure that there is no risk of damage to it from electrical apparatus.

(4) No person shall install a meter except in a readily accessible position for inspection and maintenance.

(5) Where a meter has bosses or side pipes attached to the meter by a soldered joint only, no person shall make rigid pipe connections to the meter.

(6) Where a person installs a meter and the pipes and other gas fittings associated with it, he shall ensure that—
  (a) immediately thereafter they are adequately tested to verify that they are gas tight and examined to verify that they have been installed in accordance with these Regulations; and
  (b) immediately after such testing and examination, purging is carried out throughout the meter and every other gas fitting through which gas can then flow so as to remove safely all air and gas other than the gas to be supplied.

13 Meter housings

(1) Where a meter is housed in a meter box or meter compound attached to or built into the external face of the outside wall of any premises, the meter box or meter compound shall be so constructed and installed that any gas escaping within the box or compound cannot enter the premises or any cavity in the wall but must disperse to the external air.

(2) No person shall knowingly store readily combustible materials in any meter box or meter compound.

(3) No person shall install a meter in a meter box provided with a lock, unless the consumer has been provided with a suitably labelled key to that lock.

(4) No person shall install a meter within a meter compound which is capable of being secured unless the consumer has been provided with a suitably labelled key for that compound.

14 Regulators

(1) No person shall install a primary meter or meter by pass used in connection with a primary meter unless—
  (a) there is a regulator controlling the pressure of gas supplied through the meter or the by pass, as the case may be, which provides adequate automatic means for preventing the gas fittings connected to the downstream side of the regulator from being subjected to a pressure greater than that for which they were designed;
  (b) where the normal pressure of the gas supply is 75 millibars or more at the inlet to the regulator, there are also adequate automatic means for preventing, in case the regulator should fail, those gas fittings from being subjected to such a greater pressure; and
  (c) where the regulator contains a relief valve or liquid seal, such valve or seal is connected to a vent pipe of adequate size and so installed that it is capable of venting safely.

(2) Without prejudice to the requirements of paragraph (1), no person shall cause gas to be supplied from a gas storage vessel (other than a re-fillable cylinder or a cylinder or cartridge designed to be disposed of when empty) to any service pipework or gas fitting unless—
  (a) there is a regulator installed which controls the nominal operating pressure of the gas;
  (b) there is adequate automatic means for preventing the installation pipework and gas fittings downstream of the regulator from being subjected to a pressure different from that for which they were designed; and
  (c) there is an adequate alternative automatic means for preventing the service pipework from being subjected to a greater pressure than that for which it was designed should the regulator referred to in sub-paragraph (a) above fail.

(3) No person shall cause gas to be supplied through an installation consisting of one or more re-fillable cylinders unless the supply of gas passes through a regulator which controls the nominal operating pressure of the gas.

(4) Without prejudice to paragraph (3) above, no person shall cause gas to be supplied through an installation consisting of four or more re-fillable cylinders connected to an automatic change-over device unless there is an adequate alternative means for preventing the installation pipework and any gas fitting downstream of the regulator from being subjected to a greater pressure than that for which it was designed should the regulator fail.

(5) Where a person installs a regulator for controlling the pressure of gas through a primary meter, a meter by pass used in connection with a primary meter or from a gas storage vessel, or installs a gas appliance itself fitted with a regulator for controlling the pressure of gas to that appliance, he shall immediately thereafter ensure, in either case, that the regulator is adequately sealed so as to prevent its setting from being interfered with without breaking of the seal.

(6) In relation to—
   (a) gas from a distribution main, no person except the transporter or a person authorised to act on his behalf;
   (b) gas from a gas storage vessel, no person except the supplier or a person authorised to act on his behalf,

shall break a seal applied under paragraph (5) above other than a seal applied to a regulator for controlling the pressure of gas to the appliance to which that regulator is fitted.

(7) A person who breaks a seal applied under paragraph (5) shall apply as soon as is practicable a new seal which is adequate to prevent the setting of the regulator from being interfered with without breaking such seal.

15 Meters—emergency notices

(1) No person shall supply gas through a primary meter installed after the coming into force of these Regulations or for the first time supply gas through an existing primary meter after the coming into force of these Regulations unless he ensures that a suitably worded notice in permanent form is prominently displayed on or near the meter indicating the procedure to be followed in the event of an escape of gas.

(2) Where a meter is installed or relocated in any premises in either case at a distance of more than 2 metres from, or out of sight of, the nearest upstream emergency control in the premises, no person shall supply or provide gas for the first time through that meter unless he ensures that a suitably worded notice in permanent form is prominently displayed on or near the meter indicating the position of that control.

16 Primary meters

(1) No person shall install a prepayment meter as a primary meter through which gas passes to a secondary meter.

(2) Any person—
   (a) who first provides gas through any service pipe or service pipework after the coming into force of these Regulations to more than one primary meter; or
   (b) who subsequently makes any modification which affects the number of primary meters so provided,

shall ensure that a notice in permanent form is prominently displayed on or near each primary meter indicating that more than one primary meter is provided with gas through that service pipe or service pipework.

(3) Where a primary meter is removed, the person who last supplied gas through the meter before removal shall—
   (a) where the meter is not forthwith re-installed or replaced by another meter—
      (i) close any service valve which controlled the supply of gas to that meter and did not control the supply of gas to any other primary meter; and
      (ii) seal the outlet of the emergency control with an appropriate fitting; and
      (iii) clearly mark any live gas pipe in the premises in which the meter was installed to the effect that the pipe contains gas; and
   (b) where the meter has not been re-installed or replaced by another meter before the expiry of the period of 12 months beginning with the date of removal of the meter and there is no such service valve as is mentioned in sub-paragraph (a)(i) above, ensure that the service pipe or service pipework for those premises is disconnected as near as is reasonably practicable to

the main or storage vessel and that any part of the pipe or pipework which is not removed is sealed at both ends with the appropriate fitting.

(4) Where a person proposes to remove a primary meter he shall give sufficient notice of it to the person supplying gas through the meter to enable him to comply with paragraph (3).

17 Secondary meters

(1) Any person supplying or permitting the supply of gas through a primary meter to a secondary meter shall ensure that a line diagram in permanent form is prominently displayed on or near the primary meter or gas storage vessel and on or near all emergency controls connected to the primary meter showing the configuration of all meters, installation pipework and emergency controls.

(2) Any person who changes the configuration of any meter, installation pipework or emergency control so that the accuracy of the line diagram referred to in paragraph (1) is affected shall ensure that the line diagram is amended so as to show the altered configuration.

Part D
Installation Pipework

18 Safe use of pipes

(1) No person shall install any installation pipework in any position in which it cannot be used with safety having regard to the position of other pipes, pipe supports, drains, sewers, cables, conduits and electrical apparatus and to any parts of the structure of any premises in which it is installed which might affect its safe use.

(2) Any person who connects any installation pipework to a primary meter shall, in any case where electrical equipotential bonding may be necessary, inform the responsible person that such bonding should be carried out by a competent person.

19 Enclosed pipes

(1) No person shall install any part of any installation pipework in a wall or a floor or standing of solid construction unless it is so constructed and installed as to be protected against failure caused by the movement of the wall, the floor or the standing as the case may be.

(2) No person shall install any installation pipework so as to pass through a wall or a floor or standing of solid construction (whether or not it contains any cavity) from one side to the other unless—
   (a) any part of the pipe within such wall, floor or standing as the case may be takes the shortest practicable route; and
   (b) adequate means are provided to prevent, so far as is reasonably practicable, any escape of gas from the pipework passing through the wall, floor or standing from entering any cavity in the wall, floor or standing.

(3) No person shall, subject to paragraph (4), install any part of any installation pipework in the cavity of a cavity wall unless the pipe is to pass through the wall from one side to the other.

(4) Paragraph (3) shall not apply to the installation of installation pipework connected to a living flame effect gas fire provided that the pipework in the cavity is as short as is reasonably practicable, is enclosed in a gas tight sleeve and sealed at the joint at which the pipework enters the fire; and in this paragraph a "living flame effect gas fire" means a gas fire—
   (a) designed to simulate the effect of a solid fuel fire;
   (b) designed to operate with a fanned flue system; and
   (c) installed within the inner leaf of a cavity wall.

(5) No person shall install any installation pipework or any service pipework under the foundations of a building or in the ground under the base of a wall or footings unless adequate steps are taken to prevent damage to the installation pipework or service pipework in the event of the movement of those structures or the ground.

(6) Where any installation pipework is not itself contained in a ventilated duct, no person shall install any installation pipework in any shaft, duct or void which is not adequately ventilated.

20 Protection of buildings

No person shall install any installation pipework in a way which would impair the structure of a building or impair the fire resistance of any part of its structure.

21 Clogging precautions

No person shall install any installation pipework in which deposition of liquid or solid matter is likely to occur unless a suitable vessel for the reception of any deposit which may form is fixed to the pipe in a conspicuous and readily accessible position and safe means are provided for the removal of the deposit.

22 Testing and purging of pipes

(1) Where a person carries out work in relation to any installation pipework which might affect the gastightness of any part of it, he shall immediately thereafter ensure that—
   (a) that part is adequately tested to verify that it is gastight and examined to verify that it has been installed in accordance with these Regulations; and
   (b) after such testing and examination, any necessary protective coating is applied to the joints of that part.
(2) Where gas is being supplied to any premises in which any installation pipework is installed and a person carries out work in relation to the pipework, he shall also ensure that—
   (a) immediately after complying with the provisions of sub-paragraphs (a) and (b) of paragraph (1) above, purging is carried out throughout all installation pipework through which gas can then flow so as to remove safely all air and gas other than the gas to be supplied;
   (b) immediately after such purging, if the pipework is not to be put into immediate use, it is sealed off at every outlet with the appropriate fitting;
   (c) if such purging has been carried out through a loosened connection, the connection is retested for gastightness after it has been retightened; and
   (d) every seal fitted after such purging is tested for gastightness.
(3) Where gas is not being supplied to any premises in which any installation pipework is installed—
   (a) no person shall permit gas to pass into the installation pipework unless he has caused such purging, testing and other work as is specified in sub-paragraphs (a) to (d) of paragraph (2) above to be carried out;
   (b) a person who provides a gas supply to those premises shall, unless he complies with sub-paragraph (a) above, ensure that the supply is sealed off with an appropriate fitting.

23 Marking of pipes

(1) Any person installing, elsewhere than in any premises or part of premises used only as a dwelling or for living accommodation, a part of any installation pipework which is accessible to inspection shall permanently mark that part in such a manner that it is readily recognisable as part of a pipe for conveying gas.
(2) The responsible person for the premises in which any such part is situated shall ensure that the part continues to be so recognisable so long as it is used for conveying gas.

24 Large consumers

(1) Where the service pipe to any building having two or more floors to which gas is supplied or (whether or not it has more than one floor) a floor having areas with a separate supply of gas has an internal diameter of 50 mm or more, no person shall install any incoming installation pipework supplying gas to any of those floors or areas, as the case may be, unless—
   (a) a valve is installed in the pipe in a conspicuous and readily accessible position; and

(b) a line diagram in permanent form is attached to the building in a readily accessible position as near as practicable to—
  (i) the primary meter or where there is no primary meter, the emergency control, or
  (ii) the gas storage vessel,
indicating the position of all installation pipework of internal diameter of 25 mm or more, meters, emergency controls, valves and pressure test points of the gas supply systems in the building.
(2) Paragraph (1) above shall apply to service pipework as it applies to a service pipe except that reference therein to "50 mm or more" is to be reference to "30 mm or more".
(3) In paragraph (1)(b) above "pressure test point" means a gas fitting to which a pressure gauge can be connected.

Part E
Gas Appliances

25 Interpretation of Part E

In this Part—

"flue pipe" means a pipe forming a flue but does not include a pipe built as a lining into either a chimney or a gas appliance ventilation duct;

"operating pressure", in relation to a gas appliance, means the pressure of gas at which it is designed to operate.

26 Gas appliances—safety precautions

(1) No person shall install a gas appliance unless it can be used without constituting a danger to any person.
(2) No person shall connect a flued domestic gas appliance to the gas supply system except by a permanently fixed rigid pipe.
(3) No person shall install a used gas appliance without verifying that it is in a safe condition for further use.
(4) No person shall install a gas appliance which does not comply with any enactment imposing a prohibition or restriction on the supply of such an appliance on grounds of safety.
(5) No person carrying out the installation of a gas appliance shall leave it connected to the gas supply unless—
  (a) the appliance can be used safely; or
  (b) the appliance is sealed off from the gas supply with an appropriate fitting.
(6) No person shall install a gas appliance without there being at the inlet to it means of shutting off the supply of gas to the appliance unless the provision of such means is not reasonably practicable.
(7) No person shall carry out any work in relation to a gas appliance which bears an indication that it conforms to a type approved by any person as complying with safety standards in such a manner that the appliance ceases to comply with those standards.
(8) No person carrying out work in relation to a gas appliance which bears an indication that it so conforms shall remove or deface the indication.
(9) Where a person performs work on a gas appliance he shall immediately thereafter examine—
  (a) the effectiveness of any flue;
  (b) the supply of combustion air;
  (c) its operating pressure or heat input or, where necessary, both;
  (d) its operation so as to ensure its safe functioning,
  and forthwith take all reasonable practicable steps to notify any defect to the responsible person and, where different, the owner of the premises in which the appliance is situated or, where neither is reasonably practicable, in the case of an appliance supplied with liquefied petroleum gas, the supplier of gas to the appliance, or, in any other case, the transporter.

(10) Paragraph (9) shall not apply in respect of—
   (a) the direct disconnection of the gas supply of a gas appliance; or
   (b) the purging of gas or air from an appliance or its associated pipework or fittings in any case where that purging does not adversely affect the safety of that appliance, pipe or fitting.

27 Flues

(1) No person shall install a gas appliance to any flue unless the flue is suitable and in a proper condition for the safe operation of the appliance.
(2) No person shall install a flue pipe so that it enters a brick or masonry chimney in such a way that the seal between the flue pipe and the chimney cannot be inspected.
(3) No person shall connect a gas appliance to a flue which is surrounded by an enclosure unless that enclosure is so sealed that any spillage of products of combustion cannot pass from the enclosure to any room or internal space other than the room or internal space in which the appliance is installed.
(4) No person shall install a power operated flue system for a gas appliance unless it safely prevents the operation of the appliance if the draught fails.
(5) No person shall install a flue other than in a safe position.

28 Access

No person shall install a gas appliance except in such a manner that it is readily accessible for operation, inspection and maintenance.

29 Manufacturer's instructions

Any person who installs a gas appliance shall leave for the use of the owner or occupier of the premises in which the appliance is installed all instructions provided by the manufacturer accompanying the appliance.

30 Room-sealed appliances

(1) No person shall install a gas appliance in a room used or intended to be used as a bathroom or a shower room unless it is a room-sealed appliance.
(2) No person shall install a gas fire, other gas space heater or a gas water heater of more than 14 kilowatt gross heat input in a room used or intended to be used as sleeping accommodation unless the appliance is a room-sealed appliance.
(3) No person shall install a gas fire, other gas space heater or a gas water heater of 14 kilowatt gross heat input or less in a room used or intended to be used as sleeping accommodation and no person shall install an instantaneous water heater unless (in each case)—
   (a) it is a room-sealed appliance; or
   (b) it incorporates a safety control designed to shut down the appliance before there is a build up of a dangerous quantity of the products of combustion in the room concerned.
(4) The references in paragraphs (1) to (3) to a room used or intended to be used for the purpose therein referred to includes a reference to—
   (a) a cupboard or compartment within such a room; or
   (b) a cupboard, compartment or space adjacent to such a room if there is an air vent from the cupboard, compartment or space into such a room.

31 Suspended appliances

No person shall install a suspended gas appliance unless the installation pipework to which it is connected is so constructed and installed as to be capable of safely supporting the weight imposed on it and the appliance is designed to be so supported.

32 Flue dampers

(1) Any person who installs an automatic damper to serve a gas appliance shall—
   (a) ensure that the damper is so interlocked with the gas supply to the burner that burner operation is prevented in the event of failure of the damper when not in the open position; and
   (b) immediately after installation examine the appliance and the damper to verify that they can be used together safely without constituting a danger to any person.
(2) No person shall install a manually operated damper to serve a domestic gas appliance.
(3) No person shall install a domestic gas appliance to a flue which incorporates a manually operated damper unless the damper is permanently fixed in the open position.

33 Testing of appliances

(1) Where a person installs a gas appliance at a time when gas is being supplied to the premises in which the appliance is installed, he shall immediately thereafter test its connection to the installation pipework to verify that it is gastight and examine the appliance and the gas fittings and other works for the supply of gas and any flue or means of ventilation to be used in connection with the appliance for the purpose of ascertaining whether—
   (a) the appliance has been installed in accordance with these Regulations;
   (b) the operating pressure is as recommended by the manufacturer;
   (c) the appliance has been installed with due regard to any manufacturer's instructions provided to accompany the appliance; and
   (d) all gas safety controls are in proper working order.
(2) Where a person carries out such testing and examination in relation to a gas appliance and adjustments are necessary to ensure compliance with the requirements specified in sub-paragraphs (a) to (d) of paragraph (1) above, he shall either carry out those adjustments or disconnect the appliance from the gas supply or seal off the appliance from the gas supply with an appropriate fitting.
(3) Where gas is not being supplied to any premises in which any gas appliance is installed—
   (a) no person shall subsequently permit gas to pass into the appliance unless he has caused such testing, examination and adjustment as is specified in paragraphs (1) and (2) above to be carried out; and
   (b) a person who subsequently provides a gas supply to those premises shall, unless he complies with sub-paragraph (a) above, ensure that the appliance is sealed off from the gas supply with an appropriate fitting.

34 Use of appliances

(1) The responsible person for any premises shall not use a gas appliance or permit a gas appliance to be used if at any time he knows or has reason to suspect that it cannot be used without constituting a danger to any person.
(2) For the purposes of paragraph (1) above, the responsible person means the occupier of the premises, the owner of the premises and any person with authority for the time being to take appropriate action in relation to any gas fitting therein.
(3) Any person engaged in carrying out any work in relation to a gas main, service pipe, service pipework, gas storage vessel or gas fitting who knows or has reason to suspect that any gas appliance cannot be used without constituting a danger to any person shall forthwith take all reasonably practicable steps to inform the responsible person for the premises in which the appliance is situated and, where different, the owner of the appliance or, where neither is reasonably practicable, in the case of an appliance supplied with liquefied petroleum gas, the supplier of gas to the appliance, or, in any other case, the transporter.
(4) In paragraph (3) above the expression "work" shall be construed as if, in the definition of "work" in regulation 2(1) above, every reference to a gas fitting were a reference to a gas main, service pipe, service pipework, gas storage vessel or gas fitting.

The Gas Safety (Installation and Use) Regulations 1998 (as amended)

PART F
MAINTENANCE

35 Duties of employers and self-employed persons

It shall be the duty of every employer or self-employed person to ensure that any gas appliance, installation pipework or flue installed at any place of work under his control is maintained in a safe condition so as to prevent risk of injury to any person.

36 Duties of Landlords

(1) In this regulation—
"landlord" means—
- (a) in England and Wales—
  - (i) where the relevant premises are occupied under a lease, the person for the time being entitled to the reversion expectant on that lease or who, apart from any statutory tenancy, would be entitled to possession of the premises; and
  - (ii) where the relevant premises are occupied under a licence, the licensor, save that where the licensor is himself a tenant in respect of those premises, it means the person referred to in paragraph (i) above;
- (b) in Scotland, the person for the time being entitled to the landlord's interest under a lease;

"lease" means—
- (a) a lease for a term of less than 7 years; and
- (b) a tenancy for a periodic term; and
- (c) any statutory tenancy arising out of a lease or tenancy referred to in sub-paragraphs (a) or (b) above,

and in determining whether a lease is one which falls within sub-paragraph (a) above—
  - (i) in England and Wales, any part of the term which falls before the grant shall be left out of account and the lease shall be treated as a lease for a term commencing with the grant;
  - (ii) a lease which is determinable at the option of the lessor before the expiration of 7 years from the commencement of the term shall be treated as a lease for a term of less than 7 years;
  - (iii) a lease (other than a lease to which sub-paragraph (b) above applies) shall not be treated as a lease for a term of less than 7 years if it confers on the lessee an option for renewal for a term which, together with the original term, amounts to 7 years or more; and
  - (iv) a "lease" does not include a mortgage term;

"relevant gas fitting" means—
- (a) any gas appliance (other than an appliance which the tenant is entitled to remove from the relevant premises) or any installation pipework installed in any relevant premises; and
- (b) any gas appliance or installation pipework which, directly or indirectly, serves the relevant premises and which either—
  - (i) is installed in any part of premises in which the landlord has an estate or interest; or
  - (ii) is owned by the landlord or is under his control,

except that it shall not include any gas appliance or installation pipework exclusively used in a part of premises occupied for non-residential purposes;

"relevant premises" means premises or any part of premises occupied, whether exclusively or not, for residential purposes (such occupation being in consideration of money or money's worth) under—
- (a) a lease; or
- (b) a licence;

"statutory tenancy" means—
- (a) in England and Wales, a statutory tenancy within the meaning of the Rent Act 1977 and the Rent (Agriculture) Act 1976; and

(b) in Scotland, a statutory tenancy within the meaning of the Rent (Scotland) Act 1984, a statutory assured tenancy within the meaning of the Housing (Scotland) Act 1988 or a secure tenancy within the meaning of the Housing (Scotland) Act 1987;

"tenant" means a person who occupies relevant premises being—
  (a) in England and Wales—
     (i) where the relevant premises are so occupied under a lease, the person for the time being entitled to the term of that lease; and
     (ii) where the relevant premises are so occupied under a licence, the licensee;
  (b) in Scotland, the person for the time being entitled to the tenant's interest under a lease.

(2) Every landlord shall ensure that there is maintained in a safe condition—
  (a) any relevant gas fitting; and
  (b) any flue which serves any relevant gas fitting,

so as to prevent the risk of injury to any person in lawful occupation or relevant premises.

(3) Without prejudice to the generality of paragraph (2) above, a landlord shall—
  (a) ensure that each appliance and flue to which that duty extends is checked for safety within 12 months of being installed and at intervals of not more than 12 months since it was last checked for safety (whether such check was made pursuant to these Regulations or not);
  (b) in the case of a lease commencing after the coming into force of these Regulations, ensure that each appliance and flue to which the duty extends has been checked for safety within a period of 12 months before the lease commences or has been or is so checked within 12 months after the appliance or flue has been installed, whichever is later; and
  (c) ensure that a record in respect of any appliance or flue so checked is made and retained for a period of 2 years from the date of that check, which record shall include the following information—
     (i) the date on which the appliance or flue was checked;
     (ii) the address of the premises at which the appliance or flue is installed;
     (iii) the name and address of the landlord of the premises (or, where appropriate, his agent) at which the appliance or flue is installed;
     (iv) a description of and the location of each appliance or flue checked;
     (v) any defect identified;
     (vi) any remedial action taken;
     (vii) confirmation that the check undertaken complies with the requirements of paragraph (9) below;
     (viii) the name and signature of the individual carrying out the check; and
     (ix) the registration number with which that individual, or his employer, is registered with a body approved by the Executive for the purposes of regulation 3(3) of these Regulations.

(4) Every landlord shall ensure that any work in relation to a relevant gas fitting or any check of a gas appliance or flue carried out pursuant to paragraphs (2) or (3) above is carried out by, or by an employee of, a member of a class of persons approved for the time being by the Health and Safety Executive for the purposes of regulation 3(3) of these Regulations.

(5) The record referred to in paragraph (3)(c) above, or a copy thereof, shall be made available upon request and upon reasonable notice for the inspection of any person in lawful occupation of relevant premises who may be affected by the use or operation of any appliance to which the record relates.

(6) Notwithstanding paragraph (5) above, every landlord shall ensure that—
  (a) a copy of the record made pursuant to the requirements of paragraph (3)(c) above is given to each existing tenant of premises to which the record relates within 28 days of the date of the check; and
  (b) a copy of the last record made in respect of each appliance or flue is given to any new tenant of premises to which the record relates before that tenant occupies those premises save that,

The Gas Safety (Installation and Use) Regulations 1998 (as amended)

in respect of a tenant whose right to occupy those premises is for a period not exceeding 28 days, a copy of the record may instead be prominently displayed within those premises.

(7) Where there is no relevant gas appliance in any room occupied or to be occupied by the tenant in relevant premises, the landlord may, instead of ensuring that a copy of the record referred to in paragraph (6) above is given to the tenant, ensure that there is displayed in a prominent position in the premises (from such time as a copy would have been required to have been given to the tenant under that paragraph), a copy of the record with a statement endorsed on it that the tenant is entitled to have his own copy of the record on request to the landlord at an address specified in the statement; and on any such request being made, the landlord shall give to the tenant a copy of the record as soon as is practicable.

(8) A copy of the record given to a tenant pursuant to paragraph (6)(b) above need not contain a copy of the signature of the individual carrying out the check if the copy of the record contains a statement that another copy containing a copy of such signature is available for inspection by the tenant on request to the landlord at an address specified in the statement, and on any such request being made the landlord shall make such a copy available for inspection as soon as is practicable.

(9) A safety check carried out pursuant to paragraph (3) above shall include, but shall not be limited to, an examination of the matters referred to in sub-paragraphs (a) to (d) of regulation 26(9) of these Regulations.

(10) Nothing done or agreed to be done by a tenant of relevant premises or by any other person in lawful occupation of them in relation to the maintenance or checking of a relevant gas fitting or flue in the premises (other than one in part of premises occupied for non-residential purposes) shall be taken into account in determining whether a landlord has discharged his obligations under this regulation (except in so far as it relates to access to that gas fitting or flue for the purposes of such maintenance or checking).

(11) Every landlord shall ensure that in any room occupied or to be occupied as sleeping accommodation by a tenant in relevant premises there is not fitted a relevant gas fitting of a type the installation of which would contravene regulation 30(2) or (3) of these Regulations.

(12) Paragraph (11) above shall not apply in relation to a room which since before the coming into force of these Regulations has been occupied or intended to be occupied as sleeping accommodation.

Part G
Miscellaneous

37 Escape of gas

(1) Where any gas escapes from any pipe of a gas supplier or from any pipe, other gas fitting or gas storage vessel used by a person supplied with gas by a gas supplier, the supplier of the gas shall, within 12 hours of being so informed of the escape, prevent the gas escaping (whether by cutting off the supply of gas to any premises or otherwise).

(2) If the responsible person for any premises knows or has reason to suspect that gas is escaping into those premises, he shall immediately take all reasonable steps to cause the supply of gas to be shut off at such place as may be necessary to prevent further escape of gas.

(3) If gas continues to escape into those premises after the supply of gas has been shut off or when a smell of gas persists, the responsible person for the premises discovering such escape or smell shall immediately give notice of the escape or smell to the supplier of the gas.

(4) Where an escape of gas has been stopped by shutting off the supply, no person shall cause or permit the supply to be re-opened (other than in the course of repair) until all necessary steps have been taken to prevent a recurrence of such escape.

(5) In any proceedings for an offence under paragraph (1) above it shall be a defence for the supplier of the gas to prove that it was not reasonably practicable for him effectually to prevent the gas

from escaping within the period of 12 hours referred to in that paragraph, and that he did effectually prevent the escape of gas as soon as it was reasonably practicable for him to do so.
(6) Nothing in paragraphs (1) and (5) above shall prevent the supplier of the gas appointing another person to act on his behalf to prevent an escape of gas supplied by that supplier.
(7) Nothing in paragraphs (1) to (6) above shall apply to an escape of gas from a network (within the meaning of regulation 2 of the Gas Safety (Management) Regulations 1996) or from a gas fitting supplied with gas from a network.
(8) In this regulation any reference to an escape of gas from a gas fitting includes a reference to an escape or emission of carbon monoxide gas resulting from incomplete combustion of gas in a gas fitting, but, to the extent that this regulation relates to such an escape or emission of carbon monoxide gas, the requirements imposed upon a supplier by paragraph (1) above shall, where the escape or emission is notified to the supplier by the person to whom the gas has been supplied, be limited to advising that person of the immediate action to be taken to prevent such escape or emission and the need for the examination and, where necessary, repair of the fitting by a competent person.

38 Use of antifluctuators and valves

(1) Where a consumer uses gas for the purpose of working or supplying plant which is liable to produce pressure fluctuation in the gas supply such as to cause any danger to other consumers, he shall comply with such directions as may be given to him by the transporter of the gas to prevent such danger.
(2) Where a consumer intends to use for or in connection with the consumption of gas any gaseous substance he shall—
   (a) give to the transporter of the gas at least 14 days notice in writing of that intention; and
   (b) during such use comply with such directions as the transporter may have given to him to prevent the admission of such substance into the gas supply;
and in this paragraph "gaseous substance" includes compressed air but does not include any gaseous substance supplied by the transporter.
(3) Where a direction under paragraphs (1) or (2) above requires the provision of any device, the consumer shall ensure that the device is adequately maintained.
(4) Any direction given pursuant to this regulation shall be in writing.

39 Exception as to liability

No person shall be guilty of an offence by reason of contravention of regulation 3(2) or (6), 5(1), 7(3), 15, 16(2) or (3), 17(1), 27(5), 30 (insofar as it relates to the installation of a gas fire, other gas space heater or a gas water heater of more than 14 kilowatt gross heat input), 33(1), 35 or 36 of these Regulations in any case in which he can show that he took all reasonable steps to prevent that contravention.

40 Exemption certificates

(1) Subject to paragraph (2), the Health and Safety Executive may, by a certificate in writing, exempt any person or class of persons from any requirement or prohibition imposed by these Regulations, and any such exemption may be granted subject to conditions and to a limit of time and may be revoked at any time by a certificate in writing.
(2) The Health and Safety Executive shall not grant any such exemption unless, having regard to the circumstances of the case and in particular to—
   (a) the conditions, if any, which it proposes to attach to the exemption; and
   (b) any other requirements imposed by or under any enactment which apply to the case,
it is satisfied that the health and safety of persons likely to be affected by the exemption, will not be prejudiced in consequence of it.

41 Revocation and amendments

(1) The Gas Safety (Installation and Use) Regulations 1994, the Gas Safety (Installation and Use) (Amendment) Regulations 1996 and the Gas Safety (Installation and Use) (Amendment) (No 2) Regulations 1996 are hereby revoked.

(2) Schedule 2B to the Gas Act 1986 shall be amended as follows—
   (a) In paragraph 17(1) the words "pressure fluctuation in the transporter's pipe-line system and any other" and the words "or danger" shall be deleted;
   (b) In paragraph 17(2) after the words "if so required" there shall be added "other than for the purpose of preventing danger"; and
   (c) In paragraph 17(5) and (6) after the words "this paragraph" there shall be added "or regulation 38 of the Gas Safety (Installation and Use) Regulations 1998 or directions made thereunder".

Signed by order of the Secretary of State.

Alan Meale

Parliamentary Under Secretary of State,

Department of the Environment, Transport and the Regions.

3rd October 1998

APPENDIX N

Health and Safety: The Electricity at Work Regulations 1989 (as amended)

1989 No. 635

Made   *7th April 1989*

Authority: Health and Safety at Work etc. Act 1974, ss 15(1), (2), (3)(a), (b), (4)(a), (5)(b), (6)(b), (8), (9), 82(3)(a), Sch 3, paras 1(1)(a), (c), (2), (3), 6(2), 9, 11, 12, 14, 15(1), 16, 21(b).

PART I
INTRODUCTION

1 Citation and commencement

These Regulations may be cited as the Electricity at Work Regulations 1989 and shall come into force on 1st April 1990.

2 Interpretation

(1) In these Regulations, unless the context otherwise requires—

"approved" means approved in writing for the time being by the Health and Safety Executive for the purposes of these Regulations or conforming with a specification approved in writing by the Health and Safety Executive for the purposes of these Regulations;

"circuit conductor" means any conductor in a system which is intended to carry electric current in normal conditions, or to be energised in normal conditions, and includes a combined neutral and earth conductor, but does not include a conductor provided solely to perform a protective function by connection to earth or other reference point;

"conductor" means a conductor of electrical energy;

"danger" means risk of injury;

"electrical equipment" includes anything used, intended to be used or installed for use, to generate, provide, transmit, transform, rectify, convert, conduct, distribute, control, store, measure or use electrical energy;

"firedamp" means any flammable gas or any flammable mixture of gases occurring naturally in a mine;

"injury" means death or personal injury from electric shock, electric burn, electrical explosion or arcing, or from fire or explosion initiated by electrical energy, where any such death or injury is associated with the generation, provision, transmission, transformation, rectification, conversion, conduction, distribution, control, storage, measurement or use of electrical energy;

"safety-lamp mine" means—
    (a) any coal mine; or
    (b) any other mine in which—
        (i) there has occurred below ground an ignition of firedamp; or
        (ii) more than 0.25% by volume of firedamp is found on any occasion at any place below ground in the mine;

"system" means an electrical system in which all the electrical equipment is, or may be, electrically connected to a common source of electrical energy, and includes such source and such equipment.

(2) Unless the context otherwise requires, any reference in these Regulations to—
   (a) a numbered regulation or Schedule is a reference to the regulation or Schedule in these Regulations so numbered;
   (b) a numbered paragraph is a reference to the paragraph so numbered in the regulation or Schedule in which the reference appears.

3 Persons on whom duties are imposed by these Regulations

(1) Except where otherwise expressly provided in these Regulations, it shall be the duty of every—
   (a) employer and self-employed person to comply with the provisions of these Regulations in so far as they relate to matters which are within his control; and
   [(b)
      (i) manager, in relation to a mine within the meaning of the Mines and Quarries Act 1954, and
      (ii) operator, in relation to a quarry within the meaning of regulation 3 of the Quarries Regulations 1999,
   to ensure that all requirements or prohibitions imposed by or under these Regulations are complied with in so far as they relate to the mine of which he is the manager or quarry of which he is the operator and to matters which are within his control].[1]

(2) It shall be the duty of every employee while at work—
   (a) to co-operate with his employer so far as is necessary to enable any duty placed on that employer by the provisions of these Regulations to be complied with; and
   (b) to comply with the provisions of these Regulations in so far as they relate to matters which are within his control.

Part II
General

4 Systems, work activities and protective equipment

(1) All systems shall at all times be of such construction as to prevent, so far as is reasonably practicable, danger.
(2) As may be necessary to prevent danger, all systems shall be maintained so as to prevent, so far as is reasonably practicable, such danger.
(3) Every work activity, including operation, use and maintenance of a system and work near a system, shall be carried out in such a manner as not to give rise, so far as is reasonably practicable, to danger.
(4) Any equipment provided under these Regulations for the purpose of protecting persons at work on or near electrical equipment shall be suitable for the use for which it is provided, be maintained in a condition suitable for that use, and be properly used.

5 Strength and capability of electrical equipment

No electrical equipment shall be put into use where its strength and capability may be exceeded in such a way as may give rise to danger.

---

[1] Amended by SI 1999/2024.

6 Adverse or hazardous environments

Electrical equipment which may reasonably foreseeably be exposed to—

(a) mechanical damage;
(b) the effects of the weather, natural hazards, temperature or pressure;
(c) the effects of wet, dirty, dusty or corrosive conditions; or
(d) any flammable or explosive substance, including dusts, vapours or gases,

shall be of such construction or as necessary protected as to prevent, so far as is reasonably practicable, danger arising from such exposure.

7 Insulation, protection and placing of conductors

All conductors in a system which may give rise to danger shall either—

(a) be suitably covered with insulating material and as necessary protected so as to prevent, so far as is reasonably practicable, danger; or
(b) have such precautions taken in respect of them (including, where appropriate, their being suitably placed) as will prevent, so far as is reasonably practicable, danger.

8 Earthing or other suitable precautions

Precautions shall be taken, either by earthing or by other suitable means, to prevent danger arising when any conductor (other than a circuit conductor) which may reasonably foreseeably become charged as a result of either the use of a system, or a fault in a system, becomes so charged; and, for the purposes of ensuring compliance with this regulation, a conductor shall be regarded as earthed when it is connected to the general mass of earth by conductors of sufficient strength and current-carrying capability to discharge electrical energy to earth.

9 Integrity of referenced conductors

If a circuit conductor is connected to earth or to any other reference point, nothing which might reasonably be expected to give rise to danger by breaking the electrical continuity or introducing high impedance shall be placed in that conductor unless suitable precautions are taken to prevent that danger.

10 Connections

Where necessary to prevent danger, every joint and connection in a system shall be mechanically and electrically suitable for use.

11 Means for protecting from excess of current

Efficient means, suitably located, shall be provided for protecting from excess of current every part of a system as may be necessary to prevent danger.

12 Means for cutting off the supply and for isolation

(1) Subject to paragraph (3), where necessary to prevent danger, suitable means (including, where appropriate, methods of identifying circuits) shall be available for—
    (a) cutting off the supply of electrical energy to any electrical equipment; and
    (b) the isolation of any electrical equipment.
(2) In paragraph (1), "isolation" means the disconnection and separation of the electrical equipment from every source of electrical energy in such a way that this disconnection and separation is secure.
(3) Paragraph (1) shall not apply to electrical equipment which is itself a source of electrical energy but, in such a case as is necessary, precautions shall be taken to prevent, so far as is reasonably practicable, danger.

13 Precautions for work on equipment made dead

Adequate precautions shall be taken to prevent electrical equipment, which has been made dead in order to prevent danger while work is carried out on or near that equipment, from becoming electrically charged during that work if danger may thereby arise.

14 Work on or near live conductors

No person shall be engaged in any work activity on or so near any live conductor (other than one suitably covered with insulating material so as to prevent danger) that danger may arise unless—

(a) it is unreasonable in all the circumstances for it to be dead; and
(b) it is reasonable in all the circumstances for him to be at work on or near it while it is live; and
(c) suitable precautions (including where necessary the provision of suitable protective equipment) are taken to prevent injury.

15 Working space, access and lighting

For the purposes of enabling injury to be prevented, adequate working space, adequate means of access, and adequate lighting shall be provided at all electrical equipment on which or near which work is being done in circumstances which may give rise to danger.

16 Persons to be competent to prevent danger and injury

No person shall be engaged in any work activity where technical knowledge or experience is necessary to prevent danger or, where appropriate, injury, unless he possesses such knowledge or experience, or is under such degree of supervision as may be appropriate having regard to the nature of the work.

Part III
Regulations Applying to Mines only

17 Provisions applying to mines only

(1) The provisions of regulations 18 to 28 and Schedule 1 shall apply to mines only; and the provisions of that Schedule shall have effect in particular in relation to the use below ground in a coal mine of any film lighting circuit (as defined by paragraph 1 of that Schedule) at or in close proximity to a coal face.

(2) Expressions to which meanings are assigned by the Mines and Quarries Act 1954 shall, unless the contrary intention appears, have the same meanings in regulations 18 to 27 and Schedule 1.

[18 Introduction of electrical equipment]

[Before electrical equipment (other than equipment approved for the purposes of regulation 20(1)) is first introduced into any underground part of a safety-lamp mine, the manager shall submit to an inspector a copy of the plan required to be kept for that part by regulation 29(5) of the Management and Administration of Safety and Health at Mines Regulations 1993, on which the intended locations of that equipment shall be shown together with a copy of any schematic diagram relating to that part prepared for the purposes of regulation 24(1).][2]

19 Restriction of equipment in certain zones below ground

(1) At every safety-lamp mine containing any zones below ground in which firedamp whether or not normally present is likely to occur in a quantity sufficient to indicate danger, there shall be prepared a suitable plan identifying such zones.

---

[2] Inserted by SI 1995/2005.

(2) Electrical equipment shall not be energised in such zones unless it is—
   (a) equipment of a kind approved for that purpose;
   (b) equipment approved pursuant to regulation 20(1);
   (c) equipment the use of which was lawful in such zones immediately before the coming into force of these Regulations;
   (d) equipment which has received a certificate of conformity or a certificate of inspection in accordance with Council Directive 82/130/EEC on the approximation of the laws of the Member States concerning electrical equipment for use in potentially explosive atmospheres in mines susceptible to firedamp, as adapted to technical progress by [Commission Directives 88/35/EEC, 91/269/EEC[, 94/44/EC and 98/65/EC]³];⁴
   (e) equipment such as is specified in regulation 21(2);
   (f) equipment which is not capable of producing incendive electrical sparks in normal use; or
   (g) electrically-powered equipment not permanently installed in the mine but required occasionally for monitoring, testing, recording and measurement, and used where the concentration of firedamp is 0.8% by volume or less in accordance with suitable rules drawn up by the manager to ensure that danger will not thereby arise, which rules shall in particular include provision for personal supervision of that equipment by a competent person and testing for firedamp when it is in use;
and any lights which conform with this paragraph shall be permitted lights in any mine such as is specified in paragraph (1).

20 Cutting off electricity or making safe where firedamp is found either below ground or at the surface

(1) Where any person at a mine detects firedamp in a concentration exceeding 1.25% by volume in the general body of the air either below ground at that mine or at any place on the surface thereat where any exhauster in a firedamp drainage system is installed, firedamp is monitored or its heat content measured, he shall forthwith—
   (a) cut off the supply of electricity to any electrical equipment situated at the place where the said concentration was detected; or
   (b) (where this is not possible) take all reasonably practicable steps to make such equipment safe; or
   (c) (if the taking of the measures specified in sub-paragraphs (a) and (b) above does not fall within the scope of his normal duties) report the matter to an official of the mine who shall ensure that those measures are taken;
   except that the provisions of sub-paragraphs (a) to (c) above shall not apply if the electrical equipment is approved for the purpose of remaining energised in such circumstances or (in the case of a safety-lamp mine) is electrical equipment such as is specified in regulation 21(2).
(2) If the supply of electricity to electrical equipment is cut off or the equipment made safe in accordance with paragraph (1), it shall remain in that condition until the senior official on duty at the mine having determined that it is safe to do so, directs that such precautions are no longer necessary.
(3) If the supply of electricity to electrical equipment is cut off or the equipment made safe in accordance with paragraph (1), details of the time, duration and location shall be recorded.

21 Approval of certain equipment for use in safety-lamp mines

(1) Subject to paragraph (2), no electric safety-lamp, gas detector, telephone or signalling equipment or other equipment associated therewith or required for the safety of persons shall be taken or used below ground at any safety-lamp mine unless it is equipment which has been approved

---

³ Amended by SI 1999/2550.
⁴ Amended by SI 1996/192.

pursuant to regulation 20(1) or (in the case of electric safety-lamps) is of a type for the time being approved pursuant to section 64(2) of the Mines and Quarries Act 1954.

(2) Nothing in paragraph (1) shall prevent the taking or use below ground at any safety-lamp mine of any electrical equipment which was, before the coming into force of these Regulations, approved pursuant to regulations 20 and 21A of the Coal and Other Mines (Electricity) Regulations 1956.

22 Means of cutting off electricity to circuits below ground

At every mine at which electrical equipment which may give rise to danger is installed below ground and is supplied from a power source at the surface of the mine, switchgear shall be provided at the surface for cutting off the supply of current to that equipment, and adequate provision shall be made for the operation of that switchgear, including such means of communication as will, so far as is reasonably practicable, enable the switchgear to be operated in case of danger.

23 Oil-filled equipment

Electrical equipment using oil as a means of cooling, insulation or arc suppression shall not be introduced below ground at a mine.

24 Records and information

(1) Suitable schematic diagrams of all electrical distribution systems intended to be operated at the mine (other than those operating at a voltage not exceeding 250 volts) shall, so far as is reasonably practicable—
  (a) be prepared and kept in the office at the mine; and
  (b) show the planned settings of any circuit electrical protective devices.
(2) Copies of such portions of the schematic diagrams prepared pursuant to paragraph (1) as are necessary to prevent danger and which show at least those parts of the electrical system which are served by switchgear operating at a voltage in excess of 250 volts shall be displayed at each place where such switchgear is installed.
(3) Plans on a suitable scale shall be kept in the office at the mine showing, so far as is reasonably practicable, the position of all permanently installed electrical equipment at the mine supplied at a voltage in excess of 250 volts.

25 Electric shock notices

Where, at any place at a mine, electric arc welding is taking place or electrical energy is being generated, transformed or used at a nominal voltage in excess of 125 volts ac or 250 volts dc, a notice shall be displayed in a form which can be easily read and understood and containing information on the appropriate first-aid treatment for electric shock and details of the emergency action to be taken in the event of electric shock.

26 Introduction of battery-powered locomotives and vehicles into safety-lamp mines

No locomotive or vehicle which uses an electrical storage battery, either partly or wholly, as a power source for traction purposes shall be introduced below ground at a safety-lamp mine unless it is an approved locomotive or vehicle.

27 Storage, charging and transfer of electrical storage batteries

At any mine in which electrical storage batteries are used below ground, those batteries shall, so far as is reasonably practicable, be used, stored, charged and transferred in a safe manner.

28 Disapplication of section 157 of the Mines and Quarries Act 1954

Section 157 of the Mines and Quarries Act 1954 (which provides a defence in legal proceedings and prosecutions in certain circumstances) shall not apply in relation to any legal proceedings or prosecutions based on an allegation of a contravention of a requirement or prohibition imposed by regulations 18 to 27 or by or under Schedule 1.

Part IV
Miscellaneous and General

29 Defence

In any proceedings for an offence consisting of a contravention of regulations 4(4), 5, 8, 9, 10, 11, 12, 13, 14, 15, 16 or 25, it shall be a defence for any person to prove that he took all reasonable steps and exercised all due diligence to avoid the commission of that offence.

30 Exemption certificates

(1) Subject to paragraph (2), the Health and Safety Executive may, by a certificate in writing, exempt—
   (a) any person;
   (b) any premises;
   (c) any electrical equipment;
   (d) any electrical system;
   (e) any electrical process;
   (f) any activity,
   or any class of the above, from any requirement or prohibition imposed by these Regulations and any such exemption may be granted subject to conditions and to a limit of time and may be revoked by a certificate in writing at any time.

(2) The Executive shall not grant any such exemption unless, having regard to the circumstances of the case, and in particular to—
   (a) the conditions, if any, which it proposes to attach to the exemption; and
   (b) any other requirements imposed by or under any enactment which apply to the case,
   it is satisfied that the health and safety of persons who are likely to be affected by the exemption will not be prejudiced in consequence of it.

[31 Application]

[These Regulations shall apply—
   (a) in Great Britain; and
   (b) outside Great Britain as sections 1 to 59 and 80 to 82 of the Health and Safety at Work etc. Act 1974 apply by virtue of the provisions of the Health and Safety at Work etc. Act 1974 (Application outside Great Britain) Order 1995.][5]

32 Disapplication of duties

The duties imposed by these Regulations shall not extend to—
   (a) the master or crew of a sea-going ship or to the employer of such persons, in relation to the normal ship-board activities of a ship's crew under the direction of the master; or
   (b) any person, in relation to any aircraft or hovercraft which is moving under its own power.

33 Revocations and modifications

(1) The instruments specified in column 1 of Part I of Schedule 2 are revoked to the extent specified in the corresponding entry in column 3 of that Part.

(2) The enactments and instruments specified in Part II of Schedule 2 shall be modified to the extent specified in that Part.

(3) In the Mines and Quarries Act 1954, the Mines and Quarries (Tips) Act 1969 and the Mines Management Act 1971, and in regulations made under any of those Acts, or in health and safety regulations, any reference to any of those Acts shall be treated as including a reference to these Regulations.

---

[5] Inserted by SI 1997/1993.

INDEX

absolute discharge 10.04
abuse of process 9.17–9.57
  breach of promise 9.46–9.49
  categories of abuse 9.22–9.57
  delay 9.23–9.33
  double jeopardy 9.34–9.45
  judicial review 9.01, 9.03–9.07
  jurisdiction 9.17–9.21
  magistrates 9.19–9.21
  meaning 9.17
  oppressive decision to prosecute
      contrary to policy 9.50–9.57
  prosecution decisions 9.01, 9.03–9.07
  public interest 9.20
  stay 9.18–9.19
access to legal advice
  arrest 3.35, 3.37
  causes, reports into 3.77
  improvement notices 4.06
  interviews under caution 3.35–3.38, 3.49
  PACE and codes 2.76, 3.35–3.38, 3.42–3.43
  police station, interviews at 3.35
  solicitors at interview, presence of 3.37–3.38
accident, meaning of 2.17
accidental death verdicts 11.143
ACOPs *see* **approved codes of practice (ACOPs)**
adjournments
  advance information 8.52–8.53
  committal to Crown Court for
      sentence 8.92, 8.101
  coroners' inquests 11.145–11.152
  investigations 11.146
admissibility 2.96–2.99
  answers, power to require 2.142, 2.145
  bad character evidence 8.217–8.226
  causes, reports into 3.78
  complex cases 8.153
  compulsory questioning 2.142
  hearsay 8.208–8.213
  preparatory hearings 8.151
  self-incrimination, privilege
      against 3.86–3.90
admissions 3.26, 3.42–3.43

advance information
  adjournment 8.52–8.53
  costs 8.53
  either way offences 8.48
  form of 8.50
  Friskies statements 8.66–8.68
  magistrates 8.48–8.55
  mode of trial 8.52
  protocol 8.49
  summons 8.48
  written statements 8.48
adverse inferences from silence
  conditions 3.17
  detention 3.16
  disclosure 3.11
  interviews under caution 3.13–3.19, 3.43, 3.49
  no comment interviews 3.18
  prepared statements 3.27
advice *see* **access to legal advice; advising clients in health and safety investigations**
advising clients in health and safety investigations 3.51–3.90
  causes, reports into 3.77
  conflicts of interest 3.52
  cooperation with inspectors 3.68–3.73
  Health and Safety at Work Act 1974
      section 20 powers 3.68–3.73
  improvement and prohibition notices 4.06
  interviews under caution 3.05–3.30
  legal privilege 3.74–3.85
  PACE and codes 3.53
  representation 3.55–3.67
  self-incrimination 3.86–3.90
agency workers 6.52–6.53, 12.43
aggravating factors
  committal to Crown Court for
      sentence 8.91
  death or serious personal injury 8.91, 10.29
  fines 10.23, 10.27, 10.29
  Friskies statements 8.56–8.62, 8.68
  mode of trial 8.106
  warnings, failure to heed 8.91, 10.23
agricultural land 12.56
aiding and abetting 5.27–5.29

Index

aircraft 12.56
answers, power to require
  admissibility 2.142, 2.145
  company officers, compelling answers from 2.146–2.148
  compulsory powers 2.132, 2.134, 2.142–2.143, 2.146–2.148
  directors, compelling answers from 2.146–2.148
  Environmental Protection Act 1990, powers under 2.142–2.143, 2.145, 2.147
  hearsay 2.136
  information gathering 2.140–2.141
  inspectors 2.91–2.92
  nominated person, presence of a 2.149
  self-incrimination 2.137–2.138, 2.144–2.145
  statements 2.134–2.136
appeals *see also* **appeals against improvement or prohibition notices**
  coroners' inquests 11.97
  costs 10.138, 10.150–10.151
  leaflet ITL 4.07
  preparatory hearings 8.151
appeals against improvement or prohibition notices 4.06–4.08, 4.53–4.126
  adjournments 4.79
  affirmation of notice 4.100–4.101, 4.104, 4.106–4.107
  appeals from decisions 4.126
  booklets 4.65
  burden of proof 4.84–4.87, 4.91
  cancellation 4.100, 4.104
  case management 4.72–4.74
  case stated 4.122, 4.126
  chairman 4.76
  commencement of appeals 4.62–4.67
  composition of tribunal 4.75, 4.77
  costs 4.123–4.125
  decisions 4.70, 4.120–4.122, 4.126
  defective notices, rectifying 4.103–4.109
  Employment Appeal Tribunal 4.126
  employment tribunals, appeals to 4.07–4.08, 4.53–4.126
  evidence 4.80
  example decisions 4.111–4.119
  experts, appointment of 4.81
  failure to attend 4.79
  fresh hearings 4.54
  fundamental flaws 4.100–4.110
  grounds 4.55–4.56, 4.69
  Health and Safety Act 1974 4.53, 4.69, 4.90, 4.94
  hearings 4.75–4.82
  hearsay 4.60
  insignificant contraventions 4.98–4.99
  judicial review 4.122
  legal representation 4.58, 4.79
  modification 4.56, 4.100–4.110
  notice of appeal 4.63–4.67, 4.78
  opinion, challenging the inspector's 4.88–4.97
  person, appearing in 4.79
  personal injury, risk of serious 4.88–4.90, 4.95–4.97, 4.102
  precautions 4.99
  pre-hearing orders 4.72–4.74
  preliminary arguments 4.83
  procedure 4.59–4.61
  prosecutions 4.17
  public hearings 4.75
  reasons for decisions 4.70, 4.121
  representation 4.58, 4.79
  reputation, harm to 4.56–4.57
  reviews 4.54, 4.70, 4.122
  rules 4.59–4.61
  service of notices 4.64–4.67
    date of 4.64
    methods of 4.64
  standard forms 4.65
  substitution of findings 4.105
  suspension of notices pending appeal 4.68–4.70, 4.99
  time limits 4.08, 4.17, 4.29, 4.66, 4.78
  time to correct contravention 4.100
  withdrawal 4.56, 4.71
  witnesses 4.80
  written judgments 4.120–4.121
approved codes of practice (ACOPs)
  aims of 1.10
  asbestos 12.193, 12.202–12.204
  Construction Design Management Coordinator 12.170–12.171, 12.174
  construction industry 12.146–12.148, 12.154–12.160, 12.163–12.171, 12.174
  control of substances hazardous to health 12.121
  criminal proceedings 1.11
  designers 12.178–12.179
  enforcement 1.10–1.11
  equipment 12.77, 12.79, 12.82–12.87, 12.109, 12.111, 12.119
  evidence of breach of legislation, failure to comply as 1.11
  Health and Safety Commission 1.10, 1.19
  Health and Safety Executive 1.10
  improvement notices 4.30, 4.32
  inspection 12.98–12.101, 12.103–12.104
  maintenance 12.91–12.95
  Management of Health and Safety at Work Regulations 1999 12.46
  prohibition notices 4.41

Index

regulations 12.05–12.08
risk assessment 12.20, 12.22–12.25,
 12.28–12.29
workplace 12.47
armed forces 2.16
arrest
 access to legal advice 3.35, 3.37
 arrestable offences and non-arrestable offences
  removal of distinction between 2.58, 8.03
 bail 8.47
 Crown Prosecution Service 8.04
 gross negligence manslaughter 2.65, 3.67
 indictable offences 8.03
 inspectors 2.57, 2.67–2.68, 3.03
 instituting proceedings 8.03–8.04
 interviews under caution 3.02–3.03,
  3.07, 3.19
 investigations 2.57–2.68
 magistrates 8.46
 obstruction 2.69
 police powers 2.57–2.68, 3.02
 reasonable grounds for suspicion 2.58–2.64
 search and entry, powers of 2.106
 summary arrest, powers of 2.60–2.63
 warrants 2.59, 8.46
 Work Related Death Protocol 2.65
 work related deaths 2.65–2.66
articles *see also* **dangerous articles or substances**
 definition 2.11, 6.176–6.179
 foreseeability 6.180
 Health and Safety at
  Work Act 1974 6.176–6.182
 inspectors 2.117–2.130, 2.164
 manufacturers, duties of 6.176–6.179
 plant, meaning of 6.177
 possession and detain, power to
  take 2.128–2.130
 samples, power to take 2.117–2.120, 2.164
 self-employed persons, used by 6.178–6.179
asbestos
 Approved Code of Practice 12.193,
  12.202–12.204
 clothing, cleaning and provision of 12.210
 control limit 12.206
 demolition 12.199
 duty-holders in premises 12.194–12.198
 employees' duties 12.207
 employers' duties 12.199–12.210
 equipment 12.204–12.205, 12.208–12.209
 examinations 12.209
 fines 10.107–10.112
 guidance 12.193, 12.203–12.204
 Health and Safety Executive 12.197

hierarchy of measures 12.204
instructions 12.202
leases 12.197–12.198
maintenance 12.194, 12.198–12.199
monitoring 12.206
multiple occupation, premises in 12.197
personal protective equipment 12.204–12.205,
 12.208
plans 12.196, 12.201
reasonable practicability 12.200–12.201,
 12.203–12.206
records 12.195, 12.209
regulations 12.192–12.210, App L
repair 12.194, 12.198, 12.208–12.209
respiratory protective equipment 12.204–12.205
risk assessment 12.195–12.196, 12.199–12.200
risk management 12.196
statistics 12.192
tests of equipment 12.209
training 12.202
assistance, power to require 2.157–2.158

bad character evidence 8.214–8.226
 admissibility 8.217–8.226
 meaning 8.215
 notice 8.226
 previous convictions 8.221–8.223
 procedure 8.226
 propensity 8.218–8.223
bail 8.46–8.47
Bank of England 2.71
barristers
 Bar Code of Conduct 8.138, 8.145
 charges 8.138
 costs 10.148
 pleas 8.145
bereavement expenses 10.05
biological agents 12.122
breach of general duties 6.01–6.221 *see also*
 reasonable practicability, general duties and
 charges 8.139–8.142
 Crown Court 6.03
 danger, possibility of 6.61–6.65
 employees 6.03, 6.209–6.221
 employer and employee, meaning of 6.43–6.60
 employers' duties 6.01–6.42
 exposure of risks to safety 6.61–6.65
 failure to ensure safety 6.61–6.65
 fines 6.03, 10.57–10.58
 Health and Safety at Work Act 1974 6.01–6.221
 magistrates 6.03
 manufacturers 6.165–6.208
 non-employees 6.147–6.164

breach of general duties (*cont.*)
  premises to persons other than employees, duty of persons in control of 6.147–6.164
  protective intent 6.06–6.07
  reasonable practicability 6.09
  self-employed 6.04
  service of notices 6.07
  strict liability 6.06
breach of promise
  abuse of process 9.17–9.57
  legitimate expectations 9.46
  plea bargaining 9.49
  prejudice 9.46
  prosecutions 9.46–9.47
breach of statutory duty 11.49, 11.56
building regulations 4.31, 4.40
Buncefield oil farm explosion 1.21
bundles 8.87
burden of proof
  charges 8.141
  company directors, managers and officers, secondary liability of 5.54–5.55, 5.58–5.59
  danger, possibility of 6.64, 6.66
  electricity safety 12.226
  improvement or prohibition notices, appeals against 4.84–4.87, 4.91
  installations 6.197
  limitation of liability 6.169
  Management of Health and Safety at Work Regulations 1999 6.108
  manual handling 12.64
  reasonable practicability 6.135–6.146, 7.22
  reporting incidents and offences 2.24
  research 6.195, 6.208
  reverse 6.135–6.146
case management
  case progression officers, nomination of 8.148
  complex cases 8.150–8.153
  Criminal Procedure Rules 2005 8.120
  Crown Court 8.147–8.153
  Friskies statements 8.68
  improvement or prohibition notices, appeals against 4.72–4.74
  magistrates 8.44
  plea and case management hearings (PCMH) 8.149
case progression officers, nomination of 8.148
causation
  company directors, managers and officers, secondary liability of 5.63
  corporate manslaughter 11.86
  gross negligence manslaughter 11.31

  secondary liability 5.37
  trade descriptions 5.37
causes, reports into
  admissibility 3.78
  legal advice 3.77
  production, requests for 3.78
  railway incidents, inquiries into 3.80
  self-incrimination 3.79
cautions 3.39–3.41 *see also* **formal cautions; interviews under caution**
  compulsory powers, questioning under 2.86
  enforcement 1.81–1.87
  nature of offence, being informed of 3.41
  PACE and codes 2.78–2.80, 2.85–2.86, 3.40
  questioning 2.78–2.86
  wording 3.39
censure 1.44, 1.88
character evidence *see* **bad character evidence**
charges
  Bar Code of Conduct 8.138
  burden and standard of proof 8.141
  Code of Practice for Crown Prosecutors 8.134–8.135
  Criminal Procedure Rules 2005 8.133
  fines 8.142
  general duty, breach of 8.139–8.142
  indictments 8.133–8.143
  instituting proceedings 8.05–8.06
  overriding objectives 8.133
  practice direction 8.136–8.137
  reasonable practicability 8.143
  specific duties, breach of 8.141
charitable trusts
  bare trusts 5.18–5.19
  Charitable Incorporated Organisations (CIO) 5.24
  friendly societies 5.18, 5.25
  guarantee, companies limited by 5.24
  housing associations 5.26
  incorporated trusts 5.18
  industrial and provident societies 5.18, 5.26
  legal personality 5.18–5.19, 5.24–5.26
  mutual assurance associations 5.25
  trustees, personal liability of 5.24
  unincorporated associations 5.18
civil liability and proceedings
  disclosure 7.66
  gross negligence manslaughter 11.46
  Management of Health and Safety at Work Regulations 1999 12.11
  regulations 12.05

Index

clothing, provision and cleaning of 12.210
codes *see also* approved codes of practice
    (ACOPs); PACE and codes
  Bar, Code of Conduct for the 8.138
  Code for Crown Prosecutors 8.134–8.135, 11.24
  corporate manslaughter 11.24
  disclosure 8.159, 8.170, 8.173–8.175,
      8.181, 8.189
  investigations 2.49–2.52
committal *see* committal for trial; committal to
    Crown Court for sentence
committal for trial
  arrest 8.47
  bills of indictment 8.114
  companies 8.114
  disclosure 8.162, 8.164
  exhibits 8.116
  judicial review 8.117
  magistrates 8.47, 8.100–8.101, 8.117
  material sent to Crown Court on
      committal 8.116
  mode of trial 8.102, 8.114–8.117
  orders on committal 8.115
  Plea and Case Management Hearings
      (PCMH) 8.115, 8.149
committal to Crown Court for sentence
  adjournments 8.92, 8.101
  aggravating factors 8.91
  bundles 8.87
  changing the decision to
      commit 8.99
  fact, disputed basis of 8.98
  fines 8.84, 8.91, 8.96, 8.100
  Howe factors 8.91–8.92
  mitigation 8.86–8.87, 8.91
  Mode of Trial Guidelines 8.90
  Newton hearings 8.98
  plea before venue 8.82–8.89
  powers of Crown Court 8.100–8.101
  previous convictions 8.97
companies *see also* company directors, managers
    and officers, secondary liability of; corporate
    manslaughter; company officers; directors;
    liquidation, companies in
  accounts 10.70
  bills of indictment 8.114
  committal for trial 8.114
  conflicts of interest 3.65
  delay 9.23
  fines
    groups of companies 10.63–10.64, 10.66
    instructions, employees acting outside
        their 10.41–10.43

let down by others, being 10.46–10.53
  means of companies 10.27, 10.33, 10.63,
      10.110, 10.116, 10.125, 10.128–10.129
  systemic fault 10.36–10.40
  groups of companies 10.63–10.70
  guarantee, companies limited by 5.24
  guilty pleas 8.80
  informations 8.04
  instituting proceedings 8.03
  interviews under caution 3.06
  legal personality 5.03–5.07
  let down by others, being 10.46–10.53
  limited liability 10.68
  means of companies 10.27, 10.33, 10.63,
      10.110, 10.116, 10.125, 10.128–10.129
  mode of trial 8.113–8.114
  plea before venue 8.78–8.80
  representation 8.78–8.79
  service companies 6.53–6.54
  service of notice and documents 2.28, 8.34
  summonses 8.04, 8.34
  systemic fault 10.36–10.40
  veil, piercing the 10.67–10.69
company directors, managers and officers,
    secondary liability of 5.46–5.72
  authority, persons in positions of real 5.49–5.51
  Banking Act 1987, strict liability under 5.56
  basis of liability 5.52–5.53
  burden of proof 5.54–5.55, 5.58–5.59
  causation 5.63
  co-defendants 5.70
  connivance 5.60–5.62, 5.71
  consent 5.56–5.59
  criminal liability 5.46–5.70
  Fire Precautions Act 1971 5.49
  Health and Safety at Work Act 1974 5.46–5.72
  knowledge and belief 5.53, 5.70–5.72
  manager, meaning of 5.50
  neglect 5.63–5.72
  proof of the offence 5.54–5.55
  trade descriptions 5.47
  wilful neglect 5.71
company officers
  answers, power to require 2.146–2.148
  compulsory powers 2.146–2.148
  conflicts of interest 3.65–3.66
  personal liability 3.66
compensation orders
  bereavement expenses 10.05
  employers' compulsory insurance 10.07
  funeral expenses 10.05
  personal injuries 10.05–10.06
  sentencing 10.05–10.08

Index

complex cases
  admissibility of evidence 8.151
  case management 8.150–8.153
  delay 9.25
  fraud and other complex cases, protocol for control and management of 8.152
  plea bargaining 8.154
  preparatory hearings 8.150–8.151
compulsory powers
  admissibility 2.142
  answers, power to require 2.132, 2.134, 2.142–2.143, 2.146–2.148
  cautions 2.86
  company officers 2.146–2.148
  criminal offences 7.51, 7.76
  directors 2.146–2.148
  Environmental Protection Act 1990 2.148
  false statements 7.76
  investigations 2.56
  production 3.88–3.90
  self-incrimination, privilege against 3.87–3.90
  silence, right to 3.88
computers, information held on 2.70, 2.162
conditional sale agreements 6.175
confidentiality 3.60
conflicts of interest
  advising clients 3.52
  companies 3.65
  confidentiality 3.60
  directors, advising 3.65–3.66
  disclosure 3.60, 3.62
  inspectors 3.59–3.66
  Law Society Rules and Ethics Committee Guidance 3.59–3.63
  more than one employee, representing 3.64
  public interest 3.62
  representation 3.59–3.66
  senior company officers 3.65–3.66
connivance 5.60–5.62, 5.71
Construction Design Management Coordinator 12.166–12.174
  accountability 12.169
  advice 12.169, 12.171, 12.174
  appointment 12.166–12.167, 12.171
  Approved Code of Practice 12.170–12.171, 12.174
  clients
    duties towards 12.169, 12.171
    role of 12.158, 12.166–12.168
  competence 12.154, 12.170, 12.174
  Construction (Design and Management) Regulations 2007 12.150
  construction phase plans 12.173, 12.186
  contractors 12.172, 12.174, 12.189
  creation of role of 12.150
  designers 12.174, 12.177
  designs 12.172, 12.174
  documentation 12.173
  duties 12.169–12.171
  enforcement proceedings 12.170
  files 12.173
  guidance 12.171, 12.174
  Health and Safety Executive 12.170
  information 12.166, 12.172
  notifiable projects 12.171
  planning and preparation 12.172
  planning supervisor, replacement of 12.150, 12.169
  surveys 12.171
  welfare facilities 12.173
construction industry *see also* **Construction Design Management Coordinator; contractors**
  accountability 12.158
  appointments 12.167, 12.174
  Approved Code of Practice 12.146–12.148, 12.154–12.160, 12.163–12.171, 12.174
  building regulations 4.31, 4.40
  checks 12.164
  client, role of 12.150, 12.158–12.165
  competence 12.154–12.156, 12.158, 12.168, 12.170, 12.174
  Construction (Design and Management) Regulations 2007 12.148–12.186, App J
  construction phase plans 12.166
  cooperation 12.157–12.158, 12.160–12.161
  coordination 12.150, 12.157–12.158, 12.160
  design 12.148–12.186
  designers 12.160–12.161, 12.175–12.180
  documentation 12.173
  EC law 12.147, 12.150
  enforcement action 12.158
  general provisions 12.153
  guidance 12.146, 12.155, 12.157–12.160, 12.163–12.168, 12.171, 12.174
  Health and Safety Commission 12.148
  Health and Safety Executive 12.148–12.149, 12.170
  height, work at 12.140–12.145
  high-risk projects 12.157
  information 12.162, 12.166, 12.172
  management 12.148–12.186
  notifiable projects 12.166–12.168, 12.171
  planning supervisors 12.147, 12.169
  plans 12.166
  prevention, principle of 12.153

Index

construction industry *see also* **Construction Design Management Coordinator; contractors** (*cont.*)
  project documentation 12.173
  reasonable practicability 12.159
  regulations 12.139–12.191
  standards 12.168
  statistics 12.139
  surveys 12.171
  website 12.148
  workplace 12.56
Consumer Protection Act 1987 5.40, 6.165
contractors 12.180–12.191
  appointments 12.162, 12.166–12.167, 12.174, 12.180
  Approved Code of Practice 12.181, 12.183–12.186
  competence 12.154, 12.161, 12.184–12.185
  Construction Design and Management Coordinator 12.172, 12.174, 12.189
  Construction (Design and Management) Regulations 2007 12.187
  construction phase plans 12.182, 12.186, 12.189–12.190
  design 12.182
  duties 12.180–12.191
  guidance 12.181, 12.183–12.186
  inductions 12.183, 12.188
  information 12.162, 12.180, 12.186, 12.188
  notifiable projects 12.180, 12.187, 12.189
  planning and preparation time, minimum 12.188
  plans 12.180
  prevention, principle of 12.187
  principal contractors 12.180–12.186, 12.189–12.190
  project documentation 12.180
  reasonable practicability 12.180, 12.186–12.187
  RIDDOR 12.190
  risk assessments 12.188–12.189
  site rules 12.182, 12.186
  standards 12.160
  time for planning and preparation, minimum 12.182
  training 12.183, 12.185, 12.188
  welfare facilities 12.182
contracts
  contracts for services 6.43, 6.45
  contracts of service 6.45, 6.49
  employees, meaning of 6.43–6.49, 6.53
  employment contracts 6.43–6.49, 6.53
control of substances hazardous to health
  adequate control, meaning of 12.131

  Approved Code of Practice 12.121
  biological agents 12.122
  control measures, types of 12.130
  definition 12.122
  dust 12.122
  duties 12.123–12.138
  emergency procedure 12.137
  employees' duties 12.133
  exemptions 12.138
  foreseeability 6.117–6.126
  guidance 12.121
  health surveillance 12.135
  hierarchical approach 12.129
  information 12.136
  instructions 12.136
  maintenance 12.134
  maximum exposure limits (MELs) 12.120
  monitoring of exposure 12.134
  occupational exposure standards (OESs) 12.120
  prevent or control exposure, duty to 12.129–12.132
  reasonable practicability 7.25–7.29, 12.123, 12.126, 12.129, 12.131
  regulations 12.120–12.138, App I
  reviews 12.128
  risk assessment 12.18, 12.125–12.128, 12.134
  self-employed 12.123
  specific factors 12.127
  specific measures 12.132
  toxic, corrosive, harmful or irritant substances 12.122
  training 12.136
  website 12.131
  workplace exposure limits (WELs) 12.120, 12.122, 12.131
convictions *see* **previous convictions**
cooperation with inspectors
  advising clients 3.68–3.73
  Health and Safety at Work Act 1974 section 20 powers 3.68–3.73
  information, requests for 3.68–3.72
  mitigation in sentencing 3.69
  production of documents 3.72–3.73
  reasonable excuse, defence of 3.71
CORGI installers 12.215
coroners' inquests 11.89–11.152
  accidental death verdicts 11.143
  adjournments 11.145–11.152
  appeals 11.97
  appointment of coroners 11.93
  bereaved people, involvement of 11.97, 11.109
  Chief Coroner 11.97
  compulsory statements 11.140–11.141

Index

coroners' inquests (*cont.*)
conduct can be called into question, notice to persons whose 11.122
Coronial Advisory Council 11.97
criminal or civil liability, determination of 11.118–11.121, 11.144
criminal proceedings 11.134–11.135, 11.145–11.152
current system 11.99–11.152
deaths in custody 11.95, 11.108
disclosure 11.129–11.139
documentary evidence 11.92, 11.129–11.132, 11.140
draft Bill 11.91–11.98
entry, powers of 11.94
European Convention on Human Rights 11.108–11.116, 11.121, 11.132
evidence 11.125–11.126, 11.129–11.132, 11.140
exhibits 11.136
fact, findings of 11.121
funding 11.92
gross negligence manslaughter 11.144
Health and Safety at Work Act 1974 11.139–11.141
Health and Safety Executive 11.136–11.137
inquisitorial processes 11.107
inspectors' reports 11.134
interested parties 11.107, 11.109, 11.124, 11.129
investigations 11.03, 11.125
  adjournments 11.146
  funding 11.92
  Health and Safety Executive 11.136
  independent, prompt and proper 11.109–11.111
  life, right to 11.109–11.111
  police 11.126
  powers 11.94
juries 11.95, 11.98, 11.103–11.105
legal privilege 11.134
life, right to 11.108–11.116, 11.121, 11.132
local authorities' financial responsibility for coroners 11.92
Luce Review 11.90
manslaughter 11.136, 11.144, 11.147–11.149
matters to be ascertained at inquests 11.117–11.122
Middleton-type requests 11.108, 11.116
police
  deaths caused by 11.95
  investigations, material relating to 11.126
  reports 11.134
post-mortem reports 11.129
public interest 11.137
public interest immunity 11.134
qualifications of coroners 11.93
reform 11.02, 11.89, 11.91–11.99
reports 11.134–11.135
RIDDOR (Reporting of Injuries, Diseases and Dangerous Occurrences Regulations 1995) 11.104
scope of inquests 11.106–11.116
search and seizure, powers of 11.94
Select Committee report 11.90–11.91
self-incrimination, privilege against 11.127–11.128
Senior Coroners' appointment of 11.93
Shipman Inquiry 11.90
time limits 8.15
unlawful killing 11.144
unnatural deaths 11.102
verdicts 11.142–11.144
  accidental death 11.143
  criminal or civil liability, determination of 11.118–11.121, 11.144
  expanded form of 11.114
  narrative form 11.114–11.115
  unlawful killing 11.144
violent deaths 11.101
witnesses
  advance notice of statements 11.131
  copies of statements 11.131, 11.136
  examination 11.123–11.124
  statements 11.92, 11.125–11.126, 11.131–11.132, 11.136, 11.140–11.141
Work Related Death Protocol 11.18, 11.133, 11.148
work-related deaths 11.95, 11.98, 11.133–11.139, 11.145–11.152
corporate manslaughter 11.66–11.88
causation 11.86
Code for Crown Prosecutors 11.24
common law 11.66–11.70
conflicts of interest 3.65
Corporate Homicide and Manslaughter Act 2007 11.66, 11.71
  explanatory notes 11.82
  commencement 11.71
  provisions of 11.75–11.88
  Royal Assent 11.71
Corporate Homicide and Manslaughter Bill 11.01, 11.21–11.23, 11.69–11.71
Corporate Manslaughter Bill 2.65, 11.23
Crown 11.78
Crown immunity 11.69

Index

Crown Prosecution Service 11.24, 11.73
  deaths in custody 11.82–11.83
  directing mind 11.66, 11.68
  Director of Public Prosecutions 11.24, 11.79
  directors 3.65, 5.07, 11.68, 11.77
  disasters 11.67, 11.69
  duty of care 11.81–11.84
  fines 11.78
  government bodies 11.75, 11.80
  gross breach 11.87–11.88
  gross negligence manslaughter 11.23,
      11.66–11.68, 11.77, 11.84
  identification doctrine 11.85
  impact of new offence 11.72–11.74
  investigations 11.74
  judges, existence of duty of care and 11.84
  management failure 11.86–11.87
  police 2.65, 11.73, 11.75, 11.78
  prosecution decisions 11.24
  public authorities 11.76
  publicity orders, convictions and 11.76
  rail disasters 11.67
  Regulatory Impact Assessment 11.72–11.74
  remedial orders on conviction 11.76
  secondary liability 11.78
  senior management 11.85
  Work Related Death Protocol 11.24, 11.72
corporations sole 5.04
costs 10.130–10.155
  advance information 8.53
  appeals 10.140, 10.150–10.151
  apportionment 10.152–10.155
  central funds 10.138, 10.141
  costs orders 4.123, 4.125
  Court of Appeal 10.140, 10.151
  counsel, instructing 10.148
  Crown Court 10.131, 10.140
  defence costs 10.137–10.147
  defendant's costs orders 10.139, 10.141, 10.148
  definition 4.124
  detailed assessments 4.126
  determination of amount 4.125
  discontinuance 10.138
  estimates 10.154
  European Convention on Human
      Rights 10.138
  examples 10.148–10.149
  expenses 10.139, 10.142
  fair trials 10.138
  fines 10.28, 10.155
  guilty pleas 10.153
  hourly rate 10.149

  improvement or prohibition notices, appeals
      against 3.123–4.125
  magistrates 10.131, 10.140, 10.144
  presumption of innocence 10.138
  principles 10.136
  prosecution costs 10.131–10.136
  public funding 10.137
  reasonableness 10.142, 10.146
  taxation of costs 10.141–10.145, 10.150
  time limits 10.150–10.151
court order, failure to comply with 7.90
CPS *see* **Crown Prosecution Service**
credit sale agreements 6.175
creditors' voluntary liquidation 8.24
criminal investigations 2.47–2.86 *see also* **PACE
    and codes**
  arrest 2.57–2.68
  codes of practice 2.49–2.52
  compulsory powers of questioning 2.56
  Crown Prosecution Service 2.65
  definition 2.48, 2.52
  disclosure 2.47, 2.49, 2.51
  disclosure officers, functions of 2.51
  gross negligence manslaughter 2.57
  improvement notices 2.56
  inquiry, pursuing all reasonable lines of 2.52
  inspectors, powers of 2.55–2.57
  institution of proceedings 2.54
  investigating officers, functions of 2.51
  police 2.49
  powers 2.53–2.58
  prohibition notices 2.56
  search and entry, powers of 2.57, 2.69–2.71
  seizure 2.56
criminal offences and proceedings *see also* **breach of
    general duties; criminal investigations;
    investigations of incidents and offences;
    reporting incidents and offences**
  approved codes of practice 1.11
  assistance, requirement to provide 7.51
  attending, preventing a person from 7.52–7.53
  breach of regulations 7.07–7.39
  burden of proof, reverse 6.136–6.137
  company directors, managers and officers,
      secondary liability of 5.46–5.70
  compulsory questioning 7.51, 7.76
  contravention, meaning of 7.10, 7.45
  coroners' inquests 11.118–11.121, 11.134–
      11.135, 11.144–11.152
  court orders, failure to comply with 7.89–7.90
  Crown Court 7.03, 7.09
  danger, powers to deal with immediate 7.50

675

Index

criminal investigations; investigations of incidents and offences; reporting incidents and offences (cont.)
deceive, using a document with intent to 7.83–7.87
direction, meaning of 7.47
directives 7.14
disclosure 7.61–7.69
electricity safety 12.226
employees' duties 7.05
examinations 7.48
false entries, making 7.80–7.82
false information, providing 7.74–7.79
fines 7.03, 7.09, 7.44, 7.52, 7.90
food safety 7.51
foreseeability 6.127–6.134
gas safety 12.214
Health and Safety at Work Act 1974 7.01–7.90
Health and Safety Commission 1.22–1.23, 7.59–7.60
impersonation 7.88
improvement notices 4.11, 4.13, 7.54–7.55
information 7.49, 7.51, 7.53, 7.59–7.69, 7.74
inquiries 7.40–7.43
inspection 2.152, 12.97
inspectors 2.89, 2.151
  assistance, requirement to provide 7.51
  contravening requirements by 7.44–7.51
  danger, powers to deal with immediate 7.50
  food safety 7.51
  impersonation 7.88
  information, requirement to give 7.49, 7.51, 7.53
  meaning of requirement 7.47
  obstruction 7.56–7.58
  powers 7.44–7.51
interpretation 7.13–7.14, 7.39
judges 7.13
juries 7.13
legal personality 5.01, 5.06
list of offences 7.02
magistrates 7.03, 7.09, 7.44
maintenance 12.89
Management of Health and Safety at Work Regulations 1999 6.102, 6.105, 6.110, 6.112
manufacturers, duties of 6.167, 6.181, 6.201
non-employees 6.25–6.32
obstruction 7.56–7.58
partnerships 5.11–5.15
production 2.152
prohibition notices, breach of 4.11, 4.13, 7.54–7.55

questions
  answer questions, requirement to 7.74
  preventing a person from answering 7.52–7.53
reasonable excuse, defence of 7.51
reasonable practicability 7.18, 7.39
regulations 12.02, 12.05
requirement, meaning of 7.47
research 6.208
secondary liability 5.27–5.45
self-incrimination 7.75
statistics 7.08
strict liability 7.15–7.17
unincorporated associations 5.10
Criminal Procedure Rules 2005 8.118–8.120
case management 8.120
charges 8.133
Crown Court 8.118–8.120
directions 8.120
disclosure 8.161
duplicity, rule against 8.123
expert evidence 8.190–8.204
flexibility and informality 8.120
instituting proceedings 8.02
magistrates 8.26, 8.44
overriding objectives 8.118–8.120
service of summonses 8.33
Crown
corporate manslaughter 11.78
Crown Censure 1.44, 1.88
Crown proceedings 1.43
employees 1.42, 6.56
enforcement 1.41–1.44, 1.88
Health and Safety Commission 1.13
Health and Safety Executive 1.13
immunity 11.69
improvement or prohibition notices 1.41, 1.44, 1.88
Monarch in private capacity 1.43
secondary liability 5.45
Crown Court *see also* **committal to Crown Court for sentence**
breach of general duties 6.03
case management 8.147–8.153
conducting a case in Crown Court 8.118–8.156
costs 10.131, 10.140
criminal offences 7.03, 7.09
Criminal Procedure Rules 2005 8.118–8.120
disclosure 8.161
expert evidence 8.202
fines 8.100, 10.09–10.11, 10.71–10.86, 10.123
Friskies statements 8.43
guilty pleas 8.74–8.75

676

Index

instituting proceedings 8.01–8.02
lodging the indictment 8.121–8.144
plea bargaining 8.154–8.156
plea before venue 8.69–8.81
plea negotiations 8.144–8.147
sentencing discounts 8.74–8.75
time limits 8.14
Crown Prosecution Service 1.39–1.40
  arrest 8.04
  Code for Crown Prosecutors 8.134–8.135, 11.24, 11.73
  corporate manslaughter 11.24, 11.73
  enforcement 1.39–1.40
  institution of proceedings 2.54
  investigations 2.54, 2.65
  liaison, protocol for 1.40
  manslaughter 1.39, 11.24
  pleas 8.144
  Work Related Death Protocol 11.19
custody, deaths in 11.82–11.83, 11.95, 11.108

damages 2.100
danger, possibility of
  breach of general duties 6.61–6.65
  burden of proof 6.64, 6.66
  dangerous occurrences, meaning of 2.18
  employees' duties 6.214
  Health and Safety at Work Act 1974 6.61–6.65
  non-employees 6.61–6.65
  preventive aims 6.63
  reporting incidents and offences 2.10–2.24
  risk, meaning of 6.62
danger, powers to deal with immediate
  criminal offences 7.50
  evacuation procedure 12.45
  Health and Safety at Work Act 1974 2.163–2.166
  inspectors 2.163–2.165, 7.50
  seizure 2.163, 2.165
dangerous articles or substances
  articles, definition of 2.121
  damage or destruction 2.124–2.125
  detention 2.128–2.130
  dismantle or subject to testing, power to 2.121–2.127
  examination 2.128
  Health and Safety at Work Act 1974 2.121–2.127
  immediate danger, powers to deal with an 2.163–2.164
  national security 2.126

notice 2.130
parts of machinery 12.112–12.115
possession and detain, power to take 2.128–2.130
samples 2.164
seizure 2.163, 2.165
substance, definition of 2.122
testing 2.121–2.127
dangerous parts of machinery
  accidental contact 12.113
  definition of dangerous parts 12.115
  foreseeability 12.115
  guards 12.114
  hierarchy of measures 12.114
  regulations 12.112–12.115
  risk assessment 12.114
deaths and serious personal injury *see also* **corporate manslaughter; gross negligence manslaughter; manslaughter; Work Related Death Protocol; work-related deaths**
  aggravating factors 8.91, 10.29
  Crown Prosecution Service 1.40
  custody, deaths in 11.82–11.83, 11.95, 11.108
  electricity safety 12.225
  fines 10.23, 10.29–10.32, 10.38–10.40, 10.72–10.86
  illegal immigrants 10.100, 11.38–11.44
  liaison, protocol for 1.40
  reporting incidents and offences 2.10–2.11, 2.13
  representation 3.67
deception
  calculated to deceive, meaning of 7.87
  criminal offences 7.83–7.87
  fines 7.84
  Health and Safety at Work Act 1974 7.83–7.87
  intention 7.86
declarations of truth 2.135–2.136, 7.78–7.79
delay
  abuse of process 9.23–9.33
  companies 9.23
  complex cases 9.25
  European Convention on Human Rights 9.27–9.33
  fair trials 9.27–9.33
  Human Rights Act 1998 9.30–9.31
  prejudice 9.25–9.27
  stay 9.24–9.25, 9.31
demolition 12.199
Department of Trade and Industry
  inspectors 2.75
designers *see also* **designs**
  Approved Code of Practice 12.178–12.179

designers *see also* **designs** (*cont.*)
  Construction Design and Management
      Coordinator 12.174, 12.177
  Construction (Design and Management)
      Regulations 2007 12.175
  construction industry 12.160–12.161,
      12.175–12.180
  duties 12.176
  files 12.177
  guidance 12.178–12.179
  information 12.176–12.178
  manual handling 12.61
  meaning 6.174
  notifiable projects 12.177
  prevention, principle of 12.175
  reasonable practicability 12.176
  regulations 12.175–12.179
designs *see also* **designers**
  construction industry 12.148–12.186
  contractors 12.182
  equipment 12.87, 12.118
  Health and Safety at Work
      Act 1974 6.173
  meaning 6.174
  research 6.193
detention
  Construction Design Management
      Coordinator 12.172, 12.174
  dangerous articles or substances 2.128–2.130
  interviews under caution 3.32–3.33
  PACE and codes 3.32–3.33
  silence, adverse inferences from 3.16
Director of Public Prosecutions 1.38–1.39, 8.07,
      11.24, 11.79
directors *see also* **company directors, managers and
      officers, secondary liability of**
  answers, power to require 2.146–2.148
  Companies Act 2006 5.73
  compulsory powers 2.146–2.148
  conflicts of interest 3.65–2.66
  corporate manslaughter 5.07, 11.68, 11.77
  Corporate Manslaughter and Homicide Bill 5.07
  delegation 11.63–11.65
  Directors' Duties for Health and Safety, revision
      of 5.82
  disqualification orders 10.88, 10.102–10.105
  duties 5.73–5.82
  employers' duties 11.63–11.65
  fines 10.113–10.114
  gross negligence manslaughter 11.63–11.65,
      11.77
  guidance 5.75–5.76
  inspectors 3.21–3.25

  interviews under caution 3.06, 3.21–3.25
  legal representation 3.06
  letters, interviews by 3.21–3.25
  manslaughter 3.65, 5.07, 11.63–11.65,
      11.68, 11.77
  personal liability 3.66
  prosecutions 1.95–1.99
  reform 5.76–5.82
  success of company, duty to promote the 5.73
directors' disqualification orders 10.88,
      10.102–10.105
  duration 10.105
  fines 10.105
  imprisonment 10.105
  management of company, in connection
      with 10.102–10.103
disasters *see also* **railway accidents**
  corporate manslaughter 11.67, 11.69
  fines 10.02, 10.29–10.32, 10.38–10.40,
      10.73–10.78
  Morecambe Bay disaster 10.101
disclosure 8.157–8.189
  adverse inferences from silence 3.11
  Attorney-General's guidelines 8.158,
      8.160, 8.162
  civil proceedings 7.66
  code of practice 8.159, 8.170, 8.173–8.175,
      8.181, 8.189
  committal for trial, before 8.162, 8.164
  conflicts of interest 3.60, 3.62
  consent 7.63–7.65
  coroners' inquests 11.129–11.133
  criminal offences 7.61–7.69, 8.161
  Criminal Procedure Rules 2005 8.161
  Crown Court, protocol for 8.161
  defence 8.166–8.168
  disclosure officers 2.51, 8.174–8.175, 8.183
  durable or retrievable form, material in 8.180
  expert evidence 8.201, 8.207
  fines 7.61
  fishing expeditions 8.181
  Health and Safety at Work Act 1974 7.61–7.69
  Health and Safety Commission 7.62
  Health and Safety Executive 7.61
  inquiries, failure to follow
      reasonable line of 8.177–8.178
  inspection 8.183
  inspectors 8.173–8.175
  investigation 2.47, 2.49, 2.51, 8.170–8.180
  legal privilege 8.188
  legislative framework 8.159–8.161
  magistrates 8.162
  PACE and codes 3.11–3.12

police 8.170
primary disclosure 8.163
procedure 8.163–8.165
process material 8.187
prosecutors 8.184–8.185
public interest immunity 8.188
retention 8.179–8.180
schedule of material 8.165, 8.187
secondary disclosure 8.163
third party disclosure 8.169, 8.181–8.182
time limits 8.166
unused material 8.157, 8.163, 8.186–8.188
discontinuance, costs and 10.138
documentary evidence 11.92, 11.129–11.132, 11.140
domestic servants 6.57
double jeopardy
abuse of process 9.34–9.35
special circumstances 9.40
stay 9.37
Work Related Death Protocol 9.36
work-related deaths 9.36, 11.07
due diligence
electricity safety 12.226
reasonable practicability 7.19
secondary liability 5.35
trade descriptions 5.35
duplicity, rule against 8.123–8.130
amendments of information 8.132
continuing offences 8.124, 8.128
convictions, overturning 8.131
Criminal Procedure Rules 2005 8.123
cure 8.132
effect of proceeding on duplicitous charge or count 8.131–8.132
form, as matter of 8.125–8.126
indictments 8.123–8.130
informations 8.123–8.130
quasi-duplicity 8.127
summons, separate informations in one 8.125
two or more offences, charging 8.124
dust 12.122
duty of care *see also* **breach of general duties**
asbestos 12.194–12.198
corporate manslaughter 11.81–11.84
directors 5.73–5.82
gross negligence manslaughter 11.27, 11.32–11.37, 11.62
judges 11.84
specific duties, breach of 8.141
workplace 12.49–12.52, 12.57–12.59

EC law *see also* **Framework Directive (89/391)**
construction industry 12.147, 12.150
criminal offences 7.14
directives 7.14, 12.01–12.02
enforcement 1.45
equipment 12.77
manual handling 12.60
regulations 1.08–1.09, 12.01–12.02
Single European Act 1.47
supremacy of EC law 1.46
Third Community Action Programme 1.47
workplace 12.48
economic loss 4.51–4.52
electricity safety
burden of proof 12.226
construction, meaning of 12.230
criminal offences 12.226
death or personal injury 12.225
defence 12.226, 12.238
due diligence 12.226
equipment 12.228, 12.233
general duties 12.236
inspection 12.232
live conductors 12.237–12.239
maintenance 12.228, 12.231–12.232
precautions 12.225, 12.236
protective equipment 12.233
railway tracks 12.239
reasonable practicability 12.226, 12.228, 12.233, 12.235
records 12.232
regulations 12.225–12.239, App N
self-employed 12.225
systems, definition of 12.229
use 12.228
electronic records 2.70, 2.162
emergencies
control of substances hazardous to health 12.137
evacuation procedures 12.45
powers to deal with 2.163–2.166
employees *see* **employees' duties; employees, meaning of employees' duties**
asbestos 12.207
at work, definition of 6.212
breach of general duties 6.03, 6.209–6.221
carelessness 6.217
control of substances hazardous to health 12.133
cooperate, duty to 6.216
course of employment 6.212
criminal offences 7.05

Index

employees' duties (*cont.*)
  fines 6.03, 6.210
  Health and Safety at Work
    Act 1974 6.209–6.221
  horseplay 6.219
  obvious dangers 6.214
  prosecutions 1.92–1.94, 6.217–6.221
  public interest 6.220
  reasonable care, meaning of 6.213–6.215
  safe systems of work 6.218
  supervision 6.218
  training 6.218
  violence 6.219
  warnings 6.220
employees, meaning of
  agency workers 6.52–6.53
  breach of general duties 6.43–6.60
  contracts for services 6.43, 6.45
  contracts of service 6.45, 6.49
  control test 6.49–6.51
  Crown servants 6.56
  domestic servants 6.57
  economic reality test 6.50
  employment contracts 6.43–6.49, 6.53
  independent contractors 6.43–6.49
  Health and Safety at Work Act 1974 6.43–6.52, 6.55–6.57
  multiple test 6.50
  mutuality of obligation test 6.50
  organizational test 6.50
  police officers 6.56
  proof 6.58–6.59
  prosecutions 6.43–6.45
  tax or national insurance, liability for 6.48
  work experience 6.55
employers' compulsory insurance 10.07
employers' duties
  asbestos 12.199–12.210
  continuing offences 6.14
  directors 11.63–11.65
  elements of offence 6.14–6.15
  failure to discharge duty 6.14–6.17
  Health and Safety at Work Act 1974 6.01–6.42
  non-employees, duty to 6.24–6.32
  policies 6.16
  reasonable practicability 6.11–6.12
  risk assessments 6.16, 12.12–12.35
  safety representatives and committees, appointment of 6.17
  undertakings, meaning of 6.33–6.42
  work and at work, meaning of 6.18–6.23
  written statements of policy 6.16

employers, meaning of
  breach of general duties 6.01–6.60
  Health and Safety at Work
    Act 1974 6.53–6.54, 6.60
  proof 6.60
  service companies 6.53–6.54
employment contracts 6.43–6.49, 6.53
Employment Medical Advisory Service 1.29
employment tribunals
  constitution, text of regulations on App D
  Employment Appeal Tribunal 4.126
  improvement or prohibition notices, appeals against 4.126
  rules, text of App D
enforcement framework 1.01–1.105 *see also* prosecution
  adequate arrangements for, duty to make 2.36–2.40
  approved codes of practice (ACOPs) 1.10–1.11
  cautions 1.81–1.87
  censure 1.88
  Construction Design Management Coordinator 12.170
  construction industry 12.158
  Crown 1.41–1.44, 1.88
  Crown Prosecution Service 1.39–1.40
  Director of Public Prosecutions 1.38–1.39
  discretion 1.63
  early intervention 1.76
  EC law 1.45
  Enforcement Concordat, text of App B
  Enforcement Policy Statement 1.75, 4.16, 4.18
  enforcing authority regulations, text of App C
  expert evidence 8.196–8.198
  formal cautions 1.81–1.87
  Framework Directive 1.45–1.61
  gas safety 12.213
  guidance 1.12
  Health and Safety at Work Act 1974 1.01–1.06
  Health and Safety Commission 1.13–1.14, 1.16–1.23
  Health and Safety Executive 1.13–1.14, 1.24–1.31, 1.36–1.37, 1.40, 8.196–8.198
  improvement notices 1.79, 4.18–4.22, 4.24, 4.33
  Initial Enforcement Expectation (IEE) 1.74
  inspectors 1.63
  institutions 1.13–1.44
  investigations 2.32–2.91
  letters of advice/recommendation 1.78
  licensing regimes 1.80
  local authorities 1.32–1.37
  methods of enforcement 1.75–1.88

Index

notices, breach of enforcement 10.87
Office of the Rail Regulator
    (ORR) 1.15, 1.40
permissioning and licensing
    regimes 1.80
principles of enforcement action 1.62–1.74,
    2.43–2.45, 3.28–3.29
prohibition notices 1.79, 4.18–4.22,
    4.24, 4.37
prosecutions 9.50–9.57
regulations 1.07–1.09
risk gap, concept of 4.20–4.22
verbal warnings 1.77
Work Related Death Protocol App B
engineering works 12.56
entry, inspector's powers of *see* **search and entry, powers of environment**
answers, power to require 2.142–2.143,
    2.145, 2.147
compulsory powers 2.148
criminal investigations 2.54
Environment Agency 2.54, 8.07
Environmental Health Officers (EHO) 1.34
Environmental Protection Act 1990 2.142–2.143,
    2.145, 2.147–2.148, 3.87
pollution offences 5.31
secondary liability 5.31
self-incrimination, privilege against 3.87
equipment *see* **work equipment**
ergonomics 12.87
European Convention on Human Rights
burden of proof, reverse 6.138, 6.142
coroners' inquests 11.108–11.116,
    11.121, 11.132
costs 10.138
delay 9.27–9.33
fair trials 3.86–3.88, 6.138, 6.142,
    9.09, 10.138
gross negligence manslaughter 11.26
judicial review 9.09
life, right to 2.35, 11.108–11.116,
    11.121, 11.136
self-incrimination, privilege
    against 3.86–3.88
silence, right to 3.87
strict liability 7.16–7.17
evacuation procedure 12.45
evidence *see also* **bad character evidence; expert evidence; hearsay**
admissibility 2.96–2.99
approved codes of practice 1.11
breach of legislation, failure to comply as 1.11
coroners' inquests 11.92, 11.129–11.132,
    11.140

documentary evidence 11.92, 11.129–11.132,
    11.140
excluded material 2.70
exclusion of 2.96–2.99
fairness 2.98–2.99
Health and Safety at Work Act 1974 2.96–2.99
improvement or prohibition notices, appeals
    against 4.80
inspection 2.96–2.98, 12.101
PACE and codes 2.97
special procedure material 2.70
ex turpi causa rule 11.39–11.41
examinations *see also* **inspection**
asbestos 12.209
criminal offences 7.48
dangerous articles or substances 2.128
inspectors 2.109–2.114
limitation of liability 6.172
manufacturers, duties of 6.183, 6.203
research 6.193
substances 6.198
exhibits 11.136
expert evidence 8.190–8.207
appointments 4.81
back to back, calling witnesses 8.206
calling the expert 8.205–8.206
conflicts between evidence 8.192
Criminal Procedure Rules 2005 8.19–8.204
Crown Court 8.202
disclosure 8.201, 8.207
HSE Enforcement Guide 8.196–8.198
improvement or prohibition notices, appeals
    against 4.81
independence 8.195–8.204
inspectors 8.196–8.197
instructing experts 8.207
investigators 8.198
overriding duty 8.195
pre-trial discussions 8.200
purpose 8.190
reports 8.199, 8.205
service of written statements 8.204
specialist inspectors 8.196–8.198
ultimate issue rule 8.191, 8.193–8.194

facilities, power to require 2.157–2.158
Factories Acts 1.01
fair trials
costs 10.138
delay 9.27–9.33
judicial review 9.09
magistrates 8.39
self-incrimination, privilege against 3.86–3.88

Index

fairground equipment
  installations 6.196
  manufacturers, duties of 6.188, 6.190
  research 6.193
false entries
  criminal offences 7.80–7.82
  fines 7.81
  Health and Safety at Work Act 1974 7.80–7.82
false statements 7.67–7.73
  compulsory questioning 7.76
  consents, obtaining 7.78
  costs 10.155
  criminal offences 7.74–7.79
  declarations of truth 7.74
  documents, for purpose of obtaining issue of 7.78–7.79
  fines 7.68
  Health and Safety at Work Act 1974 7.67–7.73, 7.74–7.79
  inspectors 2.89
  licences, obtaining 7.78
  knowledge or recklessness 7.69–7.71
  making a statement, meaning of 7.72
  RIDDOR 7.73
  self-incrimination 2.138, 7.75
  statutory provisions, requirement to furnish information under 7.73
fatal accidents *see* **deaths or serious personal injury**
Field Operations Directorate (FOD) 1.28–1.29
financial circumstances orders 10.14
Financial Services Authority Mutual Societies Section, registration with 5.26
fines
  accounts 10.12
  aggravating factors 10.23, 10.27, 10.29
  asbestos 10.107–10.112
  charges 8.142
  committal to Crown Court for sentence 8.84, 8.91, 8.96, 8.100
  commonly encountered situations 10.41–10.56
  companies
    directors 10.113–10.114
    groups of 10.63–10.64, 10.66
    instructions, employees acting outside their 10.41–10.43
    let down by others, being 10.46–10.53
    means of companies 10.27, 10.33, 10.63, 10.110, 10.116, 10.125, 10.128–10.129
    systemic fault 10.36–10.40
  consequences of breach 10.29–10.40
  corporate manslaughter 11.78

  costs 10.28
  court order, failure to comply with 7.90
  criminal offences 7.03, 7.09, 7.44, 7.52, 7.90
  Crown Court 8.100, 10.09–10.11, 10.71–10.86, 10.123
  death or serious injury 10.23, 10.29–10.32, 10.38–10.40, 10.72–10.86
  deception 7.84
  directors 10.105, 10.113–10.114
  disasters 10.02, 10.29–10.32, 10.38–10.40, 10.73–10.78
  disclosure 7.61
  employees 6.03, 6.210
  established principles 10.17–10.40
  examples of fines 10.71–10.86
  false entries 7.81
  false statements 7.68
  financial circumstances orders 10.14
  financial information 10.12–10.14
  foreseeability 10.43–10.44
  gas explosions 10.72, 10.75
  general duties and regulations, breach of 10.57–10.58
  good safety record, relevance of 10.54–10.56
  groups of companies 10.63–10.64, 10.66
  guilty pleas, discounts for 8.96, 10.40
  Health and Safety at Work Act 1974, breach of 10.121
  Howe principles 8.91, 10.17–10.29, 10.33, 10.122
  impersonation of an inspector 7.87
  improvement notices 4.13, 7.54
  information 7.60
  instructions, employees acting outside their 10.41–10.43
  let down by others, being 10.46–10.53
  level of fines 10.02–10.03, 10.17–10.40
  liquidation 8.19–8.20
  magistrates 10.10, 10.121–10.129
  maximum fines 10.09
  means of offenders 10.26–10.27, 10.33, 10.63, 10.110, 10.116, 10.125, 10.128–10.129
  mitigation 10.24, 10.47–10.48, 10.54–10.56
  obstruction 7.56
  prohibition notices 4.13, 7.54
  proportionate to means of offender 10.24–10.27, 10.63
  public bodies and public element 10.59–10.61
  rail disasters 10.02, 10.38–10.40, 10.73–10.74
  regulations, breach of 10.121
  section 42 orders, failure to comply with 10.119
  Sentencing Guidelines 10.122–10.125

systemic fault, importance of 10.29–10.40
tariff 10.29–10.40, 10.122
third parties, reliance on 10.46–10.53
time to pay, allowing defendants 10.106–10.117
venue for sentence 10.09–10.11
warnings, failure to heed 10.23
fire safety 5.49
floors 12.59
flues 12.220–12.223
food safety 5.31, 5.41–5.42, 7.51
foreseeability
articles 6.180
control of substances hazardous to health 6.117–6.126
criminal offences 6.127–6.134
equipment 12.81
fines 10.43–10.44
gross negligence manslaughter 11.59
Health and Safety at Work Act 1974 6.116–6.117, 6.127–6.128, 6.133
limitation of liability 6.169
maintenance 12.90
manual handling 12.65
parts of machinery, dangerous 12.115
reasonable practicability 6.101, 6.116–6.134, 7.24
forests 12.56
formal cautions
breach 1.86
guidance 1.82, 1.84
inspectors 1.81–1.87
procedure 1.87
prosecutions 1.86
public interest 1.84–1.85
purposes 1.83
repeat cautions 1.85
Framework Directive (89/391)
aims 1.48, 1.59
burden of proof, reverse 6.139–6.140
challenges to 1.53
contents 1.49–1.51
daughter directives 1.60–1.61, 12.11
enforcement 1.45–1.61
Management of Health and Safety at Work Regulations 1999 6.103–6.104, 12.11, 12.36
reasonable practicability 1.53–1.54, 1.57, 1.59
strict liability 1.55–1.59
fraud
complex cases 8.152
legal privilege 3.76
Serious Fraud Office 3.76

friendly societies 5.18, 5.25
***Friskies* statements**
advance information, relationship with 8.66–8.68
aggravating features 8.56–8.62, 8.68
case management 8.68
Crown Court 8.43
drawing up 8.59–8.61
guilty pleas 8.58
Health and Safety Executive 8.57
judicial review 8.63
magistrates 8.43, 8.52, 8.54–8.68
mitigation 8.58
mode of trial 8.108
not proving a schedule, consequences of 8.62–8.65
previous convictions 8.108
recommendations 8.57–8.58
funeral expenses 10.05

gas safety
appliances 11.50–11.56, 12.212, 12.215, 12.219–12.224
checks 12.220, 12.222
competent persons, fitting by 12.215
CORGI installers 12.215
criminal offences 12.214
definition of gas 12.211
delegation 11.50–11.56
domestic premises 12.212
enforcement 12.213
explosions 10.72, 10.75
fitting 12.215–12.218
flues 12.220–12.223
Health and Safety Executive 12.213
inspection certificates, landlord's 12.223–12.224
installation 12.211, 12.215, 12.219
landlords 12.220–12.224
leaks 12.216, 12.218
local authorities 12.213
maintenance 11.51, 12.211, 12.220–12.222
pipework 12.220
portable or mobile appliances 12.212
records 12.224
regulations 12.211–12.224, App M
safety checks 12.220, 12.222
second-hand appliances 12.219
use 12.211, 12.214, 12.219
work 12.216–12.218
general duties *see also* **breach of general duties**
construction industry 12.153
electricity safety 12.236
gross negligence manslaughter 11.47

general duties *see also* **breach of general duties** (*cont.*)
  reasonable practicability 6.66–6.100
  regulations 12.04
good safety record, relevance of 10.54–10.56
government bodies
  Agency Agreements 1.19
  corporate manslaughter 11.75, 11.80
gritting operations 12.75
gross negligence manslaughter
  absolute liability 11.47, 11.51, 11.56
  arrest 2.65, 3.67
  breach of statutory duty 11.49, 11.56
  Caparo tripartite test 11.57–11.61
  causation 11.31
  certainty, offence lacking 11.26
  civil liability, exclusion of 11.46
  contributory negligence 11.31
  coroners' inquests 11.144
  Corporate Homicide and Manslaughter Bill 11.23
  corporate manslaughter 11.66–11.68, 11.77, 11.84
  delegation 11.48, 11.50–11.56, 11.63–11.65
  directors
    corporate manslaughter 11.77
    delegation 11.63–11.65
    employers, duties as 11.63–11.65
    statutory duties owed by 11.63–11.65
  duty of care, breach of 11.27, 11.32–11.37, 11.62
  European Convention on Human Rights 11.26
  ex turpi causa rule 11.39–11.41
  fair, just and reasonable test 11.59
  foreseeability 11.59
  gas appliances 11.50–11.56
  general health and safety duties 11.47
  Health and Safety at Work Act 1974 11.45–11.56, 11.62–11.65
  homicide investigations 3.67
  illegal immigrants, death of 10.100, 11.38–11.44
  imprisonment 10.91–10.98
  individuals 11.25–11.65
  inquests 3.67
  investigations 2.57
  judicial review 9.14–9.16
  juries 11.28, 11.33–11.36, 11.56, 11.61
  jury directions 11.29, 11.34, 11.43
  legal expenses insurance 3.57
  Morecambe Bay disaster 10.101
  non-employees 11.48
  prosecution decisions 3.67, 9.14–9.15

  proximity 11.59, 11.62
  rail disasters 10.98–10.99, 11.57
  reasonable practicability 11.46
  representation 3.57
  risk of death 11.29–11.30
  seizure 3.67
  self-employed 11.48
  sentencing 10.89–10.101
  statutory duties 11.45–11.56, 11.62–11.65
  suspended sentences 10.96–10.97
  vicarious liability 11.48
  Work Related Death Protocol 2.65
guards 12.114
guilty pleas
  admissions, failure to make 3.26
  committal to Crown Court for sentence 8.93–8.96
  companies 8.80
  costs 10.153
  Crown Court 8.74–8.75
  fines 8.96, 10.40
  Friskies statements 8.58
  indication of sentence 10.15
  interviews under caution 3.26
  liquidation 8.20
  plea bargaining 8.154
  plea before venue 8.71–8.78, 8.80, 8.82–8.89
  reasonable opportunity, at first 8.94–8.95
  sentencing discounts 8.93, 8.96, 10.40

Hazardous Installations Directorate 1.30
hazardous substances *see* **control of substances hazardous to health**
Health and Safety at Work Act 1974
  articles 6.176–6.182
  breach of general duties 6.01–6.221
  burden of proof, reverse 6.135–6.136
  company directors, managers and officers, secondary liability of 5.46–5.72
  coroners' inquests 11.139–11.141
  criminal offences 7.01–7.90
  danger, possibility of 6.61–6.65
  dangerous articles or substances 2.121–2.127
  deception 7.83–7.87
  designers 6.173
  disclosure 7.61–7.69
  duties 1.05
  employees
    duties 6.209–6.221
    junior employees, isolated acts of 6.80–6.90
    meaning 6.43–6.52, 6.55–6.57
  employers
    duties 6.01–6.42

Index

meaning of 6.43–6.51, 6.58–6.60
enforcement 1.01–1.06
entry, powers of 2.101–2.104
evidence 2.96–2.99
false entries 7.80–7.82
false statements 7.67–7.73, 7.74–7.79
fines 10.17–10.29, 10.33, 10.122
foreseeability 6.116–6.117, 6.127–6.128, 6.133
gross negligence manslaughter 11.45–11.56, 11.62–11.65
Health and Safety Commission, creation of 1.01, 1.06
Health and Safety Executive, creation of 1.06
identification doctrine 6.68–6.74
improvement or prohibition notices 4.02, 4.25–4.27, 4.30–4.31, 4.46–4.47, 4.51–4.52
   appeals against 4.53, 4.60, 4.90, 4.94
inspectors 1.01, 2.87–2.166, 3.68–3.73
instituting proceedings 8.07
investigations 2.36, 2.40
junior employees, isolated acts of 6.80–6.90
legal personality 5.01–5.26
limitation of liability 6.168–6.172
Management of Health and Safety at Work Regulations 1999 6.107, 6.112–6.115
manual handling 12.61–12.62
manufacturers, duties of 6.165–6.208
non-employees 6.25–6.32, 6.147–6.164
obstruction 7.56–7.58
prohibition notices 4.02, 4.34, 4.46–4.47, 4.51–4.52
prosecutions 6.217–6.221
reasonable practicability 6.66–6.100, 7.21–7.26, 7.39
regulations 1.07–1.08, 12.02–12.05
risk assessment 12.20
Robens Report 1.01
samples 2.117–2.120
Scottish Assembly 1.02
secondary liability 5.30–5.45
section 20 powers 3.68–3.73
section 42 orders, failure to comply with 10.119
self-employed 6.29–6.31
service of notice and documents 2.25
suppliers 6.173–6.175
text App A
time limits 8.14–8.15
unincorporated associations 5.08–5.10
vicarious liability 6.76–6.79
Welsh Assembly 1.02
work and at work, definition of 6.18–6.23
work-related deaths 11.06
workplace 12.47

Health and Safety Commission (HSC)
Agency Agreements with government departments 1.19
aims 1.62
appointments 1.24
approved codes of practice 1.10, 1.19
Buncefield oil farm explosion 1.21
committees 1.19
composition 1.16
construction industry 12.148
creation 1.01, 1.06
criminal offences 1.22–1.23, 7.59–7.60
Crown 1.13
delegation 1.17
disclosure 7.62
duties and powers 1.18
early intervention 1.76
enforcement 1.13–1.14, 1.16–1.23, 2.41–2.45
Health and Safety at Work Act 1974 1.01, 1.06
Health and Safety Executive 1.17–1.20, 1.24
information, obtaining 1.23, 7.59–7.60
inquiries 1.22
investigations 2.41–2.45
local authorities 1.19, 1.32, 1.37
Newton Board 1.21
obstruction 1.22
permissions 1.80
prosecutions 1.102
regulations 1.07, 12.05
service of notice to obtain information 1.23
status of 1.13, 1.16
transfer of enforcement duties 1.37
undertakings 6.33–6.42

Health and Safety Executive (HSE)
advisory literature 1.26
approved codes of practice 1.10
asbestos 12.197
composition 1.24
Construction Design Management Coordinator 12.170
construction industry 12.148–12.149, 12.170
coroners' inquests 11.136–11.147
creation of 1.06
Crown 1.13
Director General 1.24
disclosure 7.61
early intervention 1.76
Employment Medical Advisory Service 1.29
employment sectors covered, list of 1.28

Index

Health and Safety Executive (HSE) (*cont.*)
  enforcement 1.13–1.14, 1.24–1.31, 1.36–1.37, 1.40, 8.196–8.198
  expert evidence 8.196–8.198
  Field Operations Directorate (FOD) 1.28–1.29
  Friskies statements 8.57
  gas safety 12.213
  guidance 1.12, 1.26
  Hazardous Installations Directorate 1.30
  Health and Safety at Work Act 1974 1.06
  Health and Safety Commission 1.17–1.20, 1.24
  Health and Safety Laboratory (HSL) 1.31
  Independent Legal Oversight 8.13
  industrial premises 1.36
  inspectors 1.28–1.30, 2.111, 2.150–2.156, 3.21–3.25
  investigations 2.32–2.34, 2.37–2.40, 11.136
  judicial review 2.45, 9.10–9.11, 9.14
  Legal Advisor's Office 8.13
  liaison, protocol for 1.40
  local authorities 1.32, 1.36–1.37
  Management of Health and Safety at Work Regulations 1999 12.03, 12.10
  Nuclear Safety Inspectorate (NSD) 1.30
  permissions 1.80
  prosecutions 1.25, 9.10–9.11, 9.14
  publications 1.26–1.27
  regulations 1.08, 12.03, 12.06–12.07
  reporting incidents and offences 2.05–2.07
  route map proposal 12.59
  staff 1.25
  statistics 1.25, 2.07
  status of 1.13
  transfer of enforcement duties 1.37
  website 1.27, 2.07
  Work Related Death Protocol 11.16–11.17
  work-related deaths 11.03
  workplace 12.59
  young persons 12.34
health surveillance 12.39, 12.135
hearsay 8.208–8.213
  admissibility 8.208–8.213
  answers, power to require 2.136
  factors taken into account 8.210–8.212
  improvement or prohibition notices, appeals against 4.60
  notice 8.213
  opposition to admission 8.213
height, work at
  construction industry 12.140–12.145
  control measures 12.142
  equipment 12.145
  falls, prevention of 12.143
  hierarchy of measures 12.143
  reasonable practicability 7.36–7.37, 12.141–12.143
  regulations 12.140–12.145, App K
  self-employed 12.140, 12.144
  training and instruction 12.143–12.144
hire equipment 12.96
hire purchase, conditional sale agreements or credit sale agreements 6.175
horseplay 6.219
housing associations 5.26
HSC *see* Health and Safety Commission (HSC)
HSE *see* Health and Safety Executive (HSE)
Human Rights Act 1998
  delay 9.30–9.31
  investigations 2.35
  public authorities 2.35

identification doctrine
  corporate manslaughter 11.85
  delegation 6.68
  directing minds 6.68, 6.70–6.71
  Health and Safety at Work Act 1974 6.68–6.74
  reasonable practicability 6.68–6.74, 6.98
  senior management 6.68
illegal immigrants, death of 10.100, 11.38–11.44
impersonation of inspectors 7.88
imports 6.189, 6.198
imprisonment
  court order, failure to comply with 7.90
  directors' disqualification orders 10.105
  gross negligence manslaughter 10.91–10.98
  improvement notices 4.13, 7.54
  licence, breach of 10.87
  prohibition notices 4.13, 6.54
  section 42 orders, failure to comply with 10.119
improvement notices 4.01–4.33 *see also* **appeals against improvement or prohibition notices**
  accuracy of notices, representations on 4.06
  achievable, aims must be 4.33
  amendments 4.06
  approved code of practice 4.30, 4.32
  bodies corporate 4.46
  breach 4.11–4.17, 7.54–7.55
  building regulations, work not more onerous than 4.31
  contents 4.25, 4.28, 4.30, 4.32
  criminal offences 4.11, 4.13, 7.54–7.55
  Crown 1.41, 1.44, 1.88
  decisions to issue notices 4.05, 4.18–4.24
  directions as to measures to be taken 4.30, 4.32

Index

economic loss 4.51–4.52
enforcement 1.79
   Enforcement Guide 4.24, 4.33
   Enforcement Management
      Model 4.18–4.22
   Enforcement Policy Statement 4.16, 4.18
   errors 4.52
   fines 4.13, 7.54
   Health and Safety at Work
      Act 1974 4.02, 4.25–4.27, 4.30–4.31,
      4.46–4.47, 4.51–4.52
   imprisonment 4.13, 7.54
   informal notification 4.05
   inspector's power to issue 4.01–4.02
   investigations 2.56
   leaflet ITL 19 4.07
   legal advice 4.06
   local authorities 4.51
   negligent misstatements 4.51–4.52
   notification of intent to serve notice 4.05
   partnerships 4.46
   proportionality 4.12
   prosecutions 4.15–4.17
   public interest 4.11
   public register 4.10
   purpose of enforcement 4.18
   reasonable practicability defence 7.55
   risk gap, concept of 4.20–4.22
   Robens Report 4.01
   service 4.46–4.47
   standard forms 4.32, 4.49
   standards, enforcement of 4.10
   statistics 4.09
   suspension 4.08
   time periods
      extension of 4.06, 4.48–4.50
      remedying defects 4.03, 4.29, 4.33
   visits 4.50
   withdrawal 4.06, 4.48–4.50
incapacitation 2.11
independent contractors
   contracts for services 6.43, 6.45
   employees, meaning of 6.43–6.49
   undertakings 6.36–6.38
   vicarious liability 6.77
   workplace 12.58
Independent Legal Oversight 8.13
indictable offences
   arrest 8.03
   mode of trial 8.102, 8.112
   search and entry, powers of 2.70
indictments
   bills of indictment 8.122
   charges, selecting the 8.133–8.143

   Crown Court 8.121–8.144
   duplicity, rule against 8.123–8.130
   extension of time limits 8.121
   lodging the indictment 8.121–8.144
   signing 8.122
   time limits 8.121
individuals
   person, definition of 5.02–5.07
   prosecutions 1.89–1.91, 1.100–1.101, 6.218
   sentencing 10.87–10.129
inductions 12.183, 12.188
industrial and provident societies 5.18, 5.26
information
   advance information 8.48–8.55, 8.66–8.68
   agency workers 12.43
   Construction Design Management
      Coordinator 12.166, 12.172
   construction industry 12.162, 12.166, 12.172,
      12.180, 12.186, 12.188
   control of substances hazardous to
      health 12.136
   criminal offences 7.49, 7.51, 7.53, 7.59–7.69, 7.74
   designers 12.176–12.178
   equipment 12.108–12.109
   fines 7.60
   Health and Safety Commission
      1.23, 7.59–7.60
   inspection 12.107
   inspectors 3.68–3.72, 7.49, 7.51, 7.53
   instituting proceedings 8.04–8.06,
      8.40–8.42
   magistrates 8.27–8.36, 8.38, 8.40–8.42
   Management of Health and Safety at Work
      Regulations 1999 12.41
   manufacturers, duties of 6.184–6.185,
      6.204–6.205
   notice 7.59
   self-incrimination 7.59
   temporary workers 12.43
informations
   amendments 8.40–8.42, 8.132
   duplicity, rule against 8.132
   contents 8.36
   duplicity 8.38
   errors 8.41
   laying an information 8.16, 8.26–8.35
   oral informations 8.27
   time limits 8.16
Initial Enforcement Expectation (IEE) 1.74
innocence, presumption of *see* **presumption of**
   innocence
inquiries
   criminal offences 7.40–7.43
   Health and Safety Commission 1.22

Index

Rail Safety and Standards Board 3.80
inquiry, duty to pursue all reasonable lines of 2.46–2.52
inquisitorial processes 11.107
insolvent companies, holding companies' liability for debts of 10.68
inspection *see also* **inspectors**
  Approved Code of Practice 12.98–12.101, 12.103–12.104
  arrest 2.57
  competence of persons carrying out inspections 12.106–12.107
  criminal offences 2.152, 12.97
  disclosure 8.183
  electricity safety 12.232
  equipment 12.78, 12.87–12.107
  evidence of inspection 12.101
  exceptional circumstances 12.102
  extent and frequency of 12.103–12.105
  gas safety 12.223–12.224
  guidance 12.99–12.101, 12.103
  information 12.107
  inspectors 2.111, 2.150–2.152
  installation 12.97
  instructions, 12.107
  investigations 2.55–2.57
  legal privilege 2.155–2.156
  maintenance 12.90
  records 12.100–12.102
  regulations 12.97–12.107
  training 12.107
  types of equipment 12.99, 12.105
inspectors *see also* **inspection**
  answers, power to require 2.93, 2.131–2.149
  appointment 2.91–2.92
  arrest 2.67–2.68, 3.03
  articles
    dangerous articles, power to dismantle or subject to testing of 2.121–2.127
    possession and detain, power to take 2.128–2.130
    samples, power to take 2.117–2.120, 2.164
  assistance, power to require 2.157–2.158
  cautions 1.81–1.87
  computers, information held on 2.162
  conflicts of interest 3.59–3.66
  cooperation with inspectors 3.68–3.73
  copies, power to take 2.111, 2.150–2.156
  coroners' inquests 11.134
  criminal offences 2.89, 2.151
    assistance, requirement to provide 7.51
    contravening requirements by 7.44–7.51
    danger, powers to deal with immediate 7.50

    food safety 7.51
    impersonation 7.88
    information, requirement to give 7.49, 7.51, 7.53
    meaning of requirement 7.47
    obstruction 7.56–7.58
    powers 7.44–7.51
  damages 2.100
  danger, power to deal with immediate 2.163–2.165, 7.50
  dangerous articles or substances
    dismantle or subject to testing, power to 2.121–2.127
    possession and detain, power to take 2.128–2.130
  Department of Trade and Industry 2.75
  detention of dangerous articles and substances 2.128–2.130
  disclosure 8.173–8.175
  electronic records 2.162
  enforcement 1.63
  entry, powers of 2.93, 2.101–2.108
  evidence, exclusion of 2.96–2.99
  examine and investigate, powers to 2.109–2.114
  expert evidence 8.196–8.197
  facilities, power to require 2.157–2.158
  false or reckless statements 2.89
  food safety 7.51
  Health and Safety at Work Act 1974 1.01, 2.87–2.166
  Health and Safety Executive 1.28–1.30, 2.91
  impersonation 7.88
  improper use of powers, consequences of 2.96–2.100
  improvement notices 4.01–4.02
  indemnities 2.100
  information 7.49, 7.51, 7.53
  inspection 2.111, 2.150–2.156
  instituting proceedings 8.07–8.12
  investigations 2.03
  legal privilege 2.155–2.156
  letters of advice/recommendation 1.78
  local authorities 1.33, 2.91, 2.93
  magistrates 8.11
  measurements, taking 2.115–2.116
  necessary, any other power which is 2.159–2.162
  obstruction 2.68, 2.89, 7.56–7.58
  PACE and codes 3.32–3.33
  photographs 2.115–2.116
  police officers, accompanied by 2.68

Index

possession and detain dangerous articles and substances, power to take 2.128–2.130
powers 2.87–2.166, 7.44–7.51
premises, meaning of 2.110
production, inspection and to take copies, power to require 2.111, 2.150–2.156
prohibition notices 4.01–4.02
qualifications 2.92
recordings 2.115–2.116
reporting incidents and offences 2.08
samples, power to take 2.117–2.120, 2.164
searches 2.112, 2.152
solicitors, instruction of 8.12
substances
    dangerous substances, power to dismantle or subject to testing of 2.121–2.127
    possession and detain, power to take 2.128–2.130
    samples, power to take 2.117–2.120, 2.164
    testing 2.121–2.127
undisturbed, directions to leave 2.113–2.114
verbal warnings 1.77
inquests *see* **coroners' inquests**
instituting proceedings 8.01–8.42
arrestable offences 8.03–8.04
companies, against 8.03
Criminal Procedure Rules 2005 8.02
Crown Court 8.01–8.02
Crown Prosecution Service 2.54
Director of Public Prosecutions 8.07
Environment Agency 2.54, 8.07
Health and Safety at Work Act 1974 8.07
Independent Legal Oversight 8.13
informations
    amendment 8.40–8.42
    companies 8.04
    future abolition of 8.05–8.06
inspectors 8.07–8.12
investigations 2.54
liquidation, companies in 8.17–8.25
magistrates 8.01–8.02, 8.11, 8.26–8.39
persons who may commence proceedings 8.07–8.13
requisitions 8.05
solicitors, instruction by inspectors of 8.12
summonses
    amendment of 8.40–8.42
    companies 8.04
    future abolition of 8.05–8.06
time limits 8.14–8.16
written charges 8.05–8.06

instructions
asbestos 12.202
control of substances hazardous to health 12.136
employees acting outside instructions 10.41–10.43
equipment 12.108–12.109
expert evidence 8.207
fines 10.41–10.43
height, work at 12.143–12.144
inspection 12.107
work and at work, definition of 6.22
insurance
employers' compulsory insurance 10.07
gross negligence manslaughter 3.57
legal expenses insurance 3.55–3.58
legal funding 3.55–3.58
interviews under caution 3.01–3.50
access to legal advice 3.35–3.38, 3.49
admissions 3.26, 3.42–3.43
adverse inferences
    prepared statements 3.27
    refusal to be interviewed 3.20, 3.26
    silence, from 3.13–3.19, 3.43, 3.49
advising suspects 3.05–3.30
arrest 3.02–3.03, 3.07, 3.19
corporate suspects, directors and employees, legal representation of 3.06
correspondence 3.04, 3.08–3.09, 3.21–3.25
decision to be interviewed 3.26
declining invitation to be interviewed, effect of 3.19–3.20, 3.26, 3.30
definition 2.81
detention 3.32–3.33
directors
    HSE inspectors, interviews by 3.21–3.25
    legal representation 3.06
    letter, interviews by 3.21–3.25
disclosure 3.11–3.12
guilty pleas, failure to make admissions and 3.26
investigations 2.04, 2.81
limitations on power to interview 3.46–3.49
local authority inspectors, correspondence and 3.04, 3.08, 3.21
mental disabilities, persons with 3.45
no comment interviews 3.18
oppression 3.44
PACE and codes 2.81–2.84, 3.31–3.50
Principles of Good Enforcement 3.28–3.29
prosecution decisions 3.01
records 2.83, 3.84
refusal to be interviewed 3.19–3.20, 3.26, 3.30

interviews under caution (*cont.*)
  representations in writing 3.28–3.30
  significant statements or
      silence 3.42–3.43
  solicitors
    HSE guidance to 3.06
    presence of 3.37–3.38
  statements, prepared 3.27
  taped interviews 3.04, 3.08
  vulnerable persons 3.45
  written statements under caution 3.50
  young persons 3.45
**investigations of incidents and
    offences** 2.32–2.91 *see also* **advising clients in
    health and safety investigations; criminal
    investigations**
  ambit of duty 2.34–2.52
  chief officers of police 2.39–2.40
  coroners' inquests 11.03, 11.92, 11.94,
      11.109–11.111, 11.125–11.126, 11.146
  corporate manslaughter 11.74
  decision to investigate 2.41–2.45
  disclosure 8.170–8.180
  enforcement 2.32–2.91
    adequate arrangements for, duty to
        make 2.36–2.40
    principles 2.43–2.45
  expert evidence 8.198
  gross negligence manslaughter 3.67
  Health and Safety at Work
      Act 1974 2.36, 2.40
  Health and Safety Commission Enforcement
      Policy Statement 2.41–2.45
  Health and Safety Executive 2.32–2.34,
      2.37–2.40, 11.136
  Human Rights Act 1998, public authorities
      under 2.35
  independent, prompt and proper
      investigations 11.109–11.111
  inquiry, duty to pursue all reasonable lines
      of 2.46–2.52
  inspectors 2.03, 2.109–2.114
  interviews under caution 2.04, 2.81
  Investigators' Guide 11.09
  judicial review 2.45
  life, right to 2.35, 11.109–11.111
  police 11.126
  public authorities 2.35
  public law 2.35
  social security 2.76
  statistics 2.32
  Work Related Death
      Protocol 11.09
  work-related deaths 11.01–11.07, 11.09

irrationality 9.12, 9.15

judicial review *see also* **judicial review of
    decisions to prosecute**
  committal for trial 8.117
  Friskies statements 8.63
  Health and Safety Executive 2.45
  improvement or prohibition notices, appeals
      against 4.122
  investigations 2.45
judicial review of decisions to prosecute 9.01–9.16
  abuse of process 9.01, 9.03–9.07
  European Convention on Human
      Rights 9.09
  fair trials 9.09
  gross negligence manslaughter 9.14–9.16
  Health and Safety Executive 9.10–9.11, 9.14
  irrationality 9.12, 9.15
  not to prosecute, decisions 9.09–9.16
  permission 9.08
  quashing decisions 9.07, 9.16
  Wednesbury irrationality 9.12
junior employees, isolated acts of
  directing minds 6.84
  Health and Safety at Work Act 1974 6.80–6.90
  precautions 6.88–6.90
  reasonable practicability 6.80–6.90
  vicarious liability 6.88
juries
  coroners' inquests 11.95, 11.98, 11.103–11.105
  criminal offences 7.13
  gross negligence manslaughter 11.28,
      11.33–11.36, 11.56, 11.61

knowledge
  company directors, managers and officers,
      secondary liability of 5.53, 5.70–5.72
  false statements 7.69–7.71

leases
  asbestos 12.197–12.198
  gas safety 12.220–12.224
legal advice *see* **access to legal advice;
    representation; solicitors**
legal expenses insurance 3.55–3.58
legal funding 3.55–3.58
legal personality
  charitable trusts 5.18–5.19, 5.24–5.25
  corporations 5.03–5.07
  corporations sole 5.04
  criminal offences 5.01, 5.06
  friendly societies 5.25
  Health and Safety at Work
      Act 1974 5.01–5.26

Index

industrial and provident societies 5.26
limited liability partnerships 5.16
natural persons 5.03, 5.06
partnerships 5.11–5.16
person, definition of 5.02–5.07
statute, bodies incorporated by 5.05
trusts 5.17–5.26
unincorporated bodies 5.08–5.10, 5.21–5.22

legal privilege
advice privilege 3.74, 3.85
advising clients 3.74–3.85
causes, reports into 3.77–3.80
coroners' inquests 11.134
definition of privileged material 3.75
disclosure 8.188
dominant purpose test 3.81–3.83
inspection 2.155–2.156
inspectors 2.155–2.156
litigation privilege 3.74, 3.81–3.85
legal professional privilege 3.73–3.85
pre-existing documents 3.85
production 2.155–2.156, 3.73, 3.76
search and seizure, police powers of 2.70, 3.75
search warrants 3.75
Serious Fraud Office 3.76

legal representation *see* **representation; solicitors**
legitimate expectations 9.46
letters
advice/recommendation, of 1.78
interviews by letter 3.21–3.25
local authority inspectors 3.04, 3.08, 3.21
prosecutions 1.104

liability *see* **company directors, managers and officers, secondary liability of liaison, protocol for** 1.40
licensing 1.80, 7.78
life, right to
coroners' inquests 11.108–11.116, 11.121, 11.132
investigations 2.35

lifting *see* **manual handling**
limitation of liability
burden of proof 6.169
companies 10.68
examination 6.172
foreseeability 6.169
Health and Safety at Work Act 1974 6.168–6.172
manufacturers, duties of 6.168–6.172
research 6.172
testing 6.172

limited liability partnerships 5.16
limited partnerships
bodies corporate 5.16
criminal offences, personal liability for 5.16
institution of proceedings 5.16
service 5.16

liquidation, companies in
compulsory liquidation 8.18, 8.21–8.23
creditors' voluntary liquidation 8.24
fines 8.19–8.20
guilty pleas 8.20
instituting proceedings 8.17–8.25
members' voluntary liquidation 8.24–8.25
public interest 8.19, 8.23
reports 8.20
voluntary liquidation 8.18, 8.24–8.25

litigation privilege 3.74, 3.81–3.85
local authorities
commercial premises 1.36
coroners' inquests 11.92
correspondence 3.04, 3.08, 3.21
division of responsibilities 1.32
enforcement 1.32–1.37
Environmental Health Officers (EHO) 1.34
gas safety 12.213
guidance 1.19
Health and Safety Commission 1.19, 1.32, 1.37
Health and Safety Executive 1.32, 1.36–1.37
Health and Safety/Local Authority Enforcement Liaison Committee (HELA) 1.33
improvement notices 4.51
inspectors 1.33, 2.91, 2.93, 3.04, 3.08, 3.21
reporting incidents and offences 2.06
scope of enforcement duty 1.35
transfer of enforcement duties 1.37
Work Related Death Protocol 11.17

lodging the indictment 8.121–8.144
Luce Review 11.90

machinery, dangerous parts of 12.112–12.115
magistrates
abuse of process 9.19–9.21
advance information 8.48–8.55
arrest, warrants for 8.46
bail 8.46–8.47
arrest 8.47
committal to Crown Court 8.47
conditions 8.47
breach of general duties 6.03
case management 8.44
City of London Magistrates' Court 8.29
committal to Crown Court 8.47, 8.100–8.101, 8.117

magistrates (*cont.*)
  conducting a case in magistrates'
    court 8.43–8.117
  control of substances hazardous to
    health 12.134
  costs 10.131, 10.140, 10.144
  criminal offences 7.03, 7.09, 7.44,
    8.26, 8.44
  Criminal Procedure Rules 2005 8.26, 8.44
  date for attendance 8.45
  disclosure 8.162
  fair trials 8.39
  fines 10.10, 10.121–10.129
  Friskies statements 8.43, 8.52, 8.54–8.68
  information
    amendments 8.40–8.42
    contents 8.36
    duplicity 8.38
    errors 8.41
    laying an 8.26–8.35
    orally 8.27
  initial appearance 8.45–8.47
  instituting proceedings 8.01–8.02, 8.07–8.13,
    8.26–8.39
  jurisdiction 8.26–8.39
  location 8.29–8.31
  mode of trial, procedure for
    determining 8.102–8.116
  overriding objectives 8.44
  plea before venue 8.69–8.81
  Sentencing Guidelines 10.122–10.125
  service of summons 8.32–8.35, 8.45
    companies 8.34
    contents 8.36–8.39
    Criminal Procedure Rules 2005 8.33
    methods 8.33–8.34
    single summons for more than one
      information 8.37
    time limits 8.35
  summons
    amendments 8.40–8.42
    errors 8.41
    issue 8.32
    serving 8.32–8.35, 8.45
    withdrawal 8.42
  time limits 8.14, 8.16
  venue for trial or sentence, procedure for
    determining 8.81–8.99
maintenance
  Approved Code of Practice 12.91–12.95
  asbestos 12.194, 12.198–12.199
  breakdown technique 12.95
  condition-based technique 12.95
  criminal offences 12.89
  electricity safety 12.228, 12.231–12.232
  equipment 12.78, 12.88–12.96
  foreseeability 12.90
  gas safety 11.51, 12.211, 12.220–12.222
  guidance 12.91–12.95
  hire equipment 12.96
  inspection 12.90
  latent defects 12.89
  logs 12.91
  planned preventive technique 12.95
  regulations 12.88–12.96
  safety checks 12.93
  strict liability 12.89
Management of Health and Safety at Work
    Regulations 1999 6.102–6.115,
    12.09–12.46
  agency workers, information provided
    to 12.43
  ambit 12.09
  approved codes of practice 12.46
  breach 12.11
  burden of proof 6.108
  civil liability 12.11
  criminal offences 6.102, 6.105, 6.110, 6.112
  delegation 6.100
  effect of regulations 12.46
  evacuation procedures 12.45
  fault of junior employees 6.109–6.114
  Framework Directive 6.103–6.104,
    12.11, 12.36
  guidance 12.03, 12.10, 12.46
  Health and Safety at Work Act 1974 6.107,
    6.112–6.115
  Health and Safety Executive 12.03, 12.10
  health surveillance 12.39
  information 12.41
  manual handling 12.62
  non-employees 12.41
  operation of regulations 12.46
  precautions 6.111
  prevention 12.36
  protective measures 12.36
  public policy 6.112
  reasonable practicability 6.102–6.115, 7.21,
    7.30–7.33
  review, continuing 12.37–12.40
  risk assessment 12.12–12.35, 12.41
  surveillance 12.37–12.40
  temporary agency workers, information provided
    to 12.43
  text App E
  training 12.40, 12.42, 12.44

Index

managers *see* company directors, managers and officers, secondary liability of
manslaughter 11.21–11.88 *see also* corporate manslaughter; gross negligence manslaughter
  Code for Crown Prosecutors 11.24
  coroners' inquests 11.136, 11.144, 11.147–11.149
  Crown Prosecution Service 1.39, 11.24
  Director of Public Prosecutions 11.24
  prosecution decisions 11.24
  reform 11.21–11.23
  Work Related Death Protocol 11.20
  work-related deaths 11.06–11.07, 11.20
manual handling
  burden of proof 12.64
  designers 12.61
  EC law 12.60
  foreseeability 12.65
  gritting operations 12.75
  guidance 12.74
  Health and Safety at Work Act 1974 12.61–12.62
  Management of Health and Safety at Work Regulations 1999 12.62
  manual handling operations, definition of 12.70–12.76
  manuals and specifications, obtaining 12.74
  manufacturers 12.61, 12.74
  primary duties 12.63–12.69
  reasonable practicability 12.63, 12.66
  regulations 12.60–12.76, App G
  risk assessment 12.16, 12.62, 12.66, 12.68–12.69, 12.73–12.76
  risk, meaning of 12.65
  self-employed 12.61
  statistics 12.60
  training 12.72
manufacturers, duties of
  articles for use at work 6.165–6.208
  breach of general duties 6.165–6.208
  Consumer Protection Act 1987 6.165
  criminal offences 6.167, 6.181, 6.201
  examination 6.183, 6.203
  fairground equipment 6.188, 6.190–6.193, 6.196
  fines 6.167
  foreseeability 6.180, 6.200
  Health and Safety at Work Act 1974 6.165–6.208
  imports 6.189
  information
    provision of adequate 6.184–6.185, 6.204–6.205
    revisions of 6.205
  installations 6.196–6.197
  limitation on liability 6.168–6.172
  manual handling 12.61, 12.74
  proof 6.203
  research 6.183, 6.193–6.195, 6.206–6.208
  serious risk, definition of 6.187
  standard of proof 6.186, 6.205
  substances for use at work 6.165–6.208, 6.198–6.199, 6.204, 6.206–6.208
  testing 6.183
  written undertakings 6.188–6.189
measurements, taking 2.115–2.116
medical treatment, operations and examination 2.14
members' voluntary liquidation 8.24–8.25
mental disabilities, persons with 3.45
mines 12.56
mitigation
  committal to Crown Court for sentence 8.86–8.87, 8.91
  fines 10.24, 10.47–10.48, 10.54–10.56
  Friskies statements 8.58
  good safety record, relevance of 10.54
  inspectors, cooperation with 3.69
  let down by others, being 10.47–10.48
mode of trial
  advance information 8.52
  aggravating guidelines 8.106
  committal for trial
    abolition of 8.102
    companies 8.114
    exhibits 8.116
    judicial review 8.117
    material sent to Crown Court on committal 8.116
    orders on committal 8.115
    Plea and Case Management Hearings (PCMH) 8.115
  committal to Crown Court for sentence 8.90
  companies 8.113–8.114
  either way offences 8.102–8.103
  elections
    companies 8.113–8.114
    individuals 8.111–8.112
  Friskies statements 8.108
  indictable offences 8.102, 8.112
  magistrates 8.102–8.116
  matters to take into account 8.104
  Mode of Trial Guidelines 8.90. 8.106
  plea before venue 8.90
  previous convictions 8.108–8.110
  summary offences 8.102, 8.109

Morecombe Bay disaster 10.101
mutual assurance associations 5.25

national insurance, liability for 6.48
national security 2.126
natural persons 5.03, 5.06
negligent misstatements 4.51–4.52
Newton hearings 8.96
non-employees, duties to 6.24–6.32
   breach of general duties 6.147–6.164
   criminal offences 6.25–6.32
   danger, possibility of 6.61–6.65
   domestic premises, definition of 6.149
   gross negligence manslaughter 11.48
   Health and Safety at Work Act 1974 6.25–6.32, 6.147–6.164
   Management of Health and Safety at Work Regulations 1999 12.41
   place of work, non-domestic premises used by non-employees as 6.148–6.154
   plant
      definition of 6.151, 6.154
      safety of 6.160–6.161
   premises
      common parts of blocks of flats 6.153–6.154
      definition 6.150
      duties owed by persons in control of 6.147–6.164
      safety of 6.160–6.161
   reasonable practicability 6.91–6.100
   reasonableness 6.158–6.159, 6.162–6.164
   risk assessment 12.32
   risk, meaning of 6.160
   safe and without risks to health, premises, plant or substances are 6.160–6.161
   self-employed 6.29–6.32
   substance, definition of 6.151
not guilty pleas 8.71, 8.78, 8.82
Nuclear Safety Inspectorate (NSD) 1.30

obstruction
   arrest 2.68
   criminal offences 7.56–7.58
   entry, powers of 2.105–2.106
   fines 7.56
   Health and Safety at Work Act 1974 7.56–7.58
   Health and Safety Commission 1.22
   inspectors 2.68, 2.89, 7.56–7.58
   intention 7.58
   police officers 7.57–7.58
   serious obstruction 2.105, 2.107

occupational diseases 2.19–2.20
Office of the Rail Regulator (ORR) 1.15, 1.40
officers *see* company officers; company directors, managers and officers, secondary liability of
oppression, decisions to prosecute and 9.50–9.57
overriding objective 8.44, 8.118–8.120, 8.133

PACE and codes
   access to legal advice 3.35–3.38, 3.42–3.43
   advising clients 3.53
   Bank of England 2.71
   cautions 2.78–2.80, 2.85–2.86, 3.40
   Department of Trade and Industry inspectors 2.75
   detention 3.32–3.33
   disclosure 3.11–3.12
   evidence 2.97
   inspectors 2.71–2.86
   interviews 2.81–2.84, 3.31–3.50
   legal advice, right to 2.76
   mental disabilities, persons with 3.45
   records 3.34
   searches with consent 2.71, 2.77
   seizure 2.77
   social security investigators 2.76
   vulnerable persons 3.45
   young persons 3.45
   written statements 3.50
partnerships
   criminal liability 5.11–5.15
   England and Wales 5.11–5.14
   improvement notices 4.46
   joint and several liability 5.11
   legal personality 5.11–5.16
   limited liability partnerships 5.11–5.16
   limited partnerships 5.16
   prohibition notices 4.46
   Scotland 5.15
   secondary liability 5.15
   service of notice and documents 2.28
parts of machinery, dangerous 12.112–12.115
permissioning and licensing regimes 1.80
person, definition of 5.02–5.07
personal injuries *see* deaths and serious personal injuries
personal protective equipment
   asbestos 12.208
   respiratory equipment 12.204–12.205
   risk assessment 12.17
photographs 2.115–2.116
planning supervisors 12.147, 12.150, 12.169

Index

plant
  articles 6.177
  definition 6.151, 6.154, 6.177
  non-employees 6.151, 6.154, 6.160–6.161
  safe and without risk to
      non-employees 6.160–6.161
plea and case management hearings (PCMH)
  case management 8.149
  committal for trial 8.115, 8.149
  plea bargaining 8.155
plea before venue
  benefits and drawbacks of procedure 8.75–8.77
  committal for sentence following guilty
      pleas 8.82–8.89
  companies 8.78–8.80
  Crown Court 8.69–8.81
  guilty pleas 8.71–8.78, 8.80
      committal for sentence 8.82–8.89
      companies 8.80
      Crown Court 8.74–8.75
      sentencing discount 8.74–8.77
  magistrates 8.69–8.81
  mode of trial guidelines 8.90
  no plea, indication of 8.71, 8.82
  not guilty pleas 8.71, 8.78, 8.82
  representation 8.78–8.79
  wording of indication of plea 8.81
pleas *see also* guilty pleas; plea before venue
  Bar Code of Conduct 8.145
  breach of promise 9.49
  Code for Crown Prosecutors 8.144
  complex cases 8.154
  lesser pleas, accepting 8.144–8.146
  negotiations 8.144–8.147
  plea and case management hearings 8.115, 8.149, 8.155
  plea bargaining 8.154–8.156, 9.49
  sentencing, plea bargaining and 8.154–8.156
police *see also* interviews under caution
  arrest 2.57–2.68, 3.02
  chief officers 2.39–2.40
  coroners' inquests 11.95, 11.126, 11.134
  corporate manslaughter 2.65, 11.73, 11.75, 11.78
  disclosure 8.170
  employees, meaning of 6.56
  inspectors accompanied by police 2.68, 2.105–2.107
  investigations 2.39–2.40, 2.49, 11.126
  obstruction 7.57–7.58
  search and entry, powers of 2.69–2.70, 2.105–2.107
  work and at work, definition of 6.19

Work Related Death Protocol 11.17
work-related deaths 11.03
pollution *see* environment
post-mortem reports 11.137
precautions
  electricity safety 12.225, 12.236
  improvement or prohibition notices, appeals
      against 4.99
  junior employees, isolated acts of 6.88–6.90
  Management of Health and Safety at Work
      Regulations 1999 6.111
  prohibition notices 4.45
  undertakings 6.36
pregnant women, risk assessments and 12.35
pre-hearing orders 4.72–4.74
premises *see also* workplace
  common parts of blocks of flats 6.153–6.154
  definition 2.102, 2.110, 6.150
  domestic premises, definition of 6.151, 6.154
  inspectors 2.110
  non-employees, duties to 6.06–6.07, 6.147–6.164
  safe and without risk to
      non-employees 6.160–6.161
preparatory hearings 8.150–8.151
presumption of innocence
  burden of proof, reverse 6.138
  costs 10.138
  strict liability 7.16–7.17
previous convictions
  bad character evidence 8.221–8.223
  committal to Crown Court for sentence 8.97
  Friskies statements 8.108
  mode of trial 8.108–8.110
  propensity 8.221–8.223
privilege *see* legal privilege; self-incrimination,
      privilege against production
  causes, reports into 3.78
  compulsory powers 3.88–3.90
  inspectors 2.111, 2.150–2.156, 3.72–3.73
  legal privilege 3.73, 3.76
  Serious Fraud Office 3.76
prohibition notices 4.01–4.24, 4.34–4.45 *see also*
      appeals against improvement or prohibition
      notices
  accuracy of notices, representations on 4.06
  amendments 4.06
  approved codes of practice 4.41
  deferred notices 4.04, 4.36
  bodies corporate 4.46
  breach 4.11–4.17, 7.54–7.55
  building regulations, work not more onerous
      than 4.40

Index

prohibition notices (*cont.*)
  contents 4.35, 4.42
  criminal offences 4.11, 4.13, 7.54–7.55
  Crown 1.41, 1.44, 1.88
  decisions to issue notices 4.05, 4.18–4.24
  economic loss 4.51–4.52
  enforcement 1.79
    Enforcement Guide 4.24, 4.37
    Enforcement Management
      Model 4.18–4.22
    Enforcement Policy Statement 4.16, 4.18
  errors 4.52
  fines 4.13. 7.54
  Health and Safety at Work Act 1974 4.02, 4.34,
    4.46–4.47, 4.51–4.52
  immediate effect, notices with 4.04, 4.06, 4.36
  imprisonment 4.13. 7.54
  informal notification 4.05
  inspector's power to issue 4.01–4.02
  investigations 2.56
  Ladbroke Grove rail crash 4.44
  leaflet ITL 19 4.07
  legal advice 4.06
  local authorities 4.41, 4.51
  negligent misstatements 4.51–4.52
  notification of intent to serve notice 4.05
  partnerships 4.46
  personal injury, risk of serious 4.35, 4.38, 4.43
  precautions 4.45
  proportionality 4.12
  prosecutions 4.15–4.17
  public interest 4.11
  public register 4.10
  purpose of enforcement 4.18
  reasonable practicability defence 7.55
  risk gap, concept of 4.20–4.22
  Robens Report 4.01
  service 4.04–4.05, 4.10, 4.34–4.35, 4.46–4.47
  standard forms 4.41, 4.49
  standards, enforcement of 4.10
  statistics 4.09
  suspension 4.08, 4.43
  time periods 4.36, 4.45
    extension of 4.06, 4.48–4.50
  undertakings 4.45
  visits 4.50
  withdrawal 4.06, 4.45, 4.48–4.50
proof *see* **burden of proof**
propensity 8.218–8.221
prosecutions 1.89–1.105 *see also* **Crown
  Prosecution Service; instituting proceedings;
  judicial review of decisions to prosecute**
  breach of promise 9.46–9.47

  cautions 1.86
  corporate manslaughter 11.24
  decisions 3.01, 3.29
    corporate manslaughter 11.24
    gross negligence manslaughter 3.67
    Work Related Death Protocol 11.19
  directors and managers 1.95–1.99
  disclosure 8.184–8.185
  discretion 1.102
  employees 1.92–1.94, 6.43–6.45, 6.217–6.221
  Enforcement Concordat 1.104, App B
  enforcement policy, contrary to 9.50–9.57
  gross negligence manslaughter 9.14–9.15
  guidance 1.89, 1.91, 1.102, 6.220–6.221
  Health and Safety at Work
    Act 1974 6.217–6.221
  Health and Safety Commission 1.102
  Health and Safety Executive 1.25
  improvement or prohibition notices 4.15–4.17
  individuals 1.89–1.91, 1.100–1.101, 6.218
  interviews under caution 3.01
  last resort, as 1.63
  letters to duty holders 1.104
  oppression 9.50–9.57
  policy 1.102, 9.50–9.57
  procedure 1.104–1.105
  prohibition notices 4.15–4.17
  public interest 1.94, 1.96, 1.101, 6.220,
    9.53–9.54
  recklessness 1.90
  representations 1.104–1.105, 3.29
  trading standards 9.51
  work and at work, definition of 6.18
  Work Related Death Protocol 11.19
public authorities
  corporate manslaughter 11.76
  fines 10.59–10.61
  Human Rights Act 1998 2.35
  investigations 2.35
public hearings 4.75
public interest
  abuse of process 9.20
  cautions 1.84–1.85
  conflicts of interest 3.62
  coroners' inquests, public interest immunity
    and 11.134
  disclosure 8.188
  employees' duties 6.220
  immunity 8.188, 11.134
  improvement notices 4.11
  liquidation 8.19, 8.23
  prohibition notices 4.11
  prosecutions 1.94, 1.96, 1.101, 6.220, 9.53–9.54

Index

publicity orders, convictions and 11.76

quashing decisions 9.07, 9.16

railway accidents
  causes, reports into 3.80
  corporate manslaughter 11.67
  electricity safety 12.239
  fines 10.02, 10.38–10.40, 10.73–10.74
  gross negligence manslaughter 10.98–10.99, 11.57
  inquiries 3.80
  Office of the Rail Regulator (ORR) 1.15, 1.40
  Rail Safety and Standards Board 3.80
reasonable excuse, defence of 3.71, 7.51
reasonable practicability
  1999, development to 6.66–6.100
  all reasonable steps, meaning of 7.34
  breach of general duties 6.09
  burden of proof 6.135–6.146, 7.22
  charges 8.143
  construction industry 12.159
  contractors 12.180, 12.186–12.187
  control of substances hazardous to health 7.25–7.29, 12.123, 12.126, 12.129, 12.131
  criminal offences 7.18–7.39
  defence involving reasonable practicability, regulatory duties with 7.30–7.34
  delegation to 6.93
  designers 12.176
  due diligence 7.19
  electricity safety 12.226, 12.228, 12.233, 12.235
  employers' duties 6.11–6.12
  equipment 12.81
  foreseeability 6.101, 6.116–6.134, 7.24
  Framework Directive 1.53–1.54, 1.57, 1.59
  general duties 6.66–6.100
  gross negligence manslaughter 11.46
  Health and Safety at Work Act 1974 6.66–6.100, 7.21–7.26, 7.39
  heights, falling from 7.36–7.37, 12.141–12.143
  HTM Ltd 6.101–6.146
  identification doctrine 6.68–6.74, 6.98
  improvement notices 7.55
  junior employees, isolated acts of 6.80–6.90
  Management of Health and Safety at Work Regulations 1999 6.102–6.115, 7.21, 7.30–7.33
  manual handling 12.63, 12.66
  non-employees, risks to 6.91–6.100
  prohibition notices 7.55
  qualified duties 7.23, 7.35–7.38

  regulatory duties 7.21–7.38
  research 6.194–6.195, 6.207–6.208
  reverse burden of proof 6.135–6.146
  safe system of work 6.91–6.92
  so far as is reasonably practicable 7.18, 7.27–7.28
  vicarious liability 6.75–6.79, 6.96
  weight 7.36
regulations 12.01–12.239 *see also* **Management of Health and Safety at Work Regulations 1999**
  approved codes of practice 12.05–12.08
  book, regulations in 12.07
  civil proceedings 12.05
  criminal offences 12.02, 12.05
  designers 12.175–12.179
  directives 12.01–12.02
  EC law 1.08–1.09, 12.01–12.02
  electricity safety 12.225–12.239, App N
  enforcement 1.07–1.09
  enforcing authority regulations, text of App C
  equipment 12.77–12.119, App H
  fines for breach 10.121
  gas safety 12.211–12.224, App M
  general application, regulations with 12.04
  goal-setting 12.04
  guidance 12.05–12.07
  Health and Safety at Work Act 1974 1.07–1.08, 12.02–12.05
  Health and Safety Commission 1.07, 12.05
  Health and Safety Executive 1.08, 12.03, 12.06–12.07
  height, work at 12.140–12.145, App K
  inspection 12.97–12.107
  interpretation 12.02
  maintenance 12.88–12.90
  manual handling 12.60–12.76, App C
  parts of machinery, dangerous 12.112–12.115
  prescriptive requirements 12.04
  purposes 1.07
  six pack 12.01–12.02, 12.09
  specific hazards 12.04
  workplace 12.47–12.59, App F
regulatory duties 7.21–7.38
reporting incidents and offences 2.01, 2.05–2.24
  accident, meaning of 2.17
  armed forces 2.16
  burden of proof 2.24
  contractors 12.190
  dangerous occurrences 2.10–2.24
  death, accident as cause of 2.13
  deaths and serious injury 2.10
  false statements 7.73

697

Index

reporting incidents and offences (*cont.*)
  Health and Safety Executive 2.05–2.07
  incapacitation 2.11
  inspectors 2.08
  local authorities 2.06
  medical treatment, operations or examinations 2.14
  notification 2.10–2.24
  occupational diseases 2.19–2.20
  record-keeping 2.22
  relevant enforcing authority 2.05–2.06
  responsible person, definition of 2.21
  RIDDOR 2.01, 2.05, 2.08–2.09, 2.18–2.21, 2.23–2.24, 7.73, 11.104, 12.190
  self-employed 2.12, 2.20
  serious injuries 2.11
  statistics 2.07
  vehicles on road, movements of 2.15
  visiting armed forces 2.16
reports *see also* **causes, reports into**
  coroners' inquests 11.134–11.135
  expert evidence 8.199, 8.205
  fines 8.20
  inspectors 11.134
  liquidation 8.20
  post-mortem reports 11.129
  time limits 8.15
representation *see also* **legal representation**
  conflicts of interest 3.59–3.66
  directors 3.06
  fatality, incidents involving a 3.67
  funding 3.55–3.58
  gross negligence manslaughter 3.57
  investigations, advising clients in 3.55–3.67
  legal aid 3.58
  legal expenses insurance 3.55–3.58
  plea before venue 8.78–8.79
  Work Related Death Protocol 3.67
reputation, harm to 4.56–4.57
requisitions 8.05
research
  burden of proof 6.195, 6.208
  criminal offences 6.208
  design 6.193
  examination 6.193
  fairground equipment 6.193
  limitation of liability 6.172
  manufacturers, duties of 6.183, 6.193–6.195, 6.206–6.208
  reasonable practicability 6.194–6.195, 6.207–6.208
  substances 6.206–6.208
  testing 6.193

reverse burden of proof
  criminal offences 6.136–6.137
  European Convention on Human Rights 6.138, 6.142
  fair trials 6.138, 6.142
  Framework Directive 6.139–6.140
  Health and Safety at Work Act 1974 6.135–6.146
  presumption of innocence 6.138
  statistics 6.143
RIDDOR 2.01, 2.05, 2.08–2.09, 2.18–2.21, 2.23–2.24
risk assessment
  approved codes of practice 12.20, 12.22–12.25, 12.28–12.29
  asbestos 12.195–12.196, 12.199–12.200
  contractors 12.188–12.189
  control of substances hazardous to health 12.18, 12.125–12.128, 12.134
  cooperation 12.27–12.28
  duration of validity of assessment 12.33
  employers' duties 6.16, 12.12–12.35
  equipment 12.84–12.87, 12.119
  ergonomics 12.87
  guidance 12.20, 12.22, 12.28–12.29
  hazard, definition of 12.21
  Health and Safety at Work Act 1974 12.20
  human factors 12.24
  identification of persons who might be affected 12.31–12.33
  legal expenses insurance 3.57
  Management of Health and Safety at Work Regulations 1999 12.12–12.35, 12.41
  manual handling 12.16, 12.62, 12.66, 12.68–12.69, 12.73–12.76
  model or generic assessments 12.30
  non-employees 12.32
  particular requirements 12.16–12.19
  parts of machinery, dangerous 12.114
  personal protective equipment 12.17
  pregnant women 12.35
  reviews 12.69
  self-employed 12.12
  size of undertakings, relevance of 12.29
  stress 12.31
  suitable and sufficient assessments 12.28
  training 12.26
  young persons 12.34
risk gap, concept of 4.20–4.22
Robens Report 1.01, 4.01

safe systems of work 6.91–6.92, 6.218
safety representatives and committees, appointment of 6.17

Index

samples
  articles, definition of 2.119
  dangerous articles or substances 2.164
  entry, inspector's powers of 2.117–2.120
  Health and Safety at
      Work Act 1974 2.117–2.120
  inspectors 2.117–2.120, 2.164
  substances, definition of 2.119
Scotland
  bodies corporate 5.15
  Health and Safety at Work Act 1974 1.02
  partnerships 5.15
search and entry, powers of
  arrest 2.70, 2.106
  coroners' inquests 11.04, 11.94
  electronic form, material in 2.70
  equipment, power to take 2.108
  excluded material 2.70
  gross negligence manslaughter 2.57
  Health and Safety at
      Work Act 1974 2.101–2.104
  inspectors 2.93, 2.101–2.108
  indictable offences 2.69–2.70
  investigations 2.57, 2.69–2.71
  inspectors 2.112, 2.152
  legal privilege 2.70, 3.75
  necessity 2.103–2.104
  notice 2.104
  obstruction 2.105–2.106
  other persons, power to take 2.108
  PACE and codes 2.71, 2.77
  police 2.69–2.70
    power to be accompanied by 2.105–2.107
    premises, definition of 2.102
    records 2.77
    samples 2.117–2.120
    seizure 2.70
    serious obstruction 2.105, 2.107
    special procedure material 2.70
    warrants 2.57, 2.70
secondary liability 5.27–5.45 *see also* **company directors, managers and officers, secondary liability of**
  act or default, meaning of 5.38–5.44
  aiding and abetting 5.27–5.29
  causation 5.37
  common law 5.27–5.29
  Consumer Protection Act 1987 5.40
  corporate manslaughter 11.78
  criminal offences 5.27–5.45
  Crown, liability of the 5.45
  due diligence 5.35
  Food and Drugs Act 1938 5.31, 5.41–5.42
  Health and Safety at Work Act 1974 5.30–5.45
  partnerships 5.15
  pollution offences 5.31
  trade descriptions 5.31, 5.34–5.35, 5.37, 5.39, 5.43
section 20 powers 3.68–3.73
section 42 orders, fines for failure to comply with 10.118–10.120
seizure
  coroners' inquests 11.94
  danger, dealing with immediate 2.163, 2.165
  dangerous articles or substances 2.163, 2.165
  gross negligence manslaughter 3.67
  investigations 2.56
  legal privilege 3.75
  PACE and codes 2.77
  police 3.75
self-employed
  articles 6.178–6.179
  breach of general duties 6.04
  control of substances hazardous to health 12.123
  electricity safety 12.225
  gross negligence manslaughter 11.48
  Health and Safety at Work Act 1974 6.29–6.31
  height, work at 12.140, 12.144
  manual handling 12.61
  meaning 6.30–6.31
  non-employees 6.29–6.32
  occupational diseases 2.12, 2.20
  risk assessment 12.12
  work and at work, definition of 6.18
  workplace 12.49, 12.55
self-incrimination, privilege against
  admissibility of documents 3.86–3.90
  advising clients 3.86–3.90
  answers, power to require 2.137–2.138, 2.144–2.145
  causes, reports into 3.79
  compulsory production of documents 3.88–3.90
  compulsory questioning 3.87
  coroners' inquests 11.127–11.128
  criminal offences 7.75
  Environmental Protection Act 1990, powers under 3.87
  European Convention on Human Rights 3.79, 3.86–3.88
  fair trials 3.86–3.88
  false declarations 2.138
  false statements 7.75
  information 7.59
  qualified right, as 3.87
  silence, right to 3.88

Index

sentencing 10.01–10.129 *see also* **committal to Crown Court for sentence; fines**
  absolute discharge 10.04
  compensation orders 10.05–10.08
  conditional discharges 10.04
  directors 10.88, 10.04, 10.102–10.105
  enforcement notices, breach of 10.87
  fines 10.122–10.125
  Goodyear procedure 10.15–10.16
  gross negligence manslaughter 10.89–10.101
  groups of companies 10.62–10.70
    accounts 10.70
    economic realities approach 10.69–10.70
  guilty plea 8.93, 8.96, 10.15
  imprisonment, breach of licence and 10.87
  indication of sentence, obtaining an 10.15–10.16
  individuals 10.87–10.129
  insolvent companies, holding companies' liability for debts of 10.68
  limited liability of companies 10.68
  magistrates 10.122–10.125
  plea bargaining 8.154–8.156
  range of punishments 10.04
  section 42 orders 10.118–10.120
  Sentencing Guidelines 10.122–10.125
  veil, piercing the 10.67–10.69
  venue for sentence 8.81–8.89, 10.09–10.11
Serious Fraud Office 3.76
service companies 6.53–6.54
service of notices and documents 2.02, 2.25–2.31
  address 2.27–2.29
  bodies corporate 2.28
  breach of general duties 6.07
  companies 2.28
  expert evidence 8.204
  Health and Safety at Work Act 1974 2.25
  Health and Safety Commission 1.23
  improvement or prohibition notices 4.05, 4.10, 4.03, 4.25, 4.28, 4.33, 4.45–4.47
    appeals against 4.64–4.67
  interpretation of 'serve' or 'send' 2.31
  methods 2.26–2.27, 4.64
  partnerships 2.28
  post 2.30
  prohibition notices 4.04–4.05, 4.34–4.35, 4.46–4.47
  responsible person, served on 2.30
  summons 8.32–8.39, 8.45
service of summons 8.32–8.39, 8.45
  companies 8.34
  contents 8.36–8.39

  Criminal Procedure Rules 2005 8.33
  methods 8.33–8.34
  single summons for more than one information 8.37
  time limits 8.35
Shipman Inquiry 11.90
ships 12.56
significant statements or silence 3.42–3.43
silence, right to *see also* **adverse inferences from silence**
  compulsory powers 3.88–3.90
  European Convention on Human Rights 3.87
  qualified right, as 3.87
  self-incrimination 3.88
social service investigators 2.76
solicitors
  Health and Safety Executive 3.06
  inspectors, instruction by 8.12
  instituting proceedings 8.12
  interviews under caution 3.06
special procedure material 2.70
sporting clubs and associations 5.09
statements *see also* **false statements;** *Friskies* **statements**
  advance information 8.48
  answers, power to require 2.134–2.136
  employers' duties 6.16
  interviews under caution 3.27, 3.50
  PACE and codes 3.50
  prepared statements 3.27
  silence, adverse inferences from 3.27
  witnesses 11.92, 11.125–11.126, 11.131–11.132, 11.136, 11.140–11.141
statute, bodies incorporated by 5.05
stay
  abuse of process 9.18–9.19
  delay 9.24–9.25, 9.31
  double jeopardy 9.37
stress, risk assessments and 12.31
strict liability
  Banking Act 1987 5.56
  breach of general duties 6.06
  criminal offences 7.15–7.17
  European Convention on Human Rights 7.16–7.17
  Framework Directive 1.55–1.59
  justification 7.15–7.16
  maintenance 12.89
  presumption of innocence 7.16–7.17
  prosecutions 9.51
  trading standards 9.51

Index

substances *see also* control of substances hazardous to health; dangerous articles or substances
  definition 2.119, 6.151, 6.199
  examination 6.198
  imports 6.198
  inspectors 2.121–2.130
  manufacturers, duties of 6.165–6.208, 6.198–6.199, 6.204, 6.206–6.208
  possession and detain, power to take 2.128–2.130
  research 6.206–6.208
  safe and without risk to non-employees 6.160–6.161
  samples, power to take 2.117–2.120, 2.164
  suppliers 6.198
  testing 6.198
summons
  abolition 8.05–8.06
  advance information 8.48
  amendment 8.40–8.42
  companies 8.04
  duplicity, rule against 8.125
  errors 8.41
  instituting proceedings 8.04–8.06, 8.40–8.42
  issue 8.32
  magistrates 8.32–8.35, 8.40–8.42, 8.45
  serving 8.32–8.35, 8.45
  withdrawal 8.42
supply
  Health and Safety at Work Act 1974 6.173–6.175
  hire purchase, conditional sale agreements or credit sale agreements 6.175
  substances 6.198
surveillance 12.37–12.40, 12.135

tax or national insurance, liability for 6.48
temporary workers 12.43
testing
  asbestos 12.209
  dangerous articles or substances 2.121–2.127
  inspectors 2.117–2.120, 2.124
  limitation of liability 6.172
  manufacturers, duties of 6.183
  substances 6.198
third parties *see also* non-employees, duties to
  disclosure 8.169, 8.181–8.182
  fines 10.46–10.53
  let down by others, being 10.46–10.53
time limits
  costs 10.150–10.151
  Crown Court 8.14
  disclosure 8.166
  Health and Safety at Work Act 1974, breach of 8.14–8.15
  improvement or prohibition notices 4.36, 4.45
    appeals against 4.08, 4.17, 4.29, 4.66, 4.78
    extension of time 4.06, 4.48–4.50
  indictment, trials on 8.14–8.15
  indictments 8.121
  information, laying an 8.16
  inquests 8.15
  instituting proceedings 8.14–8.16
  magistrates 8.14, 8.16
  reports 8.15
  service of summonses 8.35
  summary offences 8.16
trade descriptions 5.35, 5.37, 5.47
trading standards 9.51
traffic routes 12.59
training
  asbestos 12.202
  contractors 12.183, 12.185, 12.188
  control of substances hazardous to health 12.131
  employees' duties 6.218
  equipment 12.110–12.111
  height, work at 12.143–12.144
  inspection 12.107
  Management of Health and Safety at Work Regulations 1999 12.40, 12.42, 12.44
  manual handling 12.72
  risk assessment 12.26
  work and at work, definition of 6.20
trustees *see also* trusts
  breach of trust 5.17
  charitable trusts 5.24
  joint and several liability 5.17
  personal liability 5.24
trusts *see also* trustees
  breach of trust 5.17
  charitable trusts 5.18–5.19, 5.24–5.26
  legal personality 5.17–5.26
  person, trust as 5.20
undertakings
  Health and Safety at Work Act 1974 6.33–6.42
  independent contractors 6.36–6.38
  meaning 6.33
  precautions 6.36
undisturbed, directions to leave 2.113–2.114
unincorporated associations
  body incorporate, as 5.23
  charitable trusts 5.18
  criminal offences 5.10
  definition 5.21

Index

unincorporated associations (*cont.*)
   formation 5.22
   Health and Safety at Work Act 1974 5.08–5.10
   legal personality 5.08–5.10, 5.21–5.22
   sporting clubs and associations 5.09
unlawful killing verdicts 11.144

vehicles on road, movements on 2.15
veil, piercing the corporate 10.67–10.69
venue for trial or sentence 8.81–8.89, 10.09–10.11
verdicts
   accidental death verdicts 11.143
   coroners' inquests 11.114–11.115, 11.118–11.121, 11.142–11.144
   criminal or civil liability, determination of 11.118–11.121, 11.144
   expanded form of 11.114
   narrative form 11.114–11.115
   unlawful killing 11.144
vicarious liability
   common employment 6.75
   common law 6.75
   delegation 6.75
   gross negligence manslaughter 11.48
   Health and Safety at Work Act 1974 6.76–6.79
   independent contractors 6.77
   junior employees, isolated acts of 6.88
   reasonable practicability 6.75–6.79, 6.96
   work and at work, definition of 6.21
visiting armed forces 2.16
voluntary liquidation 8.18, 8.24–8.25
vulnerable persons
   interviews 3.45
   PACE and codes 3.45

Wales 1.02
warnings
   aggravating factors 8.91, 10.23
   employees' duties 6.220
   failure to heed warnings 8.91, 10.23
   fines 10.23
   inspectors 1.77
   verbal 1.77
warrants
   arrest 2.59
   search and entry, powers of 2.57, 2.70
welfare facilities 12.173, 12.182
winding up *see* **liquidation, companies in**
witnesses *see also* **expert evidence**
   advance notice of statements 11.131
   copies of statements 11.131, 11.136
   coroners' inquests 11.92, 11.125–11.126, 11.131–11.132, 11.136, 11.140–11.141
   examination 11.123–11.124
   improvement or prohibition notices, appeals against 4.80
   statements 11.92, 11.125–11.126, 11.131–11.132, 11.136, 11.140–11.141
work and at work, definition of 6.18–6.23
   course of employment 6.22–6.23
   employees' duties 6.212
   employers' duties 6.18–6.23
   Health and Safety at Work Act 1974 6.18–6.23
   instructions, following 6.22
   police officers on duty 6.19
   prosecutions 6.21
   self-employed 6.18
   training courses 6.20
   vicarious liability 6.21
   work experience 6.20
   workplace 12.55
work equipment
   Approved Code of Practice 12.77, 12.79, 12.82, 12.84, 12.86–12.87, 12.109, 12.111, 12.119
   asbestos 12.204–12.205, 12.208–12.209
   dangerous parts of machinery 12.112–12.115
   definition 12.78
   design 12.87, 12.118
   EC law 12.77
   electricity safety 12.228, 12.233
   ergonomics 12.87
   external advisers 12.85
   foreseeability 12.81
   guidance 12.79, 12.82, 12.84, 12.86, 12.109, 12.111, 12.119
   height, work at 12.145
   hire 12.96
   information 12.108–12.109
   inspection 12.78, 12.87–12.107
   instructions 12.108–12.109
   maintenance 12.78, 12.88–12.96
   personal protective equipment 12.17, 12.204–12.205, 12.208
   reasonable practicability 12.81
   regulations 12.77–12.119, App H
   respiratory equipment 12.204–12.205
   risk assessment 12.84–12.87, 12.119
   specified hazards, protection against 12.116–12.119
   suitability 12.78, 12.80–12.83
   training 12.110–12.111
work experience 6.20, 6.55
Work Related Death Protocol 11.08–11.20
   advice prior to charge 11.19
   arrest 2.65
   coroners' inquests 11.133, 11.148

Index

corporate manslaughter 11.24, 11.72
Crown Prosecution Service 11.19
decision-making 11.16–11.18
decision to prosecute 11.19
definition of work-related death 11.12–11.14
double jeopardy 9.36
Enforcement Concordat, text of App B
gross negligence manslaughter 2.65
Health and Safety Executive 11.16–11.17
inquests 11.18
Investigators Guide 11.09
local authorities 11.17
manslaughter 11.20
police 11.17
principles 11.10, 11.12–11.14
representation 3.67
scope 11.15–11.20
statements of intent 11.11
work-related deaths 11.01–11.152 *see also* coroners' inquests; Work Related Death Protocol
agencies, investigation by 11.03
arrest 2.65–2.66
Beedie decisions 11.05–11.07
coroners' inquests 11.95, 11.98, 11.133–11.139, 11.145–11.152
Corporate Manslaughter and Homicide Bill 11.01
double jeopardy 11.07
Health and Safety at Work Act 1974 11.06
Health and Safety Executive 11.03
investigations 11.01–11.07
manslaughter 11.06–11.07
police 11.03
workplace
agricultural land 12.56
aircraft 12.56
Approved Code of Practice 12.47
building operations 12.56
control, limited to matters within person's 12.50–12.51
definition of workplace 12.53–12.56
duty of care
extent of duty 12.50–12.51
nature of duty 12.57–12.59
to whom duty owed 12.52
who owns duty 12.49
EC law 12.48
engineering works 12.56
floors 12.59
forests 12.56
guidance 12.47
Health and Safety at Work Act 1974 12.47
Health and Safety Executive, route map proposed by 12.59
independent contractors 12.58
mines 12.56
regulations 12.47–12.59, App F
self-employed 12.49, 12.55
ships 12.56
traffic routes 12.59
work, definition of 12.55
written statements
advance information 8.48
employers' duties 6.16
interviews under caution 3.50
PACE and codes 3.50

young persons
Health and Safety Executive 12.34
interviews 3.45
PACE and codes 3.45
risk assessment 12.34